Lecture Notes in Artificial Intelligence   7713

Subseries of Lecture Notes in Computer Science

Shuigeng Zhou   Songmao Zhang
George Karypis (Eds.)

Advanced Data Mining and Applications

8th International Conference, ADMA 2012
Nanjing, China, December 15-18, 2012
Proceedings

 Springer

Series Editors

Randy Goebel, University of Alberta, Edmonton, Canada
Jörg Siekmann, University of Saarland, Saarbrücken, Germany
Wolfgang Wahlster, DFKI and University of Saarland, Saarbrücken, Germany

Volume Editors

Shuigeng Zhou
Fudan University
School of Computer Science
Handan Road 220, 200433 Shanghai, China
E-mail: sgzhou@fudan.edu.cn

Songmao Zhang
Chinese Academy of Sciences
Academy of Mathematics and Systems Science
East Road 55, 100190 Beijing, China
E-mail: smzhang@math.ac.cn

George Karypis
University of Minnesota
Department of Computer Science and Engineering
200 Union St SE, Minneapolis, MN 55455, USA
E-mail: karypis@cs.umn.edu

ISSN 0302-9743   e-ISSN 1611-3349
ISBN 978-3-642-35526-4   e-ISBN 978-3-642-35527-1
DOI 10.1007/978-3-642-35527-1
Springer Heidelberg Dordrecht London New York

Library of Congress Control Number: 2012953579

CR Subject Classification (1998): I.2, H.3, H.4, H.2.8, J.1, C.2, F.1, I.4

LNCS Sublibrary: SL 7 – Artificial Intelligence

Typesetting: Camera-ready by author, data conversion by Scientific Publishing Services, Chennai, India

Printed on acid-free paper

Springer is part of Springer Science+Business Media (www.springer.com)

Preface

We would like to welcome you to the proceedings of the 8th International Conference on Advanced Data Mining and Applications (ADMA 2012).

ADMA 2012 was held in Nanjing, China, December 15–18, 2012. Following the success of the series of seven events held from 2005 to 2011 in Wuhan (2005), Xi'an (2006), Harbin (2007), Chengdu (2008), Beijing (2009 and 2011), and Chongqing (2010), ADMA 2012 stuck to the tradition of providing a dedicated forum for researchers and practitioners to share new ideas, original research results, practical development experiences, case studies, and applications in all aspects related to data mining and applications.

The conference received 168 paper submissions from 38 countries and areas. All papers were peer reviewed by at least three members of the Program Committee (PC) composed of international experts in data mining fields. The PC, together with the PC Co-chairs, worked very hard to select papers through a rigorous review process and extensive discussion, and finally composed a diverse and exciting program including 32 regular papers and 32 short papers. The ADMA 2012 program featured five keynote speeches from prestigeous researchers in the artificial intelligence, machine learning, and data mining areas. They include the 2010 Turing Award winner Leslie Valiant from Harvard University, Bo Zhang from Tsinghua University, David Bell from Queen's University Belfast, Qiang Yang from Hong Kong University of Science and Technology, and Ester Martin from Simon Fraser University.

Without the support of several organizations, the success of ADMA 2012 would not have been possible. This includes financial support and local arrangements from Nanjing University of Finance and Economics, and Focus Technology Co., Ltd., and sponsorship from the University of Technology, Sydney, Chinese Academy of Sciences, Fudan University, and Nanjing University. We would also like to express our gratitude to the General Co-chairs for all their valuable advice and the Organizing Committee for their dedicated organizing efforts. Last but not least, we sincerely thank all PC members and authors for their contributions to ADMA 2012!

December 2012

Shuigeng Zhou
Songmao Zhang
George Karypis

Preface

Organization

Organizing Committee

International Steering Committee

Xue Li, Chair	University of Queensland, Australia
Kyu-Young Whang	Korea Advanced Institute of Science and Technology, Korea
Chengqi Zhang	University of Technology, Sydney, Australia
Osmar Zaiane	University of Alberta, Canada
Qiang Yang	The Hong Kong University of Science and Technology, China
Jie Tang	Tsinghua University, China

Honorary Chair

Ruqian Lu   Chinese Academy of Sciences, China

General Co-chairs

Chengqi Zhang   University of Technology, Sydney, Australia
Zhi-Hua Zhou   Nanjing University, China

Program Co-chairs

Shuigeng Zhou   Fudan University, China
Songmao Zhang   Chinese Academy of Sciences, China
George Karypis   University of Minnesota, USA

Workshop Co-chairs

Hong Cheng   The Chinese University of Hong Kong, China
Jianyong Wang   Tsinghua University, China

Tutorial Co-chairs

Joshua Zhexue Huang	Shenzhen Institutes of Advanced Technology, Chinese Academy of Sciences, China
Xingquan Zhu	University of Technology, Sydney, Australia

Publicity Co-chairs

Aixin Sun	Nanyang Technological University, Singapore
Bin Wang	Nanjing University of Finance and Economics, China

Local Co-chairs

Jie Cao   Nanjing University of Finance and Economics,
                                    China
Ruibo Yao   Focus Technologies Co., Ltd., China

Program Committee

Aijun An   York University, Canada
Annalisa Appice   University Aldo Moro of Bari, Italy
Yasuhito Asano   Kyoto University, Japan
Michael Bain   University of New South Wales, Australia
Keke Cai   IBM, China Research Lab, China
Rui Camacho   University of Porto, Portugal
Jiuxin Cao   Southeast University, China
Tru Hoang Cao   Ho Chi Minh City University of Technology,
                                    Vietnam
Pablo Castro   Universidad Nacional de Rio Cuarto, Argentina
Wei Chen   Microsoft Research Asia Beijing, China
Jianhui Chen   Arizona State University, USA
Ling Chen   University of Technology, Sydney, Australia
Shixi Chen   Fudan University, China
Shu-Ching Chen   Florida International University, USA
Songcan Chen   NUAA, China
Hong Cheng   The Chinese University of Hong Kong, China
Amanda Clare   Aberystwyth University, UK
Diane Cook   Washington State University, USA
Bruno Cremilleux   Université de Caen, France
Bin Cui   Beijing University, China
Peng Cui   Tsinghua University, China
Alfredo Cuzzocrea   University of Calabria, Italy
Dao-Qing Dai   Sun Yat-Sen University, China
Martine De Cock   Ghent University, Belgium
Zhihong Deng   Beijing University, China
Nicola Di Mauro   Università di Bari, Italy
Gaël Dias   University of Beira Interior, Portugal
Stefano Ferilli   University of Bari, Italy
Philippe Fournier-Viger   University of Moncton, Canada
Patrick Gallinari   LIP6-University Paris 6, France
Yanglan Gan   Tongji University, China
Jing Gao   University at Buffalo, USA
Jibing Gong   Tsinghua University, China
Manish Gupta   University of Illinois at Urbana-Champaign,
                                    USA
Qing He   Chinese Academy of Sciences, China
Tu Anh Nguyen Hoang   Vietnam National University, Vietnam

Wei Hu	Nanjing University, China
Xiaohua Hu	Drexel University, USA
Jimmy Huang	York University, Canada
Zi Huang	The University of Queensland, Australia
Huan Huo	University of Shanghai for Science and Technology, China
Akihiro Inokuchi	Osaka University, Japan
Sanjay Jain	National University of Singapore, Singapore
Licheng Jiao	Xidian University, China
Cheqing Jin	East China Normal University, China
Huidong Jin	Australian National University, Australia
Xin Jin	University of Illinois at Urbana-Champaign, USA
Panagiotis Karras	Rutgers, The State University of New Jersey, USA
Daisuke Kawahara	Kyoto University, Japan
Irwin King	The Chinese University of Hong Kong, China
Wai Lam	The Chinese University of Hong Kong, China
Sanghyuk Lee	Xi'an Jiaotong-Liverpool University, China
Gang Li	Deakin University, Australia
Chun-Hung Li	Hong Kong Baptist University, China
Xuelong Li	Chinese Academy of Sciences, China
Ninghui Li	Purdue University, USA
Tao Li	Florida International University, USA
Xue Li	School of ITEE, Australia
Jinbao Li	Heilongjiang University, China
Guozheng Li	Tongji University, China
Jianyuan Li	Tongji University, China
Li Li	Southwestern University, China
Jiye Liang	Shanxi University, China
Guoqiong Liao	Jiangxi University of Finance and Economics, China
Xide Lin	University of Illinois at Urbana-Champaign, USA
Mario Linares-Vásquez	Universidad Nacional de Colombia, Colombia
Danzhou Liu	University of Central Florida, USA
Wei Liu	University of Western Australia, Australia
Qingshan Liu	Chinese Academy of Sciences, China
Yubao Liu	Sun Yat-Sen University, China
Hongyan Liu	Tsinghua University, China
Xiaohui Liu	Brunel University, UK
Dexi Liu	Wuhan University, China
Hua Lu	Aalborg University, Denmark
Xiuli Ma	Beijing University, China
Marco Maggini	University of Siena, Italy

Guoyin Wang	Chongqing University of Posts and Telecommunications, China
Jason Wang	New Jersey's Science and Technology University, USA
Bin Wang	Nanjing University of Finance and Economics, China
Xin Wang	University of Calgary, Canada
Tim Weninger	University of Illinois at Urbana-Champaign, USA
Junjie Wu	Beihang University, China
Sai Wu	National University of Singapore, Singapore
Xintao Wu	University of North Carolina at Charlotte, USA
Zhiang Wu	Nanjing University of Finance and Economics, China
Zhipeng Xie	Fudan University, China
Wei Xiong	Fudan University, China
Sadok Ben Yahia	Campus Universitaire, Tunisia
Zijiang Yang	York University, Canada
Min Yao	Zhejiang University, China
Jian Yin	Sun Yat-Sen University, China
Tetsuya Yoshida	Hokkaido University, Japan
Wlodek Zadrozny	IBM T.J. Watson Research Center, USA
Daoqiang Zhang	NUAA, China
Shichao Zhang	University of Technology, Sydney, Australia
Mengjie Zhang	Victoria University of Wellington, New Zealand
Songmao Zhang	CAS, China
Zhongfei Zhang	Binghamton University, USA
Ying Zhang	The University of New South Wales, Australia
Harry Zhang	University of New Brunswick, Canada
Du Zhang	California State University, USA
Xiangliang Zhang	KAUST, Saudi Arabia
Zili Zhang	Deakin University, Australia
Jiakui Zhao	Beijing University, China
Xiao Zheng	Anhui University of Technology, China
Yong Zheng	DePaul University, USA
Sheng Zhong	SUNY Buffalo, USA
Xingquan Zhu	University of Technology, Sydney, Australia

Co-organized by

Nanjing University of Finance and Economics
Focus Technology Co., Ltd.
Jiangsu Provincial Key Laboratory of E-Business
Enterprise Academician Workstation at FOCUS

Sponsored by

University of Technology, Sydney, Australia
Chinese Academy of Sciences
Fudan University
Nanjing University

Supported by

Nanjing University of Finance and Economics
Focus Technology Co., Ltd.

Table of Contents

Social Media Mining

Clustering

Machine Learning: Algorithms and Applications

Classification

Prediction, Regression and Recognition

Optimization and Approximation

Mining Time Series and Streaming Data

Web Mining and Semantic Analysis

Data Mining Applications

Search and Retrieval

Information Recommendation and Hidding

Outlier Detection

Topic Modeling

Data Cube Computing

Leave or Stay: The Departure Dynamics of Wikipedia Editors

Dell Zhang[1], Karl Prior[1], Mark Levene[1], Robert Mao[2], and Diederik van Liere[3]

[1] DCSIS, Birkbeck, University of London
Malet Street, London WC1E 7HX, UK
dell.z@ieee.org, {kprior01,mark}@dcs.bbk.ac.uk
[2] Microsoft Research FUSE Labs
One Microsoft Way, Redmond, WA 98052-6399, USA
robmao@microsoft.com
[3] Wikimedia Foundation
149 New Montgomery Street, 3rd Floor, San Francisco, CA 94105, USA
dvanliere@wikimedia.org

Abstract. In this paper, we investigate how Wikipedia editors leave the community, i.e., become inactive, from the following three aspects: (1) how long Wikipedia editors will stay active in editing; (2) which Wikipedia editors are likely to leave; and (3) what reasons would make Wikipedia editors leave. The statistical models built on Wikipedia edit log datasets provide insights about the sustainable growth of Wikipedia.

1 Introduction

Wikipedia is "a free, web-based, collaborative, multilingual encyclopaedia project" supported by the non-profit Wikimedia Foundation (WMF). Started in 2001, Wikipedia has become the largest and most popular general reference knowledge source on the Internet. Almost all of its 19.7 million articles can be edited by anyone with access to the site, and it has about 90,000 regularly active volunteer editors around the world. However, it has recently been observed that Wikipedia growth has slowed down significantly[1]. It is therefore of utter importance to understand quantitatively what factors determine editors' future editing behaviour (why they continue editing, change the pace of editing, or stop editing), in order to ensure that the Wikipedia community can continue to grow in terms of size and diversity.

This paper is directly inspired by the Wikipedia Participation Challenge[2] (WikiChallenge) that was recently organised by WMF, Kaggle and IEEE ICDM-2011 to address the above problem of Wikipedia's sustainable growth by data mining. The contestants were requested to build a predictive model that could accurately predict the number of edits a Wikipedia editor would make in the next 5 months based on his edit history so far. Such a predictive model may be able to help WMF in figuring out how people can be encouraged to become,

---

[1] http://strategy.wikimedia.org/wiki/March_2011_Update
[2] http://www.kaggle.com/c/wikichallenge

S. Zhou, S. Zhang, and G. Karypis (Eds.): ADMA 2012, LNAI 7713, pp. 1–14, 2012.

and remain, active contributors to Wikipedia. Our work led to the 2nd best valid algorithm in this contest, which achieved 41.7% improvement over WMF's in-house solution. WMF is now implementing this algorithm permanently and looks forward to using it in the production environment.

The rest of this paper is organised as follows. First we investigate how long Wikipedia editors will stay active in editing through *survival analysis* in Section 2. Then, we investigate which Wikipedia editors are likely to leave (which is the task of WikiChallenge) through *predictive modelling* in Section 3. Next, we investigate what reasons would make Wikipedia editors leave through *correlation analysis* in Section 4. Finally, we review the related work in Section 5, and make conclusions in Section 6.

2   Survival Analysis

In order to analyse the expected active period that a Wikipedia editor can maintain his interest in contributing, we make an analogy to the modelling of people's expected lifetime: an editor is "born" when he starts editing (i.e., joins the community), and "dies" when he stops editing (i.e., leaves the community). Specifically, we consider an editor to be "dead" or inactive if he did not make any edit for a certain period of time. Here we set the threshold of inactivity to be 5 months, since it reflects WMF's concern as demonstrated in the WikiChallenge contest. Thus, we are able to utilise *survival analysis* [5,6] — a branch of statistics which is widely applied to modelling death in biological organisms and failure in mechanical systems — to deal with the departure dynamics of Wikipedia editors.

The dataset for this study consists of 110,383 registered editors of (English) Wikipedia that were randomly sampled. The bots (i.e., automatic agents that do maintenance task on Wikipedia) were excluded from data collection. Moreover, we removed 38,348 one-timers who had only made one edit so far, because their contribution and influence to Wikipedia are known to be negligible [11]. Finally the complete edit history of those remaining 72,035 editors were processed to extract a population of 86,468 "lives". If an editor started editing again after being "dead" (inactive), it would be considered as a new life instance, because in this study we are more interested in the continuous active period of an editor rather than his overall time in the community, and the same editor could exhibit quite different behaviour patterns when he came back after a very long break (e.g., due to the change of motivation). One prominent characteristic of lifetime data is that many samples may be *censored* [5,6] — they were still alive at the end of data collection therefore their lifetime values are only known to be longer than a certain duration. Such a censoring problem requires special treatment in probability estimation etc. when performing data analysis.

The histogram of Wikipedia editors' lifetime is shown in Figure 1a, where the lifetime is measured in days and scaled logarithmically (using natural logarithm). The lifetime distribution clearly consists of two distinct regimes separated roughly at the point of 8 hours: the left regime corresponds to occasional editors who fail to find interest in editing Wikipedia articles after the first few

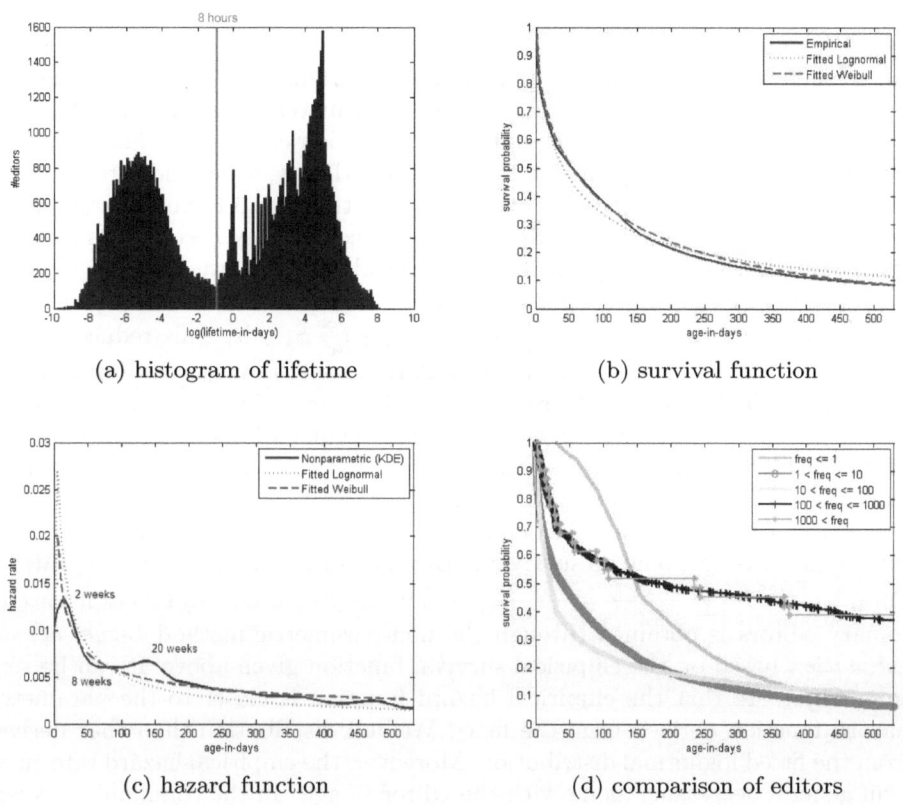

(a) histogram of lifetime

(b) survival function

(c) hazard function

(d) comparison of editors

Fig. 1. The survival analysis results for Wikipedia editors

attempts; the right regime corresponds to customary editors who stay in the community editing Wikipedia articles until they lose interest because of some reason. Let's focus on analysing the behaviour patterns of customary editors, as it is them who constitute the backbone of the community.

The objects of primary interest in survival analysis — *survival function* [5, 6] and *hazard function* [5, 6] — for customary editors are presented in Figure 1b and Figure 1c respectively.

The survival function, conventionally denoted S, is the probability that the time of death T is later than some specified time t: $S(t) = \Pr(T > t) = \int_t^\infty f(t)\,\mathrm{d}t$. The empirical survival function for customary editors is calculated using the *Kaplan-Meier estimator* [5, 6] which can handle censored data. To project out and compute editors' departure probabilities at times beyond the end of the study, we need to fit a parametric survival function to the empirical data. After trying out a number of popular probability distributions (including exponential, extreme value, lognormal, normal, Rayleigh, and Weibull), we have found that although the survival function of occasional editors roughly follows a (truncated) lognormal distribution (which confirms the finding in [2]), the survival function of customary editors can be better described by a Weibull distribution: $f(t) = \frac{\beta}{\eta} \left(\frac{t}{\eta}\right)^{\beta-1} e^{-(t/\eta)^\beta}$ when $t \geq 0$ and $f(t) = 0$ otherwise, with the scale parameter $\eta = 102.68$ and the shape parameter $\beta = 0.55$. As shown in Figure 1b, the Weibull distribution curve clearly matches the customary editors' lifetime data better than the lognormal distribution curve does. The shape parameter of the fitted Weibull distribution is less than 1, which indicates that the overall departure rate decreases over time, i.e., those who leave the community tend to leave early, and those who stay in the community become less likely to leave over time. Given the parametric survival function, we are able to make inference about the *expected future lifetime* of an editor who has stayed in the community for t_0 days: $\int_0^\infty t \frac{f(t+t_0)}{S(t_0)} dt = \frac{1}{S(t_0)} \int_{t_0}^\infty S(t) dt$. This reduces to the expected lifetime (a.k.a. mean time to failure) at birth for $t_0 = 0$. Furthermore, the age at which a specified proportion q of editors will remain can be found by solving the equation $S(t) = q$ for t. Using the fitted Weibull distribution, we estimate the *median lifetime* (at which half of the customary Wikipedia editors leave the community) to be about 53 days.

The hazard function, conventionally denoted λ, is defined as the event ("death") rate at time t conditional on survival until time t or later (that is, $T \geq t$): $\lambda(t) = \lim_{\Delta t \to 0} \frac{\Pr(t \leq T \leq T + \Delta t | T \geq t)}{\Delta t} = -\frac{dS(t)/dt}{S(t)}$. The empirical hazard function for customary editors is obtained through the non-parametric method *kernel density estimation* based on the empirical survival function given above. It can be seen from Figure 1c that the empirical hazard function is closer to the parametric hazard function derived from the fitted Weibull distribution than that derived from the fitted lognormal distribution. Moreover, the empirical hazard rate curve is in general decreasing along with the editor's "age" in the community, except for the periods of 0-2 weeks and 8-20 weeks, which suggest that these are the two critical phases to retain Wikipedia editors in the community.

In order to further understand the survival patterns of customary editors, we group them into five classes according to their monthly editing frequencies (*freq*): (i) *freq* <= 1, (ii) 1 < *freq* <= 10, (iii) 10 < *freq* <= 100, (iv) 100 < *freq* <= 1000, (v) 1000 < *freq*. The sizes of those classes are 6744, 32583, 8818, 675, and 31 respectively. The exponential scale using the powers of 10 is chosen to define the above classes because the monthly editing frequencies roughly follow the *power law*. This is the same editor classification criterion used in [11], except that the classification is not recalculated every month as we would like to analyse the relationship between an individual editor's monthly editing frequency and his whole lifetime.

Figure 1d plots the survival function for each class of customary editors separately. It is clear that low-frequency editors (class i and ii) are more likely to have a short lifetime than medium-frequency editors (class ii and class iii), and similarly we can say that medium-frequency editors (class ii and class iii) are less likely to have a long lifetime than high-frequency editors (class v). In other words, higher editing frequency implies longer lifetime — the more active an editor is, the longer he will keep active in editing.

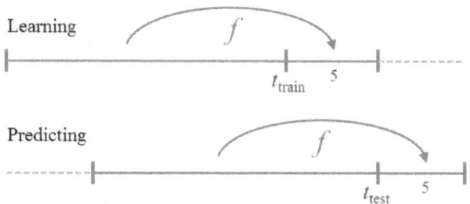

Fig. 2. Our self-supervised learning framework

3 Predictive Modelling

Now we turn to tackling the WikiChallenge task which is to predict the number of edits a Wikipedia editor would make in the next 5 months based on his edit history so far. The official dataset (for training and testing) consists of randomly sampled active editors with their full history of editing activities on the English Wikipedia (the first 6 namespaces only) in the period from 2001-01-01 to 2010-09-01. An editor is considered "active" if he or she made at least one edit in the last one year period, i.e., from 2009-09-01 to 2010-09-01. For each edit, the available information includes its user_id, article_id, revision_id, namespace, timestamp, etc. The predictive model to be constructed should predict, for each editor from the 'training' dataset, how many edits would be made in the 5 months after the end date of the 'training' dataset, i.e., from 2010-09-01 to 2011-02-01. The predictive model's accuracy is going to be measured by the Root Mean Squared Logarithmic Error (RMSLE): $\epsilon = \sqrt{\frac{1}{n} \sum_{i=1}^{n} (\log(1 + p_i) - \log(1 + a_i))^2}$, where n is total number of editors in the dataset, $\log(\cdot)$ is the natural logarithm function, p_i and a_i are the predicted and actual edit numbers respectively for editor i in the next 5 month period.

Although this problem is essentially statistical analysis of time series data, usual time series models such as autoregressive moving-average (ARMA) work poorly here, according to our initial investigation. It is probably because the departure process of editors is not really a "stationary" stochastic process.

Our basic idea is to build a predictive model f (that estimates an active editor's future number of edits based on his recent edit history) through self-supervised learning, as illustrated schematically in Figure 2. The approach is called "self-supervised" to emphasise the fact that it does not require any manual labelling of data (as in standard *supervised learning* but extracts the needed labels from data automatically.

To facilitate the description of our approach, we shall from now on talk about any time-length in the unit of months and refer to any time-point as the real number of months passed since the beginning date of the dataset. So for the official dataset 'training', the timestamp "2001-06-16 00:00:00" would be 5.5 because it is five and a half months since 2001-01-01.

Let t_{test} denote the time-point when we would like to predict each active editor's number of edits in the next 5 months. To train the predictive model, we would move 5 months backwards and assume that we were at the time-point $t_{\text{train}} = t_{\text{test}} - 5$. Thus we could know the actual number of edits made by each active editor in those 5 months after t_{train}, i.e., the label for our machine learning (regression) methods. Specifically, the target value for regression would be set as $y_i = \log(1 + a_i)$ where a_i is the actual number of edits in the next 5 months. In this way, the *squared error* loss function $L(f(\mathbf{x}), y) = (f(\mathbf{x}) - y)^2$ used by most machine learning methods (including those in our experiments and final submission) would connect the *empirical risk* directly to the evaluation metric RMSLE: $R_{emp}(f) = \frac{1}{n} \sum_{i=1}^{n} L(f(\mathbf{x}_i), y_i) = \epsilon^2$.

Given a time-point (either t_{train} or t_{test}), each active editor i would be represented as a vector $\mathbf{x}_i$ that consists of the following temporal dynamics features: (i) the number of edits in recent periods of time; (ii) the number of edited articles in recent periods of time; (iii) the length of time between the first edit and the last edit, scaled logarithmically. The periods used in our final submission for the above temporal dynamics features are: $\frac{1}{16}, \frac{1}{8}, \frac{1}{4}, \frac{1}{2}, 1, 2, 4, 12, 36$, and 108, where the length of period first doubles at each step from $\frac{1}{16}$ to 4 and then triples at each step from 4 to 108. The usage of such temporal dynamics features was inspired by the decent performance of the "most-recent-5-months-benchmark" — if using the exact number of edits in just one period (the last 5 months) for prediction could work reasonably well, we should be able to achieve a better performance by using many more recent periods. The periods were chosen to be at exponentially increasing temporal scales, because we conjecture that the influence of an editing activity to the editor's future editing behaviour would be exponentially decaying along with the time distance away from now. The process of *exponential decay* occurs in numerous natural phenomena, and it has been widely used in temporal applications where it is desirable to gradually discount the history of past events. One reason for changing from doubling to tripling midway through is to include the special period of 12-months (i.e., one year) that has been used to define the "active" editors. The periods will be capped by the time scope of the given dataset (e.g., 106 for the additional dataset 'moredata') in case they are out of range.

We have also introduced a constant *drift* term (i.e., how much the average number of edits would change after 5 months) into the formula of making final predictions, which is a crude way to cover the global shift of target values along with time. Again, its value is estimated from the situation 5 months ago.

There are three datasets available to all contestants: (i) 'training' — the official dataset for training and testing that consists of 22326031 edits; (ii) 'validation' — the official dataset for validation that consists of 274820 edits; and (iii) 'moredata' is the additional dataset generously provided by Twan van Laarhoven[3] that consists of 5717049 edits.

Since we did not have access to the true labels (target values) of the dataset 'training' during the contest, we only used it to make the final submission, but conducted our experiments (for parameter tuning etc.) on the other two datasets 'validation' and 'moredata'. It is noteworthy that these two datasets 'validation'

---

[3] http://www.kaggle.com/c/wikichallenge/forums/t/719/more-training-data

and 'moredata' had been filtered to contain only active editors (who made at least one edit in the last one year period) in order to make them exhibit the same *survivorship bias* as the dataset 'training'. This might (partially) ensure that the experimental findings on the former two datasets could be transferred to the latter one.

We have only used Python (equipped with Numpy) to write small programs for analysing data and making predictions. The machine learning methods that we have tried for our regression task all come from two *open-source* Python modules: one is scikit-learn[4], and the other is OpenCV[5].

In our experiments, we first compare different machine learning methods (with their default parameter values) in terms of their prediction performances (RM-SLE). The methods being compared include: Ordinary Least Squares (OLS), Support Vector Machine (SVM), K Nearest Neighbours (KNN), Artificial Neural Network (ANN), and Gradient Boosted Trees (GBT). The experimental results are shown in Figure 3. Gradient Boosted Trees (GBT) [3, 4] clearly outperformed all the other machine learning methods on both datasets. GBT (aka GBM, MART, and TreeNet) represents a general and powerful machine learning method that builds an ensemble of *weak* tree learners in a greedy fashion. It evolved from the application of boosting to regression trees. The general idea is to compute a sequence of very simple trees, where each successive tree is built for the prediction residuals of all preceding trees on a randomly selected subsample of the full training dataset. Eventually a "weighted additive expansion" of those trees can produce an excellent fit of the predicted values to the observed values. It allows optimisation of any differentiable loss function. Here we just use the *squared error* for the reasons given earlier. The success of GBT in our task is probably attributable to (i) its ability to capture the complex nonlinear relationship between the target variable and the features, (ii) its insensitivity to different feature value ranges as well as outliers, and (iii) its resistance to overfitting via regularisation mechanisms such as shrinkage and subsampling [3, 4].

Then, we investigate how GBT's most important parameter *weak_count* — the number of weak tree learners — affects its prediction performance for our task. Tuning *weak_count* is our major means of controlling the model complexity to avoid underfitting or overfitting. The experimental results are shown in Figure 4. It seems that on big datasets like 'moredata', a higher value of *weak_count* (i.e., more weak tree learners) would be beneficial, but on small datasets like 'validation', it might increase the risk of overfitting.

Next, we demonstrate how the prediction performance changes when we use more and more periods to generate temporal dynamics features: we start from just the shortest period ($\frac{1}{16}$) and then each time we add the next longer period to the series. The experimental results are shown in Figure 5. It seems that making use of more periods for temporal dynamics features usually helps, but the pay-off gradually diminishes.

---

[4] http://scikit-learn.org/
[5] http://opencv.willowgarage.com/wiki/

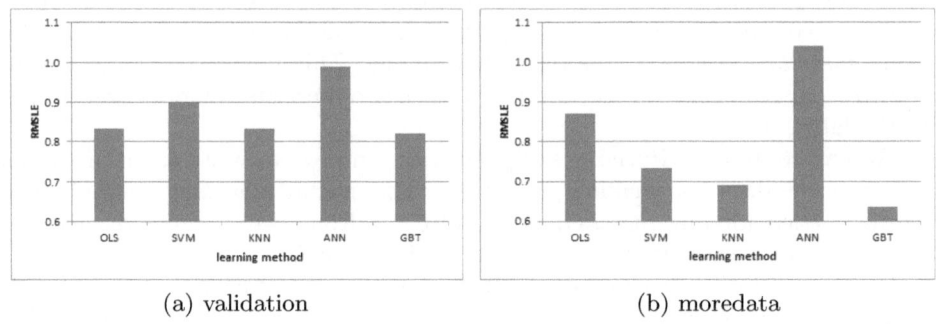

(a) validation   (b) moredata

Fig. 3. The prediction performances of different machine learning methods

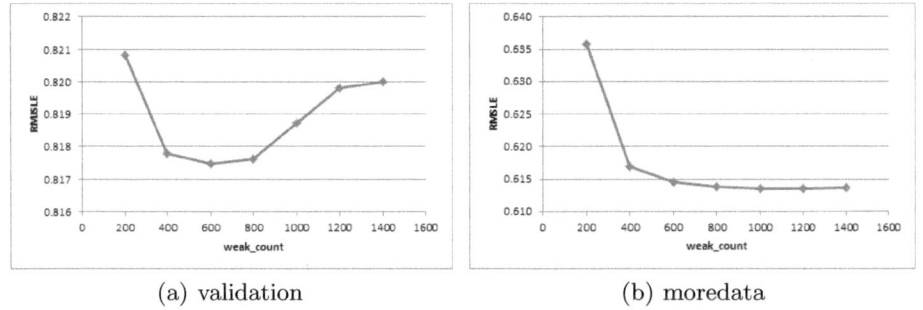

(a) validation   (b) moredata

Fig. 4. The prediction performances of GBT with different number of trees

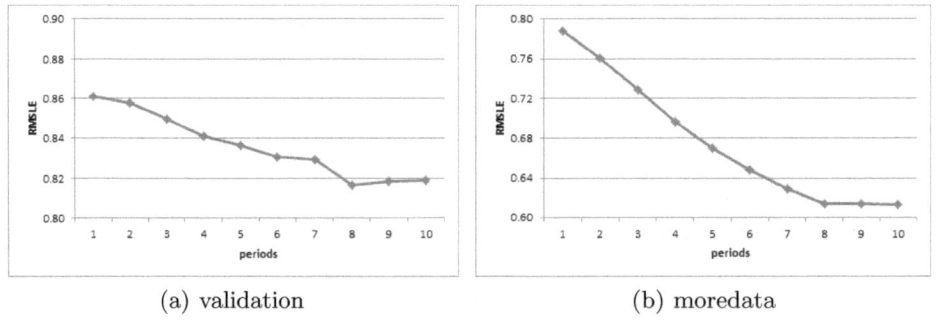

(a) validation   (b) moredata

Fig. 5. The prediction performances of GBT using different number of periods

Since 'moredata' is more similar than 'validation' to the official dataset 'training' in terms of the time scope and the number of editors, we applied the best working algorithm, GBT, with the optimal parameter setting on 'moredata' ($weak\_count = 1000$), to make the final submission based on 'training'. It got an RMSLE score of 0.862582 on the private leaderboard, which is roughly 41.7% better than WMF's baseline predictive model. The team of the first author, "zeditor", was finally ranked at the 3rd place among 96 teams. The algorithm of the top ranked team turns out to be useless, as it exploits a randomisation mistake of the dataset[6]. Therefore our approach is actually the 2nd best valid algorithm for the WikiChallenge. In comparison with the best performing valid algorithm which makes use of 206 features, our algorithm achieved very similar prediction accuracy with a much simpler model — we have only used 21 temporal dynamics features.

Our most important insight is that a Wikipedia editor's future behaviour can be largely determined by the temporal dynamics of his recent behaviour. We are a bit surprised that just temporal dynamics features (i.e., how the number of edits changes in recent periods, etc.) can go such a long way when we choose proper temporal scales and employ a powerful machine learning method. Human beings seem to be working and living in a more mechanical way than one might have thought.

Since such temporal dynamics features are actually independent of any semantics or knowledge about this specific problem, our approach could be easily generalised to other application domains, such as predicting the future supermarket spendings of shoppers, predicting the future hospital admissions of patients, and so on, based on historical behavioural data.

4   Correlation Analysis

Have we answered the question that we asked at the beginning of this paper? Yes and No.

On one hand, we have built a predictive model which can be used to identify those editors who are likely to become inactive, or in other words, who need special care and attention to be kept — if an editor is going to leave the Wikipedia community, there would probably be early signals in the temporal dynamics of his recent behaviour. On the other hand, that predictive model is pretty much a black box — it does not reveal the underlying reasons why editors become inactive, and therefore it cannot tell us how to encourage editors to remain active. For the ultimate purpose of Wikipedia's sustainable growth, we will need to investigate which character of an editor (his articles' category distribution, his relationship with other editors, etc.) and also which recent event happened to him (his articles being deleted, his revisions being reverted, unfair comments about his edits being received, etc.) could affect his behaviour.

After the WikiChallenge contest, WMF has kindly provided us the the true labels (target values) of the official dataset 'training', so we now have the complete knowledge about whether a particular editor has become inactive in the next 5 months, which enables us to look into the 'training' dataset to find possible factors relevant to an editor's departure from the community.

Intuitively, a Wikipedia editor's recent editing experience would be most informative about or have most impact upon his decision to stay or leave: if it

---

[6] http://www.kaggle.com/forums/t/980/wikipedia-participation-challenge-an-unfortunate-ending

is a good experience then he is likely to continue contributing; if it is a bad experience then he may stop contributing. Here we present the preliminary results of analysing the correlation between a Wikipedia editor's departure and the attributes of his last year edits.

The attributes that we examine for correlation analysis are listed as follows.

- namespace: the name space of the article (0: Main, 1: Talk, 2: User, 3: User Talk, 4: Wikipedia, and 5: Wikipedia Talk).
- category: a variable created as part of the Wikimedia Taxonomy Project that indicates if an article belongs to a special category (0: None, 1: Deletion, 2: Mediation, 3: Featured Lists, 4: Featured Pictures, 5: Arbitration, 6: Featured Topic, 7: Featured Portal, 8: Featured Article, 9: Featured Sounds, and 10: Good Article).
- redirect: 1 if a page redirects, 0 otherwise.
- revert: 1 if the revision was a revert, 0 otherwise.
- delta: the increase/decrease in number of characters compared with the previous revision of the article.
- cur_size: the current size in number of characters of a revision.

For each categorical attribute (namespace and category) with domain size m (i.e., m possible values), we create m binary indicator variables (one for each possible value), e.g., the variable $category_8$ would be 1 if the edit's corresponding article is in the 8-th category Featured Article, and 0 otherwise. For each numerical attribute (redirect, revert, delta, and cur_size), we simply create one numerical variable accordingly. So in the end we have 21 variables altogether. Then we calculate the average value for each such variable over the given editor's last year edits, e.g., $revert = 0.2$ means that 20% of his last year edits are revert actions. These average values, being independent from the editor's number of edits, can characterise his recent editing experience.

First, we find out the top 8 variables that are most correlated with a Wikipedia editor's future number of edits $\log(1 + a_i)$, as shown in Table 1. The strength of correlation is measured by Maximal Information Coefficient (MIC) — a measure of dependence for two-variable relationships captures a wide range of associations both functional and not, and for functional relationships provides a score that roughly equals the coefficient of determination (R^2) of the data relative to the regression function [10]. Here we choose MIC as it can measure both linear and non-linear correlations. The value of MIC is between 0 (none correlation) and 1 (perfect correlation). To know whether a correlation is positive or negative, we also calculate the Pearson product-moment correlation coefficient which ranges from -1 to $+1$, though it can only measure linear correlations. The most positively correlated variable turns out to be $revert$ (MIC = 0.1543), which implies that if an editor has made many revert edits he is likely to stay active. The most negatively correlated variable turns out to be $category_1$ Deletion (MIC = 0.1378), which implies that if an editor has many edits on articles nominated (probably by others) for deletion he is likely to leave the community. Similarly, we can make sense of the other highly correlated variables as well.

Table 1. The correlation between a Wikipedia editor's future number of edits and the attributes of his last year edits

variable	MIC	Pearson
revert	0.1543	+0.0864
namespace$_0$	0.1540	+0.0188
category$_1$	0.1378	−0.0109
namespace$_1$	0.1357	+0.0542
category$_0$	0.1355	+0.0222
redirect	0.1314	+0.0364
namespace$_3$	0.1207	+0.0030
namespace$_2$	0.1068	−0.0672
namespace$_4$	0.0955	−0.0103

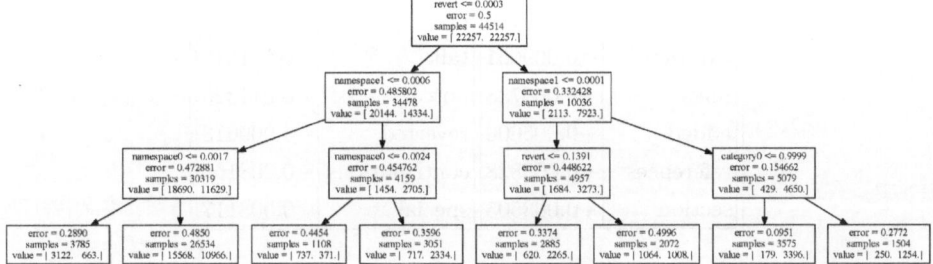

Fig. 6. The decison tree (top 3 levels) that classifies editors into **leave** or **stay**

Next, we go beyond evaluating individual variables and explore the influence of their interaction over a Wikipedia editor's departure dynamics. For this purpose, we learn a decision tree that classifies editors into 2 classes **leave** or **stay** from the 'training' dataset using the CART algorithm [1]. Since our aim now is not to make predictions but obtain interpretable knowledge, we do not apply any ensemble technique like GBT (as in Section 3): although the accuracy would be somewhat lower, the generated model would be much more readable. Figure 6 shows the top-three levels of the constructed decision tree. Such a decision tree could be easily turned into readable rules. For example, one rule we could find from the tree (more precisely the path from the root node to the second rightmost level-3 node) is as follows: editors who have a revert percentage larger than 0.03%, `Talk` namespace article percentage larger than 0.01%, and `None` category article percentage within 99.99% are very likely to stay in the Wikipedia community (179 **leave** vs 23396 **stay**).

Furthermore, we have also checked the word distributions of the article titles and edit comments. The difference between word usage frequencies in those two different class of editors — **leave** and **stay** — would be able to tell us something about their interested topics and possibly behavioural patterns as

Table 2. The difference between word usage in leaving and staying editors

title w	$p_0 - p_1$	title w	$p_0 - p_1$
school	+0.003506	wikiproject	−0.003127
italy	+0.002733	album	−0.002155
sandbox	+0.001974	season	−0.001792
high	+0.001867	station	−0.001736
episodes	+0.001621	archive	−0.001677
university	+0.001539	featured	−0.001545
iko	+0.001245	candidates	−0.001540
hot	+0.001222	administrator	−0.001498
comment w	$p_0 - p_1$	comment w	$p_0 - p_1$
aes	+0.011700	awb	−0.022537
page	+0.010565	category	−0.014285
created	+0.010170	project	−0.012324
external	+0.009651	talk	−0.011513
links	+0.006788	hotcat	−0.011326
added	+0.005600	reverted	−0.009613
references	+0.003828	contributions	−0.008165
section	+0.003505	special	−0.008117

well. Table 2 shows the words with maximum difference in $p_0(w) = \Pr(w|\textbf{leave})$ and $p_1(w) = \Pr(w|\textbf{stay})$ where w represents a word (with stopwords excluded). It can be seen that leaving editors ($p_0 - p_1 > 0$) tend to edit articles about "high schools" or "universities", "create" new "pages" or "sections", "add" "external" "links", pursue "hot" topics, leave the edit summaries blank to be filled by automatic edit summaries "aes", and experiment with "sandbox" pages; while staying editors ($p_0 - p_1 < 0$) tend to edit articles about music "albums" or "stations", get involved in "wikiproject" teams or other "projects", "archive" an article or make it "featured", 'nominate "candidates" for "administrators", use "awb" and "hotcat" tools to put articles into right "categories", "revert" (probably messy or spamming) edits, and "talk" about the "contribution" or "special" features of articles. Roughly speaking, editors who have been more active in the Wikipedia community are more likely to stay active. Therefore organising editors into "wikiproject" teams, teaching them edit assistance tools like "awb" and "hotcat", encouraging them to contribute to articles with long-term value, and letting them take part in the "administrator" election might be good ideas for retaining Wikipedia editors.

5   Related Work

The global slowdown of Wikipedia's growth rate (both in the number of editors and the number of edits per month) has been investigated by Suh et al. [11]. It is found that medium-frequency editors now cover a lower percentage of the total population while high-frequency editors continue to increase the number of their edits. Moreover, there are increased patterns of conflict and dominance (e.g., greater resistance to new edits in particular those from occasional editors), which may be the consequence of the increasingly limited opportunities in making novel contributions. An ecology inspired population model that assumes a resource limitation has been proposed to characterise the overall growth of Wikipedia. In this paper, we approach the problem from several different angles and arrive at conclusions which complement theirs.

The technique of survival analysis [5,6] has been shown to be very useful in analysing information systems. For example, the estimated lifetime of a webpage could reflect its desirability [9]. It has recently been applied to a couple of studies on Wikipedia editors' behaviour. In one work [7,8], the survival function for all Wikipedia editors is empirically estimated, but no parametric model has been produced. In another work [2], the survival function for all Wikipedia editors is fit by a mixture of two truncated lognormal distributions, but our work has revealed that the survival function for customary editors is better described by a Weibull distribution. Furthermore, those previous studies have not looked into the hazard function for Wikipedia editors.

The general problem of user participation has been investigated to various degrees in mobile telecom networks, social networks, social media, and so on. However, we have not seen any published paper about predictive modelling of Wikipedia edit number or the WikiChallenge, and the related correlation analysis of underlying reasons.

6   Conclusions

The major contributions of this paper are as follows.

– We have used the technique of survival analysis to investigate how long Wikipedia editors remain active in editing. Our results show that for customary Wikipedia editors, the survival function can be well described by a Weibull distribution (with the median lifetime of about 53 days); there are two critical phases (0-2 weeks and 8-20 weeks) when the hazard rate of becoming inactive increases; more active editors tend to keep active in editing for longer time.
– We have built a predictive model in a self-supervised learning framework using GBT with a small number of temporal dynamics features only, which can accurately forecast the number of edits a Wikipedia editor will make in the next 5 months. The best submission to the WikiChallenge from our team, "zeditor", achieved 41.7% improvement over WMF's baseline predictive model.

- We have demonstrated that the attributes of a Wikipedia editor's last year edits, particularly the percentage of reverts and the distribution of articles' namespaces as well as categories, are closely related to his departure. Some ideas about how to retain Wikipedia have also bee proposed by contrasting the word distributions of leaving editors and staying editors.

Acknowledgement. We thank WMF and Kaggle for their wonderful job in organising the Wikipedia Participation Challenge. We are grateful to Twan van Laarhoven (Radboud) for creating and sharing the additional dataset 'moredata'. Ian Soboroff (NIST) has provided insightful comments on a previous version of this paper. We also appreciate the anonymous reviewers' helpful comments.

References

1. Breiman, L., Friedman, J., Stone, C., Olshen, R.: Classification and Regression Trees. Wadsworth, Belmont (1984)
2. Ciampaglia, G.L., Vancheri, A.: Empirical analysis of user participation in online communities: the case of Wikipedia. In: Proceedings of the 4th International Conference on Weblogs and Social Media (ICWSM), Washington, DC, USA, pp. 219–222 (2010)
3. Friedman, J.: Greedy function approximation: A gradient boosting machine. Tech. rep., IMS 1999 Reitz Lecture (February 1999)
4. Friedman, J.: Stochastic gradient boosting. Tech. rep., Stanford University (March 1999)
5. Hosmer, D.W., Lemeshow, S., May, S.: Applied Survival Analysis: Regression Modeling of Time to Event Data, 2nd edn. Wiley-Interscience (2008)
6. Kleinbaum, D., Klein, M.: Survival Analysis: A Self-Learning Text, 3rd edn. Springer (2011)
7. Ortega, F.: Wikipedia: A Quantitative Analysis. Ph.D. thesis, Universidad Rey Juan Carlos (2009)
8. Ortega, F., Izquierdo-Cortazar, D.: Survival analysis in open development projects. In: Proceedings of the 2nd International Workshop on Emerging Trends in Free/Libre/Open Source Software Research and Development (FLOSS), Vancouver, Canada, pp. 7–12 (2009)
9. Pitkow, J.E., Pirolli, P.: Life, death, and lawfulness on the electronic frontier. In: CHI, Atlanta, GA, USA, pp. 383–390 (1997)
10. Reshef, D., Reshef, Y., Finucane, H., Grossman, S., McVean, G., Turnbaugh, P., Lander, E., Mitzenmacher, M., Sabeti, P.: Detecting novel associations in large data sets. Science 334(6062), 1518–1524 (2011)
11. Suh, B., Convertino, G., Chi, E.H., Pirolli, P.: The singularity is not near: Slowing growth of Wikipedia. In: Proceedings of the 2009 International Symposium on Wikis (WikiSym), Orlando, FL, USA (2009)

Cross-Modal Information Retrieval – A Case Study on Chinese Wikipedia

Yonghui Cong[1], Zengchang Qin[1,*], Jing Yu[1], and Tao Wan[2]

[1] Intelligent Computing and Machine Learning Lab
School of ASEE, Beihang University, Beijing, China
zcqin@buaa.edu.cn
[2] Department of Biomedical Engineering, Rutgers University, USA
{yonghuicong,jing.emy.yu}@gmail.com, tao.wan.wan@rutgers.edu

Abstract. Probability models have been used in cross-modal multimedia information retrieval recently by building conjunctive models bridging the text and image components. Previous studies have shown that cross-modal information retrieval system using the topic correlation model (TCM) outperforms state-of-the-art models in English corpus. In this paper, we will focus on the Chinese language, which is different from western languages composed by alphabets. Words and characters will be chosen as the basic structural units of Chinese, respectively. We also set up a test database, named Ch-Wikipedia, in which documents with paired image and text are extracted from Chinese website of Wikipedia. We investigate the problems of retrieving texts (ranked by semantic closeness) given an image query, and vice versa. The capabilities of the TCM model is verified by experiments across the Ch-Wikipedia dataset.

Keywords: Cross-modal information retrieval, topic correlation model (TCM), word-based topics, character-based topics, Ch-Wikipedia.

1 Introduction

The amount of multimedia information on the Internet is growing by an explosive rate in recent years. Much attention has been attracted to build more efficient search engines for multi-modal information including music, videos, texts, images and so on. As we know that, most of popular information retrieval systems we use at present such as Google and Baidu[1] are still uni-modal. Relations between different modalities are not well modeled. Captions or category labels are used to build information manually, which are both time-consuming and laborious. Many techniques have been proposed aimed at bridging information in different modalities [4,6,8,9,11,14]. It has been showed that multi-modal retrieval systems have made significant progress compared to uni-modal approaches [11,12,14].

Previous research in [8] maps the information in different modalities into a higher dimensional semantic space where the similarity is measured. In recent

---

[*] Corresponding author.

[1] Websites: www.google.com, www.baidu.com

S. Zhou, S. Zhang, and G. Karypis (Eds.): ADMA 2012, LNAI 7713, pp. 15–26, 2012.
© Springer-Verlag Berlin Heidelberg 2012

studies, a new probabilistic model was proposed in [14] to investigate mid-level feature correlation between texts and images. In the new model, probabilistic correlations between the mid-level "topics" are considered. Given a query image (text), a SVM classifier can be applied to compute the probability distribution over categories.

Most of the models proposed for cross-modal information retrieval are focused on English corpus and the techniques used in studying English can be easily extended to other alphabetic languages. For example, an English text can be regarded as a collection of words, which are the basic structural units in the majority of western language. However, the basic semantic units in Chinese language are not necessarily to be the Chinese words [13,17]. In this paper, we extend the TCM model to study Chinese language by employing two language models for semantic modeling. Zhao *et al.* [17] first show that the computational evidence that character-based topic models outperform the word-based topic models. The experimental results for both two models will be given in the following sections.

This paper is structured as follows: we introduce topic representation for multimedia contents in Section 2. In Section 3, topic correlation model for cross-modal information retrieval is described in details. We also create a database extracted from the Chinese Wikipedia. In Section 4, a series of experiments are conducted based on this database and the results are given and analyzed in details. The conclusions and the future work are given in the end.

2   Topic Representation

It is a significant issue to represent multimedia information by appropriate features. Low-level features have many limitations such as colors, textures for images and keywords, captions for texts. When considering the multi-modal documents such as Wiki articles, each article contains the semantically similar contents in different modalities (e.g., images and texts). However, the semantic relations cannot be well captured by low level features. Mid-level features such as visual words in the bag-of-features model and latent topics in the topic models [5] can be used to model semantic correlations between contents in different modalities. For example, given a corpus of Wiki articles with paired text and image, we may find that the texts with words like 'sky, blue, sunny' may somehow occurs more often with images containing blue colors. These latent semantic relations can be modeled by correlation between the topics of words and topics of image features (visual words in the bag-of features model). In this way, the semantic gap can be reduced by using mid-level features to model different content modalities.

2.1   Bag-of-Features Model

In this paper, scale invariant feature transform (SIFT) features are used to model the image components in a document. There have been many other low-level image features such as HOG (histograms of oriented gradients), LBP (local binary pattern). The reason for choosing SIFT is for its effectiveness and stability. SIFT features are invariant to rotation, scaling, translation and small distortions [15]. It has been empirically proven to be one of the most robust among the local invariant feature descriptors with respect to different geometrical changes [10]. The bag-of-features (BoF) model has been getting popular recently. It has two key concepts: local features and codebook. The essential aspect of local feature concept is to extract global image descriptors and represent images as a collection of local properties calculated from a set of small sub-images called patches. Codebook is a way that an image can be represented by a set of local features. The idea is to cluster the feature descriptors of all patches based on a given cluster number and each cluster represents a "visual word" that will be used to form the codebook. After obtaining the codebook, each image can be represented by the BoF histograms of the visual vocabulary of the codebook. The similarity of images can be measured by comparing between the BoF histograms. The Bag-of-features model has been well studied as one of the most effective approaches in image classification [5,15].

2.2 Word-Based Topics and Character-Based Topics

In text modeling, the bag-of-words (BoW) assumption is also well used. For example, by using topic models such as latent Dirichelet allocation (LDA) [1], a text is represented by a mixture of latent topics, and each topic is represented by a probability distribution over vocabulary. Such word topics can be used to model text components in a document [2] or used in other natural language processing applications such as Q&A [7]. Most research on topic models only concern English language. Different from western languages such as English, the morphology of Chinese language is more complex. Characters, instead of words, are the basic structure unit for Chinese language. This has been both discussed in Chinese linguistics [13] and testified using computational model [17,18]. In this research, the word-based and character-based topic models are tested, respectively. Experiments are conducted to show the differences between segmenting Chinese in words and characters. The process of modeling for the text and image components in a Wikipedia document is schematically showed in Fig. 1.

The representation of both images and texts here are not using the low-level features directly. We construct the mid-level representations for modeling the content in a two-level hierarchical structure to make them more robust and abstract. In this paper, we may use the term "topics of features" to represent the visual words in order to highlight the similarity between BoF and the topic model. Because in this research we are interested in the correlations between these topics of different modalities.

3 Topic Correlation Model

Large and heterogeneous collections of images are usually accompanied with noisy texts. The cross-modal retrieval systems are developed for users to be

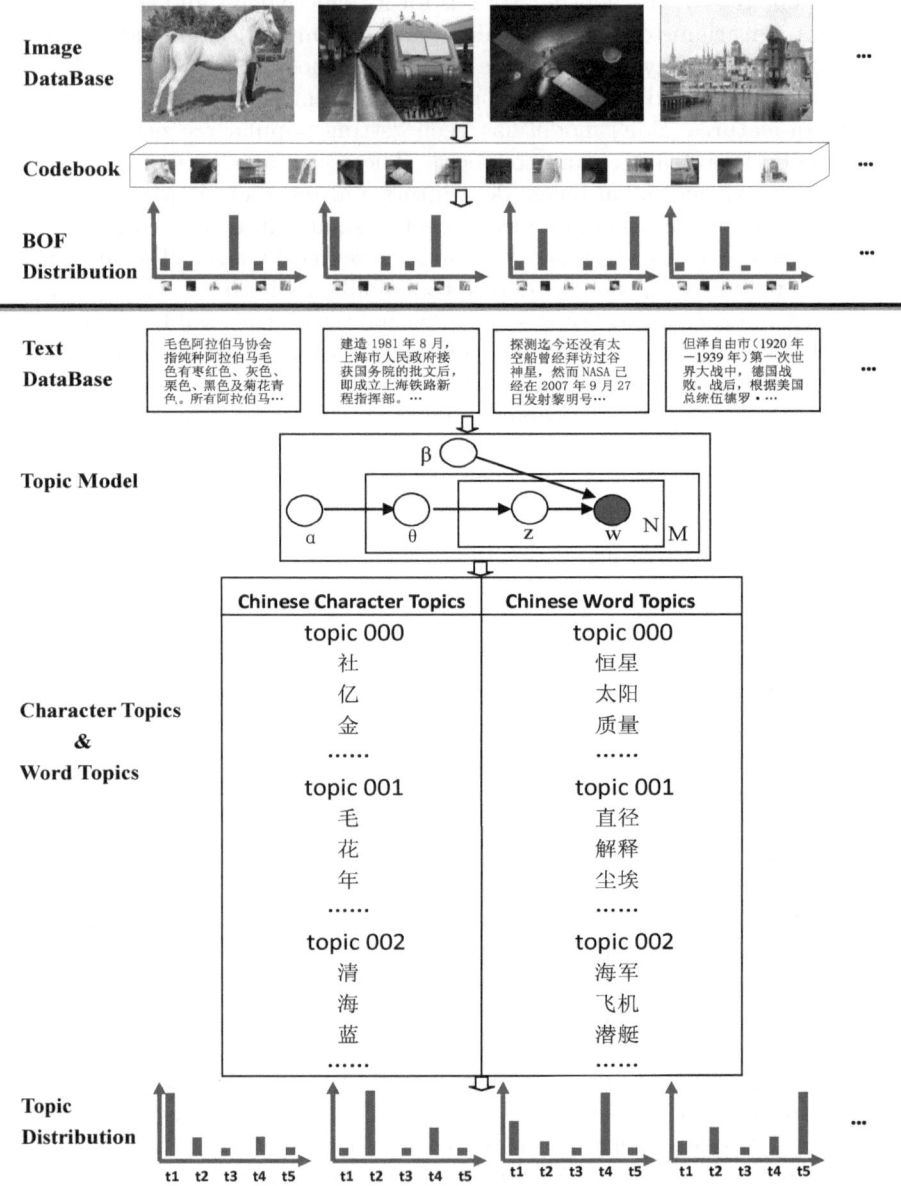

Fig. 1. Process of modeling for the text and image components for Chinese Wikipedia articles. Two modalities (image and text) are separated by a straight line. The upper part shows the procedure of extracting SIFT features from images and build distribution of BOF. The lower part shows how the topic distribution over words is computed using LDA. Chinese texts are modeled by LDAs based on words and characters, respectively. The topic list has showed that text component can be represented by a mixture of latent topics and each topic is a distribution over the vocabulary (either words or characters).

able to browse and search these collections more efficiently. Rasiwasia *et al.* [8] has demonstrated the benefits of joint model for text and image components by mapping both into one high dimensional semantic space. In recent years, the statistical correlation methods have attracted much attention. Probability models for matching words and pictures has been used to segment images with associated text [3].

The underlying relation between topics may reflect the correlation between image and text components. In [14], images are represented by distributions on topics of features and texts are represented by distributions on topics of words. In the topic correlation model, naive probabilistic correlations between image and text features are considered. Given a collection of documents is defined by $\mathbf{D} = [D_1, D_2, ..., D_K]$. We assume that there is only one-to-one mapping between image and text is acquiescent for research purpose, e.g, $D_k = [I_k, TX_k]$. where $D_k \in \mathbf{D}$, I_k and TX_k is the related image and related text in D_k respectively. Given a query I_q (or TX_q), how to find a document $D_j \in \mathbf{D}$ that has the most semantically related texts (images).

We define that $\mathbf{V} = [V_1, V_2, ..., V_M]$ is the set of visual words in the codebook where M is the codebook size. $\mathbf{T} = [T_1, T_2, ..., T_N]$ is the set of topics and N is a predefined number of topics. For a visual word V_i and a topic T_j, the underlying probabilistic relation can be computed on the training document $\mathbf{D}$,

$$P(V_i|T_j) = \sum_{k=1}^{K} P(V_i|I_k)P(I_k|TX_k)P(TX_k|T_j) \qquad (1)$$

where $P(V_i|I_k)$ is the BoF distribution over V_i of the image I_k. Since the image I_k and the text TX_k appear in the same document D_k, then

$$P(I_k|TX_k) = P(TX_k|I_k) = 1$$

For the third term $P(TX_k|T_j)$, according to the Bayes theorem, we can obtain

$$P(TX_k|T_j) = \frac{P(T_j|TX_k)P(TX_k)}{\sum_{k=1}^{K} P(T_j|TX_k)P(TX_k)} \qquad (2)$$

where $P(TX_k)$ is the prior probability of the text component in document D_k. Without any information on each document, we use the uniform distribution as the prior according to the principle of maximum entropy. Formally,

$$P(TX_k) = P(I_k) = \frac{1}{K} \qquad (3)$$

Similarly, the likelihood of topic T_j given a visual word V_i is evaluated by

$$P(T_j|V_i) = \sum_{k} P(T_j|TX_k)P(TX_k|I_k)P(I_k|V_i) \qquad (4)$$

where $P(T_j|TX_k)$ is the topic distribution over T_j of the text TX_k. According to the Bayes theorem,

$$P(I_k|V_i) = \frac{P(V_i|I_k)P(I_k)}{\sum_{k=1}^{K} P(V_i|I_k)P(I_k)} \qquad (5)$$

When the category information is available, this information can be applied to find better matching patterns between image and text components. In this framework, SVM is used as such a classifier [14]. Given a query text (image), text classifier (image classifier) is applied to compute its probability distribution over categories. C_i denotes the ith category given a set of categories $\mathbf{C} = [C_1, C_2, ..., C_n]$. Then the probability of the image I_k given a query text TX_q is computed by summing up the conditional probabilities across all the categories. Formally,

$$P(I_k|TX_q) = \sum_i P(I_k|C_i)P(C_i|TX_q) \qquad (6)$$

Based on the Bayes theorem, we can obtain

$$P(I_k|C_i) = \frac{P(C_i|I_k)P(I_k)}{\sum_k P(C_i|I_k)P(I_k)} \qquad (7)$$

Similarly, given an image query, the probability of a text component is computed by

$$P(TX_k|I_q) = \sum_i P(TX_k|C_i)P(C_i|I_q) \qquad (8)$$

where

$$P(TX_k|C_i) = \frac{P(C_i|TX_k)P(TX_k)}{\sum_k P(C_i|TX_k)P(TX_k)} \qquad (9)$$

The values of $P(C_i|I_k)$ and $P(C_i|TX_k)$ can be obtained by the predictions from trained SVM classifiers. And these two values are not necessarily the same as the classifiers are trained individually based on the contents of different modalities.

Table 1. Summary of the Ch-Wikipedia Dataset, the articles are extracted from Chinese Wikipedia website

Category	Training	Test	Total
Culture	285	71	356
Biology & Medicine	327	82	409
Natural Science	279	70	349
Geography	374	93	467
History	424	106	530
Traffic	156	39	195
Warfare & Military	206	52	258
Scholar & Occupational Figures	145	36	181
Political & Military Figures	286	72	358

4   Experimental Studies

In this section, we conduct a series of experiments to demonstrate the effectiveness of the TCM on a new database in Chinese. In these experiments, 20% of the documents were randomly chosen as the test set and the remaining as training.

4.1 Database and Preprocessing

Since there is no well accepted image-text paired Chinese corpus for cross-modal retrieval research. We create a database by our own, which is named Ch-Wikipedia Dataset[2]. It consists 3103 documents of paired texts and images from 9 categories listed in Table 1. The documents in this corpus are crawled from a collection of articles in Chinese Wikipedia, which is one of the biggest Internet information websites in Chinese language. There are 20 classes in original corpus that cover literature, media, sports, politics and other topics. Each article is split into some parts by section heading. The texts containing less than 100 Chinese characters will be ignored in our research. The first image associated with a text is chosen as its related image and other texts without images are removed. Topics of some classes are similar, such as "humanities" and "culture & society", which can be classified into one bigger category "culture". On the other hand, some classes with less than 150 documents will be abandoned as well if they can not be merged to a larger category. We list the final categories in Table 1 after merging similar ones manually.

Table 2. Accuracy rate of SVM classifiers with different kernels

Database	Linear	Polynomial	RBF	Sigmoid
Training images	0.356	0.203	0.316	0.303
Test images	0.309	0.198	0.308	310
Word-based training texts	0.642	0.548	0.663	0.627
Word-based test texts	0.654	0.385	0.665	0.668
Character-based training texts	0.820	0.639	0.811	0.776
Character-based test texts	0.712	0.533	0.721	0.704
Average	**0.582**	0.418	0.576	0.564

A score function is introduced to evaluate the likelihood of an image I_k given a text query TX_q by $S(I_k) = P(I_k|TX_q)$ and used to get a ranked list of returned data. We test the probabilistic correlation model on Ch-wikipedia dataset for the following tasks: (1) obtaining a ranked list of texts from the training database given a query image, (2) obtaining a ranked list of images from the training database given a query text. The mean average precision (MAP) is applied to measure the performance[3]. The word-based and character-based topic models are trained separately [17]. When we do the Chinese character-based topic modeling, we remove the rare characters that appear less than 3 times across the whole corpus and the terms appearing in over 50% of the documents which we consider as stop words [17]. When modeling for word-based topics, we first get a "stopwords list" from network[4], and ignore them for building the wordlist.

---

[2] http://icmll.buaa.edu.cn/zh_wikipedia
[3] http://en.wikipedia.org/wiki/Information_retrieval
[4] The stopword list of Chinese language can be downloaded from:
   http://www.byywee.com/page/M0/S639/639550.html

Table 3. The MAP value with different topic numbers for TCM. The best performance is highlighted.

Topic Number	Topic correlation model		
	Word-based	Character-based	Average
10	0.247	0.262	0.255
50	0.266	0.275	0.271
100	0.263	0.292	0.278
200	0.261	0.30	0.281
300	0.259	0.304	0.282
500	0.261	0.306	**0.284**
1000	0.264	0.301	0.282

Words appear less than 3 times through all documents are removed as well. After the above preprocessing we get 21240 unique Chinese words and 3419 unique Chinese characters. We set the topic number to 100 for the LDA and the same visual words number for the BoF model. It's showed in Table 2 that the SVM classifiers with linear kernel have the highest accuracy rate in average. So that in the following experiments, we use the SVM with linear kernel for category prediction.

4.2   Experimental Results

We vary the topic number and the empirical evaluations are shown in Fig. 2. The exact accuracy values are shown in Table 3. When the number of topics was set to 500 for TCM, we get the best performance. Thus, appropriate numbers for topics will boost the accuracy. The MAP for the TCM based on Chinese words and characters are showed in Fig. 3. Agreed with previous research in [16,17], the character-based topic model has a better performance than word-based model for any given topic number. Along with the increase of topic number from 10 to 1000, the MAP of the the word-based model has a tendency of increasing with tiny fluctuation. As for character-based model, it keeps increasing and the MAP is up to 31%, improved by 5% in average compared with the word-based. One possible reason is that the size of word vocabulary is much larger than the size of character vocabulary.

Fig. 4 and 5 show two examples of cross-modal retrieval. For Fig. 4, the given image on the top left is a picture of train and railway. The top three outputs of the TCM are texts about airport, railway system, and airport hotel which are all semantically close to travel and train. For Fig. 5, the system is given a text describe nepenthe (the corresponding image is also shown at below). The top three outputs are all nepenthe images just in different kinds.

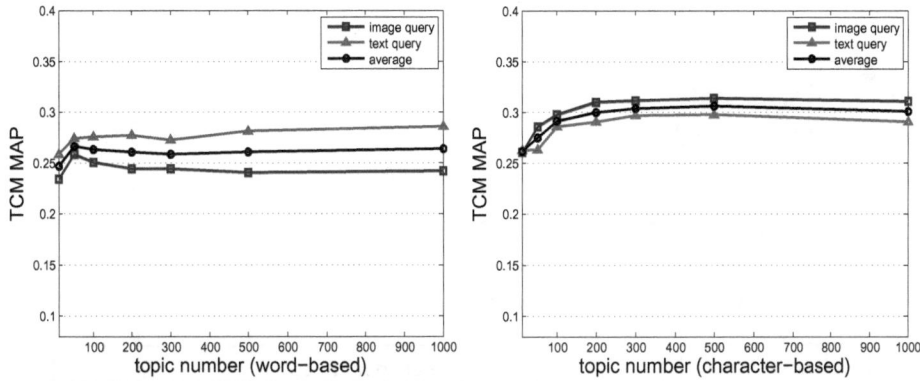

Fig. 2. The performance with increasing topic numbers. It is obvious that the model trained by character-based topic model achieved better performance than word-based one. The average retrieval accuracy for the character-based TCM model is about 31%, improved by 5% compared to the word-based model.

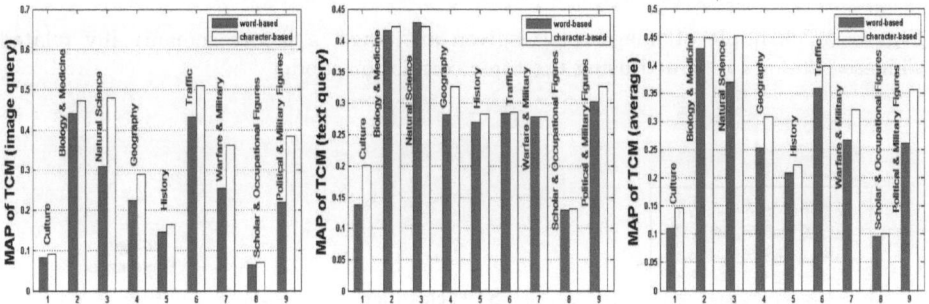

Fig. 3. Comparisons of retrieval results in each category in MAP. From the left to the right: (1) results of image retrieval; (2) results of text retrieval; (3) average results of retrieval.

Since there is no previous research about cross-modal information retrieval in Chinese language, so we cannot compare our results directly to other retrieval systems. However, pervious research [14] has shown that the TCM outperforms the other state-of-the-art cross-modal retrieval models on English Wikipedia corpus [9]. The average accuracy is up to 27%.

However, what should not be ignored is that the larger the topic number, the more time is spent to train topic distribution. The computational time is doubled when the given number of topics are doubled. But the performance can only be improved slightly. Thus, considering the time complexity, we may not like to set the number of topics too large and we can still obtain fair performance.

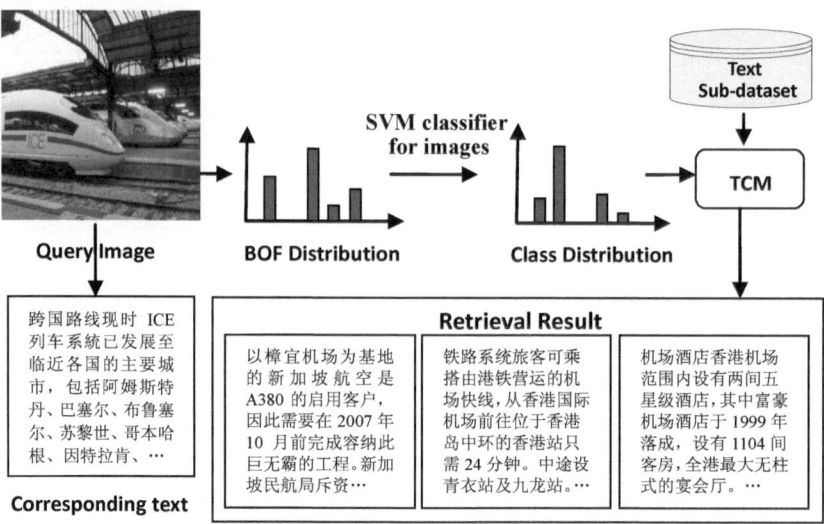

Fig. 4. An example of image query: given an image, a list of semantically related Chinese texts are returned using the topic correlation model

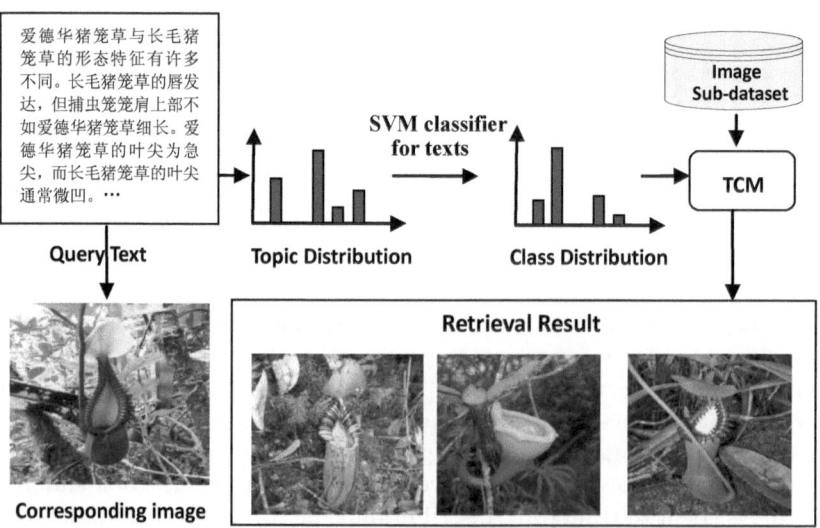

Fig. 5. An example of text query: give the description of nepenthe in Chinese, a list of images related to nepenthe are returned using the topic correlation model

5 Conclusions and Future Work

In this paper, we use a probabilistic model to study the correlations between mid-level features in different modalities. Topic correlation model is used for cross-modal information retrieval on a Chinese database Ch-Wikipedia. Comprehensive experimental results are presented to show the effectiveness. We also use the word-based and character-based topic model for text modeling. Empirical results agree with previous research that the character-based model outperforms the word-based model. We have achieved a good performance for cross-modal information retrieval in Chinese language.

In this research, the topic number is fixed for both image and text components. It is not necessarily true. The optimal topic numbers could be depend on image and text properties in the training set. How to independently find appropriate number of topics for image and text may be an interesting research problem. We can also explore more in Chinese language modeling to boost the accuracy of the TCM. For example, we can consider the word-character relations to build more accurate language model [18]. Other future work can be focused on weakening the noises when we build the correlation between different modalities.

Acknowledgments. This work is partially funded by the NCET Program of MOE, the SRF for ROCS, the Fundamental Research Funds for the Central Universities and Graduate Innovative Practice Fund of BUAA.

References

1. Blei, D.M., Ng, A., Jordan, M.: Latent Dirichlet allocation. Journal of Machine Learning Research 3, 993–1022 (2003)
2. Blei, D.M., Lafferty, J.D.: Topic models. Chapman & Hall/CRC Data Mining and Knowledge Discovery Series (2009)
3. Barnard, K., Duygulu, P., Forsyth, D., Freitas, N., Blei, D., Jordan, M.: Matching words and pictures. Journal of Machine Learning Research 3, 1107–1135 (2003)
4. Jeon, J., Lavreko, V., Manmatha, R.: Automatic image annotation and retrieval using cross-media relevance models. In: ACM SIGIR Conf. Research and Development in Information Retrieval, New York, pp. 119–126 (2003)
5. Jiang, Y., Ngo, C., Yang, J.: Towards optimal Bag-of-features for object categorization and semantic video retrieval. In: CIVR, pp. 494–501 (2007)
6. Metzler, D., Manmatha, R.: An inference network approach to image retrieval. In: Image and Video Retrieval, pp. 42–50 (2005)
7. Qin, Z., Thint, M., Huang, Z.: Ranking Answers by Hierarchical Topic Models. In: Chien, B.-C., Hong, T.-P., Chen, S.-M., Ali, M. (eds.) IEA/AIE 2009. LNCS, vol. 5579, pp. 103–112. Springer, Heidelberg (2009)
8. Rasiwasia, N., Pereira, J.C., Coviello, E., Doyle, G., Lanckriet, G.R.G., Levy, R., Vasconcelos, N.: A new approach to cross-modal multimedia retrieval. In: ACM Multimedia (MM), pp. 251–260 (2010)
9. Rasiwasia, N., Moreno, P., Vasconcelos, N.: Bridging the gap: Query by semantic example. IEEE Transactions on Multimedia 9(5), 923–938 (2007)

10. Schmid, C., Mikolajczyk, K.: A performance evaluation of local descriptors. In: ICPR, vol. 2, pp. 257–263 (2003)
11. Snoek, C., Worring, M.: Multimodal video indexing: A review of the state-of-the-art. Multimedia Tools and Applications 25(1), 5–35 (2005)
12. Westerveld, T.: Probabilistic multimedia retrieval. ACM 25, 438 (2002)
13. Xu, T.Q.: Fundamental structural principles of Chinese semantic syntax in terms of Chinese Characters. Applied Linguistics 1, 3–13 (2001) (in Chinese)
14. Yu, J., Cong, Y., Qin, Z., Wan, T.: Cross-modal topic correlations for multimedia retrieval. To appear in ICPR (2012)
15. Yuan, X., Yu, J., Qin, Z., Wan, T.: A SIFT-LBP image retrieval model based on bag-of features. In: International Conference on Image Processing (ICIP), pp. 1061–1064 (2011)
16. Zhang, Y., Qin, Z.: A topic model of observing Chinese characters. In: Proceedings of the International Conference on Intelligent Human-Machine Systems and Cybernetics (IHMSC), vol. 2, pp. 7–10 (2010)
17. Zhao, Q., Qin, Z., Wan, T.: What Is the Basic Semantic Unit of Chinese Language? A Computational Approach Based on Topic Models. In: Kanazawa, M., Kornai, A., Kracht, M., Seki, H. (eds.) MOL 12. LNCS, vol. 6878, pp. 143–157. Springer, Heidelberg (2011)
18. Zhao, Q., Qin, Z., Wan, T.: Topic Modeling of Chinese Language Using Character-Word Relations. In: Lu, B.-L., Zhang, L., Kwok, J. (eds.) ICONIP 2011, Part III. LNCS, vol. 7064, pp. 139–147. Springer, Heidelberg (2011)

Unsupervised Learning Chinese Sentiment Lexicon from Massive Microblog Data*

Shi Feng[1,2], Lin Wang[1], Weili Xu[1], Daling Wang[1,2], and Ge Yu[1,2]

[1] Northeastern University, China
{fengshi,wangdaling,yuge}@ise.neu.edu.cn
[2] Key Laboratory of Medical Image Computing (Northeastern University), Ministry
of Education, Shenyang 110819, P.R. China

Abstract. Analyzing people's feelings and emotions in social media has
become a major concern for both academic researchers and commercial
companies. The sentiment lexicon plays a crucial role in the most sentiment analysis applications. However, existing thesaurus based lexicon
building methods suffer from the coverage problems when faced with
the new words and new meanings in social media. Nowadays, millions
of users share their opinions on different aspects of life everyday in microblogs. In this paper, a novel method based on occurrence probability
with emoticons is presented to learn the candidate sentiment words from
the massive microblog data and the accuracy of the learned lexicon is
further improved by using the whole microblog space as the corpus. Extensive experiments were conducted on real world datasets with different
topics. The results show that the proposed method is able to extract the
emerging words, and learned lexicon outperforms two well-known Chinese lexicons in classifying the sentiments in microblogs.

1 Introduction

As more and more people are willing to publish their attitudes and feelings in
Web 2.0 based social media, how to provide an efficient way to analyze users'
sentiment has become a major concern for both academic researchers and commercial companies. In sentiment analysis, one fundamental problem is to recognize whether a given word expresses positive or negative meaning. Although
machine learning based algorithms have become commonplace in the related
literatures [1][2], the sentiment lexicon still plays an important role for many
sentiment analysis tasks.

* Project supported by the State Key Development Program for Basic Research of
China (Grant No. 2011CB302200-G), State Key Program of National Natural Science of China (Grant No. 61033007), National Natural Science Foundation of China
(Grant No. 61100026, 60973019), and the Fundamental Research Funds for the Central Universities (N100704001).

S. Zhou, S. Zhang, and G. Karypis (Eds.): ADMA 2012, LNAI 7713, pp. 27–38, 2012.

Some papers have been published for manual or automatic sentiment lexicon building [3][4][5], however, the challenges still remain for Web 2.0 based social media.

Emerging Words. A great number of new words are emerging in the online social media every day. Typos, ad hoc abbreviations and phonetic substitutions are common phenomena in the Web 2.0 texts. Some widely used typos and phonetic substitutions evolve into new sentiment-bearing words.

Emerging Meanings. Even the same word may have different explanations or sentiment orientations at different time periods.

Coverage. Due to the above two challenges, the traditional thesaurus and knowledge base, such as WordNet, usually suffer from the coverage problem.

Nowadays, with the fast development of mobile Internet, the microblog has become a very popular communication tool in Web 2.0 based social media. Everyday, enormous numbers of text posts that contain people's rich sentiments are published in microblogging websites such as Twitter and Weibo . By the end of May 2012, the number of registered users in the largest Chinese microblog platform Weibo.com has reached 300 million and there are more than 100 million microblog posts everyday.

The microblog is a good source for extracting sentiment lexicon. Firstly, as a convenient way to record daily personal feelings and emotions, the microblog data contains rich sentiment information. Secondly, the users usually like to employ free writing style and talk about up-to-date hot topics, so the new words and new meanings are common in the huge microblog dataset. With the microblog API, it is much easier to collect millions of posts for training. At last but not the least, many microblogs contain graphical emoticons, which can be considered as natural sentiment labels for the corresponding posts in the microblog dataset.

There are already some studies on emoticons in microblog data. Pak et al. collected tweets with happy and sad emoticons as training dataset, and built sentiment classifier based on traditional machine learning methods [6]. Davidov et al. chose 50 tags and 15 smileys as sentiment labels to classify twitter data [7]. These existing methods have verified the effectiveness of the microblog emoticons in the sentiment analysis task. However, they only focus on finding the appropriate features such as unigrams, bigrams, trigram and POS structures for sentiment classification. The basic sentiment lexicon building procedure is neglected, which may have many potential applications, such as opinion retrieval and opinion summarization.

In this paper, we propose an unsupervised sentiment lexicon learning method based on the emoticons in the microblog data. Intuitively, our basic assumption is that positive words often appear in the microblog posts with positive emoticons, and vice versa. We design appropriate rules to eliminate spam data, and collect the purified training microblog dataset with emoticons. We develop an algorithm to pick out sentiment words based on their occurrence probability in each category of the training dataset, and the accuracy of the learned lexicon is further improved by using the whole microblog space as the training corpus. Our approach can create a sentiment lexicon free of laborious efforts of the experts who must be familiar with both linguistic and psychological knowledge.

The rest of the paper is organized as follows. Section 2 introduces the related work on lexicon building and emoticon analysis. Section 3 analyzes the characteristics of the emoticons in Chinese microblogs and proposes the algorithm for sentiment lexicon building. Section 4 describes the lexicon optimization method. Section 5 provides experimental results on real world datasets. Finally we present concluding remarks and future work in Section 6.

2 Related Work

There are mainly two approaches for building sentiment lexicon. The first direction is automatic learning sentiment lexicon from thesaurus. Hu et al. resorted to the synonym and antonym relationship in WordNet to predict the orientation of the candidate words [8]. Kim and Hovy proposed two probabilistic models to estimate the strength of polarity [4]. In their models, synonyms were used as features. Their basic hypothesis was that the synonyms had the same orientation and the antonyms had the opposite orientation. Esuli et al. utilized the glosses in WordNet to represent the candidate word and based on the new representation the word was classified into positive and negative categories [5]. Esuli et al. also proposed a random walk based algorithm to rank the word polarities in Word-Net [9]. They assumed that the occurrence of the words in the glosses might be viewed as a transmitter of polarity properties.

Another direction is building sentiment lexicon based on large corpus. Turney et al. determined polarity value based on co-occurrence with seed words ('excellent' and 'poor'). The co-occurrence was measured by the number of hits returned by a search engine, i.e. the whole Web was considered as the corpus to determine the word orientation [10]. Kanayama et al used both intra- and inter-sentential co-occurrence to learn the orientation of words and phrases [11]. Kaji et al. utilized structural clues to extract sentiment words from large collection of HTML documents [12]. Velikovich et al. constructed a graph from web-computed lexical co-occurrence statistics, and employed a graph propagation algorithm to rank the words and phrases in the graph [13].

Although the thesaurus based methods can lead to higher accuracy, they also suffer from the new words and new meanings problems, which are common in Web 2.0 social media. The corpus based methods usually need laborious efforts of the experts or deep syntactic parsing. In this paper, we propose an unsupervised sentiment lexicon learning methods based on the occurrence probability of words and emoticons in microblogs, which starts from scratch and does not need the seed words

Recently, the attention of sentiment analysis researchers has gradually shifted from news articles, blogs and product reviews to microblogs. Bermingham et al. utilized the traditional machine learning based algorithms to classify the microblogs into positive and negative categories [14]. Brody et al. showed that lengthening was strongly associated with subjectivity and sentiment in tweets.

They proposed several rules to change the lengthening words into their canonical forms and classified the sentiments in Twitter based on the learned words [15]. In [7], 50 Twitter tags and 15 smileys were treated as sentiment labels and a supervised sentiment classification framework was proposed to classify the tweets. The authors evaluated diverse feature types and the experiment results validated the effectiveness of emoticons as sentiment indicators.

3   Learning Sentiment Lexicon from Massive Microblog Data

3.1   The Characteristics of Emoticons in Chinese Microblogs

Nowadays, with the help of mobile devices, people usually like to record their personal emotions and feelings in microblogs at anytime and anywhere. Due to the length limitation, users prefer to utilize emoticons to directly express their sentiments especially in Chinese microblogs. According to statistics,16.28% of the Chinese microblogs in Weibo contain emoticons, compared with 10.88% in Twitter. This is probably because Weibo has provided more convenient user interface when typing the emoticons. Fig 1 shows some examples of microblogs in Weibo.

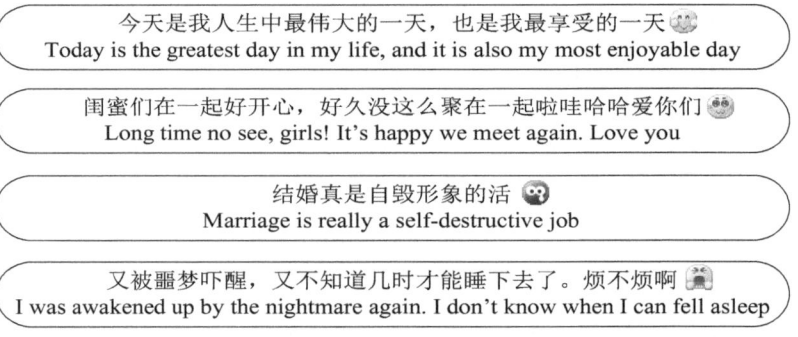

Fig. 1. Examples of Chinese microblogs with emoticons

From Fig 1 we can see that people tend to utilize emoticons to emphasize their emotion feelings. Since the microblogs have the length limitation, the sentiments expressed in the short text are usually consistent with the embedded emoticons.

3.2   Unsupervised Sentiment Lexicon Learning Based on Emoticons

From the discussion in Section 3.1 we know that the microblog is a good labeled data source for sentiment lexicon learning. However, there are also noisy data embedded in them. To address these challenges, we design strategies to eliminate the noisy microblogs as many as possible. Since it is usually convenient to access the microblog data through API, our key idea is to learn sentiment lexicon from extremely large purified microblog dataset with emoticons.

A parallel crawling system is implemented to collect the raw microblog dataset D through Weibo API. The Weibo platform has provided more than 1,000 predefined different emoticons. We manually select the most popular emoticons with obvious sentiment orientation, and group them into two categories. Finally we have the positive emoticon set PE and the negative emoticon set NE, and each of them has 15 emoticons. Here we briefly introduce our preprocessing steps as follows.

(1) Given the dataset D, we eliminate the microblogs that do not contain emoticons. We also filter out the data that have opposite emoticons. For example, if one microblog contains emoticons in PE and at the same time it has emoticons in NE, this microblog is eliminated. The result dataset is represented by D'.

(2) We segment the microblogs into words by using Chinese text processing tools. We also tag the words with the part-of-speech information for the next steps. The microblogs that contain negation words are eliminated.

(3) For each remaining microblog m, we remove the stop words. The Chinese words that only have one character are also eliminated, because the orientation of this kind of Chinese words is usually ambiguous or domain-specific.

(4) Since not all kinds of the words are good emotion indicators, we need not to reserve all words as candidate sentiment words. Here words with part-of-speech adjective, verb, noun and adverb are selected for the further detecting steps.

Although the texts are short, Figure 2 validates that the emotions can also be mixed up in one microblog. The first preprocessing step makes sure that we can get more sentiment consistent microblogs for training. After preprocessing, each microblog m in D' is represented by several words and a sentiment label SL, $m = \{w_1, w_2, \ldots, w_n, SL\}$.

To learn the sentiment lexicon from the purified training dataset, our basic assumption is that the positive sentiment word has more probability to co-occur with positive emoticons and negative sentiment word has more probability to co-occur with negative emoticons. We utilize the classic Pointwise Mutual Information (PMI) to measure the association between the candidate sentiment words and emoticons [10]. Here we have:

$$PMI(w, PE) = \log_2 \frac{p(w, PE)}{p(w)p(PE)} \tag{1}$$

$$PMI(w, NE) = \log_2 \frac{p(w, NE)}{p(w)p(NE)} \tag{2}$$

where w denote the words in the purified training set D'; $p(PE)$ is the occurrence probability of the positive emoticons in D', which is estimated by $N_{PE}/|D'|$ and N_{PE} is the number of positive emoticons; $p(w,PE)$ is the co-occur probability of w and the positive emoticons, which is estimated by $N_{PE-w}/|D'|$. N_{PE-w}

is the number of microblogs in D' that contain both w and positive emoticons. Therefore, the sentiment weight SW of w can be measured by:

$$SW(w) = PMI(w, PE) - PMI(w, NE) \tag{3}$$

Based on Formula 3, we can traverse the purified microblog dataset and rank the candidate sentiment word by their sentiment weight SW. A threshold θ is defined to control the size of the lexicon as follow.

$$w \begin{cases} add\ to\ the\ positive\ lexicon, if\ SW(w) > 0, |SW(w) > \theta| \\ add\ to\ the\ negative\ lexicon, if\ SW(w) < 0, |SW(w) > \theta| \end{cases} \tag{4}$$

We have crawled 30 million Chinese raw microblogs and after preprocessing about 1.5 million data are used for sentiment word learning. The details of the experiment setup will be discussed in the Section 5. The crawled dataset is large enough to extract the basic sentiment lexicon and each item in the lexicon is associated with a sentiment weight SW. However, compared with billions of the microblogs in the microblogging space, the crawled data is relatively small. In the next section, we will introduce a lexicon optimization algorithm which utilizes the whole microblogging space as a training corpus.

4   Sentiment Lexicon Optimization

Millions of users share their opinions on different aspects of life everyday in microblogs. Therefore the microblogging web-sites are rich sources of data for opinion mining and sentiment analysis. In the above section, we have learned a basic sentiment lexicon from the crawled microblog data with emoticons. Due to the size limitation and the bias problem of the crawled dataset, there usually exist sentiment words that are incorrectly classified.

Inspired by the work of Cilibrasi and Vitanyi, in this section we introduce a similarity metric Weibo distance to measure the semantic relationship of the words in the learned sentiment lexicon. Our basic idea is just following the Section 3: the words and emoticons with similar orientation tend to co-occur in the same microblog. Different from using the crawled data, here we employ the whole microblogging space as the training corpus. The number of hits returned by the Weibo search engine is utilized to estimate the co-occurrence of the candidate word and emoticon.

Suppose w represents a sentiment word in the learned lexicon, e is an emoticon with explicit orientation, so we give the definition of the Normalized Weibo Distance (NWD) as follows:

$$NWD(w, e) = \frac{\max\{\log f(w), \log f(e)\} - \log f(w, e)}{\log N - \min\{\log f(w), \log f(e)\}} \tag{5}$$

where $f(w)$ denotes the number of hits returned by Weibo search interface using w as keyword; $f(e)$ represents the number of hits returned by Weibo search interface using emoticon e as keyword (Note that we can use escape characters to search emoticons in Weibo); $f(w, e)$ denotes the hit number by using w and e as joint keywords; N is the number of microblogs in Weibo system. Derived from Google distance, the NWD is implemented based on theory of information distance and Kolmogorov complexity [16]. If the two items always co-occur together, we can infer that they have potential semantic consistency and their NWD value tends to be small. If the two items have opposite sentiment orientations, we can get bigger NWD value. Given a word w in the learned lexicon, we measure its NWD with a positive emoticon and a negative respectively, and the difference of NWD is used to further detect the orientation of w.

$$DD(w) = NWD(w, e_p) - NWD(w, e_n) \tag{6}$$

where e_p and e_n represent positive and negative emoticons respectively. If w has a similar distance between the positive and negative emoticons, DD will get a smaller value. Otherwise, the bigger value of DD indicates that w is more likely to belong to a predefined sentiment category. Therefore, we calculate the DD value of each word in the learned lexicon. Given a threshold , if $DD(w) < \alpha$, we eliminate w from the learned lexicon because it does not show an obvious orientation to either of the sentiment categories.

To evaluate the performance of the lexicon learning and the optimization methods, we propose a lexicon based sentiment analysis algorithm for Chinese microblog, which is given in Algorithm 1. In Algorithm 1, the number of matched positive and negative words from the lexicon is counted and the negation words are also considered during the classification.

Algorithm 1. Lexicon based sentiment analysis for Chinese microblog

---

  Input   : the microblog mb, the sentiment lexicon L, negation word set NG
  Output: sentiment score sentiscore of mb

1 Split mb into sentences;
2 **foreach** *sentence s in mb* **do**
3 | Segment s into words;
4 | **foreach** *word w of s*
5 | **do**
6 | | **if** $w \in NG$
7 | | **then**
8 | | | ng++;
9 | | **if** $w \in L$
10 | | **then**
11 | | | $sentiscore = sentiscore + SW(w)$;
12 | **if** ng is odd
13 | **then**
14 | | $sentiscore = -sentiscore/2$;

15 **return** $\Sigma sentiscore$

In Algorithm 1, firstly the microblog is split into sentences according to the punctuations. Then we calculate the sentiment score of every sentence using the lexicon L, and the word sentiment weights are summed to represent *sentiscore*. Moreover, if the number of negation words is odd, the sentiment score *sentiscore* will be decreased to $-sentiscore/2$ because the tone of the sentence can be weakened by the negation structure to certain extend. For example, "not happy" is not stronger than "sad" on expressing the negative emotions. In Algorithm 1, if the sum *sentiscore* > 0, then this microblog is regarded as positive. If the sum *sentiscore* < 0, the microblog is regarded as negative. Otherwise, it is neutral. The Algorithm 1 is easy to understand, and its performance mainly depends on the quality of the sentiment lexicon. In the next section, Algorithm 1 is used to evaluate different sentiment lexicons.

5   Experiments

5.1   Experiment Setup

Training Dataset. The Weibo API has a limit of 200 microblogs in one response for any request and also has a limit of requests per hour. To address this challenge, a parallel system with three PC nodes is designed to collect the Chinese microblogs as many as possible. For each node, we periodically send requests to public timeline Weibo API. After filtering out the duplicate and spam data, the microblogs from the three nodes are integrated together. Finally, we collected more than 30 million Chinese microblogs from October 1, 2011 to December 31, 2011.

In the 30 million crawled raw data, there are about 5.88 million microblogs containing emoticons. We preprocess the raw data using the four steps in the Section 3.2, at last we have 1,481,775 purified microblogs, of which 979,534 contain positive emoticons and 502,241 contain negative emoticons.

Testing Dataset. Since there is no benchmark dataset for the Chinese microblog sentiment analysis task, we crawled about 1,000 items on the hot topics "Pirates of the Caribbean" and the public timeline data respectively between June 10 and July 10 in 2011. Note that the time interval is different from the raw dataset used for training. For dataset annotation, we design a microblog sentiment tagging system. Firstly we load the microblog data to the annotation system, which has four basic tags: positive, neutral, negative, and unrelated. After the annotation, the unrelated, spam and advertisement microblogs are eliminated. Three graduate students who major in opinion mining annotate the tags of the microblogs. The final tag result is taken based on the majority opinions of the three students. If there is a disagreement between them, the item is eliminated.

Evaluation Measure. We use *F-Score* to measure the performance of Algorithm 1. In Formula (7), S presents the number of correct classified microblogs in one category. C is the number of the elements classified in one category. We denote the number of the manual tagged microblogs as R.

$$Precision = \frac{S}{C} \qquad Recall = \frac{S}{R} \qquad F-Score = \frac{2 \times Precision \times Recall}{Precision + Recall} \quad (7)$$

5.2   Experiment Results

Lexicon Size. Firstly we analyze the performance of the learned lexicon with different sizes. The experiment result is shown in Figure 2 and the parameter θ is used to limit the size of the lexicon. From Figure 2 we can see that when the lexicon size grows from 500 to 3,000, the classification performance increases dramatically for the lexicon has a better coverage. When the lexicon size further increases, there is no obvious corresponding growth for *F-Score*. This may because the larger size brings in more noisy data. In the further experiments, we employ the learned lexicon with about 5,000 positive words and 5,000 negative words respectively. The lexicon sizes with the number of positive and negative entries of each lexicon are shown in Table 1.

In Table 1, we denote the sentiment lexicon that learned from massive Chinese microblog data as Learned Lexicon. The Optimized Lexicon is the lexicon that utilizes Weibo distance to filter out the ambiguous words in the Learned Lexicon. The parameter is set to be 0.5 here. We can see from Table 1 that almost all the lexicons have balanced number of positive and negative words except NTUSD, which contains much more negative words [17]. About two thousand words are eliminated in the Optimized Lexicon.

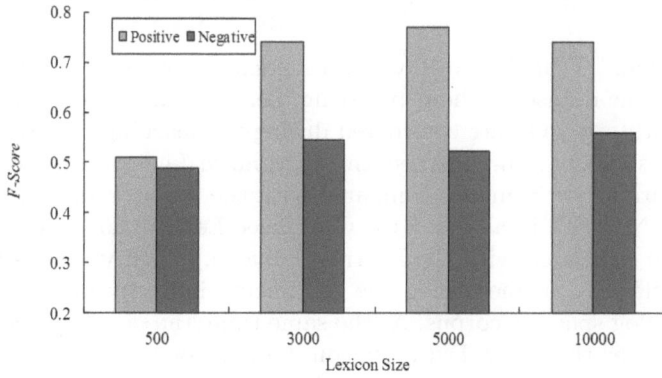

Fig. 2. The *F-Score* with different size of the learned lexicon

Table 1. Lexicon size

	HowNet	NTUSD	Learned Lexicon	Optimized Lexicon
Positive Words	4,528	2,813	5,022	4,066
Negative Words	4,320	8,277	4,986	4,122

Classification Performance. We compare our learned lexicon with two famous Chinese sentiment lexicons HowNet [18] and NTUSD. Note that there is no sentiment weight for each word in HowNet and NTUSD. Therefore, their weights are set to be 1 and -1 for positive and negative words respectively. The classification performances with different sentiment lexicon are shown in Table 2 and Table 3.

Table 2. The classification performance for the "Pirates of the Caribbean" data

Sentiment Lexicon	Positive $F\text{-}Score$	Neutral $F\text{-}Score$	Negative $F\text{-}Score$
HowNet	69.1%	42.9%	53.7%
NTUSD	75.1%	54.1%	57.2%
Raw Data Lexicon	70.2%	45.7%	50.5%
Learned Lexicon	74.9%	56.4%	52.3%
Optimized Lexicon	77.4%	57.8%	53.4%

Table 3. The classification performance for the public timeline data

Sentiment Lexicon	Positive $F\text{-}Score$	Neutral $F\text{-}Score$	Negative $F\text{-}Score$
HowNet	63.2%	53.4%	44.1%
NTUSD	66.2%	45.1%	57.0%
Raw Data Lexicon	62.1%	38.3%	42.1%
Learned Lexicon	65.6%	40.5%	45.4%
Optimized Lexicon	68.2%	45.6%	45.9%

In Table 2 and Table 3, the Raw Data Lexicon represents the lexicon that is learned from the dataset without purifying, i.e. more ambiguous emoticons and the words with any POS are considered during the learning process.

Generally speaking, our Learned and Optimized Lexicon can achieve a better or comparable performance compared with the manual constructed Lexicon HowNet and NTUSD. In all cases, the Optimized Lexicon has a better classification $F\text{-}Score$ compared with the Learned Lexicon, which validates that Weibo distance is effective in measuring the sentiment similarity of words using the whole microblog space as corpus. At the same time, the performance of Learned Lexicon is better than Raw Data Lexicon, because our preprocessing steps can eliminate ambiguous and noisy training data as much as possible. The NTUSD lexicon achieves a better negative $F\text{-}Score$. This may be because it has the largest manually selected negative words (8,277).

Another interesting observation is that the positive $F\text{-}Score$ is much better than the neutral and negative $F\text{-}Score$. This may be because people are accustomed to using more obvious sentiment words to express their positive emotions. The classification performance of the public timeline data in Table 3 is not as good as topic specified data in Table 2. After detailed analysis of the testing data, we find that the topics and words in Weibo public timeline are much more scattered, which generates less matches for all the sentiment lexicons.

Case Study. Figure 3 presents a selection of the top ranked positive and negative words in the Optimized Lexicon. We are glad to find that the emerging Internet new words such as awesome, tragic are also included in our lexicon. On the other hand, the proposed method can also find the words with new meanings. However, because of homophones it now has a new meaning "damn good" in the Internet that expresses strong emotions. Another interesting observation is that the word drainage may be a wrong segmentation result for the word "drainage oil", which means the illegally recycled cooking oil. These examples validate our basic assumption that the microblog is an effective data source for learning emerging words and new meanings for sentiment analysis task.

Positive Words		Negative Words	
动人	梦想成真	愧惜	白眼
(moving)	(dreams come true)	(regret)	(supercilious look)
过瘾	震撼	肤浅	聒噪
(have fun)	(electrifying)	(superficial)	(noisy)
可爱	碉堡	命苦	荒废
(lovely)	(damn good)	(bitter life)	(waste)
挂念	天真	悲摧	不知所措
(missing)	(naive)	(tragic)	(at a loss)
雪中送炭	佳人	挂彩	尼玛
(timely help)	(beautiful woman)	(wounded)	(damn)
美满	开小灶	冰冷	交通
(happy)	(special treatment)	(cold)	(traffic)
一路顺风	给力	伤脑筋	地沟
(bon voyage)	(awesome)	(knotty)	(drainage)

Fig. 3. Examples of positive and negative words in the Optimized Lexicon

6   Conclusion and Future Work

In this paper, we explore to use emoticons as sentiment labels to extract sentiment words from massive microblog data and the accuracy of the learned lexicon is further improved by using the whole microblog space as a corpus. The key idea is that positive words often appear in the microblog posts with positive emoticons, and vice versa. The proposed method can create a sentiment lexicon free of laborious efforts of the experts who must be familiar with both linguistic and psychological knowledge. The experiment results validate that the learned lexicon outperforms the traditional sentiment lexicons which may suffer from the coverage problem when faced with new words and new meanings in social media.

There are a wide variety of topics contained in the microblog space. For further work, we intend to combine our method with known methods to learn domain or topic specified sentiment lexicons from the massive microblog data.

References

1. Pang, B., Lee, L., Vaithyanathan, S.: Thumbs up? Sentiment Classification using Machine Learning Techniques. In: Proc. of EMNLP, pp. 79–86 (2002)
2. Jin, W., Ho, H., Srihari, R.: OpinionMiner: A Novel Machine Learning System for Web Opinion Mining and Extraction. In: Proc. of KDD, pp. 1195–1204 (2009)
3. Das, S., Chen, M.: Yahoo! for Amazon: Sentiment Extraction from Small Talk on the Web. Management Science 53(9), 1375–1388 (2007)
4. Kim, S., Hovy, E.: Determining the Sentiment of Opinions. In: Proc. of COLING, pp. 1367–1373 (2004)
5. Baccianella, S., Esuli, A., Sebastiani, F.: SentiWordNet 3.0: An Enhanced Lexical Resource for Sentiment Analysis and Opinion Mining. In: Proc. of LREC, pp. 2200–2204 (2010)
6. Pak, A., Paroubek, P.: Twitter as a Corpus for Sentiment Analysis and Opinion Mining. In: Proc. of LREC, pp. 1320–1326 (2010)
7. Davidov, D., Tsur, O., Rappoport, A.: Enhanced Sentiment Learning Using Twitter Hashtags and Smileys. In: Proc. of COLING, pp. 241–249 (2010)
8. Hu, M., Liu, B.: Mining and Summarizing Customer Reviews. In: Proc. of KDD, pp. 168–177 (2004)
9. Esuli, A., Sebastiani, F.: PageRanking WordNet Synsets: An Application to Opinion Mining. In: Proc. of ACL, pp. 424–431 (2007)
10. Turney, P.: Thumbs Up or Thumbs Down? Semantic Orientation Applied to Unsupervised Classification of Reviews. In: Proc. of ACL, pp. 417–424 (2002)
11. Kanayama, H., Nasukawa, T.: Fully Automatic Lexicon Expansion for Domain-oriented Sentiment Analysis. In: Proc. of ENMLP, pp. 355–363 (2006)
12. Kaji, N., Kitsuregawa, M.: Building Lexicon for Sentiment Analysis from Massive Collection of HTML Documents. In: Proc. of EMNLP-CoNLL, pp. 1075–1083 (2007)
13. Velikovich, L., Blair-Goldensohn, S., Hannan, K., McDonald, R.: The Viability of Web-derived Polarity Lexicons. In: Proc. of HLT-NAACL, pp. 777–785 (2010)
14. Bermingham, A., Smeaton, A.: Classifying Sentiment in Microblogs: Is Brevity an Advantage? In: Proc. of CIKM, pp. 1833–1836 (2010)
15. Brody, S., Diakopoulos, N.: Cooooooooooooooooollllllllllllllll!!!!!!!!!!!!!! Using Word Lengthening to Detect Sentiment in Microblogs. In: Proc. of EMNLP, pp. 562–570 (2011)
16. Cilibrasi, R., Vitnyi, P.: The Google Similarity Distance. IEEE Trans. Knowl. Data Eng. 19(3), 370–383 (2007)
17. Ku, L., Chen, H.: Mining Opinions from the Web: Beyond Relevance Retrieval. Journal of American Society for Information Science and Technology 58(12), 1838–1850 (2007)
18. HowNet, http://www.keenage.com/

Community Extraction Based on Topic-Driven-Model for Clustering Users Tweets

Lilia Hannachi[1], Ounas Asfari[2], Nadjia Benblidia[1], Fadila Bentayeb[2],
Nadia Kabachi, and Omar Boussaid[2]

[1] LRDSI Laboratory University of Blida, Saad Dahlab
AV. Soumaa BP 270 BLIDA, Algeria
`Hannachi.Lilia@yahoo.fr`, `Benblidia@yahoo.com`
[2] ERIC Laboratory, University of Lyon 2
5 AV. P. Mends-France 69676 Bron Cedex Lyon, France
{`Ounas.Asfari,Fadila.Bentayeb,Nadia.Kabachi,Omar.Boussaid`}`@univ-lyon2.fr`

Abstract. Twitter has become a significant means by which people communicate with the world and describe their current activities, opinions and status in short text snippets. Tweets can be analyzed automatically in order to derive much potential information such as, interesting topics, social influence, user's communities, etc. Community extraction within social networks has been a focus of recent work in several areas. Different from the most community discovery methods focused on the relations between users, we aim to derive user's communities based on common topics from user's tweets. For instance, if two users always talk about politic in their tweets, thus they can be grouped in the same community which is related to politic topic. To achieve this goal, we propose a new approach called CETD: Community Extraction based on Topic-Driven-Model. This approach combines our proposed model used to detect topics of the user's tweets based on a semantic taxonomy together with a community extraction method based on the hierarchical clustering technique. Our experimentation on the proposed approach shows the relevant of the users communities extracted based on their common topics and domains.

Keywords: Topic Model, Community extraction, Tweets, Semantic Processing, Data Mining, Social Networks.

1 Introduction

Twitter is an online social networking service that enables its users to send and read text-based posts of up to 140 characters, known as "tweets". It enables its users to communicate with the world and share current activities, opinions, spontaneous ideas and organize large communities of people. The service rapidly gained world wide popularity that led the researchers to study the characteristics of tweets content and to extract information such as opinions on a specific

S. Zhou, S. Zhang, and G. Karypis (Eds.): ADMA 2012, LNAI 7713, pp. 39–51, 2012.

topic or user's topics of interest. The tweets studies have perspectives in many domains such as, friends recommendation, opinions analysis, users topics, etc. However, the text of tweets is generally noisy, ambiguous, unstructured text data, ungrammatical, but it is a rich data set to analyze and most likely users try to pack substantial meaning into the short space, one subject by one tweet. Thus, it is important to understand the information behind the tweets and to detect the topics presented by them. To detect these topics, many applications propose to use a topic models like, PLSA (Probabilistic Latent Semantic Analysis) [1] or LDA (Latent Dirichlet Allocation) [2] which are effective and powerful approaches to detect the different latent topics. However, it presents each topic by a distribution of words. Thus we can not extract semantic concepts to the topics. In this paper, we will add concepts reconstructing the semantics of the distributed words by topic model in order to detect automatically the high level topics presented by the tweets.

In this paper, our objective is to extract users communities based on the common topics from tweets. In order to realize this objective, we propose to perform a process consisted of two phases; in the first one, we aim to discover the different treated topics by users and their different related categories by constructing a topics tree based on ODP taxonomy [1]. In the second phase, we aim to derive user's communities by grouping users according to three separated cases; topics similarity, domains similarity and topics-domains similarity. These three cases will be extracted from the first part.

Communities are a user's groups that share same characteristics and interests. It is used in many applications, such as social networks, data mining, web searching, etc. The study of community extraction generally aim to extract the communities mainly according to the relations between users. However, in the user graphs, the relations do not present the dynamics between users according to their common topics, for instance, although an existence of a friendship between users, we cannot extract their common interesting information. That's why the study of micro-blogging content attracts much attention in recent years.

We think that the relation between users, as defined in the classic methods such as [3], [4], is not enough when we look for users groups related to the same topics in order to recommend them some information. Therefore, different to these works, we aim to derive user's communities based on the common topics of user's tweets. Our contribution, in this paper, is to combine both the high-level-topics model and the user communities extraction in a unified approach called CETD: Community Extraction based on Topic-Driven-model.

For instance, if we consider a user who writes always tweets in the domain *sports* like the following real-word tweet: "*Contracts for Top College Football Coaches Grow Complicated*". Our proposed topic-driven model will assign automatically the topics "football" to this tweet. Now, if we consider another user who writes the following real-world tweet: "*Barcelona win 2-0 at Real Mallorca but Real Madrid return to form and smash Real Sociedad 5-1*". The detected topics for this tweet will be also: *football* although the word "*football*" is not

---

[1] ODP: Open Directory Project: www.dmoz.org

mentioned in this tweet. This will show the important of our semantic addition to the classic topic model. We note that the two users treat the same topic *football*. Thus we can regroup these two users in the same community based on their topics of interesting *football*.

The rest of this paper is organized as follows, Section 2 reviews the related works. In Section 3, we will analyze in details our proposed model of detection the high level topics from user's tweets. Section 4 presents the communities extraction method based on the results of the section 3. In addition, we will present the experimentation. Finally, section 5 presents the conclusion and the future works.

2   Related Works

As we use, in our proposed Topic-Driven-Model, the general probabilistic topic model LDA [2], in which the text collection is represented as a distribution of topics, and each topic is represented by a words distribution, we will present some related works which use the topic model LDA on twitter. The researchers, like [5], use the standard topic model LDA in micro-blogging environments in order to identify influential users, but the proposed influence measure is based on the number of tweets, however the twitter users usually publish a large number of noisy posts. In addition, the work of [6] compares the tweets content empirically with traditional news media by using a new Twitter-LDA model in order to detect their topics. They consider that each tweet usually treats a single topic, but this is not always the case. The works of [7] present an approach to discover a Twitter user's profile by extracting the entities contained in tweets based on Wikipedia's user-defined categories.

Moreover, as we aim to construct the user communities based on the common topics, we review some studies in this area. The traditional clustering method, such as [3], based on the arcs density in the graph. For instance, the hierarchical clustering techniques like,[8], aim to identify vertices groups with high similarity. It can be divided into two classes: Agglomerative algorithms [4], [9] and Divisive algorithms [10], [11]. In divisive algorithms technique, we do not need to specify the clusters number in advance, like Agglomerative algorithms, but the disadvantage is that many partitions are recovered. In this case we can not define the best division. In [12], the authors propose a new divisive algorithm. This algorithm is based on the concept of edge betweenness centrality. It works on moderate size networks significantly. However, the need to recompute betweenness values in every step becomes computationally very expensive. [13] propose a social topic model that incorporates both the link and content information in the social network and use this model to extract the community.

In fact, in this study, we propose an approach which is different from the existing works. We aim to construct communities by considering the semantic of tweets contents as the only one inputs data instead of being as an additional information. Also we propose a new model to cluster users tweets. We combine both semantic hierarchy presented in ODP taxonomy and topic model (LDA) in order to improve the obtained results by adding semantics relations.

3   Topic-Driven Model for Clustering Users Tweets

In this section, we present the first phase of our proposed approach which is a Topic-Driven Model to detect the high-level topics as a distribution of domains from users tweets. The architecture of this model is illustrated in Figure 1. We can divide it in four steps: Cleaning Database; ODP-Based Adapted LDA; ODP-Based Topics Semantic; n-depth High Level Topics.

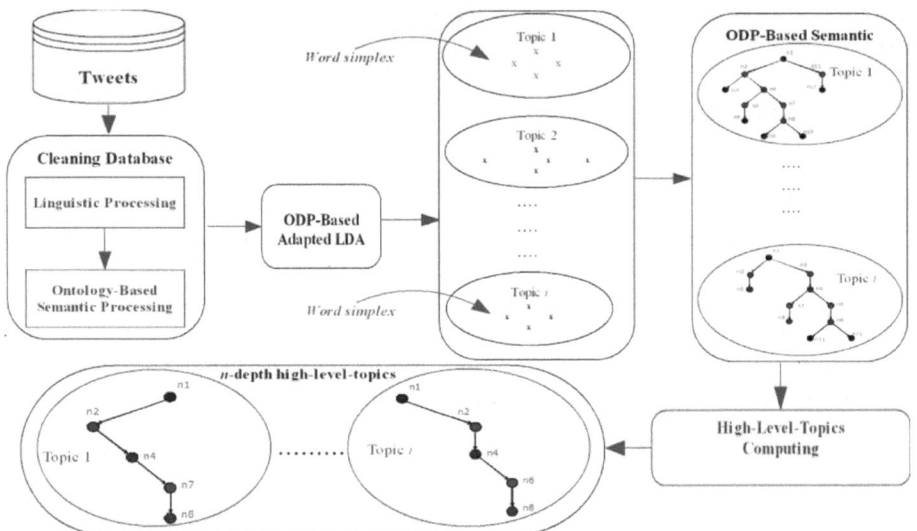

Fig. 1. System architecture

3.1   Cleaning Database

Firstly, we store the tweets corpus in a database, then, we treat tweets corpus by using both a linguistic knowledge and a semantic knowledge to clean the tweets set, because linguistic knowledge does not capture the semantic relationships between terms and semantic knowledge does not represent linguistic relationships of the terms. In the linguistic processing phase, we remove stop words, user names, hashtag and URL. Also we get the word root (streaming) for example: plays, playing, etc. ->play. Finally, we do the spelling correction.

After the first cleaning based only on linguistic processing, we note that many noisy and irrelevant words are still existed after this step. To solve this problem, we clean the corpus by using a semantic knowledge, such as, the ODP Taxonomy as an instance of a general ontology. In this semantic phase we achieve the following steps:

- We verify the existence of words in the results of the ODP indexing. If it does not exist, we remove it from the tweets. For example, A noise words such as: suprkkbwp, mirsku, etc. can be removed in this step.

– If the word exist, we will compute the number of documents (Web pages) which support it $N_D(w_i)$. In this case, we suppose a threshold: $N_D(w_i) >= 20$. Thus we remove the words that have a number of supported documents less than 20; i.e. $N_D(w_i) < 20$ because, in our experiments, we noted that the words supported by Web pages less than 20 are irrelevant words, for example, if we consider the words "awry", the number of its supported pages is 15, thus this word will be removed in this cleaning step. In this way, we can remove the irrelevant words.

3.2   ODP-Based Adapted LDA

In this step, we apply Latent Dirichlet Allocation (LDA) on the cleaned tweets data. To apply LDA, we have to define the number of iterations, the number of words allocated to each topic and the number of topics. Thus, each user tweets is represented as a distribution of topics, and each topic is represented as a words distribution. However, in order to verify the utility of using ODP Taxonomy in cleaning tweets data, LDA is applied two times: to the cleaned data by using ODP, and without using it. After comparison between the results in the two cases, we notice that the topics, sorted from the not cleaned corpus, have some noisy words with a high probability. For instance, if we specify the number of topics as five topics and five words for each topic, the words distributions for the first topic $p(w_i|T_0)$ in the two cases (by using only the linguistic phase and by using ODP) are the following:

Topic 0: sport(0.099), college(0.026), football(0.026), top(0.021), new(0.021).
Topic 0: sport(0.117), college(0.030), football(0.030), baseball(0.025), golf(0.015).

We note that the topics resulted, in the second case, are more significant and more homogeneity between words. Thus, in our model we depend on this results which is cleaned by ODP taxonomy.

3.3   ODP-Based Topics Semantic

In this phase, we construct concepts sub-trees in order to detect the semantic relations between the words of each topic resulted after the applying of LDA. For each word in the unsupervised topic, T_k, we will generate the semantic sub-tree (fragments) from the ODP containing all levels. We consider first 5 categories for each word, then we will repeat the same process for all unsupervised topics. Now we construct XML file for each topic T_k and represent each one by fragments.

3.4   N-Depth High Level Topics

As we mentioned previously, the LDA model proposes to represent a topic as a words distribution. In this case, users can not observe the difference orientations because the word is a very specific unit and connected to different topics categories. Thus, users interpret the results according to their personal background and experiences and this will decrease the performance of the model. To solve this problem, we suggest a model to present the topic in a higher level as a distribution of domains where domain is a set of topics.

Topic - Domain: In this case, we identify for each word in the topic the first five domains with their five levels at maximum from the ODP taxonomy, and then we calculate the probability of each domain in each topic based on two types of probability:

- The probability of domain j in word i which is computed by the number of the domain's occurrences of this word divided by the total number of domains for this word.

$$P(D_{j,i}) = \frac{n_j}{N} \tag{1}$$

  where n_j: the domain j occurrence in word i existed in topic k. N: the total domains occurrence for word i existed in topic k. D_j : Domain j.
- The probability of word i in topic k: the probability which was computed from the result produced by the topic model (LDA): $p(w_i|T_k)$

Thus the probability of each domain in each topic will be:

$$P(D_j, T_k) = \sum_{i=1}^{I} \frac{n_j}{N} * P(w_i|T_k) \tag{2}$$

where T_k: Topic k. I: the number of words in the Topic T_k.

For example, the top five domains associated with the topic 0, as an instance of the five used topics (presented in the LDA-Adapted step) are:

Topic 0 (T_0): Sports(0.294), Society(0.117), Football(0.098), Baseball(0.098), Colleges and Universities(0.078)

The final phase is to infer the high-level topics in the several levels. In each topic, T_k, and for each level, we will select the node which has the maximum paths (see Figure 2). Thus, to do this, we propose the following formula:

$$SD_{k,l} = ArgMax(\sum_{i=1}^{I} \frac{n_{j,l}}{N} * P(w_i|T_k)) \tag{3}$$

where $SD_{k,l}$: selected domain for Topic k in level l. $n_{j,l}$: the domain j occurrence in word i for level l.

For example, let us consider the topic k resulted after applying topic model LDA as a distribution of 10 words. Then we queried these words to ODP and we select the first five categories with their five levels. Then we build a sub-tree for each word that contains the different selected categories.

In Figure 2, we select the first four words according to their probabilities as an example. After that we combine their sub-trees in one tree. We compute the weight for each node in this tree by using our proposed formula (2). In this example, we can show the results of this step in Figure 2. Finally, to define the related high-level-topic in each level we can use the formula (3) which allow us to select the node that has the most number of paths.

The node *Arts*, for instance, will have the following weight.

Arts = (1/1 * 0.3335) + (1/3 * 0.1196) + (5/5 * 0.1196) = 0.49256

Thus, we note, from Figure 2, that the node "Arts" in level 0 have the maximum weight.

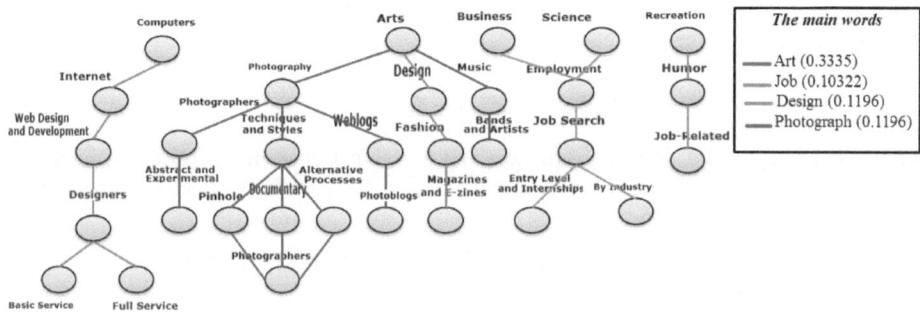

Fig. 2. Example of semantic tree

User - Domain: Usually, topics are related to different domains. However, from the topic model (LDA), we have only the probabilities of treated topics by a user. To improve this probability, we represent the users in a higher level as mixtures of domains. The following formula calculates the probability of related user s to domain *j*.

$$P(U_s, D_j) = \sum_{k=0}^{K} p(U_{s,k}) * P(D_{j,k}) \qquad (4)$$

where K: Topics number. $P(U_{s,k})$: probability of topic k for user s. $P(D_{j,k})$: probability of domain j in topic k.

4   Topic Communities Extraction

In this section, we present the second part of our proposed approach which is a method of extraction user's communities based on the topics or domains extracted from user tweets content by using our proposed Topic-Driven Model for users tweets which is divided in four steps presented in Section 3. The existing works propose to extract the communities according to the connection between users. In our study, we suggest to consider a semantic clustering to answer some queries as "what is the network structure grouped by domain?", "what is the network structure grouped by Topic?". In this method, we calculate the distance between users according to the common topics or domains and then the results will be used to construct the communities. This method consists of three steps:

calculate the distance between users; construct the graph which presents the different closeness relations between users; construct users communities based on the previous results.

4.1   Distance between Users

The researchers, in [5], define the distance between user i and user j as the Jensen-Shannon Divergence between the topics distributions on users presented by the following formula:

$$dist_T(i, j) = \sqrt{2 * D_{JS}(i, j)} \qquad (5)$$

where $D_{JS}(i, j)$: the Jensen-Shannon Divergence between the two topic distributions DT_i and DT_j. It is defined as:

$$D_{JS}(i, j) = \frac{1}{2}(D_{KL}(DT_i||M) + D_{KL}(DT_j||M)) \qquad (6)$$

where M: the average of the two probability distributions. $M = \frac{1}{2}(DT_i + DT_j)$. D_{KL}: the Kullback-Leibler Divergence which defines the divergence from distribution Q to distribution P as:

$$D_{KL}(P||Q) = \sum_i P(i)log\frac{P(i)}{Q(i)} \qquad (7)$$

Distances based on domains distributions. In our study, as each topic is related to different domains with different probabilities, there is a possibility that two users treat the same topic but not necessarily the same orientation. For example, let us consider the two users i, j which treat the same topic "*President Obama*" but they are not in the same orientation, because user i talk about the *politic* and user j talk about *health*. On the other side, they may be talk about different topics but with the same orientation. In this case, we propose two distance measures; domains and topics-domains.

- In the first measure (domains), we calculate the distance between users as the Jensen-Shannon Divergence between domains distributions over users as in formula 5 and 6.
- The second measure (topics-domains) combines the two previous measures (Topics, Domains). This new measure allows decreasing the distance between users who do not treat only same topics but also same domains. The distance between users in this measure is computed by using:

$$dist_{TD}(i, j) = \sqrt{2 * D_{JS}(i, j)} \qquad (8)$$

where $D_{JS}(i, j)$: the Jensen-Shannon Divergence between the domains distributions DD_i, DD_j, and the topic distributions DT_i and DT_j

Here, the divergence is also computed by the following formula:

$$D_{JS}(i,j) = (\frac{1}{2}(D_{KL}(DT_i\|M)+D_{KL}(DT_j\|M)))+(\frac{1}{2}(D_{KL}(DD_i\|M)+D_{KL}(DD_j\|M)))$$
(9)

Table 1 shows the distance between users based on topics-domains. We consider, in this table, only five users as an example of the test collection which will presented in the experimentation section. For instance, the distance between user 3 and user 5 according to topics-domains is 1.142907 which is the minimum distance. That means, these two users are the closest in comparison to others.

Table 1. Distance between users based on topic-domain

User - User (Topic-Domain)					
	User1	User2	User3	User4	User5
User1	0.000000	1.260869	1.163385	1.226157	1.243853
User2	1.260869	0.000000	1.231731	1.264484	1.247072
User3	1.163385	1.231731	0.000000	1.263051	1.142907
User4	1.226157	1.264484	1.263051	0.000000	1.230949
User5	1.243853	1.247072	1.142907	1.230949	0.000000

4.2   Graph Construction Based on Users Closeness Relations

The graphs have a great expressive and they are simple for modeling. They are based on two concepts nodes and edges. In the social network case, the nodes represent a set of social entities such as users or social organizations while the edges between nodes indicate that a direct relationship has been created during social interactions. In this paper, we create the graph which consists of nodes and edges as in the existing works but the nodes, in our method, represent the users and the edges represent the closeness between them according to selected topics, domains or both selected topic-domain and not the communication relationship as in the existing works. Thus, we can define three types of graphs based on the user's closeness according topics, domains and topics-domains.

Topics or domains are selected either by the choice of users or we consider all topics and domains of users which are produced by our model if he/she dose not choose any topic or domain.

In the topic graph, we create an edge from the user i to the user j, if the user j is the closest to the user i for the topic k, the weight of this link is calculated as the distance between them for this selected topic k. Moreover, if there is another edge from the user i to the user j for another topic, it is enough to choose the minimal distance between these two users i and j. For example, Table 2 shows the topics distributions for the experimented five users. We selected, as an instance, the five most related topics for these five users. Figure 3 shows the corresponded graph(topic graph). We use the same method to create the other two graphs which are based on the domains closeness or topics-domains closeness.

Table 2. Example of user distribution over topic

	Topic1 Arts	Topic2 Society	Topic3 Business	Topic4 Regional	Topic5 Sports
User1	0.016541	0.001504	0.001504	0.978947	0.001504
User2	0.001835	0.001835	0.992661	0.001835	0.001835
User3	0.952838	0.000873	0.000873	0.044541	0.000873
User4	0.000608	0.006687	0.000608	0.012766	0.979331
User5	0.079154	0.906949	0.006647	0.000604	0.006647

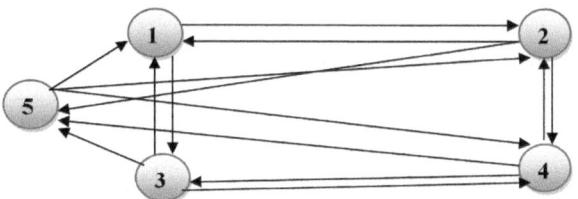

Fig. 3. Topic graph

4.3   Construct Users Communities

Based on the constructed graphs, we can extract the users communities by adapt-ing the approach of Newman [14] which is based on the divisive classification. The divisive is a top down approach which starts with all nodes as an only com-munity and applies the division method. The algorithm [14] process is based on the following steps: firstly calculate the betweenness scores for all edges in the network, secondly, find the edge with the highest score and remove it from the network, thirdly, recalculate the betweenness for all remaining edges and finally, repeat the research and the remove until get the communities.

We improve this approach to create the communities, the classic approach uses the existing graphs to calculate the betweenness by using the communications relation between users, but in our adaptation, we are based on the semantic graphs shown in the previous section. Figure 4 shows our adapted method which is based on divisive approach.

```
User0 User1 User2 User3 User4
      ┌───┴────┐
User2          User0 User1 User3 User4
  │   ┌────────┴───┐
User2 User0         User1 User3 User4
  │   ┌───┴───┐
User2 User0 User1   User3 User4
  │   ┌───────┴──┐
User2 User0 User1 User3   User4
```

Fig. 4. The adapted divisive approach

4.4   Evaluation of the Extracted Communities

Here, we will evaluate the extracted communities based on topics, domains and topics-domains. The question asked, here, is how to get the best result? [14] give the answer by his popular modularity measure that evaluates the extracted communities. It is the number of edges within community minus expected number in an equivalent network with edges placed at random. Moreover, [15] present an extension of modularity for directed graphs. The adapted formula is:

$$Q = \frac{1}{m} \sum_{i,j \in V} (A_{ij} - \frac{k_i^{out} k_j^{in}}{m}) \delta(C_i, C_j) \tag{10}$$

where A_{ij}: the elements of the adjacency matrix of G(E, V), E: edge, V: vertex. k_j, k_i: the in-degree and out-degree of nodes j,i. m: the number of edges. $\delta(C_i, C_j)$ equal 1 if i and j belong to the same community, and 0 otherwise.

However, in our case we have not the same classic graphs that use the existing links. Thus, we will compute the modularity by using the semantic graphs constructed in Section 4.2. After modularity computation for the resulted communities over the experimented users, we note that the division of the two communities where the first one which contains user1, user3, user4 and the second which contains user0, user2 have the maximum value of modularity. This division is shown in Figure 4 (the step 2).

4.5   Experimentation

We applied our proposed model on data set, which is collected by crawling one week of public tweets by using the 140dev Twitter framework[2]. Our collection is constructed from the tweets of the 15th through the 22th of January 2012 (One Week). We select the first 100 relevant users according to the follows number that have tweets number more than 20 tweets. For each user tweets, we applied

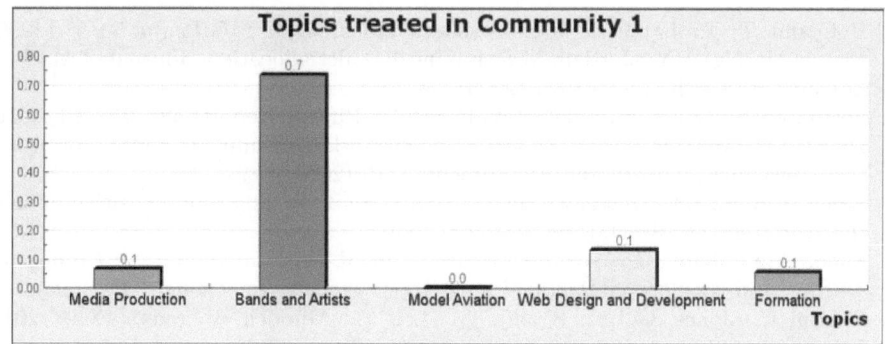

Fig. 5. Topics treated in Community 1

---

[2] http://140dev.com/free-twitter-api-source-code-library/

our proposed Topic-Driven-model in order to define the high-level-topics and their domains. We used the GibbsLDA++ C++-code[3], in the second step, to apply the topic model LDA. Then we constructed the user communities by using the two adapted algorithms based on the results of our model. We illustrate, in Figure 5, the first five most treated topics in one community. We note that the majority of users in this community treat the topic 2 "*Bands and Artists*".

5   Conclusion and Future Works

In this paper, we proposed a new approach to construct user's communities based on high-level-topics and domains which were extracted from semantic hierarchy, ODP taxonomy, to cluster users' tweets. Generally, the existing works have constructed the users communities based on the links between users. While, in our approach, the user community is based on the common topics. Thus, our method allows focusing on the relation between users according to their topics of interest. Also, we can detect the emerging topics or domains in each community.

In order to construct these communities, we proposed a model to cluster users' tweets based on ODP taxonomy as an external resource to derive high level topics and the topics domains. The existing works use a generative probabilistic model, such as LDA (Latent Dirichlet Allocation), on the tweets data to identify topics for these tweets as words distribution without considering the semantic relations. The contribution of our proposed model is using a semantic hierarchy (ODP taxonomy) to assign automatically high-level-topics to each user tweet.

This study of users' community extraction, based on the similar topics, will permit us to propose a new OLAP (On-Line Analytic Processing) operation including the semantic aspect, which is our main objective. In the future work, we will try to consider the location and the time to calculate the similarities between users.

References

1. Hofmann, T.: Probabilistic latent semantic indexing. In: SIGIR, pp. 50–57 (1999)
2. Blei, D.M., Ng, A.Y., Jordan, M.I.: Latent dirichlet allocation. Journal of Machine Learning Research 3, 993–1022 (2003)
3. Fortunato, S.: Community detection in graphs. Physics Reports 486, 75–174 (2010)
4. Donetti, L., Martinez, M.A.M.: Detecting network communities. Journal of Statistical Mechanics: Theory and Experiment, 1–15 (2004)
5. Weng, J., Lim, E.P., Jiang, J., He, Q.: Twitterrank: finding topic-sensitive influential twitterers. In: WSDM, pp. 261–270 (2010)
6. Zhao, W.X., Jiang, J., Weng, J., He, J., Lim, E.-P., Yan, H., Li, X.: Comparing Twitter and Traditional Media Using Topic Models. In: Clough, P., Foley, C., Gurrin, C., Jones, G.J.F., Kraaij, W., Lee, H., Mudoch, V. (eds.) ECIR 2011. LNCS, vol. 6611, pp. 338–349. Springer, Heidelberg (2011)
7. Michelson, M., Macskassy, S.A.: Discovering users' topics of interest on twitter: a first look. In: AND, pp. 73–80 (2010)

---

[3] http://gibbslda.sourceforge.net/

8. Hastie, T., Tibshirani, R., Friedman, J.H.: The Elements of Statistical Learning, Corrected edn. Springer (July 2003)
9. Du, H., Feldman, M.W., Li, S., Jin, X.: An algorithm for detecting community structure of social networks based on prior knowledge and modularity. Complexity 12(3), 53–60 (2007)
10. Johnson, E.L., Mehrotra, A., Nemhauser, G.L.: Min-cut clustering. Math. Program. 62, 133–151 (1993)
11. Newman, M.E.J.: Fast algorithm for detecting community structure in networks. Physical Review E 69, 66133 (2004)
12. Newman, M.E.J., Girvan, M.: Finding and evaluating community structure in networks. Phys. Rev. E 69(2), 026113 (2004)
13. Pathak, N., Delong, C., Erickson, K., Banerjee, A.: Social Topic Models for Community Extraction. Technical Report 08-005, Dept. Computer Science and Engineering, University of Minnesota (2008)
14. Girvan, M., Newman, M.E.J.: Community structure in social and biological networks. Proceedings of the National Academy of Sciences 99(12), 7821–7826 (2002)
15. Arenas, A., Duch, J., Fernandez, A., Gómez, S.: Size reduction of complex networks preserving modularity. CoRR abs/physics/0702015 (2007)

Constrained Spectral Clustering Using Absorbing Markov Chains

Jianyuan Li[1,2] and Jihong Guan[1]

[1] Department of Computer Science and Technology, Tongji University,
Shanghai 201804, China
[2] School of Mathematics and Computer Science, Shanxi Normal University,
Linfen 041000, China
jylbob@gmail.com, jhguan@tongji.edu.cn

Abstract. Constrained spectral clustering (CSC) has recently shown great promise in improving clustering accuracy or catering for some specific grouping bias by encoding pairwise constraints into spectral clustering. Essentially, the existing CSC algorithms coarsely lie in two camps in terms of encoding pairwise constraints: (1) they modify the original similarity matrix to encode pairwise constraints; (2) they regularize the spectral embedding to encode pairwise constraints. Those methods have made significant progresses, but little of them takes the extensional sense of pairwise constraints into account, e.g., respective neighbors of two musk-link points lie in a same cluster with certain high probabilities, and respective neighbors of two cannot-link points lie in different clusters with certain high probabilities, etc. In this paper, we use absorbing Markov chains to formulate the extensional sense of instance-level constraints as such, under the assumption that the formulation aids in improving the accuracy of CSC. We describe a new CSC algorithm which could propagates the extensional sense over a partly-labeled affinity graph. Experiments over publicly available datasets verify the performance of our algorithm.

1 Introduction

Clustering is an important unsupervised learning method that aims to detect structures in vector data, finding groups in which patterns of vector data are as similar as possible [1], or finding groups that statistically form smooth sub-manifolds. However, purely unsupervised learning often can not satisfy some specific grouping bias, since certain features might play noisy (or redundant) roles to the grouping bias. Motivated by this, researchers have explored a class of methods that could incorporate accessible supervision with clustering algorithms for improving clustering accuracy, as reported in [2–10].

The most accessible supervision is called pairwise constraints, meaning whether pairwise instances lie in a same group or not, and so the problem in question is called *constrained clustering*. To incorporate pairwise constraints with clustering is often practical and useful in many fields, e.g., in document clustering, whether two documents concern about a same topic or not, can be readily judged through skimming;

S. Zhou, S. Zhang, and G. Karypis (Eds.): ADMA 2012, LNAI 7713, pp. 52–63, 2012.
© Springer-Verlag Berlin Heidelberg 2012

Table 1. Notation description

Symbols	Description
X	a vector dataset $\{\mathbf{x}_i\}_{i=1}^n$ where $\mathbf{x}_i \in \mathbb{R}^d$
d	dimensionality of X
n	cardinality of X
m	the number of constrained points
G	the affinity graph over X
$\mathbf{W}$	the affinity matrix over X
$\mathbf{P}$	the original probability transition matrix
$\mathbf{F}$	the fundamental matrix of absorbing Markov chains
$\mathbf{I}$	the diagonal matrix with diagonal elements ones
$\mathcal{M}$	the set of Must-Link constraint pairs
$\mathcal{C}$	the set of Cannot-Link constraint pairs
$\mathbf{Y}$	the m-by-m constraint matrix

and in image segmentation, whether two patches represent a same meaningful entity or not, can be easily judged through observing.

The key problem to be solved is how to encode pairwise constraints into an existing clustering algorithm. In this paper, we focus on how to encode pairwise constraints into *spectral clustering*, since spectral clustering has proved to be competitive in terms of the accuracy and the wide applicability [11–17]. Existing work about this problem can be coarsely broken into two camps: (1) they modify the original similarity matrix to encode pairwise constraints, such as [5, 6, 10, 18]; (2) they regularize the spectral embedding to encode pairwise constraints, such as [8]. However, little work has ever explicitly considered the extensional sense of pairwise constraints. The aim of this paper is to explicitly characterize the extensional sense and to improve clustering accuracy. In essence, our work belongs to the first camp, using the extensional sense to modify the original affinity matrix.

The structure of this paper is introduced below. In Section 2, we review the existing constrained clustering algorithms. In Section 3 we interpret the pairwise constraints and its extensional sense, and we further characterize the extensional sense based on the theory of absorbing Markov chains. In Section 4, we describe our constrained spectral clustering algorithm. In Section 5, we evaluate the proposed algorithm over benchmark datasets. In Section 6, we draw the conclusion and mention the future work.

The notation that we used is in general as follows. Vectors are denoted by bold lower-case letters. Matrices are denoted by upper-case ones or calligraphy letters. Sets are denoted by italic upper-case letters or calligraphy letters. Scalars are denoted by italic lower-case letters. A part of symbols are introduced in Table 1.

2   Related Work

Constrained spectral clustering focuses on mitigating the blindness of unsupervised spectral clustering by incorporating few user supervision. By now, there exists a variety of algorithms which improve clustering in different ways. The typical works on CSC include [3, 5–10, 19]. The work in [5, 6, 18] enforced pairwise constraints by modifying the similarity matrix in different ways. Yu and Shi [3] formulated the problem as a subspace projection. Li et al. [8] used pairwise constraints to regularize the spectral embedding of standard spectral clustering. Lu and Carreira-Perpiñánand [7] formulated a new similarity matrix by propagating pairwise constraints into neighborhoods through the Gaussian process. Wang et al. [10] proposed a flexible framework to control the degree that pairwise constraints are respected.

3   Problem Formulation

In this section, we first interpret the extensional sense of pairwise constraints in details. And then we show how to use absorbing Markov chains to characterize the extensional sense of pairwise constraints. Finally, we introduce an approach to mixing the original probability transition matrix with pairwise constraints and their extensional sense.

3.1   Pairwise Constraints

The so-called pairwise constraints refer to Must-Links and Cannot-Links, which are first termed in [2]. A concise description for their definitions is as follows:

- If two vectors $\mathbf{x}_i$ and $\mathbf{x}_j$ lie in a same cluster, then the edge $(\mathbf{x}_i, \mathbf{x}_j)$ is called a Must-Link;
- If two vectors $\mathbf{x}_i$ and $\mathbf{x}_j$ lie in different clusters, then the edge $(\mathbf{x}_i, \mathbf{x}_j)$ is called a Cannot-Link.

If there exists at least one Must-Link or one Cannot-Link incident on data point $\mathbf{x}_i$, we say that $\mathbf{x}_i$ is a *constrained point*. Throughout this paper, we use $\mathcal{M}$ to represent the set of Must-Link constraint pairs and use $\mathcal{C}$ to represent the set of Cannot-Link constraint pairs.

3.2   The Extensional Sense of Pairwise Constraints

Let $\mathbf{a}$ and $\mathbf{b}$ denote two Must-Link points, and let $\mathbf{a}$ and $\mathbf{c}$ denote two Cannot-Link points. We have the following straightforward observations:

- For point $\mathbf{a}$, its neighbors have certain high probabilities falling into a same cluster with point $\mathbf{a}$, so is for points $\mathbf{b}$ and $\mathbf{c}$;
- Neighbors of $\mathbf{a}$ have certain high probabilities falling into a same cluster with neighbors of $\mathbf{b}$;
- Neighbors of $\mathbf{a}$ have certain high probabilities falling into different clusters with neighbors of $\mathbf{c}$;
- Neighbors of $\mathbf{b}$ have certain high probabilities falling into different clusters with neighbors of $\mathbf{c}$.

We refer to the items above as the extensional sense of pairwise constraints. We assume that this extensional sense aids in improving clustering results, if they are appropriately incorporated with spectral clustering.

3.3 Encoding the Extensional Sense Based on Absorbing Markov Chains

Graph is a powerful tool in representing the relationships between pairwise points. In a global viewpoint, an affinity graph could reflect the manifold structure in the input data. And in a local viewpoint, the similarities between pairwise neighbors reflect the neighborhood relationships. Spectral clustering is a graph-based method, which allows the similarities transmitting from neighbors to neighbors and so it could well-detect sub-manifold structures [20]. This is acted as the main advantage compared to center-based clustering methods [21]. Inspired by this, it is natural to use a graph-based method to characterize the aforementioned extensional sense of pairwise constraints. Below we use absorbing Markov chains [22] as a tool to model the extensional sense of pairwise constraints.

Let us treat all constrained points as absorbing states and treat all non-constrained points as transient states. And then we imagine that a drunken man walks randomly in the affinity graph, under the constraint that he is not allowed to leave when he arrives at an absorbing state. Starting from any transient state, the probabilities that the drunken man arrives at all absorbing states can be calculated based on the theory of absorbing Markov chains. Below is a normal description for our idea.

To be clear in description, we first give a part of notations. Let $X = \{\mathbf{x}_i\}_{i=1}^n$ denote a given vector dataset. Let us associate X with an affinity graph $G(V, E)$, in which nodes represent vectors among X and edge weights between nodes denote the similarities between pairwise vectors. In constrained spectral clustering, the graph G should be partly labeled, hence we break X into two subsets X_l and X_u where X_l consists of the vectors that the labels are known and X_u consists of the vectors that the labels are unknown. For the convenience of description, suppose that X_l comes first, i.e., $X = \{X_l, X_u\}$. Let $\mathbf{W}$ denote the weight matrix associated with the affinity graph G whose elements represent the similarities between pairwise points. For an unlabeled graph, the probability transition matrix $\mathbf{P}^u$ can be obtained from a row normalization of $\mathbf{W}$ as follows:

$$\mathbf{P}_{ij}^u = \frac{\mathbf{W}_{ij}}{\sum_{j=1}^n \mathbf{W}_{ij}}. \tag{1}$$

However, for a partly-labeled graph like G, we need modify $\mathbf{P}^u$ as a new form. Let $\mathbf{P}$ denote the probability transition matrix of G. We define its elements as

$$\mathbf{P}_{ij} = \begin{cases} \mathbf{P}_{ij}^u & \text{if } i > m \\ 0 & \text{if } i \leq m, i \neq j \\ 1 & \text{if } i \leq m, i = j, \end{cases} \tag{2}$$

where m denotes the number of constrained points.

The matrix $\mathbf{P}$ can also be written as the following block form:

$$\mathbf{P} = \begin{pmatrix} \mathbf{I} & \mathbf{O} \\ \mathbf{R} & \mathbf{Q} \end{pmatrix}, \tag{3}$$

where $\mathbf{I}_m$ represents a m-by-m diagonal matrix with all diagonal elements ones, and $\mathbf{O}$ represents a m-by-$(n-m)$ zero matrix. Hence, $\mathbf{R}$ represents the relationships between labeled points and unlabeled points, and $\mathbf{Q}$ represents the relationships between pairwise unlabeled points.

Let $\mathbf{F}=(\mathbf{I}_{(n-m)}-\mathbf{Q})^{-1}$ be the fundamental matrix of absorbing matrix chains. Then $\mathbf{F}_{ij}$ gives the average times of the random walk starting from transient states to transient states. Based on the theory of absorbing Markov chains, the absorption probabilities from each transient state to each absorbing state can be calculated as follows:

$$\mathbf{H} = \mathbf{FR}. \tag{4}$$

Here $\mathbf{H}_{ij}$ represents the probability that the drunken man starts from transient state i but is absorbed by absorbing state j. $\mathbf{H}_{ij}$ also reflects what degree unlabeled point i has the same label with labeled point j.

Above, we treat each constrained point as an absorbing state but have not used the supervision information. Intuitively, we can define co-similarities between non-constrained points (the similarities given rise to by constraints) as

$$\mathbf{S}_{ij}^{co} = \sum_{(k,l)\in\mathcal{M},k\neq l} \mathbf{H}_{ik}\mathbf{H}_{jl} - \sum_{(k,l)\in\mathcal{C},k\neq l} \mathbf{H}_{ik}\mathbf{H}_{jl} \tag{5}$$

However, the expression above has not taken into the unbalance between Must-Links and Cannot-Links. For example, if Must-Links are many but Cannot-Links are scarce, then the Eq.(5) will almost ignore the Cannot-Links so that the Cannot-Link constraints play a minor role. For this reason, we modify the co-similarity definition as follows

$$\mathbf{S}_{ij}^{co} = \delta(k,l) \max_{(k,l)\in\mathcal{M}\bigcup\mathcal{C}}\{\mathbf{H}_{ik}\mathbf{H}_{jl}\}, \tag{6}$$

where

$$\delta(k,l) = \begin{cases} 1 & \text{if } (k,l)\in\mathcal{M} \\ -1 & \text{if } (k,l)\in\mathcal{C}. \end{cases}$$

Especially, the co-similarity between a non-constrained point i and a constrained point j is given by $\mathbf{H}_{ij}\mathbf{H}_{jj}$, i.e.

$$\mathbf{S}_{ij}^{co} = \mathbf{H}_{ij}\mathbf{H}_{jj} = \mathbf{H}_{ij}. \tag{7}$$

If two points i,j are a Must-Link pair, let $\mathbf{S}_{ij}^{co}=1$; if two points i,j are a Cannot-link pair, let $\mathbf{S}_{ij}^{co}=-1$ By now, the supervision information is encapsulated in matrix $\mathbf{S}^{co}$ and we call it *co-similarity matrix*.

Below we show that this co-similarity matrix can be obtained by matrix multiplications. First, we modify matrix $\mathbf{H}$ such that the elements in each row become zeros except for the maximal row element. The obtained matrix is denoted as $\mathbf{H}^{max}$. Then, we can construct the following form of matrix $\mathcal{H}$

$$\mathcal{H} = \begin{pmatrix} \mathbf{I}_{m\times m} \\ \mathbf{H}^{max} \end{pmatrix}. \tag{8}$$

Let $\mathbf{Y}$ denote a m-by-m matrix that encapsulates all the pairwise constraints, i.e.

$$\mathbf{Y}_{kl} = \begin{cases} 1 & (k,l) \in \mathcal{M} \\ -1 & (k,l) \in \mathcal{C} \\ 0 & k = l. \end{cases}$$

Thus, we can obtain an important expression as follows

$$\mathbf{S}^{co} = \mathcal{H}\mathbf{Y}\mathcal{H}^T. \tag{9}$$

In order to incorporate the original probability transition matrix with the co-similarity matrix, we introduce a balance factor γ $(0 < \gamma < 1)$. Letting $\mathcal{P}$ denote the mixed matrix, it is formulated as

$$\mathcal{P} = \gamma\mathbf{P}^u + (1 - \gamma)\mathbf{S}^{co}. \tag{10}$$

Let $\mathbf{T}$ be a n-by-n matrix whose elements are given by $\mathbf{T}_{ij} = \max\{0, \mathcal{P}_{ij}\}$. Let $\mathcal{D}$ be a diagonal matrix whose diagonal elements are given by

$$\mathcal{D}_{ii} = \sum_j \mathbf{T}_{ij}. \tag{11}$$

Then, the mixed probability transition matrix can be obtained from

$$\mathcal{T} = \mathcal{D}^{-1}\mathbf{T}. \tag{12}$$

By now, we have renewed the original affinity matrix as a new probability transition matrix $\mathcal{T}$ in which the supervision information is included. The intensity of supervision is controlled by the parameter γ.

3.4   Our Algorithm

Based on the modeling and analysis above, we can give a new constrained spectral clustering algorithm. Since our algorithm is based on absorbing Markov chains, we write the algorithm as MAC-CSC in short.

The affinity matrix can be obtained from many methods, e.g. RBF kernel, k-nearest neighbor method [23], locally linear reconstruction [24], b-matching method [25], fitting method [26], and so forth. In our algorithm, we do not concern about how the graph construction methods influence the performance of the proposed method. We simply use the well-known RBF kernel to construct the affinity matrix, that is,

$$\mathbf{W}_{ij} = \exp(-\frac{\|\mathbf{x}_i - \mathbf{x}_j\|^2}{2\sigma^2}), \tag{13}$$

where σ is a turnable parameter. And by conventionally, we set $\mathbf{W}_{ii}=0$.

Our algorithm is suitable for the case that both Must-Links and Cannot-Links are available. In step 4, we allow the co-similarities to be negative which can make full use of the Cannot-Link constraints. In step 5, the normalization processing

---

Algorithm 1. MAC-CSC

1: Input: affinity matrix $\mathbf{W}$, constrained set, the number of clusters K
2: Compute the probability transition matrix $\mathbf{P}^u$ using Eq.(1)
3: Compute the absorbing probability matrix $\mathbf{H}$ using Eq.(3) and Eq.(4) where $\mathbf{H}_{ij}$ denotes the probability that non-constrained point i is absorbed by constrained point j
4: Compute the co-similarity matrix $\mathbf{S}^{co}$ using Eq.(9)
5: Form normalized co-similarity matrix $\mathcal{S}$ by normalizing the positive elements of $\mathbf{S}^{co}$ to row sum 1 and normalizing the remained negative elements to row sum -1
6: Form mixed matrix $\mathcal{P}=\mathbf{P}^u + \mathcal{S}$ and update all the negative elements to be zeros
7: Form matrix $\mathcal{T}$ by normalizing $\mathcal{P}$ to have row sum 1
8: Solve eigen-system $\mathcal{T}\mathbf{x}=\lambda\mathbf{x}$, and use the top K eigenvectors to form matrix $\mathbf{U} \in \mathbb{R}^{n \times K}$
9: Normalize the $\mathbf{U}$'s row to have unit length
10: Run KMeans algorithm over $\mathbf{U}$

---

aims to consider the balance between Must-Link constraints and Cannot-Link constraints. In step 6, the matrix $\mathcal{P}$ is obtained from mixing the affinity matrix and the co-similarity matrix. Considering the subsequent requirement of spectral decomposition, we eliminate the negative elements from the matrix $\mathcal{P}$ and set them to be zeros.

4 Experimental Evaluation

In this section, experiments are carried out to assess our algorithm. In the experiment, we simply set the parameter γ to be 0.5 and choose the parameter σ automatically as the average of Euclidean distances over all pairs of vector data. For each dataset, we run 50 times by choosing random constrained points and then show the average accuracies as well as the standard deviations.

4.1 Datasets

We use an artificial dataset "XOR" shown in Fig.1(a) and five real-world datasets available from UCI machine learning repository[1]. We list the fundamental information of these datasets in Table 2. The symbol # counts the number of the followed objects.

Table 2. Datasets

	XOR	Balance	Wine	Wdbc	Ionosphere	Vehicle
# Instances	400	625	178	569	351	846
# Classes	2	3	3	2	2	4
# Features	2	5	13	30	34	19

---

[1] http://archive.ics.uci.edu/ml

4.2   Evaluation Criterion on Clustering Accuracy

As used in previous studies, the most suitable evaluation criterion for constrained clustering is the Rand Index (RI). Below we introduce its definition. Let $\mathbf{y} \in \mathbb{R}^n$ be the ground-truth label vector, and let $\hat{\mathbf{y}} \in \mathbb{R}^n$ be the grouping indicator vector obtained from a clustering algorithm. The Rand Index[2] [27] is defined as

$$Rand\ Index = \frac{a + d}{a + b + c + d},\qquad(14)$$

where,

- a: the number of point pairs in X that lie in a same class in $\mathbf{y}$ and also in a same cluster in $\hat{\mathbf{y}}$;
- b: the number of point pairs in X that lie in a same class in $\mathbf{y}$ but in different clusters in $\hat{\mathbf{y}}$;
- c: the number of point pairs in X that lie in different classes in $\mathbf{y}$ but in a same cluster in $\hat{\mathbf{y}}$;
- d: the number of point pairs in X that lie in different classes in $\mathbf{y}$ and also in different clusters in $\hat{\mathbf{y}}$.

4.3   Algorithms for Comparison

Two benchmark algorithms are chosen for a comparison with our algorithm. One is the standard spectral clustering algorithm based on normalized cuts that proposed by Shi and Mailk [11], we write it as Ncut in short. The other is the constrained spectral clustering algorithm based on simple modification of the affinity matrix that proposed by [28], we write it as KKMv in short because it is a variant of KKM algorithm [5]. The KKMv algorithm is a typical method that encode constraints by modifying the similarity of each Musk-Link pair as 1 and modifying that of each Cannot-Link pair as 0, which has not explicitly consider the extensional sense of pairwise constraints.

4.4   Clustering Results

The Result over XOR Dataset. The XOR dataset shown in Fig.1(a) is a very typical dataset that has been used in the previous studies. The aim is to partition the dataset into two groups. The symbols $-$ and $+$ represent the two classes respectively, and in each class has 200 data points. The problem is linearly inseparable. We input this dataset into Ncut algorithm, KKMv algorithm, and our MAC-CSC algorithm. The results are shown in Fig.1(b). It is clearly that our algorithm outperforms the two benchmark algorithms with a very large margin. The results show that the simple modification for the similarities

---

[2] The code is available from http://www.dcorney.com/ClusteringMatlab.html

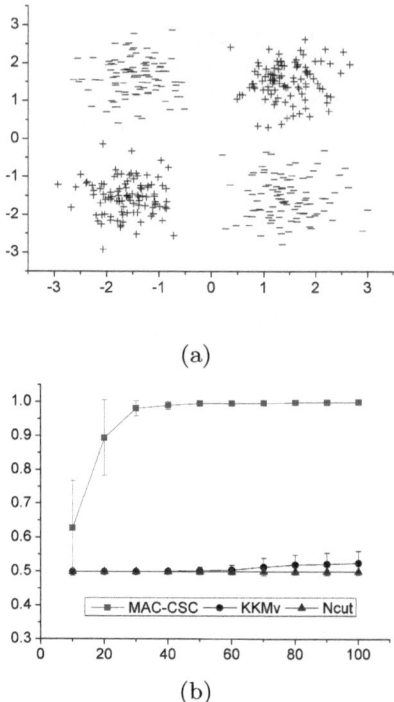

(a)

(b)

Fig. 1. (a) The XOR dataset. (b) The Rand Index versus the number of constrained points.

as [28] slightly influence the clustering accuracy. And an explicit consideration for the extensional sense of constraints could generate significant promotion in the clustering accuracy.

The Result over UCI Datasets. In order to verify the performance of our algorithm, we also carried out experiments over five real-world datasets that are available from UCI machine learning repository. These datasets are often acted as the benchmarks for verifying the performance of new clustering algorithms. We show their clustering results in Fig.2. One can see that our algorithm exhibits good performance in general. On four datasets Ionosphere, Wdbc, Balance, and Vehicle, our algorithm significantly outperforms the other two benchmark algorithms. Only on Wine dataset, our result does not have significant advantage.

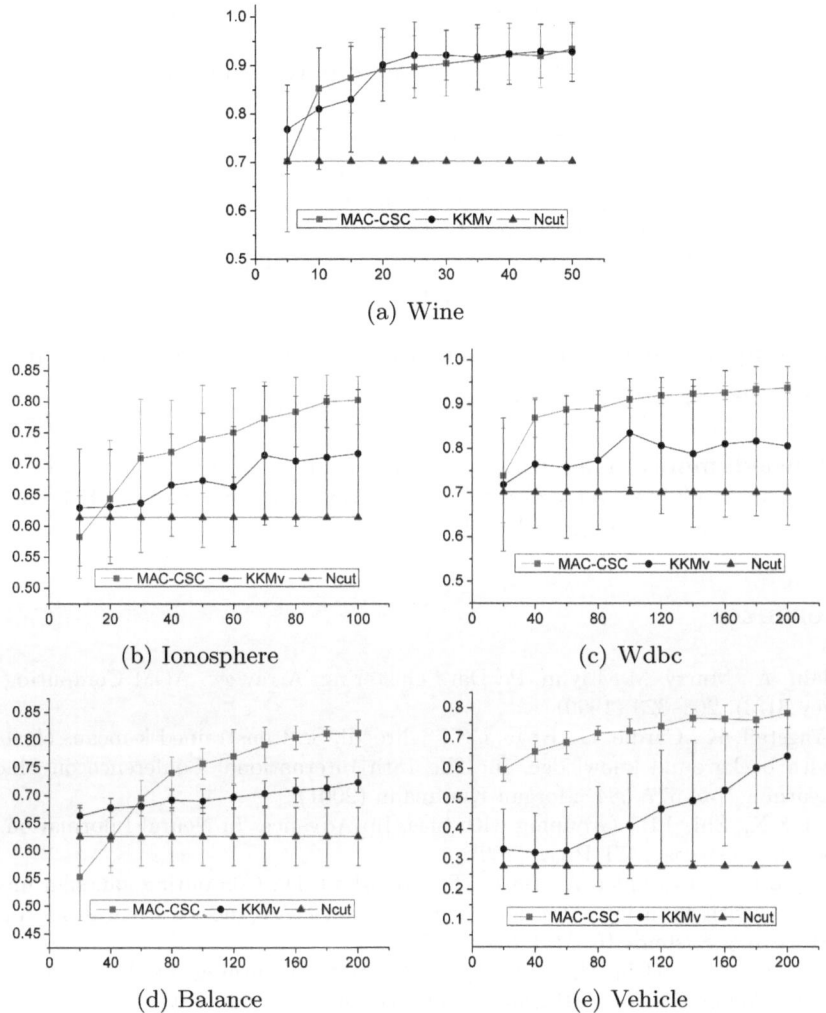

Fig. 2. The Rand Index versus the number of constrained points

5   Conclusion and Future Work

In this paper, we have shown that absorbing Markov chains can be acted as a tool to characterize the extensional sense of pairwise constraints, and that the derived constrained spectral clustering algorithm can make full use of the extensional sense of pairwise constraints. Based on experiments, we draw the following conclusions:

- The MAC-CSC algorithm can encode Must-Link constraints and Cannot-Link constraints simultaneously.
- The MAC-CSC algorithm can be applied to handle linearly-inseparable datasets.
- After we encode the pairwise constraints and their extensional sense using Markov chains, the clustering results can produce significant promotion relative to both the spectral clustering benchmark Ncut and the constrained spectral clustering benchmark KKMv.
- The MAC-CSC algorithm might obtain very good clustering results even if the available constraint information is few.

In the future, we will concentrate on how to scale our algorithm to large problem sizes.

Acknowledgment. This work was supported by National Natural Science Foundation of China under grant No. 61173118 and "Shuguang" Scholar Program of Shanghai Education Foundation.

References

1. Jain, A., Murty, M., Flynn, P.: Data clustering: A review. ACM Computing Survey 31(3), 264–323 (1999)
2. Wagstaff, K., Cardie, C., Rogers, S., Schroedl, S.: Constrained k-means clustering with background knowledge. In: The 18th International Conference on Machine Learning, pp. 577–584. Morgan Kaufmann (2001)
3. Yu, S.X., Shi, J.B.: Grouping with bias. In: Advances in Neural Information Processing Systems. MIT Press (2001)
4. Shental, N., Bar-hillel, A., Hertz, T., Weinshall, D.: Computing gaussian mixture models with em using equivalence constraints. In: Advances in Neural Information Processing Systems 16. MIT Press (2003)
5. Kamvar, S.D., Klein, D., Manning, C.D.: Spectral learning. In: The International Joint Conferences on Artificial Intelligence, pp. 561–566 (2003)
6. Ji, X., Xu, W.: Document clustering with prior knowledge. In: The 29th Annual International Conference on Research and Development in Information Retrieval, pp. 405–412. ACM, New York (2006)
7. Lu, Z.: Constrained spectral clustering through affinity propagation. In: IEEE Conference on Computer Vision and Pattern Recognition. IEEE Computer Society (2008)
8. Li, Z., Liu, J., Tang, X.: Constrained clustering via spectral regularization. In: IEEE Conference on Computer Vision and Pattern Recognition, pp. 421–428. IEEE (2009)
9. Coleman, T., Saunderson, J., Wirth, A.: Spectral clustering with inconsistent advice. In: The 25th International Conference on Machine Learning, pp. 152–159. ACM (2008)
10. Wang, X., Davidson, I.: Flexible constrained spectral clustering. In: The 16th ACM SIGKDD International Conference on Knowledge Discovery and Data Mining, pp. 563–572 (2010)

11. Shi, J.B., Malik, J.: Normalized cuts and image segmentation. IEEE Transactions on Pattern Analysis and Machine Intelligence 22(8), 888–905 (2000)
12. Ng, A.Y., Jordan, M.I., Weiss, Y.: On spectral clustering: Analysis and an algorithm. In: Advances in Neural Information Processing Systems 14, pp. 849–856. MIT Press (2001)
13. Zelnik-manor, L., Perona, P.: Self-tuning spectral clustering. In: Advances in Neural Information Processing Systems 17, pp. 1601–1608. MIT Press (2004)
14. Ning, H., Xu, W., Chi, Y., Gong, Y., Huang, T.: Incremental spectral clustering with application to monitoring of evolving blog communities. In: The SIAM International Conference on Data Mining (2007)
15. Song, Y., Chen, W.-Y., Bai, H., Lin, C.-J., Chang, E.Y.: Parallel Spectral Clustering. In: Daelemans, W., Goethals, B., Morik, K. (eds.) ECML PKDD 2008, Part II. LNCS (LNAI), vol. 5212, pp. 374–389. Springer, Heidelberg (2008)
16. Alzate, C., Suykens, J.A.K.: Multiway spectral clustering with out-of-sample extensions through weighted kernel pca. IEEE Transactions Pattern Analysis and Machine Intelligence 32(2), 335–347 (2010)
17. Rangapuram, S.S., Hein, M.: Constrained 1-spectral clustering. Journal of Machine Learning Research, W & CP 20, 1143–1151 (2012)
18. Hu, G., Zhou, S., Guan, J., Hu, X.: Towards effective document clustering: A constrained k-means based approach. Information Processing and Management 44(4), 1397–1409 (2008)
19. Xing, E.P., Ng, A.Y., Jordan, M.I., Russell, S.J.: Distance metric learning with application to clustering with side-information. In: NIPS, pp. 505–512 (2002)
20. Fowlkes, C., Belongie, S., Chung, F.R.K., Malik, J.: Spectral grouping using the nyström method. IEEE Transactions on Pattern Analysis and Machine Intelligence 26, 214–225 (2004)
21. Macqueen, J.B.: Some methods of classification and analysis of multivariate observations. In: The Fifth Berkeley Symposium on Mathematical Statistics and Probability, pp. 281–297 (1967)
22. Norris, J.R.: Markov Chains. Cambridge University Press, Cambridge (1997)
23. Luxburg, U.V.: A tutorial on spectral clustering. Statistics and Computing 17(4), 395–416 (2007)
24. Roweis, S., Saul, L.: Nonlinear dimensionality reduction by locally linear embedding. Science 290(5500), 2323–2326 (2000)
25. Jebara, T., Wang, J., Chang, S.: Graph construction and b-matching for semisupervised learning. In: ICML, p. 56 (2009)
26. Daitch, S.I., Kelner, J.A., Spielman, D.A.: Fitting a graph to vector data. In: The 26th Annual International Conference on Machine Learning, p. 26 (2009)
27. Rand, W.M.: Objective criteria for the evaluation of clustering methods. Journal of the American Statistical Association 66, 846–850 (1971)
28. Xu, Q., Desjardins, M., Wagstaff, K.: Constrained spectral clustering under a local proximity structure assumption. In: The International Conference of the Florida Artificial Intelligence Research Society. AAAI Press (2005)

Inducing Taxonomy from Tags: An Agglomerative Hierarchical Clustering Framework

Xiang Li[1], Huaimin Wang, Gang Yin, Tao Wang, Cheng Yang, Yue Yu, and Dengqing Tang[2]

[1] National Laboratory for Parallel and Distributed Processing,
School of Computer Science,
National University of Defense Technology, Changsha, China
{shockleylee,jack.nudt,taowang.2005}@gmail.com
[2] College of Mechatronics Engineering and Automation,
National University of Defense Technology, Changsha, China
http://www.nudt.edu.cn

Abstract. By amassing 'wisdom of the crowd', social tagging systems draw more and more academic attention in interpreting Internet folk knowledge. In order to uncover their hidden semantics, several researches have attempted to induce an ontology-like taxonomy from tags. As far as we know, these methods all need to compute an overall or relative generality for each tag, which is difficult and error-prone. In this paper, we propose an agglomerative hierarchical clustering framework which relies only on how similar every two tags are. We enhance our framework by integrating it with a topic model to capture thematic correlations among tags. By experimenting on a designated online tagging system, we show that our method can disclose new semantic structures that supplement the output of previous approaches. Finally, we demonstrate the effectiveness of our method with quantitative evaluations.

Keywords: social tagging, semantics, tag taxonomy, tag generality, agglomerative hierarchical clustering, topic model.

1 Introduction

Social tagging websites like Delicious[1] and Flickr[2] are becoming popular, witnessing a soaring click rate during recent years. By annotating web contents with free-form tags that they feel appropriate, users of social tagging websites are enabled to play the dual role of visitors and contributors simultaneously. Specifically, a user is allowed to search and browse resources (documents, images, URLs, etc.) through tags annotated by others and is free to add or delete tags whenever he or she wants.

---

[1] http://www.delicious.com/
[2] http://www.flickr.com/

S. Zhou, S. Zhang, and G. Karypis (Eds.): ADMA 2012, LNAI 7713, pp. 64–77, 2012.

(a) tag cloud in Freecode.com   (b) taxonomy in Amazon.com

Fig. 1. Different navigation mechanisms

For computer programs, tags are merely character strings with no meaningful relationships among them, which makes it hard to organize them into an informative structure. As depicted in Fig.1(a), most tagging websites manage tags in a flat tag cloud, where the font size of a tag is proportional to its frequency of usage, so users can have easy access to buzzword tags. Sometimes however, the desired tags are vague in users' head and may go beyond what is popular. A tag cloud could do little help in this situation, and tags should be better arranged and managed to satisfy the requirement. In retail websites, a user usually navigates through hierarchical taxonomy (as in Fig.1(b)) when her requirement is only a vague notion rather than a concrete keyword. So promisingly, taxonomy of tags will analogously facilitates navigation and searching in the now-booming social tagging websites. Besides, in the view of online data analysts, generating tag taxonomy is crucial for representing and understanding Internet folk knowledge. An ontology-like taxonomy is intrinsically a good knowledge structure which captures semantics of terms in a machine understandable way[1].

Previous literature[2][3] has addressed on how to generate tag taxonomy effectively. By using machine learning techniques, these researches achieved automated taxonomy generation, which could bootstrap and alleviate manual construction. However, new methods[4][5] keep emerging because resultant taxonomy never seemed satisfying enough. In an attempt to detect the bottleneck, we find that to deduce semantic generality of each tag is the most difficult and error prone (see Chapter 2 Related Work). To meet such challenge, we propose a novel approach of tag taxonomy construction. Our innovation lies in: i) By using an *Agglomerative Hierarchical Clustering* (AHC) framework, we only need to compute how similar two tags are and deliberately skip the calculation of tag generality. ii) We seamlessly integrate a *topic model* into the framework, so it is able to capture thematic correlations among tags. iii) We make comparative evaluation with existing approaches, the results demonstrate the usefulness and effectiveness of our method.

2   Related Work

Because of the convention of constructing taxonomies with the general tags at the top and the more specific tags below them[6], a key step in existing taxonomy construction methods is extracting generality of each tag, either by computing a generality score[2][7] or by pair-wise comparison[3][4]. As far as we know, two types of techniques are used to achieve this step.

Some research works[2][7] use *set theory* techniques to compute a tag generality. In such works, each tagged resource is treated as a distinct data item with their textual contents ignored. Each tag is presented as the collection of items it annotates. For example, Heymann et al.[7] proposed a simple yet effective way to learn a tag taxonomy. They first model each tag as a vector, with each entry being the number of times the tag annotates a corresponding resource. *Cosine metric* is then used to measure tag similarity. They come up with the *tag similarity graph* by connecting sufficiently similar (controlled with a threshold) tags. The *graph centrality* metric, which is originally proposed in social network analysis literature, is used to measure the generality of each tag. Finally, tags of higher generality are greedily placed at the upper levels of the resultant taxonomy. Liu et al.[2] used association rule mining which takes each tagged resource as a transaction and tags as items. The rules are in the form of "for an unknown resource X, if tag A appears then probably tag B will also appear", which they call "tag B subsumes tag A". The likelihood of such subsumption is naturally modeled as the confidence and support of the corresponding rule. Based on subsumption likelihood between each pair of tags they calculate an overall generality score for each tag by using a random walk process. Eventually, a taxonomy is constructed using a top-down manner.

However, such set theory based approaches share a common flaw. They only distinguish one resource from another and do not exploit tagged web documents. In light of this, some researches apply topic models to do the job. Such methods learn a latent topic distribution for each tag from web documents they annotate, tag relations are measured based on pair-wise comparison among topic distributions. Tang et al.[3] designed the *Tag-Topic model* based on the classic *LDA (Latent Dirichlet Allocation)* model[8]. They treat each word as a draw from a topic-specific word distribution, and a latent topic is in turn a draw from a tag-specific topic distribution. Based on such a generative model, Gibbs sampling is used to infer distribution parameters. They use Kullback-Leibler (KL) divergence to measure the difference between two distributions, and have proposed a relative generality score metric based on the intuition "if one tag has higher posterior probabilities on the latent topics, then it has a relatively higher generality". Wang et al.[4] merge documents annotated by the same tag (they use 'keyword' instead) as a new document which they believe can explain that tag. They learn a standard LDA model out of the underlying corpus. Those new documents are folded-in thereafter. Thus, topic distribution of each tag (new document) is obtained. Their measurement of relative generality is based on the "surprise" theory[9] plus an intuitive law that "given an anticipated tag A, the appear of a document on a more general tag B will cause less 'surprise' than if

A and B are switched". It is really difficult to come up with a tag generality score that both [3] and [4] resort to intuitions.

3 AHCTC: Agglomerative Hierarchical Clustering for Taxonomy Construction

In order to avoid computing tag generality, we borrow the idea from the classic agglomerative hierarchical clustering which only relies on how similar/distant two points are in building a hierarchy.

The term *agglomerative hierarchical clustering* is not new in the field of taxonomy construction. Brooks and Montanez[10] adopted such methods to organizing tags into a hierarchy, yet their hierarchy is not a taxonomy for lack of supertype-subtype relationships. In Liu et al.[2], authors also mention the use of agglomerative hierarchical clustering. However, their strategy is to iteratively place the most general tag remained, which is different from the classic meaning of the agglomerative hierarchical clustering framework. The brilliance of agglomerative hierarchical clustering is yet to be fully exploited in the field of taxonomy construction.

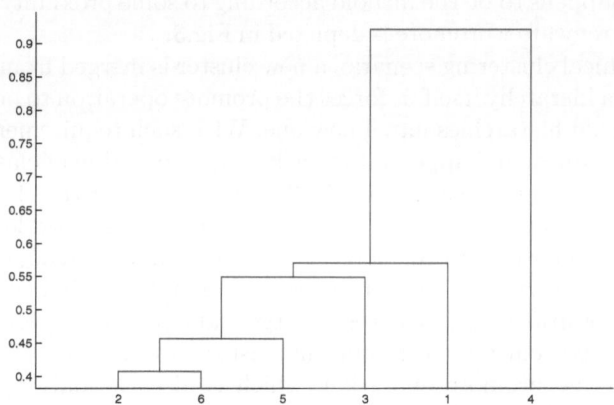

Fig. 2. Dendrogram of a typical clustering process

To illustrate the AHCTC framework in a broader picture, we eliminate any presumption on how the proximity scores are calculated, but rather assume they are known a-priori. In the next chapter, we will describe how to get a meaningful proximity score. With each data point being an initial cluster, classic agglomerative hierarchical clustering merges two closest clusters each time until only one all-inclusive cluster is left. The dendrogram of a typical AHC process consisting of 6 points are depicted in Fig.2. Sadly, the result is not a taxonomy for the absence of supertype-subtype relationships. In the dendrogram, for example,

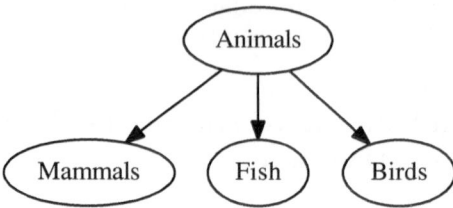

Fig. 3. A hierarchy of the cluster {*Birds, Mammals, Fish, Animals*} after promote

a supertype should be there to overarch the cluster consisting of point 2 and 6. Now that data points are tags in our problem, a tag cluster should have a direct supertype (parent tag) or the resultant tag tree is not technically a taxonomy.

Of course, a tag cluster's direct supertype should have high *semantic proximity* to each tag in the cluster. For the cluster {*Birds, Mammals, Fish Animals*}, a tag like *Animals* is more suitable to be the supertype than a *Living Things* tag. During the clustering process, tags with high proximity are assigned to the same cluster, so the supertype should better be selected from within a given cluster. Specifically, we choose the *medoid* of a cluster, i.e., the most central point. This medoid upgrades as the supertype of all other cluster members. We call this change as a **promote** operation. Now suppose the original cluster is {*Birds, Mammals, Fish, Animals*}, and *Animals* happens to be the medoid according to some proximity measure, the resultant hierarchy after promote is depicted in Fig.3.

In a hierarchical clustering scenario, a new cluster is merged from two old clusters each with a hierarchy itself, it forces the promote operation to be able to combine these two old hierarchies into a new one. With such requirements, we devise a promote mechanism and append it to each merge operation of classic AHC, the new taxonomy construction algorithm is illustrated in Algorithm 1. Two building blocks of the algorithm, $proximity(i, j)$ and $sup(m)$ are assumed known a-priori. We will illustrate how to get them later. The algorithm uses an adjacent matrix T to store resultant taxonomy. Initially, every data point is a distinct cluster. Each iteration in the **while** clause executes a merge and a promote operation. Line 4-7 finds the closest two clusters in the current cluster set to merge.

Line 8-21 is the promote operation, which is also the only operation edits T. Different from the basic promote operation we have described in Fig.3, the promote operation here will be much more intricate since a hierarchical scenario is concerned. For ease of illustration, we call the the two supertypes over the clusters being merged as **old supertypes**. Line 8-11 says if the new supertype $sup(m)$ happens to be one of the two old supertypes, an directed edge is added from it to the other old supertype (see Fig.4(a) and Fig.4(b)).

Otherwise, if $sup(m)$ comes from *the public*, the procedure will have to go through the intricacies of line 13-20. Fig.4(c) and Fig.4(d) displays them in details. Simply speaking, former subtypes of point B (the newly elected supertype), i.e. D and E, should now be A's subtypes after B's promote operation. This is because A is formerly their nearest ancestor besides B, and we want to maintain the relation that A be more general than either D or E after B's promote. To

---

Algorithm 1. AHCTC

Require: Data points $D = \{d_1, d_2, ..., d_n\}$.
Require: $T = [t_{ij}]_{n \times n}$ the adjacent matrix for the resultant taxonomy, $t_{ij} = 1$ when d_i is a direct supertype of d_j.
Require: $M = \{m_1, m_2, ...\}$, the set of all remaining clusters.
Require: $sup(m)$, index of the supertype data point for a given cluster m.
Require: $proximity(i, j)$, the proximity score between data points d_i and d_j.
Ensure: construct the resultant taxonomy T.
1: $T \leftarrow (0)_{n \times n}$
2: $M \leftarrow D$
3: **while** $|M| > 1$ **do**
4:     find m_i and m_j in M with maximum $proximity(sup(m_i), sup(m_j))$
5:     merge m_i and m_j as m
6:     add m to M
7:     delete m_i and m_j from M
8:     **if** $sup(m)$ equals $sup(m_i)$ **then**
9:         $t_{sup(m),sup(m_j)} \leftarrow 1$
10:     **else if** $sup(m)$ equals $sup(m_j)$ **then**
11:         $t_{sup(m),sup(m_i)} \leftarrow 1$
12:     **else**
13:         find k that $t_{k,sup(m)}$ equals 1
14:         $t_{k,sup(m)} \leftarrow 0$
15:         **for all** g that $t_{sup(m),g}$ equals 1 **do**
16:             $t_{sup(m),g} \leftarrow 0$
17:             $t_{k,g} \leftarrow 1$
18:         **end for**
19:         $t_{sup(m),sup(m_i)} \leftarrow 1$
20:         $t_{sup(m),sup(m_j)} \leftarrow 1$
21:     **end if**
22: **end while**

---

implement this process, the algorithm detaches $sup(m)$ from its direct supertype k (line 13-14), attaches all children of $sup(m)$ to k (line 15-18) and finally the two old supertypes are attached to $sup(m)$ (line 19-20).

If n is the number of data points, the basic AHC algorithm requires $O(n^2 \log n)$ time[11]. The only extra step of our algorithm is the promote operation (line 8-21), which at worst induces $O(n^2)$. The **while** loop will iterates for $n - 1$ times, because after each iteration the number of clusters only reduces one. At the worst case, all $n - 1$ iterations undergo line 13-20, each needs to look up and edit at most $n - 1$ edges in T. Totally speaking, our algorithm has the same time complexity as the basic AHC, $O(n^2 \log n)$.

4   Integration with Topic Models

In order to get AHCTC running, we have to mount its assemblies. Notice that $sup(m)$ finds the most central points based on $proximity(i, j)$. So essentially, we

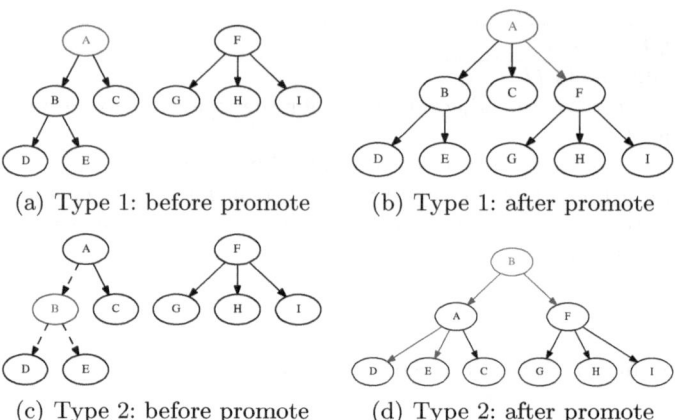

(a) Type 1: before promote   (b) Type 1: after promote

(c) Type 2: before promote   (d) Type 2: after promote

Fig. 4. Two types of promote operation during hierarchical clustering. The selected supertype and newly added edges are in red, edges to be removed are in dashed lines.

only have to implement $proximity(i, j)$ which quantifies the similarity measure between tags.

Web documents annotated by a certain tag usually contain semantic information of that tag. It will be a great loss for a tag taxonomy construction method to overlook latent semantics in tagged web documents. Recent years have witnessed drastic methodology evolution in the field of information retrieval, towards the unchanging goal of uncovering semantic information in unstructured text data. Since the publish of LDA[8], a family of algorithms[12][13] known as Probabilistic Topic Models emerge and quickly win massive adoptions. Probabilistic Topic Models are skilled at discovering hidden thematic structure of text documents. In a real corpus like New York Times, a document may be tightly related to topics like *Foreign Policy* and *Economics*, while only mentions a little on *Sports*. Topic models answer what themes or topics a document relates to and quantify how strong such relations are. Thus, a thematic similarity measure could be induced for each pair of documents. Wang et al.[4] and Tang et al.[3] each design a variant of the basic topic model able to learn thematic structure of tags from tagged corpus. Our $proximity(i, j)$ measure is built on top of Wang's model. Chapter 4.1 briefly introduces stantard LDA and Wang's topic model.

4.1   Probabilistic Topic Model

LDA is the basic Probabilistic Topic Model. In LDA, a latent topic $z = j$ is modeled as an unlabeled corpus-wide word distribution $\phi^{(j)} = P(w|z = j)$, which was drawn from a dirichlet prior distribution $Dirichlet(\beta)$. The number of topics T is specified beforehand to adjust the granularity. Each document d is a mixture of topics $\theta^{(d)} = P(z)$ with a dirichlet prior distribution $Dirichlet(\alpha)$. The generative process of each word in d is essentially a draw from the joint distribution $P(w_i) = \sum_{j=1}^{T} P(w_i|z_i = j)P(z_i = j)$. Given the observed documents, Gibbs Sampling algorithm[14] is usually used for posterior inference.

[4] modifies LDA to deal with tags (keywords). Their assumption is that documents annotated by a tag usually have thematic information of that tag. For a given tag, they merge documents annotated with it into a *new document*, removing those words occurred only once. After the standard LDA training, new documents are folded-in to the trained model by an extra run of Gibbs Sampling algorithm. Finally, a *tag-topic distribution* for each new document is estimated. Jenson-Shannon divergence or cosine similarity measure could be used to calculate the divergence between any two tag-topic distributions.

4.2   Tag Proximity Measure

Based on the above model, our tag proximity measure is chosen just as the divergence between tag-topic distributions: if we think a tag-topic distribution over T topics as the tag's coordinate in a T-dimensional *thematic space*. The divergence between two tag-topic distributions can be understood as the *thematic space coordinate distance* between the two tags. So it becomes clear that such a proximity measure exploits semantic (thematic) correlation among tags. Based on that, $sup(m)$ is defined as the most central point of cluster m in the thematic space, which ensures that a supertype can thematically represents the whole cluster. In other words, $sup(m)$ finds the data point index $\arg\min_i\{\sum_{d_j \in m} proximity(i,j)\}$.

After integrating with the topic model, AHCTC can hierarchically cluster tags that are thematically similar and pick the most central tag in the thematic space as the supertype for each cluster. Notice the difference between our method and [3][4] is that we don't have to tell which tag is more general from the topic model, which is not what a topic model good at.

5   Experiment

Ohloh[3] is a popular online open source software directory and community platform whose user number has exceeded 1,500,000. It provides information on more than 500,000 open source software projects. In the profile of each project, there is a brief description along with other valuable information like development history and technical features. Being a social tagging website, its users are given the freedom to edit this information and to tag the projects. A typical project profile in Ohloh is depicted in Fig.5. To browse projects by tags, users can either pick from the flat tag list or use keyword search for sifting. However, such a flat list does not give the conceptual scope of each tag nor the possible relationships among several tags. A tag taxonomy will apparently do a better job. Based on such a requirement, we choose Ohloh as our dataset to validate our tag taxonomy construction algorithm.

---

[3] http://www.ohloh.net/

Fig. 5. Part of project profile of the Mozilla Firefox in Ohloh

Dataset and Parameter Setting. From Ohloh, we have crawled 10,000 open source software project profiles and extracted their descriptions and tags. Suffering from the common flaws of all folk knowledge, tags might be poor phrased or too esoteric and some are even meaningless. We omit tags with less than 300 references and manually delete the meaningless tags '1', '???', '????', '?????' which somehow are widely adopted in Ohloh. The result after preprocessing is a set of 267 tags. For topic modeling, we set the number of topics at 60 and iterate 2000 times in the Gibbs Sampling process.

Baseline Methods. We compare AHCTC with two of the published methods on the same dataset. For comparison with set theory based methods, we choose the ontology induction algorithm proposed in [2] as our baseline. Since their method is based on tag subsumption, we use the shorthand SUBSUME to stand for their method. Ohloh does not provide author information of each tag, so we use $\mathbb{K}^R = (G = R, M = T, I)$ as the projection from a folksonomy onto a formal context, details are given in the original paper. We set the parameters at the best setting given by the authors, i.e., $\lambda = 0.95, \theta_s = 0.00001, \theta_c = 0.15$ and the maximum number of random walk iteration equals 1000.

For comparison with topic model based methods, we choose the LSHL algorithm proposed in [4] as the baseline. The other algorithm in [4], GSHL, is just a modified version of LSHL aiming at a different goal. Their experiment use keywords of academic paper abstracts as *Concepts* while here we use the 267 Ohloh tags to take their places. Parameters are also set at what the original paper suggests, $TH_s = 0.6, TH_d = 0.35, TH_n = 0.4$ and the number of topics for topic modeling is set to 60.

Experiment Result. The resultant taxonomies are shown in Fig.6. Since the space is limited, only the *framework* subtree of our result is depicted. For result of LSHL algorithm, the *library* subtree is chosen since it has sufficient overlapped tags with AHCTC *framework* subtree. We can see from Fig.6(a) and Fig.6(c) that despite some parts, the two trees are different in their structures.

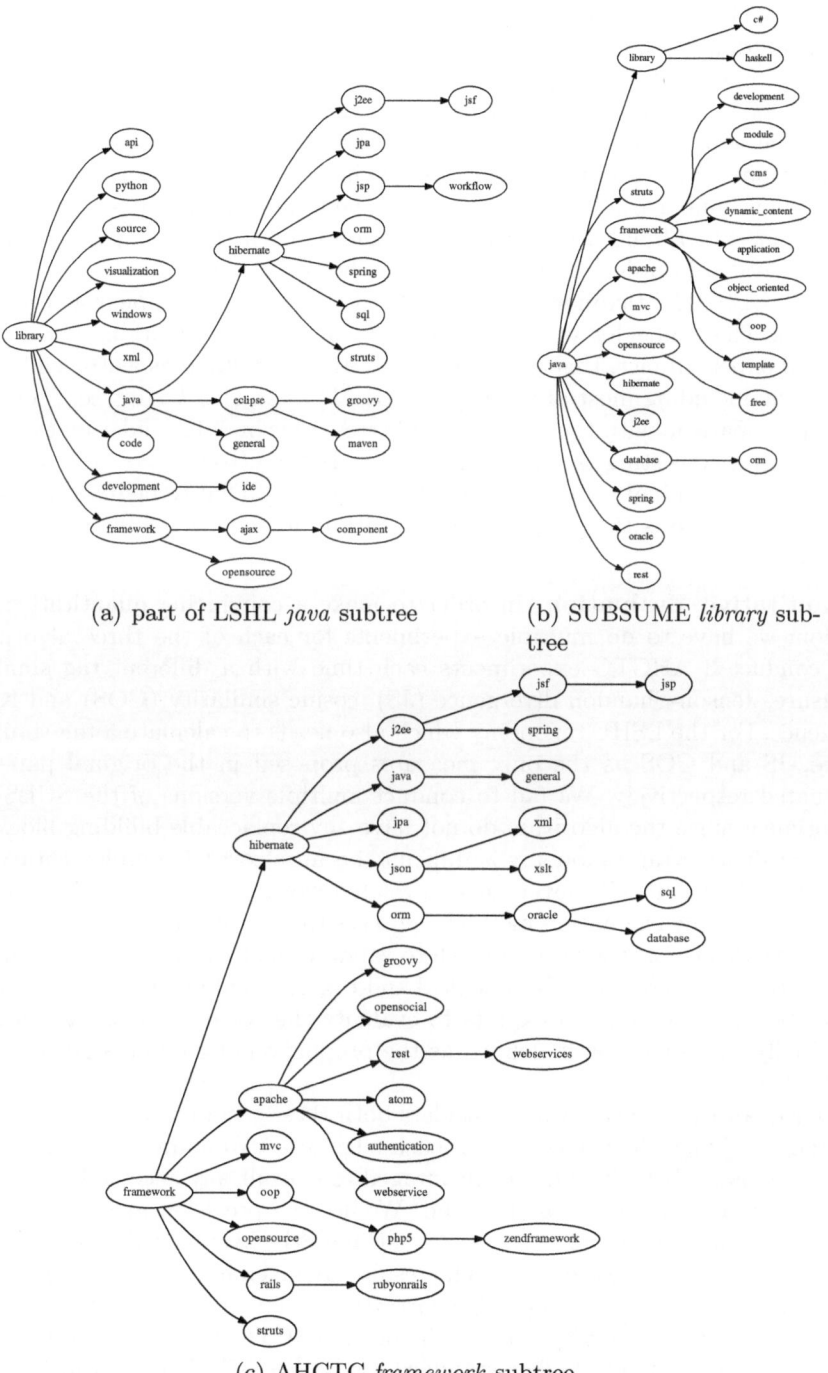

(a) part of LSHL *java* subtree

(b) SUBSUME *library* subtree

(c) AHCTC *framework* subtree

Fig. 6. Part of the resultant taxonomies

The AHCTC *framework* subtree apparently classifies the tags according to the technical framework they belong. In the LSHL *library* subtree, tags are first classified by different types of programming libraries. For example, in the AHCTC *framework* subtree *hibernate* is classified as a kind of framework but in LSHL it is deemed as a java library. The two taxonomies are of different facades. They complement one another and are both valuable in disclosing tag semantic structure. If these machine-learned taxonomies are to be used as prototypes helping ontology engineers to build a well-defined taxonomy, both the resultant taxonomies will be needed in order to provide a comprehensive picture.

The SUBSUME taxonomy seems rather flat, many tags appear as the root's children. This coincides with experiment results and illustrations given in the original paper. In fact, this taxonomy takes 91 out of 267 tags as the children of root *java*, including most of the tags in AHCTC *framework* subtree. To make a visual comparison with the other two subtrees, we can only render an excerpt of the *java* tree consisting of only pertinent tags. In Fig.6(b) the induced edges seem reasonable individually, but the whole taxonomy is too flat to be informative. We believe the cause is that SUBSUME algorithm does not exploit text of tagged documents.

Quantitative Evaluation. In order to make a convincing quantitative evaluation, we have to do multiple experiments for each of the three algorithms. We conduct 3 AHCTC experiments each time with a different tag similarity measure, Jenson-Shannon divergence (JS), cosine similarity (COS) and KL divergence. For the LSHL taxonomy which also needs to calculate a tag similarity score, JS and COS as the only measures proposed in the original paper are evaluated respectively. We fail to conduct multiple versions of the SUBSUME experiment since the algorithm do not have any replaceable building block.

To evaluate a tag taxonomy is difficult, because several complementary taxonomies might exist. However, several related works have tried in designing effective taxonomy evaluation methods. [2] uses the concept hierarchy from Open Directory Project as the ground truth to validate their learned ontology. In comparing the two hierarchies, both lexical and taxonomical precision and recall are evaluated. [4] asks domain experts to evaluate the correctness of each edge individually, and precision is defined as the proportion of correct edges in all the learned edges.

Our quantitative evaluation considers both these techniques. First, we ask 20 students to judge the correctness of each edge in a taxonomy. They are given 3 days to finish this task, and are encouraged to consult any accessible knowledge source for tags they don't understand. We use the precision measure proposed in [4] to calculate edge correctness. We calculate the average value and standard deviation of the edge correctness. The results are given in TABLE 1. All AHCTC taxonomies have high average edge correctness compared to other results, and the AHCTC-KL taxonomy has the highest. AHCTC-JS and LSHL-COS taxonomies have relatively higher standard deviations, which means many of the edges are indeterminate so judges make quite different judgements.

Table 1. Edge Correctness Judged by Human Experts

	Avg.	S.d.
LSHL-JS	0.7319	0.0935
LSHL-COS	0.6888	**0.1313**
SUBSUME	0.6897	0.0639
AHCTC-KL	**0.7940**	0.0432
AHCTC-JS	0.7895	**0.1213**
AHCTC-COS	0.7519	0.0879

To consider the correctness of each edge individually is not fully appropriate. We refer to the taxonomical precision measurement proposed in [2]. The golden truth is chosen as the *software* subtree of the Open Directory Project concept hierarchy. This subtree covers 179 out of 267 tags we are concerning, which is a 67.29% lexical precision. We also apply the taxonomical metrics defined in [2]. The taxonomical precision, recall and F-measure (TP, TR, TF) for different experiments are given in TABLE 2. AHCTC with KL divergence gets the highest value on all three metrics.

Table 2. Taxonomical Quantitative Evaluation

	TR	TP	TF
LSHL-JS	0.0194	0.0235	0.0212
LSHL-COS	0.0244	0.0296	0.0268
SUBSUME	0.0267	0.0393	0.0318
AHCTC-KL	**0.0525**	**0.0514**	**0.0519**
AHCTC-JS	0.0370	0.0334	0.0351
AHCTC-COS	0.0306	0.0328	0.0317

Discussion. The above experiments and evaluations demonstrate the usefulness and effectiveness of AHCTC algorithm, given the existence of similar methods: through qualitative evaluation, we show that AHCTC discovers valuable tag structures that are different from those discovered by any existing algorithm. Quantitative metrics suggest that AHCTC can construct better taxonomies.

As we have mentioned, parameters in SUBSUME and LSHL experiment are set at the practical best settings suggested by original papers. However, authors didn't guarantee that those settings are always the best regardless of what dataset being used. For this reason, we have actually tested several other combinations of parameter values. It turns out that the best settings still practically outrun other settings in our dataset in terms of the quality of resultant taxonomies.

It shall be noted that evidence given by our quantitative evaluations is still weak in measuring how 'good' a taxonomy is: edge correctness considers whether each individual edge is correct and can not measure the correctness of an entire taxonomy. Taxonomical metrics gained by comparing against a gold taxonomy is not persuasive, given that several complementary taxonomies might exist. It is also the reason why metrics in TABLE 2 are quite small. Taxonomy evaluation is still an open issue to be investigated in the future.

6   Conclusion

Many research efforts have been spent on inducing a taxonomy from a set of tags. This paper proposes a novel approach based on agglomerative hierarchical clustering, which can effectively skip the error prone step of calculating each tag's generality. A topic model is integrated into the AHC framework to disclose thematic correlations among tags. The experiment is built on top of data from Ohloh, an online social network software directory. With qualitative and quantitative evaluations, we demonstrate usefulness and effectiveness of the proposed method, after comparing with two representative previous works.

Acknowledgement. his research is supported by the National Science Foundation of China (Grant No.60903043), the National High Technology Research and Development Program of China (Grant No. 2012AA010101) and the Postgraduate Innovation Fund of University of Defence Technology (Grant No.120602). Our gratitude goes to authors of [4], Wang Wei and Payam Barnaghi, for their help in providing us the details of LSHL algorithm.

References

1. Staab, S., Studer, R. (eds.): Handbook on Ontologies, 2nd edn. Springer, Berlin (2009)
2. Liu, K., Fang, B., Zhang, W.: Ontology emergence from folksonomies. In: Huang, J., Koudas, N., Jones, G.J.F., Wu, X., Collins-Thompson, K., An, A. (eds.) CIKM, pp. 1109–1118. ACM (2010)
3. Tang, J., Leung, H.-F., Luo, Q., Chen, D., Gong, J.: Towards ontology learning from folksonomies. In: Boutilier, C. (ed.) IJCAI, pp. 2089–2094 (2009)
4. Wang, W., Barnaghi, P.M., Bargiela, A.: Probabilistic topic models for learning terminological ontologies. IEEE Trans. Knowl. Data Eng. 22(7), 1028–1040 (2010)
5. Navigli, R., Velardi, P., Faralli, S.: A graph-based algorithm for inducing lexical taxonomies from scratch. In: Walsh, T. (ed.) IJCAI, pp. 1872–1877. IJCAI/AAAI (2011)
6. Russell, S.J., Norvig, P.: Artificial Intelligence - A Modern Approach, 3rd internat edn. Pearson Education (2010)
7. Heymann, P., Garcia-Molina, H.: Collaborative creation of communal hierarchical taxonomies in social tagging systems. Technical report, Computer Science Department, Standford University (April 2006)

8. Blei, D.M., Ng, A.Y., Jordan, M.I.: Latent dirichlet allocation. Journal of Machine Learning Research 3, 993–1022 (2003)
9. Itti, L., Baldi, P.: Bayesian surprise attracts human attention. In: NIPS (2005)
10. Brooks, C.H., Montanez, N.: Improved annotation of the blogosphere via autotagging and hierarchical clustering. In: Carr, L., Roure, D.D., Iyengar, A., Goble, C.A., Dahlin, M. (eds.) WWW, pp. 625–632. ACM (2006)
11. Tan, P.N., Steinbach, M., Kumar, V.: Introduction to Data Mining. Addison-Wesley (2005)
12. Blei, D.M., McAuliffe, J.D.: Supervised topic models. In: Platt, J.C., Koller, D., Singer, Y., Roweis, S.T. (eds.) NIPS. Curran Associates, Inc. (2007)
13. Blei, D.M., Griffiths, T.L., Jordan, M.I.: The nested chinese restaurant process and bayesian nonparametric inference of topic hierarchies. J. ACM 57(2) (2010)
14. Griffiths, T.L., Steyvers, M.: Finding scientific topics. Proceedings of the National Academy of Science 101, 5228–5235 (2004)

Personalized Clustering for Social Image Search Results Based on Integration of Multiple Features[*]

Yi Zhuang[1], Dickson K.W. Chiu[2], Nan Jiang[3], Guochang Jiang[4], and Zhiang Wu[5]

[1] College of Computer & Information Engineering, Zhejiang Gongshang University, P.R. China
[2] Dickson Computer Systems, HKSAR, P.R. China
[3] Hangzhou First People's Hospital, Hangzhou, P.R. China
[4] The Second Institute of Oceanography, SOA, Hangzhou, P.R. China
[5] Jiangsu Provincial Key Laboratory of E-Business, Nanjing University of
Finance & Economics, P.R. China
zhuang@zjgsu.edu.cn

Abstract. The usage of Web social image search engines has been growing at an explosive rate. Due to the ambiguity of query terms and duplicate results, a good clustering of image search results is essential to enhance user experience as well as improve retrieval performance. Existing methods that cluster images only consider the image content or textual features. This paper presents a personalized clustering method called *pMFC* which is based on an integration of multiple features such as visual feature, and two conceptual features(e.g., *tag* and *title*). An unified similarity distance between two images is obtained by linearly combing the three similarity measures over three feature spaces, where three weight parameters are obtained by a multi-variable regression method. To facilitate a personalized clustering process, a user preference distribution model is introduced. Comprehensive experiments are conducted to testify the effectiveness of our proposed clustering method.

Keywords: high-dimensional indexing, probabilistic retrieval, sentiment.

1 Introduction

With the advent of Web 2.0, as one of most important new media, *social image* has played a dominated role. Many social image sharing websites such as Flickr [1] and Picasa [2], etc have been developed so far. The data size of this kind of images is exponentially increased. Retrieving and clustering of such social images has become one of the most important applications in the social media research community.

Fig. 1 shows an example of social image query results about "apple" from the Flickr [1]. In the query result, it is clear that the subjects of the images are different

---

[*] This paper is partially supported by the Program of National Natural Science Foundation of China under Grant No. 61003074, No.71072172, No.61103229, No.60903053; The Program of Natural Science Foundation of Zhejiang Province under Grant No. Z1100822, No. Y1110644, Y1110969, No.Y1090165; The Science & Technology Planning Project of Wenzhou under Grant No. G20100202.

S. Zhou, S. Zhang, and G. Karypis (Eds.): ADMA 2012, LNAI 7713, pp. 78–90, 2012.

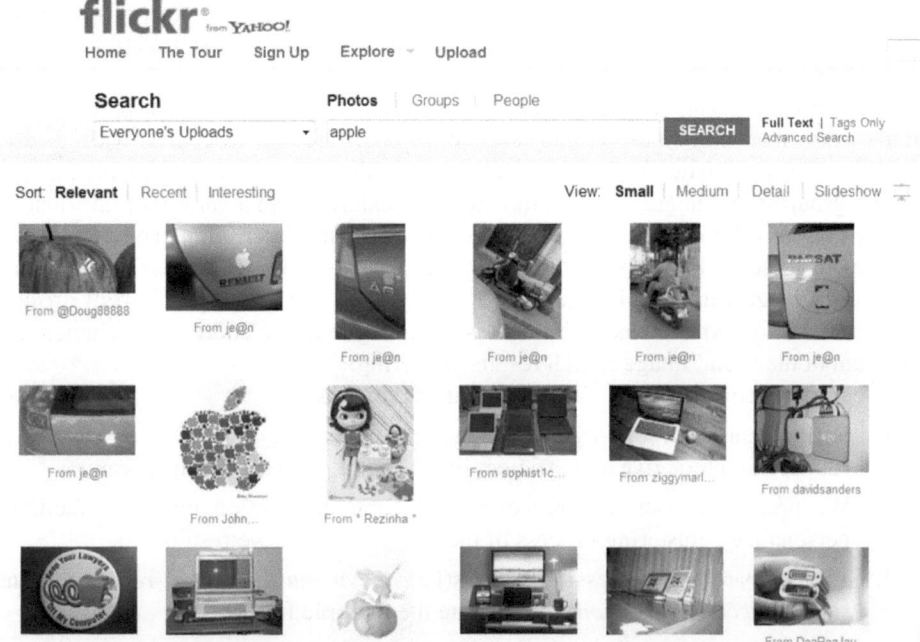

Fig. 1. An example social images for "apple" from the flickr

and orderless. To obtain a high quality query results, a clustering process of such result images is needed. In the state-of-the-art image clustering methods[3][4], color, texture *or* the surrounding texts are extracted from an image as low level visual *or* high level semantic features. The clustering effectiveness (e.g., *recall* and *precision*) of the above methods, however, is not satisfactory due to semantic gap between high level concepts and low level visual features [3]. Web social image search results clustering is clearly related to the general-purpose clustering but it has some specific requirements. Beside the above widely adopted visual features, some conceptual ones such as *tag* and *title* of a social image can affect the clustering accuracy to some extent, which have not been considered into traditional image clustering methods. For example,

— **Tag info**. Due to the characteristics of Web 2.0, users are allowed to tag the images they prefer to. Various tags are described for the image.

— **Title info**. For a social image, the title of its corresponding webpage is somehow semantically correlated to this image.

As a means of improving retrieval performance for search engines, clustering the result images into different semantic categories has been extensively investigated in text retrieval and image retrieval [3][4]. Currently available commercial social image search engines generally provide searches only based on keywords but do not fully exploit the context and content information in a natural and intuitive way.

In addition, in most cases, people would like to get some result images with some specifications they prefer to, which can not be achieved by the traditional visual feature- *or* textual-feature-based clustering methods.

This paper presents a personalized multi-feature-based social image clustering framework that combines visual and conceptual features together with a user preference probabilistic model. In this framework, the web social image search results can be grouped by integrating with the multiple features from a variety of information sources besides the image content such as title, tags and description. This framework enables us to adopt existing clustering algorithm such as the AP clustering algorithm [10] to organize returned images into semantically and visually coherent groups. To the best of our knowledge, this work is the first attempt to address this challenging problem of the social image search results clustering.

The primary contributions of this paper are as follows:

1. We present a *personalized Multi-Feature-based Clustering(pMFC)* method to support an interactive and efficient images retrieval with multiple features.

2. We propose a user preferences probabilistic distribution model to facilitate personalized clustering process of the query social image results.

3. We compare two regression models(i.e., *linear multi-variable regression* and *support vector regression*) to integrate the multiple features.

4. We perform extensive experiments to evaluate the effectiveness of our proposed clustering method.

The remainder of this paper is organized as follows. In Section 2, we provide related work. Then in Section 3, we give preliminaries of this work. In Section 4, we introduce a user preference probabilistic distribution model. Next, we propose a personalized multi-feature-based clustering method for social images in Section 5. In Section 6, we report the results of extensive experiments which are designed to evaluate the effectiveness of the proposed approach. Finally, we conclude in the final section.

2   Related Work

2.1   Web Image Clustering

Several image search result clustering (ISRC) algorithms have recently been proposed in the academic arena. Most approaches for clustering image search results exploit low-level visual features [3]. However, these approaches suffer from two problems: (a) semantic gap between the low-level features and high-level semantics and (b) low efficiency due to curse of dimensionality. Considering that global image features do not describe individual objects in the images precisely, Wang *et al* [9] proposed to use region level image analysis. They formalized the problem as a salient image region pattern extraction problem. According to the region patterns, images were assigned to different clusters. Besides visual information, textual and link information has also been used recently. A reinforcement clustering algorithm and a bipartite graph co-partitioning algorithm are proposed to integrate visual and textual features in [6].

Deng *et al*[5] first used block-level link analysis to construct an image graph. Then, spectral clustering techniques were adopted to hierarchically cluster the top image search results based on visual representation, textual representation, and graph representation. iGroup[4] took a step towards addressing these limitations by exploiting textual features such as image captions, snippets, surrounding texts. The clustering is then accomplished by combining both visual features and textual features. In the context of social tagged images, shared nearest neighbors algorithm (snn) was applied to cluster images in a collection using both tag features and visual features [8]. In our proposed *pMFC* method, we aim at clustering social images not only by using visual content but also tag and title information with a personalized probabilistic model. To be detailed in the next section, *pMFC* is flexible enough to easily accommodate different clustering algorithms

2.2 Affinity Propagation

In the *affinity-propagation*(AP)-based clustering method[10], each data point to be clustered is viewed as a node in a network which passes messages to other nodes in order to determine which nodes should be exemplars and which nodes should be associated with those exemplars. An exemplar is the point which best represents other points in its cluster.

The algorithm runs to maximize the overall similarity of all data points to their exemplars. The solution is approximated following the ideas of belief-propagation. There are two types of messages sent between data point i and candidate exemplar k: *responsibility* $r(i,k)$ and *availability* $a(i,k)$. Responsibility messages are sent from i to k and reflect how strongly data point i favors k over other candidate exemplars. Availability messages are sent from k to i and reflect how available i is to be assigned to k currently.

$$r(i,k) \leftarrow s(i,k) - \max_{k'|k' \neq k} \left\{ a(i,k') + s(i,k') \right\} \tag{1}$$

$$a(i,k) \leftarrow \min \left\{ 0, r(k,k) + \sum_{i'|i' \notin \{i,k\}} \max \left\{ 0, r(i',k) \right\} \right\} \tag{2}$$

The messages are passed during a variable number of iterations. In each iteration the evidence accumulates that some points are better exemplars. It can be seen in Eqs. (1) and (2) that there is a circular dependency between responsibility and availability. This is handled by initializing $a(i, k)$ to a zero value so that $r(i, k)$ can be calculated in the first iteration. After this the availabilities are calculated and stored to be ready for the next iteration.

3   Preliminaries

3.1 Definitions and Problem Formulation

First we briefly introduce the notations that will be used in the rest of paper.

Table 1. Meanings of Symbols Used

Symbols	Meaning
Ω	a set of social images
λ_i	the i-th social image and $\lambda_i \in \Omega$
n	the number of images in Ω
m	the number of predefined classification tags
$Sim(\lambda_i,\lambda_j)$	unified similarity distance between two images
$vSim(\lambda_i,\lambda_j)$	visual similarity distance between two images
$aSim(\lambda_i,\lambda_j)$	tag similarity distance between two images
$iSim(\lambda_i,\lambda_j)$	title similarity distance between two images
λ_q	a query image user submits
r	a query radius
$|\bullet|$	the number of objects in $\bullet$

As we know, for a social image, it is composed of two kinds of features such as objective features and conceptual ones. The objective features of an image mainly refer to visual features. For the conceptual features, the tags and webpage title of a social image need to be considered.

DEFINITION 1(Social Image). *A social image λ_i can be modeled by a four-tuple:*

$$\lambda_i ::= <i, vis, tag, title> \tag{3}$$

where

- *i refers to the social image ID;*
- *vis refers to the visual features of λ_i such as color histogram, texture, etc;*
- *tag refers to a set of tags information provided by users;*
- *title is the webpage title of the λ_i;*

Generally speaking, for a user U_i, he has different preferences in different period of time. Moreover, for his preferences, it is evident that the extent of such preferences be various during the time.

DEFINITION 2(User). *A user U_i can be modeled by a triplet:*

$$U_i ::= <i, N, P> \tag{4}$$

where

- *i refers to the user ID;*
- *N denotes the user name;*
- *P denotes the preferences of the U_i, formally represented by a triplet:*

$$P ::= <name, tem, prob> \tag{5}$$

where

- ➢ *name is the preference name of U_i;*
- ➢ *tem is the temporal information of U_i;*
- ➢ *prob is the extent of the preference of U_i and $prob \in [0,1]$;*

According to the usage of the user tags, the tag can be divided into two categories: *object tag* and *classification tag*.

DEFINITION 3(Object Tag). *An object tag(OT) of an image λ_i is a tag that describes the image.*

DEFINITION 4(Classification Tag). *A classification tag(CT) of an image λ_i is a tag that can represent the classification of the image.*

Fig. 2. An example social images for "tiger" with highlighted title, tag and comments

For example, for an image that describes an apple computer, '*apple*' belongs to the OT, and '*computer*' belongs to the CT.

3.2   Similarity of Textual Information

With the maturity of Web 2.0, the context of a social image is often very closer to its semantic meaning [4] and thus comparing textual features is beneficial. For the images, we also make comparisons for the associated metadata. The tags or the title of a social image can be regarded as a word set(*Ws*) that is composed of a couple of words.

● *Preprocessing*

The context information that accompanies an image also has to be preprocessed before any clustering process. The preprocessing step includes *normalization* and *comparison*.

The normalization process involves *stemming* and *removing stop words*. Stemming is achieved by a WordNet stemmer and can be summarized by an example: changing the word "realization" to the stem "realiz-". For comparison purposes, this is useful because the words "realize" and "realization" will now be recognized by our text comparison algorithm as a similar word through their common stem. The second part of the normalization involves eliminating stop words such as "I", "do" or "the" because they will inflate the score of the comparison algorithm and affect clustering quality. For example, two sentences being compared may have many stop words in common but this does not necessarily imply that the sentences are indeed relevant.

All punctuation symbols are removed and capitalized letters substituted for similar reasons. After normalization is complete every text can be compared using the following textual similarity measure and given a similarity score. Tags are only compared with tags, title with title, and description with description.

● *Similarity Measure*

Given two *W*s*s*(e.g., X and Y), the similarity measure of two words set X and Y can be defined in Eq.(6).

$$sim(X,Y) = \frac{\sum_{x \in X} 1_{\{y \in Y : s(x,y)=0\}} + \sum_{y \in Y} 1_{\{x \in X : s(x,y)=0\}}}{|X| + |Y|} \tag{6}$$

where $s(x,y)=0$ means word x is the same to y. This similarity is measured by the percentage of the same common words in two *W*s*s*.

For example, given two *W*s*s*: $s1$="Tiger Woods Greatest Golf Shot Ever" and $s2 =$ "Tiger Woods Amazing Golf Shot", their similarity distance is 0.73 because there are four common words between $s1$ and $s2$.

$$sim(s1,s2) = \frac{\sum_{x \in s1} 1_{\{y \in s2 : s(x,y)=0\}} + \sum_{y \in s2} 1_{\{x \in s1 : s(x,y)=0\}}}{|s1| + |s2|} = \frac{4+4}{6+5} = 0.73$$

4   Probabilistic User Preference Model

To facilitate a personalized clustering process, in this section, we propose a probabilistic user preference model.

DEFINITION 5(User Preferences Distribution Table). *A user preferences distribution table(UPDT) of the i-th user(U$_i$) can be represented by a triplet:*

$$UPDT_i ::= <i, Pref_j, Prob_j> \tag{7}$$

where

 - *i refers to user ID;*

 - *Pref$_j$ refers to the j-th preference U$_i$ has and j$\in$[1,|Pref|];*

 - *Prob$_j$ is the probability value that U$_j$ chooses the Pref$_j$ and j$\in$[1, |Pref|];*

Table 2 shows an example of the probabilistic distribution of a user's preferences. It is worth mentioning that *Pref$_j$* in definition 5 is the same to the *classification tag*(CT) in definition 4.

Table 2. An example of the probabilistic distribution of the *i*-th user's preferences

Preference(*Pref_j*)	Probability(*Prob_j*)
computer	50%
fruit	30%
history	20%

5   The Clustering Algorithm

In this section, we propose a personalized multi-feature-based social image clustering method.

5.1   Determining Features Weights

Assuming that each type of visual and conceptual features contributes equally in image recognition is not supported in human perpetual system, as different image feature plays different roles for effective image clustering.

How to obtain the relative weight of information carried by each of three kinds of features(i.e., the visual features (e.g., *color histogram*, *texture*, etc), tag and title) is critically important to the clustering effectiveness. Simply concatenating the multiple features to form a single high-dimensional feature may not be effective in terms of clustering accuracy. So in this work, we introduce two multivariable regression methods(i.e., *multivariable regression*(MVR) and *support vector regression*(SVR)) to determine the weight for each feature. The goal of our approach is to apply linear and nonlinear regression models to investigate the correlation between distance of each feature type and closeness between two social images.

In the model, distance between two image items, *Sim*, can be presented as a linear function of distance for the above three features. Symbolically, it can be written as below,

$$Sim(\lambda_i,\lambda_j)=w_v \times vSim(\lambda_i,\lambda_j)+w_a \times aSim(\lambda_i,\lambda_j)+w_i \times iSim(\lambda_i,\lambda_j) \qquad (8)$$

where $w=[w_v,w_a,w_i]$ is the vector of weight coeffcients to be determined (v, a and i denote visual feature, tag, and title feature) and $d=[vSim(\lambda_i,\lambda_j), aSim(\lambda_i,\lambda_j), iSim(\lambda_i,\lambda_j)]$ is the vector of independent distance value for each type of feature.

To determine the weight of Eq. (8), we first apply the MVR method. For training purpose, we select $2n$ image data $[m_1,m_2...m_{2n}]$ from the database, and we also have the distance value of similarity measurement $[dSim_1,dSim_2,...,dSim_n]$, e.g. Sim_i represents the distance between m_i and m_{i+1}. In this study, the distance value between two images is normalized to boolean value. For similar pair of image items, the final distance is set to be 1 and otherwise 0. The users determine whether two image objects are similar or not. After we determine the similarity between the pairs of training data, we get $n*3$

matrix. Each row has three items [*vSim, aSim, iSim*], i.e. the distance values taken by the above independent feature and *dSim* is a boolean value representing the similarity. With this matrix, we are able to calculate the weight for three features by the MVR.

Similarly, we can also obtain the corresponding weights of the above three features by adopting the SVR.

5.2   The Algorithm

Based on the above unified similarity distance, for n social images, we can obtain a distance matrix $D_{n \times n}$ as an input of the AP clustering algorithm[10] by which images are grouped into k clusters. For each cluster, there exists a set of candidate tags.

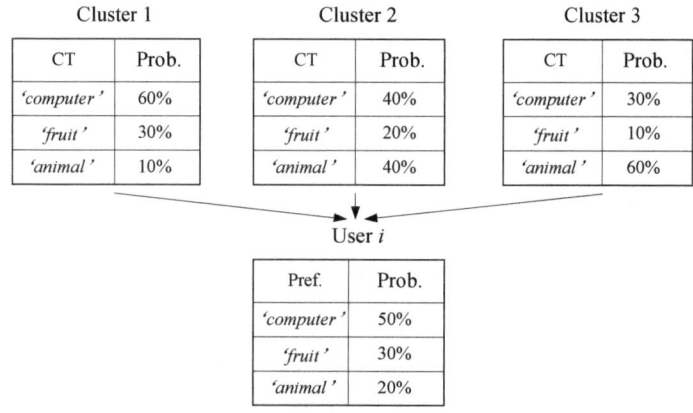

Fig. 3. An example of the final ranking score

Next, we propose a ranking method of the clusters based on users' preferences. The method is described as follows: for each cluster(C_i), we first calculate the probabilities that some predefined CTs appear in C_i. Then, a final ranking score($sClus$) between C_i and the UPDT can be derived in Eq.(9):

$$sClus(C_i) = \sum_{j=1}^{|Pref|} CT_j.Prob \times Pref_j.Prob \qquad (9)$$

where $CT_j.Prob$ denotes the probability that CT_j appears in C_i, for $i \in [1,k]$ and $j \in [1,m]$.

For example, as shown in Fig.3, suppose that there are three clusters. For each cluster, there exists a CT distribution table. Based on Eq.(9), for cluster 1, the final ranking score can be derived by: $sClus(C_1)=60\% \times 50\%+ 30\% \times 30\%+10\% \times 20\%=0.41$. Similarly, we have: $sClus(C_2)=0.34$, and $sClus(C_3)=0.3$.

For the clustering process, it is composed of two stages: 1) image clustering(lines 2-6), and 2) ranking the clustered images according to the user's preference(lines 7-9). When user U_i submits a query image λ_q, a query results(S) are returned(line 3). Its three kinds of features are extracted(line 4) and an unified similarity distance matrix is obtained(line 5). Then, the images in S are grouped by the AP clustering algorithm [10]. After that, for each cluster, we calculate the final ranking score with the U_i's preferences(lines 7-8). Therefore, the clustering result is returned to the user.

Algorithm 1. The pMFC algorithm

Input: query image λ_q, U_i, r

Output: cluster results

1. $S \leftarrow \Phi$;   // *initialization*
2. a user U_i submits a query image λ_q;
3. the image query results(S) are returned by a search engine as input images;
4. extract three kinds of features of each image in S;
5. obtain an distance matrix based on an unified similarity measure;
6. cluster the images in S by the AP clustering method;
7. **for** each cluster C_i **do**
8.   calculate the final ranking score($sClus(C_i)$) with U_i's UPDT based on Eq.(9);
9. **end for**;
10. **return** the result clusters that are ranked according to the $sClus$ descending;

Fig. 4. The *pMFC* clustering algorithm

6   Experiments

In this section, we present an extensive performance study to evaluate the effectiveness of our proposed clustering method. The image data we used are down- loaded from *Flicker.com*[1] which contains a set of the 50,000 images. All image data can be divided into 10 categories such as '*computer*', '*fruit*', '*vehicle*', '*animal*', '*bird*', '*flower*', '*people*', '*school*', '*music*' and '*food*'. We implemented the *pMFC* clustering method. All the experiments are run on a Pentium IV CPU at 2.0GHz with 2G Mbytes memory.

6.1   A Prototype Clustering System

We have implemented an online social image clustering system to testify the effectiveness of our proposed retrieval method comparing with the conventional one [3]. As shown in the right part of Fig. 5, when user submits an example image and the user's preferences as well, the clustered images are quickly obtained by the system. The right part of the figure is the clustering result in which the similarity and confidence values of the answer images are given with respective to the query one.

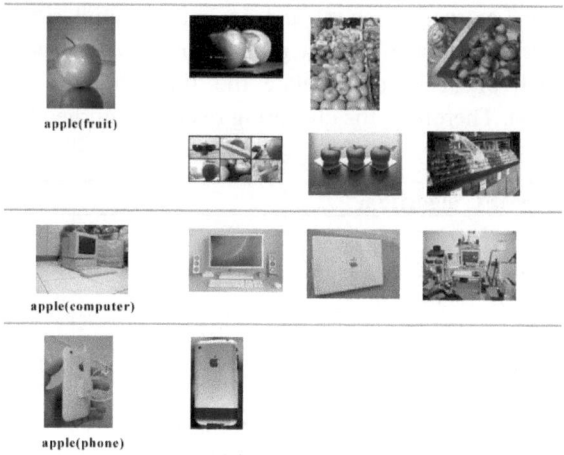

Fig. 5. One social image clustering example

6.2   Effectiveness of the Clustering Method

In the following experiments, we evaluate the effectiveness of our proposed clustering method.

Denoting the set of ground-truth as *rel*, and the set of results returned by the clustering method as *ret*, the recall and precision achieved by this clustering method is defined as:

$$recall = \frac{|rel \cap ret|}{|ret|}, \quad precision = \frac{|rel \cap ret|}{|rel|} \tag{10}$$

In the first experiment, we test the effect of two clustering methods on the clustering effectiveness. Fig. 6 illustrates a *Recall-Precision* curve for the performance comparisons of the conventional visual-feature-based approach and our proposed probabilistic clustering one. It compares the average clustering result of 20 images queries randomly chosen from the database. The figure shows that the clustering performance of the probabilistic clustering method is better than that of the conventional one by a large margin. This is because compared with the conventional similarity metric, the multi-feature-based approach can better capture the inherent similarity between two images. And the user's clustering intention can be better represented by the user preference probabilistic model.

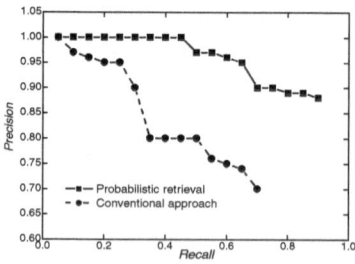

Fig. 6. Recall *vs.* precision

6.3   Comparison of Two Regression Methods

In this experiment, we compared with the two multi-variable regression methods. Fig. 7 shows the performance of query processing in terms of clustering accuracy. It is evident that the SVR outperforms the MVR. The clustering accuracy of the both regression methods increases slowly as the data size grows, then their accuracies decreased gradually.

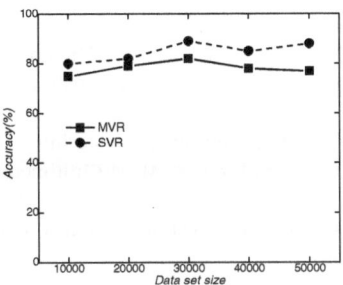

Fig. 7. Comparison of two regression methods

6.4   Effect of Data Size

In the last experiment, we proceed to evaluate the effect of data size on the clustering method. Fig. 8 indicates that with the increase of data size, the accuracy of the *pMFC* is increasing. The results conform to our expectation that the search region of *pMFC* is significantly reduced and the comparison between any two images is a CPU- intensive task. The CPU cost of sequential scan is ignored due to the expensive computation cost of it.

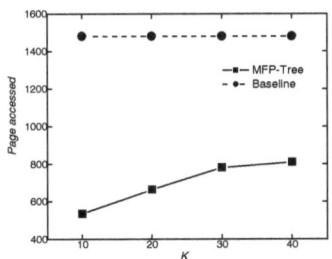

Fig. 8. Effect of data size

7   Conclusions

In this paper, we presented a novel framework for personalized probabilistic multi-feature-based clustering method for social image query results. The prototype retrieval system is implemented to demonstrate the applicability and effectiveness of our new approach to social image clustering process.

References

[1] (2011), http://www.flickr.com
[2] (2005), http://www.picasaa.google.com
[3] Chen, Y., Wang, J.Z., Krovetz, R.: Content-based image retrieval by clustering. In: Proc. of MIR, pp. 193–200 (2003)
[4] Jing, F., Wang, C.H., Yao, Y.H., et al.: IGroup: A Web image search engine with semantic clustering of search results. In: Proc. of the 14th Annual ACM Int'l Conf. on Multimedia, pp. 23–27. ACM Press, New York (2006)
[5] Cai, D., He, X., Li, Z., Ma, W.-Y., Wen, J.-R.: Hierarchical clustering of www image search results using visual, textual, and link information. In: Proc. of ACM Multimedia Conference (2004)
[6] Gao, B., Liu, T.-Y., Qin, T., et al.: Web image clustering by consistent utilization of visual features and surrounding texts. In: Proc. of ACM Multimedia Conference (2005)
[7] Rege, M., Dong, M., Hua, J.: Graph theoretical framework for simultaneously integrating visual and textual features for efficient web image clustering. In: Proc. of Int'l Conference on World Wide Web (2008)
[8] Moëllic, P.-A., Haugeard, J.-E., Pitel, G.: Image clustering based on a shared nearest neighbors approach for tagged collections. In: Proc. of CIVR, pp. 269–278 (2008)
[9] Wang, X.J., Ma, W.Y., He, Q.C., Li, X.: Grouping Web Image Search Result. In: Proc. of the 12th annual ACM int'l conference on Multimedia, pp. 436–439
[10] Frey, B.J., Dueck, D.: Clustering by passing messages between data points. Science 315(5814), 972–976 (2007)

Query Directed Web Page Clustering Using Suffix Tree and Wikipedia Links

John Park, Xiaoying Gao, and Peter Andreae

School of Engineering and Computer Science
Victoria University of Wellington, PO Box 600, Wellington, New Zealand
parkjohn@ecs.ac.nz, {xgao,pondy}@ecs.vuw.ac.nz

Abstract. Recent research on Web page clustering has shown that the user query plays a critical role in guiding the categorisation of web search results. This paper combines our Query Directed Clustering algorithm (QDC) with another existing algorithm, Suffix Tree Clustering (STC), to identify common phrases shared by documents for base cluster identification. One main contribution is the utilising of a new Wikipedia link based measure to estimate the semantic relatedness between query and the base cluster labels, which has shown great promise in identifying the good base clusters. Our experimental results show that the performance is improved by utilising suffix trees and Wikipedia links.

Keywords: Document Clustering, Semantic Distance, Semantic Relatedness.

1 Introduction

Over the past few years, there has been an exponential increase of the World Wide Web. The sheer amount of information and every genre and niche made information access difficult for everyday user. Navigation is now mainly done by search engine, such as Google, but difficulties in query construction and the long list of inaccurate results even make this facilitated approach a challenge.

One method of making the web more accessible for people is to cluster the results into clearly distinguishable topics. This allows the user to refine what would be an unclear, ambiguous query, and filter out irrelevant results relative to the topic. For example, a query "jaguar" can either mean the American big cat, the British car manufacturer, an operating system by Apple, etc. An algorithm to cluster similar documents together makes it easier to disambiguate.

Much work has been carried out in the fields of web page clustering. Almost all the data clustering algorithms have been applied on web search results clustering, including Hierarchical Agglomerative clustering, K-means, K-Medoids, Probablistic, fuzzy c-Means, Bayesian, Kohonen self-organising maps, density based, etc. The two algorithms that are closely related to this research are Suffix Tree Clustering (STC) [1], and Query Directed Web Page Clustering (QDC) [2]. STC relies on a *suffix tree* to efficiently identify common phrase shared by documents. QDC[2] takes into account the *query* that the user inputs, and our recent

S. Zhou, S. Zhang, and G. Karypis (Eds.): ADMA 2012, LNAI 7713, pp. 91–99, 2012.

research [3] has shown that it outperforms other existing algorithms including STC.

However, the query directed approach that QDC utilizes is limited by the single word labels on the clusters. This paper aims to combine the QDC with STC, so we can use suffix tree to effectively identify the phrases commonly shared by documents, and then use the phrases rather than single words to form base clusters. The key to QDC's success is a semantic distance measure that estimates the distance between a query and a base cluster label. A limitation is that the original QDC uses a semantic distance measure named normalized Google distance (NGD) [4] which requires to dynamically query the search engine many times, so it can be slow and expensive. Wikipedia as one of the largest information source on the Internet has shown promise in improving many natural language processing systems. This paper aims to use Wikipedia, especially the link structure, as the external source to provide semantic information.

This paper has two main contributions: (1) combining QDC and suffix tree clustering and building the base clusters use phrases instead of single words, and (2) using a new wikipedia-based semantic similarity measure to replace NGD.

The rest of the paper is broken down into four parts. Section 2 discusses the finer details of STC, QDC, NGD and a Wikipedia Link-Based Measure (WLM) [5], and their respective advantages and disadvantages. Section 3 details the two modifications made to the QDC algorithm via the use of suffix tree and WLM in place of NGD. Section 4 evaluates our Extended Query Directed Clustering (EQDC) algorithm, along with the two modifications separately, with the original QDC algorithm. The final section summarizes the results and possible future works.

2   Background and Related Work

There are many clustering algorithms that have been developed for web page clustering. For the purposes of this paper, we will focus on the two most relevant clustering algorithms: STC and QDC. These two clustering algorithms are broken down into several general steps to produce the output of a list of clusters that contain the set of documents relevant to the topic.

First step to any clustering algorithm is to pre-process the data. This involves removing any tags, punctuation marks and stop words, stemming words to their root form and converting all character to lowercase. After this, both STC and QDC algorithms identify and group a set of documents sharing the same word in a process called *base cluster identification*. These base clusters are then merged further to produce clusters of documents.

2.1   Suffix Tree Clustering (STC)

The unique aspect of STC [1] involves inputting sentences from the document into a suffix tree [6]. As shown in figure 1, after the documents have been read, each node of the suffix tree forms a base cluster, which are then merged if they hold more than half the documents in common for base clusters [1].

The main advantage of STC is that STC runs on linear time and has superior precision over more common algorithms such as K-means. Furthermore, the base clusters constructed use phrases instead of single words. Although STC is a very useful clustering algorithm, it is slightly outdated in comparison to algorithms such as QDC. For example, the single-link clustering that STC uses may cause cluster drifting. Such factors mean that QDC has superior precision and recall in comparison to STC overall.

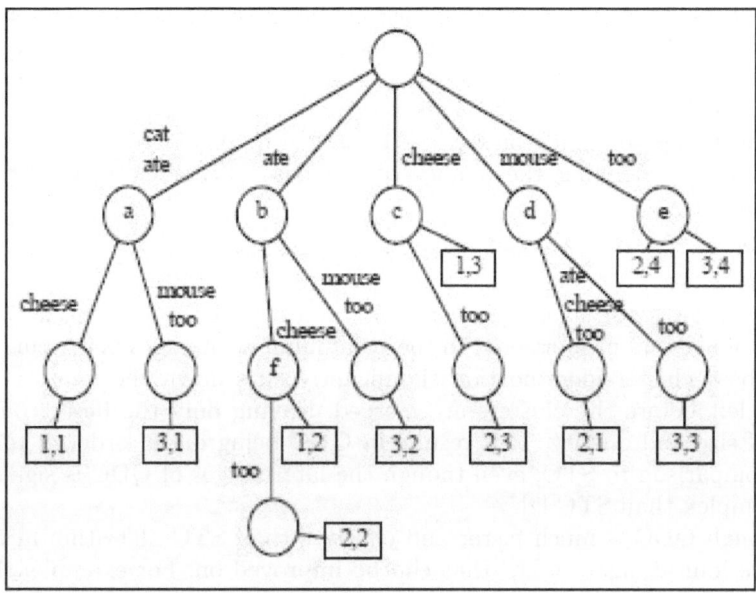

Fig. 1. Example of Suffix Tree Clustering

2.2   Query Directed Web Page Clustering (QDC)

QDC algorithm [2] consists of five major steps after pre-processing:

1. The base cluster identified and filtered out in the *base cluster identification* process.
2. The remaining base clusters are merged in a single-link clustering.
3. The merged clusters are split if necessary to prevent cluster drifting.
4. From the resultant clusters, the best quality clusters are chosen via the use of a heuristic.
5. The selected clusters are labeled, cleaned of irrelevant documents and have its pages reordered.

This clustering algorithm first introduced comparing the semantics of the query to the base clusters labels to select the most relevant clusters in the base cluster identification process. The process is broken down into three parts. First, the documents sharing a single word form a base cluster and the word is the label of the base cluster, denoted as $D(b)$. Then, only the base clusters that holds at least 4% of the total number of documents are selected. After filtering by size, the minimum query distance $QD(b)$ between each of the terms in the query and the label $D(b)$ of the base cluster b is calculated using normalized Google distance (equation 3, see 2.3 for details), as shown by the equation below:

$$QD(b) = \min_{term \in query} NGD(D(b), term) \qquad (1)$$

The cutoff point for the maximum query distance allowed for a base cluster is set at 1.5 and 2.5 for full and snippet length data respectively. Cutoff point is more lenient for snippet length data due to snippet data containing dramatically less content [2].

The next step orders the base clusters by their quality in respect to the query. It does this by defining the quality of the base cluster by a heuristic value $baseQuality(b)$.

$$baseQuality(b) = \frac{|b|}{QD(b)} \qquad (2)$$

After reordering, the algorithm removes any excess base clusters, keeping the number of clusters proportional to the total number of pages being clustered.

The base cluster identification significantly cuts down the number of base clusters left before the clusters are merged, leaving only the best grouping in terms of size and quality. This results in QDC being on an order of ten times faster comparison to STC, even though the later stages of QDC is significantly more complex than STC [2].

Although QDC is much faster and effective than STC algorithm in general, there are limitations of QDC that can be improved on. For example, the base clusters during the identification process are only represented by a single word. In most cases, this would be sufficient enough to give an accurate description. However, there are outlier cases where a phrase would be better suited than a single word. The phrase 'operating system' is a lot more useful than single words such as 'operating', 'system' or even 'computer'. In such cases, the use of a suffix tree like the STC algorithm would be a valid option.

2.3   Normalized Google Distance (NGD) and Wikipedia Link-Based Measure (WLM)

Measuring semantics is another field in computer science that is ever expanding. Nowadays, the analysis of semantics is mainly done by the use of two major fields. One is of formal semantics, where the structure of the sentence is broken into precise meaning, and the other is of statistical semantics, where the sentence is broken by the analysis of patterns of words in a large database collection. For the purposes of this paper, this discussion will be focusing only on how *distant* two words are instead of their respective meanings, and will discuss two functions used to measure the distance between the words: normalized Google distance (NGD) [4] and Wikipedia Link-based Measure (WLM) [5].

$$NGD(x,y) = \frac{\max\{logf(x), logf(y)\} - logf(x,y)}{logM - \min\{logf(x), logf(y)\}} \tag{3}$$

Normalized Google distance is a semantic distance measure that utilizes Google search engine to calculate how distant two keywords are from each other. It is derived from the number of hits for each keywords x and y, denoted $f(x)$ and $f(y)$ and the number of web pages where both x and y occur, denoted $f(x,y)$. Finally, M denotes the total number of documents indexed by the search engine.

Wikipedia Link-based Measure, on the other hand, covers two new methods of measuring semantics between two words using articles from Wikipedia. This was introduced in 2008 by David Mile and Ian H. Witten [5]. The first measure uses the angle between the vectors of the two links within each article, similar to the TF-IDF measure used frequently in information retrieval. The second measure, which was the one we use, exploits the hyperlinks shared between two Wikipedia articles to find the semantic similarity, as shown by the formula below:

$$sr(a,b) = \frac{log(\max(|A|, |B|)) - log(|A \cap B|)}{log(|W|) - log(\min(|A|, |B|))} \tag{4}$$

In the above equation, A and B are the set of articles that link to articles a and b, and W is the entire Wikipedia.

The original QDC algorithm relies on NGD to compare the distance between the query to the labels of base clusters. This requires to query a search engine multiple times which can be slow and is a major limitation. By showing that other semantic similarity measures are valid in place of NGD for the QDC algorithm, there will be other opportunities to improve the efficiency and accuracy of the original algorithm.

3   EQDC Algorithm

QDC algorithm has 5 logical steps and we have modified the first step of the algorithm on two different ways: (1) modifying the way for creating base clusters such that a phrase is used as a label, and (2) modifying similarity measure from Normalized Google Distance (NGD) to WikiMiner's semantic similarity measure.

3.1   Phrase Substitution for Single Word Labels

We define phrases as being an ordered sequence of one or more words extracted from the documents after pre-processing, and base cluster as an abstract representation documents share in common.

The algorithm uses a suffix tree structure which takes in an input of phrases of limited maximum size from the documents. The suffix tree then creates base clusters containing the set of documents which containing the same ordered sequence of words as the suffices produced. Then the procedure of filtering out base clusters in terms of size and quality is carried out, where the semantic distance of the label of the base cluster is compared with each of the query terms.

The original filter for the quality was determined by the minimum distance between a single query term and the single word label [2], using the NGD formula (3). In order to avoid the high cost of calculating NGD, another change we made is to utilize the local data set of documents instead of querying the search engine. This means that $f(x)$, $f(y)$ and $f(x,y)$ equals to the number of documents that contains x, y and documents that contains both x and y respectively, and M being the total number of documents in the data set.

To accommodate for measuring semantic distance between phrases and query for the base cluster filtering, a modification was made to the original algorithm. This distance is the minimum average distance between a query and each of the words in the phrase, where D(b) is the ordered sequence of words that labels base cluster b:

$$QD(b) = \min_{x \in query} \frac{\sum_{y \in D(b)} NGD(x,y)}{|D(b)|} \tag{5}$$

The cluster merging process in the original QDC algorithm merged the base clusters together in single-link clustering [2]. Two heuristic values are calculated and is compared against two thresholds. Both threshold must be satisfied before the clusters are merged. First compares the sets of documents in the two clusters, and the second compare the semantic similarity of the labels for the clusters. For the semantic comparison between the labels, a threshold value of $\gamma = 4.0$ is used as shown in the equation below, where $f(d_1)$ and $f(d_2)$ are the number of hits for the keyword d_1 and d_2 respectively out of M documents, and $f(d_1 \wedge d_2)$ is the number of documents containing both keywords.

$$\gamma < \frac{M f(d_1 \wedge d_2)}{f(d_1) f(d_2)} \tag{6}$$

To accommodate the use of phrases in the EQDC algorithm, we have modified the above equation to search for the number of documents in the local data set containing each of the terms of the phrases $D(c_1)$ and $D(c_2)$ from the clusters c_1 and c_2.

$$\gamma < \frac{M \prod_{d_1 \in D(c_1), d_2 \in D(c_2)} f(d_1 \wedge d_2)}{\prod_{d_1 \in D(c_1)} \prod_{d_2 \in D(c_2)} f(d_1) f(d_2)} \tag{7}$$

After these have been carried out, the clusters then undergo the same procedure as outlined in the QDC algorithm [2].

3.2   WikiMiner Semantic Similarity Measure

One of our main contributions is to investigate the effect of using a different semantic distance measure in place of the original NGD. For this, we used a new semantic relatedness measure called Wikipedia Link-based measure (WLM). Our research also finds many other methods for calculating semantic relatedness between word pairs [7,8,9]. We choose to use WLM because it is a light weighted method and it is supported by a recent data mining tool called WikiMiner [5].

The raw relatedness value returned by WikiMiner[5], however, was unsuitable to be used directly in the base cluster identification process, as NGD measures the distance between two terms and range from zero distance to infinite distance, while WLM gives a value that range from zero similarity to exact similarity. Therefore, a conversion was applied that converted the semantic relatedness value into a distance value via the formula outlined in (8):

$$dist(x, y) = ln(\frac{1}{sim(x, y)}) \qquad (8)$$

The $dist(x, y)$ value is substituted for $NGD(x, y)$ in formula (5). After applying the conversion, both NGD and WLM filter out base clusters under the same cutoff threshold.

4   Evaluation

Since EQDC focuses primary on modifying the *base cluster identification* process, we have decided to leave the rest of the algorithm out and tested only the base clusters formed. By doing this, we can get a clear idea of whether the quality of the base cluster filtering process has improved from QDC to EQDC, making further extension much more promising.

For the experiments, we evaluated QDC against three different modifications: (1) using only the phrase label for the base cluster, (2) modifying the semantic measure, and (3) using the combination of both (1) and (2) by forming the EQDC algorithm. These four algorithms were tested against three subset of the raw data set [10]. Each were matched up with the respective queries 'jaguar', 'apple' and 'victoria'. All three of the queries were suitable due to their ambiguous meaning. For example, 'jaguar' could mean either a car, an animal or a Macintosh operating system, making it ideal to separate out the documents into the respective topics. Furthermore, our test data for each data set included a manually labeled gold standard clustering, to which each of the clustering produced from the algorithms was then evaluated against. More information on the data set is discussed below.

After a few preliminary tests, we discovered that it was unnecessary to exceed the maximum phrase size of 3. This was due to the base clusters formed with phrase size of 3 had for too few documents contained within the set. This meant that these base clusters would almost immediately be filtered out.

4.1   Experimental Setup

To calculate the F-score for each of the clusters produced, we used 2 dimensional arrays of precision, recall and f-measure for every possible pair of the evaluated clusters and the gold clusters. From the 2 dimensional array the best precision, recall and f-measure out of the gold clusters for each evaluated clusters were taken. Finally, the weighted average of the precision, recall and f-measure was calculated by taking the Root-Mean-Square of the values. This is done by taking the average of the square of each entry, which then a square root is applied to the average value. Apart from the weighted measures, we also present the results of simple average Precision and Recall.

4.2   Test Data Set

The test data set used to evaluate the algorithms consists of 400 snippet data and 40 full length data [10], containing documents obtained from the web on various keywords such as 'jaguar', 'apple', 'victoria', 'football' and such. Within each keyword there are subjects that the documents related to, which forms the gold labelling for the documents. Furthermore, the documents are pre-processed by removing tags, punctuation and stop words, and stemming remaining words. This test data was formerly used by Daniel Crabtree to test his QDC algorithm [2]. This is currently the best external data available for use in query directed clustering algorithms, as other clustering data does not contain the necessary query along with the gold standard.

4.3   Results

From the results, it is clear that EQDC performs marginally better in all three modification than the original algorithm. Furthermore, we can also see a slight overall improvement when phrases are used instead of single words. As shown in Table 1, the use of WLM alone produces clusters that are almost identical in both precision and recall in comparison to the clusters produced from QDC algorithm. WLM and QDC give an overall F-scores that are almost identical. By combining phrases and WLM, the final EQDC shows an improvement.

Table 1. P(Precision), R(Recall) and F(F-measure) average over the three queries

		QDC using NGD Phrase size 1	QDC using NGD Phrase size 2	QDC using WLM Phrase size 1	EQDC (using WLM Phrase size 2)
P	Average	0.6622	0.6720	0.6644	**0.6955**
	Weighted	0.6762	0.6908	0.6837	**0.7139**
R	Average	0.7430	0.7687	0.7364	**0.7770**
	Weighted	0.7781	0.8050	0.7764	**0.8096**
F	Average	0.6488	0.6725	0.6556	**0.6924**
	Weighted	0.6828	0.7093	0.6960	**0.7282**

5   Conclusion

This paper introduces Extended Query Directed Clustering (EQDC) which combines QDC with Suffix Tree clustering (STC) and utilises a new similarity measure for estimating the distance between a query and a base cluster label. The results from the tests show that EQDC is a modification that can further improve the quality of web page clustering. Although the use of phrases and different semantic measures separately did not yield any conclusive results, using the two in conjunction can lead to improvements of the QDC algorithm.

Possible future extension may include modifying other components of the QDC algorithm. One such example is the Cluster Merging process, where the semantic similarity of the description has a large effect on whether the clusters are merged or not. In addition, the use of other better semantic similarity measure algorithm for phrases, such as ESA [11] will most likely to have a positive effect on the accuracy of the algorithm. We will also test our algorithm on larger data sets.

References

1. Zamir, O., Etzioni, O.: Web document clustering: a feasibility demonstration. In: Proceedings of the 21st Annual International ACM SIGIR Conference on Research and Development in Information Retrieval, SIGIR 1998, pp. 46–54. ACM, New York (1998)
2. Crabtree, D., Andreae, P., Gao, X.: Query directed web page clustering. In: Proceedings of the 2006 IEEE/WIC/ACM International Conference on Web Intelligence, WI 2006, pp. 202–210. IEEE Computer Society, Washington, DC (2006)
3. Crabtree, D., Gao, X., Andreae, P.: Query directed clustering. The Knowledge and Information Systems (KAIS) Journal (acepted July 29, 2012)
4. Cilibrasi, R.L., Vitanyi, P.M.B.: The google similarity distance. IEEE Trans. on Knowl. and Data Eng. 19, 370–383 (2007)
5. Milne, D., Witten, I.H.: An effective, low-cost measure of semantic relatedness obtained from wikipedia links. In: Proceedings of AAAI 2008 (2008)
6. Weiner, P.: Linear pattern matching algorithms. In: Proceedings of the 14th Annual Symposium on Switching and Automata Theory (SWAT 1973), pp. 1–11. IEEE Computer Society, Washington, DC (1973)
7. Gabrilovich, E., Markovitch, S.: Computing semantic relatedness using wikipedia-based explicit semantic analysis. In: Proceedings of the 20th International Joint Conference on Artificial Intelligence, pp. 1606–1611 (2007)
8. Strube, M., Ponzetto, S.P.: Wikirelate! computing semantic relatedness using wikipedia. In: Proceedings of Association for the Advancement of Artificial Intelligence, AAAI (2006)
9. Bu, F., Hao, Y., Zhu, X.: Semantic relationship discovery with wikipedia structure. In: Proceedings of the Twenty-Second International Joint Conference on Artificial Intelligence, IJCAI 2011, vol. 3, pp. 1770–1775. AAAI Press (2011)
10. Crabtree, D.: Raw data set (2005),
    http://www.danielcrabtree.com/research/wi05/rawdata.zip
11. Gabrilovich, E., Markovitch, S.: Computing semantic relatedness using wikipedia-based explicit semantic analysis. In: Proceedings of the 20th International Joint Conference on Artifical Intelligence, IJCAI 2007, pp. 1606–1611. Morgan Kaufmann Publishers Inc., San Francisco (2007)

Mining Fuzzy Moving Object Clusters

Phan Nhat Hai[1,2], Dino Ienco[1,2], Pascal Poncelet[1,2], and Maguelonne Teisseire[1,2]

[1] IRSTEA Montpellier, UMR TETIS - 34093 Montpellier, France
{nhat-hai.phan,maguelonne.teisseire}@teledetection.fr
[2] LIRMM CNRS Montpellier - 34090 Montpellier, France
pascal.poncelet@lirmm.fr

Abstract. Recent improvements in positioning technology have led to a much wider availability of massive moving object data. One of the objectives of spatio-temporal data mining is to analyze such datasets to exploit moving objects that travel together. Naturally, the moving objects in a cluster may actually diverge temporarily and congregate at certain timestamps. Thus, there are time gaps among moving object clusters. Existing approaches either put a strong constraint (i.e. no time gap) or completely relaxed (i.e. whatever the time gaps) in dealing with the gaps may result in the loss of interesting patterns or the extraction of huge amount of extraneous patterns. Thus it is difficult for analysts to understand the object movement behavior.

Motivated by this issue, we propose the concept of *fuzzy swarm* which softens the time gap constraint. The goal of our paper is to find all non-redundant fuzzy swarms, namely *fuzzy closed swarm*. As a contribution, we propose *fCS-Miner* algorithm which enables us to efficiently extract all the fuzzy closed swarms. Conducted experiments on real and large synthetic datasets demonstrate the effectiveness, parameter sensitiveness and efficiency of our methods.

Keywords: Fuzzy closed swarm, fuzzy time gap, frequent itemset.

1 Introduction

Nowadays, many electronic devices are used for real world applications. Telemetry attached on wildlife, GPS installed in cars, sensor networks, and mobile phones have enabled the tracking of almost any kind of data and has led to an increasingly large amount of data that contain moving object information. One of the objectives of spatio-temporal data mining [2] [4] [5] [10] [11] [12] [14] is to analyze such datasets for interesting patterns. For example, Buffaloes in South Africa (165 animals reported daily), Golden Eagles in Alaska (43 animals reported daily), and so on[1]. Analyzing this data gives insight into entity behavior, in particular and migration patterns [12]. The analysis of moving objects also has applications in transport analysis, route planning and vehicle control, socio-economic geography, location prediction, location-based services.

One of the crucial tasks for extracting patterns is to find moving object clusters (i.e. group of moving objects that are traveling together). A moving object cluster can be defined in both spatial and temporal dimensions: (1) a group of moving objects should

---

[1] http://www.movebank.org

S. Zhou, S. Zhang, and G. Karypis (Eds.): ADMA 2012, LNAI 7713, pp. 100–114, 2012.

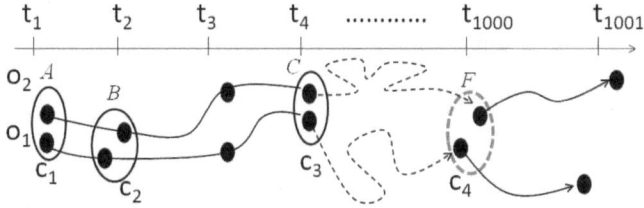

Fig. 1. An example of moving object clusters. o_1, o_2 are moving objects, $c_1, \ldots, c_4$ are clusters which are generated by applying some clustering techniques and A, B, C, F are spatial regions.

be geometrically closed to each other, (2) they should be together for at least some minimum number of certain timestamps. In this context, many recent studies have been defined to mine moving object clusters including flocks [2], moving clusters [9], convoy queries [10], closed swarms [12], group patterns [14], etc...

The common part of such patterns is that they require the group of moving objects to be together for at least min_t timestamps (i.e. could be consecutive or completely be non-consecutive), which might not be practical in the real cases. For instance, if we set $min_t = 3$ and the timestamps must be consecutive in Figure 1, no moving object cluster can be found. But essentially, these two objects, o_1 and o_2, travel together even though they temporarily leave the cluster at some snapshots. To address this issue, Zhenhui Li et al. [12] propose swarm in which moving objects are not required to be together in consecutive timestamps. Therefore, swarm can capture the movement pattern in Figure 1. The pattern is "o_1, o_2 *are moving together from A to B to C and to F at timestamps* t_1, t_2, t_4 *and* t_{1000}". This pattern could be interesting since it expresses the relationship between o_1 and o_2. However, the issue here is that it is hard to say that o_1 and o_2 moving together to F since they only meet each other at F by chance after 996 timestamps. In other words, enforcing the consecutive time constraint may result in the loss of interesting moving object clusters, while completely relaxing this constraint may generate a large number of extraneous and useless patterns.

In this paper, we propose a new movement pattern, called *fuzzy closed swarm*, which softens the consecutive time constraint without generating extraneous patterns. The key challenge is to deal with the time gap between a pair of clusters since: 1) it is difficult to recognize which size of a time gap is relevant or not, 2) we need to know when the patterns should be ended to eliminate uninteresting ones. To address these issues, we present the definition of *fuzzy time gap* and *fuzzy time gap participation index*. Obtained patterns are of the type "o_1, o_2 *are moving together from A to B to C with 60% weak, 20% medium and 20% strong time gaps*". These patterns are characterized by their time gap frequency (or support), which is by definition the proportion of time gaps involved in the patterns. As a contribution, we propose *fCS-Miner* algorithm to efficiently extract the complete set of fuzzy closed swarms. The approach shares the same spirit with the GeT_Move algorithm [5] [6] but is different in terms of goal and properties. The effectiveness as well as efficiency of our method are demonstrated on both real and large scale synthetic moving object databases.

This paper is structured as follows. Section 2 discusses the related work. The definitions of fuzzy time gap and fuzzy closed swarm are given in Section 3. fCS-Miner algorithm will be clearly presented in Section 4. Experiments testing effectiveness and efficiency are shown in Section 5. Finally, we draw our conclusions in Section 6.

2   Related Work

As mentioned before, many approaches have been proposed to extract patterns. For instance, Gudmundsson and Van Kreveld [2] define a flock pattern, in which the same set of objects stay together in a circular region with a predefined radius, Kalnis et al. [11] propose the notion of *moving clusters*. Jeung et al. [10] define a convoy pattern and propose three algorithms $CMC, CuTS, CuTS^*$ that incorporate trajectory simplification techniques in the first step. Then, the authors proposed to interpolate the trajectories by creating virtual time points and by applying density measurements on trajectory segments. Additionally, the convoy is defined as a candidate when it has at least k clusters during k consecutive timestamps.

Recently, Zhenhui Li et al. [12] propose the concept of swarm and closed swarm and the *ObjectGrowth* algorithm to extract closed swarm patterns. The ObjectGrowth method is a depth-first-search of all subsets of O_{DB} through a pre-order tree traversal. To speed up the search process, they propose two pruning rules. *Apriori Pruning* and *Backward Pruning* are used to stop traversal the subtree when we find further traversal that cannot satisfy min_t and closure property. After pruning the invalid candidates, a *ForwardClosure checking* is used to determine whether a pattern is a closed swarm. In [14], Hwang et al. propose two algorithms to mine group patterns, known as the *Apriori-like Group Pattern mining* algorithm and *Valid Group-Growth* algorithm. The former explores the Apriori property of valid group patterns and the latter is based on idea similar to the FP-growth algorithm.

The interested readers may refer to [7] where short descriptions of the most efficient approaches and interesting patterns are proposed. Nevertheless, all the work above is not able to address the problem of capturing fuzzy closed swarms.

3   Problem Statement

3.1   Preliminarily Definitions

Let us assume that we have a set of moving objects $O_{DB} = \{o_1, o_2, \ldots, o_z\}$, a set of timestamps $T_{DB} = \{t_1, t_2, \ldots, t_m\}$.

Database of Clusters. A database of clusters, $C_{DB} = \{C_1, C_2, \ldots, C_m\}$, is the collection of snapshots of the moving object clusters at timestamps $\{t_1, t_2, \ldots, t_m\}$. Note that an object could belong to several clusters at one timestamp (i.e. cluster overlapping). Given a cluster $c \in C_{DB}$ and $c \subseteq O_{DB}$, $|c|$ and $t(c)$ are respectively used to denote the number of objects belong to cluster c and the timestamp that c involved in. To make our framework more general, we take clustering as a preprocessing step. The clustering methods could be different based on various scenarios. We leave the details of this step in the Appendix *Obtaining Clusters*.

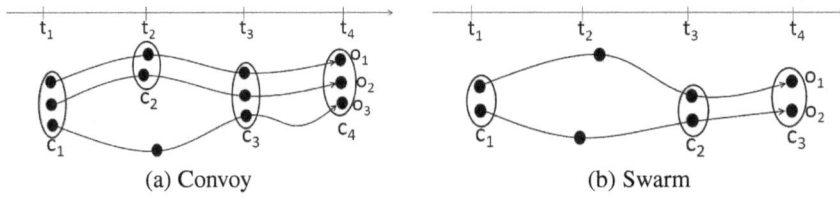

Fig. 2. An example of convoy and swarm where c_1, c_2, c_3, c_4 are clusters

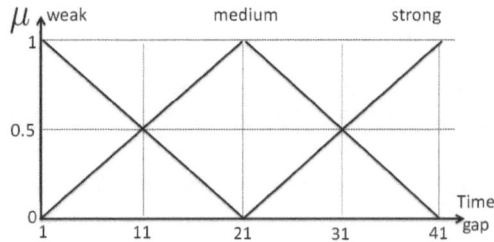

Fig. 3. Membership degree functions for fuzzy time gaps

From now, $O = \{o_{i_1}, o_{i_2}, \ldots, o_{i_p}\}(O \subseteq O_{DB})$ stands for a group of objects, $T = \{t_{a_1}, t_{a_2}, \ldots, t_{a_m}\}$ $(T \subseteq T_{DB})$ is the set of timestamps within which objects stay together.

Convoys and Closed Swarms. Informally, a *convoy* (O, T) is a group of objects O containing at least min_o individuals which are closed each other during at least min_t consecutive time points T. While, consecutive time constraint is relaxed in *swarm* in which objects in O are closed each other for at least min_t timestamps. To avoid redundancy, Zhenhui Li et al. [12] propose the notion of *closed swarm* for grouping together both objects and time. A swarm (O, T) is *object-closed* if when fixing T, O cannot be enlarged. Similarly, a swarm (O, T) is *time-closed* if when fixing O, T cannot be enlarged. A swarm (O, T) is a closed swarm if it is both *object-closed* and *time-closed*.

For instance, in Figure 2a, with $min_o = 2$, $min_t = 2$ we have two convoys $(\{o_1, o_2\}, \{t_1, t_2, t_3, t_4\})$ and $(\{o_1, o_2, o_3\}, \{t_3, t_4\})$. While, in Figure 2b, if we set $min_o = 2$ and $min_t = 2$, we can find the following swarms $(\{o_1, o_2\}, \{t_1, t_3\})$, $(\{o_1, o_2\}, \{t_1, t_4\})$, $(\{o_1, o_2\}, \{t_3, t_4\})$, $(\{o_1, o_2\}, \{t_1, t_3, t_4\})$. We can note that these swarms are in fact redundant since they can be grouped together in the following closed swarm $(\{o_1, o_2\}, \{t_1, t_3, t_4\})$.

3.2 Fuzzy Closed Swarms

As illustrated before, enforcing the consecutive time constraint or completely relaxing this constraint may result in the loss of interesting patterns or the generation of uninteresting patterns. To deal with the issue, we propose the adaptation of fuzzy logic principle in which the strength of time gaps are evaluated with a membership degree function A (see Figure 3). Given two timestamps t_1 and t_2, a time gap x between t_1

and t_2 is computed as $x = |t_1 - t_2| - 1$ (i.e. $t_1 \neq t_2$). The fuzzy time gap is defined as follows.

Definition 1. *Fuzzy Time Gap. Given two timestamps t_1 and t_2, a pair of one time gap x and one corresponding fuzzy set a, denoted by $[x, a]$, is called a fuzzy time gap if $x = |t_1 - t_2| - 1$ is involved in membership function A.*

For instance, see Figure 4, there are totally four time gaps which are $x_1 = 3, x_2 = 18, x_3 = 34$ and $x_4 = 939$. The fuzzy time gap $[x_1, weak]$, $[x_1, medium]$ and $[x_1, strong]$ respectively are $\mu_{weak}(x_1) = 0.9$, $\mu_{medium}(x_1) = 0.1$ and $\mu_{strong}(x_1) = 0$. Since x_4 is out of function A, it cannot be considered as a fuzzy time gap.

Definition 2. *Fuzzy Time Gap Set. Given an ordered list of timestamps $T = \{t_{a_1}, t_{a_2}, \ldots, t_{a_m}\}$, a set of time gaps $X = \{x_1, \ldots, x_n\}, n = m - 1$. (X, A) is a fuzzy time gap set generated from T if $\forall i \in \{1, \ldots, n\} : x_i = |t_{a_i} - t_{a_{i+1}}| - 1$ and $\forall x \in X : x$ is involved in A. Note that for any $x \in X, x = 0$ then x will be excluded from X without any affection.*

For instance, see Figure 4, a proper pattern $(\{o_1, o_2\}, \{t_1, t_2, t_6, t_{25}, t_{60}\})$ and a fuzzy time gap set is $X = \{x_1, x_2, x_3\}$ and for each time gap $x_i \in X$, there are a corresponding fuzzy set including $strong, medium$ and $weak$. Note that x_4 is out of membership function and therefore it is not included in X and $(\{o_1, o_2\}, \{t_1, t_2, t_6, t_{25}, t_{60}, t_{1000}\})$ will not be considered as a valid pattern.

To highlight the participation of time gaps given by a fuzzy set a, we further propose an adaptation of the participation index [8] which is *fuzzy time gap participation index* proposed to take into account the fuzzy time gap occurrences in the pattern.

Definition 3. *Fuzzy Time Gap Participation Ratio. Let (X, A) be a set of fuzzy time gaps and a be an item of A, the fuzzy time gap participation ratio for a in X denoted $TGr(X, a)$ can be defined as follows.*

$$TGr(X, a) = \frac{\sum_{x \in X} \mu_a(x)}{|X|} \tag{1}$$

Definition 4. *Fuzzy Time Gap Participation Index. Let (X, A) be a set of fuzzy time gaps and a be an item of A, the fuzzy time gap participation index of (X, A) denoted $TGi(X)$ can be defined as follows.*

$$TGi(X) = Max_{\forall a \in A} TGr(X, a) \tag{2}$$

For instance, see Figure 4, a fuzzy time gap set $X = \{x_1, x_2, x_3\}$ and $TGr(X, weak) = \frac{0.1+0.9+0}{3} = 0.33$, $TGr(X, medium) = \frac{0.1+0.9+0.35}{3} = 0.45$, $TGr(X, strong) = \frac{0+0+0.65}{3} = 0.22$. Thus, the fuzzy time gap participation index of $X, TGi(X) = 0.45$.

Fuzzy Swarm and Fuzzy Closed Swarm. Given a group of objects O moving together in an ordered list of timestamps T and a set of fuzzy time gaps (X, A) generated from T. (O, T, X) is a fuzzy swarm that contains at least min_o objects (resp. $|O| \geq min_o$) during at least min_t timestamps (resp. $|T| \geq min_t$) and $TGi(X) \geq \varepsilon$. The fuzzy swarm can be defined as follows.

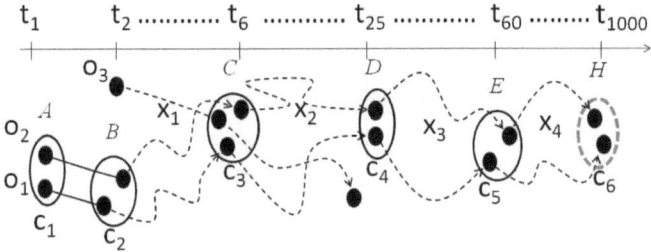

Fig. 4. A fuzzy closed swarm running example

Definition 5. *Fuzzy Swarm. Given integers min_o, min_t and a user-defined threshold ε. (O, T, X) is a fuzzy swarm if and only if:*

$$\begin{cases} (1) : |O| \geq min_o. \\ (2) : |T| \geq min_t. \\ (3) : (X, A) \text{ is a fuzzy time gap set.} \\ (4) : \forall i \in \{1, \ldots, n\}, TGi(\{x_1, \ldots, x_i\}) \geq \varepsilon. \end{cases} \quad (3)$$

Note that if $X = \{x_1, x_2, x_3\}$ then the condition (4) means that $TGi(\{x_1\}) \geq \varepsilon$, $TGi(\{x_1, x_2\}) \geq \varepsilon$ and $TGi(\{x_1, x_2, x_3\}) \geq \varepsilon$.

By definition, if we set $min_o = 2, min_t = 3$ and $\varepsilon = 0.2$ then there are totally 13 fuzzy swarms in Figure 4 such as $(\{o_1, o_2\}, \{t_1, t_2, t_6\}, \{x_1\})$, $(\{o_1, o_2\}, \{t_1, t_2, t_{25}\}, \{x = 22\})$, $(\{o_1, o_2\}, \{t_2, t_6, t_{25}, t_{60}\}, \{x_1, x_2, x_3\})$ and so on. However, it is obviously redundant to output fuzzy swarms like $(\{o_1, o_2\}, \{t_1, t_2, t_6, \})$ since it can be enlarged to $(\{o_1, o_2\}, \{t_1, t_2, t_6, t_{25}, t_{60}\}, \{x_1, x_2, x_3\})$. To avoid mining redundant fuzzy swarms, we further give the definition of fuzzy closed swarm. Essentially, a fuzzy swarm (O, T, X) is *time-closed* if fixing T, O cannot be enlarged ($\nexists O'$ s.t. (O', T, X) is a fuzzy swarm and $O \subset O'$). Similarly, a fuzzy swarm (O, T, X) is *object-closed* if fixing O then T cannot be enlarged. Finally, a fuzzy swarm (O, T, X) is a fuzzy closed swarm if it is both *time-closed* and *object-closed*. Our goal is to find the complete set of fuzzy closed swarms. The definition is formally presented as follows.

Definition 6. *Fuzzy Closed Swarm. Given a fuzzy swarm (O, T, X), it is a fuzzy closed swarm if and only if:*

$$\begin{cases} (1) : \nexists O', O \subset O' \wedge (O', T, X) \text{ is a fuzzy swarm.} \\ (2) : \nexists T', T \subset T' \wedge (O, T', X') \text{ is a fuzzy swarm.} \end{cases} \quad (4)$$

For instance (Figure 4), a closed swarm is $(\{o_1, o_2\}, \{t_1, t_2, t_6, t_{25}, t_{60}\}, \{x_1, x_2, x_3\})$.

Property 1. Anti-monotonic. For all patterns (O, T, X), if (O, T, X) is not a fuzzy swarm because of the condition (3) suffering then the following holds:

For all supersets of (O, T, X) by adding a later cluster and a fuzzy time gap in terms of time to T and X are not fuzzy swarms.

Table 1. Cluster Matrix corresponding to our running example in Figure 4

T_{DB}	t_1	t_2	t_6	t_{25}	t_{60}	t_{1000}
Clusters C_{DB}	c_1	c_2	c_3	c_4	c_5	c_6
O_{DB} o_1	1	1	1	1	1	1
o_2	1	1	1	1	1	1
o_3			1			

Proof. After construction, we have $\exists k \in \{1, \ldots, n\}$ s.t. $TGi(\{x_1, \ldots, x_k\}) < \varepsilon$. For any $X' = \{x_1, \ldots, x_n, x_m\}$, (O, T', X') is not a fuzzy swarm since $\exists k \in \{1, \ldots, m\}$ s.t. $TGi(\{x_1, \ldots, x_k\}) < \varepsilon$.

4 Discovering of Fuzzy Closed Swarms

The patterns we are interested in here, fuzzy closed swarms, is the association of a set of objects O, a set of timestamps T and a set of fuzzy time gaps X, denoted (O, T, X). As first glance, we can employ ObjectGrowth algorithm [12] to extract all closed swarms and then a post-processing step to obtain all the fuzzy closed swarms. However, moving object databases are naturally large and thus the search space of closed swarm extracting can be significantly increased (i.e. approximately $2^{|O_{DB}|} \times 2^{|T_{DB}|}$). Additionally, a huge amount of generated closed swarms (i.e. including extraneous patterns) can cause an expensive post-processing task. Furthermore, in real world applications (e.g. cars), object locations are continuously reported by using Global Positioning System (GPS). Thus, new data is always available and we need to execute again and again the algorithms on the whole database (i.e. including existing data and new data) to extract patterns. This is of course, cost-prohibitive and time consuming.

To deal with the issues, we propose *fCS-Miner* algorithm which is an adaptation of Incremental GeT_Move approach [5] [6] which has already been proved as being efficient in large moving object databases.

Basic Idea of fCS-Miner Algorithm. As in [5] [6], we first present C_{DB} in a cluster matrix (see Table 1) so that Incremental GeT_Move can be applied to extract all frequent closed itemsets (FCIs). Next, we propose an novel property which can be used to directly extract fuzzy closed swarms from generated FCIs without a post-processing step. The cluster matrix definition is as follows.

Definition 7. *Cluster Matrix [5] [6]. Given a set of clusters $C_{DB} = \{C_1, C_2, \ldots, C_m\}$ where $C_i = \{c_{i_1 t_i}, c_{i_2 t_i}, \ldots, c_{i_n t_i}\}$ is a set of clusters at timestamps t_i. A cluster matrix is thus a matrix of size $|O_{DB}| \times |C_{DB}|$. Each row represents an object and each column represents a cluster. The value of the cluster matrix cell, (o_i, c_j) is 1 (resp. empty) if o_i is in (resp. is not in) cluster c_j.*

For instance, see Table 1 and Figure 4, the matrix cell of (o_1, c_2) is 1 since $o_1 \in c_2$ and this is similar for c_1, c_3, c_4, c_6. While, the matrix cell of (o_3, c_1) is empty since $o_3 \notin c_1$.

By applying Incremental GeT_Move which mainly bases on LCM algorithm [13] on the cluster matrix, we are able to extract all FCIs. Let us denote a frequent itemset as $\Upsilon = \{c_1, c_2, \ldots, c_k\}$, O_Υ contains the corresponding group of moving objects which are closed each other in a set of timestamps $T_\Upsilon = \{t(c_1), t(c_2), \ldots, t(c_k)\}$. We can recognize that $|O_\Upsilon| = \sigma(\Upsilon)^2$, $|\Upsilon| = |T_\Upsilon|$ and X_Υ is used to denote as a fuzzy time gap set generated from T_Υ. For instance, see Table 1, a proper frequent itemset is $\Upsilon = \{c_1, c_2, c_3, c_4, c_5\}$ with $O_\Upsilon = \{o_1, o_2\}$, $T_\Upsilon = \{t_1, t_2, t_6, t_{25}, t_{60}\}$ and $X_\Upsilon = \{x_1, x_2, x_3\}$.

The following property, *f-closed swarm*, is used to verify whenever a frequent itemset Υ can be a fuzzy closed swarm or not.

Property 2. f-Closed swarm. Given a frequent itemset $\Upsilon = \{c_1, c_2, \ldots, c_k\}$, $X_\Upsilon = \{x_1, \ldots, x_n\}$. $(O_\Upsilon, T_\Upsilon, X_\Upsilon)$ is a fuzzy closed swarm if and only if:

$$\begin{cases} (1) : \sigma(\Upsilon) \geq min_o. \\ (2) : |\Upsilon| \geq min_t. \\ (3) : \forall x \in X, x \text{ is involved in } A. \\ (4) : \forall i \in \{1, \ldots, n\}, TGi(\{x_1, \ldots, x_i\}) \geq \varepsilon. \\ (5) : \nexists \Upsilon' \text{ s.t } O_\Upsilon \subset O_{\Upsilon'}, T_{\Upsilon'} = T_\Upsilon \text{ and } (O_{\Upsilon'}, T_\Upsilon, X_\Upsilon) \text{ is a fuzzy swarm.} \\ (6) : \nexists \Upsilon' \text{ s.t. } O_{\Upsilon'} = O_\Upsilon, T_\Upsilon \subset T_{\Upsilon'} \text{ and } (O_\Upsilon, T_{\Upsilon'}, X_{\Upsilon'}) \text{ is a fuzzy swarm.} \end{cases} \quad (5)$$

Proof. After construction, we have $\sigma(\Upsilon) \geq min_o$ and thus $|O_\Upsilon| \geq min_o$ since $|O_\Upsilon| = \sigma(\Upsilon)$. Additionally, $|\Upsilon| \geq min_t$ and therefore $|T_\Upsilon| \geq min_t$ since $|\Upsilon| = |T_\Upsilon|$. Furthermore, $\forall x \in X : x$ is involved in A and $\forall i \in \{1, \ldots, n\}, TGi(\{x_1, \ldots, x_i\}) \geq \varepsilon$.. Consequently, $(O_\Upsilon, T_\Upsilon, X_\Upsilon)$ is a fuzzy swarm (*Definition 5*). Moreover, if $\nexists \Upsilon'$ s.t $O_\Upsilon \subset O_{\Upsilon'}, T_{\Upsilon'} = T_\Upsilon$ and $(O_{\Upsilon'}, T_\Upsilon, X_\Upsilon)$ is a fuzzy swarm then $(O_\Upsilon, T_\Upsilon, X_\Upsilon)$ cannot be enlarged in terms of objects. Therefore, it satisfies the *object-closed* condition. Furthermore, if $\nexists \Upsilon'$ s.t. $O_{\Upsilon'} = O_\Upsilon, T_\Upsilon \subset T_{\Upsilon'}$ and $(O_\Upsilon, T_{\Upsilon'}, X_{\Upsilon'})$ is a fuzzy swarm then $(O_\Upsilon, T_\Upsilon, X_\Upsilon)$ cannot be enlarged in terms of lifetime. Therefore, it satisfies the *time-closed* condition. Consequently, $(O_\Upsilon, T_\Upsilon, X_\Upsilon)$ is a fuzzy swarm and it satisfies *object-closed* and *time-closed* conditions and therefore $(O_\Upsilon, T_\Upsilon, X_\Upsilon)$ is a fuzzy closed swarm according to the *Definition 6*.

To show the fact that from an itemset mining algorithm we are able to extract the set of all fuzzy closed swarms, we propose the following lemma.

Lemma 1. *Let $FI = \{\Upsilon_1, \Upsilon_2, \ldots, \Upsilon_l\}$ be the set of frequent itemsets being mined from the cluster matrix with $minsup = min_o$. All fuzzy closed swarms (O, T, X) can be extracted from FI.*

Proof. Let us assume that (O, T, X) is a fuzzy closed swarm. Note, $T = \{t(c_1), \ldots, t(c_k)\}$. According to the *Definition 6* we have $|O| \geq min_o$. If (O, T, X) is a fuzzy closed swarm then $\forall t(c_i) \in T, \exists c_i$ s.t. $O \subseteq c_i$ therefore $\bigcap_{i=1}^{k} c_i = O$. Additionally, we have $\forall c_i$, c_i is an item so $\exists \Upsilon = \bigcup_{i=1}^{k} c_i$ is an itemset and $O_\Upsilon = \bigcap_{i=1}^{k} c_i = O, T_\Upsilon = \bigcup_{i=1}^{k} t(c_i) = T$. Furthermore, we also have $X_\Upsilon = X$ as well. Therefore, $(O_\Upsilon, T_\Upsilon, X_\Upsilon)$ is a fuzzy closed swarm since $O_\Upsilon = O, T_\Upsilon = T$ and $X_\Upsilon = X$. So, (O, T, X) is extracted from Υ. Furthermore, $\sigma(\Upsilon) = |O_\Upsilon| = |O| \geq min_o$ then Υ is a frequent itemset

---

[2] $\sigma(\Upsilon)$ is the support value of frequent itemset Υ.

Algorithm 1. fCS-Miner

Input : double ε, int min_o, int min_t, set of items C_{DB}
1 **begin**
2 | Incremental GeT_Move(C_{DB}, min_o);
3 **PatternMining**(FCI, ε, min_t)
4 **begin**
5 | f-CS := $\emptyset$;
6 | **if** $|FCI| \geq min_t$ **then**
7 | | $\Upsilon := \emptyset$;
8 | | **for** $k := 1$ *to* $|FCI|$ **do**
9 | | | $\Upsilon' := \Upsilon \cup c_k$;
10 | | | **if** $fuzzy(X_{\Upsilon'}) = true \wedge TGi(\Upsilon') \geq \varepsilon$ **then**
11 | | | | $\Upsilon := \Upsilon'$;
12 | | | **else**
13 | | | | **if** $O_\Upsilon = O_{FCI} \wedge |\Upsilon| \geq min_t + 1$ **then**
14 | | | | | f-CS := f-CS $\cup \Upsilon$;
15 | | | | $\Upsilon := \emptyset \cup c_k$;
16 | | **return** f-CS;
17 where: $fuzzy(X_{\Upsilon'})$ returns *true* if $X_{\Upsilon'}$ is a fuzzy time gap set, otherwise returns *false*. In this function, we only need to verify that the last time gap is involved in A instead of all the time gaps in $X_{\Upsilon'}$.

---

and $\Upsilon \in FI$. Finally, $\forall (O, T)$ s.t. if (O, T, X) is a fuzzy closed swarm then $\exists \Upsilon$ s.t. $\Upsilon \in FI$ and (O, T, X) can be extracted from Υ, we can conclude that $\forall$ fuzzy closed swarm (O, T, X), it can be mined from FI.

Essentially, by scanning the FCIs from the beginning to the end with the f-closed swarm property, we are able to extract the corresponding fuzzy closed swarms. The scanning process will be ended whenever one of conditions (3), (4) is suffered (*Property 1*), after that the current frequent itemset Υ (i.e. $\sigma(\Upsilon) \geq min_o$) only need to be verified the conditions $|\Upsilon| \geq min_t$ and Υ contains the same number of objects with the FCI. This is because Υ cannot be enlarged in terms of timestamps T_Υ (i.e. *Property 1*) and objects (i.e. FCI is closed). Thus, it satisfies all the requirements to be a fuzzy closed swarm completely.

The pseudo-code of fCS-Miner is presented in the Algorithm 1. We first apply Incremental GeT_Move on cluster matrix C_{DB} with $minsup = min_o$ (line 2). Then, for each generated FCI, we directly scan it with the *f-closed swarm* property as mentioned before (lines 4-16). By using fCS-Miner, we are able to extract all fuzzy closed swarms on-the-fly without a post-processing step.

5   Experimental Results

A comprehensive performance study has been conducted on real and synthetic datasets. All the algorithms are implemented in C++, and all the experiments are carried out on a 2.8GHz Intel Core i7 system with 4GB Memory. The system runs Ubuntu 11.10 and g++ version 4.6.1. The implementation of our proposed algorithm is also integrated in a demonstration system available online[3]. As in [12] [2] [6], the following dataset[4] have been used during experiments: *Swainsoni dataset* includes 43 objects evolving over time and 764 different timestamps. It was generated from July 1995 to June 1998.

To the best of our knowledge, there is no previous work which addresses fuzzy closed swarms. Therefore, in the comparison, we employ the latest pattern mining algorithms such as $CuTS^{*5}$ [10] (convoy mining) and *ObjectGrowth* [12] (closed swarm mining). As pointed out in [12], *ObjectGrowth* outperforms *VG-Growth* [14] (group pattern mining) in terms of performance and therefore we will only consider *ObjectGrowth* and not both.

Similarly to [10] [12], we first use linear interpolation to fill in the missing data. Furthermore, as [10] [11] [12], DBScan [1] ($MinPts = 2, Eps = 0.001$) is applied to generate clusters at each timestamp. To make fair comparison, we adapt all the algorithms to accommodate clusters as input but their time complexity will remain the same. Additionally, to retrieve all the patterns including fuzzy closed swarms, convoys and closed swarms, in the reported experiments the fuzzy function in Figure 3 is applied, the default value of min_t is 1, $min_o = 1$ and $\varepsilon = 0.001$. Note that the default values are the hardest conditions for examining all the algorithms.

5.1 Effectiveness

The effectiveness of fuzzy closed swarms can be demonstrated through our online demo system. One of the extracted patterns from Swainsoni dataset is illustrated in Figure 5c. Each color represents a Swainsoni trajectory segment involved in the pattern.

To illustrate the feasibility of a fuzzy approach, we also show some of extracted closed swarms and convoys from our system[6] [3] in Figures 5a, b. We can consider that closed swarm is extraneous since the two objects meet each other at Mexico on October 1995 and after 5 months (i.e. to March 1996) for the next meeting location (i.e. Argentina). In fact, it is hard to say that they are moving together from Mexico to Argentina. While, the convoys are sensitive to time gaps and usually are short deal to the consecutive time constraint (see Figure 5a). Thus, they fail to describe the insightful relationship between objects. Either be too strict or too relaxed in dealing with time gaps may result in the loss of interesting patterns or reporting many uninteresting ones.

Distinguish from previous work, by proposing fuzzy closed swarms, we are able to reveal the relevant relationship between Swainsonies in a fuzzy point of view. Looking at the illustrated pattern in Figure 5c, we can consider that, from United States, the two objects are flying together along a narrow corridor through Central America and down to South America. Furthermore, they temporally diverge at Panama and congregate again at the Columbia central. The discovery of the fuzzy closed swarms on animal migration datasets provides useful information for biologists to better understand and

---

[3] http://www.lirmm.fr/~phan/fcsminer.jsp

[4] http://www.movebank.org

[5] The source code of $CuTS^*$ is available at http://lsirpeople.epfl.ch/jeung/source_codes.htm

[6] http://www.lirmm.fr/~phan/index.jsp

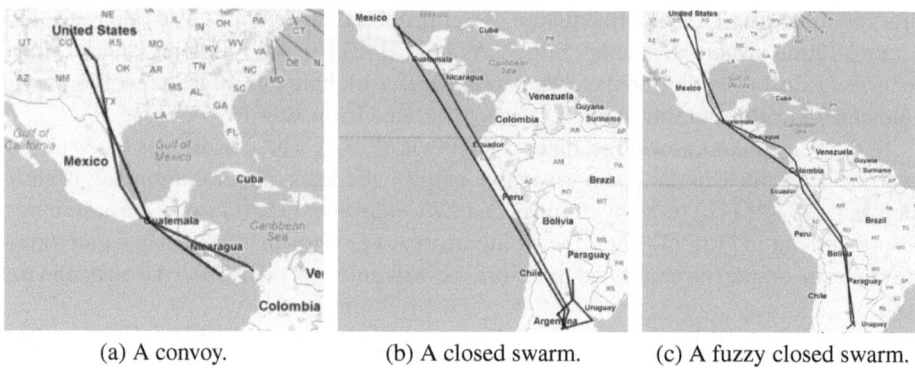

(a) A convoy.   (b) A closed swarm.   (c) A fuzzy closed swarm.

Fig. 5. An example of extracted patterns from Swainsoni dataset. The two object names are 'SW22' and 'SW40'.

examine the relationship and habits of these moving objects. Due to the space limitation, we do not provide experiments by varying the fuzzy membership function A. However, in real world context, users can express their expertise through the membership function for dealing with fuzzy approximate reasoning issues.

5.2   Parameter Sensitiveness

To show the parameter sensitiveness and efficiency of the proposed algorithm, as in [12], we also generate a large synthetic dataset using Brinkhoff's network[7]-based generator of moving objects. We generate 500 objects ($|O_{DB}| = 500$) for 10^4 timestamps ($|T_{DB}| = 10^4$) using the generator's default map with slow moving speed (5×10^6 points in total). DBScan ($MinPts = 3, Eps = 300$) is applied to obtain clusters.

Sensitiveness w.r.t ε. See Figure 6a, we can consider that fCS-Miner is not sensitive in ε. This is because ε is only used to scan the FCIs for fuzzy closed swarm extraction which is much less expensive than FCI mining task.

Sensitiveness w.r.t min_t. Figure 6b shows that ObjectGrowth is the most sensitive algorithm in min_t. This is because ObjectGrowth applies a min_t-based pruning rule, called *Apriori Pruning*, which is very sensitive in min_t. Since, it is used to limit approximately $2^{|T_{DB}|}$ candidates in total. Furthermore, with different values of min_t, there are great differences in terms of the number of extracted closed swarms (Figure 7b). Meanwhile, fCS-Miner and CuTS* only use min_t at the pattern reporting or verifying steps without any pruning rule for min_t. Additionally, as mentioned before the fuzzy closed swarm verifying task is less expensive than the FCI extraction. Consequently, be similar to CuTS*, the fCS-Miner sensitiveness in min_t is less sensitive than ObjectGrowth.

Sensitiveness w.r.t O_{DB}, T_{DB}. Figures 6c-d show the sensitiveness in the sizes of O_{DB} and T_{DB}. We can consider that all the algorithms are quite similar to each other. However, CuTS* is a little bit less sensitive than the others. This is because, in CuTS*: 1) the number of clusters at a certain timestamp is not exponentially increased due to the $|O_{DB}|$ and $|T_{DB}|$ increases, 2) for any cluster c, c can combine with the clusters

---

[7] http://iapg.jade-hs.de/personen/brinkhoff/generator/

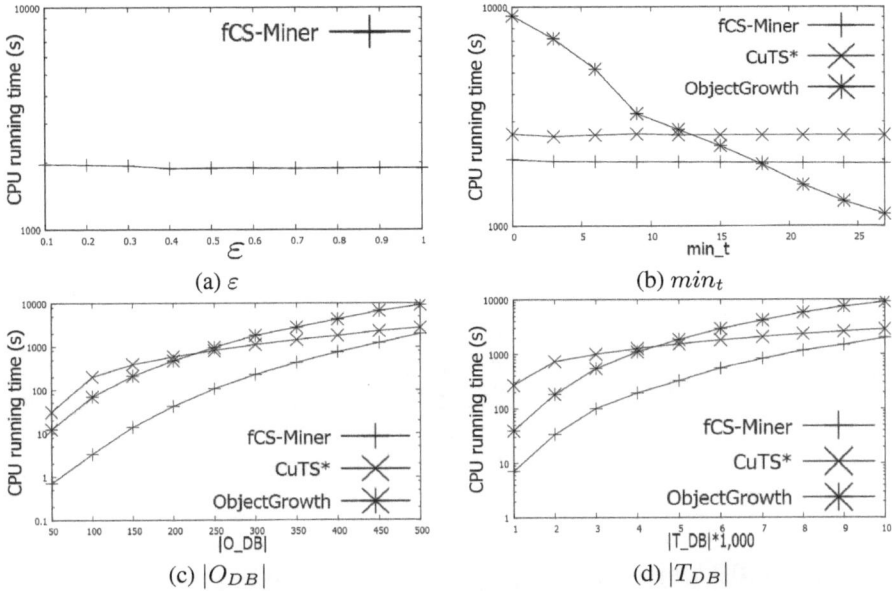

Fig. 6. Running time on Synthetic Dataset

at the next timestamp. While, for ObjectGrowth, the number of candidates is greatly increased due to the size increase of $|O_{DB}|$, $|T_{DB}|$ (i.e. approximately $2^{|O_{DB}|} \times 2^{|T_{DB}|}$ candidates). As the results, the number of closed swarms is significantly increased (see Figures 7c-d). This behavior is similar in fCS-Miner since the number of FCIs can be large. However, thanks to the fuzzy approach, there are not huge amount of generated patterns compared to ObjectGrowth. Obviously, fCS-Miner is similar to ObjectGrowth and a little bit more sensitive than CuTS* in terms of $|O_{DB}|$ and $|T_{DB}|$.

Influence of $TGi(X)$ on #f-Closed swarms. Figure 8 shows the influence of the fuzzy time gap participation index on the number of patterns that contain weak, medium and strong time gaps. We can consider that the number of patterns which have $TGi(X)$ with weak fuzzy time gaps X, medium fuzzy time gaps X and strong fuzzy time gaps X are quite distinguished from each other. Since, the number of patterns with weak X is smallest, more number of patterns with medium X and the highest number of patterns with strong X. Therefore, the $TGi(X)$ enable us to rank the fuzzy closed swarms well corresponding with the membership degree function. Furthermore, if we ignore all the fuzzy closed swarms with strong (and medium) fuzzy time gaps, a number of uninteresting patterns can be eliminated.

To summarize, fCS-Miner is effective to extract fuzzy closed swarms which are novel and useful movement patterns. By applying fuzzy function, users can express their background knowledge in order to obtain interesting patterns without generating extraneous ones. Additionally, fCS-Miner parameter sensitiveness is quite acceptable compare to the other model algorithms. Moreover, with the purpose to extract the complete set of f-closed swarms, fCS-Miner is competitive in time efficiency to state-of-the-art approaches (see Figure 6).

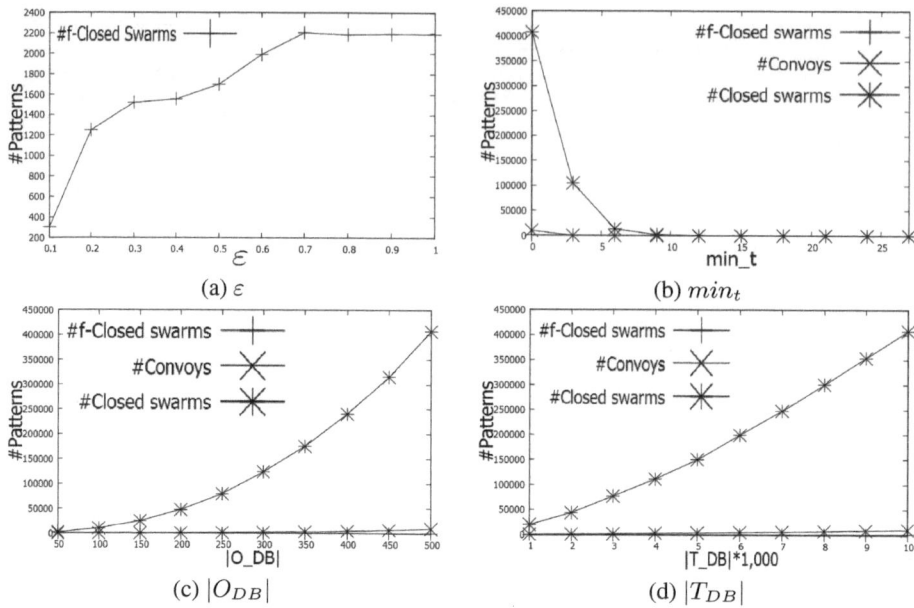

Fig. 7. Number of patterns on Synthetic Dataset

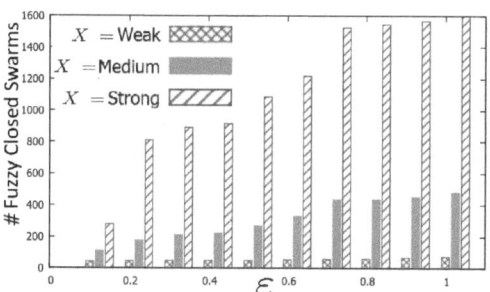

Fig. 8. Influence of $TGi(X)$ on #patterns through ε

6   Conclusions and Future Directions

In this paper, to deal with the issue that enforcing the consecutive time constraint or completely relaxing may result in the loss of interesting patterns or the generation of uninteresting patterns, we propose the concept of fuzzy swarm which softens the time gap constraint. These concepts enable the discovery of insightful movement patterns and the elimination of extraneous patterns. A new method fCS-Miner is proposed to efficiently extract all the fuzzy closed swarms. The proposed algorithm's effectiveness, and parameter sensitiveness are demonstrated using real and large synthetic datasets.

In the future work, the proposed approaches can be applied on other kinds of patterns (e.g. gradual trajectory patterns [4]). Although the number of non-interesting patterns are significantly reduced, it is still difficult to analyze the results since number of patterns is still large. Another future directions is to directly mining $top\text{-}K$ informative fuzzy movement patterns to avoid extracting redundant information.

References

1. Ester, M., Kriegel, H.P., Sander, J., Xu, X.: A density-based algorithm for discovering clusters in large spatial databases with noise. In: KDD, pp. 226–231 (1996)
2. Gudmundsson, J., van Kreveld, M.: Computing longest duration flocks in trajectory data. In: ACM GIS 2006 (2006)
3. Hai, P.N., Ienco, D., Poncelet, P., Teisseire, M.: Extracting Trajectories through an Efficient and Unifying Spatio-temporal Pattern Mining System. In: Flach, P.A., De Bie, T., Cristianini, N. (eds.) ECML PKDD 2012, Part II. LNCS, vol. 7524, pp. 820–823. Springer, Heidelberg (2012)
4. Hai, P.N., Ienco, D., Poncelet, P., Teisseire, M.: Ming time relaxed gradual moving object clusters. In: ACM SIGSPATIAL GIS (2012)
5. Hai, P.N., Poncelet, P., Teisseire, M.: GET_MOVE: An Efficient and Unifying Spatio-temporal Pattern Mining Algorithm for Moving Objects. In: Hollmén, J., Klawonn, F., Tucker, A. (eds.) IDA 2012. LNCS, vol. 7619, pp. 276–288. Springer, Heidelberg (2012)
6. Hai, P.N., Poncelet, P., Teisseire, M.: An efficient spatio-temporal mining approach to really know who travels with whom! In: BDA (2012)
7. Han, J., Li, Z., Tang, L.A.: Mining Moving Object, Trajectory and Traffic Data. In: Kitagawa, H., Ishikawa, Y., Li, Q., Watanabe, C. (eds.) DASFAA 2010. LNCS, vol. 5982, pp. 485–486. Springer, Heidelberg (2010)
8. Huang, Y., Shekhar, S., Xiong, H.: Discovering colocation patterns from spatial data sets: a general approach. IEEE Transactions on Knowledge and Data Engineering 16(12), 1472–1485 (2004)
9. Jensen, C.S., Dan, L., Ooi, B.C.: Continuous clustering of moving objects. IEEE Transactions on Knowledge and Data Engineering 19(9), 1161–1174 (2007)
10. Jeung, H., Yiu, M.L., Zhou, X., Jensen, C.S., Shen, H.T.: Discovery of convoys in trajectory databases. Proc. VLDB Endow. 1(1), 1068–1080 (2008)
11. Kalnis, P., Mamoulis, N., Bakiras, S.: On Discovering Moving Clusters in Spatio-temporal Data. In: Medeiros, C.B., Egenhofer, M., Bertino, E. (eds.) SSTD 2005. LNCS, vol. 3633, pp. 364–381. Springer, Heidelberg (2005)
12. Li, Z., Ding, B., Han, J., Kays, R.: Swarm: mining relaxed temporal moving object clusters. Proc. VLDB Endow. 3(1-2), 723–734 (2010)
13. Uno, T., Kiyomi, M., Arimura, H.: Lcm ver. 2: Efficient mining algorithms for frequent/closed/maximal itemsets. In: ICDM FIMI (2004)
14. Wang, Y., Lim, E.P., Hwang, S.Y.: Efficient mining of group patterns from user movement data. Data Knowl. Eng. 57(3), 240–282 (2006)

Appendix: Obtaining Clusters

The clustering method is not fixed in our system. Users can cluster cars along highways using a density-based method, or cluster birds in 3 dimension space using the k-means algorithm. Clustering methods that generate overlapping clusters are also applicable, such as EM algorithm or using a rigid definition of the radius to define a cluster. Moreover, clustering parameters are decided by users' requirements or can be indirectly controlled by setting the number of clusters at each timestamp.

Usually, most of clustering methods can be done in polynomial time. In our experiments, we used DBScan [1], which takes $O(|O_{DB}|log(|O_{DB}|) \times |T_{DB}|)$ in total to do clustering at every timestamp. To speed it up, there are also many incremental clustering methods for moving objects. Instead computing clusters at each timestamp, clusters can be incrementally updated from last timestamps.

Exemplars-Constraints for Semi-supervised Clustering

Hongjun Wang[1], Tao Li[2], Tianrui Li[1,*], and Yan Yang[1]

[1] School of Information Science and Technology, Southwest Jiaotong University, Chengdu 610031, China
[2] School of Computer Science, Florida International University, 33199 Miami FL, USA

{wanghongjun,trli,yyang}@swjtu.edu.cn, taoli@cs.fiu.edu

Abstract. Semi-supervised clustering aims at incorporating the known prior knowledge into the clustering process to achieve better performance. Recently, semi-supervised clustering with pairwise constraints has emerged as an important variant of the traditional clustering paradigm. In this paper, the disadvantages of pairwise constraints are analyzed in detail. To address these disadvantages, exemplars-constraints are firstly illustrated. Then based on the exemplars-constraints, a semi-supervised clustering framework is described step by step, and an exemplars-constraints EM algorithm is designed. Finally several UCI datasets are selected for experiments, and the experimental results show that exemplars-constraints can work well and the proposed algorithm can outperform the corresponding unsupervised clustering algorithm and the semi-supervised algorithms based on pairwise constraints.

Keywords: Semi-supervised Clustering, Mixture Model, Pairwise Constraints, Exemplars Constraints.

1 Introduction

In many situations when we discover new patterns using clustering, there exists known prior knowledge about the problem. We wish to incorporate the knowledge into the clustering algorithm. Recently, semi-supervised clustering (*i.e.*, clustering with knowledge-based constraints) has emerged as an important variant of the traditional clustering paradigm [2,4].

Existing semi-supervised clustering methods have been focusing on the use of background information in the form of instance level must-link and cannot-link constraints. A must-link (ML) constraint enforces that two instances must be placed in the same cluster while a cannot-link (CL) constraint enforces that two instances must not be placed in the same cluster [16]. Typical semi-supervised clustering methods based on constraints include Constrained Complete-Link [9], Constrained EM [14], HMRFKmeans [1], MPCKmeans [3], Kernel methods

---

* Corresponding author.

S. Zhou, S. Zhang, and G. Karypis (Eds.): ADMA 2012, LNAI 7713, pp. 115–126, 2012.
© Springer-Verlag Berlin Heidelberg 2012

[10,18,19,6,13,17], matrix-factorization based methods [11], and constraint projection [15,20,26,27]. Furthermore, several constraint propagation methods have been proposed by considering the neighbors around the data points constrained by the given must/cannot-links [5,7].

As far as we know, all the methods did not consider the following questions: what kind of pairwise constraints will improve the clustering accuracy; and whether any pairwise constraints improve the clustering results. In this paper, we address the two questions and illustrate exemplars constraints in detail. There are two contributions of this paper.

1. It shows that not all the constraints improve the clustering accuracy, and also studies the problem of choosing good data points to form constraints (exemplars-constraints). To the best of our knowledge, it is one of the first efforts on improving the clustering performance using exemplars-constraints.
2. Based on exemplars-constraints, a semi-supervised clustering framework is designed and especially an exemplars-constraints mixture model is proposed for semi-supervised clustering.

The rest of the paper is organized as follows. In Section 2, we introduce exemplars constraints (EC) in detail. In Section 3, a semi-supervised clustering framework based on EC is also proposed, and especially a mixture model based on EC is stated. Experimental results are presented in Section 4. The paper ends with conclusions in Section 5.

Table 1. The average accuracies of different algorithms on each dataset. Semi-supervised clustering algorithms of COP-kmeans and Constrained EM can not always obtain better results than unsupervised clustering algorithms, such as kmeans and EM.

Dataset	Kmeans	COP-Kmeans	EM	Constrained EM
haberman	0.5121±0.0254	**0.5852**±0.0216	0.6667±0.0234	**0.6729**±0.0421
iris	0.8933±0.0015	**0.9067**±0.0012	**0.9667**±0.0000	0.9660±0.0009
wdbc	**0.8541**±0.0002	0.8489±0.0023	**0.9554**±0.0008	0.9513±0.0021
wine	0.6632±0.0122	**0.7130**±0.0042	0.7528±0.0024	**0.7752**±0.0087
ionosphere	**0.7123**±0.0006	0.7068±0.0026	0.8168±0.0034	**0.8324**±0.0031
bupa	0.4840±0.0012	**0.5569**±0.0122	**0.5072**±0.0055	0.4991±0.0080
balance	0.5158±0.0048	**0.5506**±0.0016	0.5186±0.0042	**0.5280**±0.0016
heart	0.5926±0.0221	0.5926±0.0042	0.7148±0.0025	**0.7259**±0.0034

2   Exemplars-Constraints

2.1   On Pairwise Constraints

Pairwise constraints including must-link and cannot-link constraints were formally introduced by Wagstaff et al. in 2001 [16]. A must-link constraint means that two data points must be in the same cluster and cannot-link means that two data points can not be in the same cluster. It is simple but important and convenient for clustering, and therefore it attracts much attention for research. However, there are three disadvantages of pairwise constraints.

Table 2. The computation time (in seconds) of different algorithms on each dataset. It is clear that the semi-supervised clustering algorithms of COP-kmeans and Constrained EM need more computation time than unsupervised clustering algorithms.

Dataset	kmeans	COP-kmeans	EM	Constrained EM
Haberman	0.0031	**0.0375**	0.0500	**0.2781**
balance	0.0078	**0.1672**	0.1094	**0.7344**
iris	0.0047	**0.0187**	0.0437	**0.7688**
wine	0.0016	**0.0297**	0.0938	**0.7641**
bupa	0.0078	**0.0359**	0.1047	**0.4625**
sona	0.0063	**0.0219**	0.0328	**0.1516**
wdbc	0.0047	**0.0938**	0.1250	**0.3312**
wave	0.0719	**5.1219**	3.7281	**4.8172**
labor	0.0004	**0.0063**	0.0203	**0.1047**

1. They can not always improve the clustering accuracy and sometimes they even decrease the performance when compared with the corresponding unsupervised clustering alternatives.
2. It is computationally expensive for incorporating pairwise constraints.
3. Generally it needs many constraints to achieve good clustering performance. Many labeled data points are available, but some of them may misguide the results of clustering.

We use COP-kmeans [16], kmeans, Constrained EM [14] and EM algorithm on several UCI dataset for experiments. The results are shown in Tables 1 and 2. We can see that there are several times that algorithms based pairwise constraints decrease the clustering accuracy in Table 1. And for the time complexities, Cop-kmeans and kmeans are $O(nkct)$ and $O(nkt)$, respectively, where c is the number of constraints, n is the number of data points, k is the number of clusters, and t is the number of iteration. So more constraints lead to more computational time. We can see from Table 2 that algorithms based pairwise constraints need more computational time.

2.2   On Exemplars Constraints

Exemplars are the representative data points of a group in a dataset [23]. For example, there are two clusters of 1000 data points generated by four Gaussian distributions as shown in Fig 1. In Fig 1(b) there are four exemplars of data points such as A, C, D and E. In this example, totally there are $\sum_{i=1}^{1000-1} i = 499500$ pairwise constraints. If we use one data point, called exemplar, to represent all the other data points in the same Gaussian distribution, totally there are only $\sum_{i=1}^{4-1} i = 6$ pairwise constraints which are renamed as exemplars constraints. So exemplars constraints are special cases of pairwise constraints. There are two reasons why exemplars constraints are preferred. The first reason is that there are too many pairwise constraints, and it is computationally expensive to use them; The second is that semi-supervised learning based on pairwise constraints can not always improve the clustering performance (in terms of accuracy), and

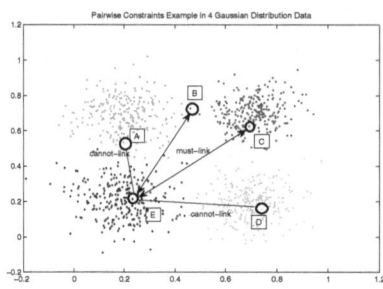
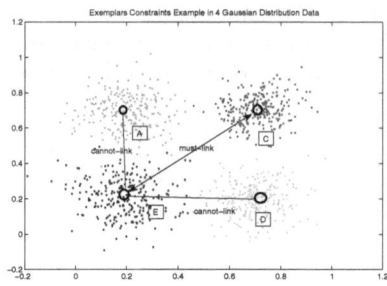

(a) Examples of pairwise constraints   (b) Examples of exemplars constraints

Fig. 1. Pairwise constraints VS exemplars constraints

sometimes it is even worse than unsupervised learning (as shown in Table 1) and it also needs more computational time (as shown in Table 2). Exemplars-constraints include two types of constraints: Exemplar-must-link (EML) means that two exemplars must be in the same cluster and Exemplar-cannot-link (ECL) means that two exemplars must be in different clusters.

2.3   Finding Exemplars

In this section, we address the problem of how to find a good exemplar which can represent the other data points in a group. Formally, we can define an objective function as follows:

$$\min F = \frac{1}{N} \sum_{e=1}^{k} \sum_{x_i \to x_e} |x_i - x_e|^2, \{i_1^n, e_1^K\}, \tag{1}$$

where x_e is an exemplar and $x_i \to x_e$ means that x_e can represent the datapoint x_i. From the objection function there are some existing methods, such as Exemplar-based clustering algorithms, to find the exemplars [21]. Exemplar-based clustering is the task of not only performing the partitioning but of identifying for each cluster its most representative member, or exemplar [21]. An important characteristic of the cluster exemplar is that its overall similarity to other data points in the cluster is maximal, which is typically done by affinity propagation algorithm [22]. Also we can employ another simple way, $i.e.$, the K-center algorithm, to find exemplars. K-center is similar to K-means and their difference is that K-center chooses k data points as the data centers instead of the cluster means.

2.4   Exemplars-Constraints Propagation

EML constraints specify that two data points have to be in the same cluster. EML constraints are also transitive as pairwise constraints [12]. Let k_i and k_j

be two connected components, and let x_i and x_j be the exemplars in k_i and k_j, respectively. Let M be the set of EML constraints, then

$$(x_i, x_j) \in M, x_i \in k_i, x_j \in k_j$$

$$\implies (a, b) \in M, \forall a, b : a \in k_i, b \in k_j.$$

ECL constraints specify that two instances must be placed in two different clusters and ECL constraints can also be entailed. Let k_i and k_j be two connected components, and let x_i and x_j be the exemplars in k_i and k_j, respectively. Let C be the set of ECL constraints, then

$$(x_i, x_j) \in C, x_i \in k_i, x_j \in k_j$$

$$\implies (a, b) \in C, \forall a, b : a \in k_i, b \in k_j.$$

Also another exemplars-constraint propagation is that exemplars constraints must influence their neighbor data points or their representative data points.

Given a set of p dimensional data $X = \{x_1, x_2, ..., x_n\}$, the corresponding pairwise EML constraint set is $M = \{(x_i, x_j) | x_i \in k_i; x_j \in k_i\}$, the pairwise ECL constraint set is $C = \{(x_i, x_j) | x_i \in k_i; x_j \in k_j; k_i \neq k_j\}$, and the neighbors set of x_i is $\mu = \{x_l | \rho \geq d(x_e); x_l \in X; l = (1, ..., n)\}$. Simply we can use Euclidean distance to choose the neighbors. So $d(x_e)$ can be $\sqrt{||x_i - x_e||^2}$, and geodesic distance can be chosen as a standard to select neighbors. Moreover, we use a Gaussian function centered at the given constraint x_A, x_B to determine the weight of x_i, x_j, because a Gaussian function can propagate constraints that are closest to the source exemplars constraint and will fall off smoothly. If the dataset is in discrete space, normalized mutual information(NMI), $NMI(x_i, x_j) = \frac{I(x_i, x_j)}{\sqrt{H(x_i)H(x_j)}}$ [24], is used for constraints propagation to select constraints neighbors.

3   A Semi-supervised Clustering Framework Based on Exemplars Constraints

3.1   Framework Description

In this subsection we design a framework based on exemplars constraints for semi-supervised clustering. The framework procedure is descried step by step as follows.

Framework 1: (A semi-supervised learning framework based on exemplars constraints [SSL-EC])

Input: Dataset, $\{x_i\}_{i=1}^n$, where x_i is the data points

1. Choose one algorithm to select exemplar in datasets:
   - Use k-centers to select exemplar in datasets;
   - Use affinity propagation to select exemplars in datasets.

2. Generate the set of exemplar-link according to the labels of exemplar.
3. Choose one constraint-based algorithm (clustering algorithms based on must-link and cannot-link) for clustering:
   - Choose COP-kmeans for clustering;
   - Choose Constrained EM for clustering;
   - Choose Constrained FCM for clustering.

Output: Cluster membership of every point.

3.2   An Exemplars-Constraints Mixture Model

A Gaussian mixture model (GMM) is a parametric statistical model which assumes that the data originates from a weighted sum of several Gaussian sources. More formally, a GMM is given by:

$$P(x|\Theta) = \sum_{i=1}^{N} \pi_i p(x|\theta_i),$$

where π_i is the weight of each Gaussian and θ_i is its corresponding parameters. Commonly EM is used to estimate the parameters as follows.

$$\pi^{t+1} = \frac{1}{N} \sum_{i=1}^{N} P(y_i = k|x_i, \Theta^t),$$

$$\mu_k^{t+1} = \frac{\sum_{i=1}^{N} \overline{X}_i P(y_i = k|x_i, \Theta^t)|x_i|}{\sum_{i=1}^{K} P(y_i = k|x_i, \Theta^t)|x_i|},$$

$$\Sigma_k^{t+1} = \frac{\sum_{i=1}^{N} \Sigma_{ik}^t P(y_i = k|x_i, \Theta^t)|x_i|}{\sum_{i=1}^{K} P(y_i = k|x_i, \Theta^t)|x_i|}.$$

Exemplar-Must-link modifies the E step in the following way: instead of summing over all possible assignments of data points to sources, only assignments which comply with the given constraints are summed over. For example, if points x_i and x_j are exemplar-must-link, assignments in which both points are assigned to the same Gaussian source are considered. On the other hand, if these points are exemplar-cannot-link, assignments in which each of the points is assigned to a different Gaussian source are considered.

3.3   EM Algorithm Procedure

With the initial guesses for the parameters (π_0, μ_0, Σ_0) of our mixture model, each data point in each constituent distribution is computed by calculating the expectation values for the membership variables. We can use an EM algorithm to find the best-fit model (π^*, μ^*, Σ^*). The EM algorithm alternates between two steps until convergence:

Exemplars-constraints EM Algorithm:

Input: Dataset and random parameters: $\{x_i\}_{i=1}^n$ and $(\pi_0, \mu_0, \Sigma_0, M, C)$

1. E-Step: Calculate the expectation of the log-likelihood over all possible assignments of data points to sources.

$$\pi^{t+1} = \frac{1}{N} \sum_{i=1}^N P(y_i = k|x_i, \Theta^t).$$

2. M-Step: Maximize the expectation by differentiating w.r.t current parameters.

$$\mu_k^{t+1} = \frac{\sum_{i=1}^N \overline{X}_i P(y_i = k|x_i, \Theta^t)|x_i|}{\sum_{i=1}^K P(y_i = k|x_i, \Theta^t)|x_i|},$$

$$\Sigma_k^{t+1} = \frac{\sum_{i=1}^N \Sigma_{ik}^t P(y_i = k|x_i, \Theta^t)|x_i|}{\sum_{i=1}^K P(y_i = k|x_i, \Theta^t)|x_i|}.$$

Output: (π^*, μ^*, Σ^*), where π^* is the probability of cluster membership of every point.

After (t) iterations, the value of expectation is π^t. In the $(t+1)^{th}$ iteration,

$$\sum_{i=1}^N L(\pi^t, \mu^t, \Sigma^t)$$

$$\leq \sum_{i=1}^N L(\pi^t, \mu^{t+1}, \Sigma^{t+1}) \qquad (2)$$

$$\leq \sum_{i=1}^N L(\pi^{t+1}, \mu^{t+1}, \Sigma^{t+1}). \qquad (3)$$

The first inequality holds because in the E-step, (2) is the maximum of $L(\pi^t, \mu^{t+1}, \Sigma^{t+1})$. The second inequality holds because in the M-step, (3) is the maximum of $L(\pi^{t+1}, \mu^{t+1}, \Sigma^{t+1})$. Therefore, the objective function is non-decreasing until convergence [25].

4   An Empirical Study

4.1   Experiment Setup

In this section, we perform experiments on 8 real-world datasets from UCI machine learning repository. The number of objects, features and classes of each

Table 3. The number of the instances, features and classes in each dataset

Dataset	Characteristic	Instances	Features	Categories
iris	real	150	4	3
wdbc	real	569	30	2
wine	real	178	13	3
ionosphere	real	351	34	2
bupa	discrete	345	6	2
balance	discrete	625	4	3
hear	real	270	13	2
haberman	discrete	306	3	2

Table 4. The average accuracies results of the experiments. Exemplars constraints EM is the proposed method.

Dataset	Kmeans	COP-Kmeans	EM	Constrained EM	Exemplars constraints EM
haberman	0.5121±0.0254	0.5852±0.0216	0.6667±0.0234	0.6729±0.0421	**0.6850±0.189**
iris	0.8933±0.0015	0.9067±0.0012	0.9667±0.0000	0.9660±0.0009	**0.9767±0.0016**
wdbc	0.8541±0.0002	0.8489±0.0023	0.9554±0.0008	0.9513±0.0021	**0.9554±0.0012**
wine	0.6632±0.0122	0.7130±0.0042	0.7528±0.0024	0.7752±0.0087	**0.8039±0.0026**
ionosphere	0.7123±0.0006	0.7068±0.0026	0.8168±0.0034	0.8324±0.0031	**0.8535±0.0042**
bupa	0.4840±0.0012	**0.5569±0.0122**	0.5072±0.0055	0.4991±0.0080	0.5154±0.0026
balance	0.5158±0.0048	**0.5506±0.0016**	0.5186±0.0042	0.5280±0.0016	0.5376±0.0017
heart	0.5926±0.0221	0.5926±0.0042	0.7148±0.0025	0.7259±0.0034	**0.7333±0.0026**

data set are listed in Table 3. For evaluation, we use micro-precision [8] to measure the accuracy of the cluster with respect to the true labels. The micro-precision is defined as $MP = \sum_{h=1}^{k} a_h/n$, where k is the number of clusters and n is the number of objects, a_h denotes the number of objects in the cluster h that are correctly assigned to the corresponding class. We identify the *corresponding class* for a cluster h as the true class with the largest overlap with the cluster, and assign all objects in cluster h to that class. Note that $0 \leq MP \leq 1$ with 1 indicating the best possible consensus clustering, which has to be in full agreement with the class labels.

In our experiments, the constraints are generated as follows: for each constraint, one pair of data points are picked out randomly from exemplars of the input data sets (The labels of which were available for evaluation purpose but unavailable for clustering). If the labels of this pair of points are the same, then an EML constraint is generated. If the labels are different, an ECL constraint is generated. The amounts of constraints are determined by the size of input data. On all the datasets, the experiments are performed 10 times and the performance measure is averaged to eliminate the difference caused by constraints.

4.2   Performance Comparison

To demonstrate how our method works for semi-supervised clustering problem and improves the clustering performance, we compare it with the following methods:

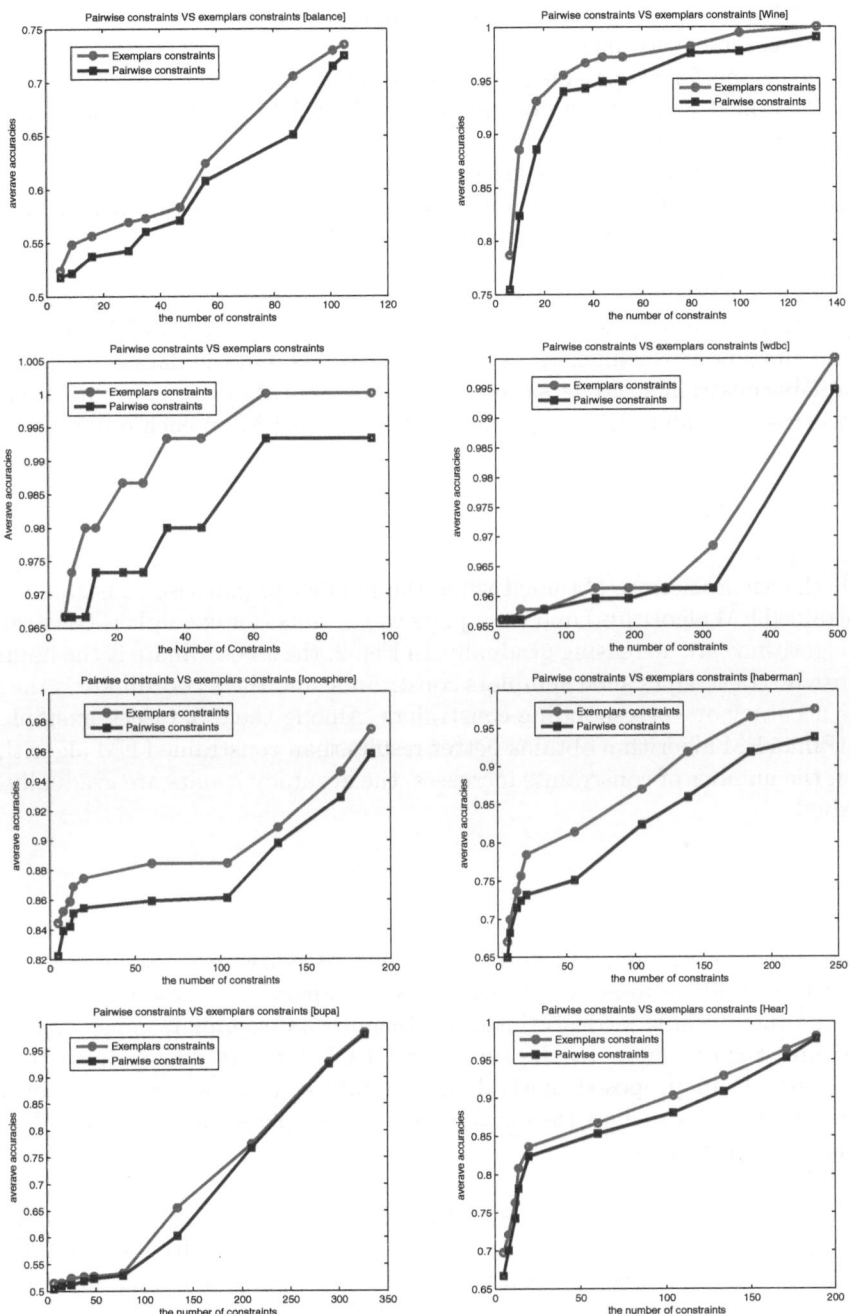

Fig. 2. The experimental results with the number of constraints increasing. In this experiment constrained EM (blue line) and exemplars-constraints EM (red line) are used for experiments. On the 8 datasets experiment constrained EM outperforms constrained EM.

1. COP-Kmeans [16]: performing COP-Kmeans on the original datasets with 5% random constraints;
2. Kmeans: performing K-means on the original datasets;
3. Constrained EM [14]: performing constrained EM on the original datasets with 5% random constraints;
4. EM: performing EM on the original datasets;
5. Exemplars-constraints EM: performing exemplars-constraints EM on the original datasets with 5% exemplars-constraints.

The performance comparison is shown in Table 4. There are total 8 datasets and 5 algorithms. The proposed algorithm of Exemplars-constraints EM achieves 7 best results among the algorithms on these datasets. On the dataset of balance, exemplars-constraints EM gets the second best result. Moreover, we can see that Exemplars-constraints EM outperforms constrained EM on each dataset.

4.3 Parameter Tuning

We also report the experiment results on constraints increasing. In this experiment, the accuracies are obtained when the number of pairwise constraints (for constrained EM algorithm) and exemplars constraints (for exemplars-constraints EM algorithm) are increasing gradually. In Fig. 2, the x coordinate is the number of pairwise constraints and exemplars constraints, and the y coordinate is the average accuracy on corresponding constraints. Among the 8 datasets, exemplars-constraints EM algorithm obtains better results than constrained EM algorithm. When the number of constraints increases, the accuracy results are gradually increasing.

5 Conclusions

In this paper, we analyzed the disadvantages of pairwise constraints. To address these disadvantages, we proposed a semi-supervised clustering framework based on the exemplars-constraints and designed an exemplars constraints EM algorithm. Experimental results on several UCI datasets demonstrate the effectiveness of our proposed method. In our future work, we will focus on the theoretical analysis such as the convergence of exemplars constraints and better selection of constraints.

Acknowledgments. This work is partially supported by NSFC (Nos. 61003142, 61175047, 61170111, 61262058), State Key Laboratory of Hydraulics and Mountain River Engineering Sichuan University (No. 1010), the Fundamental Research Funds for the Central Universities (No. SWJTU12CX092) and the construction plan for scientific research innovation groups of Leshan Normal University. We thank the anonymous reviewers for their helpful comments and suggestions.

References

1. Basu, S., Bilenko, M., Mooney, R.J.: A probabilistic framework for semi-supervised clustering. In: KDD, pp. 59–68 (2004)
2. Basu, S., Davidson, I., Wagstaff, K.L.: Constrained Clustering. CRC Press (2008)
3. Bilenko, M., Basu, S., Mooney, R.J.: Integrating constraints and metric learning in semi-supervised clustering. In: ICML, pp. 81–88 (2004)
4. Chapelle, O., Zien, A., Scholkopf, B.: Semi-supervised learning. MIT Press (2006)
5. Eaton, E.R.: Clustering with Propagated Constraints. Thesis of the University of Maryland (2005)
6. Hoi, S.C.H., Jin, R., Lyu, M.R., Wu, J.: Learning nonparametric kernel matrices from pairwise constraints. In: ICML, pp. 361–368 (2007)
7. Huang, J., Sun, H.: Lightly-supervised clustering using pairwise constraint propagation. In: Proceedings of 2008 3rd International Conference on Intelligent System and Knowledge Engineering, pp. 765–770 (2008)
8. Zhou, Z.H., Tang, W.: Clusterer ensemble. Knowledge-Based Systems 19(1), 77–83 (2006)
9. Klein, D., Kamvar, S.D., Manning, C.D.: From instance-level constraints to space-level constraints: Making the most of prior knowledge in data clustering. In: ICML, pp. 307–314 (2002)
10. Kulis, B., Basu, S., Dhillon, I., Mooney, R.J.: Semi-supervised graph clustering: a kernel approach. In: ICML, pp. 457–464 (2005)
11. Li, T., Ding, C., Jordan, M.: Solving Consensus and Semi-supervised Clustering Problems Using Nonnegative Matrix Factorization. In: ICDM, pp. 577–582 (2007)
12. Li, Z., Liu, J., Tang, X.: Pairwise constraint propagation by semidefinite programming for semi-supervised classification. In: ICML, pp. 576–583 (2008)
13. Masayuki, O., Seiji, Y.: Learning similarity matrix from constraints of relational neighbors. Journal of Advanced Computational Intelligence and Intelligent Informatics 14(4), 402–407 (2010)
14. Shental, N., Bar-Hillel, A., Hertz, T., Weinshall, D.: Computing gaussian mixture models with EM using equivalence constraints. In: NIPS, pp. 1–8 (2003)
15. Tang, W., Xiong, H., Zhong, S., Wu, J.: Enhancing semi-supervised clustering: A feature projection perspective. In: KDD, pp. 707–716 (2007)
16. Wagstaff, K., Cardie, C., Rogers, S., Schroedl, S.: Constrained k-means clustering with background knowledge. In: ICML, pp. 577–584 (2001)
17. Xuesong, Y., Songcan, C., Enliang, H.: Semi-supervised clustering with metric learning:an adaptive kernel method. Pattern Recognition 43(4), 1320–1333 (2010)
18. Yan, B., Domeniconi, C.: An Adaptive Kernel Method for Semi-supervised Clustering. In: Fürnkranz, J., Scheffer, T., Spiliopoulou, M. (eds.) ECML 2006. LNCS (LNAI), vol. 4212, pp. 521–532. Springer, Heidelberg (2006)
19. Yeung, D.Y., Chang, H.: A kernel approach for semi-supervised metric learning. IEEE Transactions on Neural Networks 18(1), 141–149 (2007)
20. Zhang, D., Chen, S., Zhou, Z., Yang, Q.: Constraint projections for ensemble learning. In: AAAI, pp. 758–763 (2008)
21. Khosla, M.: Message Passing Algorithms. PHD thesis, 9 (2009)
22. Frey, B.J., Dueck, D.: Clustering by passing messages between data points. Science 305(5814), 972–976 (2007)
23. Mzard, M.: Where are the exemplars? Science 315, 949–951 (2007)
24. Strehl, A., Ghosh, J.: Cluster Ensembles-A Knowledge Reuse Framework for Combining Multiple Partitions. Journal of Machine Learning Research 3, 583–617 (2002)

25. Neal, R., Hinton, G.: A view of the EM algorithm that justifies incremental, sparse, and other variants. In: Learning in Graphical Models, pp. 355–368 (1998)
26. Sublemontier, J.H., Martin, L., Cleuziou, G., Exbrayat, M.: Integrating pairwise constraints into clustering algorithms: optimization-based approaches. In: The Eleventh IEEE International Conference on Data Mining Workshops, Vancouver, Canada (2011)
27. Zeng, H., Cheung, Y.M.: Semi-Supervised Maximum Margin Clustering with Pairwise Constraints. IEEE Transactions on Knowledge and Data Engineering 24, 926–939 (2012)

Customer Segmentation for Power Enterprise Based on Enhanced-FCM Algorithm

Lihe Song[1], Weixu Zhan[2], Shuchai Qian[1], and Jian Yin[1,*]

[1] School of Information Science and Technology, Sun Yat-Sen University,
Guangzhou, 510006, China
[2] China Southern Power Grid Co., Ltd., Guangzhou, 510623, China
lihesong9@163.com

Abstract. Customer segmentation is an important topic of customer relationship management and FCM is a common method for customer segmentation. However, FCM is sensitive to outliers and hard to determine the number of clusters. In this paper, we define hierarchical analytical indicators for power customer segmentation, and propose a new algorithm called WKFCM_S₂ by combining enhanced-FCM algorithm with the analytical hierarchy process. Experiments on a real customer data set of a power supply enterprise show that the WKFCM_S₂ algorithm is more robust to noise and more suitable for applications.

Keywords: FCM, Clustering, Customer segmentation, AHP.

1 Introduction

With the development of technology, customer segmentation [1] becomes the key strategy for power supply enterprise to develop market and promote customer service. Clustering is one of the main methods for customer segmentation. Generally, clustering methods can be categorized into hierarchy based, density based, graph theoretic and minimizing an objective function. Fuzzy c-means (FCM) algorithm, which introduce the fuzzy set theory to improve the clustering efficiency, is the most popular method of objective function based clustering methods [2]. In this paper, we will use enhanced-FCM algorithm to perform customer segmentation.

FCM has some achievements in applications [3], but it is sensitive to outliers and hard to determine cluster number. There are some related researches about these two issues: in [4], an improved algorithm FCM_S was presented to reduce noise by adding effect of pixel's neighborhood information, but it leads to high cost of computation; on this basis, a more robust algorithm KFCM_S₂ was proposed in [5], which introduces neighborhood mean and the Gaussian kennel function; in [6], a weighted Euclidean distance strategy is introduced by considering the cohesion and coupling of features; in [7], a new cluster validation index

---

* Corresponding author.

S. Zhou, S. Zhang, and G. Karypis (Eds.): ADMA 2012, LNAI 7713, pp. 127–137, 2012.
© Springer-Verlag Berlin Heidelberg 2012

v_K was presented to eliminate the monotonically decreasing tendency with increasing number of clusters in the traditional v_{XB} index. However, they did not consider the different importance of attributes from the business view.

This paper measures customer's value from four aspects, defines a series of analysis indicators, and applies KFCM_S$_2$ algorithm and v_K index to power customer segmentation. Furthermore, a new algorithm named WKFCM_S$_2$ which combines KFCM_S$_2$ with analytical hierarchy process (AHP) is proposed to distinguish the importance of each indicator from the viewpoint of application. Finally we use the data set of a power supply enterprise to verify it.

The remaining of the paper is as follows. Section 2 introduces the analytical indicators. Section 3 develops the WKFCM_S$_2$ algorithm. In section 4, we present the experimental results. The paper concludes with Section 5.

2   Selection of Analytical Indicators

2.1   Determining Analytical Indicators

Customer segmentation proceeding from customer value is good for enterprises to allocate limited resource to high value customers. There are not uniform standards to measure customer's value. According to our investigation of power enterprise managers, we knew that their most concerns are contribution and credibility, so we determined to measure customer's value from four aspects, including credibility, contribution, growth and faithfulness, and then defined eight secondary analytical indicators, the details refer to Fig. 1.

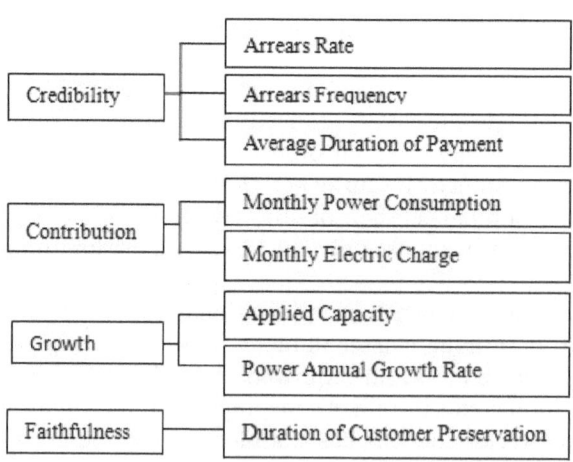

Fig. 1. Hierarchical analytical indicators

2.2   Analytical Indicators Explanation

Arrears Rate (Ar). The ratio of arrears and receivables. For the latest arrears rate best reflects customer's credibility, we use cumulation of weighted arrears rate of the latest three months, half a year, one year and history. Ar is calculated as follows.

$$Ar = \frac{1}{4} \left(\varepsilon_1 \times Ar_3 + \varepsilon_2 \times Ar_6 + \varepsilon_3 \times Ar_{12} + \varepsilon_4 \times Ar_{history} \right) \tag{1}$$

where

$$\varepsilon_i = \frac{2^{4-i}}{\sum_{k=0}^{3} 2^k} \tag{2}$$

and Ar_3 is the arrears rate of the latest three months, Ar_6 is the arrears rate of the latest half a year, Ar_{12} is the arrears rate of the latest one year, $Ar_{history}$ is the historical arrears rate.

Arrears Frequency (Af). The ratio of arrears times and receivable times.

Average Duration of Payment (Dp). The average interval between receipt of a notice and payment. It measures customers' payment initiative.

Monthly Power Consumption (Pc). Customers' average electric consumption per month.

Monthly Electric Charge (Ec). Customers' average electric charge per month.

Applied Capacity (Ac). The power rating when customer applied for installation.

Power Annual Growth Rate (Gr). Measures increase in the customers' consumption.

Duration of Customer Preservation (Dc). The number of months since the customer opened an account till now.

3   Algorithm

3.1   Enhanced-FCM (KFCM_S₂)

In [5], Songcan Chen and Daoqiang Zhang proposed an enhanced-FCM algorithm KFCM_S₂ which increases the robustness of FCM to noise by incorporating neighborhood information into the objective function of original FCM and using the Gaussian kernel function to measure the similarity between data points. This approach maps the original space of data points to high dimension feature space through nonlinear conversion while clustering is performed still in the original space and results can be interpreted intuitively. The objective function is defined as follows:

$$JS = \sum_{i=1}^{c}\sum_{k=1}^{N} u_{ik}^m \left(1 - K\left(x_k, v_i\right)\right) + \alpha \sum_{i=1}^{c}\sum_{k=1}^{N} u_{ik}^m \left(1 - K\left(\bar{x}_k, v_i\right)\right) \tag{3}$$

where x_k stands for the k-th data point, $U=\{u_{ik}\}$ is a fuzzy partition matrix and m is the exponent for the matrix U, v_i denotes centroid of the i-th cluster, the parameter α controls the effect of the penalty and $\bar{x}_k$ is the mean of neighboring data points lying within a window around x_k. $K(x, y)$ is the Gaussian kernel function defined as follows:

$$K\left(x, y\right) = \exp\left[\frac{-\left(\sum_{i=1}^{d} |x_i - y_i|^2\right)}{\delta^2}\right]. \tag{4}$$

The following solution can be obtained by minimizing (3):

$$u_{ik} = \frac{\left(\left(1 - K\left(x_k, v_i\right)\right) + \alpha\left(1 - K\left(\bar{x}_k, v_i\right)\right)\right)^{-\frac{1}{m-1}}}{\sum_{j=1}^{c}\left(\left(1 - K\left(x_k, v_j\right)\right) + \alpha\left(1 - K\left(\bar{x}_k, v_j\right)\right)\right)^{-\frac{1}{m-1}}} \tag{5}$$

$$v_i = \frac{\sum_{i=1}^{n} u_m^{ik}\left(K\left(x_k, v_i\right)x_k + \alpha K\left(\bar{x}_k, v_i\right)\bar{x}_k\right)}{\sum_{i=1}^{n} u_m^{ik}\left(K\left(x_k, v_i\right) + \alpha K\left(\bar{x}_k, v_i\right)\right)}. \tag{6}$$

3.2   Feature Weighted Algorithm

KFCM_S$_2$ uses single width Gaussian kernel function which regards all features as equally important. In fact, sample space distribution is uneven in most cases, thus single width Gaussian kernel function usually leads to partial optimization; in addition, the importance of attributes also varies in different applications. So we propose a new feature weighted algorithm (WKFCM_S$_2$), which uses AHP method to calculate the weight of features, and modifies the KFCM_S$_2$ algorithm by adding weight factors to the Gaussian kernel function.

Attribute Weight Calculation. AHP is a system analysis method proposed by A. L. Saaty in 1980 [8]. The procedure to determine the attribute weight using AHP method [9] is as follows.

1) The decision maker fills in a questionnaire like Table 1, where each grid stands for an important ratio of two indicators, thus we can construct a 8 × 8 matrix.

2) Calculate the eigenvector of the maximum eigenvalue of the above matrix.

$$\left(\omega_1, \omega_2, \omega_3, \omega_4, \omega_5, \omega_6, \omega_7, \omega_8\right). \tag{7}$$

Table 1. Questionnaire

	Ar	Af	Dp	Pc	Ec	Ac	Gr	Dc
Ar	1	5	5	5	5	5	5	5
Af	1/5	1	3	3	3	3	3	3
Dp	1/5	1/3	1	1	1	1	1	1
Pc	1/5	1/3	1	1	1	1	1	1
Ec	1/5	1/3	1	1	1	1	1	1
Ac	1/5	1/3	1	1	1	1	1	1
Gr	1/5	1/3	1	1	1	1	1	1
Dc	1/5	1/3	1	1	1	1	1	1

3) Get the weight vector by standardizing the eigenvector according to the following form.

$$\omega_1 + \omega_2 + \omega_3 + \omega_4 + \omega_5 + \omega_6 + \omega_7 + \omega_8 = 8. \tag{8}$$

Weighted Gaussian Kernel Function. The width parameter δ in Gaussian kernel function controls the radial functioning scope [10], we can amplify the linear translational distance between sample points by increasing the width parameter, so we set different δ to different attribute, give small δ to more important attributes and rewrite (4) as follows:

$$K(x, y) = \exp\left[-\sum_{i=1}^{d} \frac{|x_i - y_i|^2}{\delta_i^2}\right] \tag{9}$$

where δ_i stands for the width parameter of the i-th attribute. Then we replace δ_i by weight vector:

$$\delta_i^2 = \frac{\delta^2}{\omega_i} \tag{10}$$

and modify (9) as follows:

$$K(x, y) = \exp\left[\frac{-\left[\sum_{i=1}^{d} \omega_i |x_i - y_i|^2\right]}{\delta^2}\right]. \tag{11}$$

Thus the width parameter of each attribute can be adjusted by its weight. Finally, we deduce a new algorithm WKFCM_S$_2$ by replacing the kernel function in (3), (5), (6) with (11).

Steps of WKFCM_S$_2$ Algorithm

Step 1) Calculate attribute weights by AHP method.
Step 2) Calculate neighborhood mean of each data point.
Step 3) Update the partition matrix.
Step 4) Update the centroids and objective function.

Repeat Steps 3)-4) until the variation of the objective function less than a threshold.

3.3 Validity Index

In [7], S. H. Kwon proposed a new cluster validation index v_K, which eliminates the monotonically decreasing tendency with increasing number of clusters in the traditional index v_{XB}. We used this index to select optimal number of clusters and evaluate clustering results in our experiment.

$$v_k = \frac{\sum_{i=1}^{c} \sum_{k=1}^{n} u_{ik}^2 \left\| x_k - v_i \right\|^2 + \frac{1}{c} \sum_{i=1}^{c} \left\| v_i - \bar{v} \right\|^2}{\min_{i \neq j} \left(\left\| v_i - v_j \right\|^2 \right)} \tag{12}$$

where

$$\bar{v} = \frac{1}{n} \sum_{j=1}^{n} x_j. \tag{13}$$

4   Experimental Results

In this section, we described the experimental results based on large industrial customer data from a power supply enterprise. There are 7922 samples, all data are standardized by z-score method. We used the v_K index to select optimal number of clusters, and compared the results of FCM, KFCM_S_2 and WKFCM_S_2. We set the parameters m=2, δ=150.

4.1   Compare the Results of FCM and KFCM_S_2

We performed FCM and KFCM_S_2 with different number of clusters to compare the performance of original FCM and KFCM_S_2. Table 2 illuminates the v_K index of the two algorithms, the smaller the v_K is, the better the result is. It could be concluded that KFCM_S_2 is indeed to create better result than the original FCM, we choose KFCM_S_2 to do the following experiment.

Table 2. v_K of FCM and KFCM_S_2 with different cluster number

Cluster number	FCM	KFCM_S_2
3	11113	10965
4	11929	11847
5	23834	9690
6	18532	20266
7	12405	15981
8	20966	12728
9	20613	11102

4.2   Select Optimal Number of Clusters

As to customer segmentation, too few classifications can not match the goal of fine management, while too many classifications lead to management difficulties. We limited the customer segmentation number to 3-9 classifications in this experiment, and choosed optimal number by minimize the v_K index, as can be found in Table 2, it is 5.

4.3   Customer Segmentation Using KFCM_S$_2$

We use KFCM_S$_2$ to show how the above algorithms are used for power customer segmentation, the same process can be repeated by WKFCM_S$_2$. Table 3 shows the centroids of clusters, as the data are standardized by z-score, we can only see relative numbers which reflect the diversity of clusters. Table 4 shows averages within clusters and is not standardized.

Table 3. Centroids of clusters

Cluster	Ar	Af	Dp	Pc	Ec	Ac	Gr	Dc
1	0.97	-0.16	0.58	10.60	10.50	4.46	0.20	0.17
2	0.40	-0.27	-0.43	-0.04	-0.04	-0.06	-0.08	0.33
3	0.28	-0.17	1.27	-0.06	-0.06	-0.07	-0.04	0.26
4	0.04	2.83	0.13	-0.09	-0.08	-0.04	0.02	-2.92
5	-0.73	-0.27	-0.40	-0.13	-0.13	-0.07	-0.09	0.27

Table 4. Averages within clusters

Cluster	Ar	Af	Dp	Pc	Ec	Ac	Gr	Dc
1	1.01	8.96	9	11680	8710	99800	18	40
2	9.27	6.85	6	259	206	1316	20	45
3	0.89	7.38	12	179	141	890	40	45
4	8.36	17.40	8	149	131	1250	145	19
5	5.73	6.33	6	71	69	904	6	44

According to the result above, we analysed the characteristics of each cluster to obtain the final customer segmentation, we also designed management objective for each type according to the marketing strategy. The percentage of each type is presented in Fig. 2.

Diamond Customer. A few customers, whose power consumption is huge, although pay inactively but have low arrears rate. The management objective of them is to maintain customer loyalty.

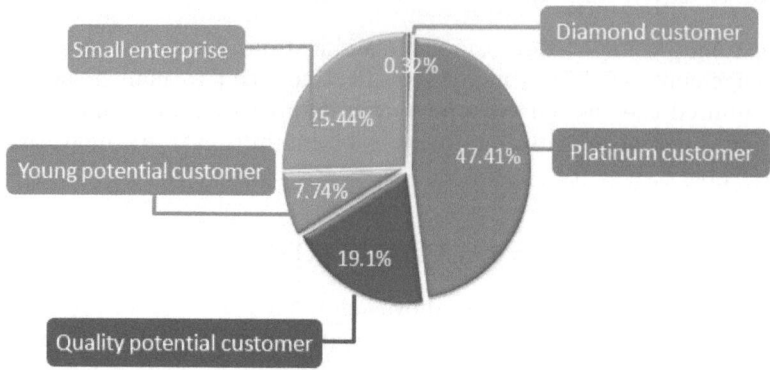

Fig. 2. Percentage of every type of customer

Platinum Customer. This type accounts for the highest dividend, almost 50%. Although their power consumption is far below diamond customers, but they are the most influential type for their big proportion. Yet in consideration of their high arrears rate, the management objective is to reduce arrears rate.

Quality Potential Customer. Big electric power consumption with low arrears rate and high consumption increase rate, implies their strong momentum of development. The management objective is to improve customer satisfaction.

Young Potential Customer. Some new customers consumption growing rapidly, causing pressure to the power load. They are mainly the new established big enterprises, the management objective is to make them long-term stability customers.

Small Enterprise. Small electric power consumption with low consumption increase rate, the enterprise should reduce the maintenance costs to them.

4.4   Compare the Result of WKFCM_S_2 and KFCM_S_2

In this experiment, we use Table 1 to create the matrix and calculate the weight vector, and conduct the customer segmentation using WKFCM_S_2 and compare its result with KFCM_S_2. Since this questionnaire regards arrears rate and arrears frequency as the two most important indicators, we speculate that the result on these two indicators could be better than unweighed algorithm KFCM_S_2.

Calculate Attribute Weights. We got the eigenvector of the maximum eigenvalue from Table 1, which is (0.86,0.38,0.14,0.14,0.14,0.14,0.14,0.14). So the standardized eigenvector (weight vector) is (3.3,1.46,0.54,0.54,0.54,0.54,0.54,0.54).

Compare the Result of WKFCM_S$_2$ and KFCM_S$_2$. From Fig. 3 we can see that the v_K index of WKFCM_S$_2$ is bigger than KFCM_S$_2$ algorithm, this does not mean the performance become poor, because the attribute weights is not based on sample space distribution, but use the expert scoring method to measure the importance of attribute. We suppose that, the performance will be better than unweighted algorithm if we use attribute weight based on sample space distribution. We choose 3 clusters according to Fig. 3.

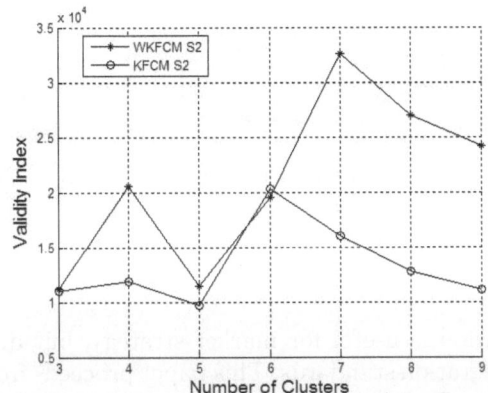

Fig. 3. v_K of WKFCM_S$_2$ and KFCM_S$_2$ with different cluster number

Fig. 4 and Fig. 5 shows the scatter diagram of WKFCM_S$_2$ and KFCM_S$_2$ on the most important two attributes, we could see the division of WKFCM_S$_2$ in Fig. 4 is much more clear and has less overlap than KFCM_S$_2$ in Fig. 5, proving that high weight attribute is strengthened in WKFCM_S$_2$.

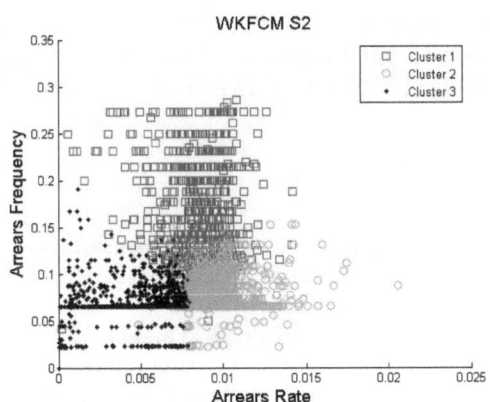

Fig. 4. Scatter diagram of WKFCM_S$_2$

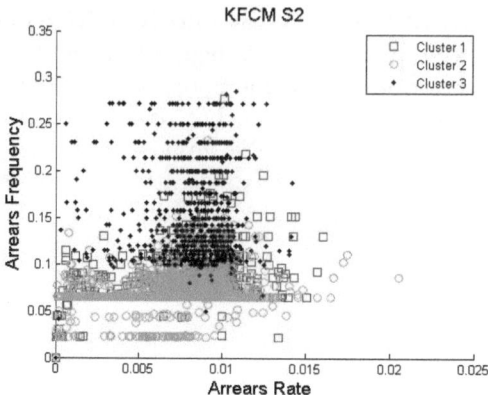

Fig. 5. Scatter diagram of KFCM_S_2

5   Conclusion

Customer segmentation is useful for market strategy, but different application have different segmentation standards, This paper proceeds from customer value, measure customer's value from four aspects, uses enhanced-FCM algorithm to perform the customer segmentation of power enterprise and achieve good results. The paper also proposes a new algorithm WKFCM_S_2 that using AHP method to add weight to attributes from the viewpoint of business. Experiment shows that the new algorithm can significantly emphasize the importance of high weighted attributes, and is more suitable for applications. For future work, we plan to develop an algorithm where the attribute weights are based on sample space distributions.

Acknowledgments. This work is supported by the Research Foundation of Science and Technology Plan Project in Guangdong Province and Guangzhou City (2010A040303004, 11A12050914, 11A31090341, 2011Y5-00004).

References

1. Shuxia, R., Qiming, S., Yuguang, S.: Customer segmentation of bank based on data warehouse and data mining. In: The 2nd International Conference on Information Management and Engineering, ICIME (2010)
2. Balafar, M.A., Ramli, A.R., Mashohor, S., Farzan, A.: Compare different spatial based fuzzy-C_mean (FCM) extensions for MRI image segmentation. In: 2010 The 2nd International Conference on Computer and Automation Engineering (ICCAE), pp. 609–611 (2010)
3. Xie, C., Xu, X., Hou, W.: Power customer credit rating based on FCM and the differential marketing strategy research. In: 2010 2nd International Conference on Information Science and Engineering (ICISE), pp. 416–418 (2010)

4. Ahmed, M.N., Yamany, S.M., Mohamed, N., Farag, A.A., Moriarty, T.: A modified fuzzy c-means algorithm for bias field estimation and segmentation of MRI data. IEEE Transactions on Medical Imaging 21, 193–199 (2002)
5. Songcan, C., Daoqiang, Z.: Robust image segmentation using FCM with spatial constraints based on new kernel-induced distance measure. IEEE Transactions on Systems, Man, and Cybernetics, Part B: Cybernetics 34, 1907–1916 (2004)
6. Jian, Y., Yufu, N.: Research on feature weights of fuzzy c-means algorithm and its application to intrusion detection. In: 2010 International Conference on Environmental Science and Information Application Technology (ESIAT), pp. 164–166 (2010)
7. Kwon, S.H.: Cluster validity index for fuzzy clustering. Electronics Letters 34, 2176–2177 (1998)
8. Huang, J., Li, C.: AHP based research on segmentation method of customer's value in aviation cargo transportation. In: IEEE International Conference on Service Operations and Logistics, and Informatics, SOLI 2006, pp. 696–698 (2006)
9. Ke, F., Jiayan, X., Qun, Z., Zhaowei, M.: An AHP-based decision support model for 3PL evaluation. In: 2010 7th International Conference on Service Systems and Service Management (ICSSSM), pp. 1–6 (2010)
10. Pu, Y.-W., Zhu, M., Jin, W.-D., Hu, L.-Z., Wang, J., Yi, Z., Zurada, J., Lu, B.-L., Yin, H.: An Efficient Similarity-Based Validity Index for Kernel Clustering Algorithm. In: Wang, J., Yi, Z., Żurada, J.M., Lu, B.-L., Yin, H. (eds.) ISNN 2006. LNCS, vol. 3971, pp. 1044–1049. Springer, Heidelberg (2006)

A MapReduce-Based Parallel Clustering Algorithm for Large Protein-Protein Interaction Networks

Li Liu[1], Dangping Fan[1,*], Ming Liu[2], Guandong Xu[3], Shiping Chen[2,4],
Yuan Zhou[1], Xiwei Chen[1], Qianru Wang[1], and Yufeng Wei[5]

[1] School of Information Science and Engineering, Lanzhou University,
Gansu 730000, P.R.China
{liliu,fandp10}@lzu.edu.cn
[2] School of Electrical and Information Engineering, The University of Sydney,
NSW 2006, Australia
{ming.liu,shiping.chen}@sydney.edu.au
[3] Advanced Analytics Institute,University of Technology Sydney,
NSW 2008, Australia
Guandong.Xu@uts.edu.au
[4] CSIRO ICT Centre, Australia
[5] The Third Peoples Hospital of Lanzhou, Gansu 730050, P.R. China

Abstract. Clustering proteins or identifying functionally related proteins in Protein-Protein Interaction (PPI) networks is one of the most computation-intensive problems in the proteomic community. Most researches focused on improving the accuracy of the clustering algorithms. However, the high computation cost of these clustering algorithms, such as Girvan and Newmans clustering algorithm, has been an obstacle to their use on large-scale PPI networks. In this paper, we propose an algorithm, called Clustering-MR, to address the problem. Our solution can effectively parallelize the Girvan and Newmans clustering algorithms based on edge-betweeness using MapReduce. We evaluated the performance of our Clustering-MR algorithm in a cloud environment with different sizes of testing datasets and different numbers of worker nodes. The experimental results show that our Clustering-MR algorithm can achieve high performance for large-scale PPI networks with more than 1000 proteins or 5000 interactions.

Keywords: PPI, Clustering, MapReduce, Edge-betweenness.

1 Introduction

Detection of physically and functionally related proteins is one of the most challeng-ing tasks for the proteomics community. Intensive computation is needed to analyze pairwise protein interaction in order to understand how proteins work together to perform their tasks. The PPI network is an important information

---

* Corresponding author.

S. Zhou, S. Zhang, and G. Karypis (Eds.): ADMA 2012, LNAI 7713, pp. 138–148, 2012.

source, which en-codes the interaction between proteins. The network contains nodes (i.e. proteins) and edges, which represent the interactions between the proteins. When proteins are expe-rienced in the same cellular process or have the same protein complex, they are ex-pected to have strong interactions with their partners [1], and vice versa. Furthermore, modularity is mostly studied by grouping physically or functionally related proteins in a cluster such that the proteins in the same cluster share common biological features [2]. Therefore, the computation objective is to discover the clustering structures in the PPI net-work, i.e. to determinate a collection of sets of nodes where each node is closer to the other nodes within the same set than to nodes outside of the set.

Several graph-based clustering algorithms have been applied to PPI network to find highly connected sub-graphs. The flow-based cluster algorithm [3] was used to cluster large-size networks and the time complexity of this algorithm was approxi-mately $O(k\,|\varepsilon|+\sum_{i=1}^{|v_c|} d_i^2)$, where the d_i is the degree of the node i in the graph, and k is a small constant which is typically set to 10. A quasi all paths-based network analysis algorithm, called CASCADE [4], was proposed to effectively detect bio-logically relevant clusters with the time complexity of $O(n^3 \log n + n^2 m)$, where n is the number of nodes and m is the number of edges. Girvan and Newmans Edge-betweeness algorithm [5] is one of the most popular clustering algorithms, which has been widely used to discover clustering structures in different domain networks, such as web, social networks, and PPI networks. The time complexity of this algorithm is $O(nm^2)$. These studies mainly focused on improving the clus-tering accuracy of Girvan and Newmans Edge-betweeness algorithm.

However, the high computation cost of using these clustering algorithms be-comes a major issue when the PPI network has thousands of nodes and millions of edges. In fact, the time complexities of these clustering algorithms for PPI networks are more than quadratic in terms of the number of nodes.

In this paper, we aim to develop a novel parallel implementation of Girvan and Newmans edge-betweenness algorithm using MapReduce to address the high com-putation cost issue. Although the edge-betweenness algorithm provides good accuracy for clustering small or medium size PPI networks, its high computation cost has become an obstacle to applying it for clustering large size PPI networks [6]. A parallel version of betweenness algorithm using the MapReduce distributed programming framework is proposed to handle large-scale PPI networks.

The rest of this paper is organized as follows: Section 2 describes background on notations and technologies used in this paper. In Section 3 we propose our MapReduce-based parallel algorithm. The experimental results are presented and discussed in Section 4. We conclude in Section 5.

2   Background and Related Work

2.1   Girvan and Newmans Edge-Betweenness Algorithm

Edge-betweenness measures the centrality of an edge within a graph. The be-tweenness is formulated as follows: Given a graph $G(V, E)$, where V is the set of n nodes and E is the set of m edges. Let $\sigma_{s,t}$ denote the total number of shortest paths from node s to node t and $\sigma_{s,t}(e)$ is the number of those paths that pass through e where $e \in E$. $\delta_{s,t}(e)$ represents the ratio of the total number of shortest paths between s and t that pass through e to the $\sigma_{s,t}(e)$, where $\delta_{s,t}(e) = \frac{\sigma_{s,t}(e)}{\sigma_{s,t}}$. Thus, the betweenness of an edge e is defined as $BC(e) = \sum_{s \neq t \in V} \delta_{s,t}(e)$.

An edge with high betweenness indicates that a large number of the shortest paths between two nodes pass through it. If the edge is removed, the graph is split into two subgraphs. Similarly, by removing the edges with the highest betweenness in descending order, we can separate the PPI graph into several subgraphs. Such subgraph contains interconnected proteins that have strongly functional relationships.

Girvan and Newmans edge-betweenness algorithm first calculates the betweenness for all edges in a graph by using breadth-first searching (BFS) method. Then it removes the edge with the highest betweenness. The algorithm repeats the calculation and removal steps until no edges remain. The total time complexity of Girvan and New-mans edge-betweenness algorithm is $O(nm^2)$, where n is the number of nodes and m is the number of edges.

2.2 PPI Modularity Evaluation

There are m collections of subgraphs produced by Girvan and Newmans edge-betweenness algorithm. In order to determine which one is the best collection in terms of functionally related proteins in a cluster, the modularity evaluation for each collection is defined as $Q = \sum_{i=1}^{k} (f_{ii} - (\sum_{j=1}^{k} f_{ij})^2)$, where k is the number of subgraphs in current collection, f_{ii} represents the fraction of all edges in the network that connect nodes in the same subgraph while f_{ij} represents the fraction of all edges in the network that connect the nodes in subgraph i to the nodes in subgraph j.

If Q approaches 1, it indicates that the collection has strong clustering structure. The higher the value of Q is, the stronger the clustering structure is. Therefore, the collection with the highest value of Q is selected as the final clustering structure in PPI network. The time complexity of calculating all the values of Q is $O(nmk^2)$.

2.3 Parallel Implementations for Computing Betweenness

Since Girvan and Newmans edge-betweenness algorithm suffers from high computa-tional cost, it is impractical for processing large-scale PPI networks. Few researches studied the parallel implementation of betweenness algorithm to handle graphs with more than hundred thousand edges. These parallel algorithms were mainly designed for the shared-memory architecture. For example, Bader and Madduri [7] implemented the first parallel algorithm for computing the betweenness based on SNA software package on the IBM p5 570 with 16-processors and 20GB-memory, and the MTA-2 with 40-processor and 160GB-memory. The algorithms time complexity and space complexity is $O(\frac{nm+n^2 \log n}{p})$

and $O(n + m)$ respectively, where p is the number of processors. Madduri et al. [8] proposed a parallel implementation based on the massively multithreaded Cray XMT system with the Threadstorm processor and 16GB memory. It also takes $O(\frac{nm+n^2 \log n}{p})$ time complexity and $O(n + m)$ space complexity to calculate the betweenness. G. Tan et al. [9] extended their work and improved the algorithm by reducing the time complexity to $O(\frac{nm}{p})$. However, usually these massive computer servers are very expensive.

2.4   MapReduce and Hadoop

MapReduce is a distributed programming model for parallel processing large-scale datasets and the computing. Compared with shared-memory architecture, MapReduce has lower economical cost and better scalability in terms of problem sizes and re-sources.

A MapReduce-based algorithm is different from the parallel algorithms based on shared-memory architecture, because MapReduce is based on message-passing architecture. The MapReduce model mainly contains two data processing phases: Map and Reduce. During the Map phase, the input data is split into smaller data segments and distributed to multiple processing units, such as virtual machines in Cloud, for parallel processing by using a map function. The intermediate output produced by the map function is a collection of key-value pair tuples. During the Reduce phase, the intermediate outputs from each map function are transferred to the machines that execute a reduce function. The key-value formed data are sorted by using the keys and aggregated by using the reduce function.

Hadoop is an open source version of Googles MapReduce and Google File System. Hadoop includes a master node and multiple worker nodes. The master node consists of a JobTracker and a NameNode. The JobTracker manages jobs scheduling and assigns the jobs to the TaskTrackers on the worker nodes, which is responsible for executing the map function and the reduce function. The NameNode is responsible for storage and management of file metadata and file distribution across several DataNodes on worker nodes, which store the data contents of these files.

3   MapReduce-Based Clustering Algorithm Implementation

In this section, we propose our MapReduce-based parallel algorithm, called *Clustering-MR*. As being seen in Fig.1, the *Clustering-MR* contains the following four steps.

Step1: An algorithm, called *forward-MR*, executes n tasks in parallel by using MapReduce. Each task has been given a node v, which is set as a root node. The object of this algorithm is to find the shortest paths by using BFS and calculate their distances from other nodes to the root node.

Step2: An algorithm, called *backward-MR*, executes m tasks in parallel by using MapReduce. Each task has been give an edge e. The object of this algorithm is to calculate the edge-betweenness $BC(e)$.

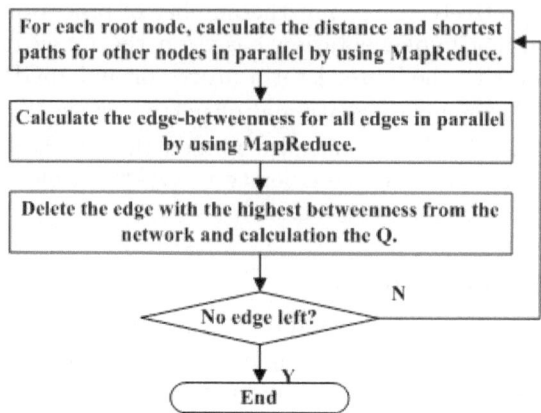

Fig. 1. The workflow of the *Clustering-MR* algorithm

Step3: The algorithm removes the edge with the highest betweenness and calculates Q value.

Step4: The algorithm repeats previous steps till no edge remains.

3.1   Input File Format for MapReduce

We use an adjacency list to represent a PPI network, which is considered as the MapReduce input file. There are totally $n \times n$ lines in the input file. The format of a line in the input file is defined as:

<*NodeId*> <*Root*> <*Neighbors*> | <*Distance*> | <*Color*> | <*Path*>

where:

. *NodeId* is the ID of a node;
. *Neighbors* is a list of nodes that are the neighbors of this node;
. *Root* is the ID of the root node;
. *Path* is the shortest path;
. *Distance* is shortest-path distance from the node to the root node;
. *Color* is the status of this node.

There are three states for each node: (1) WHITE: the node is unreached; (2) GRAY: the node is reached and going to be handled; and (3)BLACK: the node is reached and has been handled. Initially, the *Distance* for each node is set as MAX and its *color* is set as WHITE. If *NodeId* equals to *Root*, its *Distance* and *Color* are set to 0 and GRAY respectively.

3.2   Forward-MR

In *forward-MR*, the map function reads a line from the input file. If the *Color* is GRAY, it is changed to BLACK. For each node in *Neighbors*, if the *Color* is not

BLACK, the map function generates a set of *key-value* pairs as its output, where the keys are neighbors *NodeIds* and *Root*, and the values contains *Distance*+1, GRAY and *NodeId*.

Based on the map functions outputs, the reducer selects the minimum distance among the values , updates the status of this node, and records the path, then produces the output consisting of *NodeId, Root, Neighbors, Distance, Color* and *Path*, where *Distance* is the minimum distance, *Color* is the latest state and *Path* is the shortest path. If the node is not reached, then the *Distance* remains MAX and the *Color* remains WHITE. The pseudocode of the map function for *forward-MR* is shown in Algorithm 1 and reduce function is shown in Algorithm 2.

Fig.2 shows an example of the process of the *forward-MR* changing the nodes color in a network, where node 4 is the root node. Firstly, for node 2, 3 and 6, which are the *Neighbors* of the node 4, the map function emits the *key-value* pairs as <2 4 4 1 GRAY 4>, <3 4 1,4 1 GRAY 4> and <6 4 4,5,7 1 GRAY 4>. For node 4, the *key-value* pair is <4 4 2,3,6 0 BLACK >. Then the reduce function updates the *Color* of 2,3 and 6 as GRAY, the *Distance* and *Path* as 1 and 4, and the color of 4 as BLACK. Secondly, for node 1, 5 and 7 which are the *Neighbors* of the node 3 and 6, the map function emits the *key-value* pairs as <1 4 3 2 GRAY 4,3>,<5 4 6,7 2 GRAY 4,6> and <5 4 6,7 2 GRAY 4,6>. For node 2, 3 and 6, the *key-value* pairs are <2 4 4 1 BLACK 4>, <3 4 1,4 1 BLACK 4> and <6 4 4,5,7 1 BLACK 4>. Then the reduce function updates the *Color* of 1,5 and 7 as GRAY, the *Distance* and *Path* as 2 and 4,3, 4,6 and 4,6 separately, and the *Color* of node 2,3 and 6 as BLACK. Since the *Color* of node 4 is BLACK, the status of it is not changed. Thirdly, for node 1, 5 and 7, the *Color* of each neighbor *NodeId* is BLACK, the map function just emits the *key-value* pairs as <1 4 3 2 BLACK 4,3>,<5 4 6,7 2 BLACK 4,6> and <5 4 6,7 2 BLACK 4,6>. Then the reduce function updates the *Color* of node 1,5 and 7 as BLACK. At last, as each node has reached, the *forward-MR* finishes.

Algorithm 1. Mapper: Forwardmapper

Input: Each line contains NodeID, Root, Distance, Path, etc.
Output:<key, value>pair, where key contains nodeId and root while value contains node's info
1.if (line.Color is GRAY)
2.foreach node in neighbors do
3. if(the color of node is not BLACK)
4. node.distance = line.Distance + 1
5. node.color = GRAY
6. node.path = line.Path.add(line.NodeId)
7. output.write(<node,line.Root>,node.value)
8. endif
9.endfor
10.line.Color = BLACK
11.output.write(<line.NodeId, line.Root>,line.value)

3.3 Backward-MR

The final output file of *forward-MR* is used as the input file for *backward-MR*. The map function of *backward-MR* reads a line from the input file, and generates the *key-value* pairs, where the keys are the edges within the *Path*, and every value is set to 1. The reduce function calculates the edge-betweenness for each edge. Only one shortest path between two nodes is recorded in order to make computation tractable [10],thus *Clustering-MR* is an approximate algorithm.

4 Evaluation

In this section, we evaluate the performance of our *Clustering-MR* algorithm with different datasets and different numbers of worker nodes (4, 8, 12, 16 and 20). The theoretical time complexity of *Clustering-MR* algorithm is $\frac{O(nm^2)}{p}$, where n is the number of nodes, m is the number of edges and p is the number of processors. In addition, we compare *Clustering-MR* with a sequential algorithm (*Clustering-SEQ*) which executes on only one processor with single thread.

Algorithm 2. Reducer: ForwardReducer
Input: <key, list(value)>, where key contains nodeId and root while list contains the node's info.
Output:<key, value>pair, where key contains nodeId and root while value contains node's info.
1. currentNode = key.nodeId
2. foreach value in list(value) do
3. if(value.Nighbors != null)
4. currentNode.Neighbors = value.Neighbors
5. endif
6. if(value.distance <= currentNode.Distance)
7. currentNode.Distance = value.distance
8. currentNode.Path = value.path
9. endif
10. if(value.color != WHITE)
11. currentNode.Color = value.color
12. endif
13. endfor
14.output.write(key,currentvalue)

4.1 Experiment Setup

The *Clustering-MR* algorithm was tested on a Hadoop system, which consists of a single Master Node and 20 identical Worker Nodes. The Master Node contains four 64-bit Intel dual-core 3.3GHz processor, along with 4GB memory, and 500GB disk Each Worker Node contains a single 32 bit Intel 2.8GHz processor, 1GB memory and 164.7GB disk. All the nodes ran on Ubuntu 10.04 and had Hadoop v0.20.205.0 de-ployed. The network bandwidth was 12MB/s.

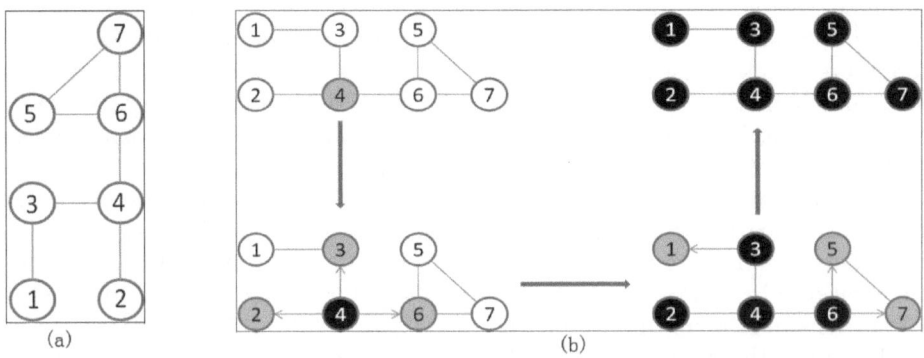

Fig. 2. (a) An example of a toy network. (b) The *forward-MR* algorithm execution process

4.2 Datasets

We evaluated the *Clustering-MR* algorithm on these seven datasets containing PPI graphs, which were downloaded from DIP database [11] on 18th, May, 2012. Table 1 shows the size of each dataset including the original size and the MapReduce input size.

Table 1. Experimental Datasets

Species	Mark	Proteins	Interactions	Size(KB) (original)	Size(KB) (MapReduce Input)
D.melanogaster	DM	7,439	22,632	227	2,637,954
S.cerevisiae	SC	2,993	7,029	70	365,973
C.elegans	CE	2,629	3,970	38	225,598
E.coli	EC	1,355	5,476	48	92,378
H.sapiens	HS	941	1,160	10	24,389
H.pylo	HP	7,01	1,358	12	15,966
M.musculus	MM	314	267	3	2,356

4.3 Experimental Results

Fig.3(a) shows the overall execution time of the Clustering-MR algorithm for processing each dataset in the Hadoop system using different worker nodes: 4, 8, 12, 16 or 20. For the DM dataset, the execution time of the algorithm with 4 worker nodes was almost 3 times longer than the runtime of the algorithm with 20 worker nodes. Similarly, for SC and CE dataset, the execution time of the algorithm with 4 worker nodes was almost 2 times longer than the runtime of the algorithm with 12 worker nodes. However, the performance didnt improve anymore when the number of worker nodes was larger than 12. For EC, HS, HP, and MM datasets, although the execution time decreased as the number of machines increases, it didnt have big improvement. These results indicate that our *Clustering-MR* algorithm with Hadoop indeed can improve performance for large size datasets. However, the improvement is not obvious for small data sizes and cannot remain linear as the number of work nodes reaches a certain threshold. This is because the system would spend more time on the management and data transferring among these machines.

Fig.3(b) shows the relationship between the number of nodes (proteins) in PPI graphs and the runtime of the *Clustering-MR* algorithm in a Hadoop system using the different number of work nodes: 4, 8, 12, 16 and 20. Similarly, Fig.3(c) shows the relationship between the number of edges (interactions between proteins) in PPI graphs and the runtime of the *Clustering-MR* algorithm. In the case of using 4 worker nodes, the execution time of the algorithm increased dramatically when the number of proteins was larger than 3000 or the number of interactions was larger than 6000. But, in the case that more worker nodes were used in the Hadoop system, the runtime slightly increased. For small size of proteins and interactions, the number of machines used in the system did not affect much the execution time of the *Clustering-MR* algorithm. But for larger size of proteins and interactions, the impact of the number of machines became big.

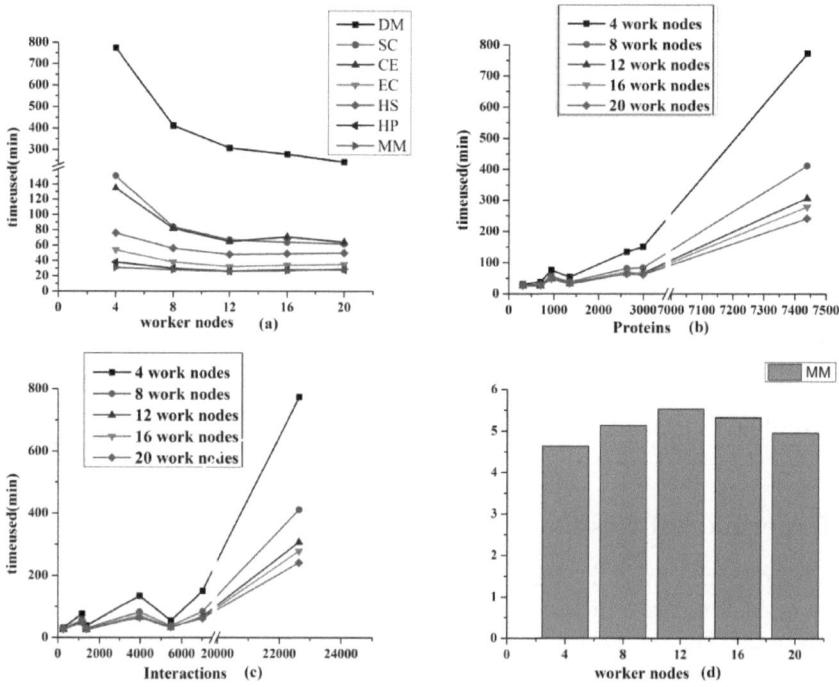

Fig. 3. The experiment results. In (a),(b),(c), the y-axis is the running time of computation, and the x-axis is the number of worker nodes, proteins and interactions respectively. In (d), the y-axis is the speedup (serial/parallel) and the x-axis is the number of worker nodes.

Fig.3(d) shows the ratio of the running time of the *Clustering-SEQ* algorithm to the *Clustering-MR* algorithm in a Hadoop system with using different worker nodes: 4, 8, 12, 16 and 20, when both algorithms are working on MM dataset. The speedup increases as the number of worker nodes increases from 4 to 12. Interestingly, the speedup decreases as the number of worker nodes increases from 12 to 20. This is because the MM dataset contain relative small PPI networks and the *Clustering-SEQ* algorithm can perform reasonably good on the small PPI networks. Furthermore, as the number of worker nodes increases, task scheduling, data transferring, large number of hard disk and memory switching took up a large portion of total running time. There-fore, given a dataset size, determining the optimal number of Hadoop worker nodes can be an interesting research work.

5 Conclusions

In this paper, we proposed a new parallel processing algorithm, called the *Clustering-MR*, applying the MapReduce model to parallelize the Girvan and Newmans edge-betweenness for large PPI networks. This new algorithm overcomes the issues of high computation cost caused by the sequential processing. Compared with the shared-memory based parallel algorithms, our algorithm is cost-effective and easy to scale. We implemented *Clustering-MR* on a real distributed system using Hadoop. Our experimental results show that: (1) *Clustering-MR* gets better performance on runtime than *Clustering-SEQ*. (2) *Clustering-MR* can efficiently handle larger-scale PPI networks with more than 1000 proteins or 5000 interactions, which would be very difficult (if not impossible) for *Clustering-SEQ*.

We observed that the experimental results in practice do not totally agree with the theoretical analysis, especially when the number of nodes or edges becomes lager and lager. The reason is that the space complexity of *Clustering-MR* algorithm that equals to $O(nm^2)$ leads to the increment of data transmission between machines which costs a large amount of time to complete. To reduce the space complexity is one of the key points in our following work. We will improve the *Clustering-MR* algorithm by changing the input data format and optimizing the Reduce step.

In our future work, we will also compare *Clustering-MR* to the MPI(Message Passing Interface)-based clustering algorithm for large PPI networks. Both of them are message-passing based algorithms. Besides, we will evaluate our *Clustering-MR* algorithm in other applications, such as social networks and linked web.

Acknowledgements. This work was partially supported by National Natural Science Foundation of China (grant no. 61003240) and Gansu Provincial Science & Technology Department (grant no. 1007RJYA010). We would also like to thank DIGICOM JAPAN Co.Ltd (http://www.digicomnet.co.jp) to provide us the Hadoop system for our experiments.

References

1. Maslov, S., Sneppen, K.: Specificity and stability in topology of protein networks. Science 296(5569), 910–913 (2002)
2. Baraba'si, A., Oltvai, Z.N.: Network Biology: Understanding the Cell's Functional Organization. Nature Reviews Genetics 5, 101–113 (2004)
3. Satuluri, V., Parthasarathy, S.: Scalable Graph Clustering Using Stochastic Flows: Applications to Community Discovery. In: Proceedings of the 15th ACM SIGKDD International Conference on Knowledge Discovery and Data Mining, KDD 2009, Paris, France, pp. 737–745 (2009)
4. Hwang, W., Cho, Y., Zhang, A., Ramanathan, M.: CASCADE: a novel quasi all paths-based network analysis algorithm for clustering biological interactions. BMC Bioinformatics 9(64) (2008)
5. Girvan, M., Newman, M.E.J.: Community structure in social and biological networks. PNAS 99(12), 7821–7826 (2002)
6. Dunn, R., Dudbridge, F., Sanderson, C.M.: The Use of Edge-Betweenness Clustering to Investigate Biological Function in Protein Interaction Networks. BMC Bioinformatics 6(39) (2005)
7. Bader, D.A., Madduri, K.: Parallel Algorithms for Evaluating Centrality Indices in Real-world Networks. In: International Conference on Parallel Processing (ICPP 2006), pp. 539–550 (2006)
8. Madduri, K., Ediger, D., Jiang, K., Bader, D.A., Chavarria-Miranda, D.: A Faster Parallel Algorithm and Efficient Multithreaded Implementations for Evaluating Betweenness Centrality on Massive Datasets. In: IEEE International Symposium on Parallel & Distributed Processing (IPDPS 2009), pp. 1–8 (2009)
9. Tan, G., Tu, D., Sun, N.: A Parallel Algorithm for Computing Betweenness Centrality. In: International Conference on Parallel Processing (ICPP 2009), pp. 340–347 (2009)
10. Maier, M., Rattigan, M., Jensen, D.: Indexing network structure with shortest-path tree. ACM Transactions on Knowledge Discovery from Data 5(3) (2011)
11. DIP Database, http://dip.doe-mbi.ucla.edu/

A New Manifold Learning Algorithm
Based on Incremental Spectral Decomposition

Chao Tan and Jihong Guan

Dept. of Computer Science and Technology, Tongji University, Shanghai, China
tanchao222@gmail.com, jhguan@tongji.edu.cn

Abstract. Manifold learning is to construct nonlinear low-dimensional manifolds from sample data points embedded in high-dimensional spaces. In streaming data applications, new data points come continually, which will change the existing data points' neighborhoods and their local distributions. Such applications call for incremental algorithms not only to deal with the adding of new data points but also to update the local neighborhoods of the existing data points. In this paper, we introduce a new manifold learning algorithm by updating the structure of eigenproblem iteratively. Incremental spectral decomposition is used in the iterative process and the resulting eigenvectors correspond to the low dimensional embedded coordinates. Experimental results show that 1) as the number of data points increases, the mapping results of the proposed approach become closer and closer to that of batch-style approaches, including LTSA and LE, and 2) the proposed approach outperforms the incremental ISOMAP (IISOMAP, a typical incremental manifold learning algorithm) in mapping accuracy. We argue that the new algorithm is suitable for incremental learning of large-scale data streams.

Keywords: Manifold learning, Incremental spectral decomposition, Incremental learning.

1 Introduction

Manifold learning [1,2] is to construct nonlinear low-dimensional manifolds from sample data points embedded in high-dimensional spaces, which has been accepted as a kind of effective nonlinear dimensionality reduction methods. In the past decade, a number of manifold learning algorithms were proposed and applied to image analysis and computer vision, information retrieval, and human gene distribution study, etc. The typical manifold learning algorithms include isometric feature mapping (ISOMAP) [1], locally linear embedding (LLE) [2], Laplacian eigenmaps algorithm (LE) [3], Hessian LLE (HLLE) [4] and local tangent space alignment algorithm (LTSA) [6].

Most existing manifold learning algorithms aim at the learning from a fixed dataset. However, for many applications, the datasets are dynamic, with new data points being added continually. To address the learning of dynamic datasets, incremental manifold learning algorithms have been developed. For example, incremental Isomap algorithm (IIsomap) proposed by Law and Jain [10] learns

S. Zhou, S. Zhang, and G. Karypis (Eds.): ADMA 2012, LNAI 7713, pp. 149–160, 2012.
© Springer-Verlag Berlin Heidelberg 2012

the input data stream incrementally, which can find the evolution process of manifold structure. The incremental learning algorithm for Laplacian eigenmaps (ILE) proposed by Jia et al. [14] computes the low-dimensional representation of a dataset by optimally preserving local neighborhood information, and a sub-manifold analysis with a formulation of linear incremental method was employed to learn the new samples incrementally. Kouropteva et al. [16] proposed the incremental LLE algorithm (ILLE), which evaluates the mapping values of the new samples and re-calculates the projections of original samples. Incremental LTSA algorithm (ILTSA) [17] can get the low-dimensional embedding coordinates of new data points by minimizing the reconstruction error.

In this paper, we propose a new manifold learning algorithm for dynamic datasets. For some existing learning methods, new points may change the current neighborhoods and local distribution of the manifold. In Isomap [1] if the neighborhood is chosen improperly, short-circuit or cavitation phenomenon will happen. Similar problems also exist in incremental methods such as IIsomap [10]. The addition of a new sample can delete critical edges in the graph and subsequently change the geodesic distances dramatically. To solve this problem, in the proposed approach we update a new point's position incrementally in the low dimensional space by updating the structure of *eigen*-problem iteratively. For this end, we employ relevance vector and matrix to deal with neighborhood modification caused by newly-coming sample points on the manifold, and establish the new points' positions in the low dimensional embedding space by updating the structure of eigen-problem iteratively, which is done by *incremental spectral decomposition*. The obtained eigenvectors correspond to the low dimensional embedded coordinates. Experimental results show that the mapping results of the proposed approach are closer to original datasets than batch-style approaches (including LE and LTSA) as the number of data points increases, and more accurate and stable than IISOMAP.

Major contribution of this paper: A new manifold learning approach for learning dynamic datasets based on incremental spectral decomposition is proposed, with its convergence proved. And an algorithm to implement the proposed approach is developed. Then we conduct extensive experiments to validate the effectiveness of the proposed approach.

The reminder of this paper is organized as follows: Section 2 introduces the preliminaries indicating notation declarations. Section 3 presents the new manifold learning approach and the algorithms. Section 4 gives the experimental results. Section 5 concludes this paper.

2   Notation Declarations

Let $\mathbb{R}$ denote the manifold. Lower case letters indicate vectors or elements, capital letters represent matrices. Subscripts are used to index the elements in vectors and matrices. For example, u_i is the i-th element of vector u. $R_{i.}$ and $R_{.j}$ indicate the i-th row and the j-th column of matrix R respectively, and R_{ij} represents the element with index (i, j) in R. $|| \cdot ||$ denotes the l_2 norm, that is, $||x|| = \sqrt{x^T x}$.

3   Manifold Learning Based on Incremental Spectral Decomposition

In this section, we introduce a new approach for incremental learning from a data stream employing relevance vector and matrix [7]. By updating the structure of *eigen*-problem incrementally, we can get the mapping of the newly added data point x_{n+1}. The relevance eigenvalue is used to represent the existing sample points $x_1, x_2, ..., x_n$ whose neighbors will be changed by the insertion of x_{n+1}.

3.1   Some Definitions

Definition 1. Given a dataset D of size n, $U_j = [U_{1j}, \cdots, U_{ij}, \cdots, U_{nj}]^T$ ($i = 1, \cdots, n$) is a *relevance vector*.

$$U_{ij} = \begin{cases} 1 \text{ if point } i \text{ is among the } k \text{ nearest neighbors of point } j \text{ or vice versa;} \\ 0 \text{ otherwise.} \end{cases}$$

Definition 2. Given a dataset D of size n, we can construct a $n \times n$ *relevance matrix* U, each column U_j is a relevance vector.

Definition 3. Given a dataset D of size n, an *incidence matrix* W is a $n \times n$ matrix whose element w_{ij}, called *incidence element*, is evaluated as:

$$w_{ij} = \begin{cases} \frac{1}{k} \text{ if point } i \text{ is among the } k \text{ nearest neighbors of point } j \text{ or vice versa;} \\ 0 \text{ otherwise.} \end{cases}$$

Definition 4. r_{ij} is a column vector with only two nonzero elements: i-th element is 1 and j-th element is -1.

3.2   Incremental Eigenmaps

In order to represent the dynamic characteristics of an eigenvalue system without changing its original representation, we propose a more effective algorithm to evaluate the incremental representation of an updated dataset based on the iteratively generated eigenvalue system. We have known the LE algorithm can preserve the local neighborhood of the low dimensional representation of a dataset [3]. Here, we use an incremental way onto the LE algorithm to learn data streams in an incremental style.

First of all we give a graph with weight: $G = G(V, W)$, where V is the vertex set and W the weight set indicating the incidence between vertex v_i and v_j. In this paper, the incidence matrix W is assumed to be the weight matrix where $w_{ij} = \frac{1}{k}$ when vertex i and vertex j are mutual k nearest neighbors, else $w_{ij} = 0$. Define diagonal matrix D as: $D = diag\{d_1, d_2, ..., d_n\}$, $d_i = \sum_j w_{ij}$. Then we introduce some propositions and theorems.

Proposition 1. Given a $n \times n$ *Laplacian* matrix $L = D - W$, there exists an relevance matrix U and R such that $L = W(RR^T)$ [5], where R contains all the column vectors r_{ij} $(1 \leq i < j \leq n)$.

Proof. According to Definition 2, $U = \begin{pmatrix} & \vdots & & \vdots & \\ \cdots & 0 & \cdots & 1 & \cdots \\ & \vdots & & \vdots & \\ \cdots & 1 & \cdots & 0 & \cdots \\ & \vdots & & \vdots & \end{pmatrix}$, which is a relevance

matrix with the element U_{ij} is 1 if points i and j are connected by an edge, otherwise it is 0.

According to Definition 4, $r_{ij}r_{ij}^T = \begin{pmatrix} & \vdots & & \vdots & \\ \cdots & 1 & \cdots & -1 & \cdots \\ & \vdots & & \vdots & \\ \cdots & -1 & \cdots & 1 & \cdots \\ & \vdots & & \vdots & \end{pmatrix}$. Since $L = D - W$,

where $D = diag\{d_1, d_2, ..., d_n\}$ and $d_i = \sum_j w_{ij}$. Add all of $w_{ij}r_{ij}r_{ij}^T$ $(1 \leq i, j \leq n)$ together, we get $L = \sum_{i,j} w_{ij}(r_{ij}r_{ij}^T + u_{ij}) - \sum_{i,j} w_{ij}u_{ij} = W(RR^T)$.

Proposition 2. $Ly = L_1 y_1 + ... + L_n y_n = W_1 R_1 R_1^T y_1 + ... + \dot{W}_n R_n R_n^T y_n$.

Proof. According to the definition of Krylov subspaces [9], an intuitive method for finding an eigenvalue (specifically the largest eigenvalue) of a given matrix L is the power iteration. Starting with an initial random vector y_0, this method calculates $Ly_0, L^2 y_0, ...$ iteratively, storing and normalizing the result into y on each turn. Ly can be computed neighborhood by neighborhood without constructing repeatedly. This suggests us to construct the Krylov form of Ly: $Ly = L_1 y_1 + ... + L_n y_n = W_1 R_1 R_1^T y_1 + ... + W_n R_n R_n^T y_n$.

For the incidence matrix W with incidence elements w_{ij}, we define $W \approx \frac{1}{k} e_i e_i^T$, where $e_i = \{1, ..., 1\}^T$, $(i = 1, ..., n)$. So, $L_i \approx \frac{1}{k} e_i e_i^T R_i R_i^T$, it fits the incremental nature of structure which can be computed neighborhood by neighborhood without explicitly forming L. We now consider the reconstructing error in low dimensional space based on the existing Laplacian framework L_i, which represents the local geometry. We assume $W_i R_i R_i^T = \frac{1}{k} e_i e_i^T R_i R_i^T + \Delta L$. Here, ΔL is the reconstruction error and we need to find optimal solutions to minimize it. We have $\Delta L = \Delta W_i R_i R_i^T - \varepsilon$. Considering **Proposition 1** and **Proposition 2**, the change of incidence matrix W can be expanded by appending the relevance vector r_i and approximate $W : \widetilde{W}_i$. The reconstruction error can be represented as

$$\Delta W_i R_i R_i^T = (I - \widetilde{W}_i) - (I - W_i) R_i R_i^T. \tag{1}$$

Substitute Eq. (1) and $\tilde{W}_i = \frac{1}{k}e_i e_i^T$ into the expression of $W_i R_i R_i^T$,

$$W_i R_i R_i^T = \frac{1}{k}e_i e_i^T R_i R_i^T + (I - \frac{1}{k}e_i e_i^T) - (I - W_i)R_i R_i^T - \varepsilon \qquad (2)$$

The error can be expressed in accordance with Eq. (2) as follows,

$$\varepsilon = \frac{1}{k}e_i e_i^T R_i R_i^T + I - \frac{1}{k}e_i e_i^T - R_i R_i^T = (I - \frac{1}{k}e_i e_i^T)(I - R_i R_i^T).$$

Let $K_i = (I - \frac{1}{k}e_i e_i^T)(I - R_i R_i^T)$. The matrix K is the projection onto the subspace spanned by the column vectors of R. For minimizing the reconstruction error, we define

$$K_i = I - E_i E_i^T \qquad (3)$$

with $E_i = [e/\sqrt{k}, r_1, ..., r_d]$. So $L_i = W_i R_i R_i^T$ can be computed as:

$$L_i = X_i(I - E_i E_i^T)X_i^T, \qquad (4)$$

where $X_i = [x_{i1}, ..., x_{ik}]$. L_i can be regarded as the *Laplacian* matrix defined on X_i during the i-th iteration. The optimal results should be constructed by resolving some *eigen*-resolvers. According to **Proposition 2**, L can be computed by partially summing as follows:

$$L(I_i, I_i) \leftarrow L(I_i, I_i) + I - E_i E_i^T \quad (i = 1, ..., n). \qquad (5)$$

Here, I_i is a set of indices for the k nearest neighbors of x_i denoted by $I_i = \{i_1, ..., i_k\}$. With respect to n, the computation cost is obviously linearly. To minimize the reconstruction error after dimensionality reduction, we compute the eigenvector matrix $[t_1, ..., t_d]$ of L corresponding to the d smallest eigenvalues to compose the data points' mapping coordinates after the new data point comes.

The Eq. (5) has constructed an incremental alignment matrix that describes the eigenvalues and eigenvectors in spectral clustering. By updating the structure of *eigen*-problem iteratively we can get the mapping of all the data points without recomputing the updated subspace.

Now we propose an alternative iterative process to compute the new point's embedding coordinate into the low dimensional space. With the definition of *Laplacian* matrix, we know $\{L_n\}$, $||L_n|| < \infty$ and $\lim_{k\to\infty} L_k = L$. We have the following lemma [13]:

Lemma 1. Let $\{M_n\}$ be a sequence of real matrices. If $\lim_{n\to\infty} M_n = M$, then

$$\lim_{n\to\infty} \frac{1}{n} \sum_{i=1}^{n} M_i = M.$$

According to **Lemma 1**, we have $\lim_{n\to\infty} \frac{1}{n} \sum_{k=1}^{n} L_k = L$.

Denote the eigenvector of L_k as v_k, λ_k as the eigenvalue. That is $L_k v_k = \lambda_k v_k$.

Define $u_k = L_k v_k$, therefore $\lim_{k\to\infty} u_k = Lv = \lambda v$. Then set $M_n = \frac{1}{n} \sum_{k=1}^{n} u_k$.

According to **Lemma 1**, we have $\lim\limits_{n\to\infty} M_n = \lim\limits_{k\to\infty} u_k = \lambda v$. Consequently, $\lim\limits_{n\to\infty} \frac{M_n}{||M_n||} = v$.

M_n can be written in a recursive form for incremental estimation as follows:

$$M_n = \frac{1}{n}\sum_{k=1}^{n} u_k = \frac{1}{n}\sum_{k=1}^{n} L_k v_k = \frac{n-1}{n}\left(\frac{1}{n-1}\sum_{k=1}^{n-1} L_k v_k\right) + \frac{1}{n}L_n v_n$$
$$= \frac{n-1}{n}M_{n-1} + \frac{1}{n}L_n v_n. \tag{6}$$

Since v_n is the eigenvector of L_n and can be estimated by $\frac{M_{n-1}}{||M_{n-1}||}$, we have $\lim\limits_{n\to\infty} v_n = v \approx \lim\limits_{n\to\infty} \frac{M_{n-1}}{||M_{n-1}||}$, thus

$$M_n \approx \frac{n-1}{n}M_{n-1} + \frac{1}{n}L_n\frac{M_{n-1}}{||M_{n-1}||}. \tag{7}$$

The convergence of the process will be given by **Theorem 1** [12] below. We can find the similar theorems with proof in Zhang and Weng's work [11].

Theorem 1. If matrices sequence $A(n)$, $||A(n)|| < \infty$ converges to a matrix $A \in \mathbb{R}^{d\times d}$, i.e., $\lim\limits_{n\to\infty} \frac{1}{n}\sum_{i=1}^{n} A(i) = A$, where A is nonnegative determined matrix and $||A|| < \infty$, the eigenvalues of A satisfies $\lambda_1 > \lambda_2 \geq ... \geq \lambda_d \geq 0$, then the iterative process converges to the maximum eigenvalue of matrix A multiplied by the corresponding eigenvector.

$$v(n) = \frac{n-1}{n}v(n-1) + \frac{1}{n}A(n)\frac{v(n-1)}{||v(n-1)||}. \tag{8}$$

According to **Theorem 1**, we can derive the convergence of (7).

Theorem 2. The iterative expression M_n converges to the maximum eigenvalue of matrix L_n multiplied by the corresponding eigenvector.

Proof. For $||L_n|| < \infty$ and $\{L_n\}$ converges to a matrix $L \in \mathbb{R}^{n\times n}$, $\lim\limits_{n\to\infty} \frac{1}{n}\sum_{k=1}^{n} L_k = L$, the eigenvalues of L are nonnegative. By **Theorem 1**, we know the iterative process (7) will converge.

3.3   Algorithm Summary

Considering the discussion above, we propose a new manifold learning method based on incremental spectral decomposition named LISD. The LISD approach can be implemented to two algorithms outlined as follows. The first one has global incremental significance, called LISD-1. The second one is a local incremental method, adding new adjacent information and revising the existing samples' low-dimensional embedding results, which is termed as LISD-2.

Algorithm LISD-1

Step 1. Solve $Ly = \lambda Dy$ as standard Laplacian Eigen-decomposition for eigen-vectors with the smallest eigenvalues. The solution and the matrix L serve as the initialization v_1 and L_1.

Step 2. With definitions in 3.1 and Proposition 1, obtain E_i on the basis of $W \approx \frac{1}{k}e_i e_i^T$ and $L_i \approx \frac{1}{k}e_i e_i^T R_i R_i^T$.

Step 3. Form the matrix L by locally summing Eq. (5). After all similarity changes occur, we compute the d smallest eigenvalues of L to get the eigenvector matrix $[t_1, ..., t_d]$ corresponding to them.

Step 4. If a data point is added, Step 2 is repeatedly conducted.

Algorithm LISD-2

Step 1. The same as the step 1 of LISD-1. Then compute the initial M: $M_1 = u_1 = L_1 v_1$.

Step 2. When a new data point arrives, use Eq. (7) to update M_n. This process will save large compute complexity when n is large.

Step 3. Towards the incremental sample point, since v_n is the Eigen-solutions of L_n and can be estimated by $\frac{M_{n-1}}{||M_{n-1}||}$, the returned vector corre-sponds to the new point's low-dimensional embedding coordinate, which is provided with incremental significance. The local linear incremental method adds new adjacent information and revises the existing samples' low-dimensional embedding results.

Step 4. If a data point is added, Step 2 is repeatedly conducted.

3.4   Complex Ananlysis

The first step of both LISD-1 and LISD-2 needs to solve a $(k \times k)$ eigenvector problem, where k is the number of the nearest neighbors. The time complexity of solving the $(k \times k)$ eigenvector problem is $O(k^2)$.

Except for step 1, the core computation process of LISD-1 and LISD-2 involves some multiplications and additions of Eq. (5) and Eq. (7). The computation of Eq. (5) involves in k nearest neighbors. So after n iterative addition we can obtain the complexity of $O(nk^2)$. The Eq. (7) is fast in convergence rate and low in the computational complexity. The computational complexity saved by the similar iterative process is $(k-1)/2$ in Weng and Zhang's work [15]. So we can derive the complexity of Eq. (7) by computing $k(k+1)/2 - (k-1)/2 = (k^2+1)/2$, in which k is the number of the nearest neighbors. After n iterative additions we can find the computational complexity is $O(nk^2)$.

Except for step 1 and step 2, the computational complexity is mainly linear in both LISD-1 and LISD-2, which involves of multiplications of determinants and some easy additions. So the core computation complexity focuses on the previous 2 steps, and we can reach a conclusion that the computation complexity of our method is $O(nk^2)$. While in ILE, ILLE and ILTSA, they all need to develop a new manifold after new points come and recompute a new alignment matrix. The time complexity of solving the $(n+1) \times (n+1)$ eigenvector problem is $O((n+1)^3)$ [14].

The complexity of incremental ISOMAP algorithm(IISOMAP) is $O(n^2 \log n + n^2 q)$ for update geodesic distance in each iteration, where n is number of points and q is the maximum degree of vertices in the graph. And the update of iteration costs $O(n^2)$ time because of the matrix multiplication [10]. So we can draw the conclusion that IISOMAP will take much longer time in updating of coordinates compared with our algorithm, which has higher efficiency.

4   Experimental Evaluation

4.1   Experiment Setting

Experimental comparison is conducted with several major manifold learning algorithms on some typical nonlinear manifold databases. The four datasets are Swiss Roll, Toroidal Helix by Coifman & Lafon [18], Twin Peaks by Saul & Roweis [8], and Punctured Sphere by Saul & Roweis [8]. Each dataset has 800 sample points that will be projected from 3-dimensional space to 2-dimensional space.

We use the error metric defined in [10] to measure the accuracy of each algorithm, which is evaluated as the mean error between the resulting coordinates of our method and the compared algorithm, further normalized by the total sample variance as follows:

$$\varepsilon_n = \sqrt{\sum_{i=1}^{n} ||X_i - \overline{X}_i||^2 / n \sum_{i=1}^{n} ||X_i||^2}. \tag{9}$$

4.2   Comparison with Batch-Style Algorithms

Fig. 1 shows the results with datasets Toroidal Helix, Twin Peaks and Punctured Sphere. For Toroidal Helix, the shape of mapping results in low dimensional space of LISD-1 are smoother and rounder than LTSA, while the results of LISD-2 are roughly similar to that of LE. For Twin Peaks, mapping results of LISD-1 is similar to LE, but the result's distribution of LISD-1 is more uniform. LISD-2 and LTSA have more evenly-distributed results, but LISD-2's results are sparser, means that the mapping results of LISD-2 are easier to discriminate. For Punctured Sphere, there are some differences between the shapes of mapping results of the four algorithms, but data distributions are quite similar. Judged from the results of the three datasets, LISD-2 has better mapping results than the other algorithms on two datasets: Toroidal Helix and Twin Peaks.

Fig. 2 shows the results on manifold Swiss Roll. We can see that: 1) when nearest neighbors' number k is small (=8), LISD-1 has quite similar result with LTSA while LISD-2's result is very similar to LE's. 2) When k gets larger ($\geq$18), LTSA fails to get acceptable dimension reduction results while the other algorithms can still work effectively. 3) Considered as a whole, LISD-2 has the best result in the sense that its mapping result is more discriminative.

In summary, the results of the four test datasets reveal that the proposed algorithms perform satisfactorily, especially LISD-2, which is the best one in the four evaluated algorithms.

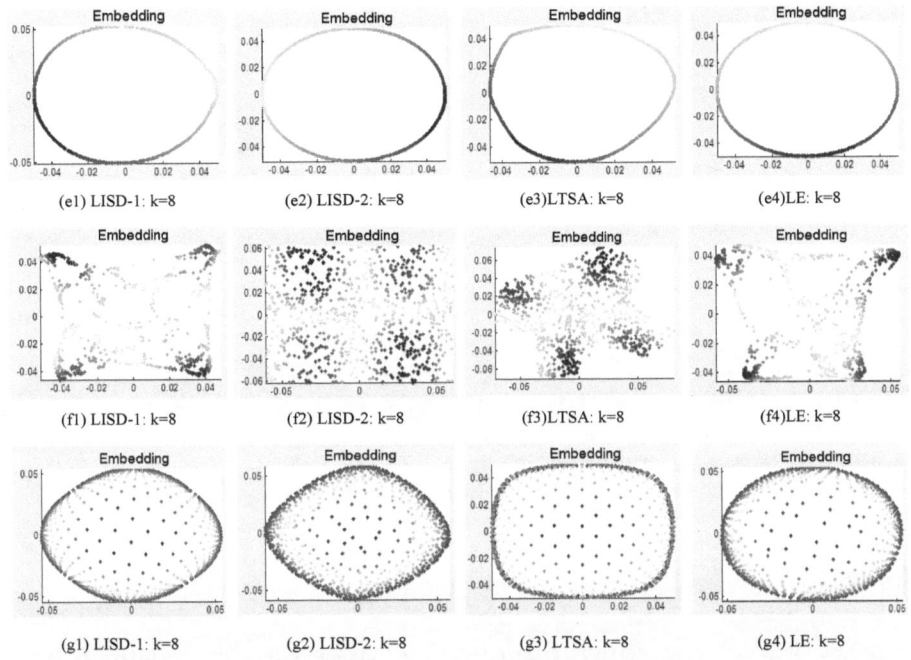

Fig. 1. Comparison of low dimensional embedding results of LISD-1, LISD-2, LTSA and LE on three manifolds: (e) Toroidal Helix, (f) Twin Peaks and (g) Punctured Sphere

4.3 Mapping Error

In this section, we compare our algorithm's mapping results in low dimensional space with LTSA, LE and incremental ISOMAP algorithm, by the error measure define in Eq. (9).

First, we take the results of LTSA and LE as references, and estimate the errors of our algorithms against LTSA and LE respectively. From Fig. 3, we can see that the error ε_n of LISD-1/LISD-2 against LE tends down gradually when n increases. And the error of LISD-2 against LTSA fluctuates around $n=800$. When n is large enough, ε_n of LISD-1 against LTSA decreases along with n slightly to disappearance. We can draw the conclusion that LISD-1 is more stable, while LISD-2 can produce good performance but is not stable enough.

Second, we take the result of LTSA as reference, and evaluate the errors of our algorithms and the incremental ISOMAP (IISOMAP) algorithm against LTSA respectively. The results are shown in Fig. 4. As more data points are projected into the low dimensional space, the errors get more stable. ε_n of our method becomes smaller and smaller, and finally close to 0; While ε_n of IIsomap against LTSA converges to 0.4. So our algorithm LISDs have smaller error than IIsomap.

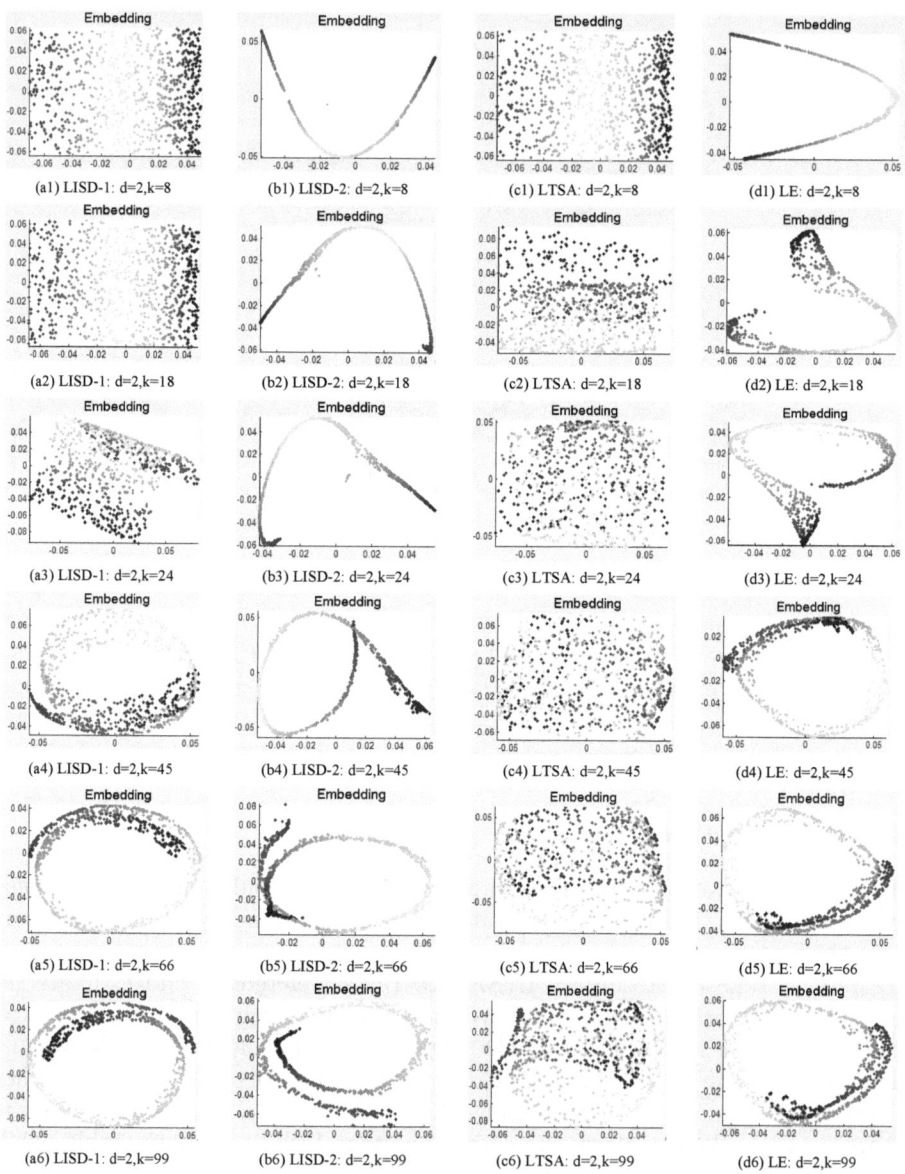

Fig. 2. Comparison of low dimensional embedding results of LISD-1, LISD-2, LTSA and LE on Swiss Roll manifold (k=8, 18, 24, 45, 66 and 99)

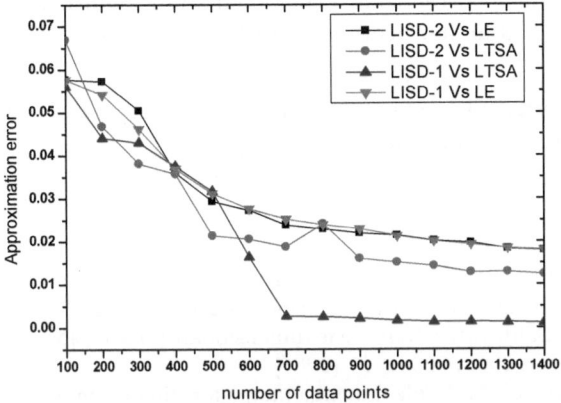

Fig. 3. Mapping error ε_n of the proposed method against LE and LTSA when the number of data points increases (Swiss Roll dataset)

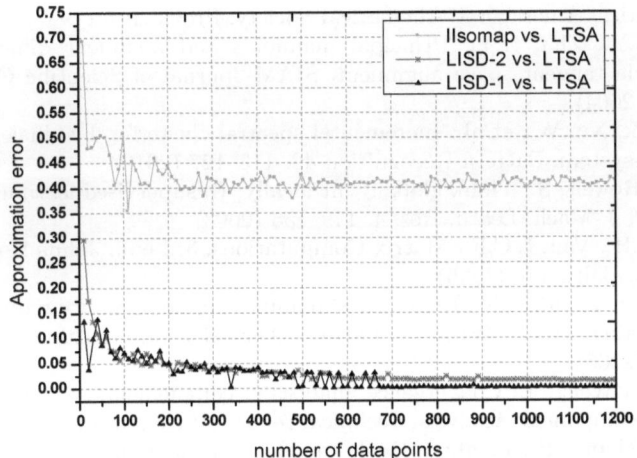

Fig. 4. Mapping error ε_n of IIsomap, LISD-1 and LISD-2 against LTSA when the number of data points increases (Swiss Roll dataset)

5   Conclusion

This paper proposes a new incremental manifold learning algorithm based on incremental spectral decomposition, which is useful to incremental learning. By updating the structure of eigen-problem incrementally, the obtained eigenvectors correspond to the low dimensional embedded coordinates. Compared with existing manifold learning dimension-reduction approaches LTSA and LE, the proposed approach LISD can obtain good or even better mapping results. When k (the number of nearest neighbors) is large, the projection results generated by LISD-2 are much better than other compared algorithms. Compared with IISOMAP, LISD has higher accuracy and lower complexity.

Acknowledgment. This work was supported "Shuguang" Scholar Program of Shanghai Education Foundation.

References

1. Tenenbaum, J.B., de Silva, V., Langford, J.C.: A global geometric framework for nonlinear dimensionality reduction. Science 290, 2319–2323 (2000)
2. Roweis, S.T., Saul, L.K.: Nonlinear dimensionality reduction by locally linear embedding. Science 290, 2323–2326 (2000)
3. Belkin, M., Niyogi, P.: Laplacian eigenmaps for dimensionality reduction and data representation. Neural Computation 15(6), 1373–1396 (2003)
4. Donoho, D.L., Grimes, C.: Hessian eigenmaps: Locally Linear embedding techniques for high-dimensional data. Proceedings of the National Academy of Sciences 100, 5591–5596 (2003)
5. Chung, F.R.K.: Spectral Graph Theory. CBMS Regional Conference Series in Mathematics. American Mathematical Society, Providence (1997)
6. Zhang, Z.Y., Zha, H.Y.: Principal manifolds and nonlinear dimensionality reduction via tangent space alignment. SIAM Journal of Scientific Computing 26, 313–338 (2004)
7. Ning, H.Z., Xu, W., et al.: Incremental spectral clustering by efficiently updating the eigen-system. Pattern Recognition 43, 113–127 (2010)
8. Saul, L., Roweis, S.: Think globally, fit locally: Unsupervised learning of nonlinear manifolds. J. Mach. Learn. Res. 4, 119–155 (2003)
9. Golub, G.H., Van, L.C.F.: Matrix Computations, 3rd edn. Johns Hopkins University Press, Baltimore (1996)
10. Law, M.H.C., Jain, A.K.: Incremental nonlinear dimensionality reduction by manifold learning. IEEE Trans on Pattern Analysis and Machine Intelligence 28, 377–391 (2006)
11. Zhang, Y., Weng, J.: Convergence analysis of complementary candid incremental principal component Analysis. Technical report, Michigan State University (2001)
12. Yan, J., Zhang, B., et al.: IMMC: Incremental maximum margin criterion. In: Proceedings of SIGKDD 2004, pp. 725–730 (2004)
13. Lu, K., He, X.: Image retrieval basedon incremental subspace learning. Pattern Recognition 38, 2047–2054 (2005)
14. Jia, P., Yin, J., et al.: Incremental Laplacian eigenmaps by preserving adjacent information between data points. Pattern Recognition Letters 30, 1457–1463 (2009)
15. Weng, J.Y., Zhang, Y.L., Hwang, W.S.: Candid Covariance-free Incremental Principal Component Analysis. IEEE Trans on Pattern Analysis and Machine Intelligence 25(8), 1034–1040 (2003)
16. Kouropteva, O., Okun, O., et al.: Incremental locally linear embedding. Pattern Recognition 38, 1764–1767 (2005)
17. Abdel-Mannan, O., Ben Hamza, A., et al.: Incremental Line Tangent Space Alignment Algorithm. In: Proceedings of CCECE 2007, pp. 1329–1332 (2007)
18. Coifman, R., Lafon, S., et al.: Geometric diffusions as a tool for harmonic analysis and structure definition of data: Diffusion maps. PNAS 102(21), 7426–7431 (2005)

Sparse Boosting with Correlation Based Penalty

Junlong Zhao

School of Mathematics and System Science, Beihang University,
Beijing,100191, China
zjlczh@126.com

Abstract. In high dimensional setting, componentwise L_2boosting method has been used to construct sparse model of high prediction, but it tends to select many ineffective variables. Several sparse boosting methods, such as, SparseL_2Boosting and Twin Boosting, have been proposed to improve the variable selection of L_2boosting algorithm. In this paper, we propose a new general sparse boosting method (GSBoosting). The relations are established between GSBoosting and other well known regularized variable selection methods in orthogonal linear model, such as adaptive Lasso, hard thresholds etc. Simulations results show that GSBoosting has good performance in both prediction and variable selection.

Keywords: boosting algorithm, sparsity, model selection, adaptive Lasso.

1 Introduction

The goal of the statistical learning is to construct the model of high prediction and to select the effective variables, particularly, in the case of high dimensional data. Boosting is an effective method for modeling. The main idea of boosting is to minimize the empirical risk and pursuing iterative steepest descent in function space ([6]). This very general and useful view of boosting has been considered by Breiman ([1],[2]) and developed further by Friedman et al. ([9]) and Friedman ([10]).

In high dimensional setting, Bühlmann and Yu ([7]) proposed componentwise L_2Boosting, which is a variant of L_2Boosting with only one covariate selected to construct the fitting function using specified base learner in each iteration. If the early stopping strategy is adopted to select the number of iteration, componentwise L_2Boosting can be used to select variables and to construct sparse model. However, besides effective variables, componentwise L_2Boosting usually selects many ineffective variables. Several sparse boosting methods have been proposed to improve the feature selection of L_2Boosting, such as, sparseL_2Boosting, Twin Boosting etc. Bühlmann and Yu ([8]) proposed sparseL_2Boosting and proved that it is equivalent to nonnegative garrote for orthogonal linear model. Bühlmann and Hothorn ([5]) proposed Twin Boosting and proved that Twin boosting with componentwise linear base learner can be equivalent to adaptive Lasso in orthogonal linear model (See the proposition 1 of [5]). The relations of Lasso type methods with sparseL_2Boosting and Twin Boosting are interesting.

S. Zhou, S. Zhang, and G. Karypis (Eds.): ADMA 2012, LNAI 7713, pp. 161–172, 2012.
© Springer-Verlag Berlin Heidelberg 2012

In this paper, we propose a new family of sparse boosting algorithm, called general sparse boosting(GSBoosting), which can build sparse model of high prediction. We show that in orthogonal linear model, GSBoosting with different parameters can be equivalent to Lasso, Adaptive lasso, ridge regression, hard threshold, OLS etc. Simulation results in different situation confirm the effectiveness of GSBoosting.

The main contents of this paper are arranged as follows. Several boosting algorithms are briefly reviewed in section 2. General sparse boosting algorithm (GSBoosting) is proposed in section 3. In section 4, the relations are established between GSBoosting and the well-known methods in orthogonal model. Section 5 presents the simulation results on the comparison of GSBoosting with other methods. And conclusions are presented in section 6.

2   Review of Generic Boosting and Sparse L_2Boosting

Boosting can be viewed as a minimization of empirical loss function by gradient descending in function space ([9],[10]). Suppose that $X = (x_1, \cdots, x_p)^T$ is the regressor and Y is responsor. $\{(X_i, Y_i)\}_{i=1}^n$ are the independent sample. Let $\rho(\cdot, \cdot)$ denote the loss function. Consider the problem of estimating a real function $f^*(\cdot) = \arg\min_{f(\cdot)} E(\rho(Y, f(X)))$. Estimation of $f^*(\cdot)$ with boosting can be done by considering the empirical risk $n^{-1} \sum_{i=1}^n \rho(Y_i, f(X_i))$ and pursuing iterative steepest descent in function space. The following generic boosting algorithm is from Bühlmann and Hothorn (2007).

1. Initialize $\hat{f}^{[0]}$. Let $\hat{f}^{[0]} \equiv \bar{Y}$. Set $m = 0$.
2. Increase m by 1. Compute negative gradient $-\frac{\partial}{\partial f}\rho(Y, f)$ and evaluate at $\hat{f}^{[m-1]}(X_i)$

$$U_i = -\frac{\partial}{\partial f}\rho(Y, f)|_{f=\hat{f}^{[m-1]}(X_i)}, \quad i = 1, 2, \cdots, n.$$

3. Fit negative gradient vector $(U_1, U_2, \cdots, U_n)^T$ by the weak learner

$$(X_i, U_i)_{i=1}^n \xrightarrow{base\ procedure} \hat{g}^{[m]}(\cdot).$$

4. Update $\hat{f}^{[m]} = \hat{f}^{[m-1]} + v\hat{g}^{[m]}$, where $0 < v \leq 1$ is a step length.
5. Iterate steps 2–4 until $m = m_{stop}$ for some stopping iteration m_{stop}.

The choice of the step length in step 4 is of minor importance. In literature, v generally takes value being sufficiently small, e.g. $v = 0.1$ ([10]). Square loss function is a special case of generic boosting algorithm with $\rho(Y, f) = \frac{1}{2}|Y - f|^2$ and $U_i = Y_i - \hat{f}^{[k-1]}(X_i)$ $(i = 1, \cdots, n)$ in step 2.

Bühlmann and Yu ([8]) proposed the sparseL_2Boosting algorithm, which leads to sparser model than componentwise L_2Boosting. They also showed that, for orthogonal linear model, componentwise L_2Boosting and sparseL_2Boosting can be equivalent to Lasso (soft threshold) ([13]) and nonnegative garrote (briefly NNG), respectively. Bühlmann and Hothorn([5]) proposed Twin boosting and proved that it is equivalent to adaptive Lasso in orthogonal linear model.

3   General Sparse Boosting

Let $\{(X_i, Y_i)\}_{i=1}^n$ be independent copy of (X, Y), where $X = (x_1, \cdots, x_p)^T$ and $Y \in R$. Consider the model

$$\mathbf{Y} = \mathbf{X}\beta + \epsilon, \tag{1}$$

where $\epsilon \sim N(0, I_n\sigma^2)$, $\mathbf{X} = (X_1, \cdots, X_n)^T = (\mathbf{x}_1, \cdots, \mathbf{x}_p)$, $\mathbf{Y} = (Y_1, \cdots, Y_n)^T$. Without loss of generality, suppose that $\mathbf{1}_p^T\mathbf{x}_j = 0, \mathbf{x}_j^T\mathbf{x}_j = 1, (j = 1, \cdots, p)$, where $\mathbf{1}_p = (1, 1, \cdots, 1)^T$.

Frank and Friedman([11]) proposed the bridge regression family, by considering the following problem

$$\arg\min_\beta\{||\mathbf{Y} - \mathbf{X}\beta||^2 + \lambda_n \cdot P_{\gamma,\mathrm{w}}(\beta)\}, \tag{2}$$

where $P_{\gamma,\mathrm{w}}(\beta) = \sum_{j=1}^p \mathrm{w}_j|\beta_j|^\gamma$. Motivated by the penalty function in bridge regression, we propose the following generalized penalty criteria to develop sparse boosting algorithm.

3.1   Generalized Sparse L_2Boosting (GSBoosting) in Linear Model

Given variable x_j $(j = 1, \cdots, p)$, let $\mathcal{H}_j$ denote the corresponding projection operator of given base learner ([7]). For example, $\mathcal{H}_j = \mathbf{x}_j\mathbf{x}_j^T/\mathbf{x}_j^T\mathbf{x}_j$ for linear base learner. Let $\hat{S}_m, m = 1, 2\cdots$, denote the index of the variable selected in m-th iteration. Assume that x_j is selected in mth iteration, and define corresponding projection as

$$\mathcal{B}_{m,j} = I_n - (I_n - v\mathcal{H}_j)(I_n - v\mathcal{H}_{\hat{S}_{m-1}})\cdots(I_n - v\mathcal{H}_{\hat{S}_1}), \quad 1 \le j \le p \tag{3}$$

where $v > 0$ is the step length. Then the residual sum of square can be denoted as ([8])

$$Rss(m, j) = || (I_n - \mathcal{B}_{m,j})Y ||^2.$$

Let $\hat{\beta}_m^{(j)}$ be the estimated coefficient with x_j been selected in m-th iteration, where $1 \le j \le p; m = 1, 2, \cdots$. Denoting by $\bar{\beta}_{m-1}$ the estimate of β after $m-1$ iteration, then we have, $\bar{\beta}_{m-1} = \hat{\beta}_{m-1}^{\hat{S}_{m-1}}$. Furthermore, denote the residual after $m-1$ iterations as

$$\mathbf{U}_{m-1} = (I_n - \mathcal{B}_{m-1,\hat{S}_{m-1}})\mathbf{Y} = \mathbf{Y} - \mathbf{X}\bar{\beta}_{m-1}.$$

where $\mathcal{B}_{m-1,\hat{S}_{m-1}}$ is defined as in (3). If x_j is selected in m-th iteration, then, by [7], the corresponding prediction of $\mathbf{Y}$ is $\widehat{\mathbf{Y}}_{m,j} = \mathcal{B}_{m,j}\mathbf{Y} = \mathbf{X}\hat{\beta}_m^{(j)}$. By the above notations, it is easy to see that $\hat{\beta}_m^{(j)} = \bar{\beta}_{m-1} + v\delta_{m,j}$, with $\delta_{m,j} = (0, \cdots, 0,$ $\mathbf{x}_j^T \mathbf{U}_{m-1}/\mathbf{x}_j^T \mathbf{x}_j , 0, \cdots, 0)^T$ and v being the step length. And the corresponding residual can be denoted as

$$\mathbf{U}_m^{(j)} = (I_n - \mathcal{B}_{m,j})\mathbf{Y} = \mathbf{Y} - \mathbf{X}\hat{\beta}_m^{(j)}.$$

Denote the sum of square residual as $Rss(m,j) = \| \mathbf{U}_m^{(j)} \|^2$. The main idea of componentwise linear L_2Boosting is to select into the model the variable that reduces the residual most, that is,

$$\begin{aligned} \hat{S}_m &= \arg\min_{1 \leq j \leq p} \|(I_n - \mathcal{B}_{m,j})\mathbf{Y}\|^2 \\ &= \arg\min_{1 \leq j \leq p} \|(I_n - v\mathcal{H}_j)\mathbf{U}_{m-1}\|^2 \\ &= \arg\min_{1 \leq j \leq p} \|(I_n - v\mathbf{x}_j\mathbf{x}_j^T/\mathbf{x}_j^T\mathbf{x}_j)\mathbf{U}_{m-1}\|^2, \end{aligned}$$

Due to the fact $\mathbf{x}_j^T\mathbf{x}_j = 1$, it holds that $\hat{S}_m = \arg\max_{1 \leq j \leq p} |\mathbf{x}_j^T\mathbf{U}_{m-1}|^2$. Let $U_m^{(j)}$ denote the population version of the $\mathbf{U}_m^{(j)}$. Let $\theta_m^{(j)} = (\theta_{m,1}^{(j)}, \cdots, \theta_{m,p}^{(j)})^T = \mathrm{cov}(X, U_m^{(j)})$, the correlation between X and the pseudo-responsor $U_m^{(j)}$, and let $\theta_0 = (\theta_{0,1}, \cdots, \theta_{0,p})^T = \mathrm{cov}(X,Y)$, the correlation between the X and Y. It should be note that θ_0 generally differs from true parameter β in (1), where we do not assume $\mathrm{cov}(X) = I_p$. The corresponding sample estimators respectively are denoted as $\hat{\theta}_m^{(j)} = \mathbf{X}^T\mathbf{U}_m^{(j)}$ and $\hat{\theta}_0 = \mathbf{X}^T\mathbf{Y}$. Recall that OLS estimator $\hat{\beta}_{OLS} = (\mathbf{X}^T\mathbf{X})^{-1}\mathbf{X}\mathbf{Y}$ is the minimizer of the problem $\arg\min_\beta \|\mathbf{Y} - \mathbf{X}\beta\|$. Smaller value of $Rss(m,j)$ means the better fitting, or equivalently $\hat{\beta}_m^{(j)}$ being closer to $\hat{\beta}_{OLS}$. On the other hand, to improve the variable selection of L_2Boosting, some penalty should be used. We use $\|\mathbf{X}^T\mathbf{U}_m^{(j)}\|^2$ as the penalty. Different from $\mathbf{U}_m^{(j)}$, $\mathbf{X}^T\mathbf{U}_m^{(j)}$ does contains information on the complexity of the model. In fact, it can be viewed as some distance between $\hat{\beta}_m^{(j)}$ and OLS/ridge estimator.

Assumed that $\mathbf{X}^T\mathbf{X} = G\Lambda G^T$, where G is the orthogonal matrix whose columns consist of eigenvectors and $\Lambda = \mathrm{diag}(\lambda_1, \cdots, \lambda_p)$ consisting of the eigenvalues. If $(\mathbf{X}^T\mathbf{X})^{-1}$ exists, then due to $\mathbf{U}_m^{(j)} = \mathbf{Y} - \mathbf{X}\hat{\beta}_m^{(j)}$ and $\mathbf{X}^T\mathbf{Y} = \mathbf{X}^T\mathbf{X}\hat{\beta}_{OLS}$, we have

$$\|\mathbf{X}^T\mathbf{U}_m^{(j)}\|^2 = \|(\mathbf{X}^T\mathbf{X})(\hat{\beta}_{OLS} - \hat{\beta}_m^{(j)})\|^2 = \|\Lambda G^T(\hat{\beta}_{OLS} - \hat{\beta}_m^{(j)})\|^2. \quad (4)$$

As $\lambda_j \equiv c > 0, 1 \leq j \leq p$, we have $\|\mathbf{X}^T\mathbf{U}_m^{(j)}\|^2 = c \cdot \|\hat{\beta}_{OLS} - \hat{\beta}_m^{(j)}\|^2$. In general setting, $\|\mathbf{X}^T\mathbf{U}_m^{(j)}\|^2$ can be viewed as a weighted distance between $\hat{\beta}_m^{(j)}$ and $\hat{\beta}_{OLS}$. If the $\mathbf{X}^T\mathbf{X}$ is ill conditioned, then $\mathbf{X}^T\mathbf{Y} = (\mathbf{X}^T\mathbf{X} + \lambda I_p)\hat{\beta}_{ridge}$, where $\lambda > 0$ and $\hat{\beta}_{ridge}$ denotes the ridge estimator. We have $\|\mathbf{X}^T\mathbf{U}_m^{(j)}\|^2 = \|(\Lambda + \lambda I_p)G^T\hat{\beta}_{ridge} - \Lambda G^T\hat{\beta}_m^{(j)}\|^2$, which can be viewed as some measurement on the difference between $\hat{\beta}_{ridge}$ and $\hat{\beta}_m^{(j)}$. In one word, $\|\mathbf{X}^T\mathbf{U}_m^{(j)}\|^2$ can be regarded as some distance between $\hat{\beta}_m^{(j)}$ and OLS/ridge estimator.

Note that OLS and ridge estimator are non-sparse solution, fitting a model of the high complexity. Smaller value of $||\mathbf{X}^T \mathbf{U}_m^{(j)}||^2$ means estimate being closer to OLS/ridge estimator. In order to get the sparse solution, $\hat{\beta}_m^{(j)}$ should not close to OLS/ridge estimator too much; hence larger value of $||\mathbf{X}^T \mathbf{U}_m^{(j)}||^2$ is preferred. Thus we construct the penalty as $P(m, j) = ||\mathbf{X}^T \mathbf{U}_m^{(j)}||^2 = \sum_{i=1}^{p} |\hat{\theta}_{m,i}^{(j)}|^2$. Furthermore, similar to adaptive Lasso ([14]), we consider further the penalty of the form $|\hat{\theta}_{m,i}^{(j)}|^\gamma / |\hat{\theta}_{0,i}^{(j)}|^\omega$. Then the penalty function can be extended as follows.

$$P_{\gamma,\omega}(m, j) = \sum_{i=1}^{p} \frac{|\hat{\theta}_{m,i}^{(j)}|^\gamma}{|\hat{\theta}_{0,i}|^\omega} = \sum_{i=1}^{p} \frac{|\mathbf{X}_i^T \mathbf{U}_m^{(j)}|^\gamma}{|\mathbf{X}_i^T \mathbf{Y}|^\omega},$$

where $\gamma > 0$ and $\omega \in R$. From the above argument, we want $Rss(m, j)$ being small and $P_{\gamma,\omega}(m, j)$ being large simultaneously. This motives us to consider minimizing the following penalized empirical risk

$$T^j(m, \lambda_n, \gamma, \omega) = Rss(m, j) - \lambda_n P_{\gamma,\omega}(m, j), \quad \lambda_n > 0.$$

The best variable selected in m-th iteration is denoted by $\hat{S}_m = \arg\min_{1 \le j \le p} T^j(m, \lambda_n, \gamma, \omega)$. For simplicity, let

$$\begin{aligned}
Rss(m) &= Rss(m, \hat{S}_m), \\
P_{\gamma,\omega}(m) &= P_{\gamma,\omega}(m, \hat{S}_m), \\
T(m, \lambda_n, \gamma, \omega) &= T^{\hat{S}_m}(m, \lambda_n, \gamma, \omega).
\end{aligned} \tag{5}$$

As m increases, $Rss(m)$ decreases and the complexity of the model increases; also the distance between estimator $\hat{\beta}_m^{(\hat{S}_m)}$ and OLS estimator decreases and consequently $-P_{\gamma,\omega}(m)$ increases. Therefore the stopping number can be taken as $\hat{m} = \arg\min_{1 \le m \le N} T(m, \lambda_n, \gamma, \omega)$, for some large integer N. We summarize the GSBoosting algorithm as follows.

1. Initialization. Let $\hat{F}_0(\cdot) \equiv 0$, $m = 0$ and $U_0 = Y$.
2. Increase m from 1. The pseudo-responsor are $U_{m,i} = Y_i - \hat{F}_{m-1}(X_i), i = 1, \cdots, n$. Let $\mathbf{U}_m = (U_{m,1}, \cdots, U_{m,n})^T$.
3. Minimize $T^j(m, \lambda_n, \gamma, \omega)$ to select the best variable, that is,

$$\hat{S}_m = \arg\min_{1 \le j \le p} T^j(m, \lambda_n, \gamma, \omega).$$

Fit the pseudo-responsor with linear base learner and construct the fitted function $\hat{f}_m(\cdot) = \mathbf{x}_{\hat{S}_m} \mathbf{x}_{\hat{S}_m}^T \mathbf{U}_m / \mathbf{x}_{\hat{S}_m}^T \mathbf{x}_{\hat{S}_m}$;
4. Let $\hat{F}_m(\cdot) = \hat{F}_{m-1}(\cdot) + v\hat{f}_m(\cdot), 0 < v \le 1$;
5. Repeat step 2–4 until some large iteration number N. The optimal iteration number can be taken as $\hat{m} = \arg\min_{1 \le m \le N} T(m, \lambda_n, \gamma, \omega)$. The final estimate is $\hat{F}_{\hat{m}}(\cdot)$.

3.2   Data-Driving GSBoosting and Parameter Selection

To avoid the selection of the tuning parameters λ_n, in this paper, we adopt the AICc criteria, the correction of AIC ([12]). Let

$$C_{AICc}(Rss, k) = \log(Rss) + \frac{1 + k/n}{1 - (k + 2)/n}. \tag{6}$$

Then the AICc based penalized loss function is defined as

$$T_{AICc}(m, \gamma, \omega) = \min_{j}\{C_{AICc}(Rss(m, j), -P_{\gamma,\omega}(m, j))\} := \min_{1 \leq j \leq p} T^j_{AICc}(m, \gamma, \omega). \tag{7}$$

Under the penalized loss function (7), the corresponding GSBoosting algorithm can be denoted as AICc-GSBoosting, a data driving method. AICc score can be used to select the variable and the optimal iteration number. AICc-GSBoosting Algorithm is as follows.

Steps 1, 2, and 4 are the same as GSBoosting algorithm. Just replace steps 3 and 5 there by the following steps, respectively.

3'. The index of the selected variable is $\hat{S}_m = \arg\min_{1 \leq j \leq p} T^j_{AICc}(m, \gamma, \omega)$. Fit the pseudo-responsor $\mathbf{U}_m$ by $\mathbf{x}_{\hat{S}_m}$ with specified base learner to obtain the estimate function $\hat{f}_m(\cdot)$.

5' The optimal stopping number is $\hat{m} = arg \min_{1 \leq m \leq N} T_{AICc}(m, \gamma, \omega)$

Two methods can be used to select the parameters. The first one is to apply multiple K-fold CV or GCV. The second one is Monte Carlo method. Separate the data at random into training data set and the testing data set and compute the prediction error on the testing data. Repeat the procedure for L times and compute the average prediction error on the testing data. The parameters which result in the smallest average prediction error can be selected as the optimal parameters.

3.3   Further Extension to General Base Learner

For nonlinear model, nonlinear base learner should be used, e.g. componentwise spline function and componentwise decision tree etc. and the penalty function should be revised. In the case of componentwise nonlinear function as base learner, let $f^{(U)}_{i,j}(\mathbf{x}_i)$ denote the function of fitting $\mathbf{U}^{(j)}_m$ by $\mathbf{x}_i$, where $f^{(U)}_{i,j}(\cdot)$ having the same complexity for $i = 1, \cdots, p$. Then, $\hat{\theta}^{(j)}_{m,i}$ can be defined as $\hat{\theta}^{(j)}_{m,i} = (f^{(U)}_{i,j}(\mathbf{x}_i))^T \mathbf{U}^{(j)}_m / \| f^{(U)}_{i,j}(\mathbf{x}_i) \|$, the correlation between the fitted value $f^{(U)}_{i,j}(\mathbf{x}_i)/ \| f^{(U)}_{i,j}(\mathbf{x}_i) \|$ and the residuals $\mathbf{U}^{(j)}_m$. And $\hat{\theta}_{0,i}$ is defined in the same way. Particularly, for spline base learner, we propose the following penalty function

$$P_{\gamma,\omega}(m, j) = \sum_{i=1}^{p} \frac{\left| \hat{g}^{(U)}_{d.f.}(\mathbf{x}_i)^T \mathbf{U}^{(j)}_m /\|\hat{g}^{(U)}_{d.f.}(\mathbf{x}_i)\| \right|^{\hat{\gamma}}}{\left| \hat{g}^{(Y)}_{d.f.}(\mathbf{x}_i)^T \mathbf{Y}/\|\hat{g}^{(Y)}_{d.f.}(\mathbf{x}_i)\| \right|^{\omega}},$$

where $\hat{g}_{d.f.}^{(U)}(\mathbf{x}_i)$ denotes the function of fitting $\mathbf{U}_m^{(j)}$ by $\mathbf{x}_i$ with spline function of fixed degree of freedom and $\hat{g}_{d.f.}^{(Y)}(\mathbf{x}_i)$ is defined similarly. Based on this extension of penalty function, AICc-GSBoosting algorithms can readily be used to build nonlinear model.

4   Relations between GSBoosting and Other Variable Selection Methods

In this section, we discuss the relations between GSBoosting method with other well known methods in the orthogonal linear model.

$$\mathbf{Y} = \mathbf{X}\beta + \epsilon, \quad \mathbf{X}^T\mathbf{X} = \mathbf{X}\mathbf{X}^T = I_n \tag{8}$$

where $\mathbf{X} = (\mathbf{x}_1, \cdots, \mathbf{x}_n)$, $\beta = (\beta_1, \cdots, \beta_n)^T$, $\mathbf{Y} = (Y_1, \cdots, Y_n)^T$ and $\epsilon = (\epsilon_1, \cdots, \epsilon_n)^T$.

For GSBoosting algorithm, we take the square loss function and the componentwise linear function as base learner. Now, we show that GSBoosting with different parameters can be equivalent to soft-threshold, hard-threshold, nonnegative garrote etc. The proofs of the following theorems are deferred to Appendix.

Let $\hat{\beta}_{GSB}^{(\hat{m})}$ denote the coefficient obtained by GSBoosting after $\hat{m}$ iterations. Denote by $\hat{\beta}_{Alasso}$, $\hat{\beta}_{hard}$, $\hat{\beta}_{ridge}$ and $\hat{\beta}_{OLS}$, the estimator of adaptive Lasso, hard-threshold, ridge regression and OLS method, respectively. The following results hold.

Theorem 1. *For model (8), let $0 < \gamma < 2$, $\omega \geq 0$ and $N < \infty$ in GSBoosting algorithm. Then, as $v \to 0$, we have $\hat{\beta}_{GSB,j}^{(\hat{m})} = \hat{\beta}_{Alasso,j}$, $j = 1, \cdots, n$.*

The proof of the following theorem 2 is quite similar to that of theorem 1, we omit it here for conciseness.

Theorem 2. *For model (8), given $\{\lambda_n > 0, n = 1, 2, \cdots\}$ where $\lambda_n < 1$ as n being large, let $v \longrightarrow 0$, the following results hold.*

(1) As $\gamma = 2$, $\omega > 0$ and $N = \infty$, we have $\hat{\beta}_{GSB}^{(\hat{m})} = \hat{\beta}_{hard}$.
(2) As $\gamma = 2$, $\omega = 0$ and $N = \infty$, we have $\hat{\beta}_{GSB}^{(\hat{m})} = \hat{\beta}_{OLS}$.
(3) As $\gamma > 2$, $\omega = \gamma - 2$ and $N < \infty$ we have $\hat{\beta}_{GSB}^{(\hat{m})} = \hat{\beta}_{ridge}$.

5   Simulation Results

In this section we compare the AICc-GSBoosting(denoted briefly as GSBoosting), L_2Boosting, SparseL_2Boosting, Twin Boosting and adaptive Lasso in their prediction and variable selection performance for linear model. The stopping number of iteration of L_2Boosting and sparseL_2Boosting are based on gMDL criteria([8]). Twin Boosting method contains two parameters, the first round iteration number m_1 and the second round iteration number m_2. We take $m_1 = 100$

and select m_2 by gMDL criteria. The step length is taken as $v = 0.1$ for all boosting methods, as usually used in literature. And for adaptive Lasso, we take the penalty function in (2) with $\gamma = 1$ and the weight function as

$$\hat{w} = 1/|\hat{\beta}_0|^{\tau} \tag{9}$$

and select the optimal τ from the set $\{0, 0.5, 1, 1.5, 2\}$ by 5-fold cross validation. As $\tau = 0$, adaptive Lasso estimator equals Lasso estimator. $\hat{\beta}_0$ is taken as either OLS estimator or ridge regression estimator and will be specified later. In application of AICc-GSBoosting method, we use Monte Carlo method in section 3.2 to select parameters γ and ω, the optimal parameter denoted as γ_{opt} and ω_{opt}.

5.1   Simulation Results

We introduce the following notations. rate_{10} denotes the ratio of the number of falsely missed effective variables over the number p_1 of the true effective variables. rate_{01} denotes the ratio of the number of falsely selected non-effective variables over the number $p - p_1$ of the non-effective variables. Repeat the simulation for L times, and let R_{10} and R_{01} denote respectively the means of rate_{10} and rate_{01} in L replications. Let Mpr and STD denote respectively the means and standard error of the prediction error in L replications. And for simplicity, in the following tables, sparseL_2Boosting, L_2Boosting, Twin Boosting, adaptive Lasso and GSBoosting are denoted briefly by spB, L_2B, TwB, Alasso, GSB, respectively.

Model 1. Consider the sparse model

$$Y_i = \beta^T X_i + \epsilon_i, \quad i = 1, \cdots, n$$

where $X_i \sim N(\mathbf{0}, \mathbf{\Sigma})$, $\epsilon_i \sim N(0, 1)$ being i.i.d. samples and $\mathbf{\Sigma} = (\sigma_{i,j})_{p \times p}$ with $\sigma_{i,j} = \rho^{|i-j|}$. $\beta = (5, 5, 5, 5, 0, \cdots, 0)^T$ with the number $p_1 = 4$ of the true effective parameters. Take $\rho = 0, 0.6$ and $p = 50, 100$.

Take sample size $n = 100$ and select at random 80% observations as training set and rest ones as testing set (that is, the size of training set is $n_1 = 80$). Repeat the simulation with $L = 50$ times. For GSBoosting method, we select the parameter γ from the set $\{0.1, 0.5 : 0.5 : 4\}$ and ω from the set $\{-4 : 1 : 4\}$, where $0.5 : 0.5 : 4$ means $[0.5, 1.0, \cdots, 3.5, 4.0]$, the the grid with width 0.5 and $\{-4 : 1 : 4\}$ is defined similarly. For adaptive Lasso, we take $\hat{\beta}_0$ in (9) being OLS estimator in the case of $p = 50$. As $p = 100$, since matrix $\mathbf{X}^T\mathbf{X}$ may be ill conditioned, we take $\hat{\beta}_0$ being ridge regression estimator.

Table 1 displays the results of sparseL_2Boosting, L_2Boosting Twin Boosting, adaptive Lasso and GSBoosting with the optimal parameter $\gamma_{opt}, \omega_{opt}$. As $p = 50, \rho = 0$, $(\gamma_{opt}, \omega_{opt}) = (4, -2)$; as $p = 50, \rho = 0.6$, $(\gamma_{opt}, \omega_{opt}) = (2, -2)$; as $p = 100, \rho = 0$, $(\gamma_{opt}, \omega_{opt}) = (4, -2)$; and as $p = 100, \rho = 0.6$, $(\gamma_{opt}, \omega_{opt}) = (1, -1)$. From Table 1, it follows that GSBoosting is better than those of sparseL_2Boosting, L_2Boosting and Twin Boosting in prediction. The

R_{01} of GSBoosting is better than those of other three boosting methods, that is, GSBoosting tends to select sparser model, especially for $\rho = 0.6$.

Table 1 also shows that GSBoosting is slightly better prediction than adaptive Lasso, especially for $p = 100$. And for $\rho = 0$, the variable selection of GSBoosting is similar to or slightly better than that of adaptive Lasso. For $\rho = 0.6$, adaptive Lasso performs best on variable selection among all methods; GSBoosting performs slightly worse than adaptive Lasso but much better than other methods.

Table 1. Comparasion of GSBoosting with other methods on model 1

p	method	$\rho = 0$				$\rho = 0.6$			
		Mpr	STD	R_{01}	R_{10}	Mpr	STD	R_{01}	R_{10}
	spB	1.401	0.517	0.301	0	1.237	0.528	0.233	0
	L_2B	1.373	0.453	0.550	0	1.211	0.493	0.505	0
50	TwB	1.312	0.454	0.206	0	1.224	0.542	0.175	0
	Alasso	1.257	0.437	0.209	0	1.259	0.422	0.022	0
	GSB	1.248	0.427	0.116	0.012	1.045	0.329	0.072	0
	spB	1.315	0.382	0.236	0	1.421	0.427	0.214	0
	L_2B	1.261	0.383	0.376	0	1.434	0.365	0.353	0
100	TwB	1.229	0.388	0.127	0	1.348	0.363	0.114	0.013
	Alasso	1.263	0.405	0.139	0	1.324	0.456	0.003	0
	GSB	1.099	0.372	0.118	0.013	1.210	0.313	0.045	0

5.2   Real Data Analysis: Ozone Data

Ozone data ([3]) contains $n = 330$ observations and 8 covariates: Height, Humidity, InversionHt, Pressure, Temp2, Temperature, Visibility, WindSpeed. We select at random 80% observations as training data and the rest as testing data. Repeat the simulation with 50 times. We consider the following case.

We add additional 20 independent variable $X_9, \cdots, X_{28}, i.i.d \sim N(0,1)$ to the original ozone data, where the original 8 variables are denoted by $X_1 \cdots, X_8$. Therefore regressor X is of dimension $p = 28$. Take the componentwise smoothing spline function as the base learner for all boosting methods. Adaptive Lasso is not applicable for this nonlinear setting. For GSBoosting method, parameter γ is selected from the set $[0.1, 1:1:4]$ and ω from the set $[-3:1:3]$.

Simulation results are presented in Table 2, where N_1 denotes the average number of variables selected from $X_9, \cdots, X_{28}$, and N_2 denotes the average number of the variables selected from $X_1, \cdots, X_8$. For GSBoosting method, the optimal parameters are $\gamma_{opt} = 2, \omega_{opt} = 0$.

From Table 2, it follows that GSBoosting is slightly better than other methods in prediction performance. Compared with L_2Boosting, values of N_1 of Twin Boosting and GSBoosting are small, which means that both methods select less ineffective variables. The N_2's of Twin Boosting and GSBoosting are

Table 2. Simulation results on Ozone data

method	Mpr	STD	N_1	N_2
spB	25.253	4.610	4.3	6.2
L_2B	23.022	4.819	11.7	6.7
TwB	23.222	4.487	2.4	4.4
GSB	22.524	4.953	2.3	5.3

smaller than those of sparseL_2Boosting and L_2Boosting, which means that the model constructed by both methods are sparser than those of L_2Boosting and sparseL_2Boostng.

6   Conclusion

In this paper, we propose a new sparse boosting algorithm, named GSBoosting, based on correlation penalty. We show in theory that in orthogonal linear model, GSBoosting with different parameter is equivalent to many existing variable selection method, such as, adaptive LASSO, hard threshold etc. Theorefore GSBoosting is very flexible as tuning parameters vary. GSBoosting can be easily extend to general base learner, such as, spline function etc. To simplify the computing, AICc criteria is used to select the tuning parameter in GSBoosting algorithm. Simulations results confirm the advantages of GSBoosting.

Appendix A: Proof of Theorem 1

Since $\| \mathbf{x}_j \| = 1$, the projection operator on $\mathbf{x}_j = (x_{1j}, \cdots, x_{nj})^T$ is $\mathcal{H}_j = \mathbf{x}_j \mathbf{x}_j^T$. Then the operator of L_2Boosting with the m iteration is

$$\mathcal{B}_m = I_n - (I_n - v\mathcal{H}_{\hat{S}_m}) \cdots (I_n - v\mathcal{H}_{\hat{S}_1}),$$

Due to the $\mathbf{x}_j^T \mathbf{x}_k = 0$, we have $\mathcal{H}_j \mathcal{H}_k = \mathcal{H}_k \mathcal{H}_j$. Therefore, we have

$$\mathcal{B}_m = I_n - (I_n - v\mathcal{H}_1)^{m_1} (I_n - v\mathcal{H}_2)^{m_2} \cdots (I_n - v\mathcal{H}_n)^{m_n},$$

where m_j is the total number that the variable x_j has been selected in m iterations. Therefore, we have $m = \sum_{j=1}^{n} m_j$. Moreover, $\mathcal{B}_m$ can be rewritten as $\mathcal{B}_m = \mathbf{X} D_m \mathbf{X}^T$, where $\mathbf{X}^T \mathbf{X} = \mathbf{X}\mathbf{X}^T = I_n$, $D_m = \text{diag}(d_{m,1}, \cdots, d_{m,n})$, $d_{m,j} = 1 - (1 - v)^{m_j}$. Therefore the residuals sum of square after m iterations is

$$Rss(m) = \|\mathbf{Y} - \mathcal{B}_m \mathbf{Y}\|^2 = \|\mathbf{X}^T \mathbf{Y} - \mathbf{X}^T \mathcal{B}_m \mathbf{Y}\|^2 = \|\mathbf{Z} - D_m \mathbf{Z}\|^2 = \|(I - D_m)Z\|^2,$$

where $\mathbf{Z} = \mathbf{X}^T \mathbf{Y} = (Z_1, \cdots, Z_p)^T$. Note that $\hat{\theta}_{0,j} = \mathbf{x}_j^T \mathbf{Y} = Z_j$ and

$$\hat{\theta}_{m,j}^{(\hat{S}_m)} = \mathbf{X}_j^T \mathbf{U}_m^{(\hat{S}_m)} = \mathbf{x}_j^T (I_n - \mathcal{B}_m) \mathbf{Y} = \mathbf{x}_j^T \mathbf{X}(I_n - \mathbf{D}_m) \mathbf{X}^T \mathbf{Y} = (1 - v)^{m_j} Z_j.$$

We have

$$T(m, \lambda_n, \gamma, \omega) = Rss(m) - \lambda_n \sum_{j=1}^{n} \frac{|\hat{\theta}_{m,j}^{(\hat{S}_m)}|^{\gamma}}{|\hat{\theta}_{0,j}|^{\omega}}$$

$$= \sum_{j=1}^{n} Z_j^2 (1-v)^{2m_j} - \lambda_n \sum_{j=1}^{n} (1-v)^{m_j \gamma} |Z_j|^{\gamma-\omega}$$

$$:= \sum_{j=1}^{n} Z_j^2 f_j(s_j),$$

where $s_j = (1-v)^{m_j}$ and $f_j(s_j) = s_j^2 - \lambda_n |Z_j|^{\gamma-\omega-2} s_j^{\gamma}$. Note that $v \longrightarrow 0$, being sufficiently small. Minimizing $T(m, \lambda_n, \gamma, \omega)$ over m is equivalent to minimizing $f_j(s_j)$ for $s_j \in [0,1]$, for $j = 1, \cdots, n$, respectively. Under the assumption that $0 \leq m_j \leq N < \infty$, minimizing $T(m, \lambda_n, \gamma, \omega)$ is equivalent to $\min\{f_j(s_j) : 0 < s_j \leq 1\}$. It is easy to obtain that

$$\frac{\partial f_j(s_j)}{\partial s_j} = 2s_j - \gamma \lambda_n |Z_j|^{\gamma-\omega-2} s_j^{\gamma-1} = s_j^{\gamma-1} [2s_j^{2-\gamma} - \gamma \lambda_n |Z_j|^{\gamma-\omega-2}].$$

Let $\frac{\partial f_j(s_j)}{\partial s_j} = 0$, then $s_j = 0$ or $s_j = (\frac{\gamma \lambda_n |Z_j|^{\gamma-\omega-2}}{2})^{\frac{1}{2-\gamma}} = (\frac{\gamma \lambda_n}{2|Z_j|^{2-\gamma+\omega}})^{\frac{1}{2-\gamma}}$. Since $0 < s_j \leq 1$, $s_j = 0$ should be deleted.

(a) As $0 < \left(\frac{\gamma \lambda_n}{2|Z_j|^{2-\gamma+\omega}}\right)^{\frac{1}{2-\gamma}} \leq 1$ or equivalently $|Z_j| \geq (\frac{\gamma \lambda_n}{2})^{\frac{1}{2-\gamma+\omega}}$. Since $\hat{\mathbf{Y}} = \mathbf{X} \hat{\beta}_{GSB}^{(\hat{m})} = \mathcal{B}_m \mathbf{Y}$, recalling $\mathcal{B}_m = \mathbf{X} D_m \mathbf{X}^T$ and the fact $\mathbf{X}^T \mathbf{X} = I_n$, we have $\hat{\beta}_{GSB}^{(\hat{m})} = D_m \mathbf{Z}$. As v being sufficiently small, minimizing $T(m, \lambda_n, \gamma, \omega)$ leads to

$$\hat{\beta}_{GSB,j}^{(\hat{m})} = (1-(1-v)^{m_j}) Z_j = (1-s_j) Z_j$$

$$= Z_j - (\frac{\gamma \lambda_n}{2|Z_j|^{2-\gamma+\omega}})^{\frac{1}{2-\gamma}} Z_j$$

$$= Z_j - \frac{(\gamma \lambda_n)^{\frac{1}{2-\gamma}}}{2|Z_j|^{\frac{\omega}{2-\gamma}}} \operatorname{sign}(Z_j).$$

(b) As $\left(\frac{\gamma \lambda_n}{2|Z_j|^{2-\gamma+\omega}}\right)^{\frac{1}{2-\gamma}} > 1$ or equivalently $|Z_j| < (\frac{\gamma \lambda_n}{2})^{\frac{1}{2-\gamma+\omega}}$, due to $v \longrightarrow 0$, then $s_j = 1$ is the minimizer of $T(m, \lambda_n, \gamma, \omega)$. Therefore, we have

$$\hat{\beta}_{GSB,j}^{(\hat{m})} = (1-(1-v)^{m_j}) Z_j = (1-s_j) Z_j = Z_j - Z_j = 0.$$

Therefore, the GSBoosting estimator is

$$\hat{\beta}_{GSB,j}^{(\hat{m})} = \begin{cases} Z_j - \frac{(\gamma \lambda_n)^{\frac{1}{2-\gamma}}}{2|Z_j|^{\frac{\omega}{2-\gamma}}} \operatorname{sign}(Z_j), & |Z_j| \geq (\frac{\gamma \lambda_n}{2})^{\frac{1}{2-\gamma+\omega}}; \\ 0, & |Z_j| < (\frac{\gamma \lambda_n}{2})^{\frac{1}{2-\gamma+\omega}}. \end{cases}$$

From the above discussion, as $0 < \gamma < 2$, $\omega \geq 0$, for model (8), GSBoosting estimator is equivalent to that of adaptive Lasso estimator. This completes the proof of theorem 1.

Acknowledgement. This project is supported by the Fundamental Research Funds for the Central Universities and National Science Foundation of China No.11101022 and No.11026049; the Ministry of Education, Humanities and Social Science Foundation Youth Project under Grant No.10YJC910013.

References

1. Breiman, L.: Arcing classifiers (with discussion). Ann. Statist. 26, 801–849 (1998)
2. Breiman, L.: Prediction games and arcing algorithms. Neural Computation 11, 1493–1517 (1999)
3. Breiman, L., Friedman, J.: Estimating Optimal Transformations for Multiple Regression and Correlation. J. Am. Statist. Ass. 80, 580–598 (1985)
4. Bühlmann, P.: Boosting for high-dimensional linear models. Ann. Statist. 34, 559–583 (2006)
5. Bühlmann, P., Hothorn, T.: Twin Boosting: improved feature selection and prediction. Statistics and Computing 20(2), 119–138 (2010)
6. Bühlmann, P., Hothorn, T.: Boosting algorithms: regularization, prediction and model fitting. Statistical Science 4(22), 477–505 (2007)
7. Bühlmann, P., Yu, B.: Boosting with the L_2 loss: Regression and classification. J. Amer. Statist.Assoc. 98, 324–339 (2003)
8. Bühlmann, P., Yu, B.: Sparse boosting. J. Machine Learning Research 7, 1001–1024 (2006)
9. Friedman, J., Hastie, T., Tibshirani, R.: Additive logistic regression: a statistical view of boosting (with discussion). Ann. Statist. 28, 337–407 (2000)
10. Friedman, J.: Greedy function approximation: a gradient boosting machine. Ann. Statist. 29, 1189–1232 (2001)
11. Frank, I.E., Friedman, J.H.: A statistical view of some chemometrics regression tools. Technometrics 35, 109–148 (1993)
12. Hurvich, C.M., Tsai, C.L.: Regression and Time Series Model Selection in Small Samples. Biometrika 76, 297–307 (1989)
13. Tibshirani, R.: Regression shrinkage and selction via the lasso. J. R. Statist. Soc. B 58, 267–288 (1996)
14. Zou, H.: The adaptive lasso and its oracle properties. J. Am. Statist. Ass. 101, 1418–1429 (2006)

Learning from Multiple Naive Annotators

Chirine Wolley and Mohamed Quafafou

Aix-Marseille University, LSIS UMR 7296
13397, Marseille, France
chirine.wolley@etu.univ-amu.fr, mohamed.quafafou@univ-amu.fr

Abstract. This paper presents a probabilistic model for coping with multiple annotators for discrete binary classification tasks. Here, annotators decline to label instances when they are unsure and therefore, ignorance and real errors are represented separately. Our model integrates both error and ignorance into a conditional Bayesian model where only the observed instance is needed to infer the label. Furthermore, we provide a more accurate study on the properties of each annotator over previous methods. Extensive experiments on a broad range of data sets validate the effectiveness of learning from multiple naive (ignorant) annotators.

Keywords: supervised learning, multiple annotators, ignorance, Bayesian analysis, properties of labelers.

1 Introduction

A classical supervised learning scenario can be formalized as the problem of inferring a function $y = f(x)$, based on a training dataset $\mathcal{D} = \{(x_i, y_i)\}_{i=1}^{N}$, containing N instances x_i with a corresponding known label y_i. x_i is a d-dimensional vector and y_i is of categorical nature. In binary classification, $y_i \in \{0, 1\}$. In real-world applications, obtaining the ground truth label y_i for each instance x_i of the training dataset may be expensive and time-consuming. Instead, recent availability of online annotation services (e.g Amazon Mechanical Turk) open up the possibility of obtaining multiple judgments $y_i^1, y_i^2, ..., y_i^T$, from T *non-expert* annotators for each data point x_i. Throughout the paper, we use the term *expert* for an annotator who can provide the correct label for all the dataset. On the other hand, the term *non-expert* is used for a random teacher (annotator) who has not the appropriate knowledge for annotating all the instances.

Acquiring labels from non-expert annotators is clearly cheaper and faster. However, this generally leads to massive databases where some annotators are more reliable and more experienced than others. Hence, in order to generate a good classifier, it is important to evaluate their reliability. Recently, several approaches have been undertaken to address this issue [4,5,10]. The common target of this family of work is both to learn a good classifier and to evaluate the truthworthiness of each annotator.

We address a new aspect of this problem, motivated by the fact that while multiple annotators may be available, we do not have control over their quality,

S. Zhou, S. Zhang, and G. Karypis (Eds.): ADMA 2012, LNAI 7713, pp. 173–185, 2012.

and their lack of knowledge may be significant. In this paper, we tackle the problem of learning from the ignorance of the annotators. Our contribution to include annotators' ignorance in the model is motivated by several points:

- Ignorance is an important aspect of real-world information. In [8] , the author shows that a proper measure of ignorance helps to avoid situations where gathering more information would be appropriate. In other words, we can learn from ignorance, and ignoring ignorance is ignorant.
- Previous works implicitly assume that annotators are always able to provide the required label. This assumption rarely holds as annotators are nonexperts: their lack of knowledge can be significant, and a teacher aware of his deficiencies may feel unconfortable at labeling an example in unsure situations.
- When labeling, annotators with the same error rate are not necessarily equally good (or bad): a labeler who is aware of his strengths and deficiencies is better than another one who isn't. Hence, real errors and ignorance should be studied separately in order to provide a more accurate analysis of the behavior of each annotator when labeling.

Our model - called Ignore - describes a Bayesian probabilistic approach for supervised learning when multiple annotators do not necessarily provide a label for each instance. Experimental results support that learning from both errors and ignorance can significantly improve the quality of the learned models. We organize the rest of the paper as follows. The next section reviews related work in the literature. Section 3 explains in detail the proposed framework, followed by the empirical evaluation in Section 4. Finally, the last section offers our conclusions.

2   Related Work

In many fields, collecting multiple annotations where gold standard labels are missing is becoming a more common practice. In the biostatistics community, authors in [9] studied the error rate estimation problem when conflicted responses are given to various medical questions. In the natural language processing community [1], the authors show that using multiple labels provided by non expert annotators can be as effective as using annotations from one expert. In a more supervised learning context, authors in [4] propose an EM-based algorithm in order to give an estimate of the actual hidden labels and to evaluate the different labelers. Many other papers aim at addressing the same problem [5,10].

All the above works aim to address the problem of multiple deficient annotators, that is, annotators who generate errors while labeling. However, they all focus on the annotators' knowledge without trying to learn from their ignorance, although ignorance is widely present in real-world applications. For this reason, a formalism dealing with real-world information should allow expressing and quantifying both aspects (knowledge and ignorance) of accurate information.

The problem of ignorance has received due attention in the statistical literature for many years. First treated as missing values, the nature of its work has considerably changed. In [8], the author shows that Dempster-Shafer theory [12] is appropriate for modeling and quantifying ignorance. In [11], the authors replace conventional point estimators with regions of estimates. Despite the extensive work on both ignorance and supervised learning with multiple annotators, we are not aware of any approach that takes into account the ignorance of annotators to generate a classifier. The closest approach in this area is a very theoretical work where the problem is analyzed in a PAC learning context [2].

Our model Ignore, which is inspired from a previous efficient algorithm [4], includes annotators' ignorance. In summary, our contributions are as follows:

- To the best of our knowledge, we are the first to present a practical and efficient algorithm dealing the ignorance issue in supervised learning from multiple annotators. These labelers are no longer forced to label all instances. In a context of binary classification, they may infer '0', '1', or '?' in case of ignorance.
- We propose a Bayesian analysis to model ignorance. We set different prior distributions depending on whether the label is given or not. Three different priors are tested on unlabeled instances: the Beta prior, and the two non-informative uniform and Jeffreys' prior.
- We study and analyse real errors and ignorance of annotators separately. This leads to a more accurate evaluation of the behaviors of the annotators when labeling, compared to previous approaches.

Extensive experiments on several different datasets validate the effectiveness of our approach against the baseline method of Raykar and Al. [4].

3   The Model

3.1   Data Rewriting Process

Given N instances $\{x_1, ..., x_N\}$, each data point x_i is described by D descriptors and labeled by T non expert annotators. We denote y_i^t the label of the i-th instance given by annotator t ($t \in \{1, ..., T\}$). As annotators are non experts, they may mark '?' in case of ignorance. Hence, in a context of binary classification, $y_i^t \in \mathcal{Y} = \{0, 1\} \cup \{?\}$. Let z_i be the true label for the i-th instance. $z_i \in \mathcal{Z} = \{0, 1\}$ is unknown. For compactness, we define the matrices $X = \left[x_1^T; ...; x_N^T\right] \in R^{N \times D1}$, $Y = \left[y_1^{(1)}, ..., y_1^{(T)}; ...; y_N^{(1)}, ..., y_N^{(T)}\right] \in R^{N \times T}$ and $Z = [z_1, ..., z_N]^T$.

In order to integrate all '?' into the model, we introduce a matrix $H \in R^{N \times T}$ of ignorance as follows:

$$h_i^t = \begin{cases} 0 \text{ if } y_i^t = \{0, 1\} \\ 1 \text{ otherwise} \end{cases} \tag{1}$$

From this idea, H can be seen as the traceability of their ignorance. Once H has been defined, all '?' can be replaced with a missing data method. While the

---

[1] $(.)^T$ is the matrix transpose.

proposed approach can use any method, we consider the random function which replaces all '?' by 0 and 1 randomly. This choice is motivated by the fact that we don't want to put any bias on the annotators' response. For compactness, we set $H = \left[h_1^{(1)}, ..., h_1^{(T)}; ...; h_N^{(1)}, ..., h_N^{(T)} \right] \in R^{N \times T}$, and our goal is to: (1) Estimate the ground truth labels Z, (2) Produce a classifier to predict the label z for a new instance x and (3) Estimate the reliability of each annotator, conditionally to the ignorance of each labeler.

3.2   Probabilistic Model

We model our problem with the joint conditional distribution $P(X, Y, Z|H, \Theta)$, where Θ is the set of parameters to be estimated (defined later in the paper). Our model can be understood as a conditional Bayesian network, as represented in Fig.1.

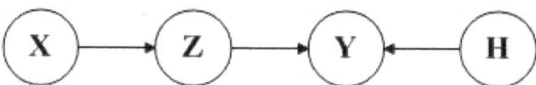

Fig. 1. Ignore model structure

Once the probability $P(X, Y, Z|H, \Theta)$ is estimated, we need to obtain an estimation of all the parameters Θ. While previous approaches maximize the log-likelihood, here, knowing the ignorance of each annotator enables us to fix priors on the parameters. Therefore, a Bayesian approach is used and our goal is to maximize the log-posterior with the maximum-a-posteriori estimator (MAP):

$$\hat{\Theta}_{MAP} = argmax_{\Theta} \{ ln Pr[X, Y, Z|H, \Theta] + ln Pr[\Theta] \} \qquad (2)$$

These estimates are obtained using a combination of the Expectation-Maximization Algorithm (EM), Newton-Raphson update and LBFGS quasi-Newton method.

3.3   Maximum a Posteriori Estimator

As the ground truth labels z are supposed unknown for all instances, we first need to estimate the probability $P(X, Y|H, \Theta)$. The variables z will be integrated as missing variables later in the model. Assuming the instances are independent from each other and considering Bayes Theorem, the likelihood of $P(X, Y|H, \Theta)$ can be expressed as follows:

$$Pr(X, Y|H, \Theta) = \prod_{i=1}^{N} Pr[y_i^1, ..., y_i^T | x_i, h_i^1, ..., h_i^T, \Theta]$$

which becomes, using Bayes theorem and including the ground truth labels Z:

$$Pr[X, Y, Z|H, \Theta] = \prod_{i=1}^{N} Pr(y_i^1, ..., y_i^T | z_i = 1, h_i^1, ..., h_i^T, \Theta_{y_1}) Pr(z_i = 1|x_i, \Theta_{z_1}) +$$

$$\prod_{i=1}^{N} Pr(y_i^1, ..., y_i^T | z_i = 0, h_i^1, ..., h_i^T, \Theta_{y_0}) Pr(z_i = 0|x_i, \Theta_{z_0})$$

with $\Theta = \{\Theta_{y_1}, \Theta_{z_1}, \Theta_{y_0}, \Theta_{z_0}\}$.

As $Pr(z_i = 0|x_i, \Theta_{z_0}) = 1 - Pr(z_i = 1|x_i, \Theta_{z_1})$, $\{\Theta_{z_1}\} = \{\Theta_{z_0}\} = \{\Theta_z\}$ and therefore $\Theta = \{\Theta_{y_1}, \Theta_{y_0}, \Theta_z\}$. We need to consider each conditional-distribution individually.

Concerning $Pr(z_i = 1|x_i, \Theta_z)$, while the proposed method can use any other classifier, we assume for simplicity a logistic regression model, i.e:

$$Pr[z_i = 1|x_i, w] = \frac{1}{1 + e^{-w^T x_i}} \tag{3}$$

where $w \in R^d$. Therefore, $\Theta_z = \{w\}$.

Concerning $Pr(y_i^1, ..., y_i^T | z_i = 1, h_i^1, ..., h_i^T, \Theta_{y_1})$, assuming that the annotators (resp. annotators' ignorance) are independent from each other, we have:

$$Pr[y_i^1, ..., y_i^T | z_i = 1, h_i^1, ..., h_i^T, \Theta_{y_1}] = \prod_{t=1}^{T} Pr(y_i^t | z_i = 1, h_i^t, \Theta_{y_1}^t)$$

As y_i^t is a binary event, a common strategy is to use a Bernoulli model. However, as h_i^t can take both modalities $\{0, 1\}$, a Bernoulli model would not be appropriate. Instead, we model y_i^t as a mixture of two Bernoulli distributions:

$$y_i^t = \begin{cases} Bernoulli(\alpha_1^t) & \text{if } h_i^t = 1 \\ Bernoulli(\alpha_0^t) & \text{otherwise} \end{cases}$$

which can also be written as follows:

$$y_i^t \sim h_i^t * B(\alpha_1^t) + (1 - h_i^t) * B(\alpha_0^t)$$

Consequently, α_1^t (resp. α_0^t) is the parameter of the Bernoulli distribution for the known (resp. unknown) labels. Therefore, $\Theta_{y_1}^t = \{\alpha_0^t, \alpha_1^t\}$ and

$$Pr(y_i^t | z_i = 1, h_i^t, \Theta_{y_1}^t) = h_i^t (\alpha_1^t)^{y_i^t} (1 - \alpha_1^t)^{1-y_i^t} + (1 - h_i^t)(\alpha_0^t)^{y_i^t} (1 - \alpha_0^t)^{1-y_i^t} \tag{4}$$

Respectively, we set $\Theta_{y_0}^t = \{\beta_0^t, \beta_1^t\}$ and we have

$$Pr[y_i^t | z_i = 0, h_i^t, \Theta_{y_0}^t] = h_i^t (\beta_1^t)^{1-y_i^t} (1 - \beta_1^t)^{y_i^t} + (1 - h_i^t)(\beta_0^t)^{1-y_i^t} (1 - \beta_0^t)^{y_i^t} \tag{5}$$

Let $p_i = Pr[z_i = 1|x_i, \Theta_z]$, $a_i = Pr[y_i^1, ..., y_i^T|z_i = 1, h_i^1, ..., h_i^T, \Theta_{y1}]$ and $b_i = Pr[y_i^1, ..., y_i^T|z_i = 0, h_i^1, ..., h_i^T, \Theta_{yo}]$, the likelihood can be written:

$$Pr(X, Y, Z|H, \Theta) = \prod_{i=1}^{N} [a_i p_i + b_i(1 - p_i)] \tag{6}$$

Next section describes which prior is assumed on each parameter.

3.4   Prior Distributions

We consider the following prior distributions :

- Prior on parameters $\{\alpha_0, \beta_0\}$
  According to equations (4) and (5), we have $\alpha_0^t = Pr[y_i^t = 1|z_i = 1, h_i^t = 0]$ and $\beta_0^t = Pr[y_i^t = 0|z_i = 0, h_i^t = 0]$. In other words, α_0^t and β_0^t are respectively the sensitivity[2] and the specificity[3] of annotator t. Since they represent the probability of a binary event, a natural choice of prior distribution is the beta prior B(a,b), where both a and b are strictly positive. Thus, we assume a beta prior on both parameters $\{\alpha_0, \beta_0\}$, i.e:

$$\begin{cases} Pr(\alpha_0^t|a_{01}^t, a_{02}^t) \sim Beta(\alpha_0^t|a_{01}^t, a_{02}^t) \\ Pr(\beta_0^t|b_{01}^t, b_{02}^t) \sim Beta(\beta_0^t|b_{01}^t, b_{02}^t) \end{cases}$$

Parameters $\{a_{01}^t, a_{02}^t, b_{01}^t, b_{02}^t\}$ are estimated as follows. Let (a,b) be the parameters of a beta distribution. (a,b) are both estimated by solving:

$$\mu = a/(a + b) \tag{7}$$

$$\int_{u1}^{u2} Beta(p|a, b)dp = C \tag{8}$$

where p is the probability to be calculated, i.e, the sensitivity and the specificity, μ the mean of the beta distribution, and C, u1, u2 three constants to initialize. The labeler provides his expectation μ of p, and equation (8) defines the percentage of chance that p is between the interval assessment [u1,u2]. As $h^t = 0$, we assume that $\mu = 0.9$, $Pr[\alpha_0^t \in [0.5 : 1]] = 0.9$, and $Pr[\beta_0^t \in [0.5 : 1]] = 0.9$. Therefore, $C = 0.9$ and [u1,u2] = [0.5,1] are the values used in our experimentations.

- Prior on parameters $\{\alpha_1, \beta_1\}$
  According to equations (4) and (5), $\alpha_1^t = Pr[y_i^t = 1|z_i = 1, h_i^t = 1]$ and $\beta_1^t = Pr[y_i^t = 0|z_i = 0, h_i^t = 1]$. In other words, $\alpha_1^t = Pr(random(0, 1) = 1|z_i = 1)$ and $\beta_1^t = Pr(random(0, 1) = 0|z_i = 0, h_i^t = 1)$. Three different priors are considered and analyzed:

---

[2] True positive rate.
[3] 1-false positive rate.

Beta(a,b) Prior: Since α_1^t and β_1^t represent the probability of a binary event, the beta distribution prior B(a,b) can be assumed for both parameters:

$$\begin{cases} Pr(\alpha_1^t|a_{11}^t,a_{12}^t) \sim Beta(\alpha_1^t|a_{11}^t,a_{12}^t) \\ Pr(\beta_1^t|b_{11}^t,b_{12}^t) \sim Beta(\beta_1^t|b_{11}^t,b_{12}^t) \end{cases}$$

And parameters $\{a_{11}^t, a_{12}^t, b_{11}^t, b_{12}^t\}$ are estimated using equations (7) and (8), with $\mu = 0.5$, C = 0.9 and [u1,u2] = [0.4,0.5].
In other words, $Pr[\alpha_1^t \in [0.4 : 0.6]] = 0.9$ (resp. $Pr[\beta_0^t \in [0.4 : 0.6]] = 0.9$).

Beta(1,1) Prior: Non-informative priors are often used to model ignorance, as they reflect a situation where there is a lack of knowledge about a parameter. In case of a binary event, many researches have advocated the use of a uniform[0,1] distribution as a non-informative prior, or equivalently, a Beta(1,1) distribution. This prior assumption can be written as follows:

$$\begin{cases} Pr(\alpha_1^t|1,1) = Beta(\alpha_1^t|1,1) \\ Pr(\beta_1^t|1,1) = Beta(\beta_1^t|1,1) \end{cases}$$

Jeffreys' Prior: Another non-informative prior widely used in Bayesian analysis and in a wide variety of applications is Jeffreys' prior [7]. For (α_1^t, β_1^t) we have:

$$\begin{cases} p(\alpha_1^t) = \frac{1}{\sqrt{\alpha_1^t(1-\alpha_1^t)}} \\ p(\beta_1^t) = \frac{1}{\sqrt{\beta_1^t(1-\beta_1^t)}} \end{cases}$$

– Prior on parameter w: For sake of completeness, we assume a zero mean Gaussian prior on the weights w with inverse covariance matrix Γ:

$$w \sim N(w|0, \Gamma^{-1})$$

3.5   Computing Issues

Our goal is to estimate $\hat{\Theta}_{MAP}$ by calculating (2). Since there are missing values z, a well-known approach to estimate the maximum log-posterior is to use the Expectation-Maximization (EM) algorithm [6]. Although it was first used to estimate the maximum likelihood, a derived algorithm has been developed for the MAP estimation.

E-Step: Given the observations X, Y, H and the current estimate of the parameters Θ, the expectation of the log-likelihood is:

$$E\left[ln\,Pr\left[Y, Z, X|H, \Theta\right]\right] \propto ln[Pr(\Theta)] + \sum_{i=1}^{N} \mu_i ln(p_i a_i) + (1 - \mu_i)ln(1 - p_i)b_i \quad (9)$$

where $\mu_i = Pr\left[z_i = 1|y_i^1, ..., y_i^T, x_i, h_i^1, ..., h_i^T, \Theta\right]$.

Using Bayes theorem and equations (3) and (4), we obtain the following expression for μ_i :

$$\mu_i \; \alpha \; Pr[y_i^1, ..., y_i^T | z_i = 1, h_i^1, ..., h_i^T, \Theta] \times Pr[z_i = 1 | x_i, \Theta] \tag{10}$$

$$= \frac{a_i p_i}{a_i p_i + b_i(1 - p_i)} \tag{11}$$

M-Step: Based on the current estimation of μ_i and on the observations X and Y, parameters Θ can be estimated by equating the gradient of (9) to 0. For both Beta(a,b) and Beta(1,1) priors, we obtain for $j \in \{0,1\}$:

$$\alpha_j^t = \frac{\sum_{i \in \{i | h_i^t = j\}} \mu_i y_i^t + a_{j1}^t - 1}{a_{j1}^t + a_{j2}^t - 2 + \sum_{i \in \{i | h_i^t = j\}} \mu_i} \tag{12}$$

$$\beta_j^t = \frac{b_{j1} - 1 + \sum_{i \in \{i | h_i^t = j\}} (1 - \mu_i)(1 - y_i^t)}{b_{j1}^t + b_{j2}^t - 2 + \sum_{i \in \{i | h_i^t = j\}} (1 - \mu_i)} \tag{13}$$

For Jeffreys' prior, it is not possible to directly equate the gradients of both parameters $\{\alpha_1^t, \beta_1^t\}$ to zero. Instead, we use the LBFGS quasi-Newton algorithm, and for convenience, the gradients with respect to both parameters are provided. We set $f_{opt} = E\left[ln\ Pr\left[X, Y, Z | H, \Theta\right]\right]$ and we obtain:

$$\frac{\partial f_{opt}}{\partial \alpha_1^t} = \sum_{i=1}^N \left[\mu_i(\alpha_1^t - y_i^t) + \frac{1 - 2\alpha_1^t}{2\sqrt{\alpha_1^t(1 - \alpha_1^t)}} \right] \tag{14}$$

$$\frac{\partial f_{opt}}{\partial \beta_1^t} = \sum_{i=1}^N \left[(1 - \mu_i)[(1 - y_i^t)(1 - \beta_1^t) - y_i^t \beta_1^t] + \frac{2\beta_1^t - 1}{2\sqrt{\beta_1^t(1 - \beta_1^t)}} \right] \tag{15}$$

Concerning the parameter w, we use the Newton-Raphson update method given by:

$$w^{q+1} = w^q - \eta H^{-1} g \tag{16}$$

where g is the gradient vector, H the Hessian matrix and η the step length (refer to [4] for more details on the calculation of g and H). Steps E and M are iterated until convergence. μ_i is initialized using $\frac{1}{|P|} \sum_{t=1}^P y_i^t$ with $P = \{t | h_i^t \neq 1\}$.

Once the parameters are estimated in the EM algorithm, a new instance x_i is classified by calculating $p(z_i = 1 | x_i) = (1 + exp(-w^T x_i))^{-1}$, the probability for x_i to have the true label z_i equals to 1.

4   Experiments

4.1   Experimental Setup and Results

In this section, we compare the performance of our Ignore approach with two methods: the baseline algorithm developed in [4], and a more classical linear regression model. For the regression model, we first encode the binary labels by -1 and 1, and the "don't know" flag, i.e. '?', by 0. Consequently, $y_i^t \in \mathcal{Y} = \{-1, 0, 1\}$. Then, we perform a linear regression model on each annotator for each instance, and majority voting is finally used to estimate the hidden true label z_i.

Simulations have been performed on eight datasets from the UCI Machine Learning Repository [3] for which binary labels are available. The data used are: Ionosphere (351,34), Cleveland Heart (297,13), Musk(version 1) (476,167), Glass (214,10), Bupa (345,7), Vertebral (310,6), Spect Heart (267,22) and Haberman (306,3) (with (number of instances, number of features)). In addition, we use the galaxy dim data described in [13], which contains 4192 samples and 14 features with also available binary labels. Before performing the simulations, each data D has been processed in four steps:

1. D is randomly divided into two folds D_{train} (training set - 80% of D) and D_{test} (testing set - 20% of D).
2. For T annotators, D_{train} is divided into T equally-sized folds $\{d_1, d_2, ..., d_T\}$ using K-means.
3. We assume annotator t is an expert for the set d_t, i.e:

$$\begin{cases} Error(y^t, d_t) = 0\% \\ Ignorance(y^t, d_t) = 0\% \end{cases}$$

4. On the rest of the data $\bar{d}_t$, we assume annotator t makes 10% of errors and $I\%$ of ignorance, $I \in \{0\%, 10\%, ..., 80\%, 90\%\}$:

$$\begin{cases} Error(y^t, \bar{d}_t) = 10\% \\ Ignorance(y^t, \bar{d}_t) = I\% \end{cases}$$

The simulations have been conducted in order to test how the proposed model behaves when confronted to ignorance, compared to the two other well-established methods. We generate T=5 annotators for each data, with the different levels of ignorance $I \in \{0\%, 10\%, ..., 80\%, 90\%\}$. Then, we simulate each model a hundred times using the bootstrap method.

To evaluate our method, we use the following two criterias, which are the AUC (Area Under roc Curves) and the classification error rate cross the different levels of ignorance. The mean of all the AUC obtained is calculated. In Figure 2, we plot the evolution of the resulted AUC for each dataset considering the different levels of ignorance. For both baseline and regression methods, the bootstrap AUC collapses when the ignorance rate increases, in opposition to our Ignore method which is clearly more robust when confronted to ignorance. In addition, we computed, for each D_{test}, the mean of the error rate classification and its

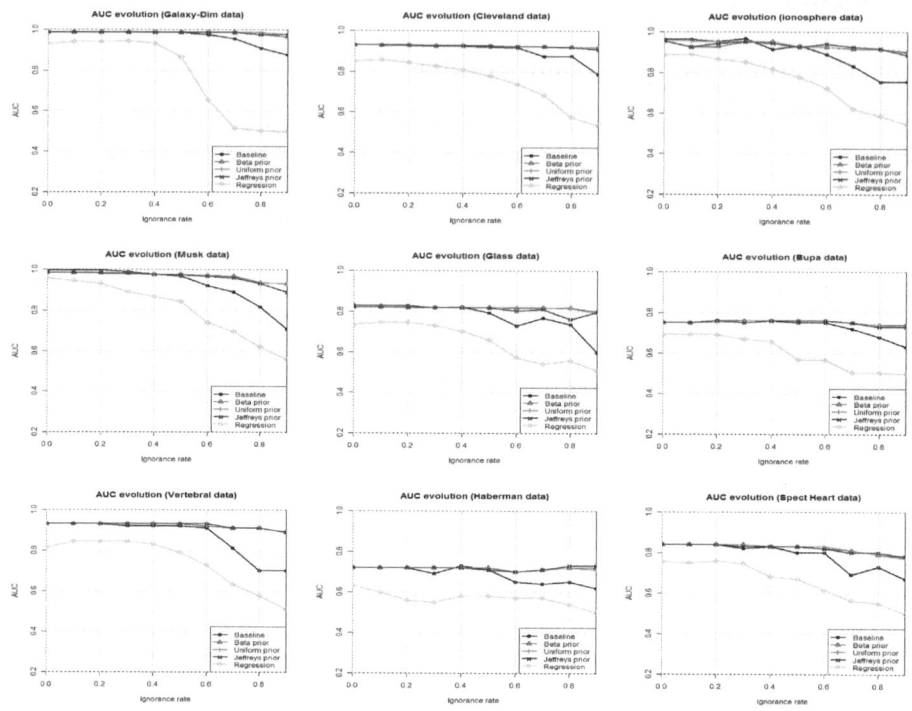

Fig. 2. Comparison of the evolution of the AUC between Ignore, Raykar's and linear Regression methods, when the ignorance rate of the annotators increases

Table 1. Experimental results of Baseline, Ignore with three priors (Beta, Uniform and Jeffrey) and Regression for all datasets. Here $E = 10\%$ and $I \in \{0\%, 10\%, ..., 80\%, 90\%\}$. (error mean $\pm$ std) are then calculated cross the different levels of ignorance.

Dataset	error rate cross different levels of ignorance				
	Baseline	Beta	Uniform	Jeffreys	Regression
Cleveland	0.164 ± 0.038	0.135 ± 0.019	0.137 ± 0.016	0.145 ± 0.016	0.313 ± 0.096
Galaxy Dim	0.336 ± 0.083	0.063 ± 0.001	0.058 ± 10^{-5}	0.060 ± 0.001	0.625 ± 0.081
Ionosphere	0.271 ± 0.088	0.215 ± 0.012	0.218 ± 0.011	0.224 ± 0.009	0.227 ± 0.044
Musk	0.331 ± 0.041	0.161 ± 0.033	0.161 ± 0.032	0.162 ± 0.023	0.234 ± 0.035
Glass	0.391 ± 0.066	0.358 ± 0.016	0.361 ± 0.017	0.359 ± 0.022	0.385 ± 0.062
Bupa	0.371 ± 0.025	0.361 ± 0.012	0.366 ± 0.016	0.352 ± 0.011	0.379 ± 0.023
Vertebral	0.429 ± 0.094	0.246 ± 0.015	0.234 ± 0.016	0.239 ± 0.022	0.478 ± 0.071
Spect Heart	0.345 ± 0.097	0.294 ± 0.015	0.301 ± 0.019	0.298 ± 0.024	0.390 ± 0.082
Haberman	0.411 ± 0.027	0.389 ± 0.021	0.379 ± 0.024	0.391 ± 0.018	0.412 ± 0.168
Mean	0.339 ± 0.066	0.247 ± 0.016	0.246 ± 0.017	0.248 ± 0.018	0.383 ± 0.074

standart deviation over the different levels of ignorance. The results are reported in table 1. We notice that the error rate prediction over the different levels of ignorance is significantly lower in our models (around 0.250 for Ignore, 0.340 for the baseline and 0.380 for the regression). Furthermore, the calculated standard deviation also confirms its stability comparing to the other two baseline methods (around 0.017 for Ignore, 0.066 for the baseline and 0.074 for the regression).

In conclusion, the obtained results for each dataset show clearly the efficiency and the stability of our approach compared to both baseline and regression methods, especially when the ignorance increases. This validates the effectiveness of learning from the ignorance of the labelers.

4.2 Annotators Properties

The proposed Ignore model considers separately the behavior of each annotator in case of errors and ignorance. In fact, for each annotator, four parameters are estimated: α_0^t and β_0^t (resp. α_1^t and β_1^t) measure the sensitivity and the specificity when annotator t is sure (resp. unsure) of the label given. Therefore, unlike previous methods, we are here able to study more precisely the behavior of each annotator when labeling, as our model estimates his performance both in sure and unsure situations. This section discusses this issue and presents our preliminary results and conclusions. In order to evaluate their performance, we assumed that annotators always label instances even if they are unsure. Therefore, they give a label in addition to the flag "?". In this context, we characterize each annotator t by its precision in both sure and unsure situations. This composed precision is represented by a pair (π_t, σ_t) where π_t is the error rate in the case the annotator t is sure, and similarly σ_t represents the error rate in unsure situations. To give a concrete example, we consider four profiles of annotators: Unconscious (π is high and σ is low), bad (both π and σ are high), expert (both π and σ are

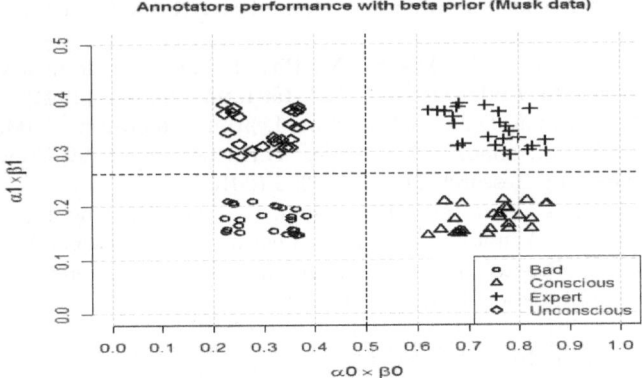

Fig. 3. Annotators properties. 100 annotators simulated, 25 for each category. Representation of the values of the parameters obtained for MUSK data.

low), and conscious (π is low and σ is high). We simulated 100 annotators, 25 for each category. In our simulation, we define a high value of π and σ when the error rate is higher than 65%. We define a low value of π and σ when the error rate is lower than 35%. Afterwards, we constructed the Ignore model for each dataset and then extracted the appropriate parameters $\alpha_0, \beta_0, \alpha_1, \beta_1$ computed for each annotator. In order to facilitate the interpretation of the parameters, we combined the two parameters of ignorance by calculating for each labeler the product $\alpha_0 \times \beta_0$ and we calculate as well the product $\alpha_1 \times \beta_1$. The result obtained for Musk data can be seen in Figure 3. The four categories of labelers are very well represented: experts labelers are opposed to bad labelers, while conscious labelers are opposed to unconscious labelers. This is a promising preliminary experimental result related to how the parameters of the Ignore models capture the properties of each annotator.

5   Conclusion

This paper addresses the challenging problem of learning from the ignorance of the annotators. The proposed framework presents a new probabilistic Bayesian approach, dealing with the presence of 'don't know' labels in the annotations of the labelers. Experiments over a broad range of datasets validate the effectiveness of our model Ignore compared to two baseline methods. Furthermore, our model Ignore enables us to study more precisely the behavior of each annotator when labeling. Due to its accuracy and efficiency, Ignore is expected to be more practical, especially in the age of the crowdsourcing services.

References

1. Snow, R., O'Connor, B., Jurafsky, D., Ng, A.Y.: Cheap and Fast - But is it Good? Evaluating Non-Expert Annotations for Natural Language Tasks. In: Proceedings of the 2008 Conference on Empirical Methods on Natural language Processing (2008)
2. Frazier, M., Goldman, S.A., Mishra, N., Pitt, L.: Learning from a Consistently Ignorant Teacher. In: Philosophical COLT 1994, pp. 328–339 (1994)
3. Asuncion, A., Newman, D.J.: UCI Machine Learning Repository (2007)
4. Raykar, V.C., Yu, S., Zhao, L.H., Valadez, G.H.: Learning from Crowds. Journal of Machine Learning Research 11, 1297–1322 (2010)
5. Yan, Y., Hermosillo, G., Rosales, R., Bogoni, L., Fung, G., Moy, L., Schmidt, M., Dy, J.G.: Modeling Annotator Expertise: Learning when everybody knows a bit of something. In: Proceedings of the 13th International Conference on Artificial Intelligence and Statistics, Italy, vol. 9 (2010)
6. Dempster, A., Laird, N., Rubin, D.: Maximum Likelihood Estimation from Incomplete Data. J. of the Royal Statistical Society (B) 39(1) (1977)
7. Jeffreys, H.: An Invariant Form for the Prior Probability in Estimation Problems. Proceedings of the Royal Society of London (1946)
8. Haenni, R.: Ignoring ignorance is ignorant. Tech. Rep., Center for Junior Research Fellows, University of Konstanz (2003)

9. Dawid, A., Skene, A.: Maximum likelihood estimation of observer error-rates using the em algorithm. Applied Statistics 28(1), 20–28 (1979)
10. Whitehill, J., Ruvolo, P., Bergsma, J., Wu, T., Movellan, J.: Whose Vote Should Count More: Optimal Integration of Labels from Labelers of Unknown Expertise. In: Advances in Neural Information Processing Systems (2009)
11. Vansteelandt, S., Goetghebeur, E., Kenward, M.G., Molenberghs, G.: Ignorance and Uncertainty Regions as Inferential Tools in a Sensitivity Analysis. Statistica Sinica 16, 953–979 (2006)
12. Shafer, G.: The Mathematical Theory of Evidence. Princeton University Press (1976)
13. Odewahn, S., Stockwell, E., Pennington, R., Hummphreys, R., Zumach, W.: Automated Star/Galaxy Discrimination with Neural Networks. Astronomical J. 103(1), 318–331 (1992)

Query by Committee in a Heterogeneous Environment

Hao Shao[1,3], Bin Tong[1], and Einoshin Suzuki[2]

[1] Graduate School of Systems Life Sciences, Kyushu University
[2] Department of Informatics, ISEE, Kyushu University
[3] School of WTO Research & Education, Shanghai Institute of Foreign Trade

Abstract. In real applications of inductive learning, labeled instances are often deficient. The countermeasure is either to ask experts to label informative instances in active learning, or to borrow useful information from abundant labeled instances in the source domain in transfer learning. Due to the high cost of querying experts, it is promising to integrate the two methodologies into a more robust and reliable classification framework to compensate the disadvantages of both methods. Recently, a few research studies have been investigated to integrate the two methods together, which is called transfer active learning. However, when there exist unrelated domains which have different distributions or label assignments, an inevitable problem named *negative transfer* will happen which leads to degenerated performance. Also, how to avoid selecting unconcerned samples to query is still an open question. To tackle these issues, we propose a hybrid algorithm for active learning with the help of transfer learning by adopting a divergence measure to measure the similarities between different domains, so that the negative effects can be alleviated. To avoid querying irrelevant instances, we also present an adaptive strategy that is able to eliminate unnecessary instances in the input space and models in the model space. Extensive experiments on both synthetic and real data sets show that our algorithm is able to query less instances and converges faster than the state-of-the-art methods.

Keywords: active learning, transfer learning, classification.

1 Introduction

Nowadays, a challenging issue in nosology is that when people encounter a new epidemic, it is crucial to classify the patients as early as possible with a high accuracy. However, even there exist thousands of suspected cases, only a few of them are labeled (diagnosed) and to label an unlabeled instance by experts is often expensive and time consuming. Active learning (AL) [3] provides a solution by selecting unlabeled examples to query, with the objective to obtain a satisfactory classifier using as few instances as possible where the labeling cost is high. In this learning scenario, a significant challenge is how to select the most informative instance in each query from the large potion of the unlabeled pool

S. Zhou, S. Zhang, and G. Karypis (Eds.): ADMA 2012, LNAI 7713, pp. 186–198, 2012.

in the target task. There have been substantial works on active learning [3], in which one representative approach is the Query by Committee (QBC) strategy which assumes a correct Bayesian prior on the set of hypotheses, and the committee members are all trained on the current labeled data set [9]. However, as pointed out in [3,15], given only a few labeled instances in the data set, the initial hypotheses sampled from these instances are not reliable since they may deviate from the optimal hypothesis with respect to the input distribution in the end of the classification procedure.

To help tackling with the problem of the lack of labeled information in the target task, transfer learning (TL) techniques aim to borrow the strength of existing data and models. In the transfer learning setting, a target data set is assumed to have only a small number of labeled instances, while abundant useful information is available in the source data sets that can be obtained with little cost. However, an essential problem in transfer learning is that, when the distributions of the source and the target domains are different, directly transferring knowledge would hurt the performance on the target task, which is also known as "negative transfer" [16]. It is likely to happen once we underestimate the side effect resulting from the distribution differences of multiple source tasks, which is common in real applications.

There exist several research works concerning integrating active learning and multi-task learning [18,22,1,23,14] with the same assumption that all tasks are similar and related. However, for our problem with existence of irrelevant domains which is common in practice, this assumption may not holdwhen two data sets have the same distribution but reversed label assignments. Moreover, informative instances are selected to improve the overall performance on multiple tasks but without a guarantee on every single task, consequently a trade-off must be made between a specific task and a set of tasks which contradicts the transfer learning setting in which the performance of the target task is most concerned. Recently, only a few research works have been done to explore the feasibility of improving active learning given the transfer learning framework [7,21]. A hybrid method is proposed in [4], however, without enough query samples on both positive and negative ones, the algorithm would fail to calculate the corresponding weights because the ratio of the positive instances are taken into the calculation form. Therefore, in the algorithm, the number of query is fixed to a large value, and it turns out to be costly in real applications. The active transfer learning method in [21] tries to use instances from the source domain to label the ones in the target domain by firstly adopting an existing active learning method to induce the initial hypothesis, then the decision function for labeling informative instances which relies on it is used to decide the instance to query. However, as we discussed above, the initial hypothesis is not reliable which may deviate from the true underlying distribution, and consequently impedes the performance on the target task.

As mentioned before, the QBC framework maintains a committee of models which consider the consensus probability based on some disagreement measure instead of an individual model. To utilize the useful information from the source domain, unlike conventional QBC methods which only consider sampled models from the target task and all members are assigned with same weights, we extend the QBC Active Learning framework in cooperation with the Transfer Learning (ALTL) setting with each member as a model from the source domain, in order to avoid training the initial hypotheses by using the re-sampled data. Models in the source tasks are assigned with different weights related to the similarities to the target task, and the weights are updated during each iteration. By exploring the advantages of KL divergence in measuring similarities between models which is proved to be useful in transfer learning, we propose a weighted KL divergence to update the weights of the initial hypotheses that can decrease the negative effects of inferior ones. We also have an adaptive strategy which could eliminate unnecessary instances in the input space and models in the model space. In such a way, our algorithm aims to query only instances of interests in order to avoid the negative transfer problem.

2   Problem Setting and the Framework of ALTL

In this paper, we deal with the classification problem on a target data set where several source data sets are available. There is a task set $\mathbb{S}$ from the source and the target domains which contains $K + 1$ data sets S_i $(i = 1, 2, ..., K, K + 1)$, where the first K data sets are in the source domain and the $(K + 1)$th data set is from the target domain. In the target data set, labeled data is regarded to be insufficient, denoted by L, while abundant unlabeled data is available which is denoted by U. Y is the set of possible labels for the instances with d dimensions and the jth class is denoted by y_j while $y_j \in Y = \{y_1, y_2, ..., y_d\}$. The objective is to obtain the class label y_j for each instance x in the unlabeled data using a d-dimensional parametric model $P_\theta(y_j|x)$ on the target data set, and we write P_θ for convenience. Note that also the model for each of the source data set S_i $(i = 1, 2, ..., K)$ is denoted by P_{θ_i}.

We are motivated to borrow the strength of transfer learning, and let each model induced from the source tasks and the target task to be a committee member, so that to avoid obtaining inferior initial hypotheses. However, an inevitable problem is that, if the discrepancy between the distributions of the source and the target domains is too large, a negative transfer is more likely to happen [16]. Therefore, given the objective to find the most informative instance, we propose a novel weighted disagreement measure to deal with the "inferior" models in the learning stage. In our framework, each model P_{θ_i} in one of the $K + 1$ tasks is regarded as a committee member. Therefore, the labeled information from multiple source data sets can be directly transferred to the target task. During the learning process, the most informative instance is selected based on the disagreement on the class label of all committee members. In the conventional QBC setting [2], all members are treated equally important. However in transfer learning setting, some models may have similar distributions with the target task, while some models have totally different distributions. This situation is also shown later in Figure 2 in Section 5. Therefore, instead of assigning equal weights to every member, a more flexible way is to use different weights based on the similarities of distributions with the target task. Therefore, we firstly consider to measure the similarities among tasks.

KL divergence was successfully implemented in active learning to measure the committee disagreement [2]. It is an information-theoretic measure which captures the expected number of extra "bits of information" required to code samples from one distribution when using a code based on the other. Therefore, we can evaluate the relevance between two models without calculating the real bits needed for coding. In our method, we add a weight w_i for each model P_{θ_i} based on the KL divergence, to measure the discrepancy between the model and the current best model P^* with the lowest error rate, which is given as follows:

$$w_i = \exp\left[-K(P_{\theta_i}(Y|X_L)||P^*(Y|X_L))\right] \tag{1}$$

where $K(Q_1||Q_2)$ denotes the K directed divergence [12] between two probability distributions Q_1 and Q_2 of a discrete random variable, which is defined as:

$$K(Q_1||Q_2) = \sum_j \left[Q_1(j)\log Q_1(j) - Q_1(j)\log\left(\frac{Q_1(j) + Q_2(j)}{2}\right)\right]$$

We adopt K directed divergence instead of KL divergence for the weights in our framework based on the following reason: for Q_1 and Q_2, $KL(Q_1||Q_2)$ is not defined when $Q_2(j) = 0$ but $Q_1(j) > 0$. In our proposal, the same technical difficulty is encountered when we try to calculate $KL(P_{\theta_i}(Y|X_L)||P^*(Y|X_L))$. There are basically two techniques to deal with this kind of problem, one is to smooth the distributions in some way such as [13], for instance with a Bayesian prior or taking the convex combination of the observation with some valid (nonzero) distribution. The second way is to employ heuristics to discard those zero frequencies. However, as pointed out by [6], these methods violate the nature of the true distributions, which may lead to unsatisfactory results. Therefore, we sidestep this problem by using the K directed divergence which inherits the good properties of KL divergence meanwhile avoids the zero frequency problems.

Note that $w_i = 1$ only when $P_{\theta_i}(Y|X_L) = P^*(Y|X_L)$. If for some y, $P^*(y|X_L) = 0$, it is easy to prove that the divergence function is still meaningful. Note that $w_i \leq 1$ all the time, thus the exponentiated function converts the number of "bits of information" into a scalar distance between 0 and 1.

Then we propose our framework, and the most informative instance selected by the committee is given as follows:

$$x^* = \arg\max_x \sum_{i=1}^{K+1} w_i KL(P_{\theta_i}(Y|x)||P_C(Y|x)) \tag{2}$$

where:

$$KL(P_{\theta_i}(Y|x)||P_C(Y|x)) = \sum_j P_{\theta_i}(y_j|x)\log\frac{P_{\theta_i}(y_j|x)}{P_C(y_j|x)} \tag{3}$$

$$P_C(y_j|x) = \frac{\sum_{i=1}^{K+1} w_i P_{\theta_i}(y_j|x)}{\sum_{i=1}^{K+1} w_i} \tag{4}$$

$$w_i = \exp\left[-K(P_{\theta_i}(Y|X_L)||P^*(Y|X_L))\right] \tag{5}$$

where $P_{\theta_i}(Y|x)$ is the class distribution of a committee member, and $P_C(y_j|x)$ is the weighted "consensus" probability that y_j is the correct label. P^* is the current best model. w_i is the weight for each model P_{θ_i} and X_L are the instances in L of the target task.

The argmax function (2) denotes that, the instance with the maximum divergence to the weighted "consensus" probability will be chosen to query. Different from the classical QBC method in the symmetric setting which only measures the disagreement between models, our method is able to balance the divergence between a single model to the current best model, and the divergence between this model to the consensus model.

3   The ALTL Algorithm and Analysis

3.1   The Procedure of ALTL

In pool-based active learning algorithms, one query is selected from the large pool of unlabeled data U in each iteration. Therefore, it is necessary to reduce the region of the instance space in U. In our algorithm, before selecting the most informative instance using the uncertainty measure, the region of the instance space is reduced by eliminating the instances with labels that all committee members agree. It means that in the eliminated region, there does not exist an instance x that, for any two committee members P_{θ_i} and P_{θ_j}, the labels predicted by the two models are identical. This is reasonable because, if all members agree on an instance x, there is no necessary to query the label of this instance as its entropy is equal to zero.

The main flow of our algorithm is summarized in Algorithm 1, and the termination condition (8) in the pseudo code will be explained in Section 5.2. Note that for each iteration, when updating the committee, we add a new committee member by building a model from the labeled data L and the instance queried. Some existing methods [4] try to reduce the model space by eliminating those models that disagree with a query in round i. However, due to the lack of the knowledge of the underlying distribution of the target task, a model that performs unsatisfactorily might possibly be the best model given enough instances, as the red real line shown in Figure 1. As pointed in [15], it is not reasonable to get rid of a model that performs badly in the current stage. In our method, we keep all the models in the committee, and the final model is determined at

---

Algorithm 1. ALTL algorithm

Input: S_i $(i = 1, 2, ..., K, K + 1)$, where the first K tasks are in the source domain
and the last one is from the target domain. Parametric models for the source tasks
$P(y|x, \theta_i)$ $(i = 1, 2, ..., K)$
Output: parametric model $P(y|x, \theta)$, where $x \in UL$
1: Build an initial model $P^0_{\theta_{K+1}}$ from the labeled data L of S_{K+1}, $P^* = P^0_{\theta_{K+1}}$
2: Create a committee C consists of P_{θ_i} $(i = 1, 2, ..., K)$ and $P^0_{\theta_{K+1}}$
3: Set $t = 1$
4: **while** $UL \neq \emptyset$ and the termination condition (8) is not satisfied **do**
5:      Eliminate the instances in UL with labels that all committee members agree
6:      **if** $UL = \emptyset$ **then**
7:          Output P^*
8:      **else**
9:          Select one instance to query by the uncertainty measure of Eq.(2)
10:         Update the current best model P^* of the model with the smallest $\epsilon(P)$
11:         $L = L \cup \{x^*\}$, $UL = UL \backslash \{x^*\}$
12:         Build a new model $P^t_{\theta_{K+1}}$ from L
13:         Update the committee C by $C \cup P^t_{\theta_{K+1}}$
14:     **end if**
15:     $t = t + 1$
16: **end while**
17: Output P^*

---

the end of the iterations. Each model is assigned a weight, and the weights of
those models deviate from the class distribution of the current best model tend
to become lower. In such a way, we try to decrease the impacts of inferior models
in each iteration instead of eliminating them from further consideration.

3.2   Termination Condition and Analysis

A simple way adopted in most existing methods is to set a number N as the
maximum number of iterations to terminate the loop of querying unlabeled
instances [2,21]. Although in such a way, it is easy to control the number of
queries in the learning procedure, we can not guarantee the performance of
the algorithm. To learn a classifier with an error less than ϵ in a noise-free
environment, passive learning requires $O(\frac{1}{\epsilon})$ labels while binary search needs
$O(\ln \frac{1}{\epsilon})$ labels [15]. That means N should not be larger than $\frac{1}{\epsilon}$ or it will be
meaningless to adopt active learning. Since ϵ is not available in the initial stage
of learning, it is not appropriate to fix the number of iterations.

   We adopt a flexible way to terminate the learning process. We firstly define
the hypothesis space $\mathcal{H}_t$ in iteration t that consists of the models with the error
rates between the lower bound LB and the upper bound UB.

$$\mathcal{H}_t = \{P_{\theta_i} : LB \leq \epsilon(P_{\theta_i}) \leq UB\} \tag{6}$$

where

$$LB = \min \epsilon(P_{\theta_i}), i \in (1, 2, ..., K + t)$$

$$UB = \frac{(|U| + |L|\epsilon_{min}(P_{\theta_i}))}{|S_{k+1}|}$$

For the termination condition, we define the Volume of the current region of Uncertainty (VOU) similar to [15] as:

$$VOU(\mathcal{H}_t) = \mathbf{Pr}_{x \in U}[\exists P_{\theta_i}, P_{\theta_j} \in \mathcal{H}_t : P_{\theta_i}(y|x) \neq P_{\theta_j}(y|x)] \qquad (7)$$

Therefore, the termination condition can be written as:

$$VOU(\mathcal{H}_t) \leq \epsilon_0 \qquad (8)$$

where ϵ_0 denotes the allowed error rate.

Our model is able to distinguish those models from the source domain which deviate from the distribution underlying the target domain. If all the models are different from the target task, the weights of these models become zero and the transfer learning problem can thereafter be regarded as a single task learning. Only if the weights of all the model are equal to 1, the asymmetric problem is regarded as the symmetric multi-task learning problem where all the tasks are treated as equally important. Therefore, our algorithm can adaptively fit to the learning scenario.

4   Experiments

We perform experiments on both the synthetic data sets [21] with specific structures to demonstrate the performance of our algorithm, and two real data sets that are the 20 Newsgroups data sets[1] and the 4 Universities data sets[2]. Our algorithm is compared with the basic QBC model [10], the state-of-the-art active learning method, as well as the random query. We also compare our method with Active Transfer [21] which is a hybrid method combine the active learning and the transfer learning. The baseline method is chosen as SVM. We follow the values of the parameters in the original papers. 10 labeled instances are chosen in the beginning for each experiment. In the following figures, we use "ALTL" to denote our algorithm, "QBC" for the QBC algorithm, "RQ" for the random query learning method, and "AT" as the active transfer algorithm.

---

[1] http://people.csail.mit.edu/jrennie/20Newsgroups
[2] http://www.cs.cmu.edu/afs/cs/project/theo-20/www/data/

4.1   Results on Synthetic Data Sets

For the synthetic data sets, we use the same two-dimensional data sets in [21] as shown in Figure 2, in which the red circles denote the positive instances and the blue crosses represent the negative instances. Note that, except for the two data sets (a) and (b), others have different distributions. For the data set (c), it shares some similarities with (a) and (b) but for the data set (d), the underlying distribution is dramatically different. The data set (e) has the reversed distribution from (a) and (b). We conduct experiments on each of the data set as the target task, with others as the source tasks.

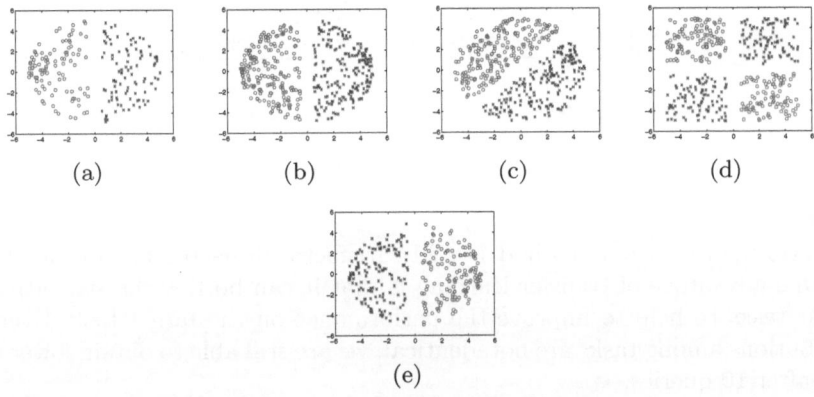

(a)   (b)   (c)   (d)

(e)

Fig. 1. Structures of the synthetic data sets

The objectives of our method is to query as few instances as possible, as well as to obtain a high accuracy in classification on the unlabeled data sets. For the experiments with the synthetic data sets, we set each of the data set in Figure 2 as a single target task, while others are treated as the source tasks. The results are presented in Figure 3. Generally, our method is the best among all the methods especially when the number of queries becomes larger than 5, and also it converges quickly than others.

For Figure 3(a) and 3(b), the performances of all the methods are obviously better than those on the other data sets. For example in Figure 3(a), the error rate for the baseline method is about 0.12, while the result of our method becomes stable and converges to 0 after about 20 queries. For RQ and QBC, the convergence is not as good as our method. The reason can be concluded as that for the data set (a) as the target task, there exists a similar source task (b) with the same underlying distribution and therefore transfer learning could bring much improvements on the accuracy. The same situation occurs in (b) in which our method converges quickly after about 10 queries.

The distribution for the target task (c) is similar but not identical with those underlying (a) and (b). However, the results shown in Figure 3(c) illustrate that

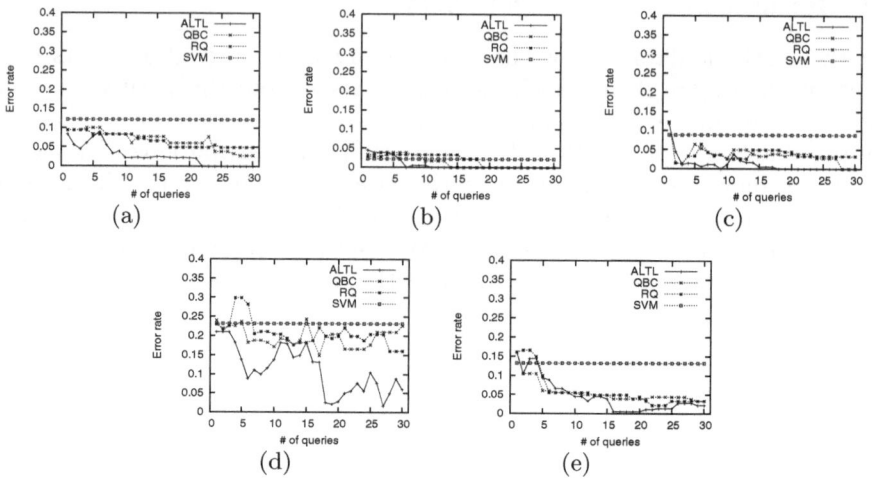

Fig. 2. Error rates on synthetic data sets

the performance of our method is still satisfactory due to the transfer learning. One advantage of transfer learning is that it can borrow the strength from similar tasks to help to improve the performance on the target task. Even the distributions among tasks are not identical, we are still able to obtain lower error rates after 10 queries.

The challenge tasks are (d) and (e), while in (d), the distribution is quite different from others but still there are some similarities, and the data set (e) has the reversed distribution with (a) and (b). The results on the two data sets show us that the performances of all the algorithms on task (d) fluctuate and our algorithm becomes much stable after 20 queries. On the task (e), the reversed distribution with (a) and (b) helps to improve the results and our method obtains an error rate nearly to 0 after 15 queries. The possible reason is that, in transfer learning, the negative information is sometimes helpful to improve the performance as shown in (e) when the target task has a totally different distribution with others but spatially similar. However in the task (d), the useful information can be transferred from the source domain to the target domain is inadequate.

4.2   Results on Real Data Sets

Firstly, we show the results on 20 Newsgroups data sets in Figure 4. Generally, our method outperforms others in most circumstances especially after 10 queries. For example in (a) rec vs sci, the performance of our algorithm becomes stable after about 15 queries. However, in the initial stage, the error rate for ALTL fluctuates and sometimes is higher than others. We believe that, in the initial stage, some committee members outfit the target task and therefore they gain higher weights, but after several iterations, our algorithm could decrease their weights and meantime more robust committee members start to dominate the learning process. Therefore, after several queries, the performance becomes stable and the informative instances could be adaptively selected.

For the results on the other categories, our ALTL is able to achieve lower error rates than the other methods. It outperforms the basic QBC algorithm by larger than 10% in most circumstances, and converges more quickly than others. The results could serve as an evidence of the effectiveness of our algorithm.

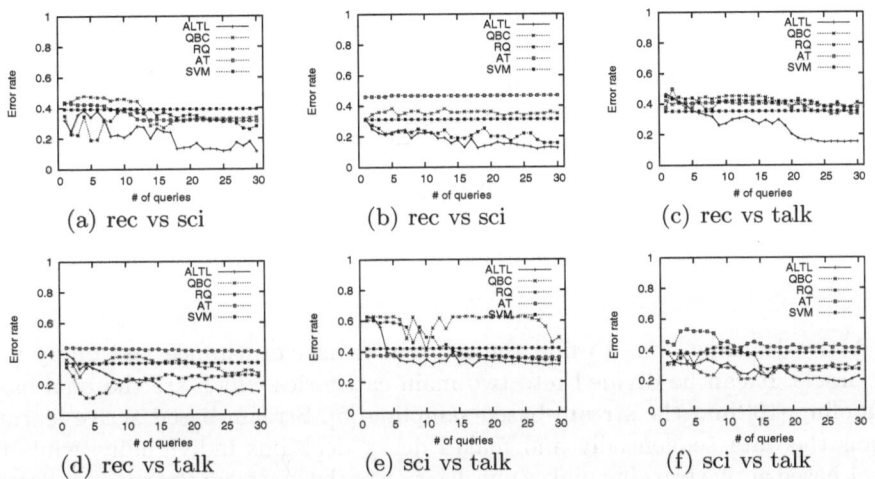

(a) rec vs sci   (b) rec vs sci   (c) rec vs talk

(d) rec vs talk   (e) sci vs talk   (f) sci vs talk

Fig. 3. Error rates for the 20 Newsgroups data set

As illustrated in Figure 5, for the 4 universities data sets, our method is still the best one among all the methods, especially after 10 queries. Note that, the error rates in Cornell are lower than those in the other figures, while in Texas, the discrepancy is much larger. Even in this kind of circumstances, our algorithm could obtain an error rate as low as 5%, which is much lower than the error rates of the other methods. It is obvious that our ALTL converges more quickly than the state-of-the-art algorithms. In each iteration, the most informative instance could be selected effectively which leads to the overall improvement of the classification accuracy.

5   Related Work

In this section, we review research works relevant to ours. Our ALTL belongs to the supervised inductive transfer learning where both the source and the target tasks contain labeled data [17,8]. For the perspective of active learning, our method is the pool-based active learning algorithm [3].

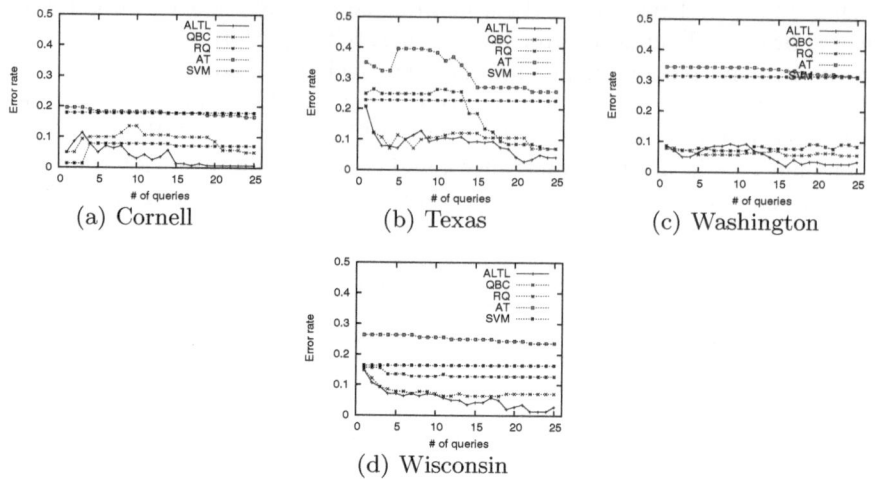

Fig. 4. Error rates for the 4 Universities data set

Active learning aims to find the most informative instance and query it to an oracle. It can be divided into two main categories which are the pool-based sampling [10] and the stream-based sampling [5]. Stream-based active learning scans the data sequentially and makes query decisions individually, while the pool-based active learning makes one decision each time from the entire collection of the unlabeled pool. [19] provided an active learner that can identify data points that change the current belief in the class distributions the most. In [11], AL had been applied in the multi-view setting. In multi-view problem, features can be partitioned into subsets each of which is sufficient for learning the mapping from the input to the output space.

Although there exist several papers concerning integrating active learning and multi-task learning [18,22,1,23,14], with the objective to improve the overall performance rather than a single task, they can not be applied to our problem where the performance of the target task is emphasized. Some research works have been done by combining the active learning and transfer learning together in order to improve the classification accuracy on the target task, however they did not consider the negative transfer problem explicitly. For example, [7] provided a hybrid method which combines the discriminative method and the generative method with SVM, but the query number is fixed to a large number and there is not an effective strategy to avoid selecting unconcerned instances. In [21], it is pointed out by the author that experts are heavily relied on, as the possibility to query an expert is set to be higher than 50%. Moreover, the decision function which relies on the initial hypothesis of an existing active learning method is not reliable due to the initial hypothesis problem. [22] proposed an active learning framework that can exploit relations among multiple tasks on their labels.

6 Conclusion

In this paper, we propose a novel active learning framework by extending the basic QBC (Query by Committee) algorithm in the target domain, but with the help of the useful information from the source domain. A weighted KL divergence measurement is adopted to evaluate the similarities between committee members, in order to decrease the negative effects of inferior models. An adaptive strategy is designed to get rid of those unnecessary instances and models. By incorporating the advantages of both active learning and transfer learning, our method is able to obtain high classification accuracy with less queries meanwhile converges faster than the state-of-the-art methods.

Acknowledgment. This work was partially supported by the grant-in-aid for scientific research on fundamental research (B) 21300053 and 24650070 from the Japanese Ministry of Education, Culture, Sports, Science and Technology, and the Strategic International Cooperative Program funded by the Japan Science and Technology Agency (JST).

References

1. Harpale, A., Yang, Y.: Active Learning for Multi-Task Adaptive Filtering. In: Proceeding of the 27th International Conference of Machine Learning (ICML 2010), pp. 431–438 (2010)
2. McCallum, A., Nigam, K.: Employing EM in Pool-based Active Learning for Text Classification. In: Proceeding of the 15th International Conference of Machine Learning (ICML 1998), pp. 350–358 (1998)
3. Settles, B.: Active Learning Literature Survey. Technical Report 1648, University of Wisconsin–Madison (2010)
4. Cohn, D., Ladner, R., Waibel, A.: Improving Generalization with Active Learning. In: Machine Learning, pp. 201–221 (1994)
5. Lewis, D.D., Gale, W.A.: A Sequential Algorithm for Training Text Classifiers. In: Proceedings of the 17th Annual International ACM SIGIR Conference on Research and Development in Information Retrieval (SIGIR 1994), pp. 3–12 (1994)
6. Pereira, F., Tishby, N., Lee, L.: Distributional Clustering of English Words. In: Proceeding of the 31st Annual Meeting of the Association for Computational Linguistics (ACL 1993), pp. 183–190 (1993)
7. Li, H., Shi, Y., Chen, M.Y., Hauptmann, A.G., Xiong, Z.: Hybrid Active Learning for Cross-domain Video Concept Detection. In: Proceedings of the International Conference on Multimedia (MM 2010), pp. 1003–1006 (2010)
8. Shao, H., Suzuki, E.: Feature-based Inductive Transfer Learning through Minimum Encoding. In: Proceeding of the SIAM International Conference on Data Mining (SDM 2011), pp. 259–270 (2011)
9. Seung, H.S., Opper, M., Sompolinsky, H.: Query by Committee. In: Computational Learning Theory, pp. 287–294 (1992)
10. Dagan, I., Engelson, S.P.: Committee-Based Sampling For Training Probabilistic Classifiers. In: Proceeding of the 23rd International Conference on Machine Learning (ICML 2006), pp. 150–157 (1995)

11. Muslea, I., Minton, S., Knoblock, C.A.: Active Semi-Supervised Learning = Robust Multi-View Learning. In: Proceeding of the 19th International Conference on Machine Learning (ICML 2002), pp. 435–442 (2002)
12. Lin, J.: Divergence Measures based on the Shannon Entropy. IEEE Transactions on Information theory 37, 145–151 (1991)
13. Church, K.W., Gale, W.A.: A Comparison of the Enhanced Good-Turing and Deleted Estimation Methods for Estimating Probabilities of English Bigrams. Computer Speech and Language 5, 19–54 (1991)
14. Li, L., Jin, X., Pan, S., Sun, J.: Multi-Domain Active Learning for Text Classification. In: Proceeding of the 18th ACM SIGKDD Conference on Knowledge Discovery and Data Mining, KDD 2012 (2012)
15. Balcan, M.F., Beygelzimer, A., Langford, J.: Agnostic Active Learning. In: Proceeding of the 23rd International Conference on Machine Learning (ICML 2006), pp. 65–72 (2006)
16. Rosenstein, M.T., Marx, Z., Kaelbling, L.P., Dietterich, T.G.: To Transfer or not to Transfer. In: NIPS 2005 Workshop, Inductive Transfer: 10 Years Later (2005)
17. Caruana, R.: Multitask Learning. Machine Learning 28, 41–75 (1997)
18. Reichart, R., Tomanek, K., Hahn, U.: Multi-task Active Learning for Linguistic Annotations. In: Annual Meeting of the Association for Computational Linguistics (ACL 2008), pp. 861–869 (2008)
19. Raj, S., Ghosh, J., Crawford, M.M.: An Active Learning Approach to Knowledge Transfer for Hyperspectral Data Analysis. In: Proceeding of the IEEE International Conference on Geoscience and Remote Sensing Symposium (IGARSS 2006), pp. 541–544 (2006)
20. Dai, W., Yang, Q., Xue, G.R., Yu, Y.: Boosting for Transfer Learning. In: Proceeding of the 24th International Conference of Machine Learning (ICML 2007), pp. 193–200 (2007)
21. Shi, X., Fan, W., Ren, J.: Actively Transfer Domain Knowledge. In: Daelemans, W., Goethals, B., Morik, K. (eds.) ECML PKDD 2008, Part II. LNCS (LNAI), vol. 5212, pp. 342–357. Springer, Heidelberg (2008)
22. Zhang, Y.: Multi-Task Active Learning with Output Constraints. In: Proceeding of the 24th AAAI Conference on Artificial Intelligence, AAAI 2010 (2010)
23. Zhu, Z., Zhu, X., Ye, Y., Guo, Y.-F., Xue, X.: Transfer Active Learning. In: Proceeding of the 20th International Conference on Information and Knowledge Management (CIKM 2011), pp. 2169–2172 (2011)

Variational Learning of Dirichlet Process Mixtures of Generalized Dirichlet Distributions and Its Applications

Wentao Fan and Nizar Bouguila

Concordia Institute for Information Systems Engineering
Concordia University, QC, Canada
wenta_fa@encs.concordia.ca, nizar.bouguila@concordia.ca

Abstract. In this paper, we introduce a nonparametric Bayesian approach for clustering based on both Dirichlet processes and generalized Dirichlet (GD) distribution. Thanks to the proposed approach, the obstacle of estimating the correct number of clusters is sidestepped by assuming an infinite number of components. The problems of overfitting and underfitting the data are also prevented due to the nature of the nonparametric Bayesian framework. The proposed model is learned through a variational method in which the whole inference process is analytically tractable with closed-form solutions. The effectiveness and merits of the proposed clustering approach are investigated on two challenging real applications namely anomaly intrusion detection and image spam filtering.

Keywords: Clustering, mixture models, Dirichlet process, nonparametric Bayesian, generalized Dirichlet mixtures, variational inference, intrusion detection, spam image.

1 Introduction

During the last decade, Bayesian nonparametric models have received significant attention from the machine learning and data mining communities [24]. As opposed to parametric approaches in which a fixed number of parameters are used, Bayesian nonparametric approaches allow the complexity of models to grow with data size. The Dirichlet process (DP) is currently one of the most popular Bayesian nonparametric models and is defined as a distribution over distributions [29,23]. Thanks to the development of Markov chain Monte Carlo (MCMC) [38] techniques, Dirichlet processes are now widely used in various domains [35,40]. One of the most common applications of the Dirichlet process is in clustering data where it is translated to a mixture model with a countably infinite number of mixture components [37,4,10,13]. The DP mixture model can then be represented as an infinite mixture model in which the difficulty of selecting the number of clusters, that usually occurs in the finite case, is removed. In DP mixture model, the actual number of clusters used to model data is not

S. Zhou, S. Zhang, and G. Karypis (Eds.): ADMA 2012, LNAI 7713, pp. 199–213, 2012.

fixed and can be automatically inferred from the data set using a Bayesian posterior inference framework via MCMC techniques. However, in practice, the use of MCMC methods is often limited to small-scale problems due to the high computational resources required. Another problem regarding MCMC approaches is the difficulty of diagnosing the convergence. An efficient alternative to MCMC techniques is a deterministic approximation approach known as variational inference or variational Bayes [28,5]. The variational approach is based on analytical approximations to the posterior distribution and has received a lot of attention recently. Compared to MCMC techniques, only a modest amount of computational time is required for variational methods. Moreover, the well-known Bayesian information criterion (BIC) for model selection can be derived as a special case of variational inference learning [5]. Due to its convincing generalization power and computational tractability, variational learning have been applied in various applications [32,22,21].

Most of the works regarding infinite mixture models make the Gaussian assumption. This assumption, however, is not realistic for non-Gaussian data and in particular normalized count data (i.e., proportion vectors) which are of particular importance in several disciplines like computer vision, data mining, machine learning, and bioinformatics and which arise in a wide variety of applications such as text and image modeling [1]. Indeed, recent works have shown that in the case of proportional data which subject to two restrictions, namely, nonnegativity and unit-sum constraint, other distributions can give better results and are more appropriate, such as the generalized Dirichlet (GD) [12,9,11,16] distribution.

The purpose of this paper is to propose a Bayesian nonparametric approach for clustering based on Dirichlet processes mixtures with GD distribution, which can be seen as an infinite GD mixture model. Our contributions are summarized as the following: first, we extend the finite GD mixture model into an infinite version using a stick-breaking construction [39,27] such that the difficulty of choosing the appropriate number of clusters is avoided. Second, we develop a variational inference framework for learning the proposed model, such that the whole inference process is analytically tractable with closed-form solutions. Finally, we apply the proposed Bayesian nonparametric approach on two challenging real-world applications involving anomaly intrusion detection and image spam filtering.

The rest of this paper is organized as follows: Section 2 presents the details of our infinite GD mixture model with the stick-breaking representation. In Section 3, we describe our variational approximation procedure for the proposed model learning. Section 4 presents the experimental results. Section 5 closes this paper with conclusions and future directions.

2   The Infinite Generalized Dirichlet Mixture Model

2.1   The Finite Generalized Dirichlet Mixture

Let us consider a D-dimensional random vector $\boldsymbol{Y} = (Y_1, \ldots Y_D)$ which is assumed to be generated from a finite generalized Dirichlet (GD) distribution [12]:

$$\text{GD}(\boldsymbol{Y}|\boldsymbol{\alpha}_j,\boldsymbol{\beta}_j) = \prod_{l=1}^{D} \frac{\Gamma(\alpha_{jl}+\beta_{jl})}{\Gamma(\alpha_{jl})\Gamma(\beta_{jl})} Y_l^{\alpha_{jl}-1} \left(1 - \sum_{k=1}^{l} Y_k\right)^{\gamma_{jl}} \tag{1}$$

for $\sum_{l=1}^{D} Y_l < 1$ and $0 < y_l < 1$ for $l = 1, \ldots, D$, where $\boldsymbol{\alpha}_j = (\alpha_{j1}, \ldots, \alpha_{jD})$, $\boldsymbol{\beta}_j = (\beta_{j1}, \ldots, \beta_{jD})$, $\alpha_{jl} > 0$, $\beta_{jl} > 0$, $\gamma_{jl} = \beta_{jl} - \alpha_{jl+1} - \beta_{jl+1}$ for $l = 1, \ldots, D-1$, and $\gamma_{jD} = \beta_{jD} - 1$. $\Gamma(\cdot)$ is the gamma function defined by $\Gamma(x) = \int_0^{\infty} u^{x-1} e^{-u} du$. Let $\mathcal{Y} = \{\boldsymbol{Y}_1, \ldots, \boldsymbol{Y}_N\}$ denotes a set of N independent and identically distributed vectors assumed to arise from a finite generalized Dirichlet mixture

$$p(\boldsymbol{Y}_i|\boldsymbol{\pi},\boldsymbol{\alpha},\boldsymbol{\beta}) = \sum_{j=1}^{M} \pi_j \text{GD}(\boldsymbol{Y}_i|\boldsymbol{\alpha}_j,\boldsymbol{\beta}_j) \tag{2}$$

where $\boldsymbol{\alpha} = (\boldsymbol{\alpha}_1, \ldots, \boldsymbol{\alpha}_M)$, $\boldsymbol{\beta} = (\boldsymbol{\beta}_1, \ldots, \boldsymbol{\beta}_M)$, and $\boldsymbol{\pi}$ is the vector of mixing coefficients, $\boldsymbol{\pi} = (\pi_1, \ldots, \pi_M)$, which are positive and sum to one.

In this paper, we exploit an interesting mathematical property of the GD distribution thoroughly discussed in [14] to transform the original data points into another D-dimensional space with independent features and rewrite the finite GD mixture model in the following form

$$p(\boldsymbol{X}_i|\boldsymbol{\pi},\boldsymbol{\alpha},\boldsymbol{\beta}) = \sum_{j=1}^{M} \pi_j \prod_{l=1}^{D} \text{Beta}(X_{il}|\alpha_{jl},\beta_{jl}) \tag{3}$$

where $X_{i1} = Y_{i1}$ and $X_{il} = Y_{il}/(1 - \sum_{k=1}^{l-1} Y_{ik})$ for $l > 1$. $\text{Beta}(X_{il}|\alpha_{jl},\beta_{jl})$ is a Beta distribution defined with parameters $(\alpha_{jl}, \beta_{jl})$.

2.2   The Infinite Generalized Dirichlet Mixture

The infinite GD mixture model proposed in this paper is constructed using a Dirichlet process with a stick-breaking representation and is defined as follows: given a random distribution G, it is distributed according to a DP: $G \sim DP(\psi, H)$ if the following conditions are satisfied:

$$\lambda_j \sim \text{Beta}(1, \psi), \qquad \theta_j \sim H, \qquad \pi_j = \lambda_j \prod_{s=1}^{j-1} (1 - \lambda_s), \qquad G = \sum_{j=1}^{\infty} \pi_j \delta_{\theta_j} \tag{4}$$

where δ_{θ_j} denotes the Dirac delta measure centered at θ_j, and ψ is a positive real number. The mixing weights π_j are obtained by recursively breaking an unit length stick into an infinite number of pieces. Now assume that we have an observed data set $\mathcal{X} = (\boldsymbol{X}_1, \ldots, \boldsymbol{X}_N)$ with an infinite number of clusters. Then, a latent variable $Z = (Z_1, \ldots, Z_N)$ is introduced as the mixture component assignment variable, in which each element Z_i takes an integer value j denoting the component from which X_i is drawn. The marginal distribution over Z can be specified in terms of the mixing coefficients π_j as

$$p(Z|\boldsymbol{\pi}) = \prod_{i=1}^{N} \prod_{j=1}^{\infty} \pi_j^{\mathbf{1}[Z_i=j]} \tag{5}$$

where $\mathbf{1}[Z_i = j]$ is an indicator function which has the value 1 when $Z_i = j$ and 0 otherwise. Based on (4), we can rewrite (5) as a function of $\boldsymbol{\lambda}$ by

$$p(Z|\boldsymbol{\lambda}) = \prod_{i=1}^{N} \prod_{j=1}^{\infty} \left[\lambda_j \prod_{s=1}^{j-1} (1 - \lambda_s) \right]^{\mathbf{1}[Z_i=j]} \tag{6}$$

According to (4), the marginal distribution of $\boldsymbol{\lambda}$ is given by

$$p(\boldsymbol{\lambda}|\boldsymbol{\psi}) = \prod_{j=1}^{\infty} \mathrm{Beta}(1, \psi_j) = \prod_{j=1}^{\infty} \psi_j (1 - \lambda_j)^{\psi_j - 1} \tag{7}$$

More flexibility is added by introducing a prior distribution over the hyperparameter $\boldsymbol{\psi}$. Motivated by the fact that the Gamma distribution is conjugate to the stick lengths [7], a Gamma prior is placed over $\boldsymbol{\psi}$

$$p(\boldsymbol{\psi}) = \mathcal{G}(\boldsymbol{\psi}|\boldsymbol{a}, \boldsymbol{b}) = \prod_{j=1}^{\infty} \frac{b_j^{a_j}}{\Gamma(a_j)} \psi_j^{a_j - 1} e^{-b_j \psi_j} \tag{8}$$

It is noteworthy that the conditional distribution of $\boldsymbol{X}_i$ given Z, $\boldsymbol{\alpha}$ and $\boldsymbol{\beta}$ is equivalent to the distribution of $\boldsymbol{X}_i$ conditioned on the Z_ith component in the mixture and is given by

$$p(\mathcal{X}|Z, \boldsymbol{\alpha}, \boldsymbol{\beta}) = \prod_{i=1}^{N} p(\boldsymbol{X}_i|\boldsymbol{\alpha}_{Z_i}, \boldsymbol{\beta}_{Z_i}) = \prod_{j=1}^{\infty} \prod_{l=1}^{D} \left\{ \left[\frac{\Gamma(\alpha_{jl} + \beta_{jl})}{\Gamma(\alpha_{jl})\Gamma(\beta_{jl})} \right]^{n_j} \right.$$
$$\left. \times \prod_{i=1}^{N} [X_{il}^{\alpha_{jl}-1} (1 - X_{il})^{\beta_{jl}-1}]^{\mathbf{1}[Z_i=j]} \right\} \tag{9}$$

where n_j is the occupation number which represents the number of observations belonging to class j and is defined as $n_j = \sum_{i=1}^{N} \mathbf{1}[Z_i = j]$.

Next, we need to introduce priors over unknown parameters $\boldsymbol{\alpha}$ and $\boldsymbol{\beta}$. Since the formal conjugate prior for the Beta distribution is intractable, we adopt Gamma priors $\mathcal{G}(\cdot)$ as suggested in [32] to approximate the conjugate priors by assuming that the Beta parameters are statistically independent, such that: $p(\boldsymbol{\alpha}) = \mathcal{G}(\boldsymbol{\alpha}|\boldsymbol{u}, \boldsymbol{v})$ and $p(\boldsymbol{\beta}) = \mathcal{G}(\boldsymbol{\beta}|\boldsymbol{s}, \boldsymbol{t})$. A directed graphical representation of this model is illustrated in Fig. 1.

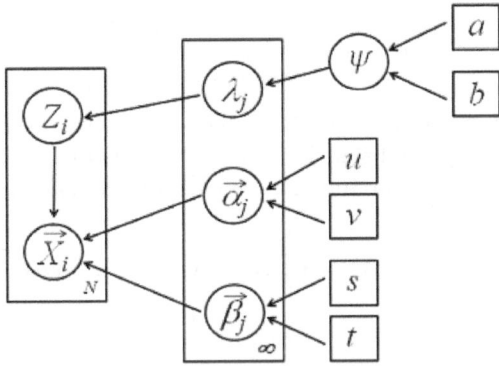

Fig. 1. Graphical model representation of the infinite generalized Dirichlet mixture model. Symbols in circles denote random variables; symbols in squares denote hyperparameters. Plates indicate repetition (with the number of repetitions in the lower right), and arcs describe conditional dependencies between variables.

3 Variational Infinite GD Mixture

In this section, we propose a truncated variational inference on the stick-breaking representation of the infinite GD mixture model. To simplify notation, we define $\Theta = \{Z, \alpha, \beta, \lambda, \psi\}$ as the set of latent variables and parameters. Similarly, the set of all observed variables is represented by $\mathcal{X}$. The central idea in variational learning is to find an approximation for the posterior distribution $p(\Theta|\mathcal{X})$ as well as for the model evidence $p(\mathcal{X})$. By applying Jensen's inequality, the lower bound $\mathcal{L}$ of the logarithm of the marginal likelihood $p(\mathcal{X})$ can be found as

$$\mathcal{L}(Q) = \int Q(\Theta) \ln[p(\mathcal{X}, \Theta)/Q(\Theta)]d\Theta \tag{10}$$

where $Q(\Theta)$ is an approximation for $p(\Theta|\mathcal{X})$. In our case, we truncated the variational distributions at a value M, such that: $\lambda_M = 1, \pi_j = 0$ when $j > M, \sum_{j=1}^{M} \pi_j = 1$, where the truncation level M is a variational parameter which can be freely initialized and will be optimized automatically during the learning process [7]. Moreover, we adopt a factorization assumptions for restricting the form of $Q(\Theta)$ [41]. Then, we have

$$Q(\Theta) = \left[\prod_{i=1}^{N} Q(Z_i)\right] \left[\prod_{j=1}^{M} \prod_{l=1}^{D} Q(\alpha_{jl}) Q(\beta_{jl})\right] \left[\prod_{j=1}^{M} Q(\lambda_j) Q(\psi_j)\right] \tag{11}$$

In order to maximize the lower bound $\mathcal{L}(Q)$, we need to make a variational optimization of $\mathcal{L}(Q)$ with respect to each of the distributions $Q_i(\Theta_i)$ in turn. For a specific factor $Q_s(\Theta_s)$, the general expression for its optimal solution can be found by

$$Q_s(\Theta_s) = \frac{\exp\langle \ln p(\mathcal{X}, \Theta)\rangle_{i \neq s}}{\int \exp\langle \ln p(\mathcal{X}, \Theta)\rangle_{i \neq s} d\Theta} \tag{12}$$

where $\langle \cdot \rangle_{i \neq s}$ is the expectation with respect to all the distributions of $Q_i(\Theta_i)$ except for $i = s$. The optimal solutions for the factors of the variational posterior can then be obtained by applying (12) to each of the factor, such that

$$Q(Z) = \prod_{i=1}^{N} \prod_{j=1}^{M} r_{ij}^{\mathbf{1}[Z_i=j]}, \quad Q(\lambda) = \prod_{j=1}^{M} \text{Beta}(\lambda_j | c_j, d_j), \quad Q(\psi) = \prod_{j=1}^{M} \mathcal{G}(\psi_j | a_j^*, b_j^*) \tag{13}$$

$$Q(\alpha) = \prod_{j=1}^{M} \prod_{l=1}^{D} \mathcal{G}(\alpha_{jl} | u_{jl}^*, v_{jl}^*), \quad Q(\beta) = \prod_{j=1}^{M} \prod_{l=1}^{D} \mathcal{G}(\beta_{jl} | s_{jl}^*, t_{jl}^*), \tag{14}$$

where we have defined

$$r_{ij} = \frac{\rho_{ij}}{\sum_{j=1}^{M} \rho_{ij}}, \quad a_j^* = a_j + 1, \quad b_j^* = b_j - \langle \ln(1 - \lambda_j)\rangle \tag{15}$$

$$c_j = 1 + \sum_{i=1}^{N} \langle Z_i = j\rangle, \quad d_j = \langle \psi_j \rangle + \sum_{i=1}^{N} \sum_{s=j+1}^{M} \langle Z_i = s\rangle \tag{16}$$

$$\rho_{ij} = \exp\left\{ \sum_{l=1}^{D} \left[\widetilde{\mathcal{R}}_{jl} + (\bar{\alpha}_{jl} - 1)\ln X_{il} + (\bar{\beta}_{jl} - 1)\ln(1 - X_{il}) \right] + \langle \ln \lambda_j\rangle + \sum_{s=1}^{j-1} \langle \ln(1 - \lambda_s)\rangle \right\} \tag{17}$$

$$u_{jl}^* = u_{jl} + \sum_{i=1}^{N} \langle Z_i = j\rangle [\Psi(\bar{\alpha}_{jl} + \bar{\beta}_{jl}) - \Psi(\bar{\alpha}_{jl}) + \bar{\beta}_{jl}\Psi'(\bar{\alpha}_{jl} + \bar{\beta}_{jl})(\langle \ln \beta_{jl}\rangle - \ln \bar{\beta}_{jl})]\bar{\alpha}_{jl} \tag{18}$$

$$s_{jl}^* = s_{jl} + \sum_{i=1}^{N} \langle Z_i = j\rangle [\Psi(\bar{\alpha}_{jl} + \bar{\beta}_{jl}) - \Psi(\bar{\beta}_{jl}) + \bar{\alpha}_{jl}\Psi'(\bar{\alpha}_{jl} + \bar{\beta}_{jl})(\langle \ln \alpha_{jl}\rangle - \ln \bar{\alpha}_{jl})]\bar{\beta}_{jl} \tag{19}$$

$$v_{jl}^* = v_{jl} - \sum_{i=1}^{N} \langle Z_i = j\rangle \ln X_{il}, \quad t_{jl}^* = t_{jl} - \sum_{i=1}^{N} \langle Z_i = j\rangle \ln(1 - X_{il}) \tag{20}$$

where $\Psi(\cdot)$ and $\Psi'(\cdot)$ are the digamma and trigamma functions, respectively. Note that, $\widetilde{\mathcal{R}}$ is the lower bounds of $\mathcal{R} = \langle \ln \frac{\Gamma(\alpha+\beta)}{\Gamma(\alpha)\Gamma(\beta)} \rangle$. Since this expectations is intractable, we use the second-order Taylor series expansion to find its lower bounds. The expected values in the above formulas are given by

$$\bar{\alpha}_{jl} = \langle \alpha_{jl}\rangle = \frac{u_{jl}^*}{v_{jl}^*}, \quad \bar{\beta}_{jl} = \langle \beta_{jl}\rangle = \frac{s_{jl}^*}{t_{jl}^*}, \quad \langle Z_i = j\rangle = r_{ij}, \quad \langle \psi_j\rangle = \frac{a_j^*}{b_j^*} \tag{21}$$

$$\langle \ln \lambda_j\rangle = \Psi(c_j) - \Psi(c_j + d_j), \quad \langle \ln(1 - \lambda_j)\rangle = \Psi(d_j) - \Psi(c_j + d_j) \tag{22}$$

$$\langle \ln \alpha_{jl}\rangle = \Psi(u_{jl}^*) - \ln v_{jl}^*, \quad \langle (\ln \alpha_{jl} - \ln \bar{\alpha}_{jl})^2\rangle = [\Psi(u_{jl}^*) - \ln u_{jl}^*]^2 + \Psi'(u_{jl}^*) \tag{23}$$

$$\langle \ln \beta_{jl} \rangle = \Psi(s_{jl}^*) - \ln t_{jl}^* \ , \qquad \langle (\ln \beta_{jl} - \ln \bar{\beta}_{jl})^2 \rangle = [\Psi(s_{jl}^*) - \ln t_{jl}^*]^2 + \Psi'(s_{jl}^*) \quad (24)$$

Since the solutions to each variational factor are coupled together through the expected values of other factors, the optimization of the model can be solved in a way analogous to the EM algorithm. The complete algorithm is summarized in Algorithm 1.[1]

Algorithm 1. Variational infinite GD mixture learning

1: Choose the initial truncation level M and the initial values for hyper-parameters u_{jl}, v_{jl}, s_{jl}, t_{jl}, a_j and b_j.
2: Initialize the value of r_{ij} by K-Means algorithm.
3: **repeat**
4:     The variational E-step: Estimate the expected values in $(21)\sim(24)$, use the current distributions over the model parameters.
5:     The variational M-step: Update the variational solutions for $Q(Z)$, $Q(\alpha)$, $Q(\beta)$, $Q(\lambda)$ and $Q(\psi)$ using the current values of the moments.
6: **until** Convergence criterion is reached.
7: Compute the expected value of λ_j as $\langle \lambda_j \rangle = \frac{c_j}{c_j + d_j}$ and substitute it into (4) to obtain the estimated values of the mixing coefficients π_j.
8: Detect the optimal number of components M by eliminating the components with small mixing coefficients close to 0.

---

4   Experimental Results

In this section, the effectiveness of the proposed approach is tested on two challenging applications: anomaly intrusion detection and image spam detection. In our experiments, the initialization value of the truncation level M is set to 15 with equal mixing coefficients. In order to provide broad non-informative prior distributions, the initial value of u and s for the conjugate priors are set to 1, and v, t are set to 0.01. The hyperparameters a and b are both initialized to 1. For comparison, we have also applied three other well-developed approaches for these two applications: the variational finite GD mixture (varGDM) model, the variational infinite Gaussian mixture (varInGM) model proposed in [7], and the variational Gaussian mixture (varGM) model proposed in [17]. Notice that, our goal is to demonstrate the advantages of using infinite mixture models over finite ones, as well as using GD mixtures over Gaussian mixtures. Comparing our results with other state of the art methods that have been applied to both applications is obviously beyond the scope of this paper.

4.1   Anomaly Intrusion Detection

Recently, an increasing number of security threats have brought a serious risk to the internet and computer networks. Therefore, Intrusion Detection Systems

---

[1] The complete source code is available upon request.

(IDSs) play a critical role in strengthening the security of information and communication systems [30,33]. Normally, IDSs are classified into two main categories: misuse (i.e. signature-based) detection and anomaly detection systems [36]. Compared to the misuse detection, anomaly detection has the advantage of being able to detect new or unknown attacks. In this experiment, we propose a novel statistical approach for unsupervised anomaly intrusion detection. In our approach, patterns of normal and intrusive activities are learned through the proposed variational infinite generalized Dirichlet mixture (varInGDM) model.

Table 1. Training and testing data set

Data Set	Normal	DOS	R2L	U2R	Probing
Training	97277	391458	1126	52	4107
Testing	60593	223298	5993	132	2377

The well known KDD Cup 1999 Data[2] is used to investigate our infinite mixture model. We use a 10 percent subset of the data in the experiments. In our case, the training data consists of 494,020 data instances of which 97,277 are normal and 396,743 are attacks. The testing set contains 292,393 data instances of which 60,593 are normal and 231,800 are attacks. All of these attacks fall into one of the following four categories: DOS: denial-of-service (e.g. *syn* flood); R2L: unauthorized access from a remote machine (e.g. guessing password); U2R: unauthorized access to local superuser (root) privileges (e.g. buffer overflow attack) and Probing: surveillance and other probing (e.g. port scanning). Therefore, we have five categories in total including the 'Normal' class. The details of the data sets can be seen in Table 1.

Since the features are on quite different scales in the data set, we need to normalize the data such that one feature would not dominate the others in our modeling approach. In the used data set, each data instance contains 41 features in which 34 are numeric and 7 are symbolic (i.e. each data instance is then represented as a 41-dimensional vector). Features represented by symbolic values are replaced by numeric values for training and testing. For example, the values of *icmp*, *tcp*, and *udp* of feature *protocol_type* are replaced by values 1, 2, and 3, respectively. By finding the maximum and minimum values of a given feature X_l in a data instance $\boldsymbol{X}$, we can transform the feature into the range $[0, 1]$ by

$$X_l = \frac{X_L - \min(X_l)}{\max(X_l) - \min(X_l)} \tag{25}$$

where X_l is set to a small value if the maximum is equal to the minimum.

Table 2 demonstrates the obtained confusion matrix with classification accuracy rate and FP rate using varInGDM. In this table, diagonal values represent the number that is correctly classified. The results of applying different

---

[2] http://kdd.ics.uci.edu/databases/kddcup99/kddcup99.html

Table 2. Confusion Matrix for Intrusion Detection using the proposed varInGDM

	Normal	DOS	R2L	U2R	Probing	Accuracy (%)	FP (%)
Normal	**60351**	141	8	4	89	99.6	10.9
DOS	7246	**215930**	26	1	95	96.7	0.8
R2L	101	1253	**4627**	2	10	77.2	0.9
U2R	26	3	8	**94**	1	71.1	10.4
Probing	29	290	3	4	**2051**	86.3	8.7
					Overall Rate	86.1	6.3

Table 3. Classification accuracy rate (Accuracy), false positive (FP) rate and the estimated number of components ($\hat{M}$) computed using different algorithms

Algorithm	$\hat{M}$	Accuracy (%)	FP (%)
varInGDM	4.98	86.1	6.3
varGDM	4.92	83.8	7.9
varInGM	4.94	78.7	12.6
varGM	4.89	76.5	14.5

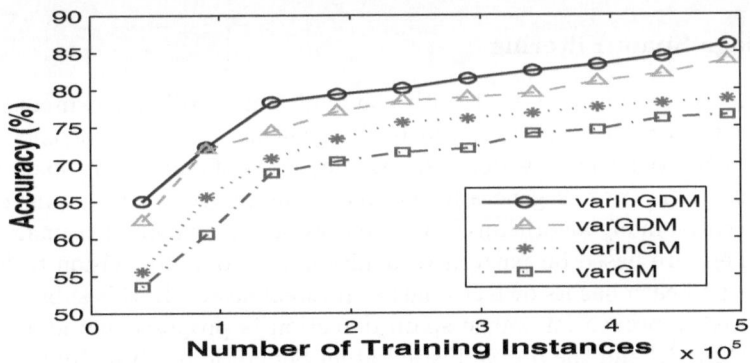

Fig. 2. The classification accuracy vs. the number of training instances

approaches on the KDD99 data set can be seen in Table 3, in terms of the average classification accuracy rate (Accuracy), false positive (FP) rate and the estimated number of components ($\hat{M}$). We can clearly observe that the proposed varInGDM algorithm outperforms the other three approaches with the highest accuracy rate (86.1%), the lowest FP rate (6.3%) and the most accurately detected number of components (4.98). It is worth mentioning that, the number of components ($\hat{M}$) in each variational algorithm is obtained by eliminating

Fig. 3. Sample spam images from Dredze spam data

components with mixing coefficients close to 0. Figure 2 shows the evolution of the classification accuracy according to the number of instances in the training data. According to this figure, the higher number of training instances, the higher the accuracy of the classification. Since the availability of many data instances for training, increases the accuracy of estimating the parameters of the varInGDM model, which then leads to an increase in the classification accuracy.

4.2   Image Spam Filtering

One of the cyber nuisances that both companies and ordinary computer users have to put up with their daily life is the growing spam phenomenon [2]. A successful new trick that is widely used now consists of embedding spam images into emails. This trick is generally referred to as image spam or image-based spam. Recently, many algorithms have been proposed for detecting image spam, most of them are based on pattern recognition and computer vision techniques [25,6,19,3] and each has its own strengths and weaknesses. In this subsection, an unsupervised approach for image spam detection is proposed by adopting the varInGDM model and the probabilistic Latent Semantic Analysis (pLSA) model [26,8] with bag-of-visual-words representation [18].

Our methodology for determining if the input image is spam or ham (i.e. legitimate email) can be summarized as follows: first, 128-dimensional Scale-Invariant Feature Transform (SIFT) vectors [31] are extracted from each input image using the Difference-of-Gaussian (DoG) interest point detector [34]. Then, these local descriptors are grouped into $\mathcal{W}$ homogenous clusters, using a clustering or vector quantization algorithm such as K-Means. Therefore, each cluster center is treated as a visual word and a visual vocabulary can then be constructed with $\mathcal{W}$ visual words. Applying the paradigm of bag-of-words, a $\mathcal{W}$ dimensional histogram representing the frequency of each visual word is calculated for each

Table 4. The average classification accuracy rate (Accuracy) and the average false positive (FP) rate of image spam detection computed using varInGDM, varGDM, varInGM and varGM over 20 random runs

	Dredze		SpamArchive		Princeton	
	Accuracy (%)	FP (%)	Accuracy (%)	FP	Accuracy (%)	FP
varInGDM	86.9	4.47	85.5	5.34	83.7	10.56
varGDM	84.2	6.21	83.6	6.93	82.4	12.75
varInGM	78.6	8.34	77.4	9.54	79.3	16.66
varGM	75.1	10.22	75.3	10.63	76.8	18.12

image. Then, the pLSA model is applied to reduce the dimensionality of the resulting histograms allowing the representation of images as proportional vectors. As a result, each image is represented now by a D-dimensional proportional vector where D is the number of latent aspects. Finally, we employ the proposed varInGDM model as a classifier to determine if an input image is spam or ham.

Three challenging real-world image spam data sets are used in our experiments: the personal spam emails collected by Dredze et al. [19], a subset of the publicly available SpamArchive corpus used by [25,19] and the Princeton spam image benchmark[3]. One common ham data set of images which is collected and used by Dredze et al. [19] is included in our experiments. In summary, there are 2,550 images in the Dredze ham data set, 3,210 images in the Dredze spam data set, 3,550 images in the SpamArchive and 1,071 images in Princeton spam image benchmark. Sample spam images are shown in Fig. 3.

In our experiments, each data set is randomly divided into two halves: one for constructing the visual vocabulary, another for testing. Subsequently, an accelerated version of the K-Means algorithm [20] is employed to cluster all the SIFT vectors into a visual vocabulary with size 800, as explained in the previous section. The pLSA model was applied by considering 45 aspects for all three data sets and each images in the data set was then represented by a 45-dimensional vector of proportions. Last, the resulting vectors were clustered by the varInGDM model. The entire procedure was repeated 20 times for evaluating the performance of our approach.

For comparison, we also applied the varGDM, varInGM and varGM models on the different data sets using the same experimental settings. Table 4 demonstrates the performances of using different algorithms for each data set in terms of the average classification accuracy rate and the average false positive rate. It is obvious that the proposed varInGDM model outperforms the other adopted approaches by providing the highest average accuracy rate and lowest false positive rate. According to our experimental results, the choice of the visual vocabulary

---

[3] http://www.cs.jhu.edu/~mdredze/datasets/image$_$spam

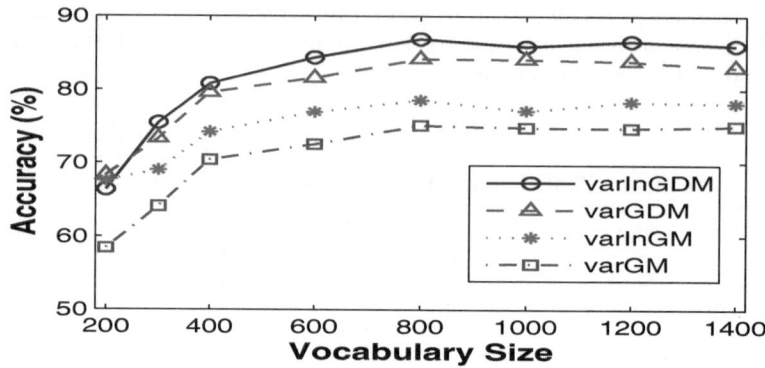

Fig. 4. Classification accuracy vs. vocabulary size for the Dredze spam data

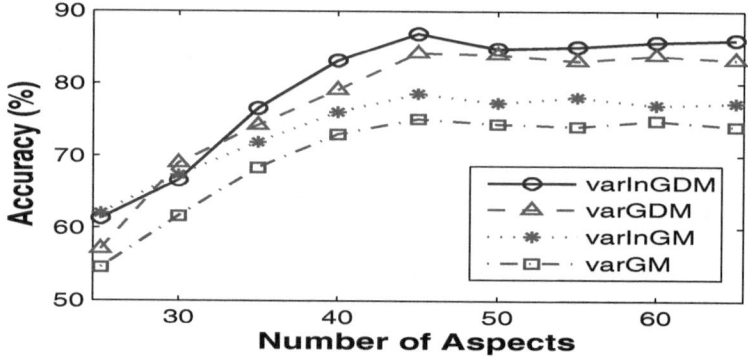

Fig. 5. Classification accuracy vs. the number of aspects for the Dredze spam data

size and the number of visual aspects may become a crucial factor for the classification accuracy. Different sizes of visual vocabulary and numbers of aspects were tested for each tested approach. As shown in Fig. 4, the highest classification rate was obtained when the vocabulary size was around 800 for the Dredze spam data set. Moreover, the optimal accuracy was reached when the number of aspects was set to 45 as illustrated in Fig. 5. Similar results were obtained for the other two data sets.

5   Conclusion and Future Work

In this paper, we have proposed a Bayesian nonparametric model namely the infinite generalized Dirichlet mixture for clustering. The proposed infinite model, which offers a practical solution to the challenging problem of model selection, is learnt within a variational framework in which the whole inference process is analytically tractable with closed-form solutions. Variational framework offers a deterministic efficient alternative to fully Bayesian inference by maximizing a lower bound on the marginal likelihood. Its main advantages are computational efficiency and guaranteed convergence that can be easily assessed. The effectiveness of the proposed approach has been tested on two challenging problems involving anomaly intrusion detection and spam image detection. Future work can be devoted to the inclusion of a feature selection component, like the one previously proposed in [15,14], within the proposed framework which may improve further modeling capabilities and clustering performance.

References

1. Aitchison, J.: The Statistical Analysis of Compositional Data. Blackburn, Caldwell (2003)
2. Amayri, O., Bouguila, N.: A study of spam filtering using support vector machines. Artif. Intell. Rev. 34(1), 73–108 (2010)
3. Amayri, O., Bouguila, N.: Content-based spam filtering using hybrid generative discriminative learning of both textual and visual features. In: ISCAS, pp. 862–865. IEEE (2012)
4. Antoniak, C.E.: Mixtures of Dirichlet processes with applications to Bayesian nonparametric problems. Annals of Statistics 2, 1152–1174 (1974)
5. Attias, H.: A variational Bayes framework for graphical models. In: Proc. of NIPS, pp. 209–215 (1999)
6. Biggio, B., Fumera, G., Pillai, I., Roli, F.: Image spam filtering using visual information. In: Proc. of the 14th International Conference on Image Analysis and Processing, pp. 105–110 (2007)
7. Blei, D.M., Jordan, M.I.: Variational inference for Dirichlet process mixtures. Bayesian Analysis 1, 121–144 (2005)
8. Bosch, A., Zisserman, A., Muñoz, X.: Scene Classification Via pLSA. In: Leonardis, A., Bischof, H., Pinz, A. (eds.) ECCV 2006. LNCS, vol. 3954, pp. 517–530. Springer, Heidelberg (2006)
9. Bouguila, N., Ziou, D.: A powreful finite mixture model based on the generalized Dirichlet distribution: Unsupervised learning and applications. In: Proc. of ICPR, pp. 280–283 (2004)
10. Bouguila, N., Ziou, D.: A Dirichlet process mixture of Dirichlet distributions for classification and prediction. In: Proc. of the IEEE Workshop on Machine Learning for Signal Processing (MLSP), pp. 297–302 (2008)
11. Bouguila, N., Ziou, D.: A dirichlet process mixture of generalized Dirichlet distributions for proportional data modeling. IEEE Transactions on Neural Networks 21(1), 107–122 (2010)
12. Bouguila, N., Ziou, D., Hammoud, R.I.: On bayesian analysis of a finite generalized Dirichlet mixture via a metropolis-within-gibbs sampling. Pattern Analysis and Applications 12(2), 151–166 (2009)
13. Bouguila, N., Ziou, D.: A Nonparametric Bayesian Learning Model: Application to Text and Image Categorization. In: Theeramunkong, T., Kijsirikul, B., Cercone, N., Ho, T.-B. (eds.) PAKDD 2009. LNCS, vol. 5476, pp. 463–474. Springer, Heidelberg (2009)
14. Boutemedjet, S., Bouguila, N., Ziou, D.: A hybrid feature extraction selection approach for high-dimensional non-Gaussian data clustering. IEEE Transactions on PAMI 31(8), 1429–1443 (2009)

15. Boutemedjet, S., Bouguila, N., Ziou, D.: Unsupervised Feature and Model Selection for Generalized Dirichlet Mixture Models. In: Kamel, M.S., Campilho, A. (eds.) ICIAR 2007. LNCS, vol. 4633, pp. 330–341. Springer, Heidelberg (2007)
16. Boutemedjet, S., Ziou, D., Bouguila, N.: A Graphical Model for Content Based Image Suggestion and Feature Selection. In: Kok, J.N., Koronacki, J., Lopez de Mantaras, R., Matwin, S., Mladenič, D., Skowron, A. (eds.) PKDD 2007. LNCS (LNAI), vol. 4702, pp. 30–41. Springer, Heidelberg (2007)
17. Corduneanu, A., Bishop, C.M.: Variational Bayesian model selection for mixture distributions. In: Proc. of AISTAT, pp. 27–34 (2001)
18. Csurka, G., Dance, C.R., Fan, L., Willamowski, J., Bray, C.: Visual categorization with bags of keypoints. In: Workshop on Statistical Learning in Computer Vision, (ECCV) (2004)
19. Dredze, M., Gevaryahu, R., Elias-Bachrach, A.: Learning fast classifiers for image spam. In: Proc. of CEAS (2007)
20. Elkan, C.: Using the triangle inequality to accelerate k-means. In: Proc. of ICML, pp. 147–153 (2003)
21. Fan, W., Bouguila, N., Ziou, D.: A Variational Statistical Framework for Object Detection. In: Lu, B.-L., Zhang, L., Kwok, J. (eds.) ICONIP 2011, Part II. LNCS, vol. 7063, pp. 276–283. Springer, Heidelberg (2011)
22. Fan, W., Bouguila, N., Ziou, D.: Variational learning for finite Dirichlet mixture models and applications. IEEE Transactions on Neural Networks and Learning Systems 23(5), 762–774 (2012)
23. Ferguson, T.S.: Bayesian density estimation by mixtures of normal distributions. Recent Advances in Statistics 24, 287–302 (1983)
24. Ferguson, T.S.: A Bayesian analysis of some nonparametric problems. The Annals of Statistics 1(2), 209–230 (1973)
25. Fumera, G., Pillai, I., Roli, F.: Spam filtering based on the analysis of text information embedded into images. J. Mach. Learn. Res. 7, 2699–2720 (2006)
26. Hofmann, T.: Unsupervised learning by probabilistic latent semantic analysis. Machine Learning 42(1/2), 177–196 (2001)
27. Ishwaran, H., James, L.F.: Gibbs sampling methods for stick-breaking priors. Journal of the American Statistical Association 96, 161–173 (2001)
28. Jordan, M.I., Ghahramani, Z., Jaakkola, T.S., Saul, L.K.: An introduction to variational methods for graphical models. In: Learning in Graphical Models, pp. 105–162 (1998)
29. Korwar, R.M., Hollander, M.: Contributions to the theory of Dirichlet processes. Ann. Probab. 1, 705–711 (1973)
30. Lippmann, R., Haines, J.W., Fried, D.J., Korba, J., Das, K.: Analysis and results of the 1999 DARPA off-line intrusion detection evaluation. In: Recent Advances in Intrusion Detection, pp. 162–182 (2000)
31. Lowe, D.G.: Distinctive image features from scale-invariant keypoints. International Journal of Computer Vision 60(2), 91–110 (2004)
32. Ma, Z., Leijon, A.: Bayesian estimation of Beta mixture models with variational inference. IEEE Transactions on PAMI 33(11), 2160–2173 (2011)
33. McHugh, J., Christie, A., Allen, J.: Defending yourself: The role of intrusion detection systems. IEEE Software 17(5), 42–51 (2000)
34. Mikolajczyk, K., Schmid, C.: A performance evaluation of local descriptors. IEEE Transactions on PAMI 27(10), 1615–1630 (2005)
35. Neal, R.M.: Markov chain sampling methods for Dirichlet process mixture models. Journal of Computational and Graphical Statistics 9(2), 249–265 (2000)

36. Northcutt, S., Novak, J.: Network Intrusion Detection: An Analyst's Handbook. New Riders Publishing (2002)
37. Rasmussen, C.E.: The infinite Gaussian mixture model. In: Proc. of NIPS, pp. 554–560. MIT Press (2000)
38. Robert, C., Casella, G.: Monte Carlo Statistical Methods. Springer (1999)
39. Sethuraman, J.: A constructive definition of Dirichlet priors. Statistica Sinica 4, 639–650 (1994)
40. Teh, Y.W., Jordan, M.I., Beal, M.J., Blei, D.M.: Hierarchical Dirichlet processes. Journal of the American Statistical Association 101, 705–711 (2004)
41. Zhang, J.: The mean field theory in EM procedures for Markov random fields. IEEE Transactions on Signal Processing 40(10), 2570–2583 (1992)

A New Multi-label Learning Algorithm Using Shelly Neighbors

Huawen Liu[1,2], Shichao Zhang[3], Jianmin Zhao[1], Jianbin Wu[1],
and Zhonglong Zheng[1]

[1] Zhejiang Normal University, Jinhua 321004, China
[2] NCMIS, Academy of Mathematics and Systems Science, CAS, China
[3] Guangxi Normal University, Guilin 541004, China

Abstract. Since multi-label data is ubiquitous in reality, a promising
study in data mining is multi-label learning. Facing with the multi-label
data, traditional single-label learning methods are not competent for the
classification tasks. This paper proposes a new lazy learning algorithm
for the multi-label classification. The characteristic of our method is that
it takes both binary relevance and shelly neighbors into account. Unlike
k nearest neighbors, the shelly neighbors form a shell to surround a given
instance. As a result, our method not only identifies more helpful neighbors for classification, but also exempts from the perplexity of choosing
an optimal value for k in the lazy learning methods. The experiments
carried out on five benchmark datasets demonstrate that the proposed
approach outperforms standard lazy multi-label classification in most
cases.

Keywords: Data mining, Multi-label learning, Shelly neighbors, kNN.

1 Introduction

Pattern classification is to train prediction models on a given dataset and then
predict the class of an unknown data using the trained models [1]. Given a dataset
$D = \{(X_i, y_i)|i = 1, ..., n\}$ consisting of n labeled instances (i.e., instances),
where X_i is a m-dimensional vector of m features (or attributes) and y_i is its
corresponding label, the common assumption of traditional learning algorithms
is that their output is a single value (i.e., label) of pre-specified classes. That is to
say, the output result of a classifier is mutually exclusive and each instance can
only be labeled as one class. This, however, is unreasonable because instances
may often belong to two or more classes at the same time in reality. Such data
is called multi-label.

Multi-label data is ubiquitous in many domains, such as text categorization
and scene image classification. For example, a movie can be classified as *action*,
horror and *science fiction* types. Facing with the multi-label data, traditional
learning algorithms are not competent for the classification tasks. To tackle with
this problem, multi-label classification derived from and deeply rooted in the
text categorization has been introduced [2]. Like traditional single-label classification, multi-label classification also aims at training classifiers for a given set of

S. Zhou, S. Zhang, and G. Karypis (Eds.): ADMA 2012, LNAI 7713, pp. 214–222, 2012.

multi-label data and predicting the classes of a query using the classifiers. The difference is that multi-label classifiers output multiple labels for an unknown instance. It is noticeable that multi-label classification is much more challenging than single-label classification.

Due to its great potential applications, in last decades multi-label classification has attracted more and more attentions from many disciplines, such as information retrieval, pattern recognition, data mining and machine learning, and now has been successfully used to text categorization [3,4], images and video annotation [13], music processing [6], bioinformatics [7], and so on. A statistical comparison of multi-label classification algorithms has been made by Demsar in [8]. Generally speaking, multi-label learning algorithms can be grouped into two major categories, i.e., problem transformation and algorithm adaption [2]. The later technique extends traditional single-label classifiers so as to they can handle multi-label data and achieve classification tasks.

The instance-based (or lazy) multi-label classification has become an important research topic since kNN was introduced to multi-label classification. The lazy multi-label learning methods usually take standard k nearest neighbors as the input and predict the labels of a new instance based on Bayesian rule or majority rule. Typical examples include MLkNN [9], BRkNN [10] and LPkNN [11]. The underlying reason of the lazy learning methods popular is that they are conceptually simple, efficient and can be easy implemented. Like the traditional kNN classifiers, this kind of learning algorithms determines the labels of a given instance in terms of its corresponding k nearest neighbors identified during the learning process.

It is worth noting that one of challenges posed by most of multi-label classifiers based on kNN is that the optimal value of k is hard to be chosen, which places a vital role to the performance. If the value of k is large, the boundaries of classes become smooth, while its efficiency is a question; if k is too small, kNN is sensitive to noisy data. To untie this knot, in this paper we introduce the concept of shelly neighbor into multi-label learning and propose a new algorithm for multi-label classification. The advantage of our method is that only the shelly nearest neighbors (SNN), which refer to those neighbors that form a shelly to encapsidate the given instance [12], are considered. Experimental results conducted on benchmark datasets downloaded from MULAN toolkit [2] demonstrates that the proposed approach outperforms standard instance-based multi-label classification.

The rest of this paper is organized as follows. Section 2 briefly recalls related work. In Section 3, some basic concepts are given and then our multi-label classification method is proposed. Simulation experiments to evaluate the proposed approach are presented in Section 4. Finally, this paper is concluded.

2   Related Work

An intuitive and naïve solution for multi-label learning is to transform the multi-label data into its corresponding single-label one, because the single-label data is a special case of the multi-label data. Tsoumakas et al. [2] discusses three different transformation strategies, i.e., *copy, selection* and *ignore*, to separate the multi-label data into several single-label ones. The *copy* method simply substitutes $|Y_i|$ single-label instances (X_i, y_i) for each multi-label instance (X_i, Y_i), where $y_i \in Y_i$, while the *selection* strategy selects only one label $y_i \in Y_i$ to take the place of Y_i. In the binary relevance (BR) learning [2], p different datasets D_k are firstly created from the original one D, where p is the number of class labels. In each new dataset D_k, the class of each instance X_i is positive if the p-th label occurs in the multi-label Y_i of X_i, and negative otherwise. These p datasets are then used to construct p different binary classifiers and the prediction results of an unknown instance are finally combined into a subset of labels, hitherto called *labelset*.

Note that the above transformation methods do not take the interrelation of labels into account. Indeed, the labels of data are relative to each other in many applications. To tackle with this issue, most of multi-label learning methods consider the whole labelset together during the learning stage. As a typical example, the label powerset (LP) algorithm [13] is such kind of learning methods. It considers each unique labelset Y_i of X_i as a new class, and the output of LP is the labelset with the highest probable over all labelsets. RAkEL (RAndom k labELsets) [11] trains several classifiers with different k, i.e., the length of labelset, and then integrates them into an ensemble one.

Unlike BR, the ranking by pairwise comparison (RPC) [14] builds a binary classifier for each pair of labels (y_i, y_j), such that each instance exclusively labeled with either y_i or y_j. Further, Furnkranz et al. [15] proposed a variation of RPC, called calibrated label ranking (CLR or CLRanking), by adding a virtual label. The function of virtual label is to determine a natural breaking point between relevant and irrelevant labelsets. On the other hand, Tsoumakas et al. [16] in MLStacking pruned the stacking models of BR by introducing correlation coefficient, which is used to estimate the correlation of each label pair. In MLStacking, the label will not be considered further if its correlation with the label being learned at the meta-level is lower than a pre-specified threshold.

From the perspective of classifier, many multi-label learning algorithms have been fulfilled by modifying some constrain conditions of traditional learning methods, such as C4.5, SVM and AdaBoost, for the multi-label data. The advantage is that the traditional classifiers can also adapt to the problem of multi-label classification by minor modifications. It is noticeable that the k nearest neighbors (kNN) learning algorithm is very popular for the multi-label classification tasks. As a typical example, MLkNN [9] predicts the labelset of an unknown instance based on the prior and posterior probabilities for the frequency of each label within its k nearest neighbors. BRkNN [10] constructs the binary relevance classifiers upon the k nearest neighbors for each unseen instance.

For the kNN classifiers, one of challenges is that the optimal value of k is hard to be chosen. If k is too large, the boundaries of classes become smooth, while its efficiency is a question; if k is small, kNN is sensitive to noisy data. To alleviate this problem, Guo et al. [17] developed a method called MkNN to determine automatically a proper value of k for different data. MkNN firstly constructs a kNN model for the data, and then takes its classification accuracy as the basis to choose an optimal value of k. It is similar to the cross-validation (CV) technique. Additionally, Zhang [12] identified shelly neighbors, rather than nearest neighbors, of a new instance to calculate its missing values. Here we also adopt the concept of shelly neighbor into our method for multi-label learning.

3   Binary Relevance Using Shelly Nearest Neighbors

Before delving into the details of our learning method, we will give some basic concepts about multi-label data and shelly neighbors at the beginning.

3.1   Basic Concepts

Let $Y = \{y_i | i = 1, ..., p\}$ be the finite set of labels. A dataset $D = \{(X_i, Y_i) | i = 1, ..., n\}$ is called a multi-label dataset, if Y_i is a labelset and $Y_i \subseteq Y$, where X_i is represented as the vector form with m features. From this definition, one may observe that for each Y_i of the i-th instance in D, if only a single label is concerned, i.e., $|Y_i| = 1$ or $Y_i \in Y$, the multi-label dataset D is degenerated into a conventional single-label dataset. This means that single-label data is a special case of multi-label data. Under the context of multi-label data, the task of learning methods is to determine an ordering or a bipartition of the set of labels for an unknown instances in terms of specific applications at hand [2].

   K nearest neighbor (kNN) is one of the most fundamental and simple learning methods in data mining, especially when there is little or no prior knowledge about the distribution of the data [1]. Given a dataset $D = \{X_i | i = 1, ..., n\}$ consisting of n instances in a m-dimensional space. The k nearest neighbors of a query instance x in D can be formally defined as

$$N(x, k) = \{X_i \in D | d(x, X_i) \leq d(x, X_j), j \neq i, i = 1, ..., k\}, \tag{1}$$

where the function d is a distance function between two instances. Given two instances x and y in a m-dimensional space, the Minkowski distance is

$$d(x, y) = (\sum_{i=1,...,m} (x_i - y_i)^p)^{\frac{1}{p}}. \tag{2}$$

Note that the Minkowski distance is a general metric. It turns out different forms if p takes different values. For example, it is Manhattan distance as $p = 1$. In a similar vein, this formula is turned out as Euclidean or Chebychev distance as $p = 2$ or $p \to \infty$, respectively. Generally, the function d is often represented as the Euclidean distance between instances in kNN. The center idea of kNN classification is to determine the label of the unclassified instance x by considering the labels of its $N(x, k)$ in a major voting strategy. As mentioned above, determining an optimal value of k is a trial thing, and kNN is sensitive to noise and its decision boundary is very sharp if k is too small. To alleviate this problem, Zhang [12] introduces a concept called shell nearest neighbor.

For a query instance X_t, its left and right nearest neighbors in the dataset D are defined as follows:

$$X_t^- = \{X_i \in D | X_{ij} \leq X_{tj}, i = 1, ..., n, j = 1, ..., m\}, \tag{3}$$

$$X_t^+ = \{X_i \in D | X_{ij} \geq X_{tj}, i = 1, ..., n, j = 1, ..., m\}, \tag{4}$$

where X_{ij} is the j-th attribute value of X_i. From this definition, one may observe X_k^- or X_k^+ may not exist in the nearest neighbors $N(X_t, k)$ of X_t. On the other hand, for a nearest neighbor X_i in $N(X_t, k)$, either $X_{ij} \leq X_{tj}$ if there is a X_t^- in $N(X_t, k)$, or $X_{ij} \geq X_{tj}$ if there is a X_t^+ in $N(X_t, k)$. The shelly neighbors $SN(X_t)$ of X_t refers to both the left neighbors X_t^- and the right neighbors X_t^+, i.e.,

$$SN(X_t) = X_t^- \cup X_t^+. \tag{5}$$

3.2 Our Method

Given a dataset D, the shelly neighbors of an unknown instance X_t are determined and can be identified in advance. One of advantages of adopting shelly neighbors for classification is that the number of the selected nearest neighbors is a variable determined by data, whereas the kNN method uses a fixed k. That is to say, it is free from the parameter k. Based on this analysis, here we take use of shelly neighbors for the purpose of multi-label classification.

As aforementioned discussion, BRkNN [10] is a typical multi-label classification algorithm, which combines the kNN learning algorithm with binary relevance together. Due to its intuitive idea and easy implement, here we incorporate the concept of shelly neighbor into BRkNN and propose a new algorithm called BRSN (Binary Relevance using Shelly Neighbors). Our method is similar to BRkNN. Unlike BRkNN, our method takes shelly neighbors of test instances, rather than k nearest neighbors, as the basis of prediction for the instances. The details of BRSN are described as follows (Algorithm 1).

Algorithm 1. BRSN: Binary Relevance using Shelly Neighbors

Input: A multi-label dataset $D = \{(X_i, Y_i)\}$ and a query instance x.
Output: The prediction labelset Y_x of x.
1).   Obtain the shelly neighbors $SN(x)$ of x in terms of Eq.(5);
2).   For each label y occurring in the labels of $SN(x)$, calculate its probability or certainty factor;
3).   Add the label y into the labelset Y_x, if it is larger than a pre-specified threshold;
4).   Return Y_x as the final results.
End

The BRSN algorithm works in a straightforward way and can be easily understood. It firstly obtain the shelly neighbors $SN(x)$ of the query instance x are identified from the training dataset D. The advantages include that all shelly neighbors of x in D can be found and none neighbor will be ignored. Additionally, from the perspective of computational cost the kNN method and the parameter k do not need be considered during the learning process. The second step aims at calculating the class distribution of each label y occurring in the shelly neighbors $SN(x)$. Subsequently, the label with large probability will be added into the labelset Y_x, and the final prediction labelset of the query instance is determined.

For the first step in the Alg. 1, we can also obtain the shelly neighbors $SN(x)$ from the its corresponding k nearest neighbors $N(x, k)$, rather than D. To achieve this purpose, the k nearest neighbors $N(x, k)$ of the given query/test x should be firstly identified from the training multi-label dataset D. It should be mention that the obtained shelly neighbors $SN(x)$ from $N(x, k)$ may be not equal to the one from D. In addition, for the parameter k of $N(x, k)$, it can be specified in advance or obtained with a cross validation manner.

4 Experiments

4.1 Datasets

In our experimental study, five benchmark datasets, including *Emotions, Medical, Scene, Yeast* and *Hierarchical*, with different types and sizes were collected from different domains. They are frequently used to validate performance of multi-label classifiers. Table 1 summarizes some of their general description information, where the second column is the name of dataset. The numbers of instances, attributes and labels in each datasets are given in the next three columns. The cardinality column refers to the average number of labels of the instances in dataset. It is used to quantify the number of alternative labels that characterize the instances in a multi-label dataset. The density column is a fraction of the cardinality by the number of labels. The purpose of introducing this concept is that it might expose some different characters between two datasets with the same label cardinality but with different number of labels. Note that these datasets have different number of labels and differ greatly in the sample size (range from 593 to 2417) and the number of attributes (from 72 to 1449).

Table 1. The brief descriptions of datasets used in our experiments

No	Dataset	Instances	Attributes	Labels	Cardinality	Density
1	Emotions	593	72	6	1.869	0.311
2	Medical	978	1449	45	1.245	0.028
3	Scene	2407	294	6	1.074	0.179
4	Yeast	2417	103	14	4.237	0.303
5	Hierarchical	1600	80	366	7.138	0.020

4.2 Experimental Settings

Currently, many outstanding learning algorithms for multi-label data have been proposed. For the reasons mentioned above, we mainly place our interesting on the kNN classifiers, especially on BRkNN [10] and MLkNN [9]. Indeed, our method (i.e., BRSN) is an extension of BRkNN by using shelly neighbors, rather than nearest neighbors. As recommended by Zhang and Zhou [9], the value of k in BRkNN and MLkNN was assigned as 10 when compared to the BRSN classifier, and the distance function was the simple Euclidean metric on the whole attribute space. To achieve impartial results, ten 10-fold cross validations had been adopted for each algorithm-dataset combination in verifying classification capability.

Unlike the single-label classification, the situation of the multi-label learning is relatively intricate and difficult, where only the classification accuracy is not enough to measure the performance of classifiers. Thus, more measures are needed to evaluate the predictive performance of classifiers for the multi-label classification. In our experiments, five popular metrics, such as *subset accuracy* (SA), *micro F-measure* (MF), *micro AUC* (MAUC), *one error* (OE) and *ranking loss* (RL), were adopted to measure the performance of classifiers.

4.3 Experimental Results

In order to validate the performance of our method, we carried out experiments on the benchmark datasets, and made a comparison BRSN to BRkNN and MLkNN. The experimental results are given in Table 2, where the bold value refers to the best one among these three classifiers on the same dataset. In addition, the notation '*' means that the value in the current entry is significantly worse than BRSN on the same dataset in a statistical t-test, where the baseline is BRSN and the confidence level is 95%. For example, the *ranking loss* of MLkNN (i.e, 0.7271 in the 2th row and the last column) on the *emotions* dataset is significantly worse than BRSN.

From the results in Table 2, one may observe that BRSN outperforms BRkNN and MLkNN in most cases. For example, the performance achieved by BRSN is the best one than others at the aspects of *micro-F1* (MF), *micor-AUC* (MAUC) and *ranking loss* (RL). This is indicates that shelly neighbor is more effective than nearest neighbor in the multi-label classifier with binary relevance.

For the subset accuracy, BRkNN achieved better performance than others on the *Yeast* datasets. Similarly, this classifier also has lowest *one error* on the *Emotions* and *Yeast* datasets, while the *one error* of BRSN is significant better than MLkNN on these two datasets. This, however, is reasonable because the label densities of these datasets are large. As a matter of fact, most of shelly neighbors can be identified from the k nearest neighbors when the dataset has relative label density and k is assigned with an appropriate value.

It is noticeable that the performance of MLkNN is relatively poor among these three classifiers in many situations, especially on the *Hierarchical* dataset, where the performance is significantly worse. For example, the *subset accuracy*

Table 2. Experimental results of BR*k*NN, ML*k*NN and BRSN

		SA	MF	MAUC	OE	RL
Emotions	BR*k*NN	0.2915	0.6487	0.8588	**0.2565**	0.1610
	ML*k*NN	0.2831	0.6598	0.2835*	1.7884*	0.7271*
	BRSN	**0.3048**	**0.6662**	**0.8667**	0.2581	**0.1567**
Medical	BR*k*NN	0.4018	0.5840	0.9470	0.3067	0.0474
	ML*k*NN	0.5061	**0.6784**	0.2403*	2.6378*	NaN
	BRSN	**0.5480**	0.6682	**0.9508**	**0.2679**	**0.0436**
Scene	BR*k*NN	0.5962	0.7012	0.9345	0.2522	0.0889
	ML*k*NN	0.6248	0.7343	0.2239*	0.4744*	0.8253*
	BRSN	**0.7046**	**0.7433**	**0.9508**	**0.2347**	**0.0727**
Yeast	BR*k*NN	**0.1982**	0.6344	0.8415	**0.2309**	0.1778
	ML*k*NN	0.1874	0.6427	0.2292*	6.2324*	0.5091*
	BRSN	0.1891	**0.6471**	**0.8484**	0.2342	**0.1687**
Hierarchical	BR*k*NN	0.0287*	0.0991*	0.9319	0.5806	0.0627
	ML*k*NN	0.0300*	0.1276*	0.5631*	26.955*	NaN
	BRSN	**0.8994**	**0.9445**	**0.9676**	**0.0738**	**0.0176**

and *micro-F1* of ML*k*NN on the *Hierarchical* dataset are only 3% and 0.1276, respectively, whereas the *one error* almost reaches to 27. The reason underlying it is that the number of labels is too large. As a result, its label density is relatively sparse and the elapsed time by ML*k*NN is high. Moreover, the *one error* and *ranking loss* achieved by ML*k*NN are significant worse than BRSN over all benchmark datasets.

5   Conclusions

In this paper we have proposed a new and efficient algorithm for lazy multi-label classification. The center idea of the proposed method is that it exploits the binary relevance of labels and the shelly neighbors, rather than k nearest neighbors, to determine the set of labels of multi-label data. As a result, it exempts from choosing an optimal value of k in the lazy learning algorithms. The experiments conducted on five benchmark datasets demonstrate that the proposed method performs better than the state-of-the-art multi-label learning algorithms.

Acknowledgments. The authors are grateful to anonymous referees for their valuable comments and suggestions. This work is supported by the National NSF of China (61170131, 61100119, 61170108, 61170109, 61272130, 61272468), Postdoctoral Science Foundation of China, and Key discipline of CST of Zhejiang Province (ZSDZZZZXK05).

References

1. Duda, R.O., Hart, P.E., Stork, D.G.: Pattern Classification, 2nd edn. Wiley, New York (2001)
2. Tsoumakas, G., Katakis, I., Vlahavas, I.: Mining Multi-label Data. In: Maimon, O., Rokach, L. (eds.) Data Mining and Knowledge Discovery Handbook, 2nd edn. Springer (2010)
3. Rousu, J., Saunders, C., Szedmak, S., Shawe-Taylor, J.: Kernel-based learning of hierarchical multi-label classification methods. Journal of Machine Learning Research 7, 1601–1626 (2006)
4. Schapire, R.E., Singer, Y.: Boostexter: a boosting-based system for text categorization. Machine Learning 39, 135–168 (2000)
5. Boutell, M.R., Luo, J., Shen, X., Brown, C.: Learning multi-label scene classiffication. Pattern Recognition 37(9), 1757–1771 (2004)
6. Trohidis, K., Tsoumakas, G., Vlahavas, I.: Multi-label classification of music into emotions. In: Proc. of International Conference on Music Information Retrieval (ISMIR 2008), Philadelphia, PA, USA, pp. 320–330 (2008)
7. Zhang, M.L., Zhou, Z.H.: Multi-label neural networks with applications to functional genomics and text categorization. IEEE Transactions on Knowledge and Data Engineering 18, 1338–1351 (2006)
8. Demsar, J.: Statistical comparisons of classifiers over multiple data sets. Journal of Machine Learning Research 7, 1–30 (2006)
9. Zhang, M.L., Zhou, Z.H.: ML-kNN: A lazy learning approach to multi-label learning. Pattern Recognition 40(7), 2038–2048 (2007)
10. Spyromitros, E., Tsoumakas, G., Vlahavas, I.P.: An Empirical Study of Lazy Multilabel Classification Algorithms. In: Darzentas, J., Vouros, G.A., Vosinakis, S., Arnellos, A. (eds.) SETN 2008. LNCS (LNAI), vol. 5138, pp. 401–406. Springer, Heidelberg (2008)
11. Tsoumakas, G., Katakis, I., Vlahavas, I.: Random k-Labelsets for Multi-Label Classification. IEEE Transactions on Knowledge and Data Engineering 23(7), 1079–1089 (2011)
12. Zhang, S.C.: Shell-Neighbor Method And Its Application in Missing Data Imputation. Applied Intelligence 35(1), 123–133 (2011)
13. Boutell, M.R., Luo, J., Shen, X., Brown, C.: Learning multi-label scene classification. Pattern Recognition 37(9), 1757–1771 (2004)
14. Hullermeier, E., Furnkranz, J., Cheng, W., Brinker, K.: Label ranking by learning pairwise preferences. Artificial Intelligence 172, 1897–1916 (2008)
15. Furnkranz, J., Hullermeier, E., Loza Mencia, E., Brinker, K.: Multi-label classification via calibrated label ranking. Machine Learning 73(2), 133–153 (2008)
16. Tsoumakas, G., Dimou, A., Spyromitros, E., Mezaris, V., Kompatsiaris, I., Vlahavas, I.: Correlation-Based Pruning of Stacked Binary Relevance Models for Multi-Label Learning. In: Proc. ECML/PKDD 2009 Workshop on Learning from Multi-Label Data (MLD 2009), pp. 101–116 (2009)
17. Guo, G., Wang, H., Bell, D.J., Bi, Y., Greer, K.: KNN Model-Based Approach in Classification. In: Meersman, R., Schmidt, D.C. (eds.) CoopIS/DOA/ODBASE 2003. LNCS, vol. 2888, pp. 986–996. Springer, Heidelberg (2003)

Kernel Mean Matching with a Large Margin

Qi Tan[1,2], Huifang Deng[1], and Pei Yang[1]

[1] South China University of Technology, Guangzhou, China
[2] South China Normal University, Guangzhou, China

Abstract. Various instance weighting methods have been proposed for instance-based transfer learning. Kernel Mean Matching (KMM) is one of the typical instance weighting approaches which estimates the instance importance by matching the two distributions in the universal reproducing kernel Hilbert space (RKHS). However, KMM is an unsupervised learning approach which does not utilize the class label knowledge of the source data. In this paper, we extended KMM by leveraging the class label knowledge and integrated KMM and SVM into an unified optimization framework called KMM-LM (Large Margin). The objective of KMM-LM is to maximize the geometric soft margin, and minimize the empirical classification error together with the domain discrepancy based on KMM simultaneously. KMM-LM utilizes an iterative minimization algorithm to find the optimal weight vector of the classification decision hyperplane and the importance weight vector of the instances in the source domain. The experiments show that KMM-LM outperforms the state-of-the-art baselines.

Keywords: Transfer learning, Instance weighting, Kernel mean matching, Large margin.

1 Introduction

In many real-world applications of data mining, it is normally expensive and time-consuming to obtain appropriate training data to learn the robust models. For example, sentiment classifiers for online reviews need to work properly on data of different types of products; search engines must provide consistent quality of service on the Web data in the markets of different languages. However, the training data commonly exist only in a limited number of domains. Collecting and annotating data for each different domain would become practically prohibitive. On the other hand, abundant labeled data may exist in some related domains. The target domain data are commonly drawn from a different feature space and follow a different distribution from that of the source domain. To bridge the domain gap is a challenging issue for the model learned from source domain to be generalized well in target domain.

Transfer learning utilizes labeled data available from some related domain in order to achieve effective knowledge transformation from source to target domain. It is of great importance in many applications, such as cross domain document classification [1], sentiment classification [2, 3], collaborative filtering [4],

S. Zhou, S. Zhang, and G. Karypis (Eds.): ADMA 2012, LNAI 7713, pp. 223–234, 2012.

online recommendation [5] and Web search ranking [6]. If done successfully, knowledge transfer can greatly improve learning performance and meanwhile avoid excessive data annotation effort.

One of the most intuitive methods for transfer learning is to identify a subset of source instances and reuse them to build the model for the target domain data. The major technique is to apply instance weighting to the source domain data according to their importance to the target domain. Huang et al. [7] presented a two-step approach called Kernel Mean Matching (KMM) which directly produces re-sampling weights without distribution estimation. The objective is to minimize the discrepancy between means of instances in a reproducing kernel Hilbert space (RKHS) between the two domains by re-weighting the instances in the source domain. Then the reweighted instances can be incorporated into a variety of classification algorithms, such as support vector machine (SVM) [8]. Experiments shown that using KMM to re-weight the instances in the source domain as a preprocessing step can improve the performance of adaptation in the target domain.

However, KMM is an unsupervised learning method which does not utilize the class label knowledge of the source data. Though the simple combination of KMM and SVM is effective, it would be expected that the adaptation performance can be further improved by leveraging the label knowledge while reweighting the instances. This paper therefore looks at the possibility of combining the two distinct stages of KMM and SVM into an unified optimization framework that will be called KMM-LM. The objective of KMM-LM is to maximize the geometric soft margin, and minimize the empirical classification error together with the domain discrepancy based on KMM simultaneously. KMM-LM utilizes a two-phase iterative minimization algorithm to find the optimal weight vector of the classification decision hyperplane and the importance weight vector of the source data. Firstly, it searches for a good weight vector of the decision hyperplane while keeping the instance importance weight vector fixed. Secondly, it re-weights the instance while keeping the decision weight vectors fixed. The two weight vectors constrain mutually with each other. The more optimal the instance importance weight, the better discriminative the decision hyperplane, and vice versa, which creates a positive cycle and can gradually improve the performance of adaptation.

The rest of the paper is organized as follows. Section 2 reviews the related work. The proposed KMM-LM model and the iterative minimization algorithm are presented in Section 3. Section 4 reports the experiments and results. Finally, in Section 5 the conclusions are drawn.

2   Related Work

Traditional machine learning methods make a basic assumption that the training and testing data should be drawn from the same feature space and the same distribution. However, in many real-world applications, this identical distribution assumption does not hold true. Transfer learning assumes that multiple tasks can benefit from certain structures of data shared among different distributions. As pointed out by Pan and Yang [9], there are three fundamental issues in transfer learning, i.e. "what to transfer", "how to transfer" and "when to transfer". "What to transfer" asks which common part of knowledge can be transferred across domains. "How to transfer" is related to the design of appropriate algorithms to extract and transfer the common knowledge. "When to transfer" is concerned with whether the transferring skills should be done or not, which is related to negative transfer. Most current works on transfer learning focused on what to transfer and how to transfer by implicitly assuming that the source and target domains are related to each other. Existing methods can be divided into instance-based approach [10, 11], feature-based approach [2, 12] and parameter-based approach [13].

Various instance weighting methods have been proposed for instance-based transfer learning [10, 11]. Instance-based approach assumes that some training examples in the source domain are similar to the data in target domain and can be reused to train the model for the target domain. Re-weighting is the major techniques in instance-transfer learning. Therefore, the problem with estimating the ratio of two probability densities has been actively explored in the literature. A naive approach to density-ratio estimation is to first separately estimate the training and testing probability densities, and then estimate the instance importance by taking into account the ratio of the estimated densities. However, density estimation is known to be a hard problem particularly in high-dimensional cases in practice [14]. The difference in distributions may occur both in marginal and conditional probabilities. Most of the existing domain adaptation works focus on the marginal probability distribution difference between the domains, assuming that the conditional probabilities are similar [7, 14–18]. Some recent work [19] focused on the estimation of both marginal and conditional probabilities.

Sugiyama [15] proposed the KLIEP approach which tries to match an importance based estimation of the test input distribution to the true test input distribution in terms of the Kullback-Leibler divergence. This approach solved this matching problem in a non-parametric fashion, i.e. the training and test input distributions are not parameterized, but only the importance is parameterized. Tsuboi et al. [16] proposed a scalable direct importance estimation method called LL-KLIEP which is an extension of KLIEP [15]. The key difference is that the KLIEP used a linearly parameterized function for modeling the importance, while LL-KLIEP adopted a log-linear model. The log-linear model only takes non-negative values, which enables reformulating the KLIEP optimization problem as an unconstrained convex problem. Then a new scalable estimation procedure is developed whose computation time is nearly independent of the number of test samples.

Kanamori et al. [17] developed a squared-loss version of the M-estimator for linear density-ratio models called unconstrained Least-Squares Importance Fitting (uLSIF), and have shown that uLSIF possesses superior computational properties. That is, a closed-form solution is available and the leave-one-out cross-validation score can be analytically computed. Kanamori et al. [14] proposed a kernelized variant of uLSIF for density-ratio estimation, which is called kernel unconstrained least-squares importance fitting (KuLSIF). They investigated its fundamental statistical properties including a non-parametric convergence rate, an analytical-form solution, and a leave-one-out cross-validation score.

LogReg [18] built a probabilistic classifier that separated training input samples from test input samples, by which the importance can be directly estimated. The maximum likelihood estimation of logistic regression models can be formulated as a convex optimization problem.

Sun et al. [19] proposed a two-stage domain adaptation methodology which combined weighted data from multiple sources based on marginal probability differences (first stage) as well as conditional probability differences (second stage), with the target domain data. The weights for minimizing the marginal probability differences are estimated independently, while the weights for minimizing conditional probability differences are computed simultaneously by exploiting the potential interaction among multiple sources.

Huang et al. [7] presented a two-step approach which directly produced resampling weights without distribution estimation. The kernel mean matching (KMM) method directly gives estimates of the importance at the training inputs by matching the two distributions efficiently based on a special property of universal reproducing kernel Hilbert spaces. Then the reweighted instances can be incorporated into a variety of classification algorithms, such as SVM.

In this research, we extended KMM by incorporating the label knowledge to estimate the instance importance. Furthermore, we integrated KMM and SVM into an unified optimization framework to find the optimal classification decision hyperplane and the instance importance vector simultaneously.

3   The Proposed Model

We first introduce the notations and the problem formulation, and then present the integrated optimization framework. Finally, an iterative algorithm is proposed to minimize the objective function.

Suppose we are given a set of labeled source domain data and unlabeled target domain data. The target domain is different but related to the source domain. The goal is to build a classifier with the help of both domain data and use it to predict the class label of the instances in the target domain as accurately as possible. For the simplicity of expression, we are focused on binary classification problem. However, the generalization of our proposed approach to multi-class classification is not difficult.

Let n and m be the number of instances in the source and target domains, respectively. Let $D_s = \{(x_i, y_i)\}_{i=1}^n$ denote data set in the source domain where x_i is the column vector for the ith instance, and y_i is the corresponding class label. The unlabeled data set in the target domain is denoted by $D_t = \{(z_i)\}_{i=1}^m$ where z_i is the column vector for the ith instance.

Let $\phi(\cdot)$ be the kernel function of mapping the instances from the original feature space to a reproducing kernel Hilbert space (RKHS). The weight vector in the mapped feature space is denoted by w. Let u be the instance importance weight vector of the source domain. The kernel matrix for the source data is denoted by K_s, and the kernel matrix for both domains data is denoted by K. The label vector of the source data is represented by $y = (y_1, \cdots, y_n)^T$.

3.1 Objective

Our proposed approach extends KMM [7] and makes use of label knowledge to estimate the instance importance. Furthermore, instead of using a two-step approach which first estimates the instance importance and then uses it to build the classifier, we integrate KMM and SVM into an unified optimization framework. The goal of KMM-LM is to maximize the geometric soft margin, while minimizing the empirical classification error together with the domain discrepancy based on KMM simultaneously. Mathematically speaking, the objective of KMM-LM is to minimize

$$\min_{w,u} J(w,u) = \min_{w,u} \|w\|^2 + \lambda_1 \sum_{i=1}^{n} u_i \xi_i + \lambda_2 \| \frac{1}{n} \sum_{i=1}^{n} u_i \phi(x_i) - \frac{1}{m} \sum_{i=1}^{m} \phi(z_i) \|^2 \quad (1)$$

$$s.t. \quad y_i(w^T \phi(x_i) + b) \geq 1 - \xi_i$$
$$\xi_i \geq 0$$
$$|\sum_{i=1}^{n} u_i - n| \leq n\varepsilon$$
$$0 \leq u_i \leq n$$
$$(i = 1, 2, \cdots, n)$$

where ξ is the soft slack variances. Here the first two terms in the right side of Eq. 1 can be regarded as the weighed version of SVM, and the last term measures the domain distance based on KMM. The coefficient λ_1 is the same as that used in SVM. Another trade-off parameter λ_2 is introduced to balance the domain distance and the structural risk. When $\lambda_2 = 0$, Eq. 1 is reduced to the weighted version of SVM.

Next, we will show that the above optimization problem can be solved efficiently by alternately finding the optimal weight vector of decision hyperplane w and the instance importance vector u.

The Computation of Optimal w^*. When u is kept fixed, the optimization problem defined in Eq. 1 is reduced to

$$\min_{w} J(w) = \min_{w} \|w\|^2 + \lambda_1 \sum_{i=1}^{n} u_i \xi_i \quad (2)$$

$$s.t. \quad y_i(w^T \phi(x_i) + b) \geq 1 - \xi_i$$
$$\xi_i \geq 0$$
$$(i = 1, 2, \cdots, n)$$

The above optimization problem can be solved in a way similar to SVM [8] as follows. We will derive a Lagrangian in order to arrive at its dual optimization problem. Introducing Lagrange multipliers α and γ, we can obtain

$$L_P = ||w||^2 + \lambda_1 \sum_{i=1}^{n} u_i \xi_i - \sum_{i=1}^{n} \alpha_i [y_i (w^T \phi(x_i) + b) - 1 + \xi_i] - \sum_{i=1}^{n} \gamma_i \xi_i \qquad (3)$$

The KKT conditions can be stated as follows

$$\frac{\partial L_P}{\partial w} = w - \sum_{i=1}^{n} \alpha_i y_i \phi(x_i) = 0 \qquad (4)$$

$$\frac{\partial L_P}{\partial b} = -\sum_{i=1}^{n} \alpha_i y_i = 0 \qquad (5)$$

$$\frac{\partial L_P}{\partial \xi} = \lambda_1 u - \alpha - \gamma = 0 \qquad (6)$$

$$\alpha_i [y_i (w^T \phi(x_i) + b) - 1 + \xi_i] = 0, \quad i = 1, 2, \cdots, n \qquad (7)$$

$$\gamma_i \xi_i = 0 \quad i = 1, 2, \cdots, n \qquad (8)$$

$$\alpha_i \geq 0, \quad i = 1, 2, \cdots, n \qquad (9)$$

$$\gamma_i \geq 0, \quad i = 1, 2, \cdots, n \qquad (10)$$

By re-substituting Eq.4, Eq. 5,and Eq. 6 into Eq.3, we can obtain the dual form of the primal problem as follows

$$\min_{\alpha} \frac{1}{2} \sum_{i,j=1}^{n} \alpha_i \alpha_j y_i y_j \phi(x_i)^T \phi(x_j) - \sum_{i=1}^{n} \alpha_i = \min_{\alpha} \frac{1}{2} \alpha^T H \alpha - p^T \alpha \qquad (11)$$

$$s.t. \quad \sum_{i=1}^{n} y_i \alpha_i = 0$$

$$0 \leq \alpha_i \leq \lambda_1 u_i, \quad i = 1, 2, \cdots, n$$

where

$$H = \Lambda \cdot K_s \cdot \Lambda$$

$$p_i = 1, \quad i = 1, 2, \cdots, n$$

Here $\Lambda = Diag\{y_1, \cdots, y_n\}$ is diagonal matrix. It is easy to show that the optimization problem of Eq. 11 is a linear constrained convex quadratic optimization problem which can be solved using quadratic problem solvers.

After solving the problem Eq. 2, we use the constraints in Eq. 2 to approximately estimate the soft margin

$$\xi_i = \max\{0, \ 1 - y_i (w^T \phi(x_i) + b)\}$$

$$= \max\{0, \ 1 - y_i (\sum_{j=1}^{n} y_j \alpha_j \phi(x_j)^T \phi(x_i) + b)\} \quad (i = 1, 2, \cdots, n) \qquad (12)$$

The Computation of Optimal u^*. Next, the obtained soft margin ξ is then incorporated into the following optimization framework. Since the class label knowledge is embedded into the soft margin ξ, it would guide the model to estimate the more optimal instance importance in comparison with the original KMM [7] approach.

When keeping w unchanged, the optimization problem of Eq. 1 can be reduced to

$$\min_u J(u) = \min_u \lambda_1 \sum_{i=1}^{n} u_i \xi_i + \lambda_2 || \frac{1}{n} \sum_{i=1}^{n} u_i \phi(x_i) - \frac{1}{m} \sum_{i=1}^{m} \phi(z_i)||^2 \tag{13}$$

$$s.t. \quad |\sum_{i=1}^{n} u_i - n| \leq n\varepsilon$$

$$0 \leq u_i \leq n$$

$$(i = 1, 2, \cdots, n)$$

Eq. 13 can be transformed into

$$\min_u \frac{1}{2} u^T Q u - q^T u \tag{14}$$

$$s.t. \quad |\sum_{i=1}^{n} u_i - n| \leq n\varepsilon$$

$$0 \leq u_i \leq n$$

$$(i = 1, 2, \cdots, n)$$

where

$$Q = \frac{2\lambda_2}{n^2} K_s$$

$$q_i = -\lambda_1 \xi_i + \frac{2\lambda_2}{nm} \sum_{j=1}^{m} \phi(x_i)^T \phi(z_j), \quad i = 1, 2, \cdots, n \tag{15}$$

Since Q is positive semi-definite, Eq. 14 is also a linear constrained convex quadratic optimization problem, which can be solved using quadratic optimization solvers.

Eq. 15 can also lead to an intuitive interpretation about the difference between KMM and our proposed approach. In KMM [7], large value of $\kappa_i = \sum_{j=1}^{m} \phi(x_i)^T \phi(z_j)$ corresponds to particularly important instance x_i and is likely to lead to large instance weight u_i. In other word, the importance weight tends to be more large for an instance which is more similar to the target domain data. But if the instance is difficult to be classified, its importance weight would be reduced. Therefore, the soft margin ξ_i is included into Eq. 15 as a discount term to penalize the misclassified instances though they are similar with the target data.

Algorithm 1. Algorithm for KMM-LM

Require:
   The source dataset $D_s = \{(x_i, y_i)\}_{i=1}^{n}$
   The target dataset $D_t = \{(z_i\}_{i=1}^{m}$
   Iteration number N_{max}
Ensure:
   Class label assigned to each instance in D_t;
 1: **repeat**
 2:     Compute the weight vector of the decision function w using Eq. 2 while keeping u fixed;
 3:     Compute the instance importance weight vector u using Eq. 13 while keeping w fixed;
 4: **until** iteration $\geq N_{max}$
 5: **for** each unlabeled $z_i \in D_t$ **do**
 6:     Assign z_i the class label based on Eq. 16;
 7: **end for**

Prediction. For an unlabeled instance in the target domain, the classification decision function is as follows

$$f(z_j) = sgn(w^T \phi(z_j) + b) = sgn\left(\sum_{i=1}^{n} y_i \alpha_i \phi(x_i)^T \phi(z_j) + b\right) \qquad (16)$$

3.2 Algorithm

We summarize the above training and testing procedure using Algorithm 1. It solves the optimization problem using an iterative method which updates w and u alternately until convergence. Then the obtained decision function is used to assign the class labels to the test data.

4 Experiments and Results

In this section, we empirically evaluate the KMM-LM algorithm for cross-domain document classification in comparison with the state-of-the-art baselines. Two popular real-world text datasets, i.e. Cora [20] and Reuters-21578 [21], are used in our experiments.

4.1 Datasets and Setup

Cora [20] is an online archive of computer science research papers. The Cora dataset contains approximately 37,000 papers, and over 1 million links among roughly 200,000 distinct documents. The documents in the dataset are categorized into a hierarchical structure. We choose a subset of Cora for our model training and test, which are contained in the five top categories and ten corresponding sub-categories (the numbers of papers are shown in the parenthesis):

- DA_1: /data_structures_algorithms_and_theory/
computational_complexity/ (711)
- DA_2: /data_structures_algorithms_and_theory/
computational_geometry/ (459)
- EC_1: /encryption_and_compression/encryption/ (534)
- EC_2: /encryption_and_compression/compression/ (530)
- NT_1: /networking/protocols/ (743)
- NT_2: /networking/routing/ (477)
- OS_1: /operating_systems/realtime/ (595)
- OS_2: /operating_systems/memory_management/ (1102)
- ML_1: /machine_learning/probabilistic_methods/ (687)
- ML_2: /machine_learning/genetic_algorithms/ (670)

Based on this data, we used a way similar to Dai et al. [12] to construct our training and test sets. For each set, we chose two top categories, one as positive class and the other as the negative. Different sub-categories were regarded as different domains. The task is defined as top category classification. For example, the dataset denoted as DA-EC consists of source domain: DA_1(+), EC_1(-); and target domain: DA_2(+), EC_2(-). The method ensures the domains of labeled and unlabeled data related due to same top categories, but the domains are different because they are drawn from different sub-categories. Such a preprocessing is a common practice for data preparation for adaptation purpose. Some previous works [9, 12] found that SVM classifier trained on in-domain data performed much worse on the out-of-domain, implying large domain gap.

We preprocessed the data for text information. The stop words and low-frequency words with count less than 5 were removed. Then the standard TF-IDF [22] technique was applied to the text datasets.

Reuters-21578 [21] is one of the most famous test collections for evaluation of automatic text categorization techniques. Reuters-21578 corpus also has hierarchical structure. It contains 5 top categories. Among these categories, orgs, people and places are three big ones. We generated three data sets, i.e. orgs vs people, orgs vs places and people vs places, for evaluation in a way similar to what we have done on the Cora datasets.

4.2 Performance Comparison

We compared KMM-LM with both SVM [8] and KMM [7]. For the implementation details, we used the default parameter $\varepsilon = (\sqrt{n} - 1)/\sqrt{n}$ and RBF kernel $k(x_i, x_j) = \exp(-\sigma||x_i - x_j||^2)$ in KMM. After obtaining the importance weight using KMM, we used the weighted version of libsvm [1] to build the classifier by incorporating the instance importance. SVM also used the RBF kernel. For both SVM and KMM, we tuned the parameters using five cross-validation on the

---

[1] http://www.csie.ntu.edu.tw/~cjlin/libsvm/

Table 1. Error rate of classification adaptation on Cora datasets

Data	SVM	KMM	KMM-LM
DA-EC	0.309	0.282	**0.241**
DA-NT	0.145	0.091	**0.082**
DA-OS	0.197	**0.108**	0.158
DA-ML	0.258	0.237	**0.204**
EC-NT	0.339	0.323	**0.253**
EC-OS	0.322	0.299	**0.267**
EC-ML	0.355	0.332	**0.299**
NT-OS	0.406	**0.367**	0.384
NT-ML	0.141	0.126	**0.099**
OS-ML	0.226	0.185	**0.157**

Table 2. Error rate of classification adaptation on Reuters datasets

Data	SVM	KMM	KMM-LM
Orgs-People	0.285	0.212	**0.209**
Orgs-Places	0.329	0.292	**0.275**
People-Places	0.387	0.334	**0.297**

source data. The classification error rate on the target data is used to evaluate classification performance, which is defined as follows:

$$\varepsilon = \frac{N_{misclassified}}{N_{total}}$$

where $N_{misclassified}$ denotes the number of misclassified test instances and N_{total} denotes the total number of test instances.

The KMM-LM algorithm used the same RBF kernel as what we used in KMM to map the data from the original feature space to the RKHS. For each dataset, we tuned the kernel parameter σ and the trade-off parameters, λ_1 and λ_2, by using five fold cross-validation on the source data. For the kernel parameter σ, the adjusting range is $\{0.125, 0.25, 0.5, 1, 2, 4, 8\}$. The adjusting ranges for both λ_1 and λ_2 are $\{0.1, 1, 10, 100, 1000\}$.

Table 1 and 2 show the test error rates for the Cora and the Reuters, respectively. SVM is a state-of-the-art algorithm for traditional classification problem, which gained superior performance on a large number of classification tasks. However, from the two tables we can see that SVM does not work well on the cross-domain text classification tasks as it did not consider the domain discrepancy. Thus SVM would not be effective for transfer learning. In contrast, both KMM and KMM-LM perform better by taking the distribution divergence between two domains into consideration than by treating them indiscriminately.

The results show that KMM-LM outperforms KMM on most datasets. Two reasons may account for the competencies of our proposed method. Firstly, KMM-LM could obtain the more appropriate instance importance than KMM by making use of the knowledge of class label in the source data. Specifically, KMM assigns bigger importance weights to the instances which are more similar to the target data. But if the instances are hard to classify, it would be expected that their importance would decrease. Hence, the soft margins are used to penalize the misclassified instances in our proposed method. Secondly, instead of using a two-step procedure in KMM, KMM-LM combines the two separate stages into an unified optimization framework. Both the classification decision weight vector and the instance importance weight vector mutually constrain with each other. The more optimal the instance importance weight, the better discriminative the decision hyperplane, and vice versa. The complementary cooperation between the two weight vectors creates a positive cycle and could gradually improve the adaptation performance.

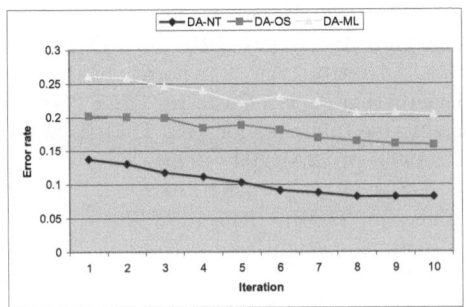

Fig. 1. Error rate curves with iterations

We also empirically analyse the convergence property of KMM-LM since it is an iterative minimization algorithm. Figure 1 shows the error rate curves with iterations for the three Cora datasets. We can see that the test error rate decreases slowly when the iteration number is close to 10. It indicates that our algorithm converges very fast. Thus, we empirically set the maximum iteration count $N_{max} = 10$.

5   Conclusions

In this paper, we presented the KMM-LM approach for transfer learning which integrated KMM and SVM into an unified optimization framework. The thrust of our technique is to incorporate the class label knowledge of the source data to guide the model to find the more appropriate instance importance in comparison with KMM. KMM-LM further combined the two separate stages of KMM and SVM into an optimization framework by leveraging the complementary cooperation between the two weight vectors. We empirically evaluated the KMM-LM algorithm for cross-domain document classification. The comparison experiments shown that it outperformed the baseline methods.

References

1. Sarinnapakorn, K., Kubat, M.: Combining Sub-classifiers in Text Categorization: A DST-Based Solution and a Case Study. IEEE Transactions on Knowledge and Data Engineering 19(12), 1638–1651 (2007)

2. Blitzer, J., Dredze, M., Pereira, F.: Biographies, Bollywood, Boom-boxes and Blenders: Domain Adaptation for Sentiment Classification. In: ACL 2007, pp. 440–447 (2007)
3. Blitzer, J., Kakade, S., Foster, D.P.: Domain Adaptation with Coupled Subspaces. In: AISTATS 2011, pp. 173–181 (2011)
4. Pan, W., Xiang, E.W., Liu, N.N., Yang, Q.: Transfer Learning in Collaborative Filtering for Sparsity Reduction. In: AAAI 2010, pp. 230–235 (2010)
5. Ma, H., Zhou, D., Liu, C., Lyu, M.R., King, I.: Recommender Systems with Social Regularization. In: WSDM 2011, pp. 287–296 (2011)
6. Gao, W., Cai, P., Wong, K.-F., Zhou, A.: Learning to Rank only using Training Data from related Domain. In: SIGIR 2010, pp. 162–169 (2010)
7. Huang, J., Smola, A.J., Gretton, A., Borgwardt, K.M., Schölkopf, B.: Correcting Sample Selection Bias by Unlabeled Data. In: NIPS 2006, pp. 601–608 (2006)
8. Joachims, T.: Transductive Inference for Text Classification using Support Vector Machines. In: IMCL 1999, pp. 200–209 (1999)
9. Pan, S.J., Yang, Q.: A Survey on Transfer Learning. IEEE Transactions on Knowledge and Data Engineering 22(10), 1345–1359 (2010)
10. Jiang, J., Zhai, C.: Instance Weighting for Domain Adaptation in NLP. In: ACL 2007, pp. 264–271 (2007)
11. Dai, W., Yang, Q., Xue, G., Yu, Y.: Boosting for Transfer Learning. In: ICML 2007, pp. 193–200 (2007)
12. Dai, W., Xue, G., Yang, Q., Yu, Y.: Co-clustering Based Classification for Out-of-domain Documents. In: KDD 2007, pp. 210–219 (2007)
13. Dayanik, A.A., Lewis, D.D., Madigan, D., Menkov, V., Genkin, A.: Constructing Informative Prior Distributions from Domain Knowledge in Text Classification. In: SIGIR 2006, pp. 493–500 (2006)
14. Kanamori, T., Suzuki, T., Sugiyama, M.: Statistical Analysis of Kernel-based Least-squares Density-ratio Estimation. Machine Learning (ML) 86(3), 335–367 (2012)
15. Sugiyama, M., Nakajima, S., Kashima, H., von Bnau, P., Kawanabe, M.: Direct Importance Estimation with Model Selection and Its Application to Covariate Shift Adaptation. In: NIPS 2007 (2007)
16. Tsuboi, Y., Kashima, H., Hido, S., Bickel, S., Sugiyama, M.: Direct Density Ratio Estimation for Large-scale Covariate Shift Adaptation. In: SDM 2008, pp. 443–454 (2008)
17. Kanamori, T., Hido, S., Sugiyama, M.: A Least-squares Approach to Direct Importance Estimation. Journal of Machine Learning Research (JMLR) 10, 1391–1445 (2009)
18. Bickel, S., Brckner, M., Scheffer, T.: Discriminative Learning for Differing Training and Test Distributions. In: ICML 2007, pp. 81–88 (2007)
19. Sun, Q., Chattopadhyay, R., Panchanathan, S., Ye, J.: A Two-Stage Weighting Framework for Multi-Source Domain Adaptation. In: NIPS 2011, pp. 505–513 (2011)
20. McCallum, A.K., Nigam, K., Rennie, J., Seymore, K.: Automating the Construction of Internet Portals with Machine Learning. Information Retrieval 3(2), 127–163 (2000)
21. Lewis, D.D.: Reuters-21578 Test Collection, http://www.daviddlewis.com/
22. Salton, G., Buckley, C.: Term-weighting Approaches in Automatic Text Retrieval. Information Processing & Management 24(5), 513–523 (1988)

Properly and Automatically Naming Java Methods: A Machine Learning Based Approach

Shusi Yu[1], Ruichang Zhang[1], and Jihong Guan[2]

[1] School of Computer Science, Fudan University, Shanghai, China
{yushusi,rczhang}@fudan.edu.cn
[2] Dept. of Computer Science and Technology, Tongji University, Shanghai, China
jhguan@tongji.edu.cn

Abstract. Method names play an important role in software maintenance. A good name explains the function of a method to developers,
while bad names mislead them. However, method naming is a complicated task. For example, Java programming specification suggests that
method names should be verbs or verb phrases. Previous research shows
that each identical verb is related to at least one name rule and each
rule is consists of 30 conditions. Since large-scale software contains hundreds of identical verbs, choosing the proper verb for a method is related
to thousands of conditions. There exists some semi-automated method
verb rule mining techniques. these rules are useful to find if a method
uses the wrong verb. However, they are not effective for general case
when no rules are applicable. In addition, none of them discusses how
to find the proper target of the verb. This paper proposes an automated
method-naming tool, based on machine learning approach. Experiments
on Eclipse and other Java projects show that our tool can successfully
predict the verbs of 70% methods and pick out the proper targets in
method contents for 90% methods.

1 Introduction

Method name is the abstraction of a method, telling what the method is and
what it can do. Proper method names help program comprehension. For example,
if a class has both Method *lock()* and *unlock()*, the class is a mutex. We can also
assert that method *lock()* shall be executed before *unlock()*. Such "lock&unlock"
rules is mined by [5] and the rules are further used to find bugs. Using method
names and method invocation rules, a Finite State Automata (FSA) model of
the behavior of the Java class can be built [4].

Software maintenance also highly depends on proper method names [9,11]. For
example, Eclipse bug #40165 shows that Eclipse users often forget to include
default tools "tools.jar" into their projects. If all methods in the class containing
the bug (*AntCorePreferences.java*) are properly named, it is not difficult for the
developer to locate the bug. Since the bug arises when something important is
not loaded into the project, methods such as *load**** and *get**** are likely to
contain it. Second, the bug deals with *tools.jar*, so the search space is further

S. Zhou, S. Zhang, and G. Karypis (Eds.): ADMA 2012, LNAI 7713, pp. 235–246, 2012.
© Springer-Verlag Berlin Heidelberg 2012

limited to the methods whose names also contain *tools* and *jar*. There is only 1 method among all 56 methods fulfill both conditions: Further debugging proofs that the method *getToolsJarURL* is the very one containing the bug.

Unfortunately, although method names are so important for software engineering, it is not an easy task to name methods properly. A proper method name means most software developers can understand the function of the method by the name. In other words, the name must follow method name conversions agree among most developers. The first step of Java method name conventions can be found in Java language specification [6], "Method names shall be verbs and verb phases." For example, *getToolsJarURL* consists of a verb *get* and its target *ToolsJarURL*. However, the specification [6] do not provide further guidance.

The developers of Eclipse (the most famous open source IDE) deployed class *NamingConventions.java* in package *jdt.core* to implement method name conventions by hard coding. One of the implemented rules is: if the method *reads* the value of an *attribute*, *AND* it *returns* an *attribute* of the class, its action name should be *get*. The hard coded rules can only check two verbs (get and set) and the rules are not at all precise. It can not go further because more precise rules are too complex to be manually coded.

[7] shows that each identical action name is related to at least one such name rules and each rule is related to at least 30 *AND* or *OR* conditions. Eclipse has 237 identical verbs, which mean the rules to choose the proper action for methods is related to at least 237*30 conditions. This is certainly too complicated for developers to learn and implement, either by source code or by natural language. Therefore an automated tool for method naming is well required.

Previous researches try to automatically discover above mentioned name rules [7,8,1]. They pre-defined a set of conditions, such as "creates regular objects" or "reads field". If most methods using the same action name fulfill the same set of conditions, the condition set is considered a rule. They assist developers by checking existing method names against such rules. Their approach is successful in rule mining and method name debugging. However, the approach has at least the following three problems.

First, they only focus on how to choose verbs. Our experiments show that more than 90% method names have both verbs and targets. Therefore choosing proper targets for method names is almost as important as choosing proper verbs. However, previous techniques provide little knowledge about target choosing.

Second, their approach only works when there are applicable rules. For the major grey zone where no rules are available, their approach does not work. For example, if 40% *check**** methods follows rule *A* and another 40% follows rule *B*, no applicable name rules for verb *check* can be found since neither rules is followed by majority.

Since automated method naming tool is well required and existing approaches are not sufficient for the requirement, we develop a method-naming tool, which solves the problems. Our tool is based on classification — a widely used machine learning technique. Given a number of predefined categories, a classifier learns from some already-categorized objects and predict the category of new objects [12].

In this paper, classification technique is used to choose both proper verb and proper target. In order to find the proper verb, each unique verb is considered as a category and each Java method as an object. Verb naming is fitted to the problem of assigning a method to a corresponding verb category. In order to find the proper target, local variables of a method is considered an object and they are categorized to being or being not the target of the method. The already categorized data is called *training set* and the un-categorized data are called *test set*.

Major contributions of this paper are as follows:

1. an vector space model to capture the function & structure of Java methods and the environment of variables.
2. A classification based approach to properly and automatically name Java methods.
3. Extensive empirical experiments on Eclipse dataset are conducted to evaluate the proposed approach. We successfully discover the proper verbs of 70% Eclipse methods and other Java projects, achieving a 70% accuracy. The experiments on predicting targets achieve a high accuracy as 90%.

The rest of this paper is organized as follows: Section 2 presents our approach to automatically naming Java methods. Section 3 gives the experiment results. Section 4 concludes the paper.

2   The Approach

Our tool consists of two namers: a namer to choose the proper verb and another to choose the proper target. Both our verb namer and target namer are classifiers. Verb naming is a multi-class categorizing, each verb is a category and each method is an object. A method belongs to a certain category if its name uses corresponding verb. The target naming is a bio-class categorization. Each variable is an object. If an variable is the target of the method using it, the variable belongs to the positive category. If not, it belongs to the negative category.

The rest of this section introduces the approach to build Java method namer. We first give the roadmap, and then present the detailed implementation techniques. Since both our namers share very similar steps, in most of the following subsections we do not explain them separately.

2.1   Roadmap

Figure 1 shows the overview of the roadmap to build the Java method namer. The *dataset* is a Java project used as training set. The *parser* extracts semantic items from dataset by parsing its abstract syntax tree (AST). The *feature matcher* extracts method names, variable names and features from the AST. It models a method or a variable as a feature vector. The *POS tagger* identifies the verb

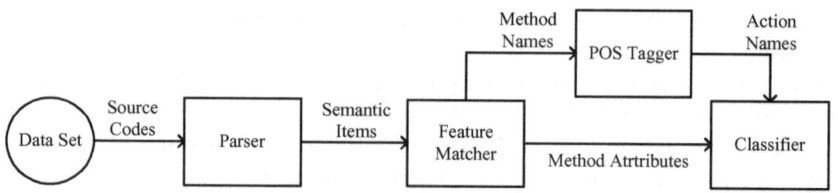

Fig. 1. The roadmap to build the Java method namer

and the target of each method. The results of feature matcher and POS tagger are used to train the *classifier*, which we used as the Java method namer.

The application process of the namer is similar to building it. A new method or variable is parsed and its feature vector is extracted. The trained classifier is used to predict its category, *i.e.*, the verb of the method.

2.2   Parser

Java source codes are text files. A parser transforms each source code file into an AST. In AST, method nodes are children of their class nodes. Method names and operation statements are children of method nodes. Smaller semantic items such as variables or operators are children of statement nodes. The features used in our approach are extracted from the AST.

2.3   Feature Matcher

Feature matcher traverses the AST and extracts method names, variable names and other required items from AST. Based on these items, a set of chosen features are used to model methods and variables. A method or a variable is modeled as a vector in the feature space. Each unique feature forms a dimension of the space. The occurrence number of the feature is the value on corresponding dimension.

```
                                    public List getURLs_2() {
                                        List result = new List();
                                        if (fDefaultURLs != null) {
                                            result.addAll(fDefaultURLs);
   public List getURLs_1() {           }
       List result = new List();       if (fCustomURLs != null) {
       if (fDefaultURLs != null) {         result.addAll(Arrays.asList(fCustomURLs));
           result.addAll(fDefaultURLs);  }
       }                                return result ;
       return result ;              }
   }
```

Fig. 2. Example methods

In Figure 2, Method *getURLs_1* creates 1 *List* object and contains 1 *if* branch. Method *getURLs_2* also creates 1 *List* object, but it contains 2 *if* branches. Considering a simple feature space model $< Create object, If branch >$, the corresponding vector of Method $getURLs\_1$ is $< 1, 1 >$. The vector of Method $getURLs\_2$ is $< 1, 2 >$.

There are three types of method features: signature features, content features and sub-verbs. Signature features represent the basic behaviors of a method, such as the parameter types and return type. Content features locate in the method body, such as the number of objects created, the number of for loops and so on. [8] also models methods by features, our signature features and content features are mainly adopted from it. Sub-verb are the verbs of methods called by the original method. A complete list of the features is shown in Table 1.

Table 1. Attributes for modeling methods

Signature features	
Returns void	Returns boolean
Returns digit	Returns string
Returns object	Has parameters
Has boolean parameters	Has digit parameters
Has string parameters	Has string parameters
	Returns field value
Content features	
Reads field	Writes field
Accesses argument	Accesses boolean
Accesses digit	Accesses string
Accesses object	Accesses array
Creates string	Creates arrary
Declares variable	Creates object
Reads constant	Casts variable
For loop	While loop
If branch	Switch branch
Return points	Throws exceptions
Catches exceptions	Exposes checked exceptions
Recursive Method call	Method call on field value
Method call on parameter value	
Sub-verb features	
Methods called with certain action names	

Most listed features in Table 1 are self-explanatory. Only sub-verbs need further explanation. Each unique verb is a unique sub-verb feature. For example, method $getURLs\_1$ calls method $addAll()$ once. Therefore its value on dimension add is 1. Signature attributes are different from the other two types of attributes because they only tell whether a certain attribute exists. In other words, it value is binary, a choice of 0 or 1. Content attributes and sub-verbs are numerical attributes. They tell how many times a certain attribute occurs in a method.

The approach to vectorize variables is quite similar to method vectorizing. Variable features are listed in Table 2.

If branch condition in Table 2 means how many times the variable is used in the condition expression of an *if* branch. *If branch body* means how many

Table 2. Attributes for modeling variables

Signature features	
Is parameter	Is attribute
Is return value	Is boolean
Is digit	Is string
Is object	
Content features	
Being assigned	Being accessed
If branch condition	If branch body
Switch branch condition	Switch branch body
For loop condition	For loop body
While loop condition	While loop body
Try branch body	Catch branch body
Sub-method parameter	Sub-method return value
Sub-verb features	
Sub-method verb, as a parameter	
Sub-method verb, as a return value	

times the variable is used in the body of an *if* branch. The definition of other content features is the same. For Sub-verb features, if the variable is used by 2 *add* methods as a parameter, its value on dimension *Sub-method add, as a parameter* is 2. The definition of the other sub-verb feature is the same.

2.4   POS Tagger

Java programmers usually use "camel case" when forming multiple-word method names. For example, method name *setCustomTypes* consists of 3 words: *set*, *custom* and *types*. In order to combine them into a single method name and keep the boundary between two words, Java programmers capitalize the fist letter of *custom* and *types*. We reverse the process and split method names and variable names into word sequences [10].

Section 1 shows that Java method names are usually leaded by verbs. The POS tagger is used to identify whether the first word in the word sequence extracted from a method name is a verb. If it is, the verb is taken as the verb of the method.

Although most method names are leaded by a verb, there exist methods not following name convention. For example, the method returning the size of a list can be *size()*. In this paper we only consider methods leading by verbs. Our approach can be easily extended to other method leading words such as *size*.

2.5   Classifier

Classification intends to group objects into a set of predefined categories. Typical classification approaches transform an object into a space vector based on the vector space model (VSM).

Formally, we denote the set of categorized method vectors as M and the not categorized vectors M'. Each vector in M or M' belongs to a certain category c of the category set C. An function $c = C(m)$ maps m to a proper category c. In this paper, the complete set of verb $Verb$ is equal to C. A classifier CF is trained by M and predicts the category of any vector $m' \in M'$. A correct prediction means $CF(m') = C(m')$.

Practically, our verb namer is trained by named and vectorized methods and predict the verb of unnamed methods. Our target namer is trained by trained by named and vectorized variables and predicts if a variable is or is not the target of the method using it. We use *precision* to evaluate the effectiveness of a classifier, which is measured as follows:

$$precision = \frac{\#of\,Hits}{\#of\,Tests}, \tag{1}$$

where *#ofHits* is the number of correctly categorized methods and *#ofTests* means the number of methods being categorized.

In this paper, we adopt support vector machine (SVM) [2] as our classification approach. A support vector machine constructs a hyperplane or set of hyperplanes in a high or infinite dimensional space, which can be used for classification, regression or other tasks. Intuitively, a good separation is achieved by the hyperplane that has the largest distances to the nearest training vectors of any class. SVM is powerful for datasets characterized by small sample size, nonlinearity, and high dimensionality.

3   Experiment Results

In this section, we present the experiment results. We first describe the setup of the experiments, and then present the results to show the effectiveness of the proposed approach in naming Java methods.

3.1   Experiment Setup

The Eclipse project 3.6[1] is used as experimental dataset. It is an open source SDK, mainly coded in Java. We use antlr 3.1[2] to parse Java codes, get each method and the corresponding AST. We develop our attribute matcher in Java. GATE5.1[3] is adopted as our POS tagger. Libsvm [3] is used to implement the classifier.

3.2   Verb Namer Effectiveness

We train the method namer with source codes of Eclipse. 237 verbs are used by the 101015 methods of Eclipse project. That is, the total number of categories is

---

[1] http://www.eclipse.org/
[2] http://www.antlr.org/
[3] http://gate.ac.uk/

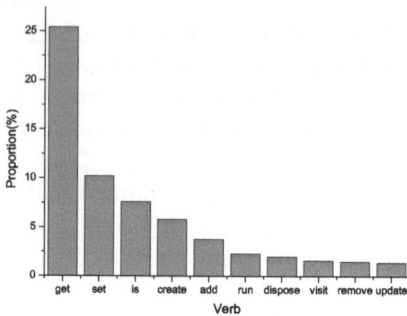

Fig. 3. Size of top 10 Eclipse verbs

237. Figure 3 shows the sizes of verbs used in Eclipse project. *get* is the biggest verb which leads about 25% methods. That means a naive namer which predicts *get* for every method reaches precision 25%.

Figure 4 show the precision results for different sizes of training set. Each training set consists of randomly chosen methods, and the rest of dataset is used to evaluate the namer's precision. Figure 4 shows that 1) as more methods are used for training, the classifier ca get higher precision. 2) our approach can get a precision of 73% when 60% methods are used for training. Compare to 25% of naive naming, it is a significant improvement.

Furthermore, we can see that our classifier can achieve good precision even with small training sets. The precision does not drop drastically when the training set goes smaller. When only 10% methods are used to train the classifier, our approach can still predict correctly the action names of 69% methods. This result suggests that our classifier can work well in real applications when software is under development and no large training datasets are available.

In the models of [7,8], the occurrence frequencies of features are not considered. All features have only 0-1 values. In this paper, we not only consider whether a method has a certain feature, but also consider how frequently the feature appears. Figure 5 compares the results of the two cases. Here, 60% Eclipse methods chosen randomly are used to train the classifier and the rest 40% for testing. The results show that the numeric model significantly outperforms the boolean model.

Here we show whether the method namer trained by one Java project is also effective over other projects. For this end, we randomly choose 60% Eclipse methods to train a classifier, and then test it on two other Java projects, TomCat 7.0.16 and Hadoop 0.20.203. The results are shown in Figure 6.

Figure 6 show that the classifier trained by Eclipse methods is almost as effective in Eclipse as in TomCat 7.0.16 and Hadoop 0.20.203. The result indicates that Java developers of at least the three projects follow the similar name conventions so that a method namer trained by one project can be applied to the others.

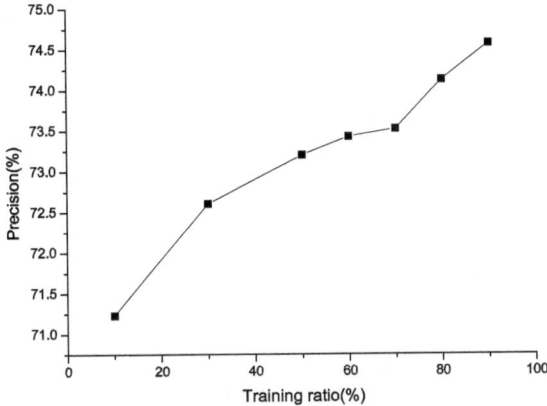

Fig. 4. Precision *vs.* the size of training dataset

3.3 Target Namer Effectiveness

In this subsection we show that our approach can not only discover the proper verb for method names but also the proper target. The target is usually a noun or a noun phrase. For Method *getProperties*, *get* is the verb and *Properties* is the target.

Not like verb choosing, which is limited to a verb list, target choosing is a much wider choice. Fortunately, most method targets can be directly found in the method itself. The target is an attribute of the class, or a parameter/variable of the method.

If we can find a variable in the method which is of the same name with the target, we call it a *full-target* method. The corresponding variable is called *full-target* variable. If only a part of the target can be found in method body, we call the method a *semi-target* one. For example, if we cannot found variable *AllProperties* in Method *getAllProperties()* but only variable *Properties*, *getAllProperties()* is a *semi-target* method. The corresponding variable *Properties* is called a *semi-target* variable. If a variable is neither a *full-target* variable nor a *semi-target* one, it is *none-target*. An attribute of the class used in a certain method or a parameter of the method is respected as an independent variable of the method.

Our investigation shows that slightly more than 10% method names are single verbs. We need not bother to find the proper target for them. Less than 10% method names have targets, but neither full-targets nor semi-targets can be found in method body. Our technique cannot be applied to these methods. Our tool focuses on the major 80% method names. Their targets can be found in method bodies.

In Eclipse, 12.7% attributes/parameters/variables are targets, 15.1% are semi-targets, the majority 70% are non-targets. Following three experiments are developed to verify the effectiveness of our target namer. A 0-1 classifier is used

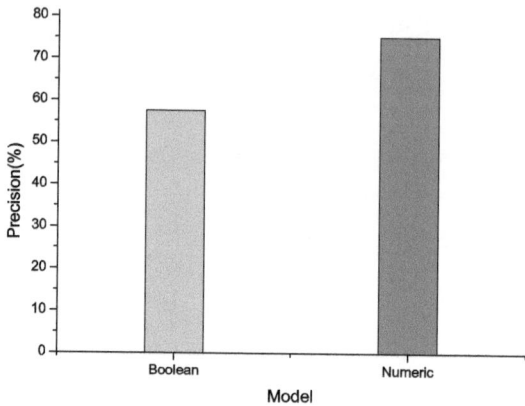

Fig. 5. Boolean model *vs.* numeric model

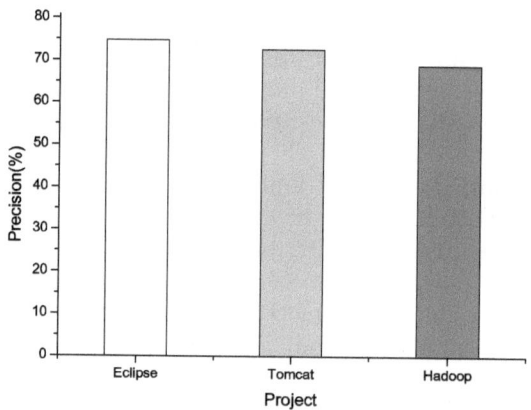

Fig. 6. Precision comparison of the method namer over three Java Projects

to identify if a variable is a target variable. If the identification is correct, our target namer find the proper target for the method containing the variable. In each experiment, 60% of the data set is used as the training set and the rest 40% the test set.

1. Experiment *Full target*. Only full-target methods are used in this experiments. All full target variables in these methods constitute the positive category. Non-target variables in full-target methods constitute the negative category. The high precision (93.6%) in Figure 7 shows that our approach is successful in discovering method targets.

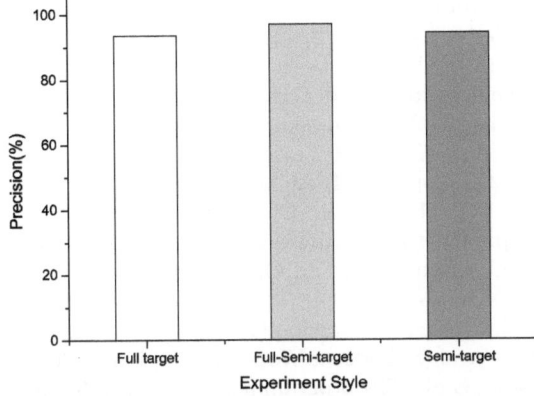

Fig. 7. Precision of target prediction

2. Experiment Full-Semi target. All full-target variables and semi-target ones from full-target methods form the positive category. All non-target variables in in the same set of methods form the negative set. The precision of the classifier is even higher (97.1%).
3. Experiment *Semi-target* is the closest to real world applications. All full-target/semi-target variables in all methods are the positive category and all non-target variables are the negative category. In this experiment, we still get a precision higher than 90%.

4   Conclusion

This paper introduces an approach to properly and automatically name Java methods. Experimental results show that the method namer trained by 60% Eclipse methods can predict the verbs of roughly 70% Java methods and more than 90% targets.

Our method naming approach is based on Java method naming rule: "Method names shall be verbs and verb phases.". Since object oriented programming languages other than Java also follow similar rules, our approach can also be extended to other programming languages such as C++ or Python with a slight modification.

Acknowledgment. This work was supported by Research Innovation Program and "Shuguang" Scholar Program of Shanghai Municipal Education Committee.

References

1. Binkley, D., Hearn, M., Lawrie, D.: Improving identifier informativeness using part of speech information. In: Proc. of the Working Conf. on Mining Software Repositories (2011)

2. Burges, C.J.C.: A tutorial on support vector machines for pattern recognition. Data Mining and Knowledge Discovery 2, 121–167 (1998)
3. Chang, C., Lin, C.: LIBSVM: A library for support vector machines. ACM Transactions on Intelligent Systems and Technology 2, 27:1–27:27 (2011)
4. Dallmeier, V., Lindig, C., Wasylkowski, A., Zeller, A.: Mining object behavior with adabu. In: Proceedings of the 2006 International Workshop on Dynamic Systems Analysis, WODA 2006, pp. 17–24. ACM, New York (2006)
5. Engler, D., Chen, D.Y., Hallem, S., Chou, A., Chelf, B.: Bugs as deviant behavior: a general approach to inferring errors in systems code. In: Proceedings of the Eighteenth ACM Symposium on Operating Systems Principles, SOSP 2001, pp. 57–72. ACM, New York (2001)
6. Gosling, J., Joy, B., Steele, G., Bracha, G.: Java(TM) Language Specification, 3rd edn. Java Addison-Wesley. Addison-Wesley Professional (2005)
7. Host, E.W., Ostvold, B.M.: The programmer's lexicon, volume i: The verbs. In: Proceedings of the Seventh IEEE International Working Conference on Source Code Analysis and Manipulation, pp. 193–202. IEEE Computer Society, Washington, DC (2007)
8. Høst, E.W., Østvold, B.M.: Debugging Method Names. In: Drossopoulou, S. (ed.) ECOOP 2009. LNCS, vol. 5653, pp. 294–317. Springer, Heidelberg (2009)
9. Lawrie, D., Morrell, C., Feild, H., Binkley, D.: What's in a name? a study of identifiers. In: Proceedings of the 14th IEEE International Conference on Program Comprehension, pp. 3–12. IEEE Computer Society, Washington, DC (2006)
10. Lawrie, D., Binkley, D.: Expanding identifiers to normalize source code vocabulary. In: ICSM, pp. 113–122. IEEE (2011)
11. Martin, R.C.: Clean Code: A handbook of agile software craftsmanship. Prentice Hall (2009)
12. Nedjah, N., de Macedo Mourelle, L., Kacprzyk, J., Frana, F.M.G., de Souza, A.F.: Intelligent Text Categorization and Clustering. Springer (2008)

A Bag Reconstruction Method
for Multiple Instance Classification
and Group Record Linkage

Zhichun Fu[1], Jun Zhou[2], Furong Peng[3], and Peter Christen[1]

[1] Research School of Computer Science
The Australian National University
Canberra, ACT 0200, Australia
{sally.fu,peter.christen}@anu.edu.au
[2] School of Information and Communication Technology
Griffith University
Nathan, QLD 4111, Australia
jun.zhou@griffith.edu.au
[3] School of Computer Science and Technology
Nanjing University of Science and Technology
Nanjing, Jiangsu 210094, China
pengfr@njust.edu.cn

Abstract. Record linking is the task of detecting records in several databases that refer to the same entity. This task aims at exploring the relationship between entities, which normally lack common identifiers in heterogeneous datasets. When entities contain multiple relational records, linking them across datasets can be more accurate by treating the records as groups, which leads to group linking methods. Even so, individual record links may still be needed for the final group linking step. This problem can be solved by multiple instance learning, in which group links are modelled as bags, and record links are considered as instances. In this paper, we propose a novel method for instance classification and group record linkage via bag reconstruction from instances. The bag reconstruction is based on the modeling of the distribution of negative instances in the training bags via kernel density estimation. We evaluate this approach on both synthetic and real-world data. Our results show that the proposed method can outperform several baseline methods.

Keywords: Multiple instance learning, bag reconstruction, instance classification, record linkage, group linkage, historical census data.

1 Introduction

The goal of record linking is to match records referring to the same entity in different datasets [7]. This is a non-trivial task because identifying the relationship between records is not straightforward due to commomly different structures in the datasets to be studied, and the possibly low quality of data in these datasets.

S. Zhou, S. Zhang, and G. Karypis (Eds.): ADMA 2012, LNAI 7713, pp. 247–259, 2012.
© Springer-Verlag Berlin Heidelberg 2012

Record linking is used in a number of applications, for example, paper citation analysis, medical record linkage, and consumer behavior mining [7,15,17].

A record linking system contains several components, which includes data pre-processing, record pair comparison, record pair classification, and result evaluation [7]. Among them, record pair classification has attracted most attention. In this task, the similarities of record pairs determine whether the pairs are matched or non-matched. Efforts in this area can be dated back to 1946, when Dunn proposed the initial idea of record linkage [8]. Since then, many approaches, either deterministic or probabilistic, have been developed [7,16]. In recent years, machine learning approaches have been widely used to improve record linking performance. These approaches include support vector machines (SVMs) [2,5], clustering [9], and graph-based methods [19,23].

In many cases, a matching decision has to be made over a collection of record pairs instead of individuals. To address this problem, Bhattacharya and Getoor proposed a collective entity resolution method [1]. In this method, an entity graph and a reference graph are used to characterize the similarity of the attributes of entities and their co-occurrence relationships. Then a relational clustering algorithm is used to compute the similarity of two clusters as a weighted sum of the attribute similarity and co-occurrence relational similarity. This allows those records that correspond to the same entity be assigned to the same cluster. Variations of this method include collective graph identification [18], transforming graph representation [21], and collective classification [24].

On et al. [20], on the other hand, proposed to link groups of record via a group linkage measure based on weighted bipartite graph matching [3]. In this method, similarities between records are computed first. Then these weights are used to compute group similarities between two collections of record pairs. In this way, the matching of two groups is determined not only by the similarity between records, but also by how many record pairs are being matched.

Although the goal of group linkage is to classify groups of records as matched or non-matched, decisions on individual record links are often necessary as well. For example, the group linkage model of On et al. [20] requires the information on the number of matched record links for group similarity computation. This has made group linkage a binary classification problem at both the group level and the record pair level. Therefore, it would be advantageous if this two-level classification process can be solved using an integrated solution. However, such a solution is often difficult to find due to the lack of labeled training data of record pairs.

Fu et al. proposed an approach that uses the multiple instance learning with instance selection (MILIS) method to solve this problem [13,14]. In MILIS, data are represented as bags, each of which contains some instances. While the group classification can be solved by a traditional Multiple Instance Learning (MIL) classifier, the instance level classification is performed by a bag reconstruction step that groups the target instance to be classified with some negative instances selected from the negative instance prototype generated by MILIS. Then the instance classification can be naturally transformed into bag classification.

Although this solution has achieved sound results [13], the bag reconstruction step is performed in a random manner, and has not taken the negative instance distribution into consideration.

In this paper, we introduce a new method for bag reconstruction. Our method explicitly measures the distribution of instances in the negative training bags using kernel density estimation (KDE). Then the negative instances that are most representative are selected for bag reconstruction. We show in the experiments that this approach delivers a more deterministic and robust solution than the earlier method proposed in [13].

The rest of the paper is organized as follows. Section 2 introduces the MILIS method reported in [14]. Section 3 describes the proposed bag reconstruction method. The experimental results and analysis are presented in Section 4. Finally, we draw conclusions in Section 5.

2   Bag Reconstruction Based on the MILIS Method

2.1   MILIS Method

The MILIS method [14] was developed to solve the low efficiency problem of multiple instance learning of the embedded instance selection (MILES) method [4]. Both MILIS and MILES map bags into a feature space defined by a few instances. This allows a bag to be converted to a vector that characterizes the bag to instance similarities. Then the bag classification is transformed into a normal vector classification problem. While MILES uses all instances in the training set to perform feature mapping, MILIS only extract one instance prototype (IP) from each training bag. This allows MILIS to generate a much lower-dimensional feature representation with comparable classification accuracy as MILES.

In a MIL setting, suppose a bag B_i contains m_i instances denoted by $\mathbf{x}_{i,j}$ for $j = 1, \ldots, m_i$. If the labels of at least one $\mathbf{x}_{i,j} \in B_i$ is positive, i.e., with $y_{i,j} = 1$, B_i is considered as a positive bag with label $Y_i = 1$. Alternatively, if B_i contains only negative instances with $y_{i,j} = -1$, it is a negative bag and is labeled as $Y_i = -1$. MIL tries to solve the following problem: given a set of positive bags $\mathcal{B}^+ = \{B_1^+, \ldots, B_{n+}^+\}$ and a set of negative bags $\mathcal{B}^- = \{B_1^-, \ldots, B_{n-}^-\}$, predict the label of a new bag B.

In the feature mapping step, MILIS selects one IP from each training bag by measuring the probability that an instance is generated from the distribution of instances in negative training bags. For a positive bag, such IPs corresponds to an instance that is most unlikely to be generated, i.e., is least negative. On the contrary, for a negative bag, an IP is selected as the most negative instance. The likelihood of an instance $\mathbf{x}$ to be negative is calculated using kernel density estimation:

$$p(\mathbf{x}|\mathcal{B}^-) = \frac{1}{Z} \sum_{j=1}^{T} \exp\left(-\tau||\mathbf{x} - \mathbf{x}_j^-||\right), \qquad (1)$$

where $\mathbf{x}_j^- \in \mathcal{B}^-$ is a negative instance, T is the total number of negative instances in $\mathcal{B}^-$, $||.||$ denotes the Euclidean norm, Z is a normalization factor which can be omitted in computation because it is the same for every instance $\mathbf{x}$, and τ is a parameter that controls the contribution of training samples.

With the IPs ready, bag-level feature representation can be created based on the similarities between a bag and the IPs. More specifically, such similarity is calculated as:

$$s(B_i, I) = \max_{\mathbf{x}_{i,j} \in B_i} \exp\left(-\eta ||\mathbf{x}_{i,j} - I||^2\right), \tag{2}$$

where I is an IP, and η is a feature mapping parameter that controls the similarity. Then a bag can be represented as an n-dimensional vector:

$$\mathbf{v}_i = [s(B_i, I_1), \ldots, s(B_i, I_i), \ldots, s(B_i, I_n)], \tag{3}$$

where I_i is the instance prototype selected from the i-th training bag, and $n = n^+ + n^-$ is the total number of training bags. With the vectorized feature representation handy, an SVM classifier Φ is used to classify bags [22].

2.2   Bag Reconstruction for Instance Classification

The MILIS algorithm has provided a sound solution for bag classification. However, it cannot classify instances. This is mainly due to the fact that instance labels are not available in the training set. As mentioned before, in the case of group record linkage, instance labels, which correspond to whether two records are matched or not, are essential in measuring the similarity between groups.

To solve this problem, a bag reconstruction method has been proposed in [13], which provides a two-step algorithm for bag and instance level classification. In the first step, unknown bags are classified as positive or negative using the MILIS algorithm. In the second step, instances in the bags are classified. According to the MIL setting, the labels of the instances in bags that have been predicted as negative can be generated directly because the negative bags only contain negative instances. Therefore, the instance classification problem reduces to predicting the labels of only these instances in the bags that have been classified as positive. The bag reconstruction step then groups each instance in a positive bag with negative IPs to form new bags. Finally, the trained bag classifier is used to classify these newly constructed bags, and the bag classification results is treated as the corresponding instance classification results. This method is based on the rationale that if an instance is negative, then the reconstructed bag only consists of negative instances, and thus will be classified as negative. Otherwise, the new bag contains one positive instance, and therefore shall be classified as positive.

In [13], two strategies have been adopted for the bag reconstruction. The first strategy randomly selects instances from the negative IPs and groups them with the instance to be classified. The second strategy adopts a greedy algorithm and performs bag reconstruction and prediction simultaneously. The instances in the testing bag are added sequentially into an initially reconstructed bag that contains randomly selected negative IPs as in the first strategy, until the new bag is classified as positive or all instance have been added. The results show that bag reconstruction using the greedy strategy slightly outperformed the Random option [13].

3 Bag Reconstruction via Negative Instance Distribution Modeling

The classification performance of the reconstructed bags is dependent on the quality of the selected negative instances. One would expect that these new bags shall be consistent with the distribution of the bags in the training set, so that the learned classification model Φ will generate good classification results. However, the random and greedy instance selection strategies in [13] have not taken this into consideration. This means the quality of the bag reconstruction is not guaranteed due to the uncertainty in negative instance selection. To solve this problem, we seek to model the distribution of the instances in the negative training bags and propose a new method to fulfill the bag reconstruction task. For convenience, from now on, we denote the selected negative instances for bag reconstruction as $\mathcal{X}$. Please note that $\mathcal{X}$ will be used for all instances to be classified.

We commence by a formal definition of the problem. Given training sets $\{\mathcal{B}^+, \mathcal{B}^-\}$ and the learned bag classification model Φ, the goal of instance classification is to predict the binary label $y_{i,j} \in \{1, -1\}$ of $\mathbf{x}_{i,j} \in B_i$, after B_i has been predicted as positive. In bag reconstruction, $\mathbf{x}_{i,j}$ is grouped with selected instances $\mathcal{X}$ in $\mathcal{B}^-$ to create a new bag $\tilde{B}$. Then the classification model Φ can be used to classify $\tilde{B}$, whose result is also considered as the label for $\mathbf{x}_{i,j}$. The goal of our method is to find the most representative $\mathcal{X}$.

Note that Equation 1 defines a kernel density estimator with an isotropic Gaussian kernel [14]. This allows the modeling of the likelihood that an instance $\mathbf{x}$ is contained in the negative bags. Based on this observation, our first solution is to select the most negative instance in the negative bags as the member of $\mathcal{X}$. Thus, $\mathbf{x}^*$ is defined by

$$\mathbf{x}^* = \arg \max_{\mathbf{x} \in \mathcal{B}^-} p(\mathbf{x}|\mathcal{B}^-) \tag{4}$$

where $p(\mathbf{x}|\mathcal{B}^-)$ is given by Equation 1.

This solution is similar to the MILIS negative IP selection process. The difference lies in that an IP is selected from a single bag in MILIS, while the $\mathbf{x}^*$ is selected from the whole negative instance pool. Such an option has three advantages. Firstly, from the data distribution point of view, $\mathbf{x}^*$ will be close to the negative IPs and far away from the positive IPs. Because the bag level feature representation step is performed using Equation 2, the similarity between a bag and an IP is based on the instance in the bag that is most similar to the IP. Thus, with the most negative instance being selected as $\mathbf{x}^*$, it is guaranteed that high similarity to negative IPs can be achieved. Secondly, the selection of $\mathbf{x}^*$ is deterministic. Unlike the random selection strategy proposed in [13], the most

---

Algorithm 1. Instance Classification

Input:
- Training set $\mathcal{B} = \{\mathcal{B}^+, \mathcal{B}^-\}$
- A testing bag B_i that contains m_i instances $\mathbf{x}_{i,j}$ for $j = 1, \ldots, m_i$
Output:
- Label $Y_i \in \{1, -1\}$ for bag B_i and labels $y_{i,j} \in \{1, -1\}$ for instances $\mathbf{x}_{i,j} \in B_i$

1. Generate IPs using Equation(1) and MILIS instance selection strategy
2. Calculate instance-based embedding for bag feature representation using Equation(3)
3. Train bag-level SVM model Φ
4. Classify B_i
5   **if** $Y_i = -1$
6.   Classify all $\mathbf{x}_{i,j} \in B_i$ as negative
7. **else**
6.   Create $\mathcal{X}$ based on the distribution of negative training bags
8.   **for** $\mathbf{x}_{i,j} \in B_i$ **do**
9.   Create a reconstructed bag $\tilde{B} = \{\mathbf{x}_{i,j}, \mathcal{X}\}$
10.   Classify $\tilde{B}$ using Φ
11.   **if** $\tilde{B}$ is negative
12. $y_{i,j} = -1$
13.   **else**
14. $y_{i,j} = 1$

---

negative instance in the negative training bags is unique. Thirdly, the reconstruction of all instances to be tested uses the same $\mathbf{x}^*$, which is not dependent on the testing data or the number of iterations to be performed. Therefore, this approach is very efficient. A summary of this instance classification method is given in Algorithm 1.

When the data is generated from a mixture of Gaussian models or from an arbitrary distribution, it may be necessary to select multiple instances for bag reconstruction. Therefore, $\mathbf{x}^*$ is expanded to a set of instances $\mathcal{X} = \{\mathbf{x}_1^*, \ldots, \mathbf{x}_k^*\}$. This leads to a larger reconstructed bag. A simple method of generating such an $\mathcal{X}$ is to iteratively search for $\mathbf{x}^*$ from the remaining negative instances in $\mathcal{B}^-$ without replacement. This guarantees the retrieval of the most negative instances based on kernel density estimation. However, there is a high possibility that several selected negative instances are very close to each other. Then the contributions of these instances to the instance embedding step are similar. This means that the $\mathcal{X}$ may contain redundant information. On the other hand, some important negative instances may be missed.

This problem can be illustrated by an example shown in Fig. 1. In this example, the data is generated from the sum the six Gaussian distributions with means -2.1, -1.3, -0.4, 1.9, 5.1, and 6.2, respectively. The standard deviation of the Gaussian distribution is set to 1. It can be seen that there are two peaks in the curve. When $\mathcal{X}$ contains only a single element, $\mathbf{x}_1^*$ will be selected due to the highest probability density at the location. If more than one elements in $\mathcal{X}$

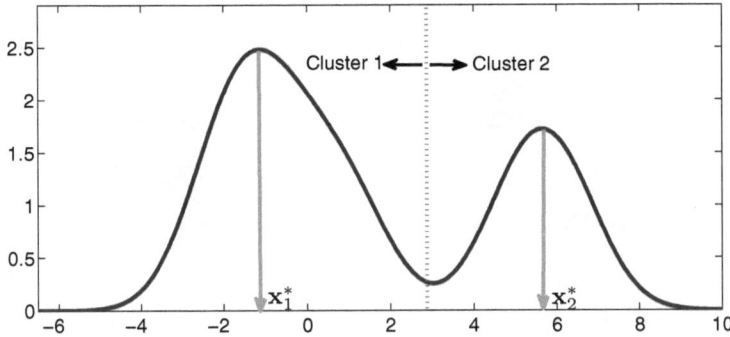

Fig. 1. Kernel density estimation for instance selection in bag reconstruction

are needed, it is most likely that points surrounding $\mathbf{x}_1^*$, will be selected, while $\mathbf{x}_2^*$ is missed.

To solve this problem, we introduce the second solution, which is based on dividing the feature space of negative instances into subspaces, and then applying kernel density estimation on each subspace. The subspace division can be performed by k-means clustering, which partitions the $\mathcal{B}^-$ into k sets $\mathcal{B}^- = \{\mathcal{B}_1^-, \ldots, \mathcal{B}_k^-\}$. For each set $\mathcal{B}_i^-$, we run kernel density estimation on all the negative instances in it. Therefore, Equation 1 is modified as

$$p(\mathbf{x}|\mathcal{B}_i^-) = \frac{1}{Z} \sum_{j=1}^{t} \exp\left(-\beta \|\mathbf{x} - \mathbf{x}_j^-\|\right), \tag{5}$$

Here, t is the total number of negative instances in $\mathcal{B}_i^-$. The negative instance selection rule in Equation 4 is updated correspondingly as

$$\mathbf{x}_i^* = \arg \max_{\mathbf{x} \in \mathcal{B}_i^-} p(\mathbf{x}|\mathcal{B}_i^-) \tag{6}$$

This allows both $\mathbf{x}_1^*$ and $\mathbf{x}_2^*$ in the above example be selected, which are the most representative instances. Note that when $k = 1$, Equations 5 and 6 reduce to the single element case in Equations 1 and 4.

4   Experiments

We have performed experiments on both synthetic data and real-world historical census datasets to demonstrate the utility of the proposed method. The implementation of MILIS method follows [14], but with the iterative model fine tune step omitted. LIBLINEAR [10] is used to train the bag classifier with all parameters set to the default values. We set $\tau = 1$ in Equation 1 and $\eta = 1$ in Equation 2. The selection of value k for the size of $\mathcal{X}$ will be discussed later.

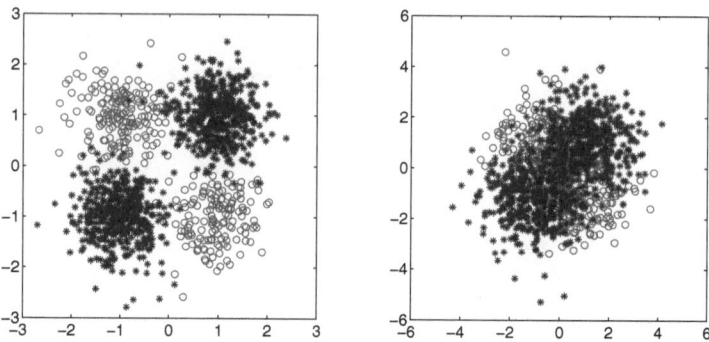

Fig. 2. Examples of synthetic data sets. Each panel contains four Gaussian distributions, whose means are $[1, 1]$, $[-1, 1]$, $[1, -1]$, and $[-1, -1]$, respectively. The standard deviation of the Gaussian distribution in the left panel is set to 0.5. The standard deviation of the Gaussian distribution in the right panel is set to 1.

4.1 Results on Synthetic Data

Two example of the synthetic datasets are shown in Fig. 2. Both sets contain 1,000 positive instances (red circles) and 5,000 negative instances (blue asteroids). These instances were randomly generated from four Gaussian distributions, two for positive instances and two for negative instances. The means of the Gaussian distributions used to generate these two datasets are identical, but their standard deviations are different. With larger standard deviation, the positive and negative instances are more overlapping with each other, and thus are more difficult to be classified. We constructed positive bags by randomly sampling from both positive and negative instances. Negative bags were constructed in a similar manner, but only from negative instances. Each bag contains a random number of instances ranging from 1 to 10. In this way, we have generated 1,000 positive bags and 1,000 negative bags. In the experiments, the instance labels are only used for evaluation purpose, without being accessed in the training stage.

Two important baseline methods to be compared are the random and greedy bag reconstruction methods proposed in [13], which are marked as "random" and "greedy", respectively. Furthermore, we have included the strategy that uses clustering centers of the negative instance prototypes for bag reconstruction. This is marked as "k-means". The proposed method based on kernel density estimation is marked as "KDE" for the option of retrieving the most negative instances in the negative bags, and "k-means+KDE" for the option of selecting multiple instances using kernel density estimation on clustered negative instances, respectively.

In the first experiment, we compare the robustness of the proposed method and the baseline methods when the difficulty level of data varies. We have randomly divided the synthetic data into a training set and a testing set with equal number of positive and negative bags. The bag level classifier was learned

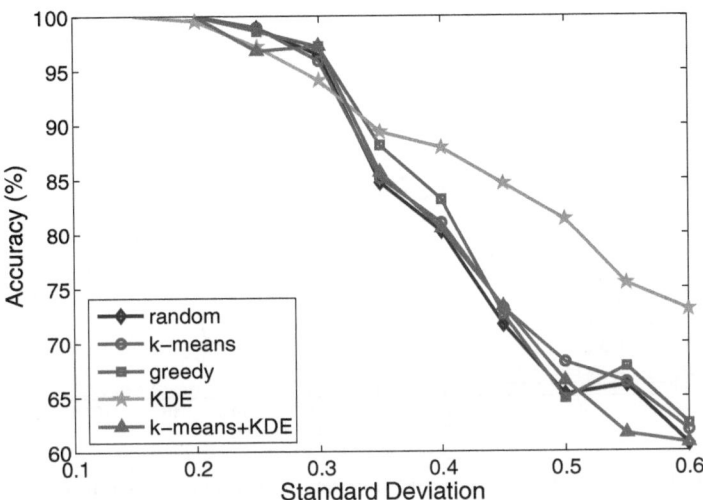

Fig. 3. Comparison of classification accuracies with data in different difficulty levels

from the training set using the MILIS algorithm [14]. The testing was only performed on the positive testing bags, which contains both positive and negative instances. This is because we are more interested in the performance of the bag reconstruction methods for instance classification. In this experiment, the number of instances selected for reconstruction is set to 5, which is the average number of instances in synthetic bags. The standard deviations of the Gaussian distributions are set from 0.1 to 0.6 with 0.1 in interval. The experiments are run for 10 times, with randomly split training and testing sets. Fig. 3 displays the mean accuracy of each method.

The results show that when the difficulty of the data is low, all methods perform similarly well. However, after the standard deviation of Gaussian distribution is set to a value larger than 0.5, the proposed method achieves much higher accuracy than the alternatives. This implies that using the most negative instances in the negative training bags for bag reconstruction is the most robust approach among all methods being compared. On the other hand, the alternative methods do not show much differences in their performance.

In the second experiment, we analyze the influence of number of negative instances in $\mathcal{X}$, i.e, the size of the reconstructed bags. Here, the standard deviation of Gaussian distribution for data generation is set to 0.5 for moderate level of data difficulty. As shown in Fig. 4, the "KDE" method performs the best among all instance classification methods in overall performance with a very flat accuracy curve. The highest accuracy is achieved at 17 negative instances, which is only slightly higher than the accuracy with 1 negative instance. This is reasonable because the selected most negative instances may be close to each other as described in Section 3. Therefore, the contribution of these instances,

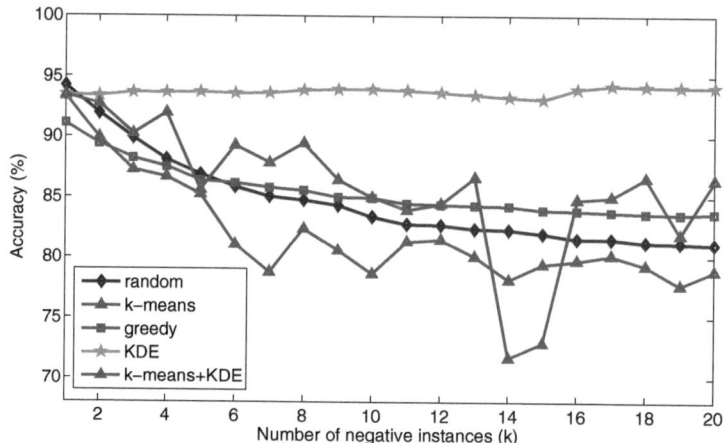

Fig. 4. Influence of reconstructed bag size to the classification accuracy

no matter how many they are, are similar in the feature embedding step of the reconstructed bags. This implies that the reconstruction with the most negative instance in the training set is sufficient to achieve good performance.

This observation can greatly simplify the $\mathcal{X}$ generation because now we only need to find the most negative instance in the negative training bags, which is already available from the MILIS IP generation step. On the other hand, the option of instance selection from clustered negative instances is not very stable due to the randomness of the initialization of k-means clustering method. The accuracy of other methods under comparison is greatly affected by the number of instances used for bag reconstruction. All of these methods achieve the highest accuracy with 1 negative instance. When the bag size increases, their accuracies drop significantly.

4.2   Historical Census Data

In this experiment, we used two historical census datasets collected in 1871 and 1881 from the United Kingdom. The 1871 dataset contains 5,575 households with 26,229 records. The 1881 dataset contains 6,025 households and 29,051 records. Each record contains the personal details of one individual, and there are no duplicates in each dataset. Each record has twelve attributes, including the address, first and family name, age, gender, relationship to head, occupation, and place of birth of each individual[1]. The goal of the historical census data linkage is to link both households and individuals in these two datasets.

In order to generate the ground truth data, we have manually labeled 2,400 household links, including 1,200 matched households and 1,200 non-matched households. The individual links in these households are also labeled. In the

---

[1] www.uk1851census.com

Table 1. Classification accuracy % on the historical census datasets

	random	greedy	k-means	KDE
Means	97.97	96.51	98.18	98.63
Stand Deviation	0.44	0.69	0.11	0.42

MIL setting, we consider household links as bags. Each bag contains several instances corresponding to individual record links. The task in this experiment is to classify unknown record links using a trained household link classifier.

Before classification, the data was cleaned and standardized using the *Febrl* data were cleaning and record linkage tool [6,11]. A variety of approximate string comparison functions were used to calculate the similarity between individual record pairs [12]. The bag level classifier was learned and used for the instance classification step following the proposed method. The experiments were ran ten times with a random split of the labeled data into equal size of training and testing sets. In the instance classification step, based on the results from the second experiment on the synthetic data, only one negative instance was used for bag reconstruction for each method.

We compare the classification accuracies of the proposed method and the baseline methods. Here, we have removed the "k-means+KDE" method because it is equivalent to "KDE" method when there is only one negative instance for bag reconstruction. The experimental results are summarized in Table 1. It can be observed that the proposed method (KDE) has generated the highest accuracy. At the same time, all alternative methods under comparison have achieved very high classification performance. This suggests that the difficulty of correctly classifying these historical census data is not high.

5   Conclusion

In this paper, we have presented a novel method of instance classification for multiple instance learning and group record linkage. This method models the distribution of the negative training bags, and groups the most representative negative instances with the target instance to be classified in order to covert instance classification to bag classification. Experimental results show that this method is very effective, and has outperformed several baseline methods. Analysis on the results also suggests that very few instances are required for the bag reconstruction purpose, which allows fast and convenient bag reconstruction. Future work will focus on the integration of the instance selection step with the multiple instance learning process, and the extension of this strategy to other multiple instance learning methods. We will also apply the proposed method to other datasets that require both group and instance level classification.

References

1. Bhattacharya, I., Getoor, L.: Iterative record linkage for cleaning and integration. In: ACM SIGMOD Workshop on Research Issues in Data Mining and Knowledge Discovery, pp. 11–18 (2004)
2. Bilenko, M., Mooney, R.J.: Adaptive duplicate detection using learnable string similarity measures. In: Proceedings of the ACM SIGKDD Conference on Knowledge Discovery and Data Mining, Washington, DC, pp. 39–48 (2003)
3. Chartrand, G.: Introductory Graph Theory. Dover Publications (1985)
4. Chen, Y., Bi, J., Wang, J.: MILES: Multiple-instance learning via embedded instance selection. IEEE Transactions on Pattern Analysis and Machine Intelligence 28(12), 1931–1947 (2006)
5. Christen, P.: Automatic record linkage using seeded nearest neighbour and support vector machine classification. In: Proceeding of the 14th ACM SIGKDD International Conference on Knowledge Discovery and Data Mining, pp. 151–159. ACM (2008)
6. Christen, P.: Development and user experiences of an open source data cleaning, deduplication and record linkage system. ACM SIGKDD Explorations 11(1), 39–48 (2009)
7. Christen, P.: Data Matching - Concepts and Techniques for Record Linkage, Entity Resolution, and Duplicate Detection. Springer (2012)
8. Dunn, H.L.: Record linkage. American Journal of Public Health 36(12), 1412–1416 (1946)
9. Elfeky, M., Verykios, V., Elmagarmid, A.: Tailor: A record linkage toolbox. In: Proceedings of the 18th International Conference on Data Engineering, pp. 17–28 (2002)
10. Fan, R.E., Chang, K.W., Hsieh, C.J., Wang, X.R., Lin, C.J.: LIBLINEAR: A library for large linear classification. Journal of Machine Learning Research 9, 1871–1874 (2008)
11. Fu, Z., Christen, P., Boot, M.: Automatic cleaning and linking of historical census data using household information. In: Proceedings of the 15th International Workshop on Domain Driven Data Mining, Vancouver, Canada, pp. 413–420 (2011)
12. Fu, Z., Christen, P., Boot, M.: A supervised learning and group linking method for historical census household linkage. In: Proceedings of the 19th Ninth Australasian Data Mining Conference, Ballarat, Australia (2011)
13. Fu, Z., Zhou, J., Christen, P., Boot, M.: Multiple Instance Learning for Group Record Linkage. In: Tan, P.-N., Chawla, S., Ho, C.K., Bailey, J. (eds.) PAKDD 2012, Part I. LNCS, vol. 7301, pp. 171–182. Springer, Heidelberg (2012)
14. Fu, Z., Robles-Kelly, A., Zhou, J.: MILIS: Multiple instance learning with instance selection. IEEE Transactions on Pattern Analysis and Machine Intelligence 33(5), 958–977 (2011)
15. Herschel, M., Naumann, F.: Scaling up duplicate detection in graph data. In: Proceedings of the ACM International Conference on Information and Knowledge Management, Napa Valley, California, pp. 1325–1326 (2008)
16. Herzog, T.N., Scheuren, F., Winkler, W.E.: Data quality and record linkage techniques. Springer ((2007)
17. Kalashnikov, D.V., Mehrotra, S.: Domain-independent data cleaning via analysis of entity-relationship graph. ACM Transactions on Database Systems 31(2), 716–767 (2006)

18. Namata, G.M., Kok, S., Getoor, L.: Collective graph identification. In: Proceedings of the 17th ACM SIGKDD International Conference on Knowledge Discovery and Data Mining, pp. 87–95 (2011)
19. Naumann, F., Herschel, M.: An introduction to duplicate detection. Synthesis Lectures on Data Management 2(1), 1–87 (2010)
20. On, B.W., Koudas, N., Lee, D., Srivastava, D.: Group linkage. In: Proceeding of the IEEE International Conference on Data Engineering, Istanbul, Turkey, pp. 496–505 (2007)
21. Rossi, R.A., KcDowell, L.K., Aha, D.W., Neville, J.: Transforming graph representations for statistical relational learning. Journal of Artificial Intelligence Research (2012)
22. Vapnik, V.: The Nature of Statistical Learning Theory. Springer (1995)
23. Winkler, W.E.: Methods for record linkage and bayesian networks. Technical report, US Bureau of the Census (2001)
24. Xiang, R., Neville, J., Rogati, M.: Modeling relationship strength in online social networks. In: Proceedings of the 19th International World Wide Web Conference, pp. 981–990 (2010)

Semi-naive Bayesian Classification by Weighted Kernel Density Estimation

Lifei Chen[1,2] and Shengrui Wang[1]

[1] Department of Computer Science, University of Sherbrooke, Quebec, J1K 2R1, Canada
{lifei.chen,shengrui.wang}@usherbrooka.ca
[2] School of Mathematics and Computer Science, Fujian Normal University, Fujian, 350108, China

Abstract. Naive Bayes is one of the popular methods for supervised classification. The attribute conditional independence assumption makes Naive Bayes efficient but adversely affects the quality of classification results in many real-world applications. In this paper, a new feature-selection based method is proposed for semi-naive Bayesian classification in order to relax the assumption. A weighted kernel density model is first proposed for Bayesian modeling, which implements a soft feature selection scheme. Then, we propose an efficient algorithm to learn an optimized set of weights for the features, by using the least squares cross-validation method for optimal bandwidth selection. Experimental studies on six real-world datasets show the effectiveness and suitability of the proposed method for efficient Bayesian classification.

Keywords: Bayesian classification, feature weighting, kernel density estimation, least squares cross-validation.

1 Introduction

Classification is a supervised learning technique aimed at assigning unlabeled samples to known classes, based on knowledge of labeled samples. The technique has been studied extensively, and a number of classification methods have been proposed in the literature [1]. Among the various existing methods, Bayesian classification, which predicts the class label using the posterior probability according to Bayes' rule, has sparked wide interest in the data mining community, due to its clear probabilistic semantics and its effectiveness [2, 3].

The existing Bayesian methods differ from each other in their ways of modeling and subsequently learning the interdependencies between attributes. In Bayesian Networks (BN for short) [2], the interdependencies are modeled as a graph in which each node corresponds to an attribute and each edge to the dependency between two attributes, represented by the conditional probability. Because learning an optimal BN is NP-hard, one has to resort to approximate algorithms such as loopy belief propagation in practice [2]. Currently, the most popular Bayesian method is Naive Bayes (NB for short), which is based on the assumption that the predictive attributes are conditionally independent given the

S. Zhou, S. Zhang, and G. Karypis (Eds.): ADMA 2012, LNAI 7713, pp. 260–270, 2012.

class attribute. Thanks to the independence assumption, NB yields incredible savings in number of model parameters and has been widely used in real-world applications [4].

Since the independence assumption adversely affects the quality of classification results when it is violated, a number of NB variants have been proposed to relax the assumption: examples include [5–7]. These variants are alternatively known as the semi-naive Bayesian methods [3, 8], among which feature-selection based methods are specially stressed because of their intuitive but effective ideas: improving NB by eliminating those noisy features that do not contribute to class prediction[9], and the redundant features that are not conditionally independent given the class [10]. Other examples are reported in [3, 11–13]. However, these latter methods usually suffer from high computational complexity because it is in general not feasible to perform an exhaustive search to find the optimal feature subsets due to the huge number of admissible subsets which is exponential in the data dimensionality.

From the perspective of dimension reduction, *feature weighting*, which closely relates to feature selection but assigning a continuous weight to each feature, can be used to address these problems. However, few attempts have been made to combine feature weighting in NB classification due to the difficulties in estimating the feature weights. Actually, in the existing feature-weighting-based methods, such as the SVM-based weighting NB [10] and the recently published FWNB [14], the weights are assigned to the features in a separated process. For these methods, feature weighting can only be regarded as a pre-processing step before NB classification, which is performed independently from the classification process.

In this paper, a new semi-naive method is proposed for Bayesian classification using an embedded feature-weighting technique, which, in effect, equates to performing a dynamic feature selection for the classes. We formulate the feature selection an integral part of the classification model, by weighted kernel density estimation such that the feature weights can be learned using an efficient optimization algorithm. The performance of the new method is evaluated on six real-world datasets, and the experimental results show its effectiveness.

The remainder of this paper is organized as follows. Section 2 presents some preliminaries and related work. Section 3 describes our semi-naive Bayesian model and the training algorithm. Experimental results are presented in Section 4. Finally, Section 5 gives our conclusion and discusses directions for future work.

2   Preliminaries and Related Work

In what follows, the training dataset is denoted by $DB=\{z_1, z_2, \ldots, z_N\}$, with $z_i=(\mathbf{x}_i, y_i)$, $i = 1, 2, \ldots, N$. Here, $\mathbf{x}_i=<x_{i1}, x_{i2}, \ldots, x_{iD}>$ is a D-dimensional input. In this paper, we will focus on categorical data. For the jth categorical attribute, where $j = 1, 2, \ldots, D$, denote the set of categories by c_j, i.e., the jth attribute takes $|c_j|(> 1)$ different values. We use y_i to denote the pre-defined class of $\mathbf{x}_i$, $y_i \in \{1,2,\ldots,K\}$, where K stands for the number of classes contained in the training dataset. The number of samples in class k is denoted by n_k.

Let $\mathbf{x}_t$ be a test sample. $p(k|\mathbf{x}_t)$ stands for the probability that $\mathbf{x}_t$ be assigned to class k. In an NB model, the probability is computed by

$$p(k|\mathbf{x}_t) = \frac{p(k) \prod_{j=1}^{D} p(x_{tj}|k)}{p(\mathbf{x}_t)} \tag{1}$$

based on Bayes' rule and the independence assumption. Subsequently, the class of $\mathbf{x}_t$, say y_t, is predicted as

$$y_t = \arg\max_k p(k) \prod_{j=1}^{D} p(x_{tj}|k) \tag{2}$$

because the denominator of Eq.(1) dose not depend on k. Since x_{tj} only takes discrete values, it is common to estimate $p(x_{tj}|k)$ by the frequency estimator

$$\bar{f}_k(x_{tj}) = \frac{f_k(x_{tj})}{n_k}$$

or, by the Laplace's law of succession [1]. Here, $f_k(x_{tj})$ denotes the number of category $x_{tj} \in c_j$ appearing in the kth training class. Moreover, the probability $p(k)$ can be estimated by $p(k) = \frac{|n_k|}{N}$, if $n_k > 0$ for $k = 1, 2, \ldots, K$, which is the case considered in this paper.

A number of semi-naive Bayes methods [3] have been proposed in recent years. They fall into two categories: structure-based and data-based methods. Those in the first category improve NB via structure extension. For example, in Tree Augmented Naive Bayes (TAN) [6] and Averaged One-Dependence Estimators (AODE) [5], the attribute dependencies are explicitly represented based on the assumption that each attribute depends upon the class and at most one other attribute. The methods in the second category aim at choosing a reduced subset of training data such that the dependencies within the subset are weaker than those in the whole dataset. Examples include the Locally Weighted Naive Bayes (LWNB) [7], which accommodates violations of the independence assumption by choosing a desired set of the training samples on which NB is applied. This is built on the observation that the independence assumption may hold or approximately hold in a subset of the training set although violated in the whole set [8].

It can be seen from Eq. (2) that the conventional NB essentially assumes that all the features are equally important for classification, which hardly holds in real-world applications. This motivates the development of another group of methods in the data-based category, called feature-selection based methods, where a feature selection algorithm is applied to remove irrelevant and redundant attributes from the data. For example, the algorithm proposed by [11] identifies the attribute whose elimination best reduces the training error in a pre-processing step. Here, a wrapper method [13] can be used to determine the attributes subset, combining with the classification algorithm. However, in practice, the number of admissible subspaces would be very large; thus such a method is time-consuming and becomes intractable, especially for high-dimensional data.

Recently, some attempts have been made to combine feature weighting for Bayesian classification. In the methods proposed by [10, 14], the posteriori probability is estimated by

$$p(\mathbf{x}_t|k) = \prod_{j=1}^{D} p(x_{kj}|k)^{w(j)}$$

where $w(j)$ denotes the weight assigned to the jth attribute. The weights are typically assigned based on some heuristic measures developed for supervised dimensionality reduction, such as feature dependency, information gain and gain ratio used for constructing a decision tree [12]. However, using these methods, the features are weighted in the pre-processing step of a classification task, where the weights are assigned independently from the classification model. In this paper, we propose a new feature-selection based Bayesian classifier, which is built on soft feature weighting techniques such that the model itself contains parameters that make variable selection as an integral part of it. Moreover, the feature weights can be optimized in linear time complexity.

3 The Proposed Method

3.1 Weighted Kernel Density Estimation

As discussed previously, $p(x_{tj}|k)$ can be conventionally estimated using the frequency estimator $\bar{f}_k(x_{tj})$. From a statistical perspective, such a non-smooth estimator may have the least sample bias; however, it may also have a large estimation variance (the finite-sample mean squared error [15]) at the same time. For the work described here, we shall employ the kernel smoothing method [15, 16] for the probability estimation, to make a good trade-off between the two conflicting factors.

Let X_{kj} be a random variable associated with the observations x_{ij} for $i = 1, 2, \ldots, n_k$ of class k, and $\kappa(X_{kj}, x_{lj}, \lambda_{kj})$ a kernel function defined for the jth attribute. Here, $x_{lj} \in c_j$ for $l = 1, 2, \ldots, |c_j|$ and λ_{kj} is the smoothing parameter called bandwidth. Aitchison and Aitken [17] proposed a widely used kernel function, defined as $\kappa_{AA}(X_{kj}, x_{lj}, \lambda_{kj}) = 1 - \lambda_{kj}$ if $X_{kj} = x_{lj}$, and $\kappa_{AA}(X_{kj}, x_{lj}, \lambda_{kj}) = \frac{1}{|c_j|-1}\lambda_{kj}$ if $X_{kj} \neq x_{lj}$. By definition, λ_{kj} of κ_{AA} is confined in interval $[0, 1 - \frac{1}{|c_j|}]$. In fact the two extreme points of the interval correspond to the two extreme cases of the smoothing estimation: In the case of $\lambda_{kj} = 0$, κ_{AA} is exactly an indicator function; and when λ_{kj} grows to $1 - \frac{1}{|c_j|}$, κ_{AA} becomes a constant $\frac{1}{|c_j|}$ for all values of X_{kj} and x_{lj}. Since the interval $[0, 1 - \frac{1}{|c_j|}]$ is related to the number of categories (say, $|c_j|$) on the individual attributes for the same class k, we propose a variation on the kernel function defined by

$$\kappa(X_{kj}, x_{lj}, \lambda_{kj}) = \begin{cases} 1 - \frac{|c_j|-1}{|c_j|}\lambda_{kj} & X_{kj} = x_{lj} \\ \frac{1}{|c_j|}\lambda_{kj} & X_{kj} \neq x_{lj} \end{cases} \tag{3}$$

with $\lambda_{kj} \in [0, 1]$ being the bandwidth.

On the other hand, for many types of real-world data, the attributes contribute unequally to classification [18, 19]. That is, the classes are only correlated with a set of relevant attributes, while the attributes relevant to classes in the same dataset are generally different from each other. In order to formalize these characteristics, we employ a locally feature-weighting scheme, where a weighting value w_{kj} is assigned to the jth attribute of class k, satisfying

$$\begin{cases} \sum_{j=1}^{D} w_{kj} = 1, & k = 1, 2, ..., K \\ 0 < w_{kj} < 1, & k = 1, 2, \ldots, K; j = 1, 2, ..., D \end{cases} \tag{4}$$

Here, the weight w_{kj} is defined to measure the contribution of the jth attribute to prediction of class k. This is inspired by the locally weighting scheme developed in projective clustering domain [19]. For a given c_k, the assignment of $w_{k1}, w_{k2}, \ldots, w_{kD}$ can be regarded as a soft feature selection procedure for the space in which c_k exists [18]. In NB, all the attributes have the same importance on classification, corresponding to the special case of $w_{k1} = w_{k2} = \ldots = w_{kD} = \frac{1}{D}$. To achieve a weighted density estimation, we then introduce a weighting coefficient in the definition of kernel bandwidth for the individual attribute, i.e.,

$$\lambda_{kj} = w_{kj}^{\beta} \lambda_k \tag{5}$$

where λ_k is a class-dependent width for each class k, and $\beta(> 1)$ a weighting exponent controlling the weights distribution. Although there might exist various relationships between λ_{kj} and λ_k, a linear projection approach as shown in Eq. (5) has been popularly adopted for feature-weighting based probability estimation [18, 19].

Based on the above definitions, we estimate $p(x_{tj}|k)$ using the weighted kernel density $\hat{p}(x_{tj}|w_{kj}, \lambda_k)$, given by

$$\begin{aligned} \hat{p}(x_{tj}|w_{kj}, \lambda_k) &= \frac{1}{n_k} \sum_{i=1}^{n_k} \kappa(x_{tj}, x_{ij}, \lambda_{kj}) \\ &= (1 - \frac{|c_j|-1}{|c_j|} w_{kj}^{\beta} \lambda_k) \bar{f}_k(x_{tj}) + \frac{1}{|c_j|} w_{kj}^{\beta} \lambda_k (1 - \bar{f}_k(x_{tj})) \\ &= \bar{f}_k(x_{tj}) + \left(\frac{1}{|c_j|} - \bar{f}_k(x_{tj}) \right) w_{kj}^{\beta} \lambda_k \end{aligned} \tag{6}$$

subject to Eq. (4). The new Bayesian classification model then can be derived. In this weighted-kernel based Bayesian classifier, the class of testing sample $\mathbf{x}_t$ is determined according to the rule of

$$y_t = \arg\max_k p(k) \prod_{j=1}^{D} \hat{p}(x_{tj}|w_{kj}, \lambda_k),$$

by replacing the probability function of Eq. (2) with the weighted kernel density function defined in Eq. (6).

3.2   Model Learning Method

Given the training dataset consisting of K classes, the goal of this subsection is to learn the set of parameters $\{\omega_k\}_{k=1}^{K}$ with $\omega_k = \{w_{kj}|1 \leq j \leq D\}$. Here, the class-dependent width λ_k is considered as a constant that depends on the training class k. Following [20], we set $\lambda_k = \frac{1}{\sqrt{n_k}}$ in this paper. Note that if $n_k \to +\infty$ then $\lambda_k \to 0$. This satisfies the general property of a kernel width, that is, the width should shrink to zero when the number of objects goes to infinity [15, 20]. Based on these, the problem of learning ω_k equates to optimization of the individual bandwidths λ_{kj} for $j = 1, 2, \ldots, D$. Then, the model parameters can be learned using a data-driven method for bandwidth selection, such as the *least squares cross-validation* (LSCV) [15, 16], from the training dataset.

The LSCV method is based on the principle of selecting a model that minimizes the total error of the resulting estimation, i.e., $\sum_{j=1}^{D} \sum_{s \in c_j} (\hat{p}(s|w_{kj}, \lambda_k) - p(s|k))^2$. Therefore, the optimized parameters should minimize the following objective function:

$$J_0(\omega_k) = \sum_{j=1}^{D} \sum_{s \in c_j} [\hat{p}(s|w_{kj}, \lambda_k) - p(s|k)]^2$$
$$\sim \sum_{j=1}^{D} \left(\sum_{s \in c_j} [\hat{p}(s|w_{kj}, \lambda_k)]^2 - 2 \sum_{s \in c_j} \hat{p}(s|w_{kj}, \lambda_k) p(s|k) \right)$$

where the term $\sum_{j=1}^{D} \sum_{s \in c_j} [p(s|k)]^2$ is omitted because it remains as a constant that is irrelevant to ω_k. Since the term $\sum_{s \in c_j} \hat{p}(s|w_{kj}, \lambda_k) p(s|k)$ in the previous equation is the expectation of $s \in c_j$, we can estimate the term by the sample mean [15]. Moreover, following [16], the leave-one-out validation method is used; then, the resulting objective function is obtained, as follows:

$$J(\omega_k) = \sum_{j=1}^{D} \sum_{s \in c_j} [\hat{p}(s|w_{kj}, \lambda_k)]^2$$
$$- \frac{2}{n_k - 1} \sum_{j=1}^{D} \left(n_k \sum_{s \in c_j} \bar{f}_k(s) \hat{p}(s|w_{kj}, \lambda_k) + \frac{|c_j| - 1}{|c_j|} w_{kj}^{\beta} \lambda_k - 1 \right) \quad (7)$$

subject to the constraints of Eq. (4).

For the optimization problem of Eq. (7), the optimal parameters can be learned by taking derivatives to the objective function. In detail, by setting the gradients of $J(\omega_k)$ with respect to w_{kj} to zero for all j, we have

$$\frac{\partial J}{\partial w_{kj}} = 2\beta w_{kj}^{\beta-1} \sum_{s \in c_j} \left(\hat{p}_{cat}(s|w_{kj}, \lambda_k) - \frac{n_k}{n_k - 1} \bar{f}_k(s) \right) \left(\frac{1}{|c_j|} - \bar{f}_k(s) \right) \lambda_k$$
$$- \frac{2}{n_k - 1} \frac{|c_j| - 1}{|c_j|} \beta w_{kj}^{\beta-1} \lambda_k = 0$$

which follows

$$w_{kj}^{\beta} = \frac{1}{\lambda_k (n_k - 1)} \frac{1 - \sum_{s \in c_j} [\bar{f}_k(s)]^2}{\sum_{s \in c_j} \left[\frac{1}{|c_j|} - \bar{f}_k(s) \right]^2}. \quad (8)$$

3.3   The Training Algorithm

The *KWNB* (denotes Kernel-Weighting for Naive Bayes) algorithm, as outlined by Algorithm 1, trains the classification model by minimizing the objective function of Eq. (7), using the methods presented in the previous subsection. Note that we have a closed-form solution to w_{kj}s, because the term $\frac{1}{\lambda_k (n_k - 1)}$ of Eq. (8)

remains the same for different attributes; thus it can be ignored when the constraints defined in Eq. (4) are applied to normalize the weights. In the algorithm, we first compute

$$\tilde{w}_{kj} = \left[\frac{1 - \sum_{s \in c_j} \left[\bar{f}_k(s) \right]^2}{\sum_{s \in c_j} \left[\frac{1}{|c_j|} - \bar{f}_k(s) \right]^2} \right]^{\frac{1}{\beta}}. \tag{9}$$

based on Eq. (8) for $j = 1, 2, \ldots, D$. In the case when the numerator or the denominator of Eq. (9) happens to be zero, two small constants ($\frac{2(n_k - 1)}{n_k^2}$ for the numerator, and $\frac{2}{n_k^2}$ for the denominator) are used for the estimation to make the weights computable in practice. Then, the weight w_{kj} is finally obtained by normalizing $\tilde{w}_{kj}$, as follows:

$$w_{kj} = \frac{\tilde{w}_{kj}}{\sum_{l=1}^{D} \tilde{w}_{kl}}. \tag{10}$$

For a training dataset consisting of K classes, $KWNB$ is called for K times, each for the kth class($k=1,2,\ldots,K$). On the whole dataset, the time complexity is $O(KND)$ while the space complexity is $O(KD)$.

Input: the training samples $\mathbf{x}_i (i = 1, 2, \ldots, n_k)$ in class k;
Output: a set of feature weights ω_k;
begin
  Compute $\tilde{w}_{kj}$ using Eq. (9) for $j = 1, 2, \ldots, D$;
  Compute w_{kj} for $j = 1, 2, \ldots, D$ by normalization using Eq. (10);
end

Algorithm 1. Outline of the $KWNB$ algorithm

4   Experimental Evaluation

In this section, we evaluate the performances of $KWNB$ on real-world categorical data. We also experimentally compare $KWNB$ with mainstream classification methods.

4.1   Datasets and Experimental Setup

The experiments were conducted on six widely used real-world datasets, all of which were obtained from the UCI Machine Learning Repository (ftp.ics.uci.edu: pub/machine-learning-databases). Table 1 lists the details of the datasets. Note that both the Mushroom and Vote datasets contain missing attribute values. The missing value in each attribute was considered as a special category in our experiments. For example, in the Mushroom dataset, the 11st attribute (named stalk-root) takes its value from $\{b, c, u, e, z, r\}$; however, there are 2480 samples miss values in this attribute. For this case, an additional category denoted ? was inserted into the original categories set; then the resulting set becomes $\{b, c, u, e, z, r, ?\}$.

Table 1. Details of the real-world datasets

Dataset	Dimension(D)	Classes(K)	Data size(N)
Balance	4	3	625
Car	6	4	1728
Vote	16	2	435
Mushroom	21	2	8124
Soybean	35	19	683
Splice	60	3	3190

Four competing classifiers: the conventional NB, Feature Weighted NB (FWNB in short)[14], Attribute Selective Bayesian Classifier (ASBC in short) [11], and Bayesian Networks (BN in short) have been chosen for comparison in our experiments. FWNB is a recently published method for semi-naive Bayesian classification, based on the locally weighted learning approach, which assigns the test sample a specified weight according to its neighborhood [14]. ASBC performs semi-naive Bayesian classification by wrapper-based feature selection [11]. We adopted the implementations of ASBC (AttributeSelectedClassifier with NaiveBayes as the classifier) and BN (Bayesian Networks) from the WEKA system [21]. For NB, the posterior probability of $s \in c_j$ is estimated using the common M-estimation, i.e., $p(s|k) = \frac{f_k(s)+1}{n_k+K}$. We set $\beta = 2$ for $KWNB$ in the experiments.

We also used the start-of-art decision tree algorithm as the baseline classifier in the experiments. The J4.8 classifier implemented in WEKA system [21] was employed. All the parameters in J4.8 were left as default values.

4.2 Experimental Results

The classification performances of the different classifiers were measured using the Micro-F1 (F1 over classes and samples) and Macro-F1 (average of within-class F1 values). Each dataset was classified by each algorithm for 10 executions using ten-fold cross validation, and the average performances are reported in the format *average* $\pm$ 1 *standard deviation*, as shown in Table 2, where the two figures in each cell represent Micro-F1 and Macro-F1, respectively. In the tables, the *better* and *worse* results are marked by symbols $\star$ and $\circ$, respectively, for the algorithms on each dataset comparing with NB, using the paired t-test with significance level 0.05.

Table 2 shows that $KWNB$ is able to obtain high-quality overall results. By examining the results in more detail, we can see that $KWNB$ outperformed the conventional NB in 7 out of 12 comparisons, whereas FWNB and BN performed poorly. ASBC achieved high classification accuracy comparable to that of *KWNB*. Compared with the other two methods, the performances of FWNB and J4.8 are unstable across the datasets. For example, J4.8 outperformed the other methods on the Mushroom dataset; however, it also performed the worst on Balance and Splice.

Table 2. Comparison of classification accuracy

Dataset	NB	*KWNB*	FWNB	ASBC	BN	J4.8
Balance	0.915 ±0.012	0.915 ±0.011	0.905 ±0.021	0.914 ±0.013	0.914 ±0.013	0.641 ±0.042 ∘
	0.635 ±0.007	0.635 ±0.007	0.628 ±0.014 ∘	0.635 ±0.008	0.635 ±0.008	0.444 ±0.029 ∘
Car	0.856 ±0.021	0.859 ±0.023	0.803 ±0.028	0.855 ±0.026	0.856 ±0.026	0.922 ±0.020 ⋆
	0.633 ±0.071	0.657 ±0.067 ⋆	0.608 ±0.060 ∘	0.632 ±0.085	0.641 ±0.086	0.811 ±0.069 ⋆
Vote	0.900 ±0.039	0.904 ±0.038 ⋆	0.929 ±0.038 ⋆	0.946 ±0.030 ⋆	0.902 ±0.039	0.966 ±0.026 ⋆
	0.896 ±0.040	0.900 ±0.040 ⋆	0.926 ±0.039 ⋆	0.943 ±0.032 ⋆	0.898 ±0.040	0.964 ±0.027 ⋆
Mushroom	0.955 ±0.007	0.991 ±0.003 ⋆	0.985 ±0.004 ⋆	0.985 ±0.005 ⋆	0.962 ±0.008 ⋆	1.000 ±0.000 ⋆
	0.954 ±.007	0.991 ±0.003 ⋆	0.985 ±0.004 ⋆	0.985 ±0.005 ⋆	0.962 ±0.008 ⋆	1.000 ±0.000 ⋆
Soybean	0.900 ±0.030	0.942 ±0.027 ⋆	0.897 ±0.030	0.922 ±0.030 ⋆	0.931 ±0.030 ⋆	0.918 ±0.032 ⋆
	0.928 ±0.028	0.964 ±0.021 ⋆	0.928 ±0.027	0.940 ±0.030 ⋆	0.955 ±0.023 ⋆	0.911 ±0.041 ∘
Splice	0.954 ±0.011	0.951 ±0.049	0.945 ±0.012	0.958 ±0.010 ⋆	0.955 ±0.011	0.941 ±0.013 ∘
	0.949 ±0.012	0.945 ±0.054	0.941 ±0.013 ∘	0.954 ±0.012 ⋆	0.950 ±0.013	0.935 ±0.014 ∘

⋆ better, and ∘ worse, comparing with NB.

KWNB owes its good performance to the embedded-in feature selection method. In effect, the soft feature weighting technique used in *KWNB* equates to performing a dynamic feature selection for the respective class during the training process. This is not the case for FWNB, which is also a feature-weighting based method. Actually, FWNB weights the features based on GainRatio of each attribute computed like in a decision tree algorithm [14]; and more importantly, here the weights are assigned in a separated pre-processing step. Such a method easily leads to the over-fitting problem especially on the datasets in relative high dimensionality, such as the Mushroom and Splice datasets. Note that the performance of J4.8, as a decision tree method, also is barely satisfactory on these two datasets. In principle, BN prevails against NB (as well as the other semi-naive Bayesian methods) due to its ability in detecting the arbitrary dependencies among attributes. Therefore, its improvement on the small datasets (such as Balance and Vote) and the datasets involving complex interdependencies like Mushroom in the experiments, may be defeated at a discount.

It is interesting to remark that both *KWNB* and ASBC, as two feature-selection based methods, yield high-quality results on most of the datasets compared with the others. This, on one hand, confirms that combination of feature selection in NB is a promising approach for semi-naive Bayesian classification. However, there is an essential difference between *KWNB* and ASBC. *KWNB* assigns a continuous value weight to each attribute; and thus is a more flexible method than hard feature selection (as used in ASBC) which can be regarded as a special case of feature weighting with the weight value being restricted to either 0 or 1. Moreover, because of its linear time complexity in training the classification model, *KWNB* is more efficient than ASBC. Due to the space limitations, evaluation on scalability of different methods is omitted.

5   Conclusion and Perspectives

In this paper, we proposed a feature-weighting based method for Bayesian classification, in order to alleviate the independence assumption made by the conventional Naive Bayes, and subsequently improve the classification performance. We proposed a weighted kernel density function to model the training data and to perform a soft feature selection scheme that adaptively identifies the different contributions of attributes to class prediction. We also proposed an efficient training algorithm, called *KWNB*, to learn an optimized set of feature weights in order to select the optimal bandwidth for kernel estimation. The experiments were conducted on UCI datasets that are widely used in real-world applications, and the results show the effectiveness of *KWNB*.

There are many directions that are clearly of interest for future exploration. One avenue of further study is to test *KWNB* on more extensive datasets, and to compare with other mainstream methods such as Minka's method [22]. Another efforts will be directed towards extending the weighted Bayesian model to mixed categorical and numeric data, incorporating with our previous work [18] aimed at numeric data classification.

Acknowledgments. The authors are grateful to the anonymous reviewers for their invaluable comments. This work was supported by the Natural Sciences and Engineering Research Council of Canada under Discovery Accelerator Supplements grant No. 396097-2010, and partially by the National Natural Science Foundation of China under Grant No. 61175123.

References

1. Hastie, T., Tibshirani, R., Friedman, J.: The elements of statistical learning: Data mining, inference, and prediction. Springer (2001)
2. Seeger, M.: Bayesian modeling in machine learning: A tutorial review. Tutorial, Saarland University (2006), http://lapmal.epfl.ch/papers/bayes-review.pdf
3. Zheng, F., Webb, G.: A comparative study of semi-naive bayes methods in classification learning. In: Proceedings of the Australalian Data Mining Workshop, pp. 141–156 (2005)
4. Wu, X., Kumar, V., et al.: Top 10 algorithms in data mining. Knowledge Information System 14, 1–37 (2008)
5. Wu, J., Cai, Z.: Learning averaged one-dependence estimators by attribute weighting. Journal of Information and Computational Science 8, 1063–1073 (2011)
6. Friedman, N., Geiger, D., Goldszmidt, M.: Not so naive bayes: Aggregating one-dependence estimators. Machine Learning 58, 5–24 (2005)
7. Frank, E., Hall, M., Pfahringer, B.: Locally weighted naive bayes. In: Proceedings of the Conference on Uncertainty in Artificial Intelligence, pp. 249–256 (2003)
8. Jiang, L., Wang, D., Cai, Z., Yan, X.: Survey of Improving Naive Bayes for Classification. In: Alhajj, R., Gao, H., Li, X., Li, J., Zaïane, O.R. (eds.) ADMA 2007. LNCS (LNAI), vol. 4632, pp. 134–145. Springer, Heidelberg (2007)
9. Fan, J., Fan, Y.: High-dimensional classification using features annealed independence rules. The Annals of Statistics 36, 2605–2637 (2008)
10. Gartner, T., Flach, P.: Wbcsvm: Weighted bayesian classification based on support vector machines. In: Proceedings of the ICML, pp. 154–161 (2001)
11. Langley, P., Sage, S.: Induction of selective bayesian classifiers. In: Proceedings of the Conference on Uncertainty in Artificial Intelligence, pp. 399–406 (1994)
12. Ratanamahatana, C., Gunopulos, D.: Feature selection for the naive bayesian classifier using decision trees. Applied Artificial Intellegence 17, 475–487 (2003)
13. Kohavi, R., John, G.: Wrappers for feature subset selection. Artificial Intelligence 97, 273–324 (1997)
14. Lee, C., Gutierrez, F., Dou, D.: Calculating feature weights in naive bayes with kullback-leibler measure. In: Proceedings of the IEEE ICDM, pp. 1146–1151 (2011)
15. Qi, L., Racine, J.: Nonparametric econometrics: Theory and practice. Princeton University Press (2007)
16. Ouyang, D., Li, Q., Racine, J.: Cross-validation and the estimation of probability distributions with categorical data. Nonparametric Statistics 18, 69–100 (2006)
17. Aitchison, J., Aitken, C.: Multivariate binary discrimination by the kernel method. Biometrika 63, 413–420 (1976)
18. Chen, L., Wang, S.: Automated feature weighting in naive bayes for high-dimensional data classification. In: Proceedings of the CIKM (2012)
19. Chen, L., Jiang, Q., Wang, S.: Model-based method for projective clustering. IEEE Transactions on Knowledge and Data Engineering 24, 1291–1305 (2012)
20. John, G., Langley, P.: Estimating continuous distributions in bayesian classifiers. In: Proceedings of the Conference on Uncertainty in Artificial Intelligence, pp. 338–345 (1995)
21. Hall, M., Frank, E., et al.: The weka data mining software: An update. SIGKDD Explorations 11 (2009)
22. Minka, T.: Estimating a Dirichlet distribution (2000), http://research.microsoft.com/en-us/um/people/minka/papers/dirichlet/minka-dirichlet.pdf

Spectral Clustering-Based Semi-supervised Sentiment Classification

Suke Li[1] and Jinmei Hao[2]

[1] School of Software and Microelectronics, Peking University, China
lisuke@ss.pku.edu.cn
[2] Beijing Union University

Abstract. This work proposes a semi-supervised sentiment classification method. Our method utilizes spectral clustering-based algorithm to improve the sentiment classification accuracy. We adopt a spectral clustering algorithm to map sentiment units in consumer reviews into new features which are extended into the original feature space. One sentiment classifier is built on the features in the original training space, and the original training features combined with the extended features are used to train the other sentiment classifier. The two basic sentiment classifiers together form the final sentiment classifier through selecting instances in the unlabeled data set into the training data set. Experimental results show that our proposed method has better performance than Self-learning SVM-based sentiment classification method.

1 Introduction

With the development of Web 2.0 technologies, many websites provide facilities to let Web users publish various consumer reviews. Lots of consumer reviews scatter in Blogs, Web forums, electronic commercial or opinion aggregation websites. Consumer reviews comprise various subjective texts that are used to express consumers' satisfaction or anger when they finish their purchasing activities for products or services. If a consumer review is not a fake review, then the sentiment polarity of the review can be very revealing the consumer's opinions.

Sentiment classification is a hot research topic in the field of opinion mining. Sentiment classification techniques are widely applied in business intelligence systems, recommendation systems, public opinion collecting and mining systems, etc. Therefore, sentiment detection and classification of consumer reviews is the main task of product opinion mining. Because there are a huge number of consumer reviews on the Web, how to automatically determine the sentiment polarity of consumer reviews has become a research problem. We can basically classify consumer reviews into three kinds of sentiment classes: positive, negative and neutral. Positive sentiment can be expressed with happy, supportive, approving language, and negative sentiment can reveal angry and depression.

The research of sentiment classification can be roughly divided into two categories. In the first category, researchers use lexical, syntactic, semantic analysis methods and nature language processing techniques, even combining with rules

S. Zhou, S. Zhang, and G. Karypis (Eds.): ADMA 2012, LNAI 7713, pp. 271–283, 2012.

or probability models to do sentiment classification. In the second category, machine learning techniques play a very important role in sentiment classification. From the perspective of machine learning, sentiment classification basically falls into three categories: supervised machine learning-based methods, semi-supervised machine learning-based methods and unsupervised sentiment classification. Unsupervised sentiment classification approaches generally need the help of an external dictionary or other language resources. On the other hand, supervised sentiment classification methods may need to manually label a number of training data in order to achieve the desired accuracy. However, a semi-supervised sentiment classification approach only needs to label a small amount of training data, and relies on proper algorithm(s) to exploit unlabeled data set to expand the training data set for improving the accuracy of sentiment classification.

According to our basic observation, we find that Web users like to use various sentiment words or expressions which may have similar meanings and the same sentiment polarity in the consumer reviews having the same overall sentiment polarity. For example, in the sentence *"The hotel staff are helpful and friendly."*, *"helpful"* and *"friendly"* are two sentiment words to describe *"staff"*. *"Helpful"* and *"friendly"* have high probability to modify *"staff"* in the reviews with positive sentiment polarity, but *"not polite"* and *"rude"* are likely to describe *"staff"* in the reviews with negative sentiment polarity. The research question is whether it is possible to improve the sentiment classification accuracy of consumer reviews through exploiting similarity of sentiment words or expressions which are used to express opinions on product features or service aspects.

The similarity of two sentiment units is measured by the frequency of the same product feature or service aspect in their context windows in the same reviews. We try to show the similarity of sentiment units can be used to conduct sentiment classification of Web reviews effectively. We firstly extract sentiment units according to simple extraction rules. (Note: a sentiment unit is different to a common uni-gram feature, and we will give a more detailed description in Section 3.) Utilizing similarity relationship of sentiment units, our method maps sentiment units into new features which are used to extend the original feature space. The uni-gram features in original training space are used to train a sentiment classifier, and the original uni-gram training features combined with extended features are used to train the other sentiment classifier. We use the two classifiers together with unlabeled data set to build the final sentiment classifier. Our method is a kind of semi-supervised sentiment classification method which only requires a small number of labeled training instances and some unlabeled instances. Experimental results show that our method has better performance than Self-learning SVM-based sentiment classification method.

The remainder of this paper is organized as follows. We firstly show some related publications in Section 2. We present our product feature ranking scheme in Section 3. In Section 4, we show our experiments that have been conducted on the extracted product reviews. Finally, in Section 5 we conclude our work.

2   Related Work

Sentiment classification is one of the main tasks of opinion mining. There are many excellent publications in the field of sentiment classification.

Lee et al. [6] divided opinion mining into three important task: (1) the development of language resources; (2) sentiment classification; and (3) opinion summarization. Pang's survey [17] gives a complete picture of recent research progress of sentiment analysis, especially the work addresses some research challenges and directions of sentiment analysis. Hatzivassiloglou and McKeown [5] adopted a regression model to predict whether conjoined adjectives had the same or different sentiment orientations. Hatzivassiloglou and McKeown' publication [5] is the early research work focusing on word level sentiment classification. Turney [21] proposed PMI (Pointwise Mutual Information) method to determine the sentiment orientation of words. PMI is a simple and practical method which can be used to infer sentiment orientation a word from its statistical association with a set of positive and negative words [21]. Based on PMI, Turney [20] also proposed an approach to document sentiment classification, namely PMI-IR (Pointwise Mutual Information and Information Retrieval) which was used to classify documents into negative and positive classes according to sentiment words appearing in the documents.

Machine learning techniques have been used for sentiment classification too. Pang [18] employed three machine learning methods, namely Naive Bayes, maximum entropy classification and support vector machines to conduct sentiment classification on movie reviews. Pang's work shows [18] that the standard machine learning techniques are promising for document sentiment classification and they outperform human-produced baselines. Machine learning techniques not only use common syntax features, but also can utilize some social information to improve the accuracy of sentiment classification. For instance, Guerra et al. [2] tried to transfer users' bias toward a topic into text features which were used to build a real-time sentiment classifiers. Consumers's sentiment towards products or services may change over time. Even for a product, the strength of sentiment can be different in different time period. Liu et al.'s work [11] considers the problem of the change in sentiments evolving over time. In this work, we don't consider the problem of sentiment evolution.

Traditional machine learning-based sentiment classification methods need training and test data sets. The challenge lies in that the user generated reviews are highly dynamic textual content in different domains. In order to construct training data sets for these domains, training and test documents usually must be labelled by humans. The labelling work is labour intensive and time consuming. In order to reduce labeling work, semi-supervised sentiment classification method came into being. For example, Zhou [22] gave the semi-supervised machine learning sentiment classification method based on the active deep network. Some studies have tried to solve imbalanced semi-supervised sentiment classification, such as Li's work [9]. Li's another article [8] describes a Co-Training-based approach to sentiment classification. In addition, there are some research publications focus on cross-domain sentiment classification, such as [16] [1].

3   Our Proposed Approach

3.1   Problem Definition

Suppose the training data set is T, and there are n training instances in T. The unlabeled data set is U, and there are m unlabeled instances in U. A training instance is expressed as an ordered pair which comprises a consumer review instance and its sentiment class label. Hence T can be shown as $T = \{(x_1, y_1), (x_2, y_2), ..., (x_n, y_n)\}$, where x_i is a training instance, y_i is sentiment class label, $1 \le i \le n$. Let all training instances and unlabeled instances be m dimensional vectors in the real space $X \subseteq \mathbb{R}^m$. Then $x_i = \{x_{i_1}, x_{i_2}, ..., x_{i_m}\}$, x_i is m dimensional input vector. U can be presented as $U = \{u_1, u_2, ..., u_m\}$. In our sentiment classification, we only consider binary classification: classifying consumer reviews into positive and negative classes. We use -1 to represent negative class, and 1 to represent positive class, so $y_i \in \{-1, 1\}$. We need extract features from the training data to form a feature vector for each training instance. To get a sentiment classifier, our task is to obtain a classification function $f(x)$:

$$f : X \longrightarrow L. \tag{1}$$

We design an algorithm to discover $f(x)$ by exploiting the training data set T and the unlabeled data set U. To predict the polarity of a consumer review x', We use function $f(x)$ to classify x' into negative class or positive class, and x' is also a m-dimension vector in the real domain. The sentiment polarity of x' is given by the function $f(x)$.

3.2   Spectral Clustering-Based Semi-supervised Sentiment Classification

A semi-supervised machine learning-based sentiment classification method only needs to manually label a small number of training instances, and then we use the training data set and the unlabeled data set to learn a sentiment classifier to carry out sentiment classification task. This work presents a spectral clustering-based sentiment classification method. Pan [16] proposed a spectral feature alignment (SFA) cross-domain sentiment classification algorithm. Pan's work [16] has proven spectral clustering can play an important role in cross-domain sentiment classification. Inspired by Pan's research [16], we try to used a spectral algorithm to do semi-supervised sentiment classification. In order to show our idea clearly, we firstly give two definitions.

Definition   (Sentiment Unit).   A sentiment unit refers to a word or expression which is subjective and has sentiment polarity.

Deninition   (Noun Unit). A noun unit is a unit which contains maximum consecutive nouns in a sentence clause. For instance, both "*staff*" and hotel "*staff room*" are noun units.

The basic idea is shown as follows:

1. Extract sentiment units from training and unlabeled consumer reviews;
2. Generate similarity matrix on the training set and the unlabeled data set according to co-occurring frequency of sentiment units.
3. Use a spectral clustering algorithm on the similarity matrix to construct sentiment unit mapping matrix (The mapping matrix is constructed by some eigenvector);
4. Map sentiment units of training instances and unlabeled instances into extended features using mapping matrix;
5. Employ SVM to build a sentiment classifier f_1 by using the original uni-gram features extracted from the training data set;
6. Employ SVM to build the other sentiment classifier f_2 using the combination of original uni-gram features and extended features mapped from training instances;
7. Use classifier f_1 and classifier f_2 to conduct sentiment classification on the unlabeled data set. When we use f_2, we also use the combination of original uni-gram features and extended features mapped from unlabeled instances; If f_1 and f_2 classify the same unlabeled instance into the same sentiment class, then put the unlabeled instance into the training data set with the predicted label;
8. Use SVM to build the final sentiment classifier f_3 on the current training data set;
9. Use f_3 to do further sentiment classification. (Note: if we want to use f_3 to conduct sentiment classification, we also need to get extended sentiment units from classified reviews using our mapping matrix.)

Some verbs and adjectives have sentiment polarity, and we look these words or expressions as "sentiment units". For example, "*love*" is a verb, and "*helpful*" and "*friendly*" are adjectives, and they are sentiment units with sentiment polarity. There are some sentiment units in the context of negation. In this case, a negative indicator and a sentiment words together can constitute a sentiment unit. For example, in the sentiment unit "*not good*", the word "*good*" is a positive sentiment word, and "*not*" is a negative indicator.

Our proposed method is based on an assumption: sentiment units appearing in the reviews which have the same overall sentiment polarity and modify the same product feature are more similar (similar polarity and meaning) than those in different reviews with different overall polarity. We try to exploit this kind of co-occurring relationship to conduct semi-supervised sentiment classification. Because semi-supervised learning-based methods only need a small amount of training data, the sentiment units in the test data set may not appear in the training data set. The disadvantageous situation may be bad for sentiment classification. In other words, the sparsity of training features will impact the sentiment classification accuracy. But if we can map common sentiment units into clustering features, we can reduce the malign influence engendered by the sparsity of training features due to the small number of training instances. We map sentiment units in the reviews into extended features in the binary values.

Algorithm 1. Unnormalized spectral clustering [13] [14] [4] [12]

Input:
   Similarity matrix $S \in \Re^{n \times n}$; the final number of clusters is k.
Output:
   k clusters;
1: Calculate the diagonal matrix D, $d_i = \sum_{1 \leq j \leq n} S_{ij}$;
2: Calculate *Laplacian* matrix L, $L = D - S$;
3: Calculate the first k eigenvectors of matrix L: $u_1, u_2, ..., u_k$. u_i is a column vector, these eigenvectors form a matrix V, $V \in \Re^{n \times k}$, y_i presents i row vector of matrix V ;
4: Use k-means on $(y_i)_{y=1..n}$ to conduct clustering, then get k classes $C = \{c_1, c_2, ..., c_k\}$;
5: **return** C;

We extract adjectives and verbs with POS (Part-of-Speech) labels of JJ, JJR, JJS, VB, VBN, VBG, VBZ, and VBP, and these words are in the context window [-3,3] of words with POS labels of NN or NNS. The context of a word window [-3,3] means the left and right distance coverage of the word in a clause is 3 words from the word. At the same time, if there has a negative indicator in the context window [-3,0] of a sentiment unit, then the negative indicator combined with the sentiment word together constitute a sentiment unit. These negative indicators including "*not*", "*no*", "*donot*", "*'do not*", "*didnt*", "*did not*", "*was not*", "*wasn't*", "*isnt*", "*isn't*", "*weren't*", "*werent*", "*doesnt*", "*doesn't*", "*hardly*", "*never*", "*neither*", and "*nor*". Co-occurrence of sentiment units means they must be in the same context scope. We provide a special co-occurring rule: two sentiment units deem as co-occurrence when there exits at least one same noun unit in their context windows([-3,3]), otherwise they don't co-occur. For example, two sentiment units appear in the same consumer review, but the two sentiment units of the context window have no one same noun unit, then the two sentiment units co-occurring frequency is 0. Co-occurring frequency refers to the frequency in the entire collection of reviews rather than just a consumer reviews.

Algorithm 2 is our proposed sentiment classification algorithm. Algorithm 2 is based on a spectral clustering algorithm as shown in Algorithm 1. Spectral clustering techniques make use of similarity matrix of the data to perform clustering tasks, and some related publications such as [13] [14] [4] [12] have given detail descriptions about these techniques. Spectral clustering techniques can be used for graph clustering. The unnormalized Spectral clustering algorithm described as Algorithm 1 shown [13][14].

According to Algorithm 1, Ng et al proposed a normalized spectral clustering algorithm [15]. Our sentiment classification is based on unnormalized spectral clustering. The reason lies in two aspects: firstly, the unnormalized spectral clustering algorithm is simpler and faster in calculation than normalized spectral clustering algorithm; secondly, the accuracy of sentiment classification is acceptable.

In Algorithm 1, S_{ij} represents a co-occurring matrix element of S, and it is the number of co-occurring times of between sentiment unit i and sentiment unit j. If $S_{ij} = 0$, then sentiment unit i and sentiment unit j don't co-occur. Diagonal matrix D is generated from S.

$$d_i = \sum_{1 \leq j \leq n} S_{ij} \tag{2}$$

So we can get Laplacian Matrix:

$$L = D - S. \tag{3}$$

Algorithm 2. Spectral clustering-based semi-supervised sentiment classification algorithm

---

Input:
    Training data set $T = \{(x_t, y_t)_{t=1}^m\}$
    Unlabeled data set to $U = \{u_1, u_2, ..., u_z\}$

Output:
    sentiment classifier f;

1: Extract uni-grams and sentiment units from T and U;
2: Construct matrix M on the training data set T, $M \in \Re^{m \times n}$, m is the instance number of the training data set, n is the number of distinct sentiment units; construct matrix N on the unlabeled data set U, $N \in \Re^{z \times n}$, z is the number of unlabeled instances;
3: Construct similarity matrix S according to the co-occurring frequency of sentiment units, $S \in \Re^{n \times n}$; {Note: In this work, if both sentiment units are in the same review, and there is at least one same noun unit in the their context windows, then the two sentiment units co-occur. }
4: Obtain a diagonal matrix D;
5: Get *Laplacian* matrix L, $L = D - S$;
6: Calculate the first k eigenvector of matrix L. These eigenvector are $l_1, l_2, ..., l_k$, l_i is column vector and they form matrix V, $V \in \Re^{n \times k}$;
7: Calculate matrix $B = M * V$, $B \in \Re^{m \times k}$;If the value of a B's element is greater than 0, its value is set to 1;
8: Calculate matrix $E = N * V$, $E \in \Re^{z \times k}$; If the value of a E's element is greater than 0, its value is set to 1;
9: Use SVM to train to get sentiment classifier f_1 on the training data set; In this case, uni-grams are extracted to train the classifier
10: Use SVM to generate sentiment classifier f_2 on $\{([x_t, B_t], y_t)_{t=1}^m\}$;
11: Use the classifier f_1 to do sentiment classification on U and the results are put into a vector V_x;
12: Use the classifier f_2 to do sentiment classification on $\{([y_t, E_t])_{t=1}^z\}$, and the results are put into a vector V_y;
13: **for all** $i = 1 : |V_x|$ **do**
14:    **if** $V_{x_i} == V_{y_i}$ **then**
15:      $T = T \cup \{u_i\}$;
16:    **end if**
17: **end for**
18: Employ SVM to get the final sentiment classifier f on T;
19: **return** f;

In algorithm 2, we use SVM [3] to do text classification. Algorithm 2 actually produces three classifiers f_1, f_2 and f. Assume the classification accuracy of f_1 is p, the classification accuracy of f_2 is q. For binary classification, a consumer reviews r has the probability $x_1 = pq$ to be classified into right class, and the probability of misclassification is $x_2 = (1-p)(1-q)$. In the algorithm 2, we actually need the classifier f_1 and the classifier f_2 to select training instances from unlabeled data set. That is we put the unlabeled instances with the same predicted sentiment polarity into the training data set.

If the classification results of the two classifier are the same, then either the two classifiers classify the review into the correct sentiment class, or the review is classified as the wrong class. Therefore, the probability of generating the same results for a review by the two sentiment classifiers is:

$$A = x_1 + x_2 = pq + (1-p)(1-q) = 1 - p - q + 2pq \tag{4}$$

The probability of inconsistent classification results is B:

$$B = 1 - A = p + q - 2pq. \tag{5}$$

Therefore, we can get the ratio of A to B:

$$l(p,q) = \frac{A}{B} = \frac{1 - (p+q-2pq)}{p+q-2pq} = \frac{1}{p+q-2pq} - 1 \tag{6}$$

Without loss of generality, we assume $0 < p < 1$ and $0 < q < 1$. Let $f(p,q) = p + q - 2pq$, then we can compute p and q when $f(p,q)$ has a local extremum value through partial differential:

$$\frac{\partial f}{\partial p} = 1 - 2q = 0, \tag{7}$$

$$\frac{\partial f}{\partial q} = 1 - 2p = 0. \tag{8}$$

But when $f(p,q)$ is an extremum, $f(p,q)$ is not the minimum. If the classifier accuracy is a continuous variable, classifier f_1 has the corresponding classification accuracy variable is p, and classifier f_2 has the variable q. Assume that p and q a uniform distribution, then mathematical expectation of X $(X = pq)$ is

$$E(X) = \int_0^1 \int_0^1 pq\rho(p,q)dpdq, \tag{9}$$

$\rho(p,q)$ is the probability density function. When $0 < p < 1$, $0 < q < 1$, $\rho(p,q) = \frac{1}{1-0} = 1$; otherwise $\rho(p,q) = 0$. Therefore, E(X) is

$$E(X) = \int_0^1 \int_0^1 pq\rho(p,q)dpdq = \int_0^1 \int_0^1 pqdpdq. \tag{10}$$

In more general case, assume that the lower limit of the accuracy of two classifiers is a and b. When $a < p < 1$ and $b < q < 1$, $\rho(x,y) = 1$; otherwise $\rho(x,y) = 0$. In this case, E(X) can be calculated by

$$E(X) = \int_a^1 \int_b^1 pq\rho(p,q)dpdq = \int_a^1 pdp \int_b^1 qdq = \frac{(1-a^2)(1-b^2)}{4} \tag{11}$$

4 Experiments

4.1 Experiments Setup

We crawled consumer reviews from **Amazon**[1] and **Tripadvisor**[2] respectively. We extracted consumer reviews about mobile phones and laptops from **Amzaon**, and hotel reviews from **Tripadvisor**. Statistics on the experimental data sets are shown in Table 1. Each category has 2000 consumer reviews which include 1000 positive reviews and 1000 negative reviews. Then we used **OpenNLP**[3] to segment review into sentences, we also use **OpenNLP** to get POS tags for these reviews. For example, our mobile phone data set has 1000 positive reviews and 1000 negative reviews, and these reviews are segmented into 28811 sentences. The the initial sentiment polarity of the reviews in the training data set is assigned according to their review ratings. A review rating is a real number ranging from 1 to 5. Actually, review ratings usually be represented as star numbers in these commercial websites. For example, if a consumer review has *"five stars"*, the review is assigned a rating value 5. When the a review rating is greater than 3, then the review is a positive review; when a review rating is less than 3, the review is a negative review.

Table 1. Experimental Statistics

data set	# of positive reviews	# of negative reviews	# of sentences
phone	1000	1000	28811
Laptop	1000	1000	14814
Hotel	1000	1000	18694

4.2 Compared Methods

Self-learning SVM (S-SVM). Support vector machine [3] is well known as one of the best text classification approach. The earliest use of SVM to determine the sentiment polarity of consumer reviews is Pang's research work [18]. Self-learning SVM-based sentiment classification is a basic semi-supervised classification method. Therefore, Self-learning SVM-based sentiment classification method is our baseline. We firstly decompose each review into uni-grams which are used as the training and test features for Self-learning SVM. The value of each feature in a review is the frequency of the item in the review. In this work, we use TinySVM[4] as an implementation of SVM classifier.

Self-learning SVM-based sentiment classification algorithm is a bootstrap approach to learning as Algorithm 3 shows. The algorithm selects the most likely correctly classified reviews from the results, and put them into the training data

---

[1] http://www.amazon.com/
[2] http://www.tripadvisor.com/
[3] http://opennlp.apache.org/
[4] http://chasen.org/~taku/software/TinySVM/

Algorithm 3. Self-learning SVM-based Sentiment Classification Algorithm

---

Input:

  Training set $L = \{l_1, l_2, ..., l_x\}$, L includes positive reviews and negative reviews;
  Unlabeled data set $U = \{u_1, u_2, ..., u_y\}$;

Output:

  Sentiment classifier C;

1: n is the number of selected reviews which are the most likely correctly classified
   reviews;
2: use SVM to get the initial sentiment classifier C on training data set L;
3: **while** unlabeled data set is not empty **do**
4:    use sentiment classifier C to classify the unlabeled instances in U: get positive
      set P and the negative set N;
5:    if $|P| >= d$, then select the most likely correctly classified d instances from P
      (the set is P_d) into L, $L = L \cup P_d$, $P = P - P_d$; otherwise put all the instances
      in the P into L, $L = L \cup P$;
6:    if $|N| >= d$, then select the most likely correctly classified d instances from N
      (the set is N_d) into L, $L = L \cup N_d$, $N = N - N_d$; otherwise put all the instances
      in the N into L, $L = L \cup N$;
7:    employ SVM to train a new sentiment classifier C on the current training data
      set L;
8: **end while**
9: **return** C;

---

set. A new sentiment classifier is built by training on the new training data set.
Repeat the above steps until there are no unlabeled review that can be added
to the training data set so far. There are a lot of SVM-based active learning
approaches such as [19]. However, our proposed method is only compared to
the basic active learning method. When a classified review has greater distance
from SVM hyperplane, the review is considered having higher probability to be
correctly classified.

4.3   Experimental Results

Table 2 gives experimental results. We randomly sample from the data set of
each product category to get the test data set firstly. A training data set is
randomly sampled from the remain data set. When use our method to classify
test instances, these instances also must be extended their original features as
Algorithm 2 does.

   Take the first line for example: the data set category is hotel, the number of training instances is 100, including 50 positive instances, and 50 negative instances. The number of unlabeled consumer reviews is 1100, the test
data set includes 800 reviews. The classification accuracy of Self-learning SVM-
based method is 68.88%. Our proposed method has the best performance with

Table 2. Classification accuracies of different sentiment classification methods

Data set	Training#	Unlabelled#	Test#	Self-learning SVM	Our method
Hotel	100	1100	800	68.88%	**83.38%**
Phone	100	1100	800	66%	**81.38%**
Laptop	100	1100	800	64.63%	**81.75%**
Hotel	200	1000	800	76.25%	**82.13%**
Phone	200	1000	800	64%	**82.5%**
Laptop	200	1000	800	70.5%	**81.25%**
Hotel	300	900	800	79.5%	**82.88%**
Phone	300	900	800	72.13%	**80.88%**
Laptop	300	900	800	67.5%	**81.25%**
Hotel	400	800	800	79.75%	**83%**
Phone	400	800	800	73.38%	**81.88%**
Laptop	400	800	800	69.13%	**81%**

accuracy value of 83.38%. In Table 2, we can see our proposed method has better performance compared to Self-learning SVM-based method on all the data sets.

5   Conclusion

This work focuses on semi-supervised sentiment classification. Based on spectral clustering, we propose a semi-supervised sentiment classification method. In this method, we use a small number of training instances and an unlabeled data set to train and build the final sentiment classifier. We construct a mapping matrix according to similarity of sentiment units which are words or expressions bearing sentiment polarity. Experimental results on empirical data sets show our method is effective and promising, and it outperforms Self-learning SVM-based method.

In the future, we will continue our research to find more effective sentiment classification methods. And we are going to compare our method with some other semi-supervised sentiment classification methods too. There are some potential applications for our proposed method in the field of business intelligence. We can use the ratings and sentiment polarity of consumer reviews to rank product features [10]. We also have known that rating inference can be used with collaborative filtering algorithms [7]. Therefore, we believe the combination of sentiment classification and product feature ranking may be useful for collaborative recommendation systems.

Acknowledgement. We thank anonymous reviewers for their constructive comments. This work is supported by Funding Project for PHR(IHLB)Academic Human Resources Development in Institutions of Higher Learning under the Jurisdiction of Beijing Municipality (PHR201108431). The work is also supported by Beijing Excellent Talents Funding(2010D005022000002).

References

1. Bollegala, D., Weir, D., Carroll, J.: Using multiple sources to construct a sentiment sensitive thesaurus for cross-domain sentiment classification. In: Proceedings of the 49th Annual Meeting of the Association for Computational Linguistics: Human Language Technologies, HLT 2011, vol. 1, pp. 132–141. Association for Computational Linguistics, Stroudsburg (2011)

2. Calais Guerra, P.H., Veloso, A., Meira Jr., W., Almeida, V.: From bias to opinion: a transfer-learning approach to real-time sentiment analysis. In: Proceedings of the 17th ACM SIGKDD International Conference on Knowledge Discovery and Data Mining, KDD 2011, pp. 150–158. ACM, New York (2011)

3. Cortes, C., Vapnik, V.: Support-vector networks. Machine Learning 20(3), 273–297 (1995), http://dblp.uni-trier.de/db/journals/ml/ml20.html#CortesV95

4. Das, K.C.: The laplacian spectrum of a graph. Comput. Math. Appl. 48, 715–724 (2004)

5. Hatzivassiloglou, V., McKeown, K.R.: Predicting the semantic orientation of adjectives. In: Proceedings of the Eighth Conference on European chapter of the Association for Computational Linguistics, EACL 1997, pp. 174–181. Association for Computational Linguistics, Stroudsburg (1997)

6. Lee, D., Jeong, O.R., Lee, S.G.: Opinion mining of customer feedback data on the web. In: Proceedings of the 2nd International Conference on Ubiquitous Information Management and Communication, ICUIMC 2008, pp. 230–235. ACM, New York (2008)

7. Leung, C., Chan, S., Chung, F., Ngai, G.: A probabilistic rating inference framework for mining user preferences from reviews. World Wide Web 14, 187–215 (2011), http://dx.doi.org/10.1007/s11280-011-0117-5, doi:10.1007/s11280-011-0117-5

8. Li, S., Huang, C.R., Zhou, G., Lee, S.Y.M.: Employing personal/impersonal views in supervised and semi-supervised sentiment classification. In: Proceedings of the 48th Annual Meeting of the Association for Computational Linguistics, ACL 2010, pp. 414–423. Association for Computational Linguistics, Stroudsburg (2010)

9. Li, S., Wang, Z., Zhou, G., Lee, S.Y.M.: Semi-supervised learning for imbalanced sentiment classification. In: Proceedings of the Twenty-Second International Joint Conference on Artificial Intelligence, IJCAI 2011, vol. 3, pp. 1826–1831. AAAI Press (2011)

10. Li, S.K., Guan, Z., Tang, L.Y., Chen, Z.: Exploiting consumer reviews for product feature ranking. Journal of Computer Science and Technology 27, 635–649 (2012), http://dx.doi.org/10.1007/s11390-012-1250-z, doi:10.1007/s11390-012-1250-z

11. Liu, Y., Yu, X., An, A., Huang, X.: Riding the tide of sentiment change: sentiment analysis with evolving online reviews. World Wide Web, 1–20 (2012), http://dx.doi.org/10.1007/s11280-012-0179-z, doi:10.1007/s11280-012-0179-z

12. Luxburg, U.: A tutorial on spectral clustering. Statistics and Computing 17, 395–416 (2007)

13. Mohar, B.: The Laplacian spectrum of graphs. Graph Theory, Combinatorics, and Applications 2, 871–898 (1991)

14. Mohar, B., Juvan, M.: Some applications of laplace eigenvalues of graphs. In: Graph Symmetry: Algebraic Methods and Applications. NATO ASI Series C, vol. 497, pp. 227–275 (1997)

15. Ng, A.Y., Jordan, M.I., Weiss, Y.: On spectral clustering: Analysis and an algorithm. In: Advances in Neural Information Processing Systems 14, pp. 849–856. MIT Press (2001)
16. Pan, S.J., Ni, X., Sun, J.T., Yang, Q., Chen, Z.: Cross-domain sentiment classification via spectral feature alignment. In: Proceedings of the 19th International Conference on World Wide Web, WWW 2010, pp. 751–760. ACM, New York (2010)
17. Pang, B., Lee, L.: Opinion mining and sentiment analysis. Found. Trends Inf. Retr. 2(1-2), 1–135 (2008)
18. Pang, B., Lee, L., Vaithyanathan, S.: Thumbs up?: sentiment classification using machine learning techniques. In: Proceedings of the ACL 2002 Conference on Empirical Methods in Natural Language Processing, EMNLP 2002, vol. 10, pp. 79–86. Association for Computational Linguistics, Stroudsburg (2002)
19. Tong, S., Koller, D.: Support vector machine active learning with applications to text classification. J. Mach. Learn. Res. 2, 45–66 (2002)
20. Turney, P.D.: Thumbs up or thumbs down?: semantic orientation applied to unsupervised classification of reviews. In: Proceedings of the 40th Annual Meeting on Association for Computational Linguistics, ACL 2002, pp. 417–424. Association for Computational Linguistics, Stroudsburg (2002)
21. Turney, P.D., Littman, M.L.: Measuring praise and criticism: Inference of semantic orientation from association. ACM Trans. Inf. Syst. 21(4), 315–346 (2003)
22. Zhou, S., Chen, Q., Wang, X.: Active deep networks for semi-supervised sentiment classification. In: Proceedings of the 23rd International Conference on Computational Linguistics: Posters, COLING 2010, pp. 1515–1523. Association for Computational Linguistics, Stroudsburg (2010)

Automatic Filtering of Valuable Features for Text Categorization

Adriana Pietramala, Veronica Lucia Policicchio, and Pasquale Rullo

University of Calabria, Rende - Italy
{a.pietramala,policicchio,rullo}@mat.unical.it

Abstract. Feature selection (FS) is aimed at reducing the size of the feature space. The advantage of FS is, in general, two-fold: improved accuracy of the learned classifiers, and efficiency of the learning process. Behind the use of a FS method there is the implicit assumption that only the selected terms are representative of the category being learned, while the rest are redundant. Thus, predicting an appropriate value of the feature space dimensionality is a crucial task, as a too aggressive feature selection might discard terms that carry essential information, while redundant features might deceive the learning algorithm. In "real life", this task is usually accomplished "manually", that is, the learning process is rerun over several vocabularies of different dimensions and the best results are eventually taken. Unfortunately, this may take very long training times.

In this paper we propose a FS technique that automatically detects an appropriate number of features for text categorization (TC) that are sufficient to learn accurate classifiers and make efficient the training process. One peculiarity of the proposed approach is that of combining both positive and negative features, the latter being considered relevant for the purpose of effective TC. The proposed approach has been tested by running three well known classifiers, notably, Ripper, C4.5 and linear SVM (the SMO implementation), over 7 real-world data sets, with varying characteristics.

1 Introduction

Text Classification (TC) is the task of assigning natural language texts to one or more thematic categories on the basis of their contents. A number of machine learning methods to automatically construct classifiers using labelled training data have been proposed in the last few years, including k-nearest neighbors (k-NN), probabilistic Bayesian, neural networks and SVMs. Overviews of these techniques can be found in [12].

In a different view, rule learning algorithms have become a successful strategy for classifier induction. Direct methods extract rules directly from data, while indirect methods extract rules from other classification models, such as decision trees (e.g., C4.5 [10]). Representative examples of direct methods include Inductive Rule Learning (IRL) systems, such as Ripper [3]. Rule learning systems are

S. Zhou, S. Zhang, and G. Karypis (Eds.): ADMA 2012, LNAI 7713, pp. 284–295, 2012.

of great practical interest in real-world TC applications because, besides providing state-of-the-art predictive capabilities, they build models that are human readable. Models that have intuitive interpretation is, indeed, a critical requirement of many applications. However, systems like C4.5 and Ripper suffer from one main drawback, i.e., they perform quite inefficiently over data sets of remarkable size (as it usual in TC applications). For an instance, the learning times of Ripper and C4.5 over the data set Market with 2000 features were around 234 and 154 hours, respectively (as reported next in this paper).

Feature Selection (FS) is aimed at reducing the size of the feature space. Potentially, FS is beneficial for TC basically for two reasons. On one hand, it removes redundant or noisy features that may deceive the learning algorithm. On the other hand, it boosts time efficiency and reduces memory space requirements. In addition, by selecting an effective lexicon, capable of expressing the essential patterns, FS will boost the induction of compact and easy to interpret classifiers.

Typically, FS is performed by picking out the N highest scoring features, according to some feature selection function, e.g., Information Gain, CHI square, etc. [4]. Behind this approach there is the implicit assumption that only the N selected terms are representative of the category being learned, while the rest are redundant. Thus, one main issue that in general arises when inducing a classifier is that of detecting the "optimal" size of the feature space, i.e., how many features (and what) the classifier can access during the learning process. Unfortunately, the number N of features needed to achieve good classification results remains unclear. The literature is lacking from this regard, and even contradictory: some studies suggest an aggressive feature selection, while others suggest that feature selection may make matters worse. For an instance, in [5] it is claimed that reducing the dimensionality of the feature space may significantly improve the effectiveness and scalability of SVMs , while other works have found that SVMs perform better when no feature selection is performed [6]. No much work, to the best of our knowledge, has been done so far as far as rule learners are concerned. Thus, in the real practice, the size N of the feature space is usually managed as a tuning parameter, that is, the learning process is rerun over several feature spaces of different dimensions and the best results are eventually taken. Unfortunately, this may require very long training times, thus loosing the advantage of dimensionality reduction as far as efficiency is concerned (with systems like C4.5 and Ripper, this iterative process might even be unfeasible on large data sets).

The aim of this work is to define a filter method for FS that is able to *automatically* select a small number of high valuable (both positive and negative) features that are sufficient to learn accurate classifiers and make efficient the learning process (especially for such slow systems as Ripper and C4.5).

In summary, the main contributions of this paper are the following. First, we provide a definition of feature space, where both features indicative of membership and of non-membership are included. Second, we define a technique for the automatic detection of an "optimal" dimensionality of the feature space. Finally, we perform a thorough experimental study to see the effect of our FS technique on both accuracy and efficiency of the learning process. To this end, we use two rule induction methods, notably, Ripper and C4.5, along with the Platt's Sequential Minimal Optimization (SMO) method for linear SVM training [7]. The experimentation was carried out over 7 real-world corpora of size ranging from around 1000 to 204,000 documents.

2   Why and What Negative Features

FS relies on the (most often) realistic assumption that there exists a subset of features selectively covering most documents under a category c. This "covering set" is made of the essential words capturing the concepts related to c.

However, detecting the set of features covering most of the positive examples may not be sufficient for the purpose of effective classification, as negative evidence is recognized to play a crucial role in TC. This is essentially because natural languages are intrinsically ambiguous, and negation helps to disambiguate concepts. For an instance, the word "ball" may ambiguously refer to either the concept "sport" or "dance" , whereas the conjunction "ball and not ballroom" much likely refers to "sport". Thus, in order to achieve high predictive capabilities, the negative features (i.e., features indicative of non-membership) are also useful and should be included in the feature space.

We note that, though systems like Ripper and C4.5 do not include an explicit mechanism to build negation into the generated rules, they can anyway deal with negative features. In the bag-of-words representation, indeed, a document is represented in an n-dimensional vector space, where n is the number of features. The value of each feature in the vector space can take on either a 0 or 1 value to indicate the presence or absence of a word in a document, or a numeric value to indicate its frequency. Thus, a valued-zero feature in a rule expresses the absence of that feature as a classification condition. More recently, a few examples of rule-based classifiers explicitly dealing with negation have been proposed (see, e.g., [1, 8, 9, 11]).

However, when FS is applied, most of the potentially negative features are automatically discarded. And, as noted by Zheng et al. in [13], even the implicit combination of positive and negative features coming from the use of two-sided scoring functions (e.g., CHI square) is not in general optimal, especially for imbalanced data sets, where the values of positive features are usually much larger than those of negative ones.

Thus, to provide a classifier with a set of candidate negative features, Zheng et al. propose a method where the negative features for a class c are the most relevant features for its complement $\bar{c}$.

As we will see in the next sections, our approach differs from the one just mentioned, in that we focus on a specific subset of features that are potentially positive for $\bar{c}$. Indeed, this set is made of those features co-occurring with positive ones (for class c) within negative examples (for class c). For an instance, if "ballroom" co-occurs with "ball" within a negative example for the category being learned, say, "sport", and "ball" is a positive feature for "sport", then

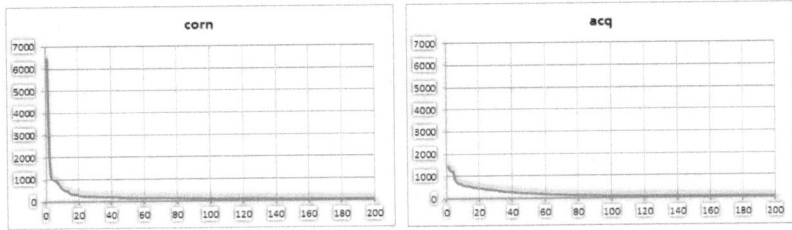

Fig. 1. Distribution of features by CHI square for two categories from R10 - "corn" (left side) and "acq" (right side). Only first 200 features are shown.

"ballroom" is a candidate negative feature for the same category. We point out that a negative feature for a category c, according to our definition, may not even be *relevant* for $\bar{c}$.

3   Combining Positive and Negative Features

Unlike in the classical definitions, where the feature space is simply a subset of positive terms from the vocabulary, in our definition the feature space consists of both a set of positive and a set of negative features.

In the following, we consider a binary classification task, and assume that all features occurring in some positive example are ranked according to some (binary) feature selection function, e.g., CC (Coefficient Correlation), CHI square or IG (Information Gain).

Figure 1 shows two empirical distributions of the CHI square scoring values of the features occurring in the positive examples of two categories from the Reuters, namely, "corn" and "acq". As it can be seen, "corn" is characterized by a few features scoring very high, while the remaining ones rapidly approach near-zero values. The sharply declining shape of this graph is indicative of an "easy" category, i.e., a category for which a high performance can be achieved with only a few (positive) discriminative words. In contrast, "acq" is a more "difficult" category. As we can see from Figure 1- right side, it has lower initial CHI square values, and the graph has a smooth (decreasing) trend. That is, no features with very high discriminative power there exist. Thus, while an aggressive reduction of the feature space should be beneficial for "corn", a more moderate policy is required for "acq".

The above observations indicate that a variable number of features should be selected, depending on the characteristics of the feature distributions of the given category. That is why we define the set of candidate positive features $Pos(k)$ as a function of k, with k denoting the size of the set.

Definition 1. (Positive candidate features) Let V_c be the set of terms occurring in the documents of category c, k a non-negative integer, and σ a feature selection function which assigns a score to every feature in V_c based on its correlation with category c (e.g., CHI Square) . Define the set $Pos(k)$ of *positive candidate features* as the set of the k highest scoring features in V_c, according to σ.

In the above definition, we intend to pick out those fetaures most indicative of membership of category c.

Now, the set $Pos(k)$ of candidate positive features induces the set $Neg(k)$ of candidate negative features as shown by the following definition.

Definition 2. (Negative candidate features) Given $Pos(k)$, consider the set N of terms co-occurring with positive features within negative examples, i.e.,

$$N = \{t \in V \mid t \notin Pos(k) \text{ and } (\Theta^+ \cap \Theta(t) \setminus T_c) \neq \emptyset\}$$

where $\Theta(t) \subseteq T$ is the set of training documents containing feature t, $\Theta^+ = \cup_{t \in Pos(k)} \Theta(t)$ and T_c is the training set of c. With each feature $t \in N$ we assign a score $\eta(t)$ as follows:

$$\eta(t) = \frac{|\Theta^+ \cap \Theta(t) \setminus T_c|}{|\Theta^+ \setminus T_c| + |\Theta(t) \cap T_c|}.$$

Then, we define $Neg(k)$ as the set of the best k elements of N according to η; we say that $t \in Neg(k)$ is a *candidate negative feature* of c.

It can be easily seen that $0 < \eta(t) \leq 1$. In particular, a term t occurring in all negative examples and in no positive one containing any positive feature has score $\eta(t) = 1$. On the other hand, $\eta(t) > 0$, $\forall t \in N$ as, by definition, t co-occurs with a candidate positive feature in some negative example.

The rationale behind the above definitions, first proposed in [9], is rather intuitive: candidate positive features are supposed to capture most of the positive examples, as they are characterized by high scoring values. On the contrary, candidate negative features, defined as terms co-occurring with positive candidate terms within negative examples, are supposed to discard most of the (potentially) false positive examples. Clearly, the higher the scoring $\sigma(t)$ (resp. $\eta(t)$) of a term t, the higher its value as a candidate positive (resp. negative) feature.

A final remark is needed here. In developing the above model of feature space, we conjectured that the amount of positive information needed to classify texts is usually much larger than that of negative one. The latter is indeed used to improve accuracy through confident rejection of non-relevant documents. This assumption is corroborated by the empirical experience showing that the number of negative features in rule-based classifiers is usually much smaller than that of positive features. Thus, the number k of candidate positive features can be regarded as an upper bound for the number of negative features. That is why we implicitly assumed that Pos and Neg are of equal size.

Definition 3. (Feature space) The feature space $\mathcal{F}(k)$ is $Pos(k) \cup Neg(k)$.

4   Detection of the Feature Space Dimensionality

Now that we have defined the notion of feature space $\mathcal{F}(k) = Pos(k) \cup Neg(k)$, we address the question of how to determine its size k. This is a key parameter and should be chosen to optimize the categorization performance.

As we have noted in the previous section, the amount of positive features to be selected strongly depends on the shape of the scoring value distributions (see Figure 1): a sharp declining curve indicates that features ranked highest contain considerable information for the purpose of discrimination among a category and its complement. On the contrary, softly declining curves indicate that even features with lower rank may be somewhat relevant.

At any rate, the score distribution of words behaves in a rather stable way: a very small number of features are highest ranked, while most features have negligible scores. Thus, we conjecture that an approximation modeling the scoring value distribution is the exponential distribution, i.e.,

$$p(x) = \lambda e^{-\lambda x}$$

where $\lambda = 1/\mu$, with μ the mean of the random variable x.

Based on the above observations, we next show how we compute the dimension k of $Pos(k)$. Let $V_c = \{t_1, \cdots, t_i, \cdots, t_n\}$ be the vocabulary of c, i.e., the set of features occurring in the positive examples of c. Further, let $S = [s_1, \cdots, s_i, \cdots, s_n]$ be the list of scoring values of the features in V_c, sorted in descending order, with s_i being the score $\sigma(t_i)$ of t_i. Further, let μ be the mean of the elements of S. By assuming an exponential distribution, the cumulative distribution function, i.e., the probability that the score s is greater or equal to a given value s_i, is given by

$$p(s \geq s_i) = e^{-\lambda s_i}.$$

Now, the method consists of the following steps:

1. *Step 1*: remove from V_c the lowest scored terms, i.e., terms $t_{q+1}, \cdots, t_n$ such that the probability that a feature has score s less or equal to the score s_{q+1} of t_{q+1} is equal to 0.25, that is, $p(s \leq s_{q+1}) = 1 - e^{-\lambda s_{q+1}} = 0.25$ (i.e., s_{q+1} is the first quartile). The aim of this step is that of cutting off the tail of the distribution, made of a usually large amount of features having very negligible scores.

2. *Step 2*: let $S' = [s_1, \cdots, s_j, \cdots, s_q]$ be the scores of the q remaining terms, and μ' their mean. Then, choose $1 \leq k \leq q$ such that $p(s \geq s_k) = e^{-\lambda' s_k} = 0.02$, where $\lambda' = 1/\mu'$. Note that, in general, $k << q$ holds.

Once k has been detected as shown above, the feature space is $\mathcal{F}(k) = Pos(k) \cup Neg(k)$, where $Pos(k)$ and $Neg(k)$ are constructed according to Definition 1 and Definition 2, respectively.

We conclude by noticing that the assumption of an exponential distribution $p(s) = \lambda e^{-\lambda s}$ is tantamount to assuming a score decay at a rate of $\lambda\%$ per features. Of course, this is not always the case, as the real distribution of the feature scores may substantially vary depending on the training data at hand. However, though there are not strong theoretical grounds for such an assumption, we will see that this familiar statistical distribution works quite reliably in practice.

Table 1. Data set description

Name	Source	Original Format	#Doc	#Feat	#Cat	Cat Size Min	Max
Oh0	Ohsumed-233445	arff	1,003	3,182	10	51	194
Oh10	Ohsumed-233445	arff	1,050	3,238	10	52	165
Ohscale	Ohsumed-233445	arff	11,162	11,465	10	709	1,621
R10	Reuters-21578	text	12,897	21,363	10	237	3,964
Ohsumed	Ohsumed-233445	text	34,389	34,359	23	427	9,611
Cade12	Gerindo Proj.	csv	40,983	69,470	12	625	8,473
Market	Rcv1	text	203,926	68,604	4	26,036	85,440

5   Empirical Study

The aim of this empirical study is to assess to what extent the proposed method affects both the accuracy and the efficiency of a number of learning algorithms. To this end, we compared the experimental results obtained by using our filter approach, with two baseline results, obtained by using 50 features (aggressive FS) and 2000 features ("lazy" FS) (the reader is referred to Section 6 for a discussion on this choice). The experimentation was performed by using a java implementation of our filter[1].

5.1   Experimental Setting

Classification methods. For this experimental study we used three most popular classifiers, namely, Ripper, C4.5, and SMO (the Platt's Sequential Minimal Optimization method for linear SVM training [7]). The former two were selected because they are of large practical interest in many real-world applications as they produce human readable classifiers. But, at the same time, they showed to be highly inefficient over large real-world data sets. SMO, in turn, was chosen as it is reported to be one of the best methods for text categorization. All such systems are available on the Weka platform (version 3.5.8).

Data sets. We carried out the empirical study on 7 real-world data sets whose properties are summarized in Table 1. As it can be seen, they span over a wide range of sizes, from a minimum of around 1000 (Oh0) to a maximum of nearly 204,000 (Market) documents. In particular, we selected 2 small data sets (Oh0, Oh10), 2 of medium size (Ohscale, R10) and 3 of large size (Ohsumed, Cade12, Market).

Pre-processing. We preliminarily pre-processed all data sets downloaded in textual format, by performing tokenization (word unigrams) and stopword removal. We used the bag-of-words representation with binary word weighting. Each feature was represented as a numerical attribute.

Experiments were performed in a binary classification setting. To this end, we binarized all data sets by performing multi-class to two-class conversion.

---

[1] Available for download at www.mat.unical.it/OlexSuite/filter/filter-download.htm

Finally, for each category, feature scoring by CHI square was performed (on the training set). Though CHI square is a "two-sided" metric, so it may select even negative features, we used it to have a realistic baseline, as it proved to be an effective and robust feature selection measure in the literature. However, the "risk" of an implicit selection of negative features is very limited in practice, as CHI square values of positive features are usually much larger than those of negative ones in imbalanced problems [13].

Performance evaluation. Due to efficiency reasons, we performed 5-fold cross validation (80% training, 20% test) on small data sets (from Oh0 up to R10), while holdout (70% training, 30% test) was applied to larger data sets, namely, Ohsumed, Cade12 and Market.

Performance was measured by the arithmetic mean PRavg of precision P and recall R (an approximation of the Precision/ Recall Break-Even Point).

5.2 Dimensionality Reduction

In Table 2, the number N of features selected by our method for each data set, averaged over all categories, is reported in the 2nd column (avg #feat). As it can be seen, N ranges from a minimum of 50 features (half positive and half negative) on average for data sets OH0 and OH10, to a maximum of 825 for Cade, with an average value, over all data sets, of 318 features. It is worth noting that $50 \leq N \leq 2000$, for all data sets. Going beyond the results given in Table 2, the minimum number of features selected for categories from R10 is 112 for *corn*, and the maximum is 354 for *acq*.

Table 2. Micro-averaged PRavg values obtained with (1) the features selected by the proposed method, (2) 2000 features and (3) 50 features

data set	avg #feat	our approach (FS-exp)			2000 features (FS-2000)			50 features (FS-50)		
		C4.5	Ripper	SMO	C4.5	Ripper	SMO	C4.5	Ripper	SMO
OH10	50	74.78	**78.25**	**77.59**	74.78	**77.82**	74.70	74.78	**78.25**	**77.59**
OH0	50	**79.85**	84.36	**85.02**	79.24	**84.37**	84.80	**79.85**	84.36	**85.02**
ohscal	186	69.84	72.51	**75.44**	70.05	**72.96**	69.52	**70.14**	72.35	74.12
R10	216	**85.16**	85.01	88.79	84.67	85.21	**88.94**	85.06	**85.30**	88.12
Ohsumed	361	**63.25**	59.29	**67.57**	61.43	**60.35**	66.94	52.25	51.84	51.49
cade	825	**48.25**	44.38	**54.60**	48,10	44,31	54,06	46.39	43.85	46.57
Market	537	95.21	**94.87**	96.02	**95.37**	94.63	**96.54**	94.83	94.30	93.33
avg	318	**73.76**	74.10	**77.86**	73.32	**74.23**	76.50	71.90	72.89	73.75

5.3 Accuracy Comparison

Table 2 compares the micro-PRavg values of classifiers obtained by using (1) our method (FS-exp), (2) 2000 features (FS-2000) and (3) 50 features (FS-50). The best accuracy values are highlighted on each data set. As it s shown, on average, FS-50 entails the worst performance, while FS-exp and FS-2000 perform quite similarly. More in details, by individually comparing our approach with each of the other two, we can observe the following.

FS-exp vs FS-50. Here, each algorithm over FS-exp achieves an average performance significantly higher than that over FS-50. In particular, SMO is the one which shows the highest improvement (over 4 points), while for the other two algorithms the improvement is of around 2 points. It should be noted that the difference in performance between F-exp and F-50 is most remarkable on larger data sets.

FS-exp vs FS-2000. When using 2000 features, the performance of both Ripper and C4.5 remains substantially unchanged on average w.r.t. FS-exp. On the contrary, the average PRavg of SMO over FS-2000 is quite lower than that over FS-exp (76.50 against 77.86). Notice that the difference in performance is most remarkable on smaller data sets.

5.4 Effect of Negative Features

Table 3 shows the effect of candidate negative features on the predictive capabilities of the induced hypotheses. Here, the micro-averaged PRavg values reported in Table 2 are compared with those obtained by using only the positive features selected by our method. As it is shown, both C4.5 and Ripper do not benefit much from negation, whereas SMO exhibits a significant accuracy improvement (around 2 points on average). Looking at each single data set, we can observe that the positive effect of negation increases as the size of the data set increases (on smaller data sets it may even have detrimental effect). For an instance, on Cade, negation entails an improvement of the SMO PRavg of around 3.5 points, while on Market the improvement is of 5 points.

Table 3. Effect of negation. For each classifier, the micro-averaged PRavg obtained by using k positive and k negative features (column "with neg") and only k positive features (column "without neg")

data set	C4.5		Ripper		SMO	
	with neg	w/o neg	with neg	w/o neg	with neg	w/o neg
OH10	**74.78**	74.31	**78.25**	77.77	77.59	**78.71**
OH0	**79.85**	79.39	84.36	**84.68**	**85.02**	84.99
ohscal	69.84	**70.14**	**72.51**	71.99	**75.44**	72.71
R10	**85.16**	84.96	**85.01**	84.81	**88.79**	87.55
Ohsumed	**63.89**	61.43	59.29	**60.00**	**67.57**	65.28
cade	**48.25**	47.79	**44.38**	43.31	**54.60**	51.14
Market	**95.21**	95.15	**94.87**	94.59	**96.02**	91.26
avg	**73.86**	73.31	**74.10**	74.02	**77.86**	75.95

5.5 Time Efficiency Comparison

Looking at Table 4, we can see that the global learning time of C4.5 passes from 238 to 44 hours and that of Ripper from 299 to 80. That is, learning times are drastically cut down. Notice that both algorithms perform faster on each single data set using our approach. In contrast, SMO shows a negligible improvement,

Table 4. Learning times (hours)

data set	our approach			2000 features		
	C4.5	Ripper	SMO	C4.5	Ripper	SMO
OH10	0.003	0.005	0.009	0.04	0.32	0.004
OH0	0.002	0.004	0.004	0.03	0.219	0.004
ohscal	0.77	0.98	4.84	5.82	10.99	3.35
R10	0.64	1.81	0.18	3.51	14.46	0.20
Ohsumed	6.90	4.14	9.31	51.08	30.05	7.83
cade	13.26	5.16	17.85	22.71	8.34	15.16
Market	22.25	68.19	29.99	154.56	234.42	36.57
total (hours)	44	80	62	238	299	63

passing from 63 to 62 hours. Even more, on some data sets, SMO performs slower (e.g., Cade).

Quite obviously, the major contribute to time reduction is provided by larger data sets. For an instance, on the Market data set (204,000 documents), the training time of C4.5 over 2000 features is around 155 hours, against 22 hours with our method, while Ripper passes from 234 to 68 hours and SMO from around 36 to 30 hours (all this with no loss in terms of accuracy - see Table 2). It is interesting noting that, with our method, C4.5 shows to be the most efficient classifier, even faster than SMO.

6 Discussion

Many previous studies claim that the best performance for linear SVM is obtained when only a small fraction of features is used [4, 5]. In particular, Gabrilovich and Markovitch have shown that, over a class of data sets characterized by many redundant features, optimal performance of linear SVM is achieved when using features between 5 and 40 [5]. Also Sebastiani claims that typical size of a local feature set is between 10 and 50 [12]. However, in contrast with these studies, other works have found that SVMs perform better when no feature selection is performed [6]. As for rule based classifiers, we are not aware of previous works.

Based on such premises, our empirical study consisted in comparing our approach with two baselines: on one side, we used a small set of 50 features (FS-50) and, on the other side, a set made of 2000 features (FS-2000)[2]. It should be noted that, for all data sets we used, the feature spaces generated by our method usually consist of a few tens/hundreds elements, thus, FS-50 ≤ F-exp ≤ FS-2000.

The analysis of the experimental results leads to some interesting observations.

First, all classifiers provide their worst performance when using 50 features. That is, in contrast with [5], we observe that text categorization does not benefit from aggressive feature selection, in particular when large data sets are involved. Quite surprisingly, SMO is the method that most suffers from aggressive dimensionality reduction.

---

[2] The attempt to use full vocabularies failed because both C4.5 and Ripper showed to be extremely slow.

Second, by comparing the results of our method with those obtained over FS-2000, we may argue that 2000 features are not only unnecessary but, in the case of SMO, even detrimental - this contrasting with findings in [6]. Indeed, while C4.5 and Ripper substantially preserve their accuracy levels when reducing the feature space from FS-2000 to FS-exp, on the contrary SMO shows an increment of predictive capability. It is noteworthy that, whereas the behavior of SMO was somewhat expected (based on previous findings), the capability of the rule induction algorithms to learn accurate classifiers using a small set of features was not so obvious.

Looking at the above results in combination, we can cautiously conclude that the proposed method is sufficiently aggressive to discard redundant terms, while not loosing informative ones.

The effect of negative features is another point that deserves some attention. As we have seen, it is negligible as far as both Ripper and C4.5 are concerned, whereas it has a meaningful impact on SMO accuracy. That is, unlike the two rule induction methods, it seems that SMO is able to exploit the potential information provided by candidate negative features (especially on larger data sets). Consistently with the findings of [13], this confirms that negative terms contain considerable information that may be essential for effective categorization.

Finally, learning times. As it was expected, C4.5 and Ripper drastically reduce training times (w.r.t. the baseline with 2000 features). As a practical implication, such slow systems can (efficiently) be used over real-world corpora of large size, without paying anything in terms of accuracy. It is worth recalling, to this end, the huge gain of training time over large data sets such as Market (86% drop in number of training hours for C4.5, and 72% for Ripper- see Table 4). In contrast, quite surprisingly, this is not the case with SMO, for which we observed only a tiny average runtime reduction. On some data sets, it was even observed an increment of the learning time (this was consistently observed also with 50 features). What is interesting in the overall is that, using our filter, C4.5 and Ripper become as efficient as SMO (actually, C4.5 even more).

7   Conclusions and Future Work

We devised a method for the automatic detection of the appropriate size of the feature space. The proposed approach selects, category by category, a suitable set of high-quality features on the basis of their statistical properties. This set is made of an equal number of positive and negative features. Experiments with three well known classifiers over 7 real-world data sets of varying characteristics have been described. The obtained results indicate that the selection performed by our method is sufficiently aggressive to eliminate redundant terms, but not so much to discard informative ones. Indeed, Ripper and C4.5 preserve their capability of learning classifiers as accurate as those induced over larger vocabularies, while for SMO an enhancement of its predictive capability was even observed. In addition, Ripper and C4.5 drastically cut down learning times (especially over data sets of remarkable size).

In conclusion, despite its empirical nature, the proposed approach seems to work quite reliably in practice: to gain time with Ripper and C4.5, thus making them usable over large corpora, and to gain accuracy with SMO (i.e., linear SVM). For this reason, we believe that it may be of real practical interest.

As a possible direction of future work, we intend to achieve a deeper understanding of the issues concerning the relationship between negative information and dimensionality reduction for rule-based systems.

References

1. Baralis, E., Garza, P.: Associative text categorization exploiting negated words. In: Proceedings of the 2006 ACM Symposium on Applied Computing, pp. 530–535 (2006)
2. http://web.ist.utl.pt/~acardoso/datasets/
3. Cohen, W.W., Singer, Y.: Context-sensitive learning methods for text categorization. ACM Trans. on Information Systems, 307–315 (1996)
4. Forman, G., Guyon, I., Elisseeff, A.: An extensive empirical study of feature selection metrics for text classification. J. of Machine Learning Research 3, 1289–1305 (2003)
5. Gabrilovich, E., Markovitch, S.: Text categorization with many redundant features: Using aggressive feature selection to make SVMs competitive with C4.5. In: Proceeding of ICML (2004)
6. Joachims, T.: Learning to Classify Text using Support Vector Machines. Kluwer (2002)
7. Platt, J.: Fast training of support vector machines using sequential minimal optimization. In: Advances in Kernel Methods. MIT Press (1998)
8. Pietramala, A., Policicchio, V.L., Rullo, P., Sidhu, I.: A Genetic Algorithm for Text Classification Rule Induction. In: Daelemans, W., Goethals, B., Morik, K. (eds.) ECML PKDD 2008, Part II. LNCS (LNAI), vol. 5212, pp. 188–203. Springer, Heidelberg (2008)
9. Policicchio, V.L., Pietramala, A., Rullo, P.: GAMoN: Discovering MofN Hypotheses for Text Classification by a Lattice-based Genetic Algorithm. In: Artificial Intelligence. Elsevier (2012), doi:10.1016/j.artint.2012.07.003
10. Quinlan, J.R.: Generating production rules from decision trees. In: Proceedings of the IJCAI, vol. 1, pp. 304–307. Morgan Kaufmann Publishers Inc. (1987)
11. Rullo, P., Policicchio, L., Cumbo, C., Iiritano, S.: Olex: effective rule learning for text categorization. IEEE Transactions on Knowledge and Data Engineering (2009)
12. Sebastiani, F.: Machine learning in automated text categorization. ACM Comput. Surv. 34, 1–47 (2002)
13. Zheng, Z., Srihari, R.: Optimally combining positive and negative features for text categorization. In: Proceedings of the ICML (2003)

A Feature Selection Method for Improved Document Classification

Tanmay Basu and C.A. Murthy

Machine Intelligence Unit, Indian Statistical Institute, Kolkata, India
mailtanmaybasu@gmail.com, murthy@isical.ac.in

Abstract. The aim of text document classification is to automatically group a document to a predefined class. The main problem of document classification is high dimensionality and sparsity of the data matrix. A new feature selection technique using the google distance have been proposed in this article to effectively obtain a feature subset which improves the classification accuracy. Normalized google distance can automatically extract the meaning of terms from the world wide web. It utilizes the advantage of number of hits returned by the google search engine to compute the semantic relation between two terms. In the proposed approach, only the distance function of google distance is used to develop a relation between a feature and a class for document classification and it is independent of google search results. Every feature will generate a score based on their relation with all the classes and then all the features will be ranked accordingly. The experimental results are presented using knn classifier on several TREC and Reuter data sets. Precision, recall, f-measure and classification accuracy are used to analyze the results. The proposed method is compared with four other feature selection methods for document classification, document frequency thresholding, information gain, mutual information and χ^2 statistic. The empirical studies have shown that the proposed method have effectively done feature selection in most of the cases with either an improvement or no change of classification accuracy.

Keywords: Feature Selection, Document Classification.

1 Introduction

Document classification is the process of automatic grouping of documents to a predefined class. It is extremely difficult to effectively retrieve some particular information from the huge online resources without good indexing of document content [1]. Document classification has become one of the key tools for automatically handling and organizing such a huge document collections [4]. The huge dimensionality of the document collections is the main difficulty of any document classification method. So effectively reduce the dimension is a key part of document classification. The standard procedure for dimensionality reduction is feature selection. Feature selection is a process that chooses a subset from the original feature set according to some criteria. The selected feature subset

S. Zhou, S. Zhang, and G. Karypis (Eds.): ADMA 2012, LNAI 7713, pp. 296–305, 2012.

retains original physical meaning and provides a better understanding for the data and learning process [12]. The feature selection for document classification task use an evaluation function that is applied to a single term [5]. Then all the terms will be sorted according to the score of the evaluation function assigned independently to each feature. Among this sorted list a predefined number of features will be selected as the best feature subset. Among various feature selection methods, Document frequency (DF) thresholding, Information Gain (IG), Mutual Information (MI), χ^2 statistic (CHI) are commonly applied techniques in document classification [1]. Yang et. al. [1] investigated five feature selection methods. They reported that good feature selection methods improve the categorization accuracy with an aggressive feature removal using DF, IG and CHI methods.

A simple technique for vocabulary reduction in document classification [1] is *DF thresholding*. Document frequency refers to the number of documents in which a term occurs. The document frequency of each term in the training documents will be computed and the terms with a document frequency less than a predefined threshold will be discarded from the vocabulary. *Mutual information* based feature selection assumes that the term with higher class ratio is more effective for classification. On the other hand rare terms will have a higher score than common terms for those terms with equal conditional probability. Hence MI might perform badly when a classifier gives stress on common terms. For a training corpus those terms whose MI score is less than a predefined threshold will be removed from the vocabulary. It is to be mentioned here that the same choice is made for other two methods IG and CHI. *Information gain*, measures the number of bits of information obtained for class prediction by knowing the presence or absence of a term in a document [1]. It gives more weight to common terms rather than the rare terms. Hence IG might perform badly when there are scarcity of common terms between the documents of the training corpus. Supervised feature selection using χ^2 *statistic* measures the association between the term and the class. In our experiments we have used the maximum CHI score for comparison. Forman [11] shows a comparative study of twelve feature selection methods on 229 text classification problem instances and proposed a new method called *bi-normal separation*. Their experiments show that bi-normal separation can perform very well in the evaluation metrics of recall rate and f-measure. But for precision, it often loses to IG. Liu et. al. [12] proposed two new unsupervised feature selection methods. First one is *term contribution* which ranks the feature by its overall contribution to the documents similarity in a data set. Another is, *iterative feature selection*, which utilizes some successful feature selection methods, such as IG and CHI, to iteratively select features and perform text clustering at the same time.

Cilibrasi et. al. [3] developed normalized google distance to measure the semantic relation between two terms/phrases using google page counts. Google page count is the number of hits of a term displayed by google. The google distance determines how close a term x is to another term y on a zero to infinity scale. A distance zero indicates two terms are exactly same. The distance will be infinity when the terms never occur together. Here we shall use this distance to develop a relation between a term and a class for document classification. Then every feature will be assigned a score depending on their relation with all the classes. All the terms will be ranked according to their individual score. Then a predefined number of terms will be selected as very important features which have high weights. The idea is new in feature selection literature and no google search results are required for the proposed method.

We have applied the method on several text data sets. Knn classifier is used to judge the effectiveness of the proposed feature selection method and precision, recall, f-measure and classification accuracy are used for experimental analysis. The empirical studies show that the proposed method performs better than the other conventional feature selection methods even after 90% terms removal in most of the cases. The proposed method is an alternative approach of the existing feature selection methods and in various situations the proposed method is showing either an improvement or no change in classification performance, when compared to the other methods.

The paper is organized as follows. The proposed method is described in section 2. The experimental results are shown and discussed in section 3. Section 4 presents some conclusions on the proposed method.

2   Feature Selection by Term Relevance

Normalized google distance (NGD) was introduced by Cilibrasi et. al. [3] to utilize the vast knowledge available on the web by using the google page counts. NGD determines the semantic distance between two terms using the number of hits returned by Google search engine as search terms. The value of NGD lies between 0 and ∞. The NGD value 0 indicates that the terms are exactly same. If two terms never appear together then the value will be ∞. This weighting scheme of NGD will be used to derive a relation between a term and a class in the proposed feature selection method. In this method we shall measure the relevance of a term with a class. The class for which the term has the highest score will be the ultimate score of the term over all the classes. The relevance of a term t with a particular class c can be derived from the following distance function *term relevance* (TR).

$$TR(t,c) = \begin{cases} -1, & \text{if } f(t,c) = 0 \\ \frac{\max\{logf(t),logf(c)\}-logf(t,c)}{logN-\min\{logf(t),logf(c)\}}, & \text{otherwise} \end{cases} \tag{1}$$

Here f(t) is total number of documents containing term t and f(c) is total number of documents in class c. f(t, c) denotes the number of documents belonging to class c and containing the term t, and N is the total number of documents. To measure the relevance of a term in global feature space, all the class specific scores of a term are to be taken into consideration. The score of a term over all the m classes can be obtained in the following way:

$$TR_{max}(t) = \max_{i=1}^{m}\{TR(t, c_i)\}$$

All the features will be ranked in decreasing order according to their $TR_{max}(t)$ value and then a predefined number of terms will be selected for classification which have high weights. The following are the main properties of the proposed term relevance.

Properties of TR

- The value of TR will be 0, when $f(t) = f(c) = f(t, c)$, for any term t and a class c.

- From the right hand side of equation 1 we have

$$\max\{log f(t), log f(c)\} - log f(t, c) - log N + \min\{log f(t), log f(c)\}$$
$$= log f(t) + log f(c) - log f(t, c) - log N$$
$$= log \frac{f(t) * f(c)}{f(t, c) * N}$$

Now $f(t) * f(c)$ may be greater than $f(t, c) * N$. Hence the value of TR may be greater than 1 for any pair of class c and term t. But the value of TR will always be finite.

- $TR(t, c) \geq 0$, if a term t appears in the class c. The only negative value of TR is -1 when t does not belong to c. Hence TR is not a metric.

- TR is symmetric. For every pair of a term t and a class c, we have $TR(t, c) = TR(c, t)$.

3   Empirical Evaluation

This section describes the performance of various feature selection methods in document classification using several document collections. The description of the data sets and performance measures are given below. The effectiveness of different feature selection algorithms is evaluated using the performance of knn classifier. The knn classifier is chosen since it shows very good performance in a previous study by Yang et. al. [2]. 10-fold cross validation method has been used to fix the value of k from the training set. The range of k has been set from 1 to 20. For all the data sets except Reuter have no separate test and training sets. So 10 fold cross validation is performed on the entire data set to split it into training and test set. Knn has been executed for 10 times to reduce the effect of random selection of the documents by cross validation. The average results of this 10 executions is reported in Table 2.

Table 1. Data Sets Overview

Data Set	No. of Documents	No. of Terms	No. of Classes
la1	3204	31472	6
la2	3075	31472	6
la12	6279	31472	6
Reuter	13324	29944	90
tr21	336	7902	6
tr41	878	7454	10

3.1   Document Collections

Reuters-21578[1] is a collection of documents that appeared on Reuters newswire in 1987. The documents were originally assembled and indexed with categories by Carnegie Group, Inc. and Reuters, Ltd. The corpus contains 21578 documents in 135 categories. In this version of Reuter the documents with multiple class labels were discarded and the largest 90 categories were selected. The corpus was divided into 9583 training documents and 3741 test documents according to the Apte split [6].

The rest of the text data sets have been developed by Karypis and Han[2] [7]. Data sets tr21, tr41 are derived from TREC-5, TREC-6, and TREC-7 collections [8]. Data sets la1, la2 and la12 are from the Los Angeles Times data of TREC-5 [8]. The classes of the tr21 and tr41 data sets were generated from the relevance judgment provided in the TREC collections. The class labels of la1 and la2 were generated according to the name of the news-paper sections that these articles appeared, such as Entertainment, Financial, Foreign, Metro, National, and Sports. The documents that have a single label are selected for the la1, la2 and la12 data sets.

The number of documents, number of terms and number of classes of all the data sets can be found in Table 1. For the above data sets, the stop words have been extracted using the standard English stop word list[3]. Then, by applying the standard porter stemmer algorithm for stemming, the inverted index is developed. We have removed those words which occurred only once, twice or thrice in the corpus at first.

3.2   Performance Measures

In order to evaluate the effectiveness of class assignments, the standard recall, precision and f-measure are used. Precision and recall for a particular class are defined as:

$$\text{recall} = \frac{\text{number of documents found which are correct}}{\text{number of actual documents}}$$

$$\text{precision} = \frac{\text{number of documents found which are correct}}{\text{number of documents found}}$$

---

[1] http://www.daviddlewis.com/resources/testcollections/reuters21578/
[2] http://www-users.cs.umn.edu/~han/data/tmdata.tar.gz
[3] http://www.textfixer.com/resources/common-english-words.txt

The f-measure combines recall and precision with an equal weight in the following form:

$$\text{f-measure} = \frac{2 \times \text{recall} \times \text{precision}}{\text{recall} + \text{precision}}$$

These scores are computed for the binary decisions on each individual class and then be averaged over all classes. The closer the values of precision and recall, the higher is the f-measure. The value of f-measure lies between 0 and 1. Classification accuracy is also measured for performance evaluation.

Table 2. Comparison of Various Feature Selection Methods for Document Classification

Data	Vocabulary Size(%)	F-measure					Accuracy (%)				
		CHI	DF	IG	MI	TR	CHI	DF	IG	MI	TR
la1	10%	0.816	0.649	0.815	0.296	**0.817**	81.55	66.15	81.54	34.06	**82.51**
	20%	0.811	0.670	0.822	0.580	**0.830**	81.12	67.24	82.30	58.22	**83.67**
	30%	0.816	0.734	0.820	0.682	**0.834**	81.71	74.10	82.05	68.66	**83.77**
	50%	0.821	0.789	0.814	0.777	**0.836**	82.24	79.13	81.51	78.15	**83.94**
	100%	0.809	0.809	0.809	0.809	0.809	80.97	80.97	80.97	80.97	80.97
la2	10%	0.844	0.660	0.834	0.512	**0.854**	84.52	66.81	83.64	52.77	**85.68**
	20%	0.835	0.731	0.835	0.738	**0.864**	83.65	73.38	83.53	73.64	**86.85**
	30%	0.832	0.742	0.838	0.785	**0.858**	83.12	74.10	83.89	78.77	**86.16**
	50%	0.835	0.798	0.833	0.824	**0.859**	83.55	79.84	83.31	82.78	**86.09**
	100%	0.831	0.831	0.831	0.831	0.831	83.09	83.09	83.09	83.09	83.09
la12	10%	0.841	0.653	0.835	0.402	**0.847**	84.15	66.62	83.56	41.62	**84.83**
	20%	0.838	0.711	0.843	0.654	**0.859**	83.87	72.13	84.34	65.28	**86.22**
	30%	0.841	0.754	0.842	0.748	**0.859**	84.15	75.96	84.23	75.20	**86.27**
	50%	0.842	0.786	0.841	0.825	**0.860**	84.24	78.60	84.14	82.89	**86.35**
	100%	0.840	0.840	0.840	0.840	0.840	84.06	84.06	84.06	84.06	84.06
Reuter	10%	0.642	0.500	0.659	0.306	**0.660**	68.05	54.10	**69.09**	34.96	**69.09**
	20%	0.644	0.573	**0.656**	0.482	**0.656**	68.29	60.94	**68.72**	52.12	68.56
	30%	0.645	0.586	0.656	0.539	**0.659**	67.76	61.96	68.75	57.84	**69.17**
	50%	0.633	0.600	0.656	**0.672**	0.640	66.02	62.76	68.16	**69.60**	66.77
	100%	0.634	0.634	0.634	0.634	0.634	65.97	65.97	65.97	65.97	65.97
tr21	10%	0.873	0.781	**0.894**	0.817	0.881	88.12	79.78	**89.87**	83.28	88.27
	20%	0.875	0.812	0.886	0.859	**0.913**	88.42	84.20	90.29	87.18	**91.93**
	30%	0.876	0.831	0.877	0.866	**0.910**	88.81	84.43	89.35	88.54	**91.51**
	50%	0.867	0.851	0.862	0.866	**0.880**	88.59	85.74	88.85	86.91	**89.33**
	100%	0.857	0.857	0.857	0.857	0.857	87.80	87.80	87.80	87.80	87.80
tr41	10%	0.919	0.761	0.924	0.399	**0.931**	92.23	76.27	92.82	43.44	**93.46**
	20%	0.926	0.855	0.921	0.666	**0.927**	92.78	86.10	92.13	66.91	**92.93**
	30%	0.922	0.877	**0.924**	0.789	0.923	92.50	87.94	**92.76**	79.03	92.49
	50%	0.919	0.904	**0.925**	0.921	0.923	92.18	90.54	92.82	92.43	**92.71**
	100%	0.922	0.922	0.922	0.922	0.922	92.60	92.60	92.60	92.60	92.60

3.3   Analysis of Results

Table 2 shows the performance of various feature selection methods in document classification using f-measure and classification accuracy. The results are shown using several thresholds on the total number of unique terms i.e., the performance of the knn classifier is reported after removing 50%, 70%, 80%, and 90% unique terms. The vocabulary size in Table 2 indicates that the experiments are performed when there are 10%, 20%, 30%, 50% and 100% unique terms in the vocabulary. The proposed TR is compared with CHI, DF, IG and MI methods.

If significant amount of terms were removed from the vocabulary at high levels (after 90% terms removal) by TS then knn would not provide improved classification performance which becomes clear from the experimental results. But the experimental results show that TR is able to detect significant terms even after 90% or more terms removal. Let us take the example of la2 data set using TR thresholding when the total number of unique terms are reduced from 100% to 10%, the f-measure is improved from 0.831 to 0.854. The same thing can be observed in all the data sets for TR. The performance of CHI and IG are also not degraded after 90% or more terms removal in most of the data sets. The DF and MI methods are not able to provide better classification performance even after removal of more than 50% terms in most of the cases.

The proposed method is compared with other four methods and six data sets have been used for analysis. So there are 192 comparisons in Table 2 for the proposed method for 50% to 90% removal of terms. TR performed better than the other methods in 179 cases and in the rest 13 cases other methods (e.g., CHI, IG) have an edge over TR. Consider two such cases where the other methods have an edge over TR. For data set reuter, when there are 20% terms in vocabulary, IG (68.72%) has an edge over TR (68.56%) in classification accuracy and for data set tr41, when there are 30% terms in vocabulary, IG (0.924) has an edge over TR (0.923) in f-measure.

A statistical significance test is needed to check whether these differences are significant, e.g., whether 0.924 is significantly different from 0.923. A generalized version of paired *t-test* is used for testing the equality of means when the variances are unknown. This problem is the classical Behrens-Fisher problem in hypothesis testing and a suitable test statistic[4] is described and tabled in [9] and [10], respectively.

It has been found using t-test that out of those 179 cases where TR performed better than the other methods, in 164 cases the difference was statistically significant for the level of significance 0.05. Out of the rest 13 cases where other methods have an edge over TR, all the differences were statistically significant for the same level of significance. Thus from the experimental results it can be observed that the proposed TR will be very useful for document classification.

Figure 1 show the precision-recall curve of CHI, IG and TR for all the data sets. CHI and IG have comparable performances with TR for precision and

---

[4] The test statistic is of the form $t = \frac{\bar{x}_1 - \bar{x}_2}{\sqrt{s_1^2/n_1 + s_2^2/n_2}}$, where $\bar{x}_1, \bar{x}_2$ are the means, s_1, s_2 are the standard deviations and n_1, n_2 are the number of observations.

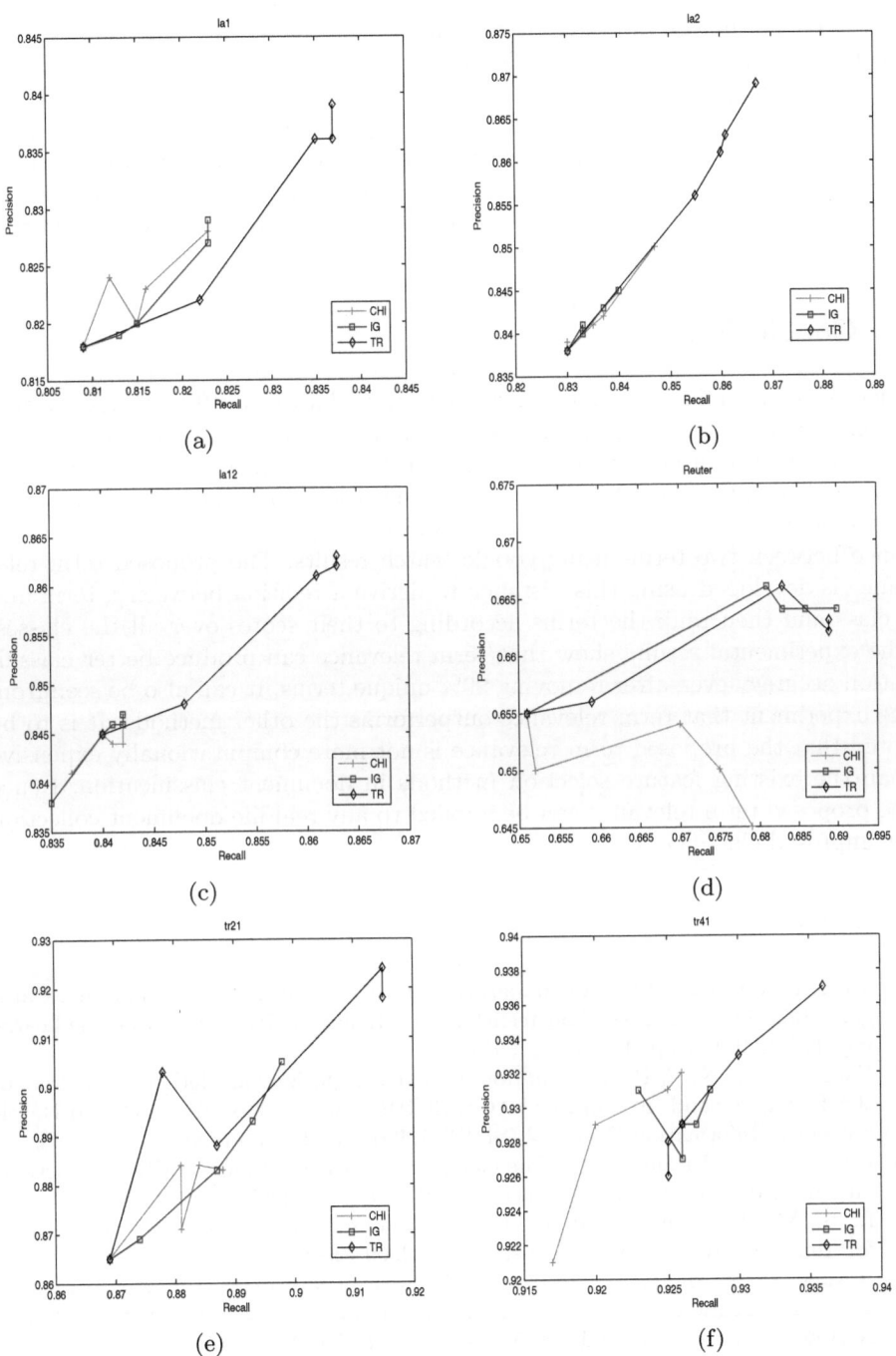

Fig. 1. Precision Recall Curve by Varying the Number of Terms for all the Data Sets

recall. DF never performed better than TR and MI performed better than TR only once. So that DF and MI are not included in the figures and the precision-recall curves are developed for CHI, IG and TR only. It is observed from figure 1(a), figure 1(b), figure 1(c) and figure 1(e) that TR is the best performer. TR shows better performance than CHI, but IG performs comparably better than TR in figure 1(d). In figure 1(f), for recall rate below 93% CHI and IG perform better than TR, but for rest of the cases TR performs better than CHI and IG. Thus the superiority of TR can be observed from the precision-recall curves of all the data sets.

4 Conclusions

Effectively managing the high dimensionality of data is a difficult task for document classification. An efficient feature selection method is needed to improve the performance of document classification. In this article a new feature selection method is proposed for document classification using the distance function of google distance. The google distance was developed to extract semantic distance between two terms using google search results. The proposed term relevance is developed using this distance to derive a relation between a term and a class and then rank the terms according to their scores over all the classes. The experimental results show that term relevance can produce better classification accuracy even after removing 90% unique terms. It can also be seen from the experiment that term relevance outperforms the other methods. It is to be noted that the proposed term relevance is not more computationally expensive than the existing feature selection methods in document classification. Hence the proposed term relevance can be applied to any real life document collection for improved classification.

References

1. Yang, Y., Pedersen, J.O.: A Comparative Study on Feature Selection in Text Categorization. In: Proc. of the Fourteenth International Conference on Machine Learning (ICML 1997), pp. 412–420 (1997)
2. Yang, Y., Liu, X.: A Re-examination of Text Categorization Methods. In: Proc. of the Twenty-Second International ACM SIGIR Conference on Research and Development in Information Retrieval (SIGIR 1999), pp. 42–49 (1999)
3. Cilibrasi, R.L., Vitanyi, P.M.: The Google Similarity Distance. IEEE Transactions on Knowledge and Data Engineering 19(3), 370–383 (2007)
4. Li, S., Xia, R., Zong, C., Huang, C.: A Framework of Feature Selection Methods for Text. In: Proceedings of ACL-IJCNLP 2009 (2009)
5. Novovicova, J., Malik, A.: Information-Theoretic Feature Selection Algorithms for Text Classification. In: Proceedings of International Joint Conference on Neural Networks, Montreal, Canada, July 31-August 4 (2005)
6. Joachims, T.: Text Categorization with Support Vector Machines: Learning with Many Relevant Features. In: Nédellec, C., Rouveirol, C. (eds.) ECML 1998. LNCS, vol. 1398, pp. 137–142. Springer, Heidelberg (1998)

7. Karypis, G., Han, E.H.: Centroid-Based Document Classification: Analysis and Experimental Results. In: Zighed, D.A., Komorowski, J., Żytkow, J.M. (eds.) PKDD 2000. LNCS (LNAI), vol. 1910, pp. 424–431. Springer, Heidelberg (2000)
8. TREC, Text REtrieval Conference, http://trec.nist.gov
9. Lehmann, E.L.: Testing of Statistical Hypotheses. John Wiley, New York (1976)
10. Rao, C.R., Mitra, S.K., Matthai, A., Ramamurthy, K.G. (eds.): Formulae and Tables for Statistical Work. Statistical Publishing Soc., Calcutta (1966)
11. Forman, G.: An Extensive Empirical Study of Feature Selection Metrics for Text Classification. The Journal of Machine Learning Research 3(1), 1289–1305 (2003)
12. Liu, T., Liu, S., Chen, Z., Ma, W.: An Evaluation on Feature Selection for Text Clustering. In: Proc. International Conference on Machine Learning (ICML 2003) (2003)

An Ensemble Approach to Multi-label Classification of Textual Data*

Karol Kurach, Krzysztof Pawłowski, Łukasz Romaszko, Marcin Tatjewski,
Andrzej Janusz, and Hung Son Nguyen

Faculty of Mathematics, Informatics and Mechanics, The University of Warsaw,
Banacha 2, 02-097, Warsaw Poland
(kkurach,kpawlowski236,lukasz.romaszko,marcin.tatjewski)@gmail.com,
andrzejanusz@gmail.com, son@mimuw.edu.pl

Abstract. In this paper, we investigate different approaches to multi-label classification of textual data, with a special focus on ensemble techniques. Commonly used classifier ensembles combine outputs of base learning models in order to enhance the learning results. The multi-label classification problem introduces some new challenges to the ensemble learning methods. For instance, one needs to decide in which order is it better to aggregate the base learners – on a level of individual labels and then for the whole label sets, or the other way around. We discuss this issue and experimentally compare selected approaches. In the experiments, we use data from JRS'2012 Data Mining Competition, whose scope was topical classification of biomedical research papers, and as the base learners we utilize the models employed by the winners of this contest.

Keywords: Data Mining, Topical Classification, Multi-label Classification, Ensemble Learning.

1 Introduction

We present a review of approaches towards developing multi-label classifiers, with a special interest in ensemble techniques. Classifier ensembles combine several different classifiers into a single one. In multi-label learning, each data instance is associated with a set of labels that are binary classification decisions for this instance. In order to obtain the best set of decision labels for a single object, common techniques employ a two-layer learning strategy. Firstly, they construct a set of binary classifiers and subsequently utilize another learning algorithm to compute optimal multi-label decision on the basis of previous binary decisions or scores assigned to individual labels [1,2,3]. Nevertheless, in this paper we show that a successful methodology may use even more than two layers. In this

---

* This research was supported by the National Centre for Research and Development (NCBiR) under grant SP/I/1/77065/10 by the strategic scientific research and experimental development program: "Interdisciplinary System for Interactive Scientific and Scientific-Technical Information".

S. Zhou, S. Zhang, and G. Karypis (Eds.): ADMA 2012, LNAI 7713, pp. 306–317, 2012.

context, the first-layer binary classifiers are commonly called base learners, while the higher-layer techniques for combining the lower-layer results are knows as meta-learning algorithms, such as voting, thresholding or stacking.

This paper is especially focused on presenting and analyzing an impact of meta-learning techniques on the multi-label classification. We investigate different compositions of learning layers, both in terms of number of layers and in terms of characteristics of methods used on each particular layer. In our scope of interest are also techniques for selecting the best parameters for the blending algorithms. In particular, we are trying to answer a question whether it is better to firstly combine predictions of different classifiers for individual labels and then use a meta-learning approach to select the final set of labels, or to use a meta-learning algorithm to discover dependencies between the labels in the first step, and then blend the results.

Our hereby analysis of ensemble methods is based on implementations that were submitted to the JRS'2012 Data Mining Competition [4]. In particular, we have reimplemented all the base learners and blending algorithms used by the winners of this contest [5]. In the experiments, we test our implementations on the contest data to show stability of this approach, as well as to check how different parameter settings and ensemble compositions influence the quality of the results.

The task for the participants of the JRS'2012 Data Mining Competition [6] was to devise algorithms capable of accurately predicting topics related to articles from PubMed Central [7], based on the association strengths of automatically generated document tags. Each document could be labelled with several topics and this number was not fixed, so the challenge could be regarded as a multi-label classification of textual data. The best results of the contest, including the result of several of the top teams, were achieved with ensemble methods.

Multi-label classification is one of the main issues in semantic indexing and semantic search systems, where the problem can be understood as assigning to each document a subset of concepts from a given thesaurus or ontology. This topic is the main interest of an ongoing project, called SONCA (Search based on ONtologies and Compound Analytics) [8], conducted at the Faculty of Mathematics, Informatics and Mechanics of the University of Warsaw. It is a part of a larger SYNAT project: "Interdisciplinary System for Interactive Scientific and Scientific-Technical Information" (http://www.synat.pl/). SONCA is a hybrid database framework application, wherein scientific articles are stored and processed in various forms. SONCA is expected to provide interfaces for intelligent algorithms identifying relations among various types of objects. It extends the typical functionality of scientific search engines by more accurate identification of relevant documents and more advanced synthesis of information. To achieve this, concurrent processing of documents needs to be coupled with an ability to produce collections of new objects using queries specific to analytic database technologies. This paper may be seen as a summary of our research on applications of the multi-label classification for semantic indexing and topical classification of textual data.

2 Multi-label Classification Problem

In this section we formalize the multi-label classification problem over a set of scientific documents D. We assume that each document $d \in D$ is represented by a vector

$$\mathcal{F}(d) = \langle f_1(d), \ldots, f_k(d) \rangle$$

in a feature space. In practice, $\mathcal{F}(d)$ can be the TFxIDF representation of the document d or it can be another semantic representation of d, see [4].

Let $Q = \{q_1, \cdots, q_n\}$ be a set of n labels describing documents from the set D. We can then specify a function $K : D \times Q \to \{0, 1\}$, such that $K(d, q) = 1 \Leftrightarrow q$ is a label of the document d. Let $K(d)$ be the set of labels assigned to d.

A solution to the multi-label classification task is to create a classification algorithm (or a classifier) $K' : D \times Q \to \{0, 1\}$, such that the values of $K'(d)$ depend on $\mathcal{F}(d)$ only. The function $K'(d)$ is required to be as close to $K(d)$ as possible. Usually, the closeness between two sets of labels is defined by one of the well known evaluation methods from literature, e.g. [9].

The prediction quality evaluation should be performed on a set of new unseen documents, which is denoted by D_{test}. Let us recall some of the existing evaluation methods.

1. Accuracy (defined in equation (1)): measures the classification quality, not distinguishing errors resulting from choosing too many labels from errors resulting from not choosing the label that should be chosen.

$$Accuracy(K, K', D_{test}) = \frac{1}{|D_{test}|} \sum_{d \in D_{test}} \frac{|K(d) \cap K'(d)|}{|K(d) \cup K'(d)|} \tag{1}$$

2. Precision, Recall and F-score: these measures are very popular in the information retrieval theory. Precision is a ratio of correct decisions made by a classifier to all the labels that have been chosen. Recall is a ratio of correct decisions made by a classifier to all the labels that were truly assigned to the document.

$$P(K, K', d) = \frac{|K(d) \cap K'(d)|}{|K'(d)|} \quad \text{and} \quad R(K, K', d) = \frac{|K(d) \cap K'(d)|}{|K(d)|} \tag{2}$$

F-score is a popular multi-label classification quality measure which deals with imbalanced label representation problem. F-score for a single document is defined as a harmonic mean of Precision and Recall:

$$F\text{-}score(K, K', d) = \frac{2}{\frac{1}{P} + \frac{1}{R}} = \frac{2 \cdot P(K, K', d) \cdot R(K, K', d)}{P(K, K', d) + R(K, K', d)} \tag{3}$$

3. Average Precision, Recall and F-score: the analogous evaluation measures to P and R, but defined for the whole test data set. These measures are the arithmetic means of measures calculated for single documents (equations: (4) and (5)).

$$Precision(K, K', D_{test}) = \frac{1}{|D_{test}|} \sum_{d \in D_{test}} P(K, K', d) \tag{4}$$

$$Recall(K, K', D_{test}) = \frac{1}{|D_{test}|} \sum_{d \in D_{test}} R(K, K', d) \tag{5}$$

In equation (6) F-score is defined as an arithmetic mean of such variables calculated for each of the documents.

$$F\text{-}score(K, K', D_{test}) = \frac{1}{|D_{test}|} \sum_{d \in D_{test}} \frac{2 \cdot P(K, K', d) R(K, K', d)}{P(K, K', d) + R(K, K', d)} \tag{6}$$

Multi-label classification is one of the most important issues in semantic indexing and text categorization systems. In the SONCA project [8], the labels are defined by concepts from a given ontology, and the multi-label classification problem can be understood as a task of assigning a set of adequate concepts to documents from a given collection of texts. Multi-label learning has also proven to be a successful method for solving classification problems in other domains, ranging from scene, image and video annotation [10] to multiple applications in bioinformatics, e.g. functional genomics [11,12].

There has been a voluminous work done on the multi-label classification problem. Classification models adapted to this specific task include decision trees, neural networks, nearest neighbor methods, the support vector machine, linear least squares optimization, Naive Bayes and more. McCallum [13] defines a Bayesian approach for the document classification, in which the multiple classes that comprise a document label set are represented by a mixture model. The parameters of the model are learned by the maximum a posteriori estimation from labeled training documents using Expectation Maximization. Given a new document, the most likely label set is selected with the Bayes rule.

Prof. Zhi-Hua Zhuo developed a multi-label lazy learning method named ML-kNN [14], which is derived from the traditional k-Nearest Neighbor (kNN) algorithm enhanced by a Naive Bayes classifier. In this approach, for each unseen instance, its k nearest neighbors in the training set are firstly identified. After that, based on statistical information gained from the label sets of these neighbor instances, i.e. the number of neighbor instances belonging to each possible class, maximum a posteriori (MAP) principle is utilized to determine the label set for the unseen instance. ML-kNN may achieve performance superior to some well-established multi-label learning algorithms, which was showed by empirical experiments on three different real-world multi-label learning problems, i.e. yeast gene functional analysis, natural scene classification and automatic web page categorization [14].

3 Ensemble Approach to Multi-label Classification

Blending, or in other words ensemble construction, is a machine learning technique in which multiple base predictors are combined into a single classification model [1,15]. Ensembles are known to be able to produce better results than any of the individual base models [16]. This approach has been successfully used in many data mining competitions, including the Netflix Prize [17] and JRS'2012 [18].

This section describes several approaches to multi-label classification using ensemble algorithms. We present the solution of the winners of JRS'2012 Data Mining Competition, which contains different types of ensembles - both on a level of finding relations between labels and improving base classifiers. Inspired by their methods, we explore more deeply neural-network based blendings.

3.1   JRS'2012 Data Mining Competition Winner's Solution

Team ULjubljana, the winners of JRS'2012 Data Mining Competition[5], used neural-network methods to find relation between labels and to combine outputs from several base learners into prediction for one label. They provided a detailed description of their approach which allowed us to re-implement their methods. In this way, we were able not only to confirm its top quality, but we also gained a better insight into how it works. We present their solution in this section.

Base Learners. The Team ULjubljana's winning solution consists of three main stages.

1. 14 base learners are used to generate predictions for each of 83 labels.
2. For each label, 14 base predictions are combined using the stacking(3.1) method into one meta-prediction.
3. A thresholding(3.1) technique, which takes 83 meta-predictions for every row, is applied to choose the final set of labels for the row.

Every base learner takes 25640 attributes and generates predictions for each of 83 labels. A prediction is a real number between 0 and 1, where value close to 1 indicates a high likelihood of having this label. There are 5 types of base learners that are used with different parameters, making a total of 14 distinct base learners.

The only parameter to each base learner type except Random Forest is λ. It controls the regularization. All logistic regression and neural network algorithms are optimized with L-BFGS [19]. We describe each base learner type below.

- Logistic Regression (lr) - run once for each of 83 labels. Four instances with different λ parameter are used.
- F-score logistic regression (flr) - modification of logistic regression which attempts to predict all the 83 labels in a single run, optimizing the F-score directly. Four instances with different λ parameter are used.
- Log F-score logistic regression (flr_log) - modification of the flr model, where cost function is changed by applying the logarithm. Three model instances are used.

– Random forest (rf) - one instance, trained only on attributes having at least 50 non-zero entries in the training set.
– Logistic Regression with Neural Network (lr_nn) - model described in more details in 3.1.

For evaluation of base learner's results, thresholding (3.1) technique is used with threshold value equal to 0.25. However, the stacking algorithm 3.1 uses raw, continuous predictions of the base learners as an input.

Logistic Regression with Neural Network (lr_nn). This model is slightly different than all other base learners, since it is an ensemble already. The drawback of Logistic Regression learner was the assumption of independence of attributes, whereas real-world values are not completely independent. To overcome this obstacle, stacking algorithm is used. Neural network is trained on 83 inputs that are equal to the predictions generated by linear regression instances. There are 100 hidden layer neurons and 83 outputs. Each output is a real value between 0 and 1 that corresponds to the likelihood that given label shall be included in the answer for current row.

Logistic regression with neural network achieves substantially better results than any other base learner algorithm.

A Stacking Technique. Stacking is a type of ensemble learning, in which predictions from base learners are used as input for one meta-learner. Algorithm 1 shows a pseudo code for this method.

In the winner's solution, neural network with 14 input units, 20 hidden units and one output unit is used as the meta-learner algorithm. Ensemble model is trained for each of 83 labels using five cross-validation folds.

Assigning Final Labels by Thresholding. The output of ensemble method described above are 83 real numbers for every row. In order to predict final answer for a given row, we need to choose subset of labels. Simple method which takes all labels with score greater than 0.25 would be enough to score 0.53378 and take 2nd place in the competition. To improve this result, Team ULjubljana adjusts thresholds for every label separately using a greedy algorithm. In the first step, each of the 83 thresholds is initialized to 0.25. Then, for each label, a threshold is chosen to maximize the average F-score on training set, assuming that all other thresholds are fixed.

3.2   Label/Learner Ensembles

Inspired by the Team ULjubljana's methods, we explore more deeply neural-network based blends on the data set from the JRS competition. We reimplemented all their methods in order to conduct the experiments with blending. We use a subset of those base learners (3.1) for our ensembles.

Input:
Let a = number of attributes (25640 in this task)
Let k = number of algorithms used in blending (14 in this task)
Let m = number of rows in training set (10000 in this task)

Data set $D = \{(x_1, y_1), \ldots, (x_m, y_m)\}$, $x_i \in \mathbb{R}^a$, $y_i \in \{0, 1\}$
Base learners: $\{L_1, L_2, \ldots, L_k\}$
L_i = function(x, y) $(x \in \mathbb{R}^a, y \in \{0, 1\})$ which returns model (function) $\mathbb{R}^a \to \mathbb{R}$
Meta learner: $Meta$: function(x, y) $(x \in \mathbb{R}^k, y \in \mathbb{R})$ which returns model $\mathbb{R}^k \to \mathbb{R}$

Output: Ensemble model (function $\mathbb{R}^k \to \mathbb{R}$)

```
 1 for f ← 1 to folds do
 2       // Every train set consist of m * (1 − 1/folds) elements.
         D^f_train ← {D_i ∈ D | i ≢ f (mod folds)};
 3       for i ← 1 to k do
 4            baseModel^f_i ← L_i(D^f_train.x, D^f_train.y);
 5       end
 6 end
 7 meta_input ← ∅;
 8 for i ← 1 to m do
 9       for j ← 1 to k do
10            // baseModel^i mod folds was NOT trained on i-th row (D.x_i)
11            // so it can be used to predict the result of i-th row. This predicted
12            // value will be one of the inputs to the blending algorithm (Meta).
13            pred_j = baseModel^i mod folds_j (D.x_i);
14       end
15       meta_input ← meta_input.append((pred_1, …, pred_k), D.y_i);
16 end
17 ensembleModel ← Meta(meta_input.pred, meta_input.y);
18 return ensembleModel;
```

Algorithm 1. Stacking for one label

Label Ensemble. As a *label ensemble* we define type of blending that takes as input all 83 label predictions generated by a single base learner, possibly to find relations between them (e.g. some pairs of labels cannot be present at the same time). This can be done by using neural network with number of inputs and outputs equal to number of labels. Various base learners are tried as neural network's inputs.

Learner Ensemble. As a *learner ensemble* we define type of blending which takes as input many base learner predictions for one label, and combines them into one value for this label. For example, this can be a neural network with one input per every base learner and one output.

Label Ensemble Followed by Learner Ensemble. We can chain two ensembles together to achieve even better results. First, we apply label ensemble to base models, and then combine results for every label using learner ensemble. That is the type of blending which Team ULjubljana used in their final winning solution to JRS competition.

Learner Ensemble Followed by Label Ensemble. We reverse the order of blending from the previous algorithm. First, for every label we blend output from base learners. It gives us 83 values, on which we can run label ensemble to find relations between labels.

4   Experimental Evaluation

In this section we describe the data and the evaluation method used in our experiments, as well as present the results of different approaches to construction of multi-label classifier ensembles.

4.1   The Data

The data set used for evaluation was taken from JRS'2012 Data Mining Competition [4]. It consists of 20,000 rows, 25,640 conditional attributes and 83 target attributes (labels). The conditional attributes are real integers in $0-1000$ range. Only about 0.4% of them are non-zero, making the data very sparse. The target attributes are either 0 or 1. The names of attributes and labels were anonymized to promote domain-independent solutions.

The original source of data is PubMed Central – a full-text archive of biomedical and life sciences journal literature at the US National Institutes of Health's National Library of Medicine (NIH/NLM). Medical Subject Headings (MeSH) is a controlled vocabulary thesaurus created and maintained by NIH/NLM. It defines over 26,000 descriptors (headings) and 83 sub-headings that are used for indexing the journal articles.

Each row in the data set describes one article from PubMed Central open-access subset. The conditional attributes correspond to the MeSH headings. They were generated by our tagging algorithm and their values represent a strength of association between an article and medical terms. The target labels are set to 1 if the corresponding MeSH subheading was assigned by the experts and 0 otherwise – they can be interpreted as presence of particular context or topic in the article. The quality of the classifiers was evaluated using the average F-score measure [4].

In the described experiment we train the classifiers on the first 10,000 rows and evaluate their performance on the remaining 10,000 rows. Although the same train/test split was used during the competition, the presented result may slightly differ because we do not make a distinction between the documents from the preliminary and final test sets [4].

4.2 Results of the Evaluation

The quality estimates for described base learners and ensemble methods can be found in Tables 1 and 2. Additionally, Figures 1 and 2 visualize these results. In the plots, *best.base* learner denotes the result of the best base learner, *best.label* is the result of the best label-ensemble, *learner* is the result of the learner-ensemble classifier and *label.learner* corresponds to the result of the label-ensemble followed by the learner-ensemble. Finally, *learner.label* is the result of the learner-ensemble applied first and followed with label-ensemble.

All the presented ensemble methods improved the quality of predictions over the base learners. The difference between the best base model and the weakest ensemble (best.label) is very similar to the difference between the weakest and the strongest ensemble (learner.label). In other words, the best ensemble helps twice as much.

Table 1. Performance of: best base learner; best label-ensembled base learner; label-ensemble followed by learner-ensemble method; learner-ensemble followed by label-ensemble method

learner	F-score
1 best.base	0.5167
2 best.label	0.5225
3 learner	0.5239
4 label.learner	0.5256
5 learner.label	0.5289

Table 2. Performance of the base learners and the corresponding label-ensembles

learner	lambda	base	ensemble
1 flr	0.1	0.5159	0.5143
2 flr	0.2	0.5150	0.5158
3 flr	0.3	0.5132	0.5180
4 flr	0.4	0.5093	0.5164
5 flr.log	0.4	0.5167	0.5202
6 flr.log	0.5	0.5165	0.5201
7 flr.log	0.6	0.5146	0.5196
8 lr	2.0	0.5061	0.5185
9 lr	3.0	0.5096	0.5214
10 lr	4.0	0.5112	0.5216
11 lr	5.0	0.5117	0.5225

It is worth noting that two ensembles chained together always achieved better results than any single ensemble. This observation was not obvious, since the more blending layers we have, the less data can be used for training the lowest layer.

We were also interested in finding out what is the best order for combining ensembles of multi-label classification models. It seems that chaining ensembles

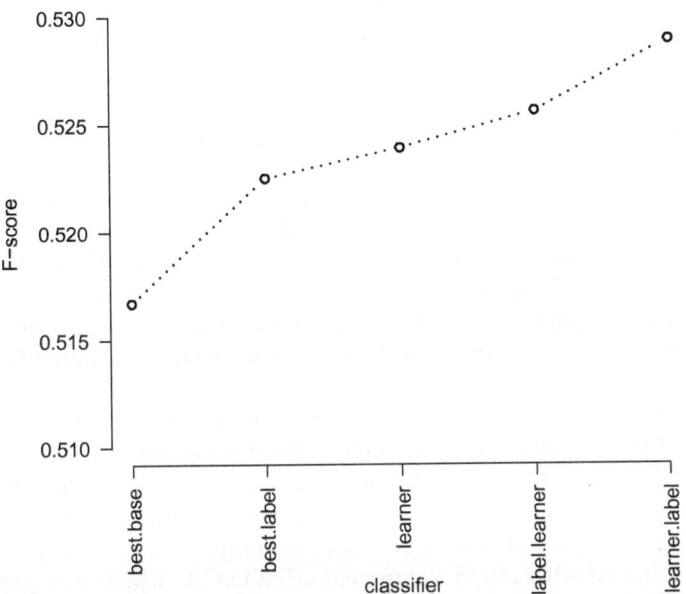

Fig. 1. The F-scores of the five approaches to construction of multi-label classifier ensembles

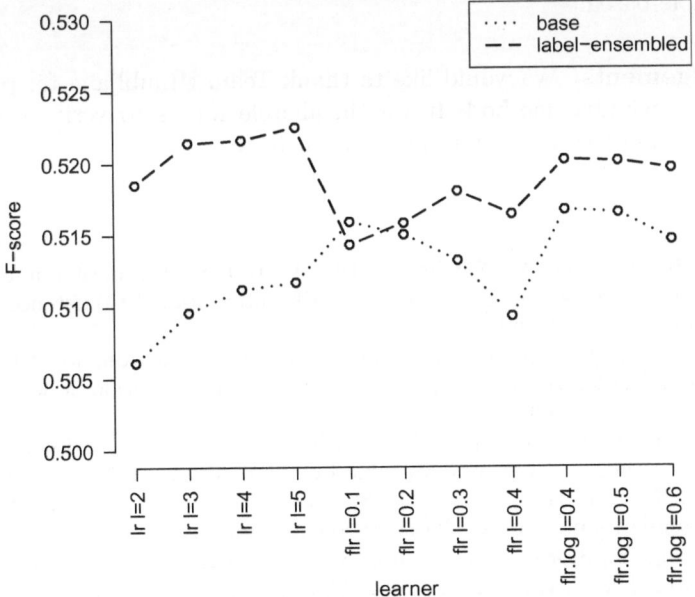

Fig. 2. Scores of the base learners in comparison to the scores of their learner-ensemble versions. l denotes a value of the regularization parameter λ.

in order: learner-ensemble, then label-ensemble outperforms the opposite order. We conclude that ensembles and chained ensembles are a promising framework for solving multi-label classification problems.

5   Conclusions and Plans for the Future

In this paper we presented a summary of our research on ensemble approaches to multi-label classification of textual data. In particular, we discuss our experiments that measured and compared the effectiveness of different blending techniques described in [18]. One of the outcomes of our work is the observation that higher classifier performance can be obtained when learning the relations between labels is conducted after the base learners results on individual labels are merged.

The origin of our research was the JRS'2012 Data Mining Competition. Our team organized this event in order to review the state-of-the-art methods in the domain of multi-label classification. We wanted to perform such a survey in order to implement the best classifiers in the SONCA system [8]. As we now analyzed, re-implemented or improved several solutions submitted to the competition, we are ready to advance with our development of SONCA, which is a part of our contribution to the SYNAT project.

To further improve our classification methods, we are going to investigate different approaches to textual document representation [6]. We would like to test their impact on results and behaviour of different learning algorithms, especially those ensemble-based.

Acknowledgements. We would like to thank Team ULjubljana for providing code for their winning method. It was invaluable for us to verify the result, deeply understand the solution and re-write it in R.

References

1. Caruana, R., Munson, A., Niculescu-Mizil, A.: Getting the most out of ensemble selection. In: Proceedings of the 6th IEEE International Conference on Data Mining, pp. 828–833 (2006)
2. Dieterich, T.G.: An experimental comparison of three methods for constructing ensembles of decision trees: Bagging, boosting, and randomization. Machine Learning 40(2), 139–157 (2000)
3. Bauer, E., Kohavi, R.: An empirical comparison of voting classification algorithms: Bagging, boosting, and variants. Machine Learning 36(1-2), 105–139 (1999)
4. Janusz, A., Nguyen, H.S., Ślęzak, D., Stawicki, S., Krasuski, A.: JRS'2012 Data Mining Competition: Topical Classification of Biomedical Research Papers. In: Yao, J., Yang, Y., Słowiński, R., Greco, S., Li, H., Mitra, S., Polkowski, L. (eds.) RSCTC 2012. LNCS, vol. 7413, pp. 422–431. Springer, Heidelberg (2012)
5. Žbontar, J., Žitnik, M., Zidar, M., Majcen, G., Potočnik, M., Zupan, B.: Team ULjubljana's Solution to the JRS 2012 Data Mining Competition. In: Yao, J., Yang, Y., Słowiński, R., Greco, S., Li, H., Mitra, S., Polkowski, L. (eds.) RSCTC 2012. LNCS, vol. 7413, pp. 471–478. Springer, Heidelberg (2012)

6. Janusz, A., Świeboda, W., Krasuski, A., Nguyen, H.S.: Interactive document indexing method based on explicit semantic analysis. In: Yao, J., Yang, Y., Słowiński, R., Greco, S., Li, H., Mitra, S., Polkowski, L. (eds.) RSCTC 2012. LNCS, vol. 7413, pp. 156–165. Springer, Heidelberg (2012)
7. Beck, J., Sequeira, E.: PubMed Central (PMC): An archive for literature from life sciences journals. In: McEntyre, J., Ostell, J. (eds.) The NCBI Handbook. National Center for Biotechnology Information, Bethesda (2003)
8. Bembenik, R., Skonieczny, L., Rybiński, H., Niezgódka, M.: Intelligent Tools for Building a Scientific Information Platform, vol. 390. Springer-Verlag New York Inc. (2012)
9. Tsoumakas, G., Katakis, I.: Multi-label classification: An overview. IJDWM 3(3), 1–13 (2007)
10. Zhou, Z., Zhang, M.: Multi-instance multi-label learning with application to scene classification. In: Advances in Neural Information Processing Systems 19, p. 1609 (2007)
11. Barutcuoglu, Z., Schapire, R.E., Troyanskaya, O.G.: Hierarchical multi-label prediction of gene function. Bioinformatics 22(7), 830–836 (2006)
12. Zhou, Z., Zhang, M., Huang, S., Li, Y.: Multi-instance multi-label learning. Artificial Intelligence 176(1), 2291–2320 (2012)
13. McCallum, A.: Multi-label text classification with a mixture model trained by em. In: Proceedings of AAAI 1999 Workshop on Text Learning (1999)
14. Zhang, M.L., Zhou, Z.H.: Ml-knn: A lazy learning approach to multi-label learning. Pattern Recognition 40(7), 2038–2048 (2007)
15. Caruana, R., Niculescu-Mizil, A., Crew, G., Ksikes, A.: Ensemble selection from libraries of models. In: Proceedings of the 21st International Conference on Machine Learning, pp. 137–144. ACM Press (2004)
16. Janusz, A.: Combining Multiple Classification or Regression Models Using Genetic Algorithms. In: Szczuka, M., Kryszkiewicz, M., Ramanna, S., Jensen, R., Hu, Q. (eds.) RSCTC 2010. LNCS, vol. 6086, pp. 130–137. Springer, Heidelberg (2010)
17. Bennett, J., Lanning, S.: The netflix prize. In: KDD Cup and Workshop in Conjunction with KDD (2007)
18. Kurach, K., Pawłowski, K., Romaszko, Ł., Tatjewski, M., Janusz, A., Nguyen, H.S.: Multi-label classification of biomedical articles. In: Bembenik, R., Skonieczny, Ł., Rybiński, H., Kryszkiewicz, M., Niezgódka, M. (eds.) Intelligent Tools for Building a Scientific Information Platform: Advanced Architectures and Solutions. Springer (2012)
19. Byrd, R.H., Lu, P., Nocedal, J., Zhu, C.Y.: A limited memory algorithm for bound constrained optimization. SIAM Journal on Scientific Computing 16(6), 1190–1208 (1995)

Hierarchical Text Classification for News Articles Based-on Named Entities

Yaocheng Gui[1,2], Zhiqiang Gao[1,2], Renyong Li[3], and Xin Yang[3]

[1] Institute of Web Science, School of Computer Science and Engineering,
Southeast University, Nanjing 211189, P.R. China
[2] Key Laboratory of Computer Network and Information Integration
(Southeast University), Ministry of Education,
Nanjing 211189, P.R. China
{yaochgui,zqgao}@seu.edu.cn
http://iws.seu.edu.cn
[3] Focus Technology Co., Ltd., Nanjing 210061, P.R. China
{lirenyong,yangxin}@made-in-china.com

Abstract. There exist a range of hierarchical text classification approaches that classify text documents into a pre-constructed hierarchy of categories. In these approaches, feature selections are often based on terms (words or phrases), which are unsuitable for hierarchically classifying news articles. Named entities are informative features in news articles which have not been studied seriously in previous hierarchical text classification approaches. This paper utilizes named entities as features for classifying news articles into a pre-constructed hierarchy about international relations. The feature selection is implemented based on named entities associated with local categories. Documents are then represented by the selected features using two types of models, which are Boolean model and Vector model. We train SVMs corresponding to both types of models based-on local information. The experimental results show that the use of named entities improves the performance of hierarchical text classification for news articles.

Keywords: Hierarchical Text Classification, Feature Selection, Named Entity, Support Vector Machine.

1 Introduction

There are millions of news articles about international relations available on the Web. Applications with the capability of analyzing or understanding news articles about international relations are developed for the purpose of helping people to manage the information. Hierarchical classification is at the heart of such applications, which categorized documents into a pre-constructed hierarchy – typically a tree or a direct acyclic graph (DAG). This paper aims at classifying Chinese news articles about international relations into a hierarchy with hundreds of categories, named international relation taxonomy (IRT). The IRT

S. Zhou, S. Zhang, and G. Karypis (Eds.): ADMA 2012, LNAI 7713, pp. 318–329, 2012.

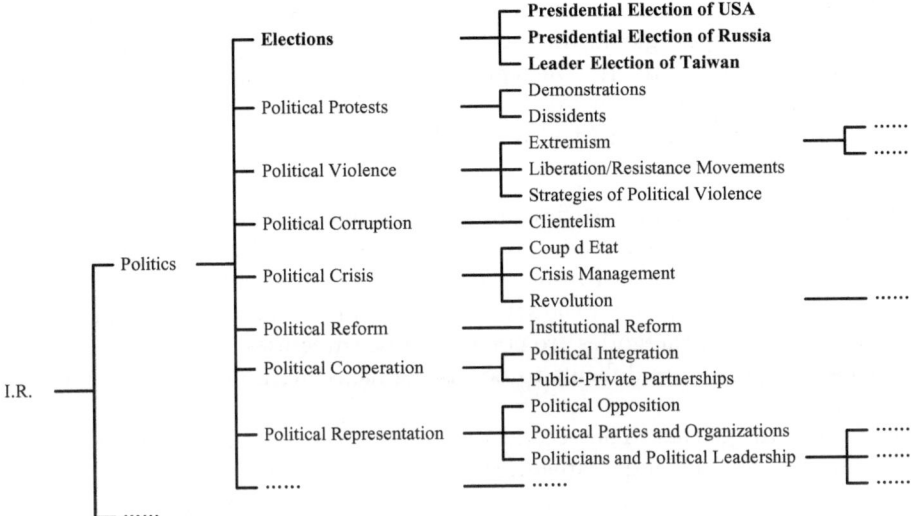

Fig. 1. Part of the International Relation Taxonomy

is constructed with a few specified categories which do not exist in other general taxonomies. Figure 1 illustrates a part of the IRT hierarchy.

The IRT is a hierarchy provided by experts of international relation domain, containing 321 categories distributed in 6 levels. A set of 5,758 Chinese news articles, which are crawled from news portals such as Sina[1] and Sohu[2], are classified into these categories manually. All the ambiguous documents are voted by three experts. Thus, 260 of these categories are related to at least one document and each category consists of 22 documents in average. Figure 2 illustrates a paragraph of a news article from the category *Presidential Election of USA* that consists of a few tagged named entities.

中新网（China News）2月29日电 据外电报道，美国（USA）密歇根州（Michigan）与亚利桑那州（Arizona）当地时间28日同时举行共和党（Republican Party）总统初选，共和党（Republican Party）总统候选人罗姆尼（Romney）已经稳赢亚利桑那州（Arizona）初选，得票率遥遥领先主要对手桑托勒姆（Santorum）；罗姆尼（Romney）还同时有望赢得密歇根州（Michigan）的初选，7成计票结果显示，罗姆尼（Romney）得票率领先桑托勒姆（Santorum）4个百分点。外电普遍称，罗姆尼（Romney）连赢两州初选，再次夺得领先优势。

Fig. 2. A paragraph of a news article from the category *Presidential Election of USA* that consists of a few tagged named entities. Three types of named entities are tagged, which are person names in blue, organization names in red, and location names in yellow. All these named entities are interpreted into English.

---

[1] http://www.sina.com.cn/

[2] http://www.sohu.com/

Previous approaches to solve the hierarchical classification problem can be divided in two types: global (or big-bang) approaches and local (or top-down) approaches according to [1]. The features in both types of approaches are often based-on terms, which are words and phrases. There exist two major difficulties when implementing hierarchical text classification approaches based-on terms on the IRT hierarchy:

- **Close Categories.** The first difficulty is the high similarity of categories with the same parent node in the IRT hierarchy. For example, the documents from categories *Presidential Election of USA* and *Presidential Election of Russia* talk about the same topic, i.e. presidential election, in different countries. Such categories are noted as close categories. The documents from close categories are difficult to be distinguished with features selected from terms [2].
- **Rare Categories.** The second difficulty is the sparsity of document distribution on leaf node categories. If we only consider documents assigned by human editors directly to those 260 categories (without counting in the documents assigned to their child categories), most categories have very few labeled documents. For instance, over 50% of the categories in the IRT hierarchy have fewer than 10 labeled documents. Such types of categories also exist in the Yahoo! categories, which are denoted as rare categories [1]. For hierarchical text classification with rare categories, there are not enough training examples for most learning algorithms to train an effective classifier in a high dimensionality feature space.

Named entity recognition (NER) is a subtask of information extraction that seeks to locate and classify atomic elements in text into predefined categories such as the names of persons, organizations, locations, expressions of times, quantities, monetary values, percentages [3]. For example, three types of named entities are illustrated in Figure 2. Through a pre-process of named entity recognition on the set of 5,758 Chinese news articles, we found that 99.4% of these articles contain named entities. Generally, each article contains a few number of named entities, 14.7 per article in average.

Intuitively, a few informative named entities, such as the candidate's name or the location name, are often sufficient to distinguish news articles about presidential election of USA from those of Russia. For example, supposing that we are trying to classify a news article into the two forementioned categories, if we have known that it is about presidential election, then the appearance of a special named entity, e.g. *Putin* or *Mitt Romney*, may directly impact the decision of classification. Given such types of named entities, other sequences of words add little differentiation power, and are therefore redundant. This type of categorization problem is characterized with redundant features [2]. It is claimed that a particular concept can be learned with a small number of features, while the rest of the features do more harm than good in close categories [2].

According to this intuition, we assume that named entities are distinguishable features for news articles about close categories. Because the named entities distribute in different categories of the IRT hierarchy, we implement feature selection method based on local collections of named entities with respect to each non-leaf categories. Furthermore, we represent text document using Boolean model as well as Vector model.

In rare categories, the dimension of the feature space consists of terms can be high for the number of training examples. The use of named entity as feature may help to reduce the dimension of the feature space. However, it can be further reduced by some aggressive feature selection strategies, e.g. select only 500 features for each category. We assume that the hierarchical text classification about rare categories will benefit from the reduced feature space brought by feature selection methods. Because it decreases the requirements of training examples .

SVM is suggested to be very robust in the presence of numerous features, according to a wide range of studies about hierarchial text classification which implement SVM as the basic classification model [1, 2, 4, 5]. We trained SVM for classification and regression according to the local training examples with respect to the category of the IRT hierarchy. Some previous work showed that the accuracy of SVM in hierarchical text classification can also be improved by local feature selection [1].

This paper aims at providing a hierarchical text classification approach for news articles by using named entities as features, implementing local feature selection with an aggressive strategies, training SVM with local information, and predicating classes with top-down approach.

The rest of the paper is organized as follows. In Section 2, we give a brief overview of related works. In Section 3, we describe the models and strategies of proposed approaches. The experimental results are shown in Section 4. Section 5 concludes with a summary and plans for future work.

2   Related Works

In hierarchical text classification, many algorithms have been proposed. According to how the hierarchical structure is explored [6], the current literature often refers to local approaches, when the system construct classifiers on each level of the hierarchy where each classifier works as a flat classifier on that level; and global approaches, when a single classifier coping with the entire class hierarchy is used. It is infeasible to directly build a classifier for a large-scale hierarchy [1]. We review the local classifier approaches and the related work on using named entity in classification in this section.

The first study of the hierarchical text categorization problem is carried out in [7]. It proposed a local approach with divide-and-conquer principle, the system first classified a document into high-level categories and then proceeded iteratively dealing only with the children of the categories selected on the previous level. This study experimentally showed that hierarchical information can be extremely benefit for text categorization by improving the classification performance over the "flat" technique, which does not use the hierarchical information.

The hierarchical SVM is first used in [4, 8] to deal with the hierarchical text classification problem. In order to assign a test example to a class, the probabilities for the predicted class were used. The first method uses a boolean condition where the posterior probability of the classes on the first and second levels must be higher than a user specified threshold. The second method uses a multiplicative threshold that takes into account the product of the posterior probability of the classes on the first and second levels.

The evaluation of SVM over the full taxonomy of the Yahoo! categories, is reported in [1]. The scalability and effectiveness of SVM and two threshold tuning methods were theoretically analyzed. A distributed classification system was developed for experiments. In implementation of SVM, the sequential minimal optimization (SMO) algorithm with a linear kernel was used for the binary version of the SVM classifier and 4,000 features were selected for each binary classification task. It is claimed that the hierarchical use of SVM is efficient enough for very large-scale classification and the effectiveness is still far from satisfactory.

Recent research has focused on two issues: increasing sparsity of training data at deeper nodes in the taxonomy, and error propagation where a mistake made high in the hierarchy cannot be recovered. The approach for classifying large scale text hierarchy at deep-nodes introduced in [5] consists of a search stage and a classification stage. In the search stage, a subset of categories from the large scale hierarchy is formed related to a given document by a category-search algorithm. In the classification stage, a specific classifier is trained for each given document. This approach prunes the large-scale hierarchy to a small subset of flat categories, which decomposes the difficult global problem into simpler local ones. Other studies suggested techniques for the error propagation issue, such as the isotonic smoothing approach provided by [9] and the Refined Experts model provided by [10].

There are a range of hierarchical text classification approaches that using terms (words or phrases) as features. A feature selection method that both consider general terms and named entities is proposed in [11]. The linear-chain CRFs is used to train named entity recognition models based on named entities extracted from the corpus. Then, a popular feature selection method – information gain (IG) – is implemented to select Chinese phrases as terms. The experimental results showed that the general feature selection method combining with named entity information performed better than traditional feature selection methods. However, it is implemented for flat text classification instead of hierarchical text classification.

The hierarchical text classification approach proposed in this paper uses named entity as feature and information gain as feature selection method according to [11]. However we consider only named entities as features instead of term features and implement information gain based on local categories of the IRT hierarchy. In addition, we compare two document representation models which are Boolean model and Vector model using named entities. In the implementation of SVM, we utilize the SMO algorithm with a linear kernel for the binary version of SVM according to [1]. Besides, we also compare the SVM for regression with the binary version of SVM. The dimension of the feature space in our approach is lower than it is reported in [1]. Finally, we implement the top-down approach for class prediction according to [7].

3   Hierarchical Text Classification Approach

In this section, we first introduce the formulation of hierarchical classification for documents. Then, we present the representation models used in this paper. Finally, we propose the hierarchical classification approach for news articles based-on named entity.

3.1   Problem Formulation

We introduce the definition of hierarchical text classification presented by [6, 12, 13].

Definition 1. *Hierarchical Text Classification is the task of assigning a boolean value to each pair $\langle d_j, c_i \rangle \in D \times C$ with a given structure $\mathcal{H} = \langle C, \prec \rangle$, where D is a domain of documents, $C = \{c_1, \ldots, c_{|C|}\}$ is a set of predefined categories and $\prec \subseteq C \times C$ is an asymmetric, anti-reflexive, transitive binary relation on C.*

The binary relation $\prec$ represents the "IS-A" relationship. "IS-A" relationship is defined as asymmetric, anti-reflexive and transitive in [6]:

- $\forall c_i, c_j \in C$, if $c_i \prec c_j$ then $c_j \not\prec c_i$
- $\forall c_i \in C, c_i \not\prec c_i$
- $\forall c_i, c_j, c_k \in C, c_i \prec c_j$ and $c_j \prec c_k$ imply $c_i \prec c_k$

For any hierarchy $\mathcal{H}$, it is assumed that the existence of the root (or top) category $Root(\mathcal{H})$ which is an ancestor of all other categories in the hierarchy. The IRT hierarchy in our application is exactly a tree.

3.2   Representation Models

Vector space model is an algebraic model for representing text documents as vectors of identifiers, such as terms. For example, document d_j is represented by vector $\langle w_{1j}, w_{2j}, \ldots, w_{ij}, \ldots, w_{mj} \rangle$, where w_{ij} is a value associate with the i-th feature in document d_j. We present two types of models to represent documents, which are Boolean model where w_{ij} is a binary value which denotes the appearance of the i-th feature in document d_j, and Vector model where w_{ij} is a real value which denotes the importance of the i-th feature in document d_j. Particularly, we implement the *term frequency/inverse document frequency* (TF-IDF) score as the value of w_{ij} in Vector model.

Corresponding to the two types of models, we build SVM for classification with respect to the Boolean model and SVM for regression with respect to the Vector model at each node of the hierarchy. The performance of these two types of SVM models are compared in Section 4.

3.3   Hierarchical Text Classification Based-on Named Entity

We implement the "local classifier per parent" [6] strategy in this paper. For each parent category (non-leaf category) in the hierarchy, a multi-class classifier is trained to classify documents into its child categories. To distinguish Feature Selection, Training and Testing from the content, we propose our hierarchical text categorization approach in three phases. Details of each phase is described as follows.

Feature Selection. In this phase, terms and named entities are considered as different types of features for the task of hierarchical text classification. We extract terms by constraining the length less than 6 words and the frequency more than 5 in the corpus. As a result, each news article contains 310 terms in average. On the other hand, we employ ICTCLAS[3] to recognize three typical types of named entities, which are the names of persons, organizations and locations, from each news article. Totally, each news article contains 14.7 entities in average. However, there exist a few named entities, such as the location name, e.g. China, and the organization name, e.g. Sina, that are common in news articles. These named entities are less informative and provide little power of differentiation.

Information gain is frequency employed as a termgoodness criterion in the field of machine learning [14]. We implement information gain as feature selection method based on named entities as well as terms in our approach to select a set of features for each category. The information gain of a feature t (named entity or term) with respect to category c_k is defined as follows [12]:

$$P(t) \sum_{k=1}^{|C|} P(c_k|t) \cdot \log_2 \frac{P(t, c_k)}{P(t) \cdot P(c_k)} + P(\bar{t}) \sum_{k=1}^{|C|} P(c_k|\bar{t}) \cdot \log_2 \frac{P(\bar{t}, c_k)}{P(\bar{t}) \cdot P(c_k)} \quad (1)$$

where $|C|$ denotes number of categories in the category set C, $c_k \in C$, $P(t)$ denotes the probability that a random document contains feature t, $\bar{t}$ represents the absence of feature t, $P(\bar{t})$ denotes the probability that a random document does not contains feature t. We compare the performance of hierarchical text classification approaches with different size of feature sets.

Comparatively, we propose a naive method, which calculating TF-IDF score for each feature and selecting the most informative features from each document. The TF-IDF score of the i-th feature in the j-th document is defined as follows [15]:

$$w_{ij}(c) = tf_{ij} \cdot \log_2 \left(\frac{|Doc(c)|}{df_i(c)} \right) \quad (2)$$

where $Doc(c)$ denotes a subset of articles that labeled by category c and $|Doc(c)|$. is the size of the subset, tf_{ij} denotes the frequency that the i-th feature is mentioned in the j-th document, and $df_i(c)$ is the number of documents that mentioned the i-th feature in set $Doc(c)$. The feature set selected by naive method

---

[3] http://ictclas.org/

consists of one or two most informative features, with the highest TF-IDF scores, from each document.

Training. In this phase, we train multi-class classifier for each non-leaf category of the hierarchy. The multi-class classifier decomposes into a set of binary classifiers or predictors corresponding to its descendant categories. We first represent training examples by feature vectors in the local feature space associated with the current category. Then we define the set of positive and negative examples for each descendant category of the current category. Finally, we train SVM for classification and regression for these descendant categories.

There are different ways to define the set of positive and negative examples for training the binary classifiers. It is suggested that the siblings policy is in advance of using considerable less data and obtain no bad accuracy [6]. The siblings policy is defined as follows:

$$Tr^+(c) = *(c) \cup \Downarrow (c) \tag{3}$$

$$Tr^-(c) = \leftrightarrow (c) \cup \Downarrow (\leftrightarrow (c)) \tag{4}$$

where $Tr^+(c)$ denotes the set of positive training examples of category c, $Tr^-(c)$ denotes the set of negative training examples of category c, $*(c)$ denotes the set of examples whose categories is c, $\Downarrow (c)$ denotes the set of descendant categories of c, and $\leftrightarrow (c)$ denotes the set of sibling categories of c. It means that the positive examples of category c are the documents belong to category c and its descendant categories, and the negative examples are the documents belong the siblings of category c and their descendant categories.

We build SVM for classification and regression in this training phase. The SVM for classification is trained by the SMO algorithm with polynomial kernel [16], and the SVM for regression is trained by the RegSMOImproved algorithm with polynomial kernel [17]. Both of the two algorithms are provided by Weka[4] using the default parameter setting.

Testing. In this phase, a top-down classification policy is employed to classify the unlabeled news articles. For each multi-class classifier, we make decision about which descendant category are classified or predicted. This process is repeating until we reach a leaf category or can not classify the test example to any descendant. There are two tasks corresponding to the SVM for classification and SVM for regression.

In classification task, the multi-class classifier returns the positive results of its binary classifiers. For example, if the binary classifier at *Presidential Election of USA* returns true for a news article, then the multi-class classifier at *Election* will return *Presidential Election of USA* as its result. In regression task, the multi-class classifier considers the average prediction value of its predictors. For example, if the predicator at *Presidential Election of USA* returns a value higher than the average of its siblings, then the multi-class classifier at *Election* will return *Presidential Election of USA* as its result.

---

[4] http://www.cs.waikato.ac.nz/ml/weka/

4   Experimental Results

In this section, we present the experimental results of named entity based hier-archical text classification approach. To exam the utility of named entity based representation models in the first assumption, we analyze the performance of named entity features compared with term features in both close categories and rare categories. Two types of feature selection methods, the naive method and information gain method, are compared with different number of features. Be-sides, the comparisons of SVM for classification and regression are provided to exam the Boolean model and Vector model.

The labeled documents that related to the IRT hierarchy are divided into two sets: the training set and the testing set. For each category, the labeled documents are randomly divided into 5 folds: 4 folds for training and 1 fold for testing. In addition, we remove the categories containing only one labeled document. This partition guarantees that the testing documents are with the same distribution of the data. Further, we select a set of categories that consists less than 10 labeled documents as the rare categories, and manually choose a set of categories that contain similar topics as the close categories. Thus, we have close categories testing set, rare categories testing set and hybrid category testing set which contains both close categories and rare categories. The micro-averaged F1 scores [18] as well as the micro-averaged precision and recall are utilized as metrics.

Comparison of Features. The effectiveness of named entity feature and term feature are compared with different feature numbers on close categories testing set. In this experiment, we implement information gain method, both Boolean model and Vector model of feature vectors are tested. All results are presented in Figure 3. It is illustrated that named entity features performs better than term features when a few number, e.g. 500, of features are selected. Because the named entities are informative features for close categories, and the use of named entity deduces the dimension of the feature space.

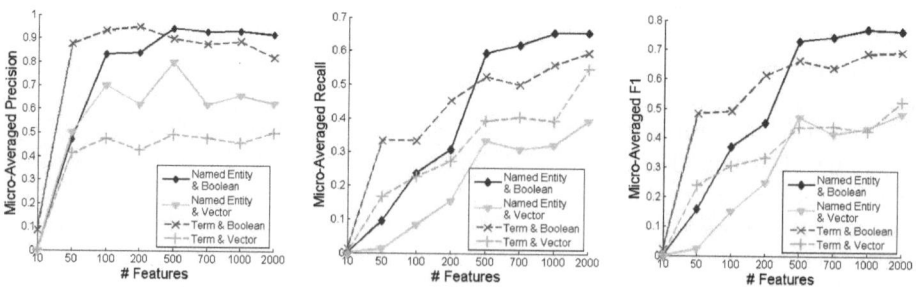

Fig. 3. Micro-averaged precision, recall and F1 score (y-axis) for hierarchical text clas-sification using named entity and term features on close categories testing set. The x-axis denotes the number of features range from 10 to 2000.

Previous research which studied document representation using named entities in new event detection, found that the named entities does not always effective[19]. Our experiments show that the effectiveness of such document representation is also associated with the number of named entities.

Comparison of Feature Selection Methods. We compare two feature selection methods, the naive method and the information gain method. In this experiment, we implement the SVM for classification on three testing sets. The naive method selects one most informative feature from each document, and the information gain method selects 2,000 features from the category. It is illustrated in Table 1 that the information gain method performs better for both named entity features and term features.

Table 1. Micro-averaged F1 score for hierarchical text classification using information gain method and naive method

Feature Selection Methods		Testing Set		
		Close Categories	Rare Categories	Hybrid Categories
Naive Method for	Named Entity	0.705	0.309	0.407
	Term	0.549	0.323	0.352
IG Method for	Named Entity	0.763	0.266	0.405
	Term	0.689	0.292	0.393

Comparison of SVM. We compare the SVM for classification and regression in this experiment. The results are illustrated in Figure 4. It is illustrated that the SVM for regression associated with the Vector model performs better than the SVM for classification associated with Boolean model on close categories testing set, while the Boolean model performs better on rare categories and hybrid categories testing sets. Because the SVM for regression requires more training examples than the SVM for classification. Hence, the performance of SVM for regression decreases on rare categories due to the lack of training examples.

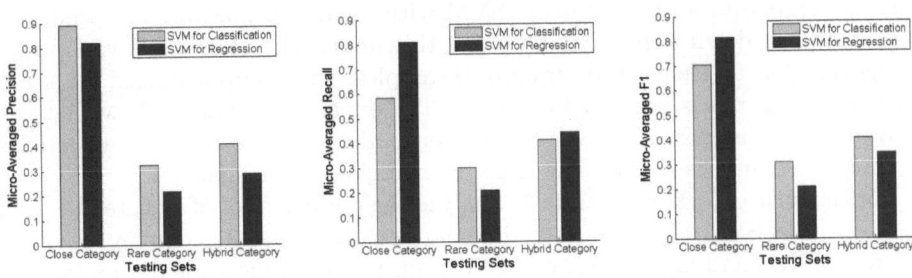

Fig. 4. Micro-averaged precision, recall and F1 score (y-axis) for hierarchical text classification with SVM for classification and regression, using named entity features on testing sets. The x-axis denotes different testing sets.

The performance of Vector model is relevant with the effectiveness of SVM for regression, it also causes that the Boolean model performs better than Vector model as illustrated in Figure 3.

Overall Performance. In this experiment, we test SVM for classification using named entity features on three testing sets, which are close categories testing set, rare categories testing set and hybrid categories testing set. The experiment results are illustrated in Figure 5. It is shown that the bad performance of hierarchical text classification on rare categories decreased the overall performance of the hierarchy. Such types of categories are suffered from the skewed distribution of data, since the use of named entity cannot improve the performance of hierarchical text classification on the rare categories.

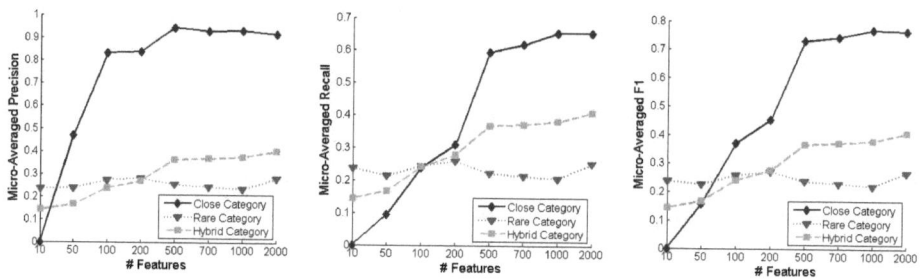

Fig. 5. Micro-averaged precision, recall and F1 score (y-axis) for hierarchical text classification using named entity features on close categories, rare categories and hybrid categories testing set. The x-axis denotes the number of features range from 10 to 2000.

5   Conclusions and Future Works

We introduce a hierarchical text classification approach for news articles about international relations, focusing on the difficulties of close categories and rare categories. We use named entities as features, implement local feature selection with an aggressive strategies, train SVM with local information, and predicate class with top-down approach. Although this approach is not effective for rare categories due to the lack of training examples, the micro-averaged F1 score for close categories is improved to 81.4%. The experiment results show that the Boolean model is more effective than Vector model for representing news articles using named entities when a few training examples are available.

For the future work, we plan to 1) analyze the performance of feature selection methods and SVM algorithms for imbalanced data sets, 2) compare to other related models and algorithms, and 3) experiment on large-scale data sets.

Acknowledgement. We gratefully acknowledge funding from the National Science Foundation of China under grants 61170165.

References

[1] Liu, T., Yang, Y., Wan, H., Zeng, H., Chen, Z., Ma, W.: Support vector machines classification with a very large-scale taxonomy. ACM SIGKDD Explorations Newsletter 7, 36–43 (2005)

[2] Gabrilovich, E., Markovitch, S.: Text categorization with many redundant features: Using aggressive feature selection to make svms competitive with c4. 5. In: Proceedings of the 21th International Conference on Machine Learning, p. 41 (2004)

[3] Nadeau, D., Sekine, S.: A survey of named entity recognition and classification. Lingvisticae Investigationes 30, 3–26 (2007)

[4] Dumais, S., Chen, H.: Hierarchical classification of web content. In: Proceedings of the 23rd ACM SIGIR Conference, pp. 256–263. ACM (2000)

[5] Xue, G., Xing, D., Yang, Q., Yu, Y.: Deep classification in large-scale text hierarchies. In: Proceedings of the 31st ACM SIGIR Conference, pp. 619–626. ACM (2008)

[6] Silla, C., Freitas, A.: A survey of hierarchical classification across different application domains. Data Mining and Knowledge Discovery 22, 31–72 (2011)

[7] Koller, D., Sahami, M.: Hierarchically classifying documents using very few words. In: Proceedings of the Fourteenth International Conference on Machine Learning, pp. 170–178. Morgan Kaufmann Publishers Inc. (1997)

[8] Chen, H., Dumais, S.: Bringing order to the web: Automatically categorizing search results. In: Proceedings of the SIGCHI Conference on Human Factors in Computing Systems, pp. 145–152. ACM (2000)

[9] Punera, K., Ghosh, J.: Enhanced hierarchical classification via isotonic smoothing. In: Proceedings of the 17th International Conference on World Wide Web, pp. 151–160. ACM (2008)

[10] Bennett, P., Nguyen, N.: Refined experts: improving classification in large taxonomies. In: Proceedings of the 32nd ACM SIGIR Conference, pp. 11–18. ACM (2009)

[11] Liu, B., Li, C.: An efficient feature selection method using named entity recognition for chinese text categorization. In: Proceedings of the International Conference on Machine Learning and Cybernetics, pp. 3527–3531 (2009)

[12] Kiritchenko, S.: Hierarchical text categorization and its application to bioinformatics. PhD thesis, Citeseer (2005)

[13] Sebastiani, F.: Machine learning in automated text categorization. ACM Computing Surveys (CSUR) 34, 1–47 (2002)

[14] Yang, Y., Pedersen, J.: A comparative study on feature selection in text categorization. In: Machine Learning International Workshop, pp. 412–420. Morgan Kaufmann Publishers, Inc. (1997)

[15] Baeza-Yates, R., Ribeiro-Neto, B.: Modern information retrieval, vol. 82. Addison-Wesley, New York (1999)

[16] Platt, J.: Fast training of support vector machines using sequential minimal optimization. In: Advances in Kernel Methods, pp. 185–208. MIT Press (1999)

[17] Shevade, S., Keerthi, S., Bhattacharyya, C., Murthy, K.: Improvements to the smo algorithm for svm regression. IEEE Transactions on Neural Networks 11, 1188–1193 (2000)

[18] Yang, Y., Liu, X.: A re-examination of text categorization methods. In: Proceedings of the 22nd ACM SIGIR Conference, pp. 42–49. ACM (1999)

[19] Kumaran, G., Allan, J.: Text classification and named entities for new event detection. In: Proceedings of the 27th ACM SIGIR Conference, pp. 297–304. ACM (2004)

Document-Level Sentiment Classification Based on Behavior-Knowledge Space Method

Zhifei Zhang, Duoqian Miao, Zhihua Wei, and Lei Wang

Department of Computer Science and Technology,
Tongji University, Shanghai 201804, China
Key Laboratory of Embedded System and Service Computing,
Ministry of Education, Tongji University, Shanghai 201804, China
zhifei.zhang@gmail.com

Abstract. There are mainly two kinds of methods for document-level sentiment classification, unsupervised learning and supervised learning. When ensemble learning is introduced, existing methods only combine unsupervised learning algorithms or supervised learning algorithms. To overcome each other's flaws, a novel sentiment classification method based on behavior-knowledge space is proposed, in which two unsupervised and two supervised learning algorithms are utilized. The experiment results not only explain the effectiveness by diversity measure, but also show that the proposed method is significantly superior to each individual classifier. In addition, our method is better than the other two common ensemble methods.

Keywords: sentiment classification, ensemble learning, behavior-knowledge space.

1 Introduction

More and more people express their attitudes and opinions about products, persons or events using the Internet. These user-generated texts are unstructured or semi-structured, which contain subjective information, such as attitudes, opinions, and emotions. Sentiment classification aims to determine the polarity of a given text at document, sentence or word level, i.e., whether the expressed opinion is positive, negative or neutral. Sentiment classification is a research hotspot of web mining, which has been widely used in many domains [1], e.g., commerce product recommendation, unhealthy information filtering, public opinion monitoring, and stock trend prediction.

Depending on the granularity of text, sentiment classification can be divided into three levels: word, sentence and document. Word-level sentiment classification is the basis of the other two tasks, whose methods are dictionary-based or corpus-based [2]. Dictionary-based methods utilize dictionaries, lexicons or their extensions to obtain the polarity of a word, e.g., WordNet [3], HowNet [4], while corpus-based methods fully utilize corpus to obtain the polarity of a word, e.g., adjectives' clustering [5], point-wise mutual information [6]. Sentence-level

S. Zhou, S. Zhang, and G. Karypis (Eds.): ADMA 2012, LNAI 7713, pp. 330–339, 2012.

sentiment classification is to decide the polarity of a sentence and also extracts sentiment-related elements, such as opinion holders and opinion aspects. As the key of sentiment classification, sentence-level sentiment classification can support fine-grained sentiment analysis. One method is to use sentiment dictionaries and domain dictionaries, extract subjective elements and calculate the polarity with weighted sum [7]. Another method is to construct classifiers based on machine learning, e.g., CRFs-based sentiment classification with redundant features [8], subjectivity summarization based on minimum cuts [9]. Document-level sentiment classification is to give the overall polarity of a document, by unsupervised learning [10][11] or supervised learning [12][13].

Ensemble learning is to combine outputs of multiple classifiers and can gain better results. It has been successfully applied in text classification [14], but less in sentiment classification, which can be regarded as the special case of text classification. Xia et al. [15] focus on ensemble of feature sets and classifiers, in which all classifiers belong to supervised learning. Wan [16] combines unsupervised learning methods using bilingual knowledge. In term of the relation among basic classifiers, ensemble learning can be classified into two groups, homogeneous ensemble learning and heterogeneous ensemble learning [17]. The common heterogeneous ensemble strategies are Voting, Bayes' Rule and Behavior-Knowledge Space [18]. Voting is simple but designs equality of basic classifiers. Bayes' Rule is limited to the assumption of independence among basic classifiers. Behavior-Knowledge Space (BKS, for short) solves their defects. In this paper, we focus on ensemble of unsupervised learning and supervised learning algorithms based on Behavior-Knowledge Space method for document-level sentiment classification, and verify the effectiveness of the proposed method on real sentiment corpus. Our contributions are summarized as follows.

- Propose an ensemble of unsupervised learning methods and supervised learning methods for sentiment classification.
- Explore the effectiveness of Behavior-Knowledge Space ensemble method for document-level sentiment classification.

The remainder of this paper is organized as follows. Section 2 introduces ensemble learning briefly and BKS method in detail. Our method is described in Section 3. The experiment results are illustrated in Section 4. Section 5 concludes this paper.

2   Ensemble Learning

2.1   Main Idea

A topic within machine learning where there has been a lot of recent activity is ensemble learning, which is to improve classification accuracy by learning ensembles of classifiers [19].

Given a set of training examples, a classifier is obtained by a learning algorithm. Given a new example, each classifier predicts the corresponding output.

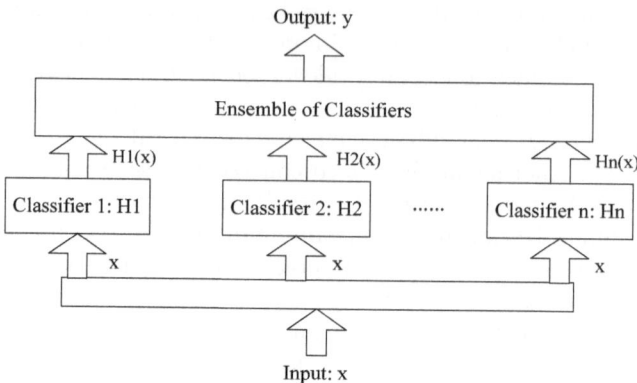

Fig. 1. Main Idea of Ensemble Learning

An ensemble of classifiers is multiple classifiers whose individual outputs are combined in some way to classify a new example. The main idea of ensemble learning is described in Fig.1.

An ensemble is more accurate than each basic classifier only if the component classifiers disagree with one another and the accurate rate of each basic classifier is not below 0.5 [19]. Diversity is the basis of the ensemble learning, which can be measured by Q-static. Given two classifiers H_i and H_j, Table 1 is the contingency table between each pair of classifiers.

Table 1. Contingency Table between Classifiers H_i and H_j

	H_j True	H_j False
H_i True	A	B
H_i False	C	D

A is the probability that both classifiers truly label input data. B is the probability that H_i truly labels but H_j falsely labels input data. C is the probability that H_i falsely labels but H_j truly labels input data. D is the probability that both classifiers falsely label input data. By this definition, $A + B + C + D = 1$ is hold. Then the Q-static measure of diversity for these two classifiers is calculated by Eq.1 [20].

$$Q_{ij} = \left| \frac{AD - BC}{AD + BC} \right| \tag{1}$$

We can see that $0 \le Q_{ij} \le 1$, when Q_{ij} is approaching 0, the diversity of two classifiers is bigger, otherwise the diversity is smaller.

2.2   Behavior-Knowledge Space Method

Behavior-Knowledge Space method [18] is proposed by Huang and Suen in 1993, which is for combination of multiple classifiers. Denote the number of basic classifiers K, the number of decision classes M, the class set $\Lambda = \{1, 2, ..., M\}$. If a classifier rejects an input, the output class is set to be $M+1$. The decision of each classifier H_i is marked by $e(i)$, $e(i) \in \Lambda \cup \{M+1\}$. A behavior-knowledge space BKS is a K-dimensional space. The intersection of the decisions of classifiers is one unit of the BKS, denoted by $BKS(e(1), e(2), ..., e(K))$. For example, $K = 2$, $M = 3$, the BKS is constructed, as presented in Table 2.

Table 2. An Example of 2-Dimensional BKS

$e(1)$ \ $e(2)$	1	2	3	4
1	(1,1)	(1,2)	(1,3)	(1,4)
2	(2,1)	(2,2)	(2,3)	(2,4)
3	(3,1)	(3,2)	(3,3)	(3,4)
4	(4,1)	(4,2)	(4,3)	(4,4)

Each unit of BKS contains three kinds of data: (1) the total number of input examples with each class $m \in \Lambda$, denoted by $n_{e(1),...,e(K)}(m)$, (2) the total number of input examples $T_{e(1),...,e(K)}$, and (3) the best representative class $R_{e(1),...,e(K)}$. The last two can be calculated using the first one.

$$T_{e(1),...,e(K)} = \sum_{m=1}^{M} n_{e(1),...,e(K)}(m) \tag{2}$$

$$R_{e(1),...,e(K)} = \{l | n_{e(1),...,e(K)}(l) = \max_{1 \leq m \leq M} n_{e(1),...,e(K)}(m)\} \tag{3}$$

For each unit, the best representative class is unique. Given such a BKS, the belief value of an input example belonging to one class is computed by Eq.4.

$$BEL(m) = \frac{n_{e(1),...,e(K)}(m)}{T_{e(1),...,e(K)}} \tag{4}$$

There are two stages in BKS method: knowledge modeling and operation. The knowledge modeling stage is to construct a BKS using training examples, mainly compute the values of T and R for each unit. The operation stage is to make final decisions for new examples according to the decision classes of basic classifiers and the following decision rule,

$$D(x) = \begin{cases} R_{e(1),\ldots,e(K)} & \text{when } T_{e(1),\ldots,e(K)} > 0 \text{ and } BEL(R_{e(1),\ldots,e(K)}) \geq \alpha \\ M+1 & \text{otherwise} \end{cases} \quad (5)$$

where $\alpha(0 \leq \alpha \leq 1)$ is the threshold which controls the reliable degree of the decision.

3 Our Method

3.1 Problem Statement

Here, document-level sentiment classification can be regarded as two-class text classification, with positive and negative labels. Four basic classifiers are used: the simple weighted sum of sentiment words (called SWS) [10], the weighted sum of sentiment words based on concepts (called WSC) [11], support vector machine (SVM) [12] and k-nearest neighbors (KNN)[13]. The former two are unsupervised learning algorithms, and the latter two are supervised learning algorithms. Thus, we get $K = 4$ and $M = 2$ in BKS method. Our ensemble method is indicated in Fig.2.

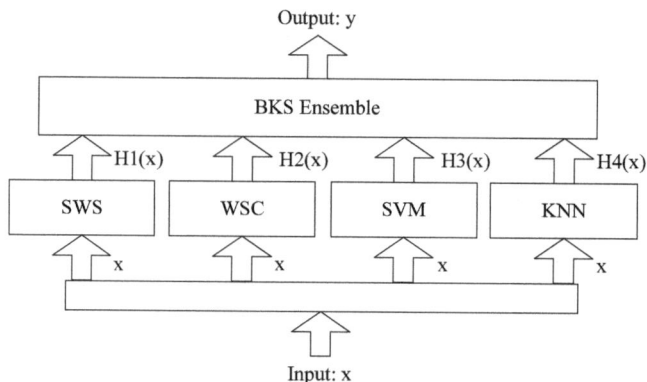

Fig. 2. BKS Ensemble Method for Document-level Sentiment Classification. Here, $H1(x)$, $H2(x)$, $H3(x)$, $H4(x)$ and y are all from {Positive, Negative, Reject}.

3.2 Method Illustration

Our ensemble method is roughly illustrated in Fig.3. In this diagram, the top half is about training and the bottom half is about testing. In training procedure, a BKS ensemble classifier is established based on a training set. In testing procedure, the BKS ensemble classifier makes decisions for a testing set. It is necessary to state that, SWC and WSC directly give the outputs of training examples or testing examples, but SVM and KNN give the outputs of testing examples after learning from training examples.

Algorithm 1 describes the proposed method in detail. The settings of these four basic classifiers are similar to that described in their corresponding references.

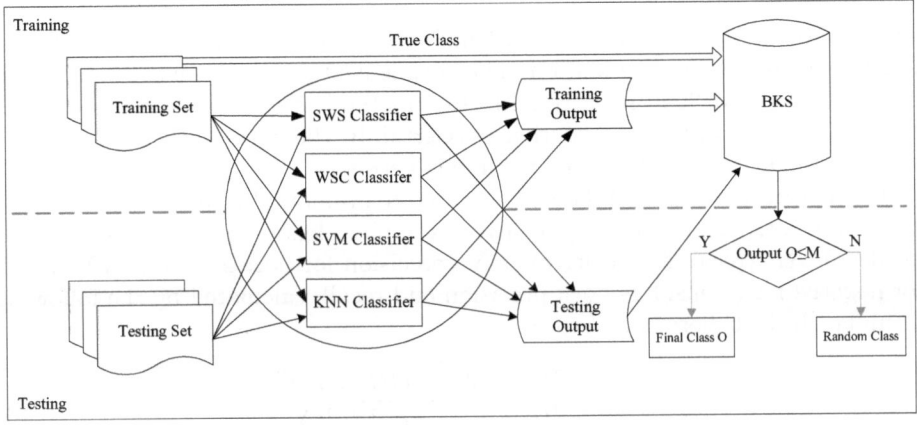

Fig. 3. Diagram of BKS Ensemble for Document-level Sentiment Classification

---

Algorithm 1. Document-level Sentiment Classification Based on BKS

Input:
> Set of training examples, TS;
> Set of testing examples, NS;
> Reliability threshold for decision, α;

Output:
> Decision classes of testing examples, D;
> 1: Building up two supervised learning classifiers, SVM and KNN, using the training set TS;
> 2: Obtaining the decision classes by four individual classifiers for each example in TS, denoted by $TID_{|TS|\times 4}$;
> 3: From $TID_{|TS|\times 4}$ and the true classes in TS, computing T and R for each unit of BKS by Eq.2 and Eq.3;
> 4: Obtaining the decision classes by four individual classifiers for each example in NS, denoted by $NID_{|NS|\times 4}$;
> 5: From $NID_{|NS|\times 4}$ and BKS, making decisions for all testing examples by Eq.5, denoted by D';
> 6: For all the i-th testing example in NS, if $D'(i) < M + 1$, then $D(i) = D'(i)$, otherwise $D(i)$ is randomly generated from $\{1, ..., M\}$; ($M = 2$)
> 7: **return** D.

4   Experiments

4.1   Experiment Settings

The experiments are carried out on ChnSentiCorp-Htl-ba-4000 from ChnSenti-Corp[1] plus 5-fold cross validation. The corpus contains 4000 texts about hotel, 2000 positive and 2000 negative. In BKS method, the reliability threshold is set to be 0.55 according to experiences. Certainly, the optimal threshold can be found automatically [18], but more time is needed.

Two metrics, precision and recall, are used to evaluate the performance respectively in positive examples and negative examples, denoted by PP (precision for positive), RP (recall for positive), PN (precision for negative) and RN (recall for negative). F_1 considers both precision and recall, calculated by the following formula. (In fact, here is Macro F_1)

$$F_1 = \frac{(PP + PN) \times (RP + RN)}{PP + PN + RP + RN} \tag{6}$$

Our experiments include two parts. First, we explain the effectiveness of BKS ensemble on the basis of the performances of four basic classifiers and their diversity. Second, the performance of our method is demonstrated via comparison with four basic classifiers and the other two ensemble methods.

4.2   Effectiveness of BKS Ensemble

As known, there are two conditions to guarantee the effectiveness of ensemble learning. The experiment results of four basic classifiers are shown in Table 3.

Table 3. Results of Four Basic Classifiers

Classifier	PP	RP	PN	RN	F_1
SWS	0.850	0.852	0.851	0.850	0.851
WSC	0.681	0.778	0.748	0.631	0.709
SVM	0.898	0.826	0.839	0.905	0.867
KNN	0.857	0.878	0.875	0.854	0.866

Obviously seen in Table 3, the accurate rate of each basic classifier is higher than 0.5, even higher than 0.6. One condition of effective ensemble learning is satisfied.

The diversity is measured by Eq.1. Table 4 displays the diversity measure of four basic classifiers used in our method. Because the diversity measure of two same classifiers is equal to 1, all values of the diagonal direction are useless and marked as "—".

---

[1] http://www.searchforum.org.cn/tansongbo/corpus-senti.htm

Table 4. Diversity Measure of Four Basic Classifiers

	SWS	WSC	SVM	KNN
SWS	—	0.51	0.61	0.45
WSC	0.51	—	0.24	0.26
SVM	0.61	0.24	—	0.90
KNN	0.45	0.26	0.90	—

SVM classifier and KNN classifier are close, because their diversity measure is up to 0.9. WSC classifier is significantly different from the other three classifiers. The difference between SWS classifier and the others is medium. In a word, the diversity of these four classifiers is guaranteed so that our ensemble method is effective.

4.3   Performance of BKS Ensemble

Document-level sentiment classification based on BKS is compared with four basic classifiers, as shown in Fig.4.

In general, BKS ensemble method outperforms four individual classifiers, F_1 is up to 92.5%. Besides, the difference of performances in positive and negative examples is not significant, which is wanted. BKS ensemble method can overcome its basic classifier's flaws and improve the accuracy of classification system.

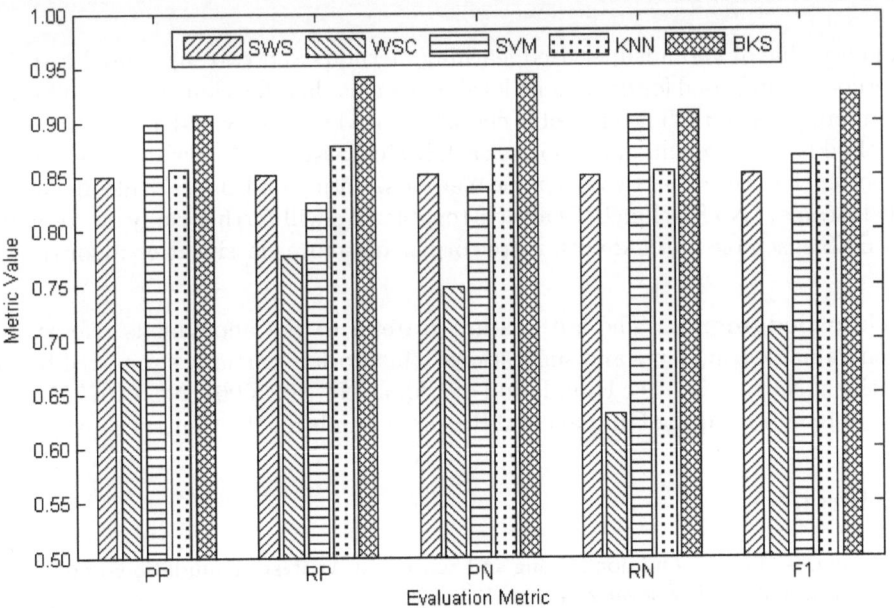

Fig. 4. Comparison Results between BKS Ensemble and Four Basic Classifiers

BKS ensemble method is also compared with the other two ensemble methods, Voting and Bayes' Rule (see Table 5).

Table 5. Comparison Results of Three Ensemble Methods

Ensemble Method	PP	RP	PN	RN	F_1	Random Times
Voting	0.884	0.883	0.883	0.884	0.884	162
Bayes' Rule	0.88	0.909	0.911	0.884	0.896	0
BKS	0.906	0.941	0.943	0.909	0.925	14

BKS is better than the other two ensemble methods. The reason is that BKS does not need the assumption of independence among basic classifiers, while Bayes' Rule needs. BKS does not treat basic classifiers equally, but Voting does. The last column in Table 5 indicates the average times of random decisions. Voting needs more times to decide the classes of examples randomly. Bayes' Rule can make decisions definitely. BKS needs few times to make random decisions, which is acceptable. If the number of basic classifiers increases, the times of random decisions will become more. In this situation, some units of BKS are not assigned more training examples and it is not useful to predict the class of a new example. Thereby, this phenomenon tells us BKS ensemble method is more suitable for document-level sentiment classification.

5 Conclusions

An ensemble of two unsupervised learning methods and two supervised learning methods is proposed for document-level sentiment classification, which is different from previous work. The ensemble classifier based on Behavior-Knowledge Space method is effective and outperforms each basic classifier. Meanwhile, our method is better than the other two ensemble methods, Voting and Bayes' Rule. In ensemble learning, "No free lunch" is always tenable. We will further carry out research on BKS ensemble method with more corpus and more classification algorithms.

Acknowledgments. The authors are grateful to the anonymous referees for their valuable comments and suggestions. This work is partially supported by the National Natural Science Foundation of China (No. 60970061, No. 61075056 and No. 61103067) and the Fundamental Research Funds for the Central Universities.

References

1. Pang, B., Lee, L.: Opinion Mining and Sentiment Analysis. Foundations and Trends in Information Retrieval 2(1-2), 1–135 (2008)
2. Rao, D., Ravichandran, D.: Semi-supervised Polarity Lexicon Induction. In: EACL 2009, pp. 675–682. ACL, Morristown (2009)

3. Kim, S.M., Hovy, E.: Automatic Detection of Opinion Bearing Words and Sentences. In: IJCNLP 2005, pp. 61–66. ACL, Morristown (2005)
4. Fu, X., Liu, G., Guo, Y., Guo, W.: Multi-aspect Blog Sentiment Analysis Based on LDA Topic Model and Hownet Lexicon. In: Gong, Z., Luo, X., Chen, J., Lei, J., Wang, F.L. (eds.) WISM 2011, Part II. LNCS, vol. 6988, pp. 131–138. Springer, Heidelberg (2011)
5. Wiebe, J.: Learning Subjective Adjectives from Corpora. In: AAAI 2000, pp. 735–740. AAAI Press, Menlo Park (2000)
6. Turney, P., Littman, M.L.: Measuring Praise and Criticism: Inference of Semantic Orientation from Association. ACM Trans. on Information Systems 21(4), 315–346 (2003)
7. Hu, M.Q., Liu, B.: Mining and Summarizing Customer Reviews. In: KDD 2004, pp. 168–177. ACM Press, New York (2004)
8. Zhao, J., Liu, K., Wang, G.: Adding Redundant Features for CRFs-based Sentence Sentiment Classification. In: EMNLP 2008, pp. 117–126. ACL, Morristown (2008)
9. Pang, B., Lee, L.: A Sentimental Education: Sentiment Analysis Using Subjectivity Summarization Based on Minimum Cuts. In: ACL 2004, pp. 271–278. ACL, Morristown (2004)
10. Turney, P.: Thumbs Up or Thumbs Down? Semantic Orientation Applied to Unsupervised Classification of Reviews. In: ACL 2002, pp. 417–424. ACL, Morristown (2002)
11. Chen, Y.F., Miao, D.Q., Li, W., Zhang, Z.F.: Semantic Orientation Computing Based on Concepts. CAAI Trans. on Intelligent Systems 6(6), 489–494 (2011) (in Chinese)
12. Pang, B., Lee, L., Vaithyanathan, S.: Thumbs Up? Sentiment Classification Using Machine Learning Techniques. In: EMNLP 2002, pp. 79–86. ACL, Morristown (2002)
13. Tan, S.B., Zhang, J.: An Empirical Study of Sentiment Analysis for Chinese Documents. Expert Systems with Applications 34, 2622–2629 (2008)
14. Dong, Y.S., Han, K.S.: A Comparison of Several Ensemble Methods for Text Categorization. In: SCC 2004, pp. 419–422. IEEE Computer Society, Washington (2004)
15. Xia, R., Zong, C.Q., Li, S.S.: Ensemble of Feature Sets and Classification Algorithms for Sentiment Classification. Information Sciences 181, 1138–1152 (2011)
16. Wan, X.J.: Using Bilingual Knowledge and Ensemble Techniques for Unsupervised Chinese Sentiment Analysis. In: EMNLP 2008, pp. 553–561. ACL, Morristown (2008)
17. Dietterich, T.G.: Ensemble Methods in Machine Learning. In: Kittler, J., Roli, F. (eds.) MCS 2000. LNCS, vol. 1857, pp. 1–15. Springer, Heidelberg (2000)
18. Huang, Y.S., Suen, C.Y.: The Behavior-Knowledge Space Method for Combination of Multiple Classifiers. In: CVPR 1993, pp. 347–352. IEEE Computer Society, Washington (1993)
19. Dietterich, T.G.: Machine Learning Research: Four Current Directions. AI Magazine 18(4), 97–136 (1997)
20. Kuncheva, L.I., Whitaker, C.J.: Measures of Diversity in Classifier Ensembles. Machine Learning 51, 181–207 (2003)

NAP-SC: A Neural Approach for Prediction over Sparse Cubes

Wiem Abdelbaki[1], Sadok Ben Yahia[1,2], and Riadh Ben Messaoud[3]

[1] Faculty of Sciences of Tunis, El-Manar University, 2092 Tunis, Tunisia
[2] Institut Mines-TELECOM, TELECOM SudParis, UMR CNRS Samovar, 91011 Evry Cedex, France
[3] Faculty of Economics and Management of Nabeul, Carthage University, 1054 Tunis, Tunisia

Abstract. OLAP techniques provide efficient solutions to navigate through data cubes. However, they are not equipped with frameworks that empower user investigation of interesting information. They are restricted to exploration tasks.

Recently, various studies have been trying to extend OLAP to new capabilities by coupling it with data mining algorithms. However, most of these algorithms are not designed to deal with sparsity, which is an unavoidable consequence of the multidimensional structure of OLAP cubes.

In [1], we proposed a novel approach that embeds Multilayer Perceptrons into OLAP environment to extend it to prediction. This approach has largely met its goals with limited sparsity cubes. However, its performances have decreased progressively with the increase of cube sparsity.

In this paper, we propose a substantially modified version of our previous approach called NAP-SC (Neural Approach for Prediction over Sparse Cubes). Its main contribution consists in minimizing sparsity effect on measures prediction process through the application of a cube transformation step, based on a dedicated aggregation technique.

Carried out experiments demonstrate the effectiveness and the robustness of NAP-SC against high sparsity data cubes.

Keywords: Data Warehouse, Data Mining, Principal Component Analysis, Machine Learning, Multilayer Perceptron, Prediction.

1 Introduction

A Data Warehouse (DW) is a corner stone in the Business Intelligence (BI) process. It is implemented to store analysis contexts within multidimensional data structures, referred to as *Data Cubes* and usually manipulated by using On-line Analytical Processing (OLAP) applications to enable senior managers exploring information and getting BI reportings through interactive dashboards.

By definition, a DW should fundamentally contain integrated data [2]. However, generally, exploring a data cube disclose a sparse structure within several empty measures. In DW models, empty measures correspond to non-existent facts reflecting either out-of-date events that did not happen or upcoming events

S. Zhou, S. Zhang, and G. Karypis (Eds.): ADMA 2012, LNAI 7713, pp. 340–352, 2012.

that have not yet occurred and may happen in the future. We argue that predicting these measures could consolidate BI reportings and provide new opportunities to DW customers by enlarging their dashboard picture. For instance, it will be extremely useful to a retailer chain to predict the potential sale amount of ice cream in January in some particular agencies. This indicator will definitely help the company to optimise the number of ice cream freezers to install in that period.

So far, non-existent measures in a data cube may potentially be learned from its neighborhood. Agarwal and Chen state that making future decisions over historical data is one crucial goal of OLAP [3]. However, OLAP is restricted to exploration and not equipped with a framework to empower user investigation of interesting information. In fact, despite the fundamental Cood's statement of goal seeking analysis models (such as "What if" analysis) required in OLAP applications since the early 90's [4], most of today's OLAP products still lack an effective implementation of this feature.

On the other hand, data mining is a mature, robust field that have proven its efficiency in dealing with complex data sets [5]. Recently, several approches have been attempting to perform data mining techniques on OLAP cubes. They tackled several issues like cube exploration [6], association rules mining [7] and prediction [8,3]. In [1], we adopted this approach while attempting to predict non-existent measures over OLAP cubes. Thus, we attempted to adapt neural networks, which are among the most popular machine learning techniques that have been explored to solve data mining problems [5], to OLAP environment. To that end, we proposed a neural based approach to predict measures over high-dimensional data cubes that we consider as a Neural Approach to Prediction over High-dimensional Cubes (NAP-HC).

The experimental study showed that NAP-HC has largely met its goals in the case of limited sparsity data cubes. However, its performances decrease within sparse data cubes. This deterioration is justifiable, since various researches affirm that sparsity affects the performances of any approach trying to combine OLAP with data mining methods [9,10]. Moreover, sparsity is an unavoidable consequence of the multidimensional structure of data cubes. It is generated by relationships between dimension attributes. For example, while investigating some product sales in a retail chain according to time and store location, drilling down the location dimension to departments level will disclose many empty cells, generated by the departments that do not sell that product from the first. Kang et al. argue that this case appears very often due to OLAP applications' nature of business [11].

In this paper, we introduce a novel Neural Approach for Prediction over Sparse Cubes (NAP-SC). We stress that our current proposal does not upgrade NAP-HC, which we still recommend for limited sparsity cubes. However, it is an alternative version designed for the particular case of high sparsity data cubes, which makes the following contributions:

- Getting more value out of our recently proposed approach [1], by further exploring its framework and techniques.
- Minimizing sparsity effect on analysis by embedding in a cube transformation step, based on a dedicated aggregation technique.
- Involving the hierarchical structure of data cubes in the analysis to enable the prediction system to deal with multiple hierarchical levels at once.

This paper is organized as follows. In Section 2, we expose a state of the art of works related to predictions in data cubes. In Section 3, we present a reminder of NAP-HC and define our analysis context. Section 4 details the method formalization that we followed in our proposal. In Section 5, we carry out experiments investigating the effectiveness of NAP-SC. Finally, Section 6 summarizes our findings and addresses future research directions.

2 Related Work

Performing data mining tasks on DWs represents an important topic in DW technology. Goil and Choudhary argue that data mining automated techniques further empower OLAP and make it more useful [12]. Several researches were proposed under different motivations (discovery-driven exploration of cubes, cube mining, cube compression, and so on). In line with our concern, we focus on those having a close linkage with prediction in DWs.

We summarize in Table 1 proposals attempting to extend OLAP to prediction. They are detailed according to seven main criteria: (1) What is the overall goal of the proposal? (2) Does it include an algorithmic optimization? (3) Does it use a reduction technique? (4) Does it introduce new classes of measures? (5) Does it provide explicit predicted values of non-existent measures? (6) Does the proposal involve the hierarchical structure of data cubes in the analysis? and(7) Does it deal explicitly with sparsity? We note (+) if the proposal meets the criteria, and (-) if not.

Palpanas *et al.* used the principle of information entropy to build a probabilistic model capable of detecting measure deviations to compress data cubes [13].

Table 1. Proposals integrating prediction in data cubes

Proposal	Goal	Optimization	Reduction	Measures	Values	Hierarchies	Sparsity
[13]	Compression	-	+	-	-	+	-
[14]	Compression	-	+	+	-	+	-
[6]	Exploration	+	-	-	-	+	-
[15]	Prediction	+	-	+	-	+	-
[8]	Prediction	+	-	-	-	-	-
[3]	Prediction	+	-	+	-	+	-
[1]	Prediction	-	+	+	+	-	-
NAP-SC	Prediction	+	+	+	+	+	+

Their approach predicts low-level measures from high-level pre-calculated aggregates. Chen *et al.* introduced a new class of data cubes, called *Regression Cubes* [14]. They contain compressible measures indicating the general tendency and variations compared to original ones. Sarawagi *et al.* proposed a log linear model to assist DW users when exploring data cubes by detecting exceptions [6]. Chen *et al.* introduced the concept of *Prediction Cubes*, where a score or a distribution of probabilities of measures are fetched beside their original values [15]. They are used to build prediction models. Bodin-Niemczuk *et al.* proposed to equip OLAP with a regression tree to predict measures of forthcoming facts [8]. Agarwal and Chen introduced a new data cube class called *Latent-Variable Cube* [3]. It is able to compute aggregate functions, such as mean and variance, over latent variables detected by a statistical model.

In [1], we proposed NAP-HC that predicts measures over high-dimensional data cubes. It introduces a new class of cubes, called PCA-cubes, integrating customized measures referring to predictors stored in an external database. The approach operates on two main stages. The first is a pre-processing one that makes use of the Principal Component Analysis (PCA) to reduce data cube dimensionality. As for the second stage, it introduces an OLAP oriented architecture of Multilayer Perceptrons (MLP)s that learns from multiple training-sets without merging them, and yet comes out with unique predicted value for each targeted measure.

From the above cited references, an outstanding common observation is dealing with data dimensionality. Indeed, the multidimensional structure of data and the usual huge facts' volumetry in DWs represent one of the most challenging issues of integrating predictive models into OLAP environment. This could be of a negative effect on prediction performance, which is supposed to provide BI reporting costumers with fast and accurate results in line with OLAP applications. Thus, some of the above proposals rely on heuristics to optimize implemented algorithms [6,15,3]. Some others rather consider a pre-processing stage to reduce dimensionality effect on their algorithms [13,14,1].

One of the most fundamental challenges of associating OLAP with a predictive model concerns involving the hierarchical structure of data cubes to improve analysis performances. While some approaches employ low level model to constitute higher aggregation level models as [6], other approaches inverse this methodology to derive, low-level facts from their existent aggregates as [13]. NAP-HC does not handle multiple hierarchical levels. It explores one level per dimension during all the analysis. Nevertheless, NAP-SC explores multiple hierarchical levels during the dimension reduction and the prediction stages.

On the other hand, sparsity, which is the case of most of OLAP cubes, remains a very serious issue. Thomsen defines it as the degrees to which cells contain invalid values instead of data [16]. In addition of increasing the access time, it degrades most of analysis techniques applied on data cubes [9,10]. For instance, the conducted experiments in [1] showed clearly that the performances of NAP-HC decreased progressively while increasing the sparsity level of the treated cube. Therefore, we argue that handling sparsity represents a fundamental challenge for any approach trying to extend data mining algorithms to OLAP environment. Despite this, all of the above cited researches do not provide explicit solutions of their confrontations with sparsity.

We affirm that the solutions reached in [1] still hold a lot of promises and major aspects of their potential contributions remain unexplored. Therefore, we propose to study them further by proposing NAP-SC, which is a novel approach for predicting non-existent measures over sparse OLAP cubes. It has the general overview of NAP-HC since the two approaches operate according to the same global stages. Nevertheless, a closer look reveals substantial differences that we will expose in the following sections of this paper.

3   General Notations

In this section we present the general notations that we use in this paper. We also present the definitions introduced in [1] accompanied with a short reminder of NAP-HC. We entend to reuse the same data cube definition provided in [7].

Let $\mathcal{C}$ be a data cube having the following proprieties:

- $\mathcal{C}$ has a non empty set of d dimensions $\mathcal{D} = \{D_i\}_{(1 \leq i \leq d)}$;
- $\mathcal{C}$ contains a non empty set of m measures $\mathcal{M} = \{M_q\}_{(1 \leq q \leq m)}$;
- Each dimension D_i contains l_i categorical attributes;
- H_i is the set of hierarchical levels of the dimension D_i. H_j^i is the jth hierarchical level in D_i.

We also use the concept of *cube level*, proposed by Agarwal and Chen in [3]. It defines a vector of distinct dimensions levels.

NAP-HC embeds MLPs, which had proven their performances in prediction tasks [17,18], within OLAP environment in a two stage proposal. The first stage consists in generating reduced information preserving training sets over the original cube while preserving the measure variations and the semantics linking attributes and dimensions. In order to do so, we resorted Principal Component Analysis (PCA) as a dimensions reduction technique [19]. We exploited its orthogonal transformation to convert the correlated dimension attributes into smaller sets of principal components.

As PCA is not designed for multidimensional structures, we introduced the concept of cube-face to identify all the possible configurations of the data cube and cover all the measure's variations.

Definition 1 (Cube-face). *Let* $\{D_k, D_v, D_{s_1}, \ldots, D_{s_f}, \ldots, D_{s_{d-2}}\}_{(1 \leq f \leq d-2)}$ *be a non-empty subset of d distinct dimensions.*

We denote by $Cf(D_k, D_v, (D_{s_1}, \ldots, D_{s_f}, \ldots, D_{s_{d-2}}))$ *a cube-face Cf of a data cube $\mathcal{C}$. It is a data view of a data cube identifiable by the geometrical positions of its dimensions that we call: Key dimension D_k, Variant dimension D_v and a set of $(d-2)$ Slicer dimensions $D_{s_{d-2}}$.*

The number, n, of extractable cube-faces over a data cube is equal to the number of its geometrical faces.

To preserve the semantics linking attributes and dimensions, we introduced the concept of PCA-slice.

Definition 2 (PCA-slice). $P(D_k, D_v, (a_{t_1}^o, \ldots, a_{t_f}^p, \ldots, a_{t_{d-2}}^q))$ *is the PCA-slice obtained by applying the OLAP* Slice *operator on* Cf; *with* $a_{t_1}^o a_{t_f}^p$ *and* $a_{t_{d-2}}^q \in D_{s_1}, D_{s_f}$ *and* $D_{s_{d-2}}$, *respectively.*

The coordinate factors are generated by iteratively applying PCA on the extracted PCA-slices and stored in external tables that we called PCA-tables. In order to track the membership of a measure and its corresponding coordinate factors, we introduced the concept of PCA-cube.

Definition 3 (PCA-cube). *A PCA-cube is a data cube that contains, beside its original measures, a new type of measures consisting of references to the sets of coordinate factors associated to each cell.*

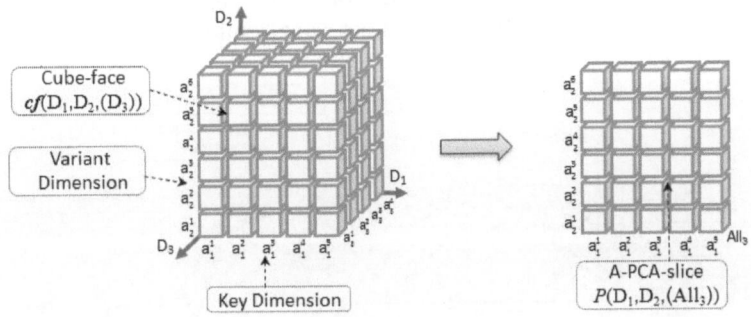

Fig. 1. Cube-face transformation

As for the second stage, we designed a new MLPs architecture to overtake the multidimensional structure of data cubes. It consists of an interconnection of $n+1$ sub-networks. Firstly, n child-networks, each one associated to a distinct cube-face. It gets that cube-face coordinate factors as inputs and provides the targeted measures as output. Then, a combinator-network that receives the outputs of all child-networks as inputs and comes out with the targeted measures. The innovative aspect of this architecture consists in involving multiple training-sets in the same learning process without having to merge them and yet generating a unique predicted value for each targeted measure.

Like NAP-HC, NAP-SC is not a cube completion technique. It is not supposed to fill all empty measures of a data cube. Its main objective is to promptly come out with prediction of any empty measure upon the request of the user.

4   Formalization of Our Proposal

Most of machine learning algorithms are not designed to deal with missing values. Many researches apply deletion techniques, which provide trivial solutions that enable data mining algorithm application. However, they may seriously affect data quality and lead to non representative data set. Other researches use imputation methods like multiple imputation [20], regression, mean imputation, etc. However, these methods are designed for bi-dimensional data and are not adapted to multidimensional structures.

In [21], Ben Messaoud *et al.* proposed a cube reorganization approach that generates more dense cubes from spare ones. While facing the sparsity issue, the authors proposed to explore OLAP aggregation operators to minimize sparsity effect. They transformed a data cube into a complete disjunctive table by fixing two dimensions, treated as instance and variable dimensions, and aggregating the remaining ones to the highest level, which is naturally the *All* level.

We affirm that aggregating some dimensions to higher levels enable the analysis to avoid many empty cells. However, we notice that fixing one specific combination of dimensions may promote some dimensions on the expense of others. This may cause a loss of the information extractable over the unexplored combinations. That is to say, any transformation of the dimensions combination will certainly lead to a whole different data set that remain unexplored in the case of [21]'s approach.

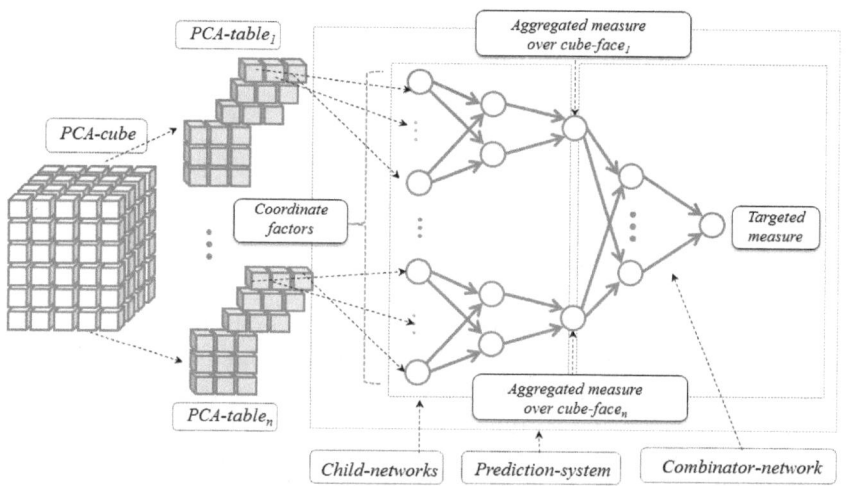

Fig. 2. Prediction system

We intend to revise this transformation technique by coupling it with NAP-HC cube decomposition technique. This will enable it to consider all the extractable dimension combinations over the data cube and thus to treat equitably all dimensions. Then, we will explore the revised version to minimize sparsity effect on cube-faces.

On the other hand, we must consider that the aggregation step causes loss of precision in term of dimension attributes. Nevertheless, we intend to make up for this by revising the forthcoming prediction system and enabling it to involve multiple hierarchical levels in the same prediction process.

4.1 Dimensions Reduction

The main purpose of the dimensions reduction stage is to generate information preserving, concentrated, decorrelated training sets over the original data cube. Like NAP-HC, NAP-SC uses PCA as a dimensions reduction technique and explore its orthogonal transformations to extract smaller sets of principal components. They serves as predictors of the forthcoming prediction system. The sheer novelty of NAP-SC dimensions reduction stage consists in including a revisited version of Ben Messaoud et $al.$'s proposal to minimize sparsity effect.

Actually, the reduction process starts by identifying all the cube-faces from the data cube following the initially selected cube level. At this point, NAP-HC extracts all the PCA-slices from the cube-faces and iteratively applies PCA on them. PCA-slices are generally sparse, what affects the reduction and thus the prediction process. To minimize sparsity effect, we propose to reuse the transformation proposal of Ben Messaoud et $al.$ after coupling it with our cube decomposition proposal to preponderate equitability among all the cube dimensions.

Algorithm 1. Training algorithm

Input: $\mathcal{P}cc, cube\_faces' set, RMSE\_max$
Output: $child, combinator, RMSE$
1 **foreach** $cube\_face$ in $cube\_faces' set$ **do**
2     $initialize(child)$;
3     $converged \leftarrow false$;
4   **while** $converged = false$ **do**
5        $agg\_m \leftarrow extract\_agg(cube\_face.A\_PCA\_slice)$;
6        $pc[] \leftarrow extract\_comp(\mathcal{P}cc, agg\_m)$;
7        $propagate(pc, agg\_m, child)$;
8        $adjust(child)$;
9   **if** $RMSE(child) < RMSE\_max$ **then**
10 | $converged \leftarrow true$;
11

12 $initialize(combinator)$;
13 $converged \leftarrow false$;
14 **while** $converged = false$ **do**
15   **foreach** $child$ **do**
16       $agg\_m \leftarrow extract\_agg(cube\_face.A\_PCA\_slice)$;
17       $pc[] \leftarrow extract\_comp(\mathcal{P}cc, agg\_m, cube\_face)$;
18       $combinator\_input[] \leftarrow propagate(pc, agg\_m, child)$;
19    $m \leftarrow extract\_measure(Pcc)$;
20    $propagate(combinator\_input[], m, combinator)$;
21    $adjust(combinator)$;
22   **if** $RMSE(combinator) < RMSE\_max$ **then**
23       $converged \leftarrow true$;
24       $RMSE \leftarrow RMSE(combinator)$;
25   **return** $child$ $networks'$ $set, combinator, RMSE$;

Thus, we transform each cube-face as follows: We keep the *key* and the *variable* dimensions still. Then, using the aggregation function that had been used in the initial cube computation, we totally aggregate all the *slicer* dimensions to the *All* level, as mentioned in Figure 1. This operation generates one single large PCA-slice per cube-face. To distinguish it from the classical PCA-slice, we call it *Aggregated-PCA-slice* (A-PCA-slice). The generated A-PCA-slices are much less sparse than the classical PCA-slices due to the aggregation step that enable the analysis to avoid many empty cells by aggregating them according to wisely selected dimensions.

Finally, we apply PCA on each A-PCA-slice. This operation generates sets of coordinate factors, which we store in PCA-tables. Then, we reuse the concept of PCA-cube to track the membership of measures and their coordinate factors.

Like most conventional OLAP pre-processing phases, the reduction stage is a time consuming process. Therefore, we believe that it should be executed in back-stage, on a regular basis, by the end of each periodic data loading of the DW.

4.2   Measure Prediction

The main goal of this stage is to apply MLPs, which can not handle multidimensional structures, on OLAP cubes. NAP-HC prediction system consists of an OLAP oriented MLPs architecture that consider multiple disjoint training-sets without having to merge them. And yet, it comes out with a unique predicted value of each targeted measure. We intend to reuse the general aspects of this system with NAP-SC. Nevertheless, we entend to empower it further, in order to make up for the loss of precision caused by the aggregation step. Thus, we preserve the general overview of the MLPs architecture and modify its training algorithm to enable the prediction system to handle multiple cube levels to restore the initial cube level targeted by the user .

As shown in Figure 2, NAP-SC prediction system consists of an interconnection of multiple sub-networks; $n - 1$ child-networks and one single combinator-network. Each child-network is associated to one distinct A-PCA-slice, and thus to one cube-face. In addition, for each cube-face, we intentionally emphasized the same dimensions in both of measures concentration and cube-face transformation steps. In such a way, the reduction stage preserves the relationships of original and aggregated measures. Thus we affirm that an appropriate prediction system can come out with the original values from the aggregated ones. In our case, it is NAP-SC's prediction system, whose child-networks consider aggregated measures as target instead of initially selected levels measures. As for its combinator-network, it targets the initially selected levels measures. These transitions between correlated dimensions' levels enable the prediction system to deal with different cube levels and to restore the lower cube-level values targeted by the user from the higher cube levels exploited in the reduction stage.

The training algorithm of the prediction system is provided in Algorithm 1. It uses Root Mean Squared Error (RMSE) as stopping criteria. It requires the PCA-cube, the set of cube-faces and a maximum tolerable value of RMSE as inputs.

Each child-network is initialized by randomizing its internal weights and setting-up its structure according to its associated cube-face dimensions. Then, it is trained using a randomly selected set of cells from the PCA-cube $\mathcal{P}cc$. It gets the coordinate factors referenced by the PCA-cube as inputs and the aggregated measures extracted from the cube-face's A-PCA-slice as output. The training process is repeated iteratively, until convergence.

After all the child-networks are trained, the combinator-network is to be initialized. It gets a number of input nodes equal to the number of child-networks and one single output node. It is trained in a batch mode with the trained child-networks' outputs. Unlike child-networks that consider aggregated measures as outputs, the combinator-network consider the measure derived from the initially targeted cube-level as output. It provides the measures's values targeted by the user.

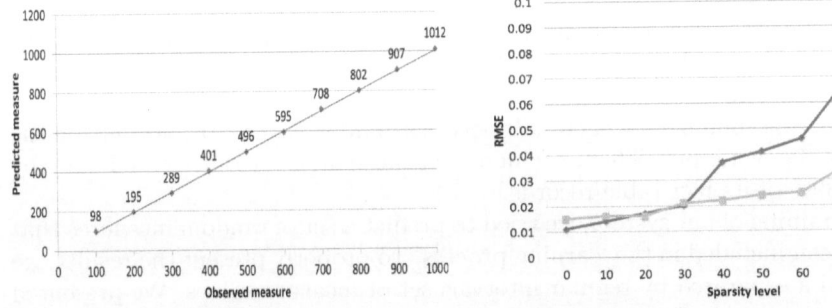

Fig. 3. Prediction quality   **Fig. 4.** Performances against sparsity

As several theoretical and empirical studies show that a single hidden layer is sufficient to achieve a satisfactory approximation of any nonlinear function [17], we restrict the architecture of all sub-networks to three layers, including a hidden one. We also use the gradient back-propagation algorithm [22] that has proved its usefulness in several applications [18,17]. We associate it with the conjugate gradient learning method and the sigmoid activation function.

5   Experimental Study

To test our approach, we implemented an experimental prototype of our system. We adapted the *American Community Surveys 2000-2003* [1] database to DW context and exploited it in our experimental study. It is a real-life, derived from the U.S.A census database. It contains samples of the American population treated between 2000 and 2003.

---

[1] American Community Surveys is accessible from the official site IPUMS-USA (Integrated Public Use Microdata Series); `http://sda.berkeley.edu`

5.1   Analysis Context

We consider a 4 dimensions data cube; *Location, Origin, Education* and *Time*, with 3.8 million facts. We aim at predicting the number of people of a certain race, according to their cities and their levels of education. To compute the initial cube level, we focus on the *person count* measure and select the hierarchical levels; *State, Race* and *Education*. These levels contain, respectively, 51, 10 and 14 dimension attributes. After the application of our reduction approach, we end up with 6 cube-faces that generate 10, 12, 4, 3, 4 and 3 principal components. For the prediction phase, we use the 10-fold cross-validation technique and the RMSE as a quality indicator. Our experimental study is spread over two experiments to investigate the following aspects.

5.2   Prediction Quality

The purpose of the first experiment is to investigate the performances of NAP-SC in the case of real-life high sparsity data cube. We elaborated a predictive system that faithfully represents our proposed architecture. We have set the hidden neurons number of each sub-network hidden layer, to one half of the number of its own input. Then, we increased the sparsity of our treated cube by deleting 30% of its fact table records.

After training of our system, we tried to predict a set of random measures that had not been included in the learning process. To properly present the results, we considered a separated by regular intervals set of measure values. We presented the resulting curve in Figure 3. We note that the predicted values have minimum distances from the line *observed measure = predicted measure*. Furthermore, the correlation coefficient of these values is equal to 0.96, what indicates an accurate prediction.

5.3   Efficiency against Sparsity

In order to investigate NAP-SC efficiency against sparsity, we compared its performances with those of NAP-HC, while varying the sparsity level. Thus, we started with our initial data cube and increased the sparsity of the data cube by 10% at each time.

We present the results in Figure 4, $S1$ and $S2$ translate, respectively, the performances of NAP-HC and NAP-SC. We notice that NAP-HC outperformed NAP-SC for a level of sparsity between 0% and 25%. From 30%, NAP-SC takes the lead and outperforms NAP-HC. Then, it preserves its robustness until a percentage of 70% of non-existent facts. However, from 70%, NAP-SC's RMSE evolution becomes important and the prediction quality decreases remarkably. This is explained by the fact that the minimum number of valid instances becomes insufficient from a level of sparsity of 70% .

Through this experiment, we highlighted the usefulness of both NAP-HC and NAP-SC. The BI customer can use either system depending on his treated cube.

We affirm that he can even apply individually both systems on the same data cube according to the sparsity level of the manipulated dimensions' hierarchical levels.

6 Conclusion and Perspectives

In this paper, we explored neural networks, which have been proven their efficiency in several data mining techniques, to empower OLAP and extend it to prediction capabilities.

We proposed NAP-SC that predicts non-existent measures over sparse data cubes. It is a substantially modified version of our previously proposed approach NAP-HC [1], designed to solve the same problem in the case of high dimensional non sparse data cubes. The main contribution of NAP-SC consists in minimizing sparsity effect on analysis. It is ensured through a cube transformation step based on a dedicated aggregation technique. In addition, the prediction stage of NAP-SC involves multiple cube levels in the same prediction process, what embeds further our proposed MLPs architecture into OLAP environment.

The experimental study showed the improved accuracy of NAP-SC and its robustness against sparsity. Notwithstanding, NAP-HC outperformed NAP-SC in the case of limited sparsity data cubes. Thus we conclude that both system are useful for different scenarios. It is up to BI customer to choose between them according to the nature of his treated cubes. He can even apply individually both systems on the same data cube in different cube levels.

As part of future work, we plan to formalize explicit criterion indicating which system to apply between NAP-HC and NAP-SC. In addition, we plan to include a framework that explains the reasons of non-existent measures occurrences, similarly to that of [23]. Finally, we aim at modeling a theoretical relation between the reduction and the prediction stage to optimize our model.

References

1. Abdelbaki, W., Ben Messaoud, R., Ben Yahia, S.: A Neural-Based Approach for Extending OLAP to Prediction. In: Cuzzocrea, A., Dayal, U. (eds.) DaWaK 2012. LNCS, vol. 7448, pp. 117–129. Springer, Heidelberg (2012)
2. Inmon, W.H.: Building the Data Warehouse. John Wiley & Sons (1996)
3. Agarwal, D., Chen, B.C.: Latent OLAP: Data Cubes Over Latent Variables. In: Proceedings of the 2011 International Conference on Management of Data, SIGMOD 2011, pp. 877–888. ACM, New York (2011)
4. Codd, E.F., Codd, S.B., Salley, C.T.: Providing OLAP (on-line Analytical Processing) to User-analysts: An IT Mandate, vol. 32. Codd & Date, Inc. (1993)
5. Olson, D., Delen, D.: Advanced Data Mining Techniques. Springer (2008)
6. Sarawagi, S., Agrawal, R., Megiddo, N.: Discovery-Driven Exploration of OLAP Data Cubes. In: Schek, H.-J., Saltor, F., Ramos, I., Alonso, G. (eds.) EDBT 1998. LNCS, vol. 1377, pp. 168–182. Springer, Heidelberg (1998)
7. Ben Messaoud, R., Loudcher-Rabaseda, S.: OLEMAR: An On-Line Environment for Mining Association Rules in Multidimensional Data. In: Advances in Data Warehousing and Mining, vol. 2. Idea Group Publishing (2007)

8. Bodin-Niemczuk, A., Ben Messaoud, R., Rabaséda, S.L., Boussaid, O.: Vers l'intégration de la prédiction dans les cubes OLAP. In: EGC, pp. 203–204 (2008)
9. Niemi, T., Nummenmaa, J., Thanisch, P.: Normalising OLAP Cubes for Controlling Sparsity. Data & Knowledge Engineering 46(3), 317–343 (2003)
10. Kriegel, H.P., Borgwardt, K.M., Kröger, P., Pryakhin, A., Schubert, M., Zimek, A.: Future Trends in Data Mining. Data Min. Knowl. Discov. 15(1), 87–97 (2007)
11. Juyoung, K., Hwanseung, Y., Yoshifumi, M.: Classification of Sparsity Patterns and Performance Evaluation in OLAP Systems. IPSJ SIG Notes 67, 449–455 (2002)
12. Goil, S., Choudhary, A.: High Performance Multidimensional Analysis and Data Mining. In: Proceedings of the High Performance Networking and Computing Conference (SC 1998), Orlando, Florida, USA (November 1998)
13. Palpanas, T., Koudas, N., Mendelzon, A.: Using Datacube Aggregates for Approximate Querying and Deviation Detection. IEEE Trans. on Knowl. and Data Eng. 17, 1465–1477 (2005)
14. Chen, Y., Dong, G., Han, J., Pei, J., Wah, B.W., Wang, J.: Regression Cubes with Lossless Compression and Aggregation. IEEE Trans. on Knowl. and Data Eng. 18 (December 2006)
15. Chen, B.C., Chen, L., Lin, Y., Ramakrishnan, R.: Prediction Cubes. In: Proceedings of the 31st International Conference on Very large Data Bases, VLDB 2005, pp. 982–993 (2005)
16. Thomsen, E.: Olap Solutions: Building Multidimensional Information Systems. Wiley Computer Publishing (2002)
17. Hornik, K., Stinchcombe, M., White, H.: Multilayer Feedforward Networks are Universal Approximators. Neural Networks 2(5), 359–366 (1989)
18. Haykin, S.: Neural Networks: a Comprehensive Foundation. Prentice Hall International Editions Series. Prentice Hall (1999)
19. Hotelling, H.: Analysis of a Complex of Statistical Variables into Principal Components. Journal of Educational Psychology 24(7), 498–520 (1933)
20. Rubin, D.B.: Multiple Imputations in Sample Surveys: a Phenomenological Bayesian Approach to Nonresponse. In: Proceedings of the Survey Research Methods Section, pp. 20–28 (1978)
21. Ben Messaoud, R., Boussaid, O., Rabaséda, S.L.: Using a Factorial Approach for Efficient Representation of Relevant OLAP Facts. In: Proceedings of the 7th International Baltic Conference on Databases and Information Systems (DB&IS 2006), pp. 98–105. IEEE Communications Society, Vilnius (2006)
22. Rumelhart, D., McClelland, J., University of California, S.D.P.R.G: Parallel Distributed Processing: Explorations in the Microstructure of Cognition. Foundations. Computational Models of Cognition and Perception. MIT Press (1986)
23. Ben Othman, L., Rioult, F., Ben Yahia, S., Crémilleux, B.: Missing Values: Proposition of a Typology and Characterization with an Association Rule-Based Model. In: Pedersen, T.B., Mohania, M.K., Tjoa, A.M. (eds.) DaWaK 2009. LNCS, vol. 5691, pp. 441–452. Springer, Heidelberg (2009)

Semi-supervised Gaussian Process Regression and Its Feedback Design

Xinlu Guo[1], Yoshiaki Yasumura[2], and Kuniaki Uehara[1]

[1] Graduate School of System Informatics, Kobe University
guo@ai.cs.scitec.kobe-u.ac.jp,
uehara@kobe-u.ac.jp
[2] College of Engineering, Shibaura Institute of Technology
yasumura@shibaura-it.ac.jp

Abstract. Semi-supervised learning has received considerable attention in the machine learning literature due to its potential in reducing the need for expensive labeled data. The majority of the proposed algorithms, however, have been applied to the classification task. In this paper we present a graph-based semi-supervised algorithm for solving regression problem. Our method incorporates an adjacent graph, which is built on labeled and unlabeled data, with the standard Gaussian process (GP) prior to infer the new training and predicting distribution for semi-supervised GP regression (GPr). Additionally, in semi-supervised regression, the prediction of unlabeled data could contain some valuable information. For example, it can be seen as labeled data paired with the unlabeled data, and under some metrics, they can help to construct more accurate model. Therefore, we also describe a feedback algorithm, which can choose the useful prediction of unlabeled data for feedback to re-train the model iteratively. Experimental results show that our work achieves comparable performance to standard GPr.

Keywords: Semi-supervised learning, Gaussian Process, Graph Laplacian, Regression, Feedback.

1 Introduction

By using enough labeled training data supervised learning algorithm can learn reasonably accurate model. However, in many machine learning domains, such as bioinformatics and text processing, labeled data is often difficult, expensive and time consuming to obtain. Meanwhile unlabeled data may be relatively easy to collect in practice. For this reason in recent years semi-supervised learning has received considerable attention in the machine learning literature due to its potential in utilizing unlabeled data to improve the predictive accuracy.

There are various algorithms to implement semi-supervised learning task[17]. However, most of work in this field consider classification problem, while regression remains largely under studied. Generally, the regression problem is more general than the classification problem since the outputs in the latter case are

S. Zhou, S. Zhang, and G. Karypis (Eds.): ADMA 2012, LNAI 7713, pp. 353–366, 2012.

constrained to have only a finite number of possible values whereas in regression they are assumed to be continuous. Thus, we have a lot of interest in regression algorithm that can make use of unlabeled data.

In this paper, we propose an inductive semi-supervised regression algorithm. Our algorithm firstly builds a adjacent graph by using the distances between labeled and unlabeled data. In regression, such distance information can be used to reflect the similarity between the output values. Then we use the graph information within the standard Gaussian process (GP) probabilistic framework to infer a new GP prior and a graph-based covariance function. From the new prior and the graph-based covariance function, we derive the key training model and predicting distribution for the semi-supervised GP regression (GPr). Since the predictions from the GP model take the form of a full predictive distribution, the unseen data can also be predicted easily by the distribution. That is to say our method is inductive. Our approach is similar to the method proposed by Sindhwani et. al.[12]. But the method is related to classification task, and we focus on solving regression problem.

Furthermore, we also present a feedback algorithm to pick up the helpful predictions of unlabeled data for feeding back and re-train the model iteratively. Most existing semi-supervised methods do not utilize the information available through the predictions of the system feedback. However, the prediction may contain some valuable information. For example, the unlabeled data and its prediction can be seen as labeled data, and under some metrics, they can help to construct more accurate model. In other words, when a learning process is performed repeatedly, we gain extra information from a new source: past unlabeled examples and their predictions, which can be viewed as a kind of experience. This kind of experience serves as a new source of knowledge related to the prediction model. The new knowledge provides the possibility of improving the performance of semi-supervised GPr. Therefore, in this paper, we exploit the feedback algorithm to improve the predictive accuracy in an iterative training setting. The framework picks up helpful predictions of unlabeled data, and add them into the labeled dataset and re-train the regression model iteratively.

2   Related Work

Zhou *et al.* [14] proposed a co-training style semi-supervised regression algorithm. The algorithm employs two k-nearest neighbor regressors as the base learners, each with a different distance metric. Each regressor labels the unlabeled data, and the most confident labeled instances are used for the other learner. The final prediction is made by averaging the regression estimates generated by both regressors. Similarly Brefeld *et al.* [3] performed another co-training style semi-supervised regression algorithm by employing multiple learners. These approaches do not modify the algorithm framework. Indeed, they just keep the supervised algorithm and only change the form of the labels of data. In other words, the co-training method do not take full advantage of the inherent structure between labeled and unlabeled data points.

Besides co-training, regularization based method has also been widely employed in semi-supervised regression. This method combines a regularization term of all data, with the predictive error of labeled data into a criterion. By using such criterion, the unlabeled data can help to get a better knowledge for what parts of the input space that the predictive function varies smoothly in. A variety of approaches using the regularization term have been proposed. Some well known regularization terms are graph Laplacian regularizer[15], Hessian regularizer [5], parallel field regularizer [6], and so on. These methods have enjoyed a great success. However, they are transductive. Transductive learning only works on the observed labeled and unlabeled training data. Although transductive task is simpler than inductive inference, it can not handle the unseen data. Therefore, semi-supervised induction has been attracting more attention recently.

3   An Overview of Gaussian Process Regression

GP has been proved to be useful and powerful tool for the purpose of regression. The important advantage of GP is the explicit probabilistic formulation. This not only provides probabilistic predictions but also gives the ability to infer model parameters. Here, we will give the key GP training and prediction distribution for supervised learning (for more details see [9]). We assume that the input training data is given as $X_D = \{X_L, X_U\} = \{x_1, \ldots, x_\ell, x_{\ell+1}, \ldots, x_N\}$, where $x_i \in R^m$, N is the total number of input data and ℓ is the number of labeled data. X_L and X_U denote the inputs of labeled and unlabeled dataset respectively. We use $y = \{y_1, \ldots, y_\ell\}$ to represent the corresponding outputs of labeled data X_L.

In the GP methodology, the corresponding output label y is assumed relating to an latent function $f(x)$ through a Gaussian noise model: $y = f(x) + \mathcal{N}(0, \sigma^2)$, where $\mathcal{N}(m, c)$ is a Gaussian distribution with mean m and covariance c. The regression task is to learn a specific mapping function $f(x)$, which maps an input vector to a label value. Usually, a zero-mean multivariate Gaussian prior distribution is placed over f. That is:

$$p(f|X_L) = (2\pi)^{-\frac{\ell}{2}} |K_L|^{-\frac{1}{2}} \exp\left(-\frac{1}{2} f^T K_L^{-1} f\right) \tag{1}$$

where K_L is an $\ell \times \ell$ covariance matrix. In particular, the element of K_L is built by means of a covariance function (kernel) $k(x, x')$. A simple example is the standard Gaussian covariance defined as:

$$k(x, x') = c \cdot \exp\left[-\frac{1}{2} \sum_{j=1}^{d} b_j \left(x^j - x'^j\right)^2\right], \theta = \{c, b\} \tag{2}$$

where $b = \{b_j\}_{j=1}^d$ plays the role of characteristic length-scales. c is the kernel over scale. The parameters b, c are initially unknown and are added to a parameter set θ, which is defined as containing all such hyper-parameters.

Given some observations and a covariance function, we wish to find out the most approximate θ, and make a prediction on the test data. There are various methods for determining the parameters. A general one is the gradient ascent, which seeks an optimal θ by maximizing the marginal likelihood. For a GP model, the marginal likelihood is equal to the integral over the product of likelihood function $p(y|f) = \mathcal{N}(f, \sigma^2 I)$ and the prior $p(f|X_L)$ (Eq.(1)), given as:

$$p(y|X_L) = (2\pi)^{-\frac{\ell}{2}} |K_L + \sigma^2 I|^{-\frac{1}{2}} \exp\left(-\frac{1}{2} y^T (K_L + \sigma^2 I)^{-1} y\right) \tag{3}$$

which is typically thought as the training probability of GP.

Given the observations and optimal θ, the prediction distribution of the target value f_* for a test input x_* can be expressed as [9]:

$$p(f_* \mid x_*, X_L, y) = (2\pi)^{-\frac{\ell}{2}} |c_*|^{-\frac{1}{2}} \exp\left(-\frac{1}{2}(f_* - m_*)^T c_*^{-1}(f_* - m_*)\right) \tag{4}$$

where the predictive mean and variance are:

$$m_* = k_*^T \left(K_L + \sigma^2 I\right)^{-1} y \tag{5}$$
$$c_* = k_{**} - k_*^T \left(K_L + \sigma^2 I\right)^{-1} k_*$$

where k_* is a matrix of covariances between the training inputs and test points. The matrix k_{**} consists of the covariances between the test points.

4   Semi-supervised Gaussian Process Regression

As we can see in standard GPr, neither the prior of latent function f (Eq.(1)) nor the predictive distribution (Eq.(4)) does not contain any information of the un-labeled data. Evidently, to train a accurate GP model, we need to get sufficient training data (labeled data). However, the training data is often difficult and expensive to obtain, while the unlabeled data is relatively easy to collect. So it appears necessary to modify the standard GP model to make it capable of learning from unlabeled data, and thereby improve the performance of prediction. In this section we present how to effectively use the information of unlabeled data to extend the standard GP model into the semi-supervised framework.

According to semi-supervised smoothness assumption, if two points are close, then so should be the corresponding outputs. Based on this assumption, the un-labeled data should be helpful in regression problem. They can help explore the nearness or similarity between outputs. And the output should vary smoothly with this distance. So, to utilize the unlabeled data, we consider building an ad-jacent graph to define the nearness between labeled and unlabeled data. Then we attempt to incorporate the graph information into the standard GP probabilistic framework to generate a new probability model for semi-supervised GPr.

4.1   Prior Condition on Graph

In order to take advantage of the information of unlabeled data, we build an adjacent graph $\mathcal{G} = (V, E)$ on all observed data points $X_D = \{X_L, X_U\}$, to find the adjacent relationship between labeled and unlabeled data, where V is the set of nodes composed by all data points, E is the set of edges between nodes. The graph can be represented by a weight matrix $W = \{w_{ij}\}_{i,j=1}^{N}$, where $w_{ij} = \exp\left(-\frac{\|x_i - x_j\|^2}{2\delta}\right)$ is the edge weight between nodes i and j, with $w_{ij} = 0$ if there is no edge.

From the previous section, we can see that regression by GP is a probabilistic approach. Probabilistic approaches to regression attempt to model $p(y|X_D)$. In this case, to make the unlabeled data affect our predictions, we must make some assumptions about the underlying distribution of input data. In our work we attempt to combine the graph information with the GP. Thus, we focus on incorporating a prior of $p(\mathcal{G})$ with the prior of $p(f)$ to infer a posterior distribution of f condition on the graph $\mathcal{G}$.

Here, we consider the graph $\mathcal{G}$ itself as a random variable. There are many ways to define a probability of the variable $\mathcal{G}$. Sindhwani et al. [12] provides a simple prior of observing the graph $\mathcal{G}$:

$$p\left(\mathcal{G}|f\right) \propto \exp\left(-\frac{1}{2}f^T \Delta f\right) \tag{6}$$

where Δ is a graph regularization matrix, which is defined as the graph Laplacian here. We can derive Δ as follows: let $\Delta = \lambda L^v$, where λ is a weighting factor, v is an integer, and L denotes the combinatorial Laplacian of the graph $\mathcal{G}$. Let $D_{ii} = \sum_j w_{ij}$, the combinatorial Laplacian is defined as $L = D - W$.

Combining the Gaussian process prior $p(f)$ with the likelihood function Eq.(6), we can obtain the posterior distribution of f on the graph $\mathcal{G}$ as follows:

$$p(f|\mathcal{G}) = \frac{1}{p(\mathcal{G})}p(\mathcal{G}|f)p(f) \tag{7}$$

Observably, the posterior distribution Eq.(7) is proportional to $p(\mathcal{G}|f)p(f)$, which is a multivariate Gaussian as follows:

$$p(f|\mathcal{G}) = (2\pi)^{-\frac{N}{2}}|K_{DD}^{-1} + \Delta|^{\frac{1}{2}} \exp\left(-\frac{1}{2}f^T(K_{DD}^{-1} + \Delta)f\right) \tag{8}$$

The posterior distribution Eq. 8 will be used as the prior distribution for the following derivation. To proceed further, we have to derive the posterior of f_X independent of graph G. Here X denotes the more general dataset, which contains observed dataset X_D and a set of unseen test data X_T, i.e., $X = \{X_D, X_T\}$. In standard GP, the joint Gaussian prior distribution of f_X can be expressed as follows:

$$p\left(f_X\right) = \mathcal{N}\left(\begin{bmatrix} 0 \\ 0 \end{bmatrix}, \begin{bmatrix} K_{DD} & K_{DT} \\ K_{DT}^T & K_{TT} \end{bmatrix}\right) \tag{9}$$

Then the same as above, the posterior distribution of f_X conditioned on $\mathcal{G}$ is proportional to $p(\mathcal{G}|f_X)p(f_X)$, and it is explicitly given by a modified covariance function defined in the following:

$$p\left(f_X|\mathcal{G}\right) = \mathcal{N}\left(0, \tilde{K}_{XX}\right) \tag{10}$$

where

$$\tilde{K}_{XX}^{-1} = \begin{bmatrix} K_{DD} & K_{DT} \\ K_{DT}^T & K_{TT} \end{bmatrix}^{-1} + \begin{bmatrix} \Delta & 0 \\ 0 & 0 \end{bmatrix} \qquad (11)$$

Eq.(10) gives a general description that for any finite collection of data X, the latent random variable f_X conditioned on graph $\mathcal{G}$ has a multivariate normal distribution $\mathcal{N}(0, \tilde{K}_X)$, where $\tilde{K}_X$ is the covariance matrix, whose elements are given by evaluating the following kernel function:

$$\tilde{k}(x, z) = k(x, z) - k_x^T (I + \Delta K)^{-1} \Delta k_z \qquad (12)$$

where K is a $N \times N$ matrix of $k(\cdot, \cdot)$, and k_x and k_z denote the column vector $(k(x_1, x), \ldots, k(x_{l+u}, x))^T$.

We notice that by incorporating the graph information Δ with the standard GP prior $p(f)$, we infer a new prior condition on the graph $\mathcal{G}$ and a graph-based covariance function $\tilde{k}$. In fact this semi-supervised kernel (covariance function) is first proposed by Sindhwani et al. [11] from the Reproducing Kernel Hilbert Space view, and is used for the semi-supervised classification task. In our work, we mainly focus on how to utilize the new prior and the graph-based covariance function to derive the training and predicting distributions for semi-supervised GPr.

4.2   Objective Functions

Our objective training function for semi-supervised GPr is the marginal likelihood $p(y|X, \mathcal{G})$, which is the integral of the likelihood times the prior $p(y|X, \mathcal{G}) = \int p(y|f)p(f|X, \mathcal{G})df$. As the same with standard GP, the term marginal likelihood refers to the marginalization over the latent function value f. But the difference is that the prior of semi-supervised GP is the posterior obtained by conditioning the original GP with respect to graph $\mathcal{G}$.

According to Eq. 8 and the likelihood $p(y|f) = \mathcal{N}(f, \sigma^2 I)$, the marginal likelihood of the observed target values y is:

$$p(y|X, \mathcal{G}) = (2\pi)^{-\frac{N}{2}} |\Sigma|^{-\frac{1}{2}} \exp\left(-\frac{1}{2} y^T \Sigma^{-1} y\right) \qquad (13)$$

where $\Sigma = \left(K_{DD}^{-1} + \Delta\right)^{-1} + \sigma^2 I$. This formulation can be seen as the training model of our proposed method. We can select the appropriate values of hyper-parameters $\Theta = \{\theta, \sigma\}$ by maximizing the log marginal likelihood $\log p(y|X, \mathcal{G})$. The goal is to solve $\hat{\Theta} = \arg\max \log p(y|X, \mathcal{G})$. In learning process we seek the partial derivatives of the marginal likelihood, and use them for the gradient ascent to maximize the marginal likelihood with respect to all hyper-parameters.

After learning the model parameters, we are now confronted with the prediction problem. In the prediction process, given a test data x_*, we are going to infer f_* given the observed vector y. According to the prior Eq.(10) and Eq.(13), the joint distribution of the training output y and the test output f_* is

$$\begin{bmatrix} y \\ f_* \end{bmatrix} \sim N\left(\begin{bmatrix} 0 \\ 0 \end{bmatrix}, \begin{bmatrix} \Sigma & \tilde{k}_* \\ \tilde{k}_*^T & \tilde{k}_{**} \end{bmatrix}\right) \tag{14}$$

Then we can use this joint probability and Eq.(13) to compute the Gaussian conditional distribution over f_*:

$$p\left(f_* | x_*, X, y, \mathcal{G}\right) \propto \exp\left(-\frac{1}{2}[y, f_*]\begin{bmatrix} \Sigma & \tilde{k}_* \\ \tilde{k}_*^T & \tilde{k}_{**} \end{bmatrix}^{-1}\begin{bmatrix} y \\ f_* \end{bmatrix}\right) \tag{15}$$

By using the partitioned inverse equations, we can derive the Gaussian conditional distribution of f_* at x_*:

$$p\left(f_* | x_*, X, y, \mathcal{G}\right) = \frac{1}{Z}\exp\left[-\frac{1}{2}(f_* - \hat{\mu})^T C^{-1}(f_* - \hat{\mu})\right] \tag{16}$$

where

$$\hat{\mu} = \tilde{k}_*^T \Sigma^{-1} y \tag{17}$$
$$C = \tilde{k}_{**} - \tilde{k}_*^T \Sigma^{-1} \tilde{k}_*$$

This is the key predictive distribution for our proposed semi-supervised GPr method. $\hat{\mu}$ is the mean prediction at the new point and C is the standard deviation of the prediction. For fixed data and fixed hyper- parameters of the covariance function we can predict the test data from the labeled data and a large amount of unlabeled data.

Note that the graph $\mathcal{G}$ contains the adjacent information of labeled and unlabeled data, and it is helpful for regression according to the smoothness assumption of supervised learning. Then, the knowledge on $p(\mathcal{G})$ that we gain through the unlabeled data carries information that is useful in the inference of $p(y|X, \mathcal{G})$ and $p(f_*|x_*, X, y, \mathcal{G})$, which is the training probability and predictive distribution for semi-supervised GP regression. Thus, our semi-supervised GPr method can be expected to yield an improvement over supervised one.

5   GPr with Feedback

In the semi-supervised regression, we learn a predictive model from labeled and unlabeled data. Then the output of the unlabeled data can be predicted through the model. In this process, predictive output can be viewed as a kind of experience. Such experience provides the possibility of improving the performance of semi-supervised GPr. Therefore, in this paper, we describe a feedback algorithm, which can pick up the useful prediction of unlabeled data for feeding back into the labeled dataset and re-train the model iteratively.

In a predictive system, we can not affirm that all the predictions of unlabeled data could be correctly predicted. For this reason, not all the predictions are helpful for re-training and we need to pick up the useful one from them. Here we call such useful prediction a confident prediction. Now we have a problem that what is the confident prediction on unlabeled data of a regressor. Intuitively, if a labeled example can help to decrease the error of the regressor on the labeled data set, it should be the confident labeled data. Therefore, in each learning iteration of feedback, the confidence of unlabeled data point x_u can be evaluated using a criterion as:

$$E_{x_u} = \sum_{x_i \in X_L} \left((y_i - M(x_i))^2 - (y_i - M'(x_i))^2 \right) \tag{18}$$

here, M is the original semi-supervised regressor trained by the labeled dataset (X_L, y_L) and unlabeled dataset X_U, while M' is the one re-trained by the new labeled dataset $\{(X_L, y_L) \cup (x_u, \hat{y}_u\}$ and unlabeled dataset $\{X_U - x_{u'}\}$. Here x_u is an unlabeled data point while $\hat{y}_u$ is the real-valued output predicted by the original regressor M, i.e. $\hat{y}_u = M(x_u)$. The first term of Eq. 18 denotes the mean squared error (MSE) of the original semi-supervised regressor on labeled dataset, and the second term is expressed the MSE of the regressor utilizing the information provided by $(x_u, \hat{y}_u)$ on the labeled dataset. Thus, $(x_u, \hat{y}_u)$ associated with the biggest positive E_{x_u} can be regarded as the most confident labeled data. In other words, If the value of E_{x_u} is positive, it means utilizing $(x_u, \hat{y}_u)$ is beneficial. So we can use this unlabeled data paired with its prediction as labeled data in the next round of model training. Otherwise, $(x_u, \hat{y}_u)$ is not helpful to train models, and will be omitted. Then the x_u should remain in the unlabeled dataset X_U.

Table 1. Algorithm of feedback

Input: Labeled dataset(X_L, y_L), Unlabeled dataset X_U,
       Learning iterations T, Initial parameters set $init\_para$
Output: Prediction model M
Step1: Training model
$M \leftarrow Semi\_train(X_L, y_L, X_U, inti\_para)$
Step2: Choosing and feedback
for $t = 1 : T$ **do**
    Create pool $X_{U'}$ by randomly picking data points from X_U
    for each $x_u \in X_{U'}$ **do**
        $\hat{y}_u \leftarrow M(x_u)$
        $M' \leftarrow Semi\_train((X_L, y_L) \cup (x_u, \hat{y}_u), \{X_U - x_u\}, init\_para)$
        $E_{x_u} \leftarrow \sum_{x_i \in X_L}((y_i - M(x_i))^2 - (y_i - M'(x_i))^2)$
    end for
    for each $E_{x_u} > 0$ **do**
        $(X_L, y_L) \leftarrow (X_L, y_L) \cup (x_u, \hat{y}_u)$
        $X_U \leftarrow \{X_U - x_u\}$
    end for
    $M \leftarrow Semi\_train(X_L, y_L, X_U, inti\_para)$
end

The pseudo code of our feedback framework is shown in Table 1, where the function $Semi\_train$ returns a semi-supervised GP regressor. The learning process stops when the maximum number of learning iterations, T, is reached, or there is no unlabeled data.

6   Experiments

In this section, we firstly evaluate the performance of proposed semi-supervised GPr (SemiGPr) on some regression datasets, and make a direct comparison to its standard version (GPr). Then we show the experimental results of SemiGPr extension by feedback (named FdGPr).

There are $D + 4$ hyper-parameters in SemiGPr: kernel length-scales $\{b_i\}_{i=1}^D$, where D is the dimension of input X, kernel over scale c, noise σ and edge weight length-scale δ. In our experiment, we select the appropriate values of $\{b, c, \sigma\}$ by maximizing the marginal likelihood. To reduce the computing complexity, we fix $\delta = 10$ for all datasets. 4-fold cross validation is performed on each dataset and all the results are averaged over 40 runs of the algorithm.

The datasets used to evaluate the performance of our method are summarized in Table 2. In the experiment, the examples contained in artificial dataset Friedman is generated from the function: $y = \tan^{-1} \frac{x_2 x_3 - \frac{1}{x_2 x_4}}{x_1}$. The constraint on the attribute is: $x_1 \sim U[0, 100]$, $x_2 \sim U[40\pi, 560\pi]$, $x_3 \sim U[0, 1]$, $x_4 \sim U[1, 11]$. Gaussian noise terms is added to the function. The real-world data sets are from the UCI machine learning repository and StatLib. In our experiment, for each dataset, we randomly choose 25% of the examples to be test data, while the remaining are training data. We take 10% of the training data as labeled examples, and the remaining are used as the set of unlabeled examples. Note that all the datasets are normalized to the range $[0, 1]$.

Table 2. Datasets used for SemiGPr. D is the features; N denotes the size of the data.

Dataset	Friedman	wine	chscase	no2	kin8nm	triazines	pyrim	bodyfat
D	4	11	6	7	8	60	27	14
N	3000	1599	400	500	2000	186	74	252
Source	Artificial	UCI	Statlib	Statlib	UCI	UCI	UCI	Statlib

6.1   Algorithmic Convergency

In this paper, we estimate the hyper-parameters by using the gradient descent method to minimize the following log marginal likelihood.

$$-\log p\left(y|X, \mathcal{G}\right) = \frac{1}{2} y^T \Sigma^{-1} y + \frac{1}{2} \log |\Sigma| + \frac{N}{2} \log 2\pi \qquad (19)$$

Firstly, we discuss the convergence of above training objective function. In Figure 1, we show how the objective function value decrease as a function of the iterations on triazines (left) dataset and no2 (right) dataset. The result of triazines shows a typical convergence process. As the number of iterations is increasing, the objective function value is decreasing smoothly. Meanwhile, the objective function value of no2 is converged in two stages. From the results, we can see that the objective function value decreases with the increase of the number of iterations and the iterative procedure guarantees a local optimum solution for the objective function in Eq.(19). According to our offline experiments, generally, the objective function converges after about 30-40 iterations for the datasets in Table 2.

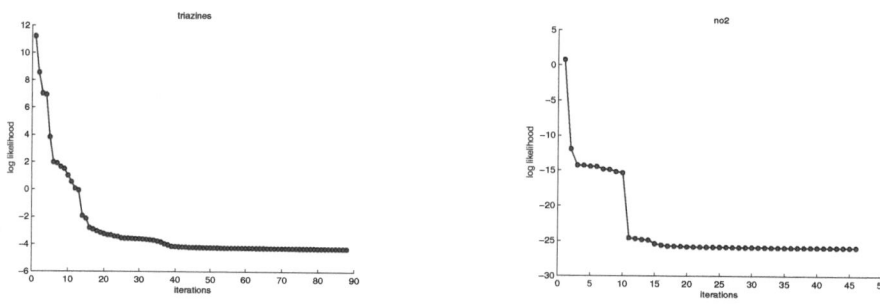

Fig. 1. Log likelihood decreases along with the increase of the iteration No. for the triazines (left) and the no2 (right)

6.2   Efficiency of Unlabeled Data

To verify the SemiGPr model can take advantage of unlabeled data, for a fixed number of labeled data, we vary the number of unlabeled examples, and plot the mean squared error (MSE) for dataset triazines and no2. The corresponding curves are shown in Fig. 2, where the dotted line and solid line indicate the predictive errors on unlabeled dataset and test dataset respectively. Note that when the proportion of unlabeled data is 0%, the result denotes the MSE of standard GPr. The figure shows that the proposed semiGPr algorithm have lower MSE compared to the standard GPr both on unlabeled and test dataset. Moreover, as the proportion of unlabeled examples increases, the advantage of semiGPr increase further. From this result, we can conclude that SemiGPr may bring extra advantage by utilizing the unlabeled data for model training. In other words, the unlabeled data provide some useful information, and our semi-supervised algorithm can make use of this information to improve predictive accuracy.

While we observe a significant performance improvement of the proposed algorithm by using unlabeled examples, the unlabeled examples are not always helpful. For example, for data no2 (right figure of Fig. 2), when the proportion

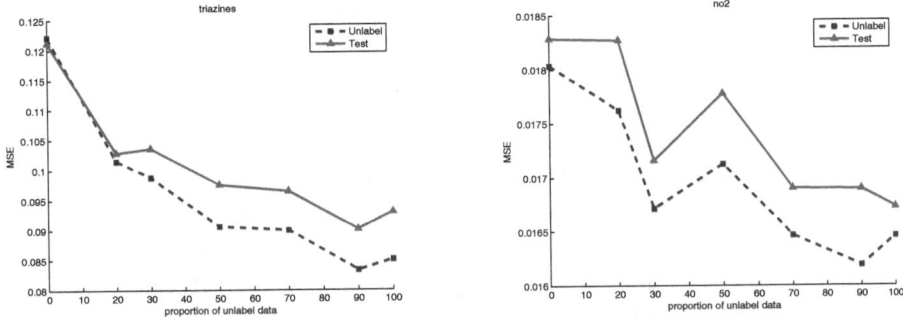

Fig. 2. Performance of SemiGPr as a function of number of unlabeled examples

of unlabeled data goes from 30% to 50%, the error rates are not reduced. On the contrary, they are increased. The same as triazines when the size of unlabeled dataset goes from 90% to 100%. The possible reason for this result is that a part of the unlabeled examples contain some noises or something else, which would be a negative effect on the training predictive model.

6.3   Evaluation of Regression Accuracy

To further clarify the effect of the proposed method, we compare the MSE between SemiGPr and GPr. The comparative results are summarized in Table 3. The above value is the performance on unlabeled dataset, and the following value is the one on test dataset. In this experiment, we consider GPr as the baseline and compare the performance of SemiGPr with it. The improvements are also listed in the table. In addition to the average MSE, we test the significance of the performance difference between SemiGPr and GPr using a paired t-test on the MSE values. The differences are significant with a paired t-test at the 0.05 level, and the results with significant improvement in the table are bold-faced.

The result in Table 3 shows that our method SemiGPr performs as well as or better than the standard GPr in terms of regression accuracy. We can observe that SemiGPr leads to improvements in most of the datasets, and the differences are significant in about half of the datasets. From the comparison, we can conclude that using the unlabeled training data with our semi-supervised regression framework, the GP regression accuracy can be improved. On some of the datasets like chscase, the precision of SemiGPr did not have a significant improvement over the standard one. As our previous analysis, the unlabeled examples sometimes had a negative effect on the regression prediction. Consequently, the possible reason for this result is that there are some noises in unlabeled data or poor hyper-parameter choices made in optimization process.

Table 3. Comparison of SemiGPr with the standard GPr on different datasets

Dataset	Friedman	Wine	chscase	no2	kin8nm	bodyfat	pyrim	triazines
GPr	0.0113	0.0196	0.0273	0.0180	0.0136	0.0026	0.0524	0.1215
	0.0114	0.0205	0.0268	0.0183	0.0134	0.0061	0.0544	0.1205
SemiGPr	0.0101	0.0190	0.0264	0.0161	0.0131	0.0026	0.0359	0.0843
	0.0102	0.0199	0.0265	0.0164	0.0132	0.0027	0.0495	0.0925
Improv.	10.62%	**3.06%**	3.30%	**10.56%**	**3.68%**	0%	**31.49%**	**30.62%**
	10.53%	**2.93%**	1.12%	**10.38%**	1.49%	55.74%	9.01%	**23.24%**

6.4 Extension by Feedback

In this part, two of the datasets used in SemiGPr are presented to demonstrate the effectiveness of the SemiGPr extended by Feedback algorithm, which is denoted by FdGPr. Experimental setting is the same as the previous subsection.

To clarify unlabeled examples and their predictions really contain some valuable information and our feedback algorithm can utilize such information to improve the predictive accuracy, we plot the MSE of FdGPr for different iteration numbers. The results are shown in Fig. 3. The dot line denotes the MSE on the unlabeled dataset, and the solid line is the result of the test dataset. The left figure is the result of dataset *no2* and the right one is that for *chscase*. Note that when the feedback iteration is 0, the result denotes the MSE of SemiGPr.

From the figures we can see that when the iteration number is increased, the feedback algorithm cut the error rate drastically over SemiGPr. The results show clearly that the unlabeled examples and their predictions have a beneficial effect on model learning. From the experiment, larger iteration number almost always produces better results, while considering the computational cost, the iteration T should be set to 20. Although FdGPr achieves comparable performance to a

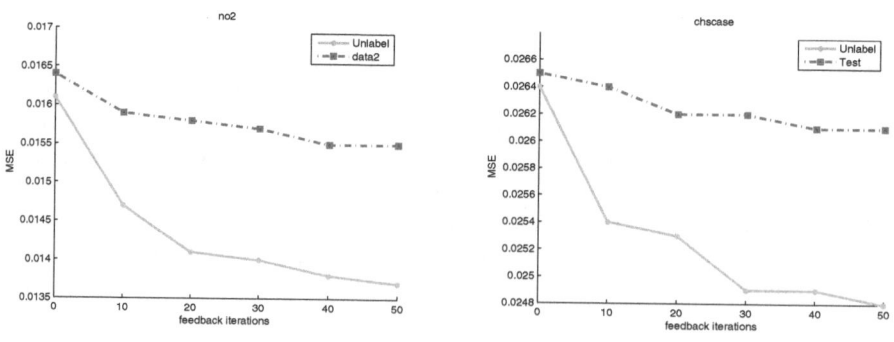

Fig. 3. The effect of different feedback iterations on unlabeled and test dataset

non-feedback baseline on unlabeled dataset, it does not have a significant improvement over the other ones on test dataset. Therefore we should point out that the feedback algorithm makes our work transductive, and we should find a new metric to pick up prediction of unlabeled examples to improve the performance on test dataset in future work. From this result, we can make a conclusion that by utilizing feedback information, FdGPr makes significant performance improvements over all other methods, especially on unlabeled data.

7 Conclusion

In this paper we presented and evaluated a semi-supervised GPr by incorporating a adjacent graph within the standard GP probabilistic framework. Through exploring the standard GP to semi-supervised setting, we can learn a regression model from only partially expensive labeled data and a large amount easy obtaining unlabeled data. Moreover, we presented a feedback algorithm, which can choose the confident prediction for feedback, to further improve the performance. The experimental results indicated that our semi-supervised regression approach can improve the prediction accuracy. Besides, by choosing the confident prediction for feedback, it brought significant improvement in prediction accuracy over a non-feedback baseline. All the experimental results showed that our method achieves comparable performance to standard GPr.

In our experiment, we compared SemiGPr with standard GPr. There exist some other semi-supervised regression methods, such as transductive regularization regression [6], co-training [14], and so on. However, because of the different experimental setting, we could not compare the proposed method with them. In the future work we will implement these methods and make a compare with SemiGPr. Moreover, although our method is considerable for achieving a comparable performance in prediction accuracy, we have to admit that it is time complexity, especially in feedback. In the future work, we plan to do some further research on fast learning to reduce the computational complexity.

References

1. Belkin, M., Niyogi, P., Sindhwani, V.: Manifold Regularization: A Geometric Framework for Learning from Labeled and Unlabeled Examples. The Journal of Machine Learning Research 7, 2399–2434 (2006)
2. Blum, A., Mitchell, T.: Combining Labeled and Unlabeled Data with Co-training. In: Proc. of the 11th Annual Conference on Computational Learning Theory, pp. 92–100 (1998)
3. Brefeld, U., Gärtner, T., Scheffer, T., Wrobel, S.: Efficient Co-regularized Least Squares Regression. In: Proc. of the 23rd International Conference on Machine Learning, pp. 137–144 (2006)
4. Brouard, C., d'Alché-Buc, F., Szafranski, M.: Semi-supervised Penalized Output Kernel Regression for Link Prediction. In: Proc. of the 28th International Conference on Machine Learning, pp. 593–600 (2011)

5. Kim, K.I., Steinke, F., Hein, M.: Semi-supervised Regression using Hessian Energy with an Application to Semi-supervised Dimensionality Reduction. In: Advances in Neural Information Processing Systems, vol. 22, pp. 979–987 (2009)
6. Lin, B.B., Zhang, C.Y., He, X.F.: Semi-supervised Regression via Parallel Field Regularization. In: Advances in Neural Information Processing Systems, vol. 24, pp. 433–441 (2011)
7. Liu, Y., Liu, Y., Zhong, S., Chan, K.C.C.: Semi-supervised Manifold Ordinal Regression for Image Ranking. In: Proc. of the 19th ACM International Conference on Multimedia, pp. 1393–1396 (2011)
8. Nigam, K., McCallum, A.K., Thrun, S., Mitchell, T.: Text Classification from Labeled and Unlabeled Documents using EM. Machine Learning 39(2), 103–134 (2000)
9. Rasmussen, C.E., Williams, C.K.I.: Gaussian Process for Machine Learning. MIT Press (2006)
10. Rwebangira, M.R., Lafferty, J.: Local Linear Semi-supervised Regression. Technical Report CMU-CS-09-106, School of Computer Science, Carnegie Mellon University (2009)
11. Sindhwani, V., Niyogi, P., Belkin, M.: Beyond the Point Cloud: from Transductive to Semi-supervised Learning. In: Proc. of the 22nd International Conference on Machine Learning, pp. 824–831 (2005)
12. Sindhwani, V., Chu, W., Keerthi, S.S.: Semi-supervised Gaussian Process Classifiers. In: Proc. of the 20th International Joint Conference on Artificial Intelligence, pp. 1059–1064 (2007)
13. Sindhwani, V., Niyogi, P., Belkin, M.: A Co-regularized Approach to Semi-supervised Learning with Multiple Views. In: Proc. of ICML Workshop on Learning with Multiple Views, pp. 74–79 (2005)
14. Zhou, Z.H., Li, M.: Semi-supervised Regression with Co-training Style Algorithms. IEEE Transactions on knowledge and Data Engineering 19(11), 1479–1493 (2007)
15. Zhu, X., Ghahramani, Z., Lafferty, J.: Semi-supervised Learning using Gaussian Fields and Harmonic Functions. In: Proc. of the 20th International Conference on Machine Learning, pp. 912–919 (2003)
16. Zhu, X., Lafferty, J.D., Ghahramani, Z.: Semi-Supervised Learning: from Gaussian Fields to Gaussian Processes. Technical Report CMU-CS-03-175, School of Computer Science, Carnegie Mellon University (2003)
17. Zhu, X.: Semi-supervised Learning Literature Survey. Technical Report 1530, Department of Computer Sciences, University of Wisconsin, Madison (2005)

A Graph-Based Churn Prediction Model for Mobile Telecom Networks

Saravanan M.[1] and Vijay Raajaa G.S.[2]

[1] Ericsson R & D, Chennai, India
{m.saravanan@ericsson.com}
[2] Thiagarajar College Of Engineering, Madurai, India
{gsvijayraajaa@gmail.com}

Abstract. With the ever-increasing demand to retain the existing customers with the service provider and to meet up the competition between various telecom operators, it is imperative to identify the number of visible churners in advance, arbitrarily in telecom networks. In this paper, we consider this issue as a social phenomenon introduced to mathematical solution rather than a simple mathematical process. So, we explore the application of graph parameter analysis to the churner behavior. Initially, we try to analyze the graph parameters on a network that is best suited for node level analysis. Machine learning and Statistical techniques are run on the obtained graph parameters from the graph DB to select the most significant parameters towards the churner prediction. The proposed novel churn prediction methodology is finally perceived by constructing a linear model with the relevant list of graph parameters that works in a dynamic and a scalable environment. We have measured the performance of the proposed model on different datasets related to the telecom domain and also compared with our earlier successful models.

Keywords: Call Graph, Churn Prediction, Hadoop Framework, Graph DB, Graph Parameters, Dynamic and Scalable environment.

1 Introduction

Churn in the telecom industry refers to the movement of customers from one operator network to the other. It is an interesting social problem which relates not only to surviving the competition among telecom service providers but also to better understanding of their own customers [1, 2]. It is all the more important as customer churn leads to diminished profits for the operator and enhanced business for the telecom operator's competitor. Moreover, it is more important for the operator to retain an existing customer than to get a new one. With the continuous addition of new operators in the market and with the availability of mobile number portability service, churners are increasing at a higher rate than before. Churn being a predictive model, there is no generalized scalable approach to capture the probable churners effectively in the telecom data.

Several methods and machine learning algorithms have been proposed for predicting churners in different domains [2,3,4]. Existing approach to churn prediction

S. Zhou, S. Zhang, and G. Karypis (Eds.): ADMA 2012, LNAI 7713, pp. 367–382, 2012.

pertains to attribute based analysis which has proven to be relatively time consuming because the process has to be rerun every time the dataset is fed or updated. Moreover the classification model proposed related to this has been proved to face issues with respect to skewness of the churn data [1]. The churn data tends to be imbalanced because the churners tend to be far less in number in the order of (2% - 5%) compared to the non-churners. Due to the existence of the class imbalance problem [5], the high accuracy value derived from a model in churn prediction analysis provides no useful result in real time. Also it has difficulty in parallelizing certain aspects of the traditional algorithms, poses difficulty in applying them over large telecom dataset. Another interesting aspect is that certain attribute based analysis was found to be specific to a particular dataset such as data from a developed country where in the same model failed miserably for the developing country [7].

In order to tackle the above core problems, this paper examines the close relationship between the graph parameter analysis and in understanding the churn behavior. In this study the telecom data is visualized in the form of graph and several graph parameters are inferred from the same. The graph parameters are computed from the vertex and edge pairs ((V (G), E (G)), visualized from the telecom dataset stored in a scalable graph DB framework. The graph DB falls in the class of NOSQL database technologies. The idea of using a NOSQL DB rather than the traditional relational DB is that the NOSQL data technologies supports scalable and schema less structure that helps in analyzing and storing huge datasets [10].

The graph parameters considered for node level analysis are as follows: In-Degree, Out-Degree, Closeness centrality, Call weight, Proximity prestige, Eccentricity centrality, Clustering coefficient, In Degree and Out degree prestige. In addition to this, Game theory approach using Shapley value is calculated to find influential members (most important members) in the network. The graph parameters chosen specifically indicate the active participation of a customer and thus aid in studying the churn behavior over a period of time. The existing call graph implementation for churn analysis related to telecom network behavior study was confined only to the degree module and participation coefficient in a network [2, 3]. It is difficult to predict the churners accurately using a confined set of parameters chosen arbitrarily. Thus we propose a novel idea to analyze graph parameters exhaustively during the training phase of the model, which can help in understanding the factors contributing to churn behavior.

Even though the considered graph parameters hold close relationship to the churn behavior analysis, evaluation of all the graph parameters over the huge dataset on a dynamic environment tends to be a costly process. Thus we need to run predictive machine learning methods like multivariate Discriminant and Regression Analysis that can aid in finding out specific graph parameters that contribute significantly to the discrimination of churn behavior. The corresponding analysis can help in bringing down the list of graph parameters for identifying the visible churners quickly.

Call Detail Record (CDRs) is generated for every transaction made in the telecom domain. With billions of CDR records to be processed, it's virtually impossible to manipulate the data over a single machine. Therefore a map reduce based parallelized framework using HADOOP architecture [11] is employed for pre-processing of CDRs over a cluster environment. The efficiency of combining the map reduce based framework with NOSQL based storage makes this innovative model work in a scalable and a dynamic platform with ease and minimal cost. Eventually, the telecom service providers can use the proposed model in identifying churners efficiently on a streaming environment and in launching retention campaign based on their priorities.

1.1 Our Specific Contributions

- **Geo-spatial data processing on a distributed environment:** The huge CDR data set is pre-processed by splitting them based on specific locations. It is further processed by splitting them into periodic chunks for graph parameter analysis using Hadoop based Map-Reduce framework.
- **Usage of predictive machine learning models to identify specific graph parameters for churn behavior analysis:** We have written the code for predictive models such as: Multivariate Discriminant Analysis and Logistic Regression to extract specific graph parameters for churn analysis that can aid in reducing the computational cost and thus it improve the effectiveness of probable churner identification on a distributed environment.
- **Proposal of a final model that can work in a scalable and dynamic environment for churn prediction:** The proposed graph-oriented model which is a replacement to the traditional approaches can be easily extended to any other domains for probable churner prediction that can work in a scalable and dynamic environment.

2   Related Work

Prediction of probable churners was analyzed by different levels of various studies relevant to the domains such as Telecom service providers, Insurance companies [12], Pay - TV service providers [13], banking and other financial service companies [14], Internet service providers [15], newspapers and magazines [16]. Existing models for churn prediction pertains to supervised and semi supervised methods. Such methods have been designed using different data mining techniques [17]. For instance, the predictive performance of the Support Vector Machine method is benchmarked to Logistic Regression and Random Forest in a newspaper subscription context for constructing a churn model [16]. Another study dealt with the prediction model built for a European pay-tv company by using Markov chains and a Random Forest model benchmarked to a basic logistic model [13]. The general issue with these approaches is that the models don't scale well in a dynamic environment and they tend to be relatively time consuming. Few studies such as [18] employ more than one method based on cluster analysis and classification but they failed to present a standardized model for churn prediction that can be applied to a generalized telecom dataset. The other issue with respect to the classification models is that they face class imbalance problem due to skewedness in the telecom data [1]. Churners correspond to a minor class and therefore building the classification model would bias the model trained towards the majority or the non-churner class [5]. The similarity in these approaches is that the model measures the overall prediction accuracy instead of measuring the accuracy of predicting churners separately. The accuracy can be improved by predicting the non-churners with a high degree of correctness. This is possible because the models trained will be good at predicting non-churners because of the relatively huge number of non-churner samples.

The usage of graph-based techniques for data analysis has been employed in social network analysis [19], analyzing the network structure in the telecom circle and World Wide Web Hyperlink graph analysis [8]. One of the first studies on graph for telecommunication was performed on a graph of land line phone calls made on a single day data. The generated graph consisted of approximately 53 million nodes and 170 million edges [20]. The graph inferred that most of nodes being pairs of telephones that called only each other. Most of the existing graph models are based on the node distributions [8]. Analysis based on confined set of parameters would lead to diminished results. The usage of graph as a means to analyze *big data* has been a successful entity in studying the usage of websites in the internet which employs the ingestion of massive data feed from World Wide Web [8]. Interesting analysis such as Page Rank for search results has been performed out of the same [9]. The telecom data holds close resemblance to the internet feed where in the real time data generated trails a power law graph and the size tends to be huge.

In this paper, we have considered exhaustive list of graph parameters that suits for node level churn analysis which includes centrality and prestige measures. Centrality is based on the choices or participation made by a user whereas prestige depends on the choices that a given user received from others. Churn is a specific business case wherein the telecom carrier would like to identify chunks of users who are likely to churn. We have analyzed the call graph properties specific to customer churn behavior on a telecom domain.

Two of the most widely used statistical methods for analyzing categorical outcome variables related to consumer behavior analytics are linear discriminant analysis and logistic regression [23]. The goal of Logistic Regression is to find the best fitting and most parsimonious model to describe the relationship between the outcome or response variable and a set of independent variables [23]. Linear discriminant analysis can be used to determine which variable discriminates between two or more classes, and to derive a classification model for predicting the group membership of new observations. These methods are more suitable to be employed for graph parameter analysis in our approach. Also one of them can be extended to generate a final linear model for predicting visible churners.

3   Graph Parameter Analysis

The structural properties of call graph are calculated for every node in the network which are analyzed over two time frames in the churn window namely: before the period of churning and during the period of churning. It is noted that the customer is likely to churn out after the second time frame. We try to understand the churn behavior pattern over different phases of the churn window and use the corresponding results for further analysis. The graph parameters chosen for the study depicts different aspects of participation by a customer in the given network. The call graph G is generated by ingesting the CDRs to create ((V(G), E(G)) pairs, where V(G) represents the vertices in the call graph and E(G) represents the edge connecting two vertices. The edge weight represents the number of calls made in the CDR. Fig 1 illustrates the nodes with specific graph parameter measures. The filled vertex (in red color) signifies that the node has high influence in the network which is determined by the Shapley value metrics. The graph parameters considered for the node level analysis are described here.

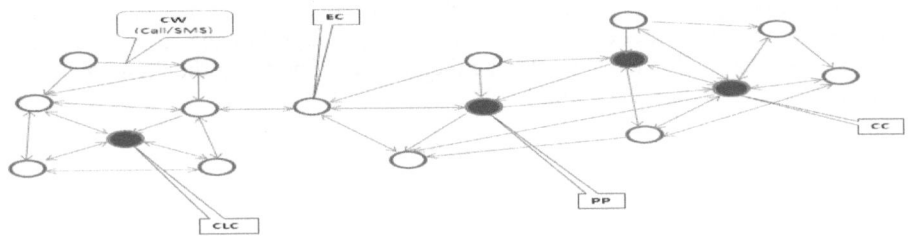

Fig. 1. Sample network with vertices with high graph measures are marked

In-Degree (ID): In-degree measures the number of incoming connections to a given user [19]. The incoming connections can represent the incoming calls or SMS received. To measure the in-degree of a given user (v_i), we count the number of unique users who have communicated to a given user in the network.

Out-Degree (OD): Out-degree measures the number of outgoing connections from a given user (v_i) [19]. The out-degree signifies the active participation of a customer in the network. We find the measure by counting the number of unique users that the user v_i has communicated to.

Closeness Centrality (CC): Closeness centrality measures the importance of a user in a network based on their location in the call graph [21]. A central user will tend to have a high closeness centrality; i.e. if a central user was thought of an information spreader, then rumors initiated by him will spread to the whole network quicker [21]. Let $d_{i,j}$ be the length of the shortest path between vertex v_i and other vertices v_j. Then the average distance between vertex v_i and all other vertices v_j which is given by:

$$l_i = \frac{1}{|v|}\sum_{j\in v}d_{i,j} \tag{1}$$

The closeness centrality is defined as the inverse of l_i.

$$cc_i = \frac{1}{l_i} \tag{2}$$

Degree Prestige (DP): It is based on the in-degree (ID) and the out-degree (OD) of a node in the graph, which takes into account the number of nodes that are adjacent to a particular node in the graph [19]. Prominent customers in the network can be found using this factor.

$$DP_i = \frac{f_i}{|V|-1} \qquad (3)$$

where f_i- is the number of first level neighbors adjacent to node v_i.

Proximity Prestige (PP): Reflects how close all the nodes are present in the graph with respect to a given node x in the network [19]. It signifies the ease of reaching a specific customer in the network. If k_i be the number of nodes in the network who can reach member v_i then PP is given as

$$PP_i = \frac{\dfrac{k_i}{|V|-1}}{\dfrac{1}{k_i}\sum_{j=1,j\in V}^{k_i} d_{i,j}} \qquad (4)$$

Eccentricity Centrality (EC): It states the most central node in the network [21]. The node with high EC value is the one that minimizes the maximum distance to any other node in the network. It signifies the closeness of the neighbor's to a given customer in the network.

$$EC(x) = \frac{1}{\max\{d_{i,j} : j \in V\}} \qquad (5)$$

Clustering Coefficient (CLC): The clustering coefficient represents the density of community accruing from a given node n in the network [19]. When a customer from a highly clustered community is likely to churn then there is a possibility that he will induce other members in the community to churn as well. It is represented as:

$$CLC = \frac{|\text{Actual edges between neighbors' of n}|}{|\text{Possible edges between neighbors' of n}|} \qquad (6)$$

Shapley Value (SV): The Shapley value represents the influential score for a given node in the network [22]. Influential nodes are the one who are not only active in participation but also holds strong influence among their neighboring nodes. The telecom carriers must target the influential churners with their retention scheme first to prevent them from becoming an influential churn spreader. It is represented as:

$$SV_i = \sum_{v_j \in v_i \cup N(v_i,d)} \frac{1}{1+\deg(v_j)} \qquad (7)$$

where, $N(v_i,d)$ represents nodes with d degree of separation from node v_j.

There exists several other graph parameters for graph analysis but we have restricted our analysis to specific parameters that is applicable to node level analysis which have closed association to the events happening in telecom domain. The algorithmic representation of Shapley value is given in Section 4.3.

4   System Overview

The CDR data is visualized as a call graph which consists of vertices and edges based on the activities of individual customers in the network. Exhaustive graph parameter analysis is inferred from the ingested graph database (InfiniteGraph [25]). The visualization and the corresponding analysis are made over a period of time. Specific graph parameters are chosen by employing two different multivariate methods that contributes more to extract churn behavior. Finally, we arrive at a linear model with more specific graph parameters using logistic regression to be employed for probable churner prediction on a dynamic environment. The overall system throws light on a novel way of churn prediction with ease and minimal cost as illustrated in Fig 2. The detailed description of individual components of the system is discussed in the following sections.

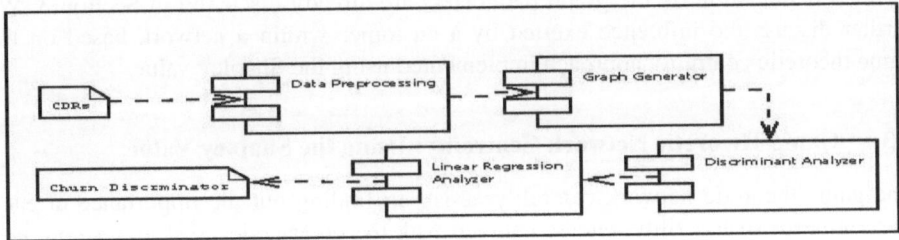

Fig. 2. Overall System Flow Diagram

4.1   Data Pre-processing

In a typical billing system of a mobile operator, for every operation performed by the customer varying from Voice, SMS usage to GPRS usage and each individual event is recorded and stored as Call Detail Records (CDRs).The dataset initially used for the model employs CDRs from leading telecom operators of a developing and a developed country respectively. The time span of the churn window is taken for a period of 3 months. One of the datasets is initially preprocessed to generate urban and rural region datasets to explore deeper analysis of telecom data related to churn problem. The CDR data is later processed by splitting them into weekly window chunks. We split the dataset to analyze the behavior of a customer over a period of time to comprehend the churn behavior. In earlier churn prediction models the attributes were aggregated over entire time period [22]. The disadvantages of other approaches is that in case of new user who has joined in later part of the month may be classified as churner due to his low usage and also it's difficult to generate a predictable pattern from those data, as differences in usage pattern cannot be derived [22]. The dataset generated in the telecom industry tends to be of huge size and hence processing those takes a lot of computational time. Thus we employ Hadoop-based Map Reduce framework to preprocess the CDR data by converging them to location-wise details and use them for graph generation and parameter computations.

4.2   Data Ingestion and Graph Generation

The graph is generated using a distributed graph database implemented in java. It is from a class of NOSQL (or Not Only SQL) data technologies focused on graph data structure. Graph data typically consist of objects (nodes) and various relationships (edges) that may connect two or more nodes. The graph is generated by ingesting the CDRs to create vertex and edge pairs. The edge weight represents the number of calls made or SMS sent. The connections among the node are represented using directed edges. Location wise call graphs are generated for every split window of data and each graph is stored in a graph DB for further graph parameter analysis. The graph parameter chosen ranges from simple computations such as degree calculation to understand the incoming and outgoing activities from a customer to complex centrality evaluation to comprehend the closeness or a closed community formation. The detailed explanation on the graph parameters are already discussed in Section 3. We further discuss the influence exerted by a customer within a network based on the game theoretic centrality approach implemented using the Shapley value.

4.3   Game Theoretic Network Centrality: Using the Shapley Value

The game theoretic network centrality assists in finding out the importance of each node in terms of its utility when combined with the other nodes [22]. In telecom network it wouldn't suffice to find the importance of a node as a mere standalone entity as in other centrality measures. Other works related to the finding of influential nodes in the network [7] didn't address the influence of a node as a combination of several nodes in a network. Given a telecom network, the game theoretic network centrality indicates the *coalition value* of every combination of nodes in the network. We have introduced *Dijkstra's algorithm* to efficiently track the shortest distance between a given node and its neighbor's in calculating Shapley value for each node.

Program: Computing SV by running a game.
Input: Graph Network ingested from CDR.
Output: SVs of all nodes in network

```
foreach node v in Network do
    DistanceVector D = Dijkstra(v,Network);
    kNeighbours(v) = null;
    kDegrees(v) = 0;
    foreach node u<= v in Network do
        if D(u) <= k then
            kNeighbours(v).push(u);
            kDegrees(v)++;
    end
end
```

```
foreach node v in Network do
   ShapleyValue[v] = 1
      kDegrees(v)++;
      foreach node u in kNeighbours(v) do
         ShapleyValue[v]+= 1
         kDegrees(u)++;
      end
   end
   return ShapleyValue;
end
```

4.4   Linear Discriminant Analysis

Linear discriminant analysis is used to determine which attribute discriminates between two or more naturally occurring groups [23]. The linear step-wise discriminant
model is used to extract graph parameters that contribute more to the churn behavior.
A discriminant function that is a linear combination of the components of graph parameters x can be written as:

$$g(x) = w^T x + w_0 \qquad (8)$$

where w is the weight vector for different graph parameters and w_0 is the threshold
weight. In our analysis we define two groups namely: Probable churners and non-
churners. In case of the LDA, a linear discriminant function is of the form

$$y = a_0 + a_1 x_1 + a_2 x_2 + ... + + a_n x_n \qquad (9)$$

with x_i being the graph parameter derived from the CDR data set. The parameters a_i has
to be determined in such a way that the discrimination between the groups is at its best.

4.5   Modified Logistic Regression Model

The usage of logistic regression model in our study is applicable as a two folds
process: First, the logistic regression aids in listing out the graph parameters based on
their significance towards the contribution of churn behavior. Next, we derive a linear
model using the selected list of graph parameters for the churn analysis. The linear
model derived using the logistic regression y is usually defined as similar to Eqn. (9).

The intercept is a constant value of y, when the value of all graph parameters is
taken to be zero. Each of the regression coefficients describes the size of contribution
of the graph parameter towards the churn behavior. The logistic regression is a useful
way of describing the relationship between the extracted graph parameters and for
predicting the churners. Using the linear model we derive a logistic function which
takes on values between zero and one [23]:

$$f(y) = \frac{e^y}{e^y + 1} = \frac{1}{1 + e^{-y}} \qquad (10)$$

The input is y and the output is $f(y)$. The logistic function is useful because it can take
a graph parameter input value ranging from negative infinity to positive infinity, whereas the output is confined to values between 0 and 1 which indicates the probability
to churn or not .

5   Analysis and Results

5.1   Datasets

The graph parameters are examined over three different datasets obtained from the leading telecom service providers of two different countries. The first dataset corresponds to a rural base whereas the second one corresponds to an urban region of a particular country and the third set corresponds to a data from a developed country. The idea of using three different datasets helps to make the model more generic and a standard one. The dataset spans to a time period of three months record. Hence during this period, who ever disconnected from the service is considered as churners. The results given in this section are related to third set for example. Since the evaluation presented in section 6 not shows much difference on the accuracy of all three models generated.

5.2   Weekly Analysis of Graph Data

Fig 3 illustrates the windowing frame used to analyze the churn behavior over a period of time. The windowing frame currently shows a spread of 6 week splits.

- **First time frame:** Time frame before the period of churning.
- **Second time frame:** Time frame during the period of churning.
- **Churn Window:** The customer is likely to churn out after the second time frame.
- **Churn Slider:** The slider moves over the windowing frame over different period of time. The size of the churn slider is a period of one week. It captures the variation in graph parameters. Significant variation in the graph parameters shows the presence of churn behavior.

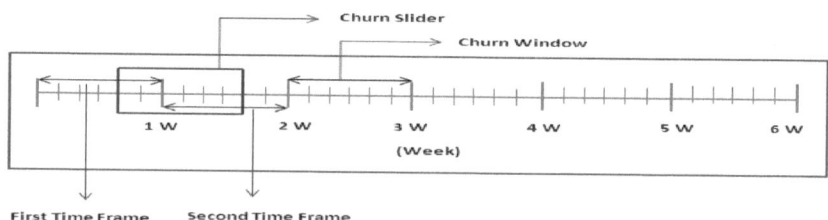

Fig. 3. Analyzing the churn behavior pattern over a period of time using the Windowing frame

The windowing frame is initially split into 3 weeks. The probable churner exists in the first two weeks and churns out in the 3rd week or in churn window. Once the slider crosses the first churn window, the three frames present in the windowing frame progresses by one frame in the forward direction leaving the first week behind. As the slider progresses in the windowing frame, graph parameter variation is captured from one time frame to the other and radical changes in graph parameters are marked as probable churners in the churn window.

5.3 Selection of Best Graph Parameters for Churn Prediction

The idea of selecting specific graph parameters contributing effectively to the churn behavior rather than considering all the parameters can be achieved by running the machine learning algorithms such as Logistic regression and Multivariate Discriminant analysis. These machine learning approaches are used to analyze the graph parameters over the two time frames to highlight a specific list of graph parameters that contribute significantly for discriminating churners from non-churners. The analysis results using the logistic regression and multivariate discriminant analysis are described here.

5.3.1 Implementation of Predictive Data Mining Model: Logistic Regression Model and Multivariate Discriminant Analysis

Logistic regression and Multivariate Discriminant analysis are run over the graph parameters calculated for the different datasets as a training model. Discriminant analysis is used to find out the canonical weighted score for the graph parameters that contribute to the churn behavior. Based on the step wise analysis the graph shown in Fig 4 shows the significant contribution of few graph parameters for churn behavior.

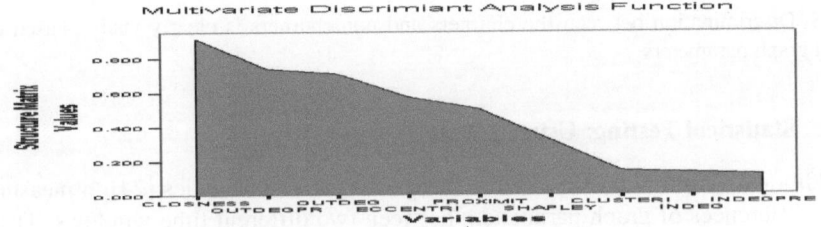

Fig. 4. Multivariate Discriminant analysis Function for graph parameters

The Logistic regression model is first run to calculate the W*ald statistics measure* with its significance for the best contributing graph parameters as illustrated in Table 1. The Wald statistic for a coefficient is the square of the result of dividing the coefficient by its standard error. The logistic regression is later used to derive a linear model for predicting churners for a specific threshold level (0.60) as elaborated in Section 6.

Table 1. Wald Statistics Scores for top four parameters

Variables	Wald	Sig
OD	9.246	0.002
SV	8.354	0.005
CC	23.550	0.000
PP	39.703	0.000

Fig 5 illustrates the discrimination of churners and non-churners based on the selective graph parameters derived from the multivariate discriminant and logistic regression models. The variation is studied by analyzing the Out degree, Shapley value, closeness centrality and proximity prestige before the period of churning. The discriminating functions based on the selected variables clearly segregate the churners and non-churners separately.

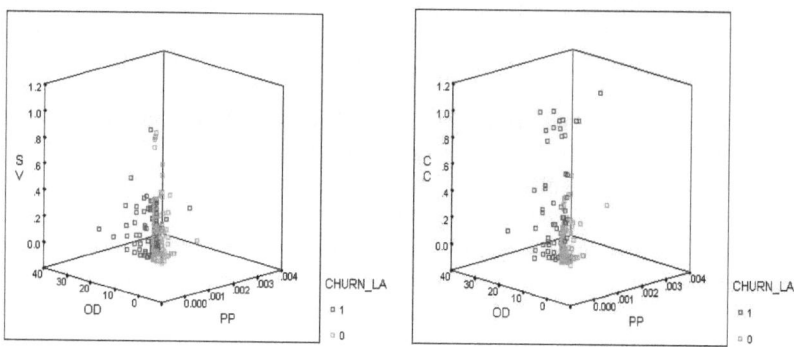

Fig. 5. Discrimination between the churners and non-churners is clearly visible based on selected graph parameters

5.4   Statistical Testing: Using T Test

In addition to Multivariate model, we have used paired t statistics [24] to measure the mean differences of graph parameters between two different time windows. The customer churns out in the third slot or window are considered as real churners. The variation in graph parameter is carefully analyzed over the period of time. The mean value is calculated to give an accumulated score for the graph parameters in terms of churners and non-churners separately. The standard deviation finds the variation in graph parameter from the mean score. The mean and the standard deviation value for the first time frame are compared with the second one for the churners and non-churners respectively. The gradual reduction in specific graph parameters shows that the churners are slowly losing interest in using the corresponding network. This is verified with paired t statistics.

In Table 2, the values given in the bracket denotes that the analysis is carried out during the second time frame in the windowing frame. We find that there is a significant variation for certain graph parameters as in the case of churners. The variation for the non-churners is minimal which proves that the usage by non-churners in the network is almost constant. Based on the t statistics we found that the out degree, closeness, proximity, eccentricity and shapely value has showed relatively high significance for probable churner identification. The drop in out degree of churners clearly illustrates their intention and changes in other important parameters depict the value of their presence in the network. Losing some of the influential users will create rickety in the present network. This result cross verifies the predictive model outputs.

Table 2. Mean and Standard deviation scores for churners and non churners

	CHURNERS		NON-CHURNERS	
	MEAN	**STD. DEV**	**MEAN**	**STD. DEV**
ID	0.70000(0.682)	1.10000(0.95)	1.00000(1.00200)	1.60000(1.65500)
OD	**2.26848(1.8571)**	**3.33069(2.7708)**	1.00000(0.99195)	3.00000(2.98535)
IDP	0.00001(0.00001)	0.00002(0.00001)	0.00002(0.00002)	0.00002(0.00002)
ODP	0.00004(0.00003)	0.00006(0.00006)	0.00002(0.00002)	0.00005(0.00004)
CC	**0.68345(0.74389)**	**0.39080(0.37275)**	0.87667(0.87579)	0.29137(0.29301)
PP	**0.00006(0.00003)**	**0.00018(0.00010)**	0.00002(0.00002)	0.00011(0.00010)
EC	**0.91679(0.93018)**	**0.21142(0.19249)**	0.96912(0.96784)	0.13192(0.13512)
CLC	0.00863(.00865)	0.06270(0.06724)	0.00211(0.00222)	0.02654(0.02797)
SV	**0.30590(0.32772)**	**0.21750(0.22551)**	0.45759(0.46464)	0.21243(0.22581)

6   Evolving a New Methodology for Churn Prediction

The Venn diagram representation given in the Fig 6 illustrates the effectiveness of graph parameters contributing for the churn behavior over a period of time. We found that the graph parameters contributing commonly from all the three datasets mention in Section 5, based on the analysis inferred from the predictive machine learning models and statistical models are: Out Degree, Shapley Value, Proximity Prestige, and Closeness Centrality. We propose that when there is significant variation in the above parameters as a whole then there is a high probability that the customer will churn out. On the other way, the considered parameters will run much faster to generate model and even it can process the data on a streaming environment.

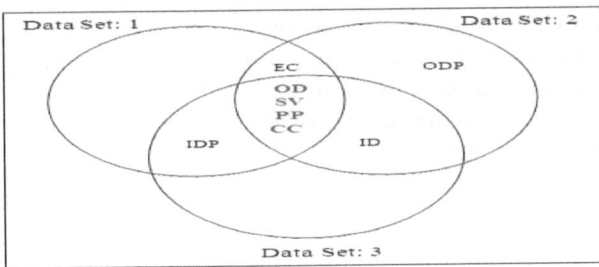

Fig. 6. Significant graph parameters contribution to churn behavior in various datasets

The proposed linear model is used for predicting the probable churners in dynamic environment. We compare the results of predicted churners with the actual churners in the test datasets to find the accuracy of the proposed model. The model was tested for three different datasets over a period of three month time scale. The average accuracy for churn prediction using the proposed model was found to be 81.67 % as shown in Fig 7. The maximum accuracy reached in our previous model using hybrid learning is 72.18% for the same dataset used in this study [22]. The improvements in results highlight the significance of the proposed graph-based model for the churn prediction on mobile telecom networks. Also we have seen the decent improvement in F-measure value from 0.46 to 0.55.

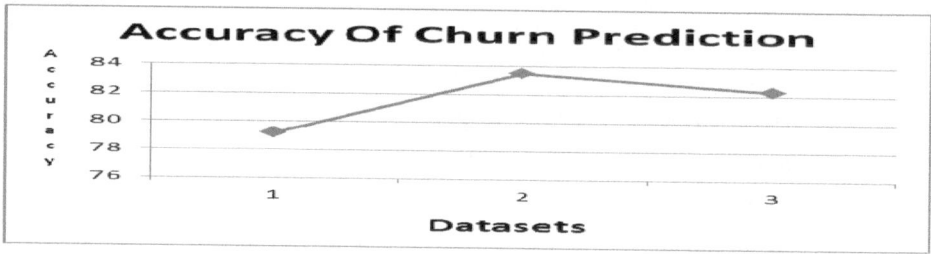

Fig. 7. Testing the accuracy of the churn prediction model

7   Conclusion

The Graph-based analysis for churn prediction is a novel idea proposed for efficient churn prediction in the telecom domain. The graph-based visualization aids in better understanding of the behavior of the customers. Also the simplicity in graph traversal aids in quick manipulation of the graph parameters. The machine learning methods and statistical test chosen for analysis have effectively worked out in choosing specific graph parameters contributing to churn behavior analysis. This approach has helped to make the study cost effective with respect to time and computation. The model also works on a distributive, parallel and scalable platform. The proposed model has overcome the class imbalance problem with graph-based solution which is prevalent in the other existing models. The new model is tuned to work on a generalized dataset. It is framed by collecting the graph parameters over a period of time, analyzing the variation in the parameters over the time period of churning and finally extracting the graph parameters that contribute more to churn behavior. Thus the new model proposed has opened a new way for finding visible churners with ease and cost effectively on a streaming environment.

References

1. Hung, S.-Y., Yen, D.C., Wang, H.-Y.: Applying data mining to telecom churn management. Expert System Applications 31(3), 515–524 (2006)
2. Dasgupta, K., Singh, R., Viswanathan, B., Chakraborty, D., Mukherjea, S., Nanavati, A.A., Joshi, A.: Social ties and their relevance to churn in mobile telecom networks. In: Proceedings of the 11th International Conference on Extending Database Technology, EDBT 2008, New York, USA, pp. 668–677 (2008)

3. Lazarov, V., Capota, M.: Churn Prediction in the Business Analytics Course. TUM Computer Science (2007)
4. Lu, J.: Predicting Customer Churn in the Telecommunications Industry – An Application of Survival Analysis Modeling Using SAS. In: Berry, M.J.A. (ed.) Data Mining Techniques (2004)
5. Burez, J., Vandenpoel, D.: Handling class imbalance in customer churn prediction. Expert Systems with Applications 36(3), 4626–4636 (2009)
6. Kurucz, M., Benczúr, A., Csalogány, K., Lukács, L.: Spectral Clustering in Telephone Call Graph. In: Joint 9th WEBKDD and 1st SNA-KDD Workshop 2007 (WebKDD/SNA.KDD 2007), San Jose, California, USA (2007)
7. Yeshwanth, V., Saravanan, M.: Churn Analysis in Mobile Telecom Data using Hybrid Paradigms. In: Second Conference on the Analysis of Mobile Phone Datasets and Networks, NetMob 2011. MIT, Cambridge (2011)
8. Broder, A.Z., Kumar, R., Maghoul, F., Raghavan, P., Rajagopalan, S., Stata, R., Tomkins, A., Wiener, J.L.: Graph structure in the web. The International Journal of Computer and Telecommunications Networking 33(1-6), 309–320 (2000)
9. Wicks, J., Greenwald, A.R.: Parallelizing the Computation of PageRank. In: Bonato, A., Chung, F.R.K. (eds.) WAW 2007. LNCS, vol. 4863, pp. 202–208. Springer, Heidelberg (2007)
10. DeCandia, G., Hastorun, D., Jampani, Kakulapati, M., Lakshman, A., Pilchin, A., Sivasubramanian, S., Vosshall, P., Vogels, W.: Dynamo: Amazon's highly available key- value store. In: Proceedings of Twenty-First ACM SIGOPS Symposium on Operating Systems Principles (SOSP 2007), pp. 205–220. ACM, New York (2007)
11. Apache Hadoop, http://hadoop.apache.org
12. Morik, K., Kopcke, H.: Analysing customer churn in insurance data a case study. In: Proceedings of the 8th European Conference on Principles and Practice of Knowledge Discovery in Databases, New York, USA, pp. 325–336 (2004)
13. Burez, J., Van den Poel, D.: CRM at a Pay –TV Company: Using analytical models to reduce customer attrition by targeted marketing for subscription services. Expert Systems with Applications 32(2), 277–288 (2007)
14. Halling, M., Hayden, E.: Bank failure prediction: A two-step survival time approach (2006), http://ssrn.com/abstract=904255
15. Khan, A.A., Jamwal, S., Sepehri, M.M.: Applying Data Mining to Customer Churn Prediction in an Internet Service Provider. In: IACSIT Hong Kong Conferences, IPCSIT, vol. 30. IACSIT Press, Singapore (2012)
16. Coussement, K., Van den Poel, D.: Churn prediction in subscription services: An application of sup-port vector machines while comparing two parameter-selection techniques. Expert Systems with Applications 34(1), 313–327 (2008)
17. Wei, C.-P., Chiu, I.-T.: Turning telecommunications call details to churn prediction: a data mining approach. Expert Systems with Applications 23(2), 103–112 (2002)
18. Tsai, C.-F., Lu, Y.-H.: Customer churn prediction by hybrid neural networks. Expert Syst. Appl. 36, 12547–12553 (2009)
19. Wasserman, S., Faust, K.: Social network analysis: Methods and applications. Cambridge University Press, New York (1994)
20. Abello, J., Pardalos, P., Resende, M.: Maximum clique problems in very large graphs. DIMACS Series, vol. 50, pp. 119–130. American Mathematical Society (1999)
21. Karnstedt, M., Rowe, M., Chan, J., Alani, H., Hayes, C.: The Effect of User Features on Churn in Social Networks. In: WebSci 2011, Koblenz, Germany (2011)

22. Yeshwanth, V., Vimal Raj, A., Saravanan, M.: Evolutionary Churn Prediction in Mobile Networks using Hybrid Learning. In: Proceedings of 24th International Florida Artificial Intelligence Research Society Conference (FLAIRS-24), Palm Beach, Florida, USA (2011)
23. Pohar, M., Blas, M., Turk, S.: Comparison of Logistic Regression and Linear Discriminant Analysis: A Simulation Study. Metodološki zvezki 1(1), 143–161 (2004)
24. Donal Zimmerman, W.: A note on interpretation of the Paired-Samples t Test. Journal of Educational and Behavioral Statistics 22(3), 349–360 (1997)
25. InfiniteGraph, http://www.infinitegraph.org

Facial Action Unit and Emotion Recognition with Head Pose Variations

Chadi Trad[1], Hazem Hajj[1], Wassim El-Hajj[2], and Fatima Al-Jamil[3]

[1] American University of Beirut, Faculty of Engineering and Architecture,
Department of Electrical and Computer Engineering
[2] American University of Beirut, Faculty of Arts And Sciences,
Department of Computer Science
[3] American University of Beirut, Faculty of Arts and Sciences,
Department of Psychology
{cht02,hh63,we07,fa25}@aub.edu.lb
http://www.aub.edu.lb/

Abstract. Facial expression recognition has been an active research topic for many years, with Facial Action Coding Systems (FACS) being among the widely used methods. FACS is a well-established scheme in psychology to annotate facial muscle contractions and relaxations, also called Action Units (AUs). Previous work on FACS-based methods focused on frontal or near-frontal head poses. In this work, we propose a method to recognize expressions in side head poses. This method builds one classifier for each possible group of occlusions. Facial expression recognition of a side facial pose is then based on a boosting approach of the different classifiers. The method is first tested with frontal and near-frontal head poses, and the results are shown to be comparable to state of the art work for AU and emotion detection. The method is then tested with a small training set for various orientations and AUs, and shown to be accurate.

Keywords: Facial Action Coding System (FACS), facial expressions, head pose, non-frontal, Action Unit (AU).

1 Introduction

Traditional Human Computer Interface (HCI) designs have focused on interface devices that convey explicit messages from the user, such as touch screens and keyboards, while the affective state has been ignored. However, recent research has demonstrated the importance of affective state in human-human communication [1]. Therefore, recognizing the user's affect can make HCI more natural.

In facial expression detection, the research areas are generally divided into two main streams: facial expression measurement and facial muscle action detection, which fall under the categories known as message-judgment and sign-judgment approaches, respectively [2]. While the aim of the former is to detect the affect underlying the facial expression, the latter aims to purely describe the state of facial components such as their movements or shapes, leaving affect judgment to a higher level process. Facial Action Coding System (FACS) is considered as

S. Zhou, S. Zhang, and G. Karypis (Eds.): ADMA 2012, LNAI 7713, pp. 383–394, 2012.

a sign-judgment approach. FACS, which was introduced by Ekman et al. [3], is the most used coding scheme that describes the muscular activity of a face. Indeed, FACS describes visually discernible facial movements in terms of action units (AUs). Ekman and Friesen identified 44 AUs, which were associated with the contraction of facial muscles. They also provided rules for recognition of the onset (start), apex (peak) and offset (end) of the AUs. After detecting the AUs in the face, the universal emotions of Ekman, such as anger, disgust, and happiness, can be recognized by using the Emotional FACS rules (EMFACS). In addition, more affective states can be also recognized using FACS Affect Interpretation Database (FACSAID). Consequently, using this representation simplifies the affect recognition problem to the muscular structure of the face, thus, reducing thousands of facial expressions to a combination of few AUs.

Generally, facial expression recognition consists of two phases: feature extraction and classification stage. A large number of methods have been developed for each phase. For instance, feature extraction is divided into two types of approaches: holistic techniques (e.g. [4]) and local techniques (e.g. [5] and [6]). While holistic approaches use the entire face as input, the local approaches only consider specific features to be used such as geometrical features and appearance features or a combination of both. On the other hand, the classification techniques can be divided in two groups: spatial approaches [4] and spatio-temporal approaches [5]. While spatial approaches consist of analyzing the features on a frame by frame basis, spatio-temporal methods consider the evolution of the features in time. Finally, we note that improving feature extraction and classification techniques can improve the accuracy of automatic AU recognition systems, independently of the application. Furthermore, systems that can consider realistic situations, such as various head poses and occlusions, are more applicable to real life applications, since such difficulties occur in naturalistic settings. Most of the current methods proposed are robust to small head pose variations, such as the work of [5]. Few researches have worked on 3D databases, such as [7] and [8], but have focused on identifying the universal emotions rather than the FACS AUs. In this study, we propose a system that can recognize FACS AUs for various poses. Rather than training one classifier per AU for the whole face, we train multiple classifiers for each part of the face. The final decision combines the classification results while considering which parts of the face were occluded. The rest of this paper is organized as follows. In Section 2, we present related work and background on our proposed system. In Section 3, we introduce our proposed method. Experimental results are shown in Section 4 before drawing conclusions in Section 5.

2   Related Work and Preliminary Concepts

In this section, we first present related work for the particular area of facial expression recognition for various poses as well as geometrical-feature-based methods. Then, a background on the methods used is provided. In AU feature extraction, two types of local features are generally used: geometrical and appearance based features, or a combination of both. In this work, we focus our experiments on geometrical features to demonstrate the overall performance of our system. However, the same proposed approach can be used for a combination of geometrical and appearance based method.

2.1 Related Work

Virtually most of the AU methods reported so far have been based on near frontal views data [5] or on data with moderate head pose variation [9]. For instance, the work of Pantic et al. [5] investigated facial AU recognition from near-frontal views using geometrical features, and modeled the temporal phases of AUs. One of the weaknesses reported is that significant out-of-plane rotations affected the recognition accuracy. Other work such as the work of Tyan et al. [6] used a combination of geometrical and appearance based features. This method was reported to be robust for moderate face rotations, but no direct measure for the accuracy/angle dependency was given.

To extract facial expressions in less-constrained environments, such as different head poses, Pantic and Patras [10] investigated facial AU recognition from profile views. Also, several experiments were done on the BU-3DEF [11] database, which is a facial expression database that consists of 3D shapes and 2D facial textures from 100 subjects. One of these works is the work of Hu et al. in [7] where they studied facial expression recognition for different head orientations, i.e. yaw and pitch angles, by extracting appearance based features such as Histogram of Oriented Gradients (HoG), Local Binary Patterns (LBP) and Scale Invariant Feature Transform (SIFT). These descriptors are well known in the area of computer vision for object and texture classification. In [8], Hu et al. use the displacement of manually selected feature points from the face for classification. However, all of these methods were tested on detecting the basic emotions of Ekman and not on detecting FACS AUs.

In the work of Valstar et al. [5], the authors extracted the facial points using a tracking scheme based on particle filtering using factorized likelihoods (PFFL). Affine transformation was then performed on the obtained coordinates to reverse the effect of scaling and small head orientations. Geometrical features as well as temporal features were extracted from the image sequences. Finally, a combination of Gentle-Boost and Support Vector Machine (SVM) was used in the classification stage. The system was further extended to detect the temporal activation model (neutral, onset, apex and offset). The main disadvantages of their system in detecting AUs for pose variations can be summarized by the following: The affine transformation cannot model out-of-plane rotations assigned with the head pose, and (2) PFFL cannot handle facial point occlusions associated with head pose variation.

2.2 Preliminary Concepts

In this sub section, we describe the general parts in a geometrical based AU recognition system and introduce the methods used in our system. First, a facial tracker is employed to detect and track the facial points. One of the most used models for facial tracking is the Deformable Model Fitting (DMF). DMF is a classic problem formulation in which the shape of object deformations is modeled using the Point Distribution Model (PDM) founded by Taylor [12]. In this model, the facial points' positions are calculated using the following equation:

$$\mathbf{x}_i = s\mathbf{R}(\bar{\mathbf{x}}_i + \boldsymbol{\Phi}_i \mathbf{q}) + t , \tag{1}$$

where $\mathbf{x}_i$ denotes the location of the ith landmark, s denotes a scale, $\mathbf{R}$ a rotation matrix, $\bar{\mathbf{x}}_i$ the mean location of the ith landmark, q a set of non-rigid parameters, and $\boldsymbol{\Phi}$ a submatrix of basis variations. The aim is to determine the landmarks positions x_i, by determining the set of shape parameters (shape, rotation, translation and non-rigid parameters). Particularly, the Regularized Deformable Model Fitting (RDMF) tracker [13] follows the DMF model. RDMF uses a logistic regressor function to determine the likelihood of a facial point position, given an input image. The values of the landmark positions are determined by minimizing the misalignment error according to the PDM model as well as maximizing the new position likelihood. The search space of an optimal solution is minimized using hill climbing methods. The advantage of this tracker is that it is robust to multiple occlusions since it leverages the relationship among the facial points in the PDM model. After extracting features from the obtained facial points' positions, a machine learning algorithm such as SVM can be used to classify an AU. However, if the feature dimension is greater than the training data, overfitting to the training data is rather probable. Many feature reduction techniques can be used at this stage. Gentle-Boost combines a weighted vote of weak classifiers in the final classification. In the next section, we propose how to combine RDMF with Gentle-Boost to detect the AUs for various orientations.

3   Proposed Method

In this section, we describe our proposed method illustrated in Fig. 1. The first step is to detect and track a set of facial point coordinates using the RDMF tracker proposed in [13]. These coordinates are then separated into two groups: left-face points and right-face points. The features extracted are the distances' variation among the points. Finally, we describe the model for detecting the activation status of each AU. In the following subsections, we first describe how the features are extracted for each face region, and then we explain the AU classification scheme. Finally, we describe a classification system for emotion detection.

3.1   Feature Extraction from Left and Right Face Regions

Each image sequence is first processed using the RDMF tracker [13] obtain facial points coordinates across all frames. We note the coordinates of these points as:

$$X = ((x_1, y_1), (x_2, y_2), ..., (x_n, y_n)) . \tag{2}$$

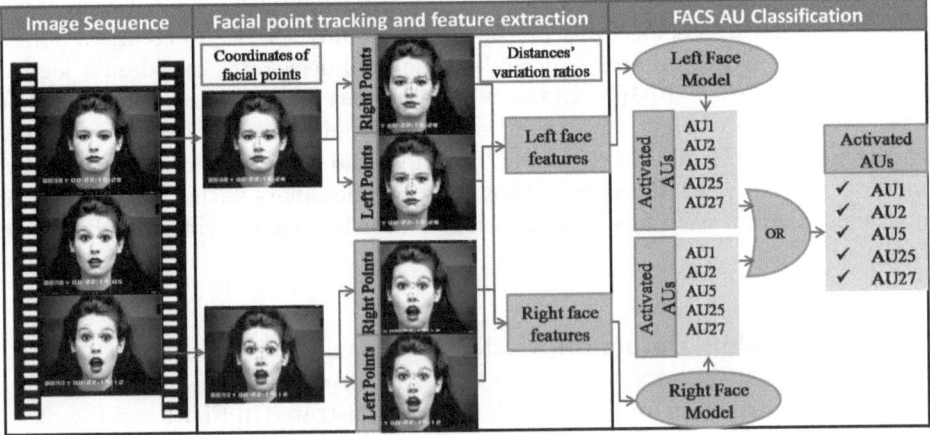

Fig. 1. AU detection algorithm

In this work, the number of facial points is 66, which is based on the RDMF implementation. We note at this point that not all of these points can be extracted when an area from the face is occluded, such as the right or the left facial area. After detecting the facial points, a set of features are then extracted. Euclidean distances d_{ij} among the points are calculated, such that: $d_{ij} = \sqrt{(x_i - x_j)^2 + (y_i - y_j)^2}$. Then, the set of features are extracted from the specified facial area. The features are defined by the ratios:

$$R_{ij} = \frac{d_{ij}}{d_{ij}^r} \tag{3}$$

where d_{ij}^r is the distance between the facial points i and j in a reference frame: a frame where the facial expression is neutral. The choice of these features is suitable when the head poses in the neutral frame and the tracked frame are relatively close. Rather than employing all the points in the AU classification, we propose to train multiple classifiers for each facial area (in this case left or right). It follows that only pair of points within one part of the face are used as features in each classifier. For instance, consider a facial tracker that can track three disjoint sets of facial points: $\mathbf{A}$, $\mathbf{B}$ and $\mathbf{C}$, and assume that either one of the set of points $\mathbf{A}$ or $\mathbf{C}$ can be occluded at once (for example the left or right facial area). Rather than training a single classifier $\mathbf{M_{ABC}}$ on all the landmarks that belong to $\mathbf{A}$, $\mathbf{B}$ or $\mathbf{C}$, we propose to train two classifiers: $\mathbf{M_{AB}}$ and $\mathbf{M_{BC}}$, that is for all possible combinations of landmarks being present/absent. For instance, in the testing phase, in the case where $\mathbf{A}$ (respectively $\mathbf{C}$) is occluded, $\mathbf{M_{BC}}$ classifier should be used (respectively $\mathbf{M_{AB}}$). The occlusion status of the facial area can be directly extracted from the RDMF tracker by checking the point coordinates. On the other hand, if no landmarks from $\mathbf{A}$ or $\mathbf{B}$ are occluded, the decision should be weighted between $\mathbf{M_{BC}}$ and $\mathbf{M_{AC}}$. The same procedure applies for the training phase, where the classifiers can be trained only when their corresponding sets are not occluded. In our implementation, we consider three sets of landmarks: left-face-only set $\mathbf{L}$, right-face-only set $\mathbf{R}$ and common points $\mathbf{J}$. In the case where no area is occluded, the final decision is based on a logical OR between $\mathbf{M_{LJ}}$ and $\mathbf{M_{RJ}}$. In the remaining part of the paper, the term "left face" (respectively "right face") will refer to the points in $\mathbf{L}$ and $\mathbf{J}$ (respectively to the points in $\mathbf{R}$ and $\mathbf{J}$). It is worth noting that the difference between this method and conventional ones is that multiple classifiers with various features are being used for each face region rather than using one classifier for the whole face.

3.2   Action Unit Detection Model

The training algorithm for each AU classifier is illustrated in Fig. 1 and Fig. 2. In the training phase, the neutral and apex frames are extracted from each video. The features from the left and right areas of the face are collected separately. One classifier is trained to classify the activation state of each AU (activated or not) and for each area of the face, using the features collected. The activation state of each AU should be available in the database or manually annotated by a FACS coder. We employ Gentle-Boost algorithm to avoid data overfitting on one hand, since the number of features is higher than the number of training data, and to make our work more comparable with other works in the literature. In the testing phase, if a facial area is occluded, the classifier of the other area will be used for classification. In the case where no facial area is occluded, the activation state of the AU is calculated by performing a logical OR on both left and right classifications. In fact, the FACS manual states that if an AU is activated in one part of the face, e.g. left eye brow raiser, the AU is annotated to be activated.

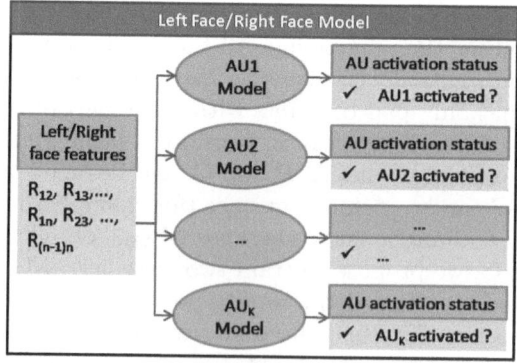

Fig. 2. Left/Right face model

3.3   Emotion Detection Model

In order to detect emotions, the same features extracted from the face are used to detect the various emotions expressed by the subjects: anger, disgust, fear, happiness, sadness, surprise or neutral. A number of classifier are trained using the Gentle-Boost algorithm to differentiate between each pair from the set of emotion (one-versus-one approach). The detected emotion in the left and right parts of the face cannot be combined together in the same way the detected AUs were combined in the previous section (using the logical OR). In the experiments section, we show the results of each region separately.

4   Experiments and Evaluation

In our system, we employed the author's implementation of the RDMF facial tracker in our system [13]. The code executes in real time and its output ranges from 20 - 30 fps based on the processor and the compiler used. In order to evaluate our method, we perform three experiments. In the first one, we test our system on the Cohn-Kanade (CK) database [12] which contains 480 gray scale videos that were made public. The head orientation of subjects in the recorded videos is near-frontal. This database was collected for the purpose of facial expression recognition and this is currently the most used database in this research area. We study our system on this database in order to validate our results by comparing them with a state-of-the-art geometrical approach for small head orientations. In this comparison, we use the results obtained in the frame-based experiments by Valstar et al. in [5]. In the second experiment, the Emotion detection system is tested against the one proposed also in [5]. Finally, we validate our system in the third experiment for various poses of the head, ranging from 0 degree to 90 degrees. For this case, multiple videos were recorded featuring different orientations.

4.1   Benchmarking for Action Unit Detection

In the CK database, for each sequence of images, the facial landmarks were detected using the RDMF tracker. The coordinates of the 66 facial landmarks are tracked through each image sequence in the database. The left face area consists of 37 points while the right one contains 38. The numbers were determined experimentally and depend on the training of the facial tracker.

For each facial area, the ratios of the distances, described in section 3, were extracted from the neutral face and the apex frames. In the CK database, rather than manually labeling the apex frames for each AU in each video, which is time consuming, we considered the neutral and the apex frames to be the first and the last frames of the image sequence, respectively. We note that this assumption is fair for most AUs in this database, since most image sequences were recorded till the beginning of the apex stage for all AUs. However, we are aware of this assumption's limitation since it is not always valid, especially for certain AUs like AU45 (blink). As per features and class labels, the left face features and the right face features were extracted for each image as specified in the previous section. In summary, the size of the feature sets are presented in Table 1. Finally, we mention that we trained each of the Gentle-Boost classifiers for 10 rounds.

Table 1. Size of the facial landmark sets and their corresponding features

Size of landmark sets			Size of feature vector		
Full-Face	Left	Right	Full-Face	Left	Right
66	37	38	2145	666	703

We conduct our experiment using the leave-one-subject-out strategy. In each fold, one subject is left out of the database, and all classifiers are trained on the remaining subjects and then tested on the subject that was left out. Binary confusion matrices are then summed together for all the experiments, i.e. for each subject in the database. Table 2 shows a comparison between the results obtained in our work, named Distance Ratio Classifier (DRC), and the one in the paper [5] by Valstar et al., illustrated as (TMP). In the third column, the number of positive examples for each AU is illustrated. All AUs that were previously studied, except for AU10, are also studied in this work. In our experiment, all 500 image sequences from the CK database were used, whereas the number of image sequences used in the TMP algorithm is 153. Four measures were calculated: accuracy, recall, precision and F1 measure. While the accuracy measure is a highly biased measure due to the unbalanced nature of the data, precision and recall are a better approximation of the data. The F1 measure combines the two latter measures by favoring them equally. The table is interpreted as follows. For each AU, compare the F1 column of the DRC and TMP algorithms. Precision and recall can be used for further investigation on the property of the classifier used. When needed, we refer to the accuracy of column. However, we note again that the latter measure is not very significant since the number of negative examples for each AU is much bigger than the ones with positive examples. Although, the results obtained in our method are highly optimistic, no direct conclusion on the superiority of our algorithm over the temporal based algorithm can be made since the selection of videos used is not the same. As can be seen, AU1 (inner brow raisers) and AU24 (lip pressor) show very close results with superior measures for the DRC method. Additionally, our method is also superior for other AUs such as AU2, AU4, AU5, AU7 and AU9. We believe that the reason for this improvement is behind the DMF model used by the Facial Tracker. On the other hand, other AUs such as AU6 (cheek raiser) and AU12 (lip corner puller) show that the method proposed is not accurate. In fact, it can be observed that the landmarks of the lip corners are not tracked effectively using the RDMF when performing AU12. Lastly, the poorest result achieved by DRC is for AU45 (blink). The lack of precision for this AU detector is mostly attributed to the preprocessing assumption that we made. In fact, most image sequences end after

Table 2. Comparison between the work in this paper (DRC) and the one in [5] (TMP). The number of videos used in the test is specified in the third column.

AU	Meth.	Videos	Acc.	Recall	Prec.	F1
1	DRC	144	0.910	0.809	0.864	0.835
	TMP	68	0.918	0.808	0.844	0.826
2	DRC	97	0.964	0.871	0.946	0.907
	TMP	50	0.939	0.791	0.879	0.833
4	DRC	156	0.896	0.755	0.864	0.806
	TMP	54	0.870	0.604	0.658	0.630
5	DRC	78	0.926	0.708	0.761	0.734
	TMP	37	0.904	0.566	0.629	0.596
6	DRC	111	0.870	0.713	0.694	0.703
	TMP	39	0.930	0.789	0.811	0.800
7	DRC	108	0.862	0.685	0.679	0.682
	TMP	31	0.870	0.268	0.315	0.290
9	DRC	50	0.972	0.864	0.826	0.844
	TMP	30	0.928	0.676	0.497	0.573
12	DRC	113	0.904	0.780	0.780	0.780
	TMP	42	0.930	0.827	0.844	0.836
15	DRC	81	0.910	0.609	0.661	0.634
	TMP	19	0.969	0.500	0.283	0.361
20	DRC	70	0.924	0.638	0.772	0.698
	TMP	34	0.908	0.466	0.582	0.517
24	DRC	43	0.928	0.421	0.533	0.471
	TMP	17	0.935	0.395	0.497	0.440
25	DRC	303	0.888	0.917	0.905	0.911
	TMP	19	0.851	0.717	0.782	0.748
26	DRC	39	0.926	0.175	0.636	0.275
	TMP	27	0.902	0.336	0.380	0.357
27	DRC	77	0.972	0.919	0.895	0.907
	TMP	30	0.964	0.836	0.873	0.854
45	DRC	19	0.954	0.091	0.400	0.148
	TMP	23	0.943	0.584	0.408	0.480
DRC Avg.			0.920	0.664	0.748	0.689
TMP Avg.			0.917	0.611	0.619	0.609

the offset of this AU has occurred, i.e. the final frame does not generally contain the apex for AU45. Any accurate result for this AU is due to the mere correlation between AUs in the database.

4.2 Benchmarking for Emotion Detection

In this section, the same features extracted previously are used in the emotion detection. We compare our method to the one proposed by Valstar et al. in [5] on the CK database. Note that the annotation for the emotions is provided in the database. The confusion matrix obtained in [5] is illustrated in Table 3(a).

The results obtained using our method are illustrated in Table 3(b). We test our system by using the points from left face only. The classification accuracy in our method is very comparable to the one in [5]. The classification rate for the emotion anger and surprise is much better in our method. On the other hand, the sadness classification rate is lower in our case. We note that the database subsets used are not the same in our experiment and the one in [5]. Thus, we don't elaborate more on the comparison, and simply state that the results obtained in our method are comparable to the state of the art approach in [5].

Table 3. (a) Confusion matrix for emotion classification using the method in [5]. (b) Confusion matrix for emotion classification using the features from the left face.

	An.	Di.	Fe.	H.	Sad.	Sur.	Rate
Ang.	2	3	2	0	9	1	0.118
Disg.	1	19	1	1	4	1	0.704
Fear	1	4	15	5	2	1	0.536
Hap.	1	0	3	33	0	1	0.868
Sad.	4	2	1	0	16	1	0.667
Sur.	0	1	1	1	0	34	0.919

	An.	Di.	Fe.	H.	Sad.	Sur.	N.	Rate
Ang.	19	3	0	2	4	0	1	0.655
Disg.	1	32	0	0	0	0	1	0.941
Fear	0	1	14	1	1	0	0	0.824
Hap.	0	0	1	60	0	0	0	0.984
Sad.	4	0	2	0	9	0	1	0.563
Sur.	0	0	8	0	0	63	0	0.887
Neu.	0	0	0	0	0	0	228	1

4.3   Various Pose Evaluation for Brow Raisers

In this section, we test our method on various orientations. For this purpose, a training set was created featuring two subjects in 45 videos in total. The sequences were recorded for three discrete yaw orientations (horizontal rotations), namely no yaw, moderate yaw and extreme yaw, approximated by: 0 degree, 45 degrees and 90 degrees angles to the camera imager. All videos were taken using a DMC-F3 Panasonic camera at a resolution of 1280x720 pixel^2, a rate of 30 fps and a distance of 2 meters from the subject.

The subject was asked to stand in a frontal pose, and then to rotate his head by a specific angle until facing a marker on the wall and to perform an AU or a combination of AUs. Afterwards, we manually annotated the neutral frames and apex frames of each sequence. We note that only the neutral frame preceding the onset phase is considered. A sample recorded set of images is shown in Fig. 3. The subjects were asked to perform AU1 and AU2 (brow raisers). As a matter of fact, it is essential to assess the validity of any FACS system for the most common AUs (AU1 and AU2 consists about 20% of the CK database).

In the testing phase, only the tracked points from the neutral and apex frame were extracted from the video. A previously trained DRC classifier set from the CK database was used on the data set. Table 4 illustrates the evaluation of our method for the three yaw intensities, and for the two AUs. We note that the tracker failed to track some videos for extreme face orientations. These videos are excluded from the final statistics. The second column shows the intensity of the head orientation. The numbers of positive and negative examples are illustrated

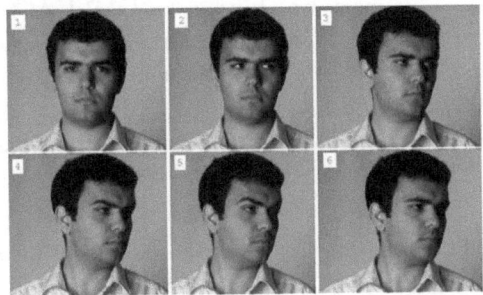

Fig. 3. Samples in an images sequence from the special database. The subject is first asked to rotate their face, then to perform the AU. The third image is annotated as the neutral face, and the fifth is annotated as the apex frame.

Table 4. Evaluating algorithm on the special database for 3 head orientation intensities and two AUs

AU	Int.	P	N	Acc.	Recall	Prec.	F1
1	1	4	10	0.929	1.000	0.800	0.889
	2	14	22	0.861	0.786	0.846	0.815
	3	9	11	0.950	0.889	1.000	0.941
2	1	4	10	0.643	0.500	0.400	0.444
	2	7	29	0.889	0.571	0.800	0.667
	3	5	15	0.850	0.600	0.750	0.667

in the third and the fourth column. The same measures from experiment 1 are used in this experiment. Not surprisingly, the F1 increased when the orientation intensity was stronger. In fact, this is consistent with the work in [10] which concluded that profile views are better than frontal views for AU detection.

5   Conclusion

Finally, we conclude our work with the following analysis. We have developed a working system that can detect AUs for various head poses without any prior training on these poses. Our method has comparable results with the state-of-the-art geometrical algorithm in [5] for near-frontal head orientation. Moreover, we realize that our model was able to generalize to a new database and that AUs were still detectable at various pose orientations. The main issue that we wish to address in our future work is to extract features that are more robust to pose variations. Lastly, using the tracker from [13] had its disadvantages, since some AUs couldn't be detected accurately, such as lip corner pullers. We wish to evaluate our work on other trackers that were previously used in the literature.

Acknowledgments. This work was funded by Intel-KACST Middle East Energy Efficiency Research (MER) and the American University of Beirut (AUB) University Research Board (URB). Special acknowledgment is due to Lama Nachman for her valuable inputs.

References

1. Russell, J., Dols, J.: The psychology of facial expression. Studies in emotion and social interaction. Cambridge University Press (1997)
2. Tian, Y.L., Kanade, T., Cohn, J.F.: Facial expression recognition. In: Li, S.Z., Jain, A.K. (eds.) Handbook of Face Recognition, 2nd edn., pp. 487–519. Springer (2011)
3. Ekman, P., Friesen, W.V.: The facial action coding system: A technique for the of facial movement. In: Consulting Psychologists (1978)
4. Lien, J., Kanade, T., Cohn, J., Li, C.: Detection, tracking, and classification of action units in facial expression. Robotics and Autonomous Systems 31(3), 131–146 (2000)
5. Valstar, M., Pantic, M.: Fully automatic recognition of the temporal phases of facial actions. IEEE Transactions on Systems, Man, and Cybernetics, Part B: Cybernetics 42(1), 28–43 (2012)
6. Tian, Y., Kanade, T., Cohn, J.: Recognizing action units for facial expression analysis. IEEE Transactions on Pattern Analysis and Machine Intelligence 23(2), 97–115 (2001)
7. Hu, Y., Zeng, Z., Yin, L., Wei, X., Zhou, X., Huang, T.S.: Multi-view facial expression recognition. In: FG, pp. 1–6 (2008)
8. Hu, Y., Zeng, Z., Yin, L., Wei, X., Tu, J., Huang, T.: A study of non-frontal-view facial expressions recognition. In: 19th International Conference on Pattern Recognition, ICPR 2008, pp. 1–4. IEEE (2008)
9. Tong, Y., Liao, W., Ji, Q.: Facial action unit recognition by exploiting their dynamic and semantic relationships. IEEE Transactions on Pattern Analysis and Machine Intelligence 29(10), 1683–1699 (2007)
10. Pantic, M., Patras, I.: Dynamics of facial expression: Recognition of facial actions and their temporal segments from face profile image sequences. IEEE Transactions on Systems, Man, and Cybernetics, Part B: Cybernetics 36(2), 433–449 (2006)
11. Yin, L., Wei, X., Sun, Y., Wang, J., Rosato, M.: A 3d facial expression database for facial behavior research. In: 7th International Conference on Automatic Face and Gesture Recognition, FGR 2006, pp. 211–216. IEEE (2006)
12. Cootes, T., Taylor, C.: Active shape models–smart snakes. In: Proc. British Machine Vision Conference, vol. 266275. Citeseer (1992)
13. Saragih, J., Lucey, S., Cohn, J.: Deformable model fitting by regularized landmark mean-shift. International Journal of Computer Vision 91(2), 200–215 (2011)

Use of Supervised Learning to Predict Directionality of Links in a Network

Sucheta Soundarajan and John E. Hopcroft

Cornell University, Ithaca NY 14853, USA

Abstract. Often, the information contained in network data is incomplete. Many avenues of research are aimed at addressing this incompleteness. For example, the link prediction problem attempts to identify which missing links are most likely to exist in the complete network. In this paper, we consider a related, but different, problem: predicting the directions of links in a directed network. We treat this problem as a supervised learning problem in which the directions of some edges are known. We calculate various features of each known edge based on its position in the network, and use a Support Vector Machine to predict the unknown directions of edges. We consider four networks, and show that in each case, this method performs significantly better than other compared methods.

1 Introduction

Over the past few decades, network analysis has become an increasingly important way for scientists to study the relationships, interactions, and roles of individuals in groups. Although scientists and engineers have developed sophisticated, valuable techniques for studying networks, any such methods are inherently limited by the quality of the data.

Much of available data is incomplete. In many cases, actual nodes or edges may be missing from a network. For example, it is believed that in some genetic networks, where links are experimentally determined, fewer than 1% of the edges have been discovered. Even in cases where information is not gathered experimentally, networks may be incomplete due to other factors, such as lack of participation (e.g., individual people may choose not to participate in Facebook, and even if they do participate, they may not connect to all of their actual friends). In other cases, while the existence of nodes or links may be known, we may be missing information about their characteristics.

This type of information is valuable to many social network analysis techniques: for instance, community detection algorithms will work more accurately on a more complete network, and algorithms to predict the flow of information through a network rely on accurate knowledge of link directionality. Thus, inferring this sort of missing information is an especially active and important research area.

In this paper, we consider the problem of predicting the direction of links. In much of network data, links may be reported as undirected, and yet the underlying network may actually consist of directed links. For example, in Facebook, all

S. Zhou, S. Zhang, and G. Karypis (Eds.): ADMA 2012, LNAI 7713, pp. 395–406, 2012.

links (friendships) must be reciprocated (in contrast with a network like Twitter, where one user may follow another without reciprocation). However, even in a network such as Facebook, a link was likely initiated by one person and may be stronger in that direction. Although a researcher studying a Facebook network is given undirected data, information about the underlying direction of the links may help, for instance, to more accurately predict the spread of news or trends through the network.

We treat this problem as a supervised learning problem. We assume that we are given the directions of some links, and attempt to use this information to predict the directions of other links. In the example of Facebook, such a situation may arise if one can use information about the timing of similar posts, and then infer the direction of some friendships (e.g., if one user consistently posts articles, news, or events before another user, we might infer that the second user is following the first user, and so the edge goes from the second user to the first user). See Adamic, et al. [2] for an analysis of such a situation. Although one might use this type of information to predict the directions of some links, it is unlikely that similar information will be available for every pair of linked users. In such cases, one can apply supervised learning techniques to predict the unknown directions of these links.

In our experiments, we consider four networks from different domains, each containing directed links. We train a Support Vector Machine (SVM) classifier on some of these links, and use this model to predict the directions of other links (the directionality of which is withheld, but the existence is assumed). For each link, we create a feature vector that includes information about its position in the network. Some of these features are obtained from other algorithms intended to solve this same problem, while other features represent more general node, edge, and network features. Each edge (a, b) is labeled with one of three possible class memberships, indicating whether the edge is from a to b, b to a, or both. We compare our algorithm to four other methods, and show that it is the best performer on every network.

This paper begins with an overview of related work, including work from the areas of link prediction, as well as detailed discussions of two other methods for predicting link directionality. We then discuss the methodology used in our work, beginning with a description of our datasets and calculations used to produce the feature vectors, and continuing with detailed information on our experiments. We next discuss and analyze our results, and conclude with a discussion of future work.

2   Related Work

Much work has been done in the related area of link prediction, and some of these methods have proven valuable for the problem of predicting link direction. In this section, we will begin with a description of relevant link prediction methods. We then describe various methods for predicting link directions, including some that rely on external information (such as timing of blog posts), and others that use only the structure of the network.

2.1   Link Prediction

Many popular link prediction methods use information about the structure of the network to determine the probability that two nodes are connected. One simple method, the Common Neighbors metric, simply calculates the number of neighbors that two nodes share, with the assumption that nodes with many shared neighbors are likely to be connected. This method and its variants are fast, simple, and reasonably accurate [3]. Although these methods are not directly useful for our task in this paper, they demonstrate that simple, local metrics can be valuable in inferring missing information in networks.

These metrics and others can be incorporated into a supervised learning framework, as we do in this paper. For example, Al Hasan et al. approach this problem by applying SVM and k-Nearest Neighbors methods to feature vectors that include both topological features and external features (such as keywords in papers), with good results [1].

2.2   Predicting Link Direction

Here, we discuss three methods for predicting link direction. The first method resembles ours, in that it uses a supervised learning method; however, it uses external, non-topological information. The other two methods are ranking methods, intended to place the nodes into a hierarchy. Under these methods, link directionality is assumed to go from lower-ranked nodes to higher-ranked nodes.

Supervised Learning. Adar and Adamic consider the problem of tracking information flow through blogspace [2]. Bloggers online often see posts or information in another blog, and then repost that information to their own blog. However, this information is often not cited, and so it may be unclear where the blogger learned the information.

In some cases, bloggers do cite the source of their posts, or explicit links may be declared. Adar and Adamic use these cases as positive examples in their classifier models. The features they consider include textual similarity between blog posts, similarity between posted URLs, and relative timing of blog posts. A two-class SVM trained on these features resulted in high cross-validation scores, demonstrating the value of a supervised learning framework to this problem. This work differs from ours in that it relies on external information beyond the topology of the network.

PageRank. PageRank is a very popular, effective algorithm that uses the topology of a directed graph to rank nodes according to their importance in the network [7]. Although this algorithm is probably best known for its use in online search engines, it can also be used as a predictor of link directionality. The output of the algorithm represents the probability that a random walk on the network will arrive at a particular node. The algorithm also includes a 'restart value,' or probability that the random walk will restart back to its first node on any given step of the walk.

A node's PageRank depends on the PageRank of other nodes: if many highly ranked nodes point to some node, then that node is believed to be of high importance, and so will have a high PageRank score. A PageRank score can be used to predict link direction by assuming that an edge between a node with a low PageRank and a node with a high PageRank is directed from the former to the latter.

Leader-Follower Ranking. We also consider another algorithm for ranking nodes, again with the assumption that edges go from lower-ranked nodes to higher-ranked nodes. This algorithm, by Guo et al. [4], is based on a recursive algorithm that partitions the network into 'leaders' and 'followers.' The user specifies a value α between 0 and 1 that defines the fraction of leaders at each step. In each step of the algorithm, for each node n, the algorithm calculates δ_n, defined as the difference between node n's in-degree and out-degree. Nodes with a high δ_n are considered leaders, whereas nodes with a low δ_n are considered followers.

After partitioning the graph into leaders and followers, the algorithm recursively calls itself again on each of these two sets. At the lowest level, if the number of nodes is less than $\frac{1}{\alpha}$, the algorithm returns the nodes ranked in order of their δ_n. When joining together two sets, the algorithm places the followers below the leaders.

Formally,

$$R_G(x) = \begin{cases} D_x & \text{if } |G| < 1/\alpha \\ R_{G_L(x)} & \text{if } |G| \geq 1/\alpha \text{ and } x \in L \\ |L| + R_{G_F(x)} & \text{if } |G| \geq 1/\alpha \text{ and } x \notin L \end{cases}$$

where G is the graph currently being considered, $R_G(x)$ is the ranking of node x, D_x is the ranking of node x induced by the δ_n values, L is the set of leaders in G (and G_L is the subgraph of G induced by L), and F is the set of followers in G (and G_F is the subgraph of G induced by F).

The output of this algorithm is a list of nodes, where the index of the node corresponds to its ranking. Because no two nodes can share the same position in a list, it is impossible for two nodes to have an equal rank. Because in our problem of predicting link directions, we want to allow for reciprocated links, we thus modify the algorithm slightly to allow nodes to share a rank (we accomplish this by allowing leaders who share a δ_n to have the same rank, rather than requiring a completely ordered list).

To determine the optimal α value, one can consider several different values, and determine which value produces a ranking that conforms best to the actual directions of edges in the graph.

3   Methodology

We treat the problem of predicting link direction as a classification problem. To accomplish this, we collect a set of four directed networks. For each network we create a set of training edges, for which directionality is known, and a set of test edges, for which directionality is withheld. For edges in the test set, we assume that the existence of that edge is known, but the direction is unknown. For each edge, we calculate a feature vector that contains information about that edge's location in the network, and then determine how well a classifier can use this feature vector to predict edge directionality. In this section, we first describe the network datasets used in our experiments, giving background information and statistics on each. We then discuss the various network, node, and edge features used to characterize edges in the networks. We conclude with a description of our experimental methodology, including information on the creation of test and training sets, as well as a brief background on the classification method used.

3.1 Datasets

In this section, we describe the datasets used in our experiments. We consider 4 directed networks from different domains. Three of these networks (Amazon, Epinions, and Slashdot) are quite large, and so for these networks we considered a 5000 node subgraph of each, where the 5000 nodes were chosen using a breadth-first search beginning from a randomly chosen node. All datasets were downloaded from the online dataset collection at SNAP, the Stanford Network Analysis Project.

Amazon. "Amazon" is a product co-purchasing network obtained from the online retailer Amazon.com [9]. For each item, Amazon.com reports up to five other items that were frequently bought with that item. An edge from node A to node B indicates that customers who bought item A also frequently bought node B. The original Amazon network contained nearly 300,000 nodes, and as described above, we consider a 5000 node subset of this full network.

Epinions. "Epinions" is a social network from the review website Epinions.com [12]. Each node represents a user of the website, and a link from user A to user B indicates that user A 'trusts' user B's reviews. This network contains approximately 75,000 nodes, and as with Amazon, we consider a 5000 node subgraph of the larger network.

Slashdot. "Slashdot" is a social network from the technology website Slashdot.com [11]. In this website, users submit news and reviews, and are able to mark other users as friends or foes. Each node represents one user, and a directed link from A to B indicates that user A has tagged user B as either a friend or foe. This network originally contained approximately 80,000 nodes, so we again consider a 5000 node subgraph.

Wiki. "Wiki" is a social network from the free online encyclopedia Wikipedia.com, which contains content created and edited by users [10]. Some of the users are elected to be administrators with special privileges. In this network, each node is one user, and a directed link from user A to user B indicates that user A participated in user B's election. This network is fairly small, containing only 7115 nodes, and so we do not need to consider a subgraph.

3.2   Network Features

To produce features for each edge, we consider two graphs: N, the directed network formed from the edges in the training set, and G, an undirected network formed from edges in both the training and the test set (we include edges from the test set because in our problem, the directionality of the edge is unknown, but its existence is given). To create G, we simply ignore the directions on edges, and convert all of them to undirected edges. For edge (a, b) in the training or test sets for each network, we calculate 9 features, listed below. To calculate the degree and betweenness features, we consider the undirected graph G that is produced by converting each directed edge into an undirected edge, and for the various ranking features, we consider the directed graph N induced by the edges in the training set.

The first 5 features are calculated using the undirected graph G. These features have traditionally not been used to order nodes, but we consider them here in order to learn whether they can be useful in predicting the direction of an edge. For example, for an edge (a, b), we consider the degrees of both nodes a and b. It may be the case that nodes tend to be directed from low degree nodes to high degree nodes (or vice versa), and so this feature might increase prediction accuracy. The node betweenness features might play a similar role.

1. Degree of node a: The degree of a node is simply the number of edges adjacent to it. To calculate this value, we use undirected graph G.
2. Degree of node b
3. Node Betweenness for node a: The node betweenness feature, proposed by sociologist Linton Freeman, is a measure of a node's centrality in the network [5]. There are different forms of this metric: the one that we use is simply the fraction of all shortest paths between all pairs of nodes that pass through the node in question. That is, for a given node a, we consider all other pairs of nodes b, c, and determine how many of the shortest paths from node b to node c pass through node a. We then normalize this by the total number of shortest paths between all pairs of nodes b, c. To determine node betweenness values, we use undirected graph G.
4. Node Betweenness for node b
5. Edge Betweenness for edge (a, b): The edge betweenness feature is defined similarly to the node betweenness factor, except it indicates the fraction of shortest paths that use a particular edge. For edge betweenness, we again use undirected graph G.
6. Leader-Follower Ranking for node a: To calculate the rank of a node, we apply the algorithm described in [4] to directed graph N. The original algorithm does not allow for two nodes to have equal rank (rather, it produces a list of nodes, where the index of a node is its rank), so as described earlier,

we modify the algorithm slightly by allowing two nodes to share an index in the ordered ranking list. In this metric, a node that is highly ranked receives a low score (that is, it appears early in the list).

7. Leader-Follower Ranking for node b
8. PageRank of node a: To calculate a node's PageRank, we apply the well-known algorithm that calculates the probability that a random walk will include that particular node. In our calculations, we apply a restart factor of 0.15, representing the probability that the random walk returns to its starting node at any given step.
9. PageRank of node b

3.3 Experiments

To predict directionality of links, we use a Support Vector Machine (SVM) classifier. In our experiments, we assume that the existence of each edge is known, and attempt to predict its direction. For each network, we create a set of training links, for which the direction is known, and test links, for which direction is unknown. The training links constitute 90% of the edges in the network, while the other 10% of edges are withheld for testing.

For some of the features described above, we calculate feature values using a graph G containing undirected links from both the training and test sets; for the other features, we create a graph N that uses only directed links from the training set. Because we need every node to appear in N, some caution is necessary in partitioning the edges into training and test sets; in particular, we must ensure that the edges in the training set describe a graph that has only one component, and that we do not create 'orphan' nodes by placing all of a node's edges into the test set. Thus, to produce the training set, we first find a minimum spanning tree on network G, and then continue adding edges to this tree until 90% of the edges from the network are present.

For each edge (a, b) in the training set, we assign one of three class labels: Class 0 indicates that the edge is reciprocated (both nodes a and b link to each other), Class 1 indicates that the edge is directed from node a to node b, and Class 2 indicates that it is directed from node b to node a. From this set of edges, we randomly sample a maximum of 1000 elements from each class to train the SVM classifier. Network Amazon had a small number of reciprocated edges, and so to maintain class balance, each class contains only 400 elements.

For the test set, we again sample up to 1000 elements from each class (again, less for Amazon). In both the training and test sets, it is possible that some edges appear twice, as both (a, b) and (b, a). These two elements would have different class labels (unless the edge is reciprocated, in which case both would be labeled as Class 0), and the feature vector for element (b, a) would simply be a reordering of the feature vector for element (a, b).

We then apply the SVM classifier from the LibSVM software package [6] to produce a classifier model for these training sets. A SVM classifier accepts as input a set of data points (feature vectors) labeled with classes, and then projects the feature vector into a higher dimensional space and attempts to divide the classes with a hyperplane. Because data is typically noisy, the method that we use allows for a 'soft margin,' in which points may appear on the wrong side of the dividing hyperplane, with a penalty. In order to allow for multiple classes, LibSVM creates many two-class models, one for each pair of classes, and then combines these models into a single multi-class model. In this software, a cross-validation grid search is used to select optimal parameters (such as for the soft margin penalty) for the SVM.

When the model is applied to the test set, each element in the test set is assigned a probability vector, indicating the probability that the element belongs to each of the three classes. An element is then assigned to whichever class receives the bulk of its probability mass. However, because each element is assigned a probability vector, we are able to identify relationships between the three different classes: for example, we might observe that elements in class 1 (edges directed from node a to node b) are also similar to elements in class 0 (reciprocated edges), but are quite different from elements in class 2 (edges from node b to node a).

We perform 4 sets of experiments. First, we create a SVM model using the full feature vectors (SVM-Full). Next, we consider only those features that are obtained from the Leader-Follower Ranking algorithm (SVM-LFRank). Then, we consider only those features obtained from the PageRank algorithm (SVM-PR). Finally, we measure accuracy by using both PageRank and Leader-Follower (SVM-PR-LFRank). Although both PageRank and our modified version of the Leader-Follower Ranking algorithm allow for two nodes to have equal rank (thus predicting a reciprocated edge between the two nodes), it is likely that there is some small, non-zero difference between two nodes' rankings that would still indicate a reciprocated edge; that is, if two nodes have very similar, but slightly different, rankings, we may still wish to predict that each is linked to the other. Training a SVM model on these features allows us to discover this similarity threshold, and so allows for more accurate predictions.

We additionally calculate accuracy using PageRank and Leader-Follower without an SVM.

4   Results

Our results show that using a SVM classifier framework to predict directions of links is successful. Our first experiment, in which we used the entire feature vector, resulted in higher accuracy scores than using feature vectors containing either PageRank or the Leader-Follower Ranking scores alone, and also higher than the accuracy obtained by using PageRank and Leader-Follower together. As expected in all of these results, Classes 1 and 2 behave close to symmetrically.

Table 1 contains the results of our experiments. The columns correspond to the accuracy scores from 6 different experiments- the four SVM experiments described above, as well as using PageRank and Leader-Follower Ranking scores alone, without a SVM. The percentage represents the fraction of elements from the test set that were correctly classified into the proper class.

Table 1. Accuracy scores for each method and network

	SVM-Full	SVM-PR	SVM-LFRank	SVM-PR-LFRank	PageRank	Leader-Follower
Amazon	68%	67%	64%	65%	61%	58%
Wiki	78%	60%	76%	77%	47%	59%
Epinions	69%	59%	65%	65%	53%	60%
Slashdot	73%	44%	67%	68%	41%	62%

Notably, SVM-Full outperforms every other method for each of the 4 networks. Although in some cases, either SVM-PR, SVM-LFRank, or SVM-PR-LFRank perform nearly as well, these other three methods are much less consistent (for example, SVM-PR scores nearly as well as SVM-Full on network Amazon, but much worse on network Slashdot). SVM-PR-LFRank also performs fairly well, but is typically several percentage points behind SVM-Full. The two non-SVM methods tended to perform poorly. Further examination of the data shows that these methods tend to misclassify elements from Class 0 (reciprocal links) as one of the other two classes. This is to be expected, because, as described earlier, for these methods to classify a link as reciprocal, the two nodes must have exactly the same rank. Using a SVM allows some variability in this rank comparison.

Table 2 describes how the elements of each class were classified. For example, 54.2% of elements from Class 0 in Amazon were classified as Class 0, 19.7% as Class 1, and 26.1% as Class 2. We see that, typically, the SVM achieves very high accuracy rates for the directed classes, but lower accuracy for the undirected class. When a directed edge is misclassified, it is usually classified as undirected, and only very rarely is it classified as belonging to the other class of directed edges.

Table 3 contains the average feature values for each class. Notably, the $LF_b - LF_a$ and $PR_b - PR_a$ features tend to show high variance between the different classes: Class 1 always has a low value of $LF_b - LF_a$, and a high value of $PR_b - PR_a$, while the opposite holds true for Class 2. Class 0 typically has average values close to 0 for these scores. However, recall from Table 1 that these metrics alone gave poor accuracy. This occurs for two reasons: first, these methods will only predict a reciprocal link if the two nodes have identical ranks, and second, because although they perform well on average, there are many elements where they fail.

These results demonstrate two key points. First, although ranking methods such as PageRank and Leader-Follower Ranking score reasonably well on their own, they are vastly improved with use of a supervised classifier. Second, relatively simple metrics such as degree, node betweenness, and edge betweenness can further boost the accuracy of methods for predicting the direction of a link.

Table 2. SVM-Full Classification. The value in row R, column C is the fraction of elements in class R that were classified as class C.

		Class 0	Class 1	Class 2
Amazon	Class 0	54.2%	19.7%	26.1%
	Class 1	22.1%	73.1%	4.7%
	Class 2	16.2%	7.0%	76.8%
Wiki	Class 0	81.0%	10.4%	8.6%
	Class 1	23.0%	76.3%	0.6%
	Class 2	19.4%	2.7%	77.9%
Epinions	Class 0	57.4%	21.1%	21.4%
	Class 1	19.8%	75.8%	4.5%
	Class 2	21.3%	3.7%	75.0%
Slashdot	Class 0	53.5%	19.3%	27.1%
	Class 1	15.3%	81.2%	3.5%
	Class 2	11.1%	4.1%	84.9%

Table 3. Average Feature Value per Class: D_a is degree of node a, NB_a is node betweenness of node a, EB is edge betweenness, LF_a is leader-follower ranking of node a, PR_a is PageRank of node a

		D_a	D_b	NB_a	NB_b	EB	$LF_a - LF_b$	$PR_a - PR_b$
Amazon	Class 0	7.8	8.1	0.0015	0.0018	2.78e-05	10.4	4.5e-0.5
	Class 1	6.3	26.3	0.0013	0.0142	9.40e-05	-1550.0	0.0015
	Class 2	25.8	6.2	0.0146	0.0019	10.0e-05	1532.0	-0.0016
Wiki	Class 0	219.2	214	0.0044	0.0041	1.01e-07	-214.1	7.3e-05
	Class 1	195.4	124.4	0.0040	0.0022	1.21e-07	-2905.6	0.0003
	Class 2	120.5	191.1	0.0020	0.0040	1.25e-07	3656.7	-0.0003
Epinions	Class 0	169.0	180.3	0.0021	0.0033	2.40e-08	-3.7	2.45e-05
	Class 1	146.6	210.1	0.0043	0.0028	3.96e-08	-1762.4	0.0007
	Class 2	213.5	164.4	0.0029	0.0058	3.92e-08	1741.1	-0.0007
Slashdot	Class 0	91.8	95.4	0.0018	0.0020	1.38e-07	-98.9	-6.69e-07
	Class 1	83.4	84.8	0.0015	0.0014	1.28e-07	-2335.4	0.0002
	Class 2	84.7	81.7	0.0013	0.0014	1.23e-07	2264.1	-0.0002

5   Conclusion and Future Work

In this paper, we have considered the problem of predicting the directions of links in a network, motivated by the frequent lack of complete network data. Solutions to this problem may be useful for networks such as Facebook, which represents every link as undirected, even though the underlying social network is likely to be directed. Determining the directions of links will allow researchers to more accurately perform network analysis tasks, such as tracing the spread of information through the network.

We treat this problem as a supervised learning problem in which the directions of some edges are known and the directions of other edges are unknown. We considered 4 network datasets from different domains, each containing both directed and reciprocated edges. For each network, we partitioned the edges into a training set, containing 90% of the edges, and a test set containing the remaining 10% of the edges. For each edge, we calculated several features related to the edge's position in the network topology. To calculate these features, we used two graphs: an undirected version of the complete graph (including edges in the test set), and a directed version of the graph represented by the training set. We compared our methods to PageRank and Leader-Follower Ranking, two other methods for predicting link directionality. Both of these methods rank nodes based on their importance in their networks, with the assumption that links are likely to go from low-value nodes to high-value nodes. In our experiments, we saw that the SVM trained on the full feature vector produced results substantially better than other methods. A detailed analysis of our results showed that directed links most often tended to be confused with undirected links, rather than links of the other direction. Our method achieved very high accuracy on the directed links.

Interestingly, although the two ranking methods showed high inter-class variability, they performed much more poorly than the other methods at classification. When the scores from these methods were used in a classifier, we were able to achieve much higher accuracy results, even though there was no additional information. This is likely due to these methods' poor ability to identify reciprocated links. This issue can be remedied through use of a SVM; however, adding further network topological information to these features provided an additional gain in accuracy. Of the additional features we considered, node degree seemed the most valuable in general, though the node betweenness and edge betweenness features were occasionally valuable on individual networks.

This area of research has many interesting potential future directions. Most obviously, it is worthwhile to determine which features (including ones not considered in this paper) are most valuable for this task. We are also interested in incorporating external metadata (as in [2]), in order to learn how much additional accuracy such metadata can provide when used in conjunction with the features in this paper.

Acknowledgements. Funding for this work is provided by grant AFOSR FA9550-09-1-0675.

References

1. Al Hasan, M., Chaoji, V., Salem, S., Zaki, M.: Link prediction using supervised learning. In: SIAM Workshop on Link Analysis, Counterterrorism and Security with SIAM Data Mining Conference (2006)
2. Adar, E., Adamic, L.: Tracking information epidemics in blogspace. In: Proceedings of the 2005 IEEE/WIC/ACM International Conference on Web Intelligence (2005)
3. Lu, L., Zhou, T.: Link prediction in complex neworks: a survey. Phys. A: Stat. Mech. 390(6), 1150–1170 (2011)
4. Gup, F., Yang, Z., Zhou, T.: Predicting link directions via a recursive subgraph-based ranking (2012), http://arxiv.org/abs/1206.2199
5. Freeman, L.C., Borgatti, S.P., White, D.R.: Centrality in valued graphs: a measure of betweenness based on network flow. Social Networks 13, 141–154 (1991)
6. Chang, C.-C., Lin, C.-J.: LIBSVM: a library for support vector machines. ACM Transactions on Intelligent Systems and Technology 2 (2011)
7. Page, L., Brin, S., Motwani, R., Winograd, T.: The PageRank citation ranking: bringing order to the Web. Technical Report, Stanford Digital Libraries (1998)
8. Stanford Network Analysis Project, http://snap.stanford.edu/index.html (accessed 2012)
9. Leskovec, J., Adamic, L., Adamic, B.: The dynamics of viral marketing. ACM Transactions on the Web 1(1) (2007)
10. Leskovec, J., Huttenlocher, D., Kleinberg, J.: Signed networks in social media. In: CHI 2010, pp. 1361–1370 (2010)
11. Leskovec, J., Lang, K., Dasgupta, A., Mahoney, M.: Community structure in large networks: natural cluster sizes and the absence of large well-defined clusters. Internet Mathematics 6(1), 29–123 (2009)
12. Richardson, M., Agrawal, R., Domingos, P.: Trust Management for the Semantic Web. In: Fensel, D., Sycara, K., Mylopoulos, J. (eds.) ISWC 2003. LNCS, vol. 2870, pp. 351–368. Springer, Heidelberg (2003)

Predicting Driving Direction
with Weighted Markov Model

Bo Mao[1,*], Jie Cao[1], Zhiang Wu[1], Guangyan Huang[2], and Jingjun Li[3]

[1] Jiangsu Provincial Key Laboratory of E-Business,
Nanjing University of Finance and Economics,
Nanjing, China
maoboo@gmail.com
[2] Centre for Applied Informatics,
Victoria University, Melbourne, Australia
[3] Yurun Meat Industry Company Limited, Nanjing, China

Abstract. Driving direction prediction can be useful in different applications such as driver warning and route recommendation. In this paper, a framework is proposed to predict the driving direction based on weighted Markov model. First the city POI (Point of Interesting) map is generated from trajectory data using weighted PageRank algorithm. Then, a weighted Markov model is trained for the near term driving direction prediction based on the POI map and historical trajectories. The experimental results on real-world data set indicate that the proposed method can improve the original Markov prediction model by 10% at some circumstances and 5% overall.

Keywords: driving direction prediction, trajectory mining, weighted PageRank.

1 Introduction

Along with the development of GPS and wireless communication technologies, the driving trajectory data is increased dramatically. Basically ever vehicle can be easily equipped with GPS device and becomes a sensor on the road. As the most driven type of all vehicles, taxi is excellent for road condition detection since there is not too much privacy issue compared with other cars. The trajectory data from taxi has been applied in driving route recommendation, road map generation, map matching and driving direction prediction [1]. It is important to predict the near term direction for the car drivers, which can be useful to provide information about upcoming road situation and is the basic for destination prediction.

History driving trajectories have been proved to be an efficient basis to predict the near-term driving direction [2]. Meanwhile, the city itself contains many "points of interesting (POIs)" that attract more traffic. It is useful to detect these POIs and apply them into driving direction prediction. Currently, POI is

---

* Corresponding author.

S. Zhou, S. Zhang, and G. Karypis (Eds.): ADMA 2012, LNAI 7713, pp. 407–418, 2012.
© Springer-Verlag Berlin Heidelberg 2012

usually collected manually, and many places may be missed. In this paper, we generate city POI map based on taxi trajectory data, since more getting on or off events happened near the POIs. However, the identified POI is not an explicit point but an area with description of its interesting level. Then combining the created POI map and trajectory data, driving directions are predicted. We select weighted PageRank algorithm to generate the POI map and Markov Model to predict the direction.

The rest of paper is structured as follows. Section 2 lists the related work. Section 3 provides the problem statement. The proposed methodology is described in section 4. Experimental results are given in section 5 and Section 6 concludes the whole paper.

2 Related Work

Driving route prediction has been intensively studied in data mining and GIS fields. Patterson et al. [3] have used a dynamic Bayes net to detect a moving person's mode of transportation (i.e. bus, foot, car) based on GPS trajectories and predicted their route. Li et al. [4] have analyzed the periodic behaviors of animal based on their traces. In [5], trajectories of road network are classified by mining the discriminative patterns. These models and patterns can be used for direction prediction. Krumm [2] proposed a Markov model, trained from past behavior, to predict the next few road segments that a driver would take. Froehlich and Krumm [6] predicted entire routes from GPS traces by looking at which previous route a driver appeared to lock onto partway into the trip. In [7], the author developed a basic algorithm, and variations, to predict the aggregate turn behavior of drivers at intersections. Besides trajectory data, geographic features are also considered in the direction prediction [3], such as land covered by water or ice made for less popular destinations than commercial and residential areas. Also Jiang [8] suggested street-based topological representations for prediction traffic flow in GIS. Therefore, it is necessary to combine the city features or POIs into the prediction.

To detect POIs, Internet and mobile phone data are analyzed [9, 10] but it may arise the privacy concerns. Lee et al. [11] tried to turn the geo-tagged photo into list of POIs. Brouwers [12] discussed three different sensor sources (GPS, Wifi and Geolocation) and their idiosyncrasies when used for dwelling detection. Jiang [8] ranked the city space based on road network using weighted PageRank. It found that the PageRank scores are better correlated to human movement rates than the space syntax metrics. However, the geo-related data is not allows available, so trajectory based POI detection method are employed.

In this work, we present a driving direction prediction framework based on just trajectory data using Weighted PageRank algorithm. In the framework, city POI map is generated from trajectories and the driving prediction is performed based on the POI map and history trajectories.

3   Problem Statement

In this paper, near-term driving direction prediction is based on the city POI map and trajectory data. This section will give the definition of POI map and other related concept we use.

Definition 1. City POI map is a function $f(c_P)$, in which $c_P = (x_P, y_P)$ is the coordinate of location P and $f(c_P)$ returns the attraction level in location P.

Definition 2. Trajectory TR=$\{c_1, c_2, \cdots, c_n\}$ represents the coordinates in a time sequence from t_1 to t_n, in which $t_i < t_j$ if $i < j$.
A trajectory can be also viewed as a set of line segments TR=$\{c_1 c_2, \cdots, c_{n-1} c_n\}$. Since no road graph is generated, we approximately define a coordinate $c_x \in TR$ (c_x belongs to TR), *iff* $\min(\text{dist}(c_x, c_i c_{i+1}) | 0 < i < n) < d$, which means c_x in a buffer of TR as shown in Fig. 1a. There are two trajectories: $TR_1 = \{c_0, c_1, c_2, c_3\}$ and $TR_2 = \{c_x, c_y, c_z\}$. Because the points in TR_2 are all contained by TR_1, then we define that TR1?TR2. Assume two trajectories TR_i and TR_j, if $TR_i \subset TR_j$ or $TR_j \subset TR_i$, then they are the same route or $TR_i = TR_j$.

At some circumstance, there are certain points that not belong to another trajectory even if they are along the same road e.g. point cy in Fig. 1b. To overcome the problem, we add the condition that if c_i and c_{i+1} belong to TR, then $c_i \in$TR.

In this paper, we try to predict the near term driving direction of a vehicle based on the history trajectories. Suppose $TR_c = c_{-n}, c_{-n+1}, c_{-1}, c_0$, we will find out the possible options of the next direction that contains c1 and their corresponding probabilities. The one with the biggest possibility will be assigned as the predicted direction.

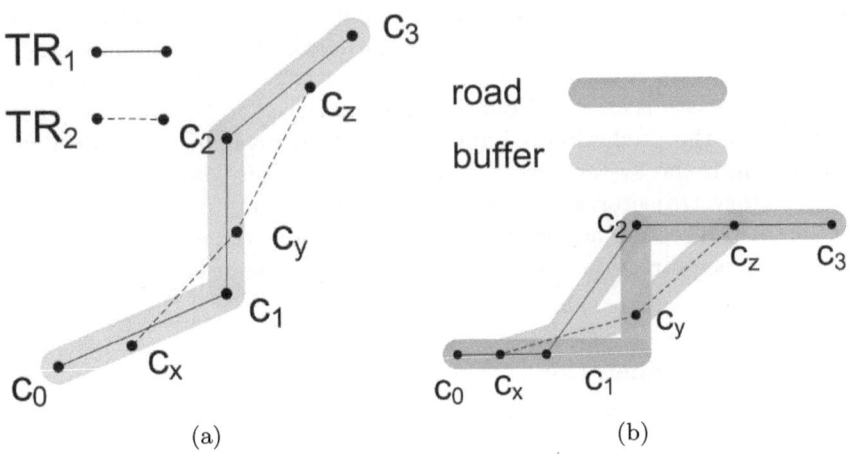

(a)   (b)

Fig. 1. Example of belong to relationship

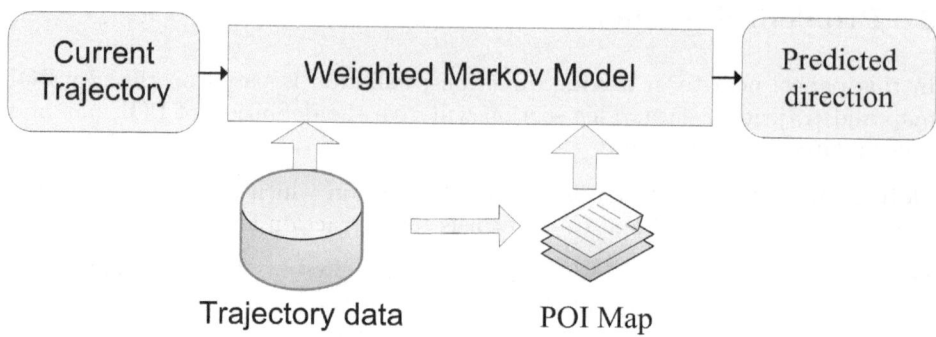

Fig. 2. Prediction Framework

4   Methodology

In this section, we present the framework of weighted Markov model based prediction. This framework consists of two steps: 1) POI map generation, and 2) weighted Markov prediction. We describe the framework in Section 4.1 and these two steps through Section 4.2 and 4.3 respectively.

4.1   Driving Direction Prediction Framework

The proposed prediction framework contains three main parts: trajectory dataset, POI map and weighted Markov model. Initially, trajectory data is collected from different cars with GPS tracker. The data is structured according to **Defination 2**. To separate trajectories from GPS points, we can connect the tracker with some devices such as taximeter or just apply data mining methods [13]. The POI map is created based on trajectory dataset using weighted PageRank algorithm. Each trajectory can be regarded as an edge that link two parts of the city, based on which a graph about city is generated. With weighted PageRank method, all parts of the city are ranked according to their attractions to the drivers, and a POI map of the city is created from the ranking. Based on the trajectory dataset and POI map, the prediction is implemented. When a real time trajectory TR is inputted into the system, we first test its nth-order Markov Model to find out the history trajectories that contain n latest points of TR and select the most weighted direction according to the POI map as the output result instead of most frequent direction. Rest of the section will describe the framework in details.

4.2   POI Map Generation

As defined, POI map $f(c_x)$ is the attractive level to the drivers or passages. PageRank [14] has been proved as an efficient method to define the "attractiveness" of a web page and city morphology [15]. Therefore, we employ PageRank to calculate the POI map. To take the number of trajectories into consideration, weighted PageRank [16] is selected.

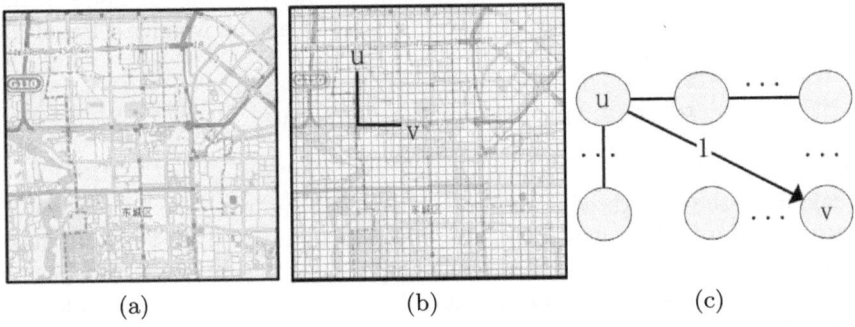

Fig. 3. POI map generation

The analysis in this paper is performed on the city raster map. We first divide the city (Fig. 3a) into square areas (Fig. 3b). Every square area is represented by a node in the directed city graph (Fig. 3c). For each square area or node such as u and v in Fig. 3b, if there is a trajectory connected from u to v, the number of outlinks in u and number of inlinks in are both increased by 1 as shown in Fig. 3c.

The weighted PageRank of each node in graph G is calculated according to [17]. The original PageRank algorithm [14] defined the pagerank of a node, as formula 1:

$$PR_u = 1 - d + d \sum_{v \in B(u)} \frac{PR(v)}{N_v}. \tag{1}$$

where u represents an area or node. B(u) is the set of nodes that point to u. PR(u) and PR(v) are rank scores of node u and v, respectively. N_v denotes the number of outgoing links of node v. d is a dampening factor that is usually set to 0.85.

The original PageRank is proposed for web pages, but in city transportation scenario the number of trajectory from one area to another is critical. For example, the train station trend to be an important POI since it is the destination of many trajectories, which can be reflected well in the weighted PageRank algorithm. The Weighted PageRank Algorithm assigns larger rank values to more important (popular) nodes instead of dividing its rank value evenly among its outlink nodes. The weight from the number of inlinks and outlinks is recorded as W_{in} and $W_{out}(v,u)$, defined as follows respectively.

$$W_{(v,u)}^{in} = \frac{I_u}{\sum_{p \in R(v)} I_p}. \tag{2}$$

$$W_{(v,u)}^{out} = \frac{O_u}{\sum_{p \in R(v)} O_p}. \tag{3}$$

Where I_u/O_u and I_p/O_p is the number of inlinks/outlinks of node u and p. R(v) denotes all nodes that are pointed from v. Then the weighted PageRank is defined in formula (4)

$$PR(u) = (1 - d) + d \sum_{v \in B(u)} PR(v) W_{(v,u)}^{in} W_{(v,u)}^{out}. \tag{4}$$

To calculate the Weighted PageRank, PR(u) is set to 1 for all nodes. Then iteratively compute the new PR(u) based on formula (4) until the difference between two rounds are smaller than a threshold (e.g. ¡10_{-5}). Previous experiments [14, 17] showed that the PageRank gets converged to a reasonable tolerance in the roughly logarithmic (logn), which is also applicable for Weighted PageRank.

4.3   Weighted Markov Model

According to [2], the prediction of near-term driving direction can be based on its past route. Therefore, Markov model is employed. A trajectory can be defined as $\{\cdots, c_{-2}, c_{-1}, c_0, c_1, c_2, \cdots\}$, where c_0 is the current position that can be updated, and $c_1, c_2, \cdots$ are the unknown future position of the vehicle. We try to predict the direction that contains both c_0 and c_1 as defined section 2. At anytime, we have the information $c_0, c_{-1}, \cdots$ back to the beginning of the trip. The Markov model is built on these known positions, and gives a probabilistic prediction over the future directions. The first order Markov model is only based on current position c_0, and the second order Markov model is based on c_0 and c_{-1}. In general, n-th order Markov is based on $c_0, \cdots, c_{-n+1}$ as shown in formula (5). Note that $P[c_1]$ is not the probability of the vehicle appearing at c_1 but its direction to c_1.

$$P[c_1] = P[c_1|c_0, c_{-1}, \cdots, c_{-n+1}]. \tag{5}$$

In original Markov model, the probability of each direction ci is calculated as formula (6), in which $N[c_p \cdots c_q]$ is the number of trajectories containing position $c_p \cdots c_q$.

$$P[c_i|c_0, c_{-1}, \cdots, c_{-n+1}] = \frac{N[c_i, c_0, c_{-1}, \cdots, c_{-n+1}]}{N[c_0, c_{-1}, \cdots, c_{-n+1}]}. \tag{6}$$

Based on formula (6), we can select the direction with biggest probability as the next possible direction.

The original Markov model only takes the number of trajectories into consideration, but the POI distribution along trajectories could also affect the driving direction as people trend to go to these "interesting spots". Therefore, in this paper we improved the original Markov model with the POI map and proposed the weighted Markov model in formula (7):

$$P[c_i|c_0, c_{-1}, \cdots, c_{-n+1}] = \frac{\sum f(c_k\ c_{k_e})}{\sum f(c_l\ c_{l_e})}. \tag{7}$$

In formula (7) $f(c_p\ c_q)$ is the maximum value of POI map along the route from c_p to c_q. k_e and l_e are the end point of k and l respectively. If POI map is a

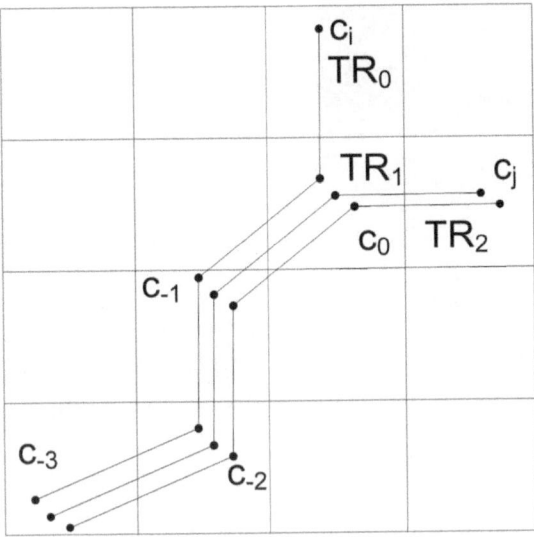

Fig. 4. Weighted Markov model

constant value, then formula (7) will be the same as (6). Fig. 4 illustrates an example of original and weighted Markov model.

In Fig. 4, three history trajectories are listed. For trajectory $[c_0, c_{-1}, c_{-2}, c_{-3}]$, we can predict its next direction will be c_j with probability 2/3. However, suppose the attractive value in area ci is 3 and attractive value in area c_j is 1, then the predicted next direction will be ci with probability 3/5.

5   Performance Evaluation

In this section, we evaluate the performance of the near-term driver direction prediction based on weighted Markov model. We describe the experimental data and environment in Section 5.1. We demonstrate the POI map generated from trajectories in Section 5.2 and report the prediction results comparing with original Markov model in Section 5.3.

5.1   Experimental Setting

All experiments are conducted on an Intel Core2 Duo 2.4GHz PC with 3.25GB of RAM, running on Windows XP SP3. Eclipse 3.5.2 is selected as the IDE. All programs are written in Java.

The test dataset came from a taxi company in San Francisco area. The collected trajectories contain 536 taxis driving in three months from April to July, 2008. Each trajectory is consisted of several records that have latitude, longitude, flag bit (1 for with passenger and 0 for not), and the current time. The

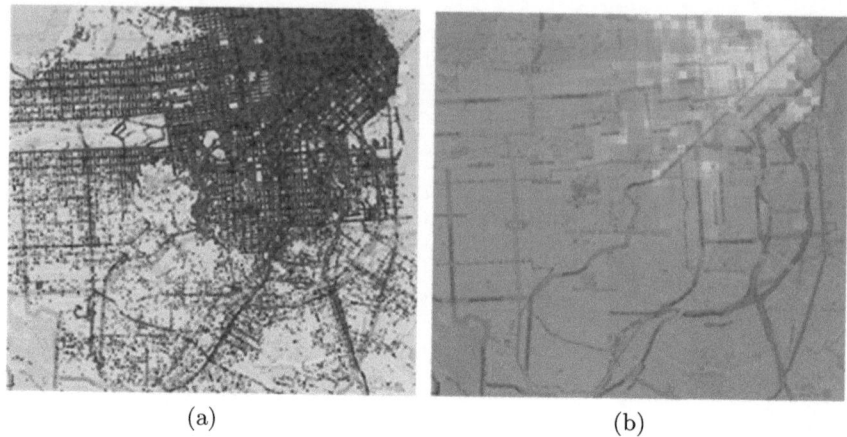

(a)   (b)

Fig. 5. Snapshots of the start/end point distribution of training trajectories

Table 1. Generative models of shilling attacks

Label	Location	Label	Location
1	San Francisco National Cemetery	6	Golden Gate Heights Park
2	Downtown, Nob Hill	7	School of the Arts
3	Golden Gate Bridge	8	Center BART station
4	Golden Gate Park	9	West Portal
5	University of San Francisco	10	Train Station and AT&T Park

duration between two records is one minute, which reduces the accuracy of the prediction but still preserve the repre-sentativeness of the methods comparison.

357 taxis out of total 536 are selected as the training data (2/3) and others (1/3) are test data. Only the trajectories that are sending passengers are taken into consideration, and there are totally 250,438 trajectories used as history data. The POI map will also be generated from these train trajectories. Fig. 5 gives the snapshots of the distribution of training trajectories. Fig. 5a shows the start pint (red) and end point (blue), and Fig. 5b illustration their density map.

5.2 Generated POI Map

The generated POI map based on different number of trajectory data is given in Fig. 6. The test area is divided into 50*50 sub areas. The weighted PageRank algorithm identifies many local attractive centers (POIs) compared with the density distribution map in Fig. 5b. We can see that the detected POIs are gathered into a stable location along with the increasing of taxi number. These POIs are distributed in highway junc-tion, residential area and some park areas. The experimental result indicates the effectiveness of the proposed method to detect the POIs in the city. Based on the POI map, we will predict the near term driving direction of vehicle.

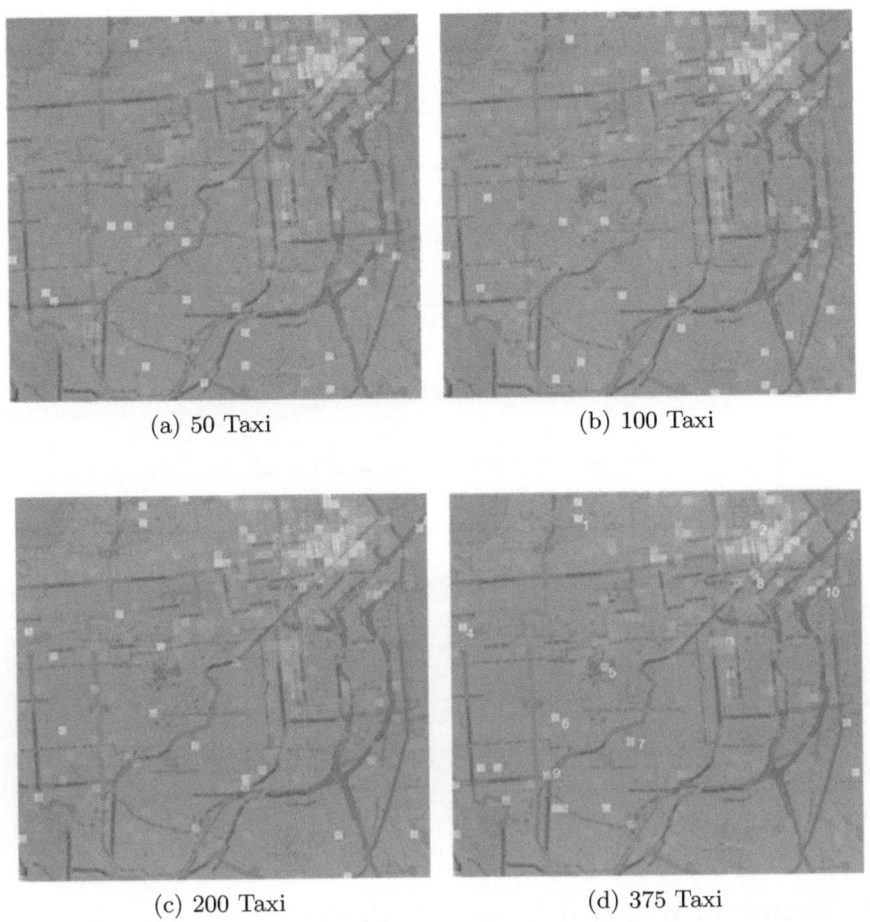

(a) 50 Taxi   (b) 100 Taxi

(c) 200 Taxi   (d) 375 Taxi

Fig. 6. Generated POI map from different number of Taxis data

Finally, 2/3 of the whole dataset are select to generate the POI map used for pre-diction as Fig. 6d. We identify some of the POIs according to the local map and shown in Table 1. It is shown that the weighted PageRank is effective in POI detec-tion. For example, the most weighted parts are located in the downtown area (Lable 2); some other detected POI areas are scenic spots such as Golden Gate Bridge (Lable 3) and Golden Gate Park (Lable 4); Universities like University of San Francisco (Lable 5) and School of the Arts (Lable 6) are also identified as POI area; Finally, the transportation centers e.g. Center BART station, West Portal and Train Station and AT&T Park (Lable 8,9,10 respectively) are correctly detected as POIs.

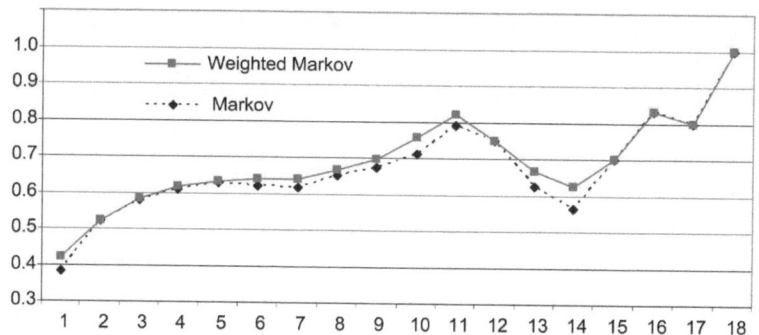

Fig. 7. Prediction accuracy of the original and Weighted Markov Models

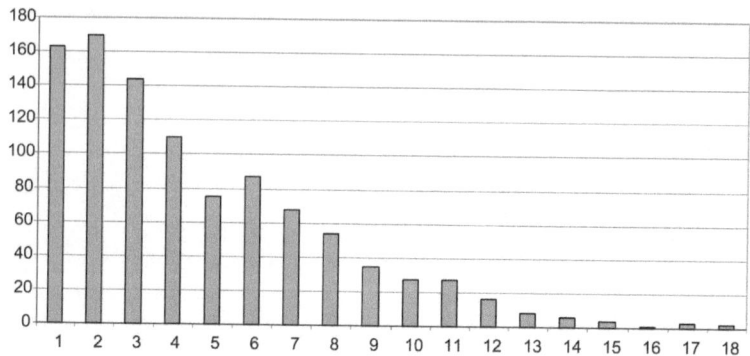

Fig. 8. Number of matched points

5.3 Results of Weighted Markov Prediction

In the prediction experiment, 250,438 trajectories from 357 taxis are used as train data. Set P which contains1000 trajectories from remain taxis is used as test data. We first generate the POI map of the area as shown in Section 5.2. For each trajectory TR_i in P, we randomly set a position c_0^i as the current location, and we will predict the next driving direction $c_0^i c_1^i$ with the proposed weighted Markov method.

The next direction is determined by $HT(k)$ the historical trajectories which have the same direction with TR_i in the last k points as defined in Section 4.3. Therefore, we can calculate the accuracy of prediction by testing if HT(k,1), the historical trajectories which have the same direction with TR_i in the last k points and the next 1 point, have the majority weight (over 50%) of $HT(k)$. For example, in Fig. 4, $HT(4) = TR_0, TR_1, TR_2$, and $HT(4,1) = TR_0$ if $c_1 = c_i$ or $HT(4,1) = TR_1, TR_2$ if $c_1 = c_j$. For the original Markov model, if $c_1 = c_j$ the prediction will be correct since TR_1, TR_2 takes 2/3 of $HT(4)$; for the weighted Markov model, if $c_1 = c_i$, the prediction will be correct, because TR_0 takes 3/5 of $HT(4)$; otherwise the prediction will be wrong.

In the prediction, we calculate the prediction accuracy of original Markov model and the proposed weighted Markov model and the results are given in Fig 7. The proposed method increases the prediction accuracy more than 10% for the trajectories with only one point match. The overall accuracy is increased around 5% by applying the weighted Markov model.

The accuracy improvement is mainly gained from the POI map included by weighted Markov model, which actually reflect the attractiveness of the potential destinations according to the experimental results.

We also notice that the prediction accuracy is increasing along the number of matched points. However, as the total number of matched trajectories decreases, the random factor increases. Therefore, we can see the accuracy drop when number of matched points over 12. Fig. 8 demonstrates the number of matched points of the 1000 trajectories (set P) for prediction. For example, we can see that around 160 trajectories in P only have one matched point with the base dataset. Meanwhile there are fewer than 20 trajectories when number of matched points is 12, which is the reason for the drop of prediction accuracy shown in Fig. 7.

6   Conclusions

In this paper, we have presented a framework of weighted Markov model based driving direction prediction with trajectory data. We have performed the analysis about the driving behavior, which leads to the conclusion that POI map can improve the predic-tion accuracy.

We have conducted experiments by comparing prediction accuracy of original Markov model and the proposed weighted Markov model. The experimental results show that weighted PageRank algorithm is effective to compute the city POI map and the POI map can improve the driving direction prediction accuracy.

Overall, we believe that we have provided a method to analysis distribution of POIs in city area and a framework to better predict the near term driving direction. In future research, we will focus on the time related POI map considering different time people may go to different places.

Acknowledgments. This research is supported by National Natural Science Foundation of China under Grants No. 41201486, 61103229 and 71072172, Jiangsu Provincial Colleges and Universities Outstanding S&T Innovation Team Fund under Grants No. 2011013, Key Project of Natural Science Research in Jiangsu Provincial Colleges and Universities under Grants No. 12KJA520001, Key Technologies R&D Program of China under Grants No. SQ2013BAJY4130, National Key Technologies R&D sub Program in 12th five-year-plan under Grants No. SQ2011GX07E03990, the Natural Science Foundation of Jiangsu Province of China under Grant BK2012863, International S&T Cooperation Program of China under Grants No. 2011DFA12910, and Transformation Fund for Agricultural Science and Technology Achievements under Grants No. 2011GB2C100024. Jiangsu Province demonstration project of Internet of things (Yurun Group).

References

[1] Zheng, Y., Zhou, X.: Tri-Training: Exploiting Unlabeled Data Using Three Classifiers. Spinger (2011)

[2] Krumm, J.: A markov model for driver turn prediction. In: World Congress on Society of Automotive Engineers (SAE), Detroit, MI, USA (2008)

[3] Patterson, D.J., Liao, L., Fox, D., Kautz, H.: Inferring High-Level Behavior from Low-Level Sensors. In: Dey, A.K., Schmidt, A., McCarthy, J.F. (eds.) UbiComp 2003. LNCS, vol. 2864, pp. 73–89. Springer, Heidelberg (2003)

[4] Li, Z., Han, J., Ding, B., Kays, R.: Mining periodic behaviors of object movements for animal and biological sustainability studies. Data Min. Knowl. Discov. 24(2), 355–386 (2012)

[5] Lee, J.G., Han, J., Li, X., Cheng, H.: Mining discriminative patterns for classifying trajectories on road networks. IEEE Trans. Knowl. Data Eng. 23(5), 713–726 (2011)

[6] Froehlich, J., Krumm, J.: Route prediction from trip observations. In: 2008 World Congress on Society of Automotive Engineers (SAE) (2008)

[7] Krumm, J.: Where will they turn: predicting turn proportions at intersections. Personal Ubiquitous Comput. 14(7), 591–599 (2010)

[8] Jiang, B.: Ranking spaces for predicting human movement in an urban environment. Int. J. Geogr. Inf. Sci. 23(7), 823–837 (2009)

[9] Schonland, A., Williams, P.: Using the internet for travel and tourism survey research: Experiences from the net traveler survey. Journal of Travel Research 35(2), 81–83 (1996)

[10] Ahas, R., Aasa, A., Roose, A., Ülar, M., Silm, S.: Evaluating passive mobile positioning data for tourism surveys: An estonian case study. Tourism Management 29(3), 469–486 (2008)

[11] Lee, C., Greene, D., Cunningham, P.: Detecting grand tours of europe with geotags. In: 2nd Workshop on Computational Social Science and the Wisdom of Crowds at NIPS 2011 (2011)

[12] Brouwers, N., Woehrle, M.: Detecting dwelling in urban environments using gps, wifi, and geolocation measurements. In: Proc. 2nd Int'l Workshop on Sensing Applications on Mobile Phones, pp. 1–5 (November 2011)

[13] Zhu, Y., Zheng, Y., Zhang, L., Santani, D., Xie, X., Yang, Q.: Inferring taxi status using gps trajectories. CoRR abs/1205.4378 (2012)

[14] Page, L., Brin, S., Motwani, R., Winograd, T.: The pagerank citation ranking: Bringing order to the web. Technical Report 1999-66, Stanford InfoLab (November 1999); Previous number = SIDL-WP-1999-0120

[15] Jiang, B., Liu, C.: Street-based topological representations and analyses for predicting traffic flow in gis. International Journal of Geographical Information Science 23(9), 1119–1137 (2009)

[16] Xing, W., Ghorbani, A.: Weighted pagerank algorithm. In: CNSR, pp. 305–314. IEEE Computer Society (2004)

[17] Ridings, C., Shishigin, M.: PageRank Convered. Technical report (2006)

Pattern Mining, Semantic Label Identification and Movement Prediction Using Mobile Phone Data

Rong Xie[1], Jun Luo[2,3], Yang Yue[4,5], Qingquan Li[4,5], and Xiaoqing Zou[6]

[1] International School of Software, Wuhan University,Wuhan 430079, China
[2] Shenzhen Institutes of Advanced Technology, CAS, Shenzhen 518055, China
[3] Shenzhen Key Laboratory of High Performance Data Mining, Shenzhen 518055, China
[4] Shenzhen University, Shenzhen, China
[5] State Key Lab of Information Engineering in Surveying, Mapping and Remote Sensing, Wuhan University, Wuhan 430079, China
[6] Faculty of Land Resource Engineering, Kunming University of Science and Technology, Kunming, China
{xierong,yueyang,qql}@whu.edu.cn, jun.luo@siat.ac.cn, zuoxq@163.com

Abstract. Data collected from mobile phones have potential knowledge to provide with important behavior patterns of individuals. In this paper, we present approaches to discovering personal mobility and characteristics based on mobile phone location information and semantic analysis. We discuss three aspects related to very common mobile phone-related applications such as pattern mining, semantic label identification and movement prediction. We use real mobile phone data to perform functions of discovering these behavior patterns and demonstrate effectiveness of our approaches.

Keywords: Mobile phone log data, Mobility pattern, Frequent pattern mining, Semantic labels, Movement prediction.

1 Introduction

Understanding and extracting meaningful information from trajectory data is important to many applications, such as location-based services/location-based advertising (LBS/LBA), context-aware services, route prediction and social network analysis. Using these patterns, it is helpful to provide users with more intelligent and customized services, and benefit service providers as well. Mobile phones record trajectories of our life in the form of call logs, and the data is a rich information source for discovering personal behavior characteristics, particularly mobility patterns.

Different from those continuous and fine-grained GPS-enabled data, most of the existing mobile phone location data are generated only when users make a phone call or use a data communication service, such as sending short message service (SMS) and browsing website based on cell tower locations. So, it usually records a user's trajectories in a discrete, less precise, and uncompleted way [1]. On the other hand, telecommunication providers record mobile phone locations based on their connected

S. Zhou, S. Zhang, and G. Karypis (Eds.): ADMA 2012, LNAI 7713, pp. 419–430, 2012.
© Springer-Verlag Berlin Heidelberg 2012

cell towers. Mobile phone location data is thus coarse in space at the granularity of cellular tower coverage radius and sparse in time only when an event happens [2]. In summary, some challenges exist in the mobile phone data analysis and mining as follows. 1) It is difficult to directly access meaningful information about mobility patterns of users since all these location logs are in low level data unit [3]. To make mobile phone data more readily accessible to related applications, higher level data abstractions are needed. 2) In the traditional data mining, some classical algorithms, for example, *Apriori* method [4] or pattern growth-based method [5], ignoring spatial and temporal characteristics, often lead to incomplete, even wrong knowledge, and cannot be fully used in discovering mobility patterns.

The aim of our research is to overcome the limitations of mobile phone log data, and discover some personal mobility and behavior patterns. In the paper, we study processing and mining methods of personal behavior patterns from mobile phone location information. Based on these methods, some improved data mining algorithms are developed, including frequent pattern mining, semantics label extraction and movement prediction.

The rest of the paper is organized as follows. Approaches and algorithms are proposed to handle frequent pattern mining, semantics label extraction and movement prediction in section 2-4, respectively. Section 5 shows some visualization results. We also present the related work in Section 6. Conclusion and future work are finally discussed in Section 7.

2   Trajectory Pattern Mining

2.1   History Trajectory Modeling

For handling pattern mining from trajectory, it is firstly required to model history trajectory of each individual. For the case of GPS data, a record can be returned every few seconds, containing accurate location information. It is thus convenient to generate a complete trajectory from GPS data. For the original mobile phone log data, a record represents information that a user stops in a place at a specific time, containing user ID, time, and cell tower ID. In our research, this information of single stop point is inadequate, so it is necessary to define and generate trajectory by relying on these original data. A trajectory of a mobile phone user is a sequence of cell tower locations. To account for approach to modeling, we introduce two basic mobility concepts as follows.

Definition 1 (Stop): A *stop* is a position that a user k stays within the range of a cell tower (*cell_id*) at time t. Let s_k denote a collection of stop points of the user k such that $s_k = \{s_{k1}, s_{k2}, ..., s_{kn}\}$, where each s_{ki} contains identifier of cell tower ($s_{ki}.cell\_id$) and timestamp ($s_{ki}.t$).

Definition 2 (Trajectory): We can sequentially connect some stop points of a user k into a trajectory. Thus, a trajectory tr is defined as equation (1), where tr_id stands for trajectory identifier of user k, s_k is a stop point of user k.

$$tr :=< tr\_id, \{s_k\} >\qquad(1)$$

According to definition 2, trajectory can be generated with the following rules. 1) Frequent staying points can be discovered, where calls are made frequently, such as home, place of work etc. 2) We can merge some points with the same location and adjacent time, resulting in start time st_i and end time et_i. 3) According to rules of trajectory segment, add the start point and the end point into trajectory. 4) If there is no information on record, trajectory can be directly truncated. Based on these rules, a user's history trajectory can be expressed as a sequence of stop point with the format tr_id: $<p_i, st_i\text{-}et_i>$, where tr_id is identifier of trajectory, p_i stands for position that user is located at a place (i.e. a cell tower) at time t_i. st_i represents start time when user enters a specific place; while et_i represents end time when user leaves the place. For example, a trajectory can be generated and described as shown in Figure 1.

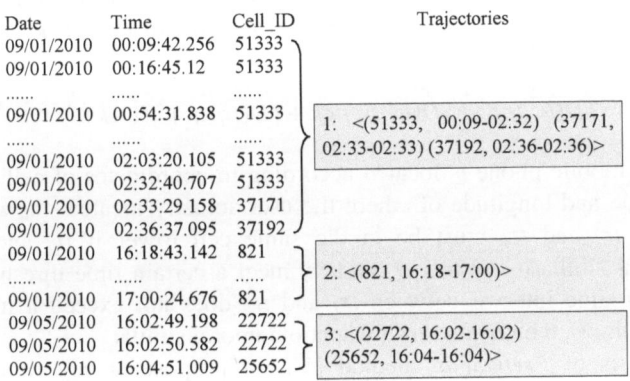

Fig. 1. An example of history trajectory modeling

2.2   Frequent Pattern Mining

Under modeling historical trajectory, we can use algorithms to discover frequent item sets in data, for example, *"User A usually goes to supermarket in the morning on weekend"*; or *"User A always goes to cinema on Friday night"*. Frequent pattern mining is aim to analyze large number of historical trajectory of a user in order to discover frequent spatiotemporal patterns that meet the minimum frequency degrees.

Definition 3 (Frequent Pattern): Frequent pattern *FP* is defined as equation (2), where ci_i is identifier of cell tower, t_i is the time when user enters the cell tower, and f is represented as frequent degree of pattern, which meets the requirements of minimum frequent degree threshold δ_f.

$$FP :=< (ci_1, t_1)(ci_2, t_2)...(ci_n, t_n) : f \mid f \geq \delta_f >\qquad(2)$$

Trajectory is naturally represented as a sequence of stop points. Spatiotemporal frequent pattern not only meets the requirements of general sequential pattern, but also has the characteristics as follows. 1) Each point in the sequence has the strict time characteristics, such as entering time, departure time and duration. 2) Each stop point in the sequence is a separate event. 3) When two stop points in sequence match with each other, traditional similarity standard is no longer suitable, thus a new ranging method must be defined. 4) Different from general sequential pattern, it is necessary to rely on background information for trajectory pattern mining, and filter and screen on the patterns.

Definition 4 (Spatiotemporal Similarity): If two trajectories pass through the same place at a certain time or during a certain time period, they are often with spatial and temporal similarity. The degree of spatiotemporal similarity is a kind of transformation from spatiotemporal distance, which can be directly calculated from spatiotemporal distance STD between trajectory tr_1 and trajectory tr_2 as indicated in equation (3), where $0 \leq k \leq 1$, SD stands for spatial distance between tr_1 and tr_2. TD is time difference between tr_1 and tr_2. In the specific instances, k can be adjusted according to the actual situation.

$$STD(tr_1, tr_2) := SD(tr_1, tr_2) \times k + TD(tr_1, tr_2) \times (1-k) \qquad (3)$$

In our study, mobile phone is located according to coordinates of cell tower, not the precise latitude and longitude of where the user actually located. So we can directly suppose that tr_1 and tr_2 must be in the same cell tower if tr_1 and tr_2 are with spatiotemporal similarity. When tr_1 and tr_2 meet a certain time threshold δ_t, that is, occurrence of time interval between tr_1 and tr_2 does not exceed a maximum time interval, as follows, then tr_1 and tr_2 are spatiotemporal similar.

On the basis of *PrefixSpan* method [5], we propose an algorithm of frequent pattern mining as follows, which re-defines measure method of spatiotemporal similarity to make trajectory mining possible. Similar to *PrefixSpan*, our algorithm does not produce candidate sequences in the mining process. However, it may generate many projected databases, corresponding to each frequent prefix sequence. If we have to physically produce the projection databases, it recursively increases cost to build a large number of databases. An optimization can be pseudo-projection, which records index of corresponding sequence and starting position of projection suffix, rather than a physical projection. In other words, physical projection of sequence is replaced by identifier of recorded sequence and index of projection location. When pseudo-projection is implemented in memory, cost of projection can be obviously reduced. However, if pseudo projection is used as hard disk-based access, it may be less effective, because random access of hard disk space is expensive. One solution is that physical projection is used if original trajectory database or projection database is too large to put in memory; while pseudo-projection is used if projection database can be placed in memory. This idea is used in our study.

Trajectory pattern mining algorithm based on pattern growth *ProcessProjectedDB* is described as follows. The results of the algorithm are stored in a pattern set *PatternSet*, whose format of each pattern is as defined in Definition 2.

In the following algorithm, we use logical structure of *PrefixSpan* algorithm to ensure the order of stop points in trajectory pattern. We also define similarity measurement method to ensure efficiency of the algorithm.

Algorithm 1. ProcessProjectedDB(sequenceIndex, frequency, preStr)

for i:=1 to sequenceIndex.Count
  do if sequenceIndex[i]<sequenceSet[i].Count
    then for j:=sequenceIndex[i] to sequenceSet[i].Count
      do Count number of each space-time point. Compare distance with a
threshold to know if two points are similar.
  for each term in termCount
  do if number(term)>frequency
    then if frequentSet.ContainsKey(preStr+term)=FALSE
      then frequentPatternSet.Add(preStr+term, number(term))
      for k:=1 to sequenceIndex.Count
      do for l:=sequenceIndex[k] to sequenceSet[k].Count
        do Find the index of current term in sequenceSet[k] and get the
new sequenceIndex to newIndex
        ProcessProjectedDB(newIndex, frequency, preStr)

---

3   Identifying User Semantic Labels

The basic assumption of this study is that, where people live, work and hangs out are those regions that can meet their preferences. Therefore, the characteristics of those regions, to a great extent, also reflect the characteristics of the person. We first identify regions closely attached to a user, i.e., ROI (Region of interests), by spatially aggregating the mobile phone traces. Then, we extract the features of these ROIs using data crawling from POI reviewing websites and real estate websites, together with map data. Last, the regional features are associated to the user as his/her personal labels. This process is shown in Figure 2.

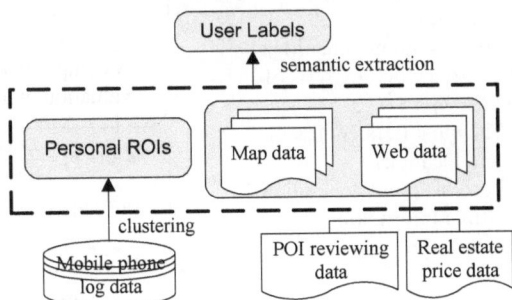

Fig. 2. Approach to semantic label generation

We use clustering approach *ST-DBSCAN* [6] to identify ROIs attached to a user considering time of day, day of week, and public holidays. 500m is used as the threshold radius. Since ROI represents an area that a person spends an amount of time and/or an area visited frequently, a cluster cannot be generated in this study unless the person had at least a call, a text message, or a data communication within the region. To generate semantic features for each ROI, frequency analysis is performed on POI category data to extract the features of these regions, such as shopping and park etc. Real estate price data is also used to label the positioning of the region, by assuming regions with high price are associated with high-expenditure, and vice verses.

Figure 3 illustrates the semantic label extraction for an identified place that a user frequently visited. As shows in the figure, the top three categories of POIs in this region are restaurant, recreation, and educational service. Besides, by referring the average housing price of where the user lives, then a label marking his/her family income/expenditure is "*middle-high*".

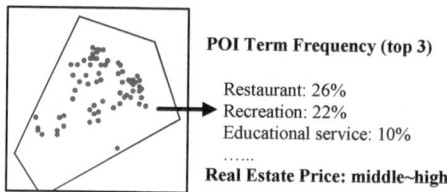

Fig. 3. Semantic label generation

4   Movement Prediction

According to the available relevant background information, location P_1 of user A at time T_1, denoted as (A, P_1, T_1), can be inferred where the user is the most likely appear at location P_2 at the next time T_2, for example (A, P_2, T_2). The process of prediction pattern mining is presented in Figure 4 which depends on background information (i.e. location, predicted point generation), history trajectory analysis (i.e. the most frequent traces and locations), and real-time information (i.e. time) etc.

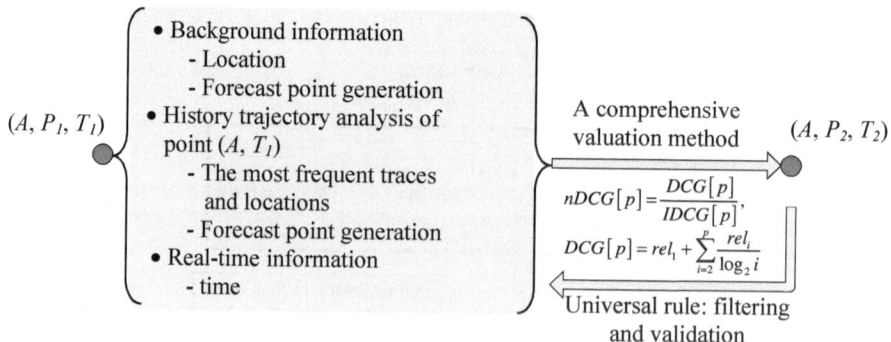

Fig. 4. Movement prediction

We firstly analyze the background information and extract the required characteristics for evaluation. These values are sorted using comprehensive valuation method of mathematical modeling. The sequence of the predicted location is given according to the descending order of the value of these possibilities.

In the paper, classical *DCG* (Discounted Cumulative Gain) is used to measure the quality of screening sorting results. The evaluation factors are *nDCG* and *DCG*, defined in equations of (4) and (5), where, $rel_i=-lgK_i$ (unit of K_i:$\mu mol/L$). The sort of K_i is as the ideal sort result (*IDCGp*). *nDCGp* is to be normalized as *DCG*. If *nDCGp* is closer to 1, then the sort of result is closer to the desired result (i.e. sort of K_i).

$$nDCG[p] = \frac{DCG[p]}{IDCG[p]}; \tag{4}$$

$$DCG[p] = rel_1 + \sum_{i=2}^{p} \frac{rel_i}{\log_2 i}; \tag{5}$$

This is a gradual process to get precise values, the parameters are needed to adjust and improve. The results are filtered and validated by universal rules.

Further, the implementation of algorithm *GetPrediction* is described as follows. The algorithm extracts subscripts of the current stop point in the sequence, and determines whether the next stop point meets the predicted conditions. If so, the scenario is stored as a candidate. After getting all frequent degree of possible scenarios, the likelihood of each scenario is then calculated.

Algorithm 2. GetPrediction(number, ci, interval, frequency, startTime, endTime, nextTime)

---

  patternSet:=GetFrequentSet(number, interval, frequency)
  possiblePointSet:=NIL
  startTime(CurrentPoint):=startTime
  endTime(CurrentPoint):=endTime
  ci(CurrentPoint):=ci, j:=0, sum:=0
  for each pattern in patternSet
    do patternFrequency:=Frequency(pattern)
      places:=SpatioTemporalSequence(pattern)
      for i:=1 to places.Count
        do if SpatioTemporaDistance(CurrentPoint, placesi)<=threshold
          then if startTime<=startTime(placesi+1)<=endTime
              &&endTime<=endTime(placesi+1)<=nextTime
            then possiblePointSet.Add(placesi+1, patternFrequency)
              sum:=sum+patternFrequency)
  for i:=1 to possiblePointSet.Count
    do possibility(possiblePointSet[i]):=patternFrequency/sum

---

Under the implementation of movement prediction, we have several next stop points with a certain extent of possibility. We can further make labels for a user, which can be summarized as *"high expenditure"*, *"night life"* for example, who frequently visited some regions at night (6pm-11pm), so we can give the user LBA services about "KTV" or "pub".

5   Case Study

In this section we summarize some results of our implementation in mining some individual mobility patterns obtained from mobile phone data as below.

5.1   Raw Data Set

The dataset for our work is collected by telecommunications service provider involving calling log files of 14 users during September, 2010 in the city of Kunming, China.

Raw data are described as follows. 1) Map data. The whole area is divided into several *LAC*s (i.e. large area). *LAC* contains several cell towers. Each cell tower can be divided into several sectors. It includes some data of large area (*LAC*), cell tower (*cell_id*), and sector (*site_id*). 2) User information. It includes data of calling or message, International Mobile Subscriber Identity (*IMSI*), normal location update, and normal cycle update. 3) Mobility log. It includes phone number (*number*), time (*time*), region number (*LAC*), cell tower (*cell_id*), sector (*site_id*).

5.2   Framework of Mobility Mining

Figure 5 illustrates the general architecture of our framework in which four steps need to be performed, such as data preprocessing, history trajectory modeling, mobility pattern mining and mining visualization based on raw data, like map data, user information and mobile phone log. Here, we focus on modeling and describing basic pattern types related to personal mobility, such as frequent pattern mining, semantic label generation, and movement prediction.

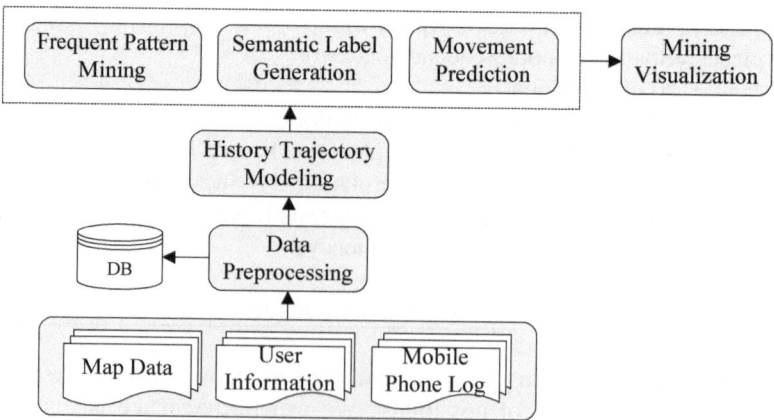

Fig. 5. A framework of mining personal mobility patterns

5.3   Results

The visualization result of frequent pattern mining is shown in Figure 6 of a user identified by 91961, which invokes algorithms of trajectory generation and frequent pattern mining which are discussed in Section 2.

Figure 7 presents the spatial distribution of user 91861's mobile phone traces (red points), and three ROIs (in orange) attached to him/her. Although this user's trace spread over a wide area, he/she only has three important regions attached.

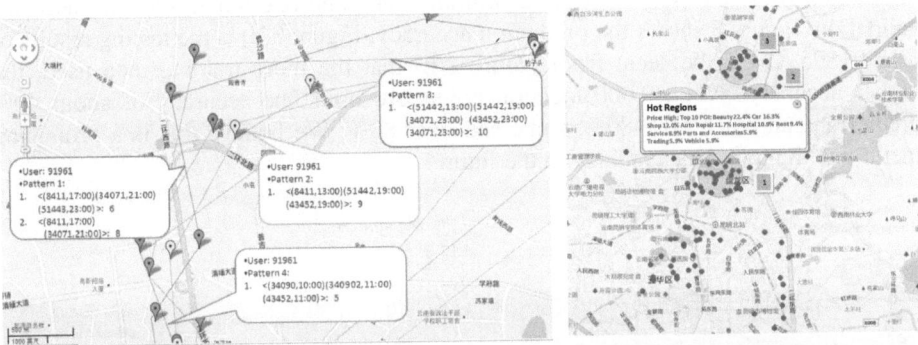

Fig. 6. Visualization of trajectory pattern mining (Taking an example of user 91961)

Fig. 7. User 91861's mobile phone trace and ROIs attached

We consider time stamp associated with each mobile phone location data in the three clusters (as shown in Table 1).

Table 1. Time distribution of user 91861's location on workday

Time period	Cluster 1 (1.50km^2)		Cluster 2 (0.31km^2)		Cluster 3 (0.15km^2)	
	n	%	n	%	n	%
2am-5am	0	0	36	22%	0	0
5am-8am	0	0	33	21%	15	68%
8am-11am	73	28%	0	0	0	0
11am-2pm	97	38%	1	0	1	4%
2pm-5pm	51	20%	2	1%	0	0
5pm-8pm	32	12%	24	15%	4	18%
8pm-11pm	0	0	30	19%	2	9%
11pm-2am	0	0	31	19%	0	0

Here, an assumption is that people work at day time and go home at night. Judged by the time period when the call events occur, it is not difficult to infer that the user works at Cluster 1 and lives at Cluster 3. And most of the time, this person goes to work and back home very regularly. He/she seldom works over-time and has any night life. We estimate the housing price of Cluster 3 where the person lives; it can be found that 75% of the house price is between 6000-8000RMB/m^2 which is the average price of the study area. Then a label possibely associated with the region is "*middle income*". The top three categories of POI in Cluster 2 are restaurant, recreation, and educational service. Based on these information, it can be inferred that Cluster 2 is possibly a leisure place. Similarly, labels associated with Cluster 1 where the user works are: "*government office*", "*university*", "*park*". Hence, some labels about the users can be summarized as: *middle income, civil servant/university staff,* and *regular life*.

Using such semantic lables as input, movement prediction are conducted. 5-month log data of fourteen users are used as training data, and one-month data are used for validation. Figure 8 shows the prediction accuracy. Figure 8(a) is the testing results of user 91895. It can be seem from Figure 8(b) that the more training data used, the higher accuracy. When 9,000 location points are used, the accuracy of about 40% prediction can reach 50%-80%, some can over 80%. We believe this is a promising method with more data available in the future.

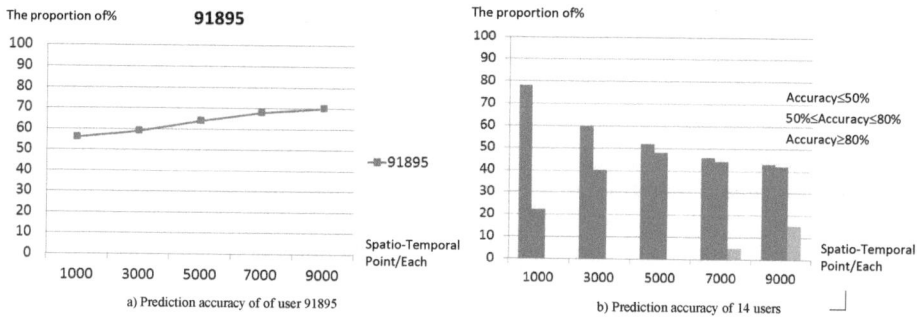

Fig. 8. Accuracy of movement prediction

6   Related Work

To study pattern mining in human mobility, [7] studies the trajectory of 100,000 anonymized mobile phone users of six-month period and find a high regularity degree in human trajectories contrasting with estimation by prevailing Lévy flight and random walk models. People tend to return a few frequent locations and follow simple repeated patterns despite the diversity of their travel history. [8] introduces a system for sensing complex social systems with data collected from 100 mobile phones over 9 months and demonstrate the ability to use standard Bluetooth-enabled mobile telephones to measure information access and use in different contexts, recognize social patterns in daily user activity, infer relationships, identify socially significant locations, and model organizational rhythms. [3] presents formal definitions to capture the cellphone users' mobility patterns and profiles, and provide a complete framework Mobility Profiler, for discovering mobile user profiles starting from cell based location log data. They use real-world cellphone log data to demonstrate their framework and experiments.

As to semantic information, [1] uses ZIP codes to define ROI. But it is still very limited to understanding users' characteristics for their studies on extracting personal features from the important places, such as home, work locations and other ROI. [9] develops an activity-aware map that describes the most probable activity associated with a specific area of space based on POIs information from a large mobile phone data of nearly one million records of the users in the central Metro-Boston area. They find a strong correlation in daily activity patterns within the group of people who share a common work area's profile.

To study call prediction, [10] proposes a Call Predictor (CP) that computes the probability of receiving calls and makes call prediction based on caller's behavior and reciprocity. They also propose a novel concept of Call Predicted List (CPL) that provides phone user an ability to predict future incoming calls as well as an improvement over the "last received calls" functionality [11].

7   Conclusion and Future Work

In this paper, we propose approaches to mining personal mobility patterns from mobile phone log data. Based on these approaches, some typical data mining algorithms are developed, including frequent pattern mining, semantic label extraction and movement prediction. We improve the existing PrefixSpan approach to better meet the characteristics of spatial and temporal data, and propose a method for measuring spatial-temporal similarity. Identifying frequently visited places, a user's characteristics are generalized by extracting the semantic labels of these places. According to spatial and temporal correlation, we can predict the location of a user who may arrive at the next moment. Combined with these user labels, a more accurate LBA services can be provided to the user.

In particular, the contributions of this paper are listed as follows. 1) In order to capture mobility behavior of a mobile phone user for the purpose of analysis and mining mobility patterns, we introduce definitions for the concepts of mobility pattern, i.e. stop and trajectory. Considering time stamp associated with the stops and trajectories, it is possible to find out where the person lives, works and goes for leisure. Semantic analysis is performed on the POIs and real estate price data crawling from websites, to refer personal characteristics which are valuable for LBA etc. services. 2) We design and implement approaches for discovering mobility patterns based on mobile phone log data, containing frequent pattern mining, semantic label generation and movement prediction. We define the mobility pattern concept for mobile phone environment and present a trajectory modeling method. 3) We demonstrate our approaches by using real mobile phone data set. Using this dataset, visualization results are presented to show trajectory pattern mining, semantic label generation, and movement prediction, proving our approaches are capable of discovering individual mobility pattern that can be used for wide applications.

As future work, we are going to work on a similar framework that uses mobile phone data to discover individual behaviors in some new applications, such as periodical pattern and social networking etc.

Acknowledgements. The authors would like to thank Yanbing Ji, Wei Feng, Jia Chen, Peng Zhou at Wuhan University for their helpful algorithm development.

References

1. Isaacman, S., Becker, R., Cáceres, R., Kobourov, S., Martonosi, M., Rowland, J., Varshavsky, A.: Identifying Important Places in People's Lives from Cellular Network Data. In: Lyons, K., Hightower, J., Huang, E.M. (eds.) Pervasive 2011. LNCS, vol. 6696, pp. 133–151. Springer, Heidelberg (2011)
2. Licoppe, C., Diminescu, D., Smoreda, Z., Ziemlicki, C.: Using Mobile Phone Geolocalisation for 'Socio-Geographical' Analysis of Co-ordination, Urban Mobilities, and Social Integration Patterns. Journal of Economic & Social Geography 99, 584–601 (2008)
3. Bayir, M.A., Demirbas, M., Eagle, N.: Discovering Spatiotemporal Mobility Profiles of Cellphone Users. In: Proceedings of IEEE International Symposium on a World of Wireless, Mobile and Multimedia Networks & Workshops, pp. 1–9. IEEE Press (2009)
4. Agrawal, R., Srikant, R.: Fast Algorithms for Mining Association Rules. In: Proceedings of 20th International Conference on Very Large Data Bases (VLDB), pp. 487–499. Morgan Kaufmann Press (1994)
5. Pei, J., Han, J., Mortazavi-Asl, B., Pinto, H.: PrefixSpan: Mining Sequential Patterns Efficiently by Prefix-Projected Pattern Growth. In: Proceedings of the 17th International Conference on Data Engineering (ICDE), pp. 215–224. IEEE Computer Society (2001)
6. Birant, D., Kut, A.: ST-DBSCAN: An Algorithm for Clustering Spatial-Temporal Data. Data & Knowledge Engineering 60, 208–221 (2007)
7. González, M.C., Hidalgo, C.A., Barabási, A.L.: Understanding Individual Human Mobility Patterns. Nature 453(5), 779–782 (2008)
8. Eagle, N., Pentland, A.S.: Reality Mining: Sensing Complex Social Systems. Personal and Ubiquitous Computing 10(4), 255–268 (2006)
9. Phithakkitnukoon, S., Dantu, R.: Mobile Social Closeness and Similarity in Calling Patterns. In: Proceedings of IEEE Conference on Consumer Communications & Networking Conference (CCNC 2010), pp. 1–5. IEEE Press (2010)
10. Phithakkitnukoon, S., Dantu, R.: Predicting Calls – New Service for an Intelligent Phone. In: Krishnaswamy, D., Pfeifer, T., Raz, D. (eds.) MMNS 2007. LNCS, vol. 4787, pp. 26–37. Springer, Heidelberg (2007)
11. Phithakkitnukoon, S., Dantu, R.: CPL: Enhancing Mobile Phone Functionality by Call Predicted List. In: Meersman, R., Tari, Z., Herrero, P. (eds.) OTM 2008 Workshops. LNCS, vol. 5333, pp. 571–581. Springer, Heidelberg (2008)

Using Partially-Ordered Sequential Rules to Generate More Accurate Sequence Prediction

Philippe Fournier-Viger[1], Ted Gueniche[1], and Vincent S. Tseng[2]

[1] Dept. of Computer Science, University of Moncton, Canada
[2] Dept. of Computer Science and Inf. Eng., National Cheng Kung University, Taiwan
philippe.fournier-viger@umoncton.ca, ted.gueniche@gmail.com,
tsengsm@mail.ncku.edu.tw

Abstract. Predicting the next element(s) of a sequence is a research problem with wide applications such as stock market prediction, consumer product recommendation, and web link recommendation. To address this problem, an effective approach is to mine sequential rules from a set of training sequences to then use these rules to make predictions for new sequences. In this paper, we improve on this approach by proposing to use a new kind of sequential rules named partially-ordered sequential rules instead of standard sequential rules. Experiments on large clickstream datasets for webpage recommendation show that using this new type of sequential rules can greatly increase prediction accuracy, while requiring a smaller training set.

Keywords: symbolic sequence prediction, sequential rules, partial order.

1 Introduction

Predicting the next element(s) of a sequence is an important research problem with wide applications such as stock market analysis, consumer product recommendation, weather forecasting, text generation and web link recommendation. Techniques for sequence prediction can be categorized according to the types of sequences on which they are applied. There are two main types. On one hand, time series are sequences of numeric data typically recorded at an equal time interval (e.g. sale data and temperature data). On the other hand, symbolic sequences are sequences of events or nominal data generally recorded at unequal time intervals (e.g. customer shopping sequences, program execution sequences and web click streams). Previous research on sequence prediction has mainly focused on predicting time-series. This is usually done by applying statistical methods to find mathematical functions that fit the data. These functions are then used to make predictions [3]. In this paper, we are interested by the case of symbolic sequences, which has many applications [13]. Because the data in symbolic sequences is not numeric, techniques for time-series forecasting cannot be applied to this problem.

To predict symbolic sequences, researchers have used techniques such as recurrent neural networks [11] and Markov models [2] (see [13] for a comprehensive

S. Zhou, S. Zhang, and G. Karypis (Eds.): ADMA 2012, LNAI 7713, pp. 431–442, 2012.
© Springer-Verlag Berlin Heidelberg 2012

survey). Limitations of those techniques are that they (1) require labeled training data, (2) require defining a reward function and/or (3) assume that the next element of a sequence only depends on the previous element or requires building a model that is exponentially large if more than one element has to be considered [13], [2]. These assumptions are unrealistic or unpractical for many real applications [13]. To perform symbolic sequence prediction without making these assumptions, an effective solution is to use algorithms to discover sequential rules occurring in a set of training sequences. These rules are then used to make predictions [14], [8], [9], [12]. This approach has the advantage of being unsupervised, scalable, and to generate accurate predictions [10], [9], [12]. In this paper, we improve on this approach by proposing to use a new type of sequential rules named partially-ordered sequential rules [5], [6], [7] that we have proposed in previous works. We compare the prediction accuracy of these rules with standard sequential rules [14], [10], [9], [12]. for the task of webpage recommendation for large clickstream datasets. Experimental results show that using the new type of sequential rules can greatly improve prediction accuracy, while requiring a smaller training set.

The rest of this paper is organized as follows. Section 2 defines the problem of sequence prediction. Section 3 defines the two types of sequential rules that are compared. Section 4 presents our comparison framework that we have named the Predictor. Section 5 reports experimental results. Finally, section 6 draws a conclusion and discusses future works.

2   The Problem of Sequence Prediction

To define the problem of sequence prediction, we first need to define what are a sequence and a sequence database.

An *itemset* $I = \{i_1, i_2, ..., i_m\}$ is an unordered set of items, where an item is a symbolic value. For example, $\{a, b, c\}$ represents the sets of items a, b and c.

A *sequence database* SDB is a set of sequences $S = \{s_1, s_2...s_s\}$ and a set of items $I = \{i_1, i_2, ..., i_m\}$ occurring in these sequences [3], [4], [14].

A *sequence* is an ordered list of itemsets $s = \langle I_1, I_2, ...I_n \rangle$ such that $I_k \subseteq I$ for all $1 \leq k \leq n$. For instance, the sequence $s1 = \langle \{a, b\}, \{c\}, \{f\}, \{g\}, \{e\} \rangle$ represents that items a and b occurred at the same time, and were followed successively by c, f, g and lastly e. For example, consider the sequence database SDB depicted in Figure 1. It contains four sequences named s_1, s_2, s_3 and s_4. In this example, each single letter represents an item. Items between curly brackets represent an itemset. The sequence of SDB could encode, for example, the sequence of webpages visited by four users or transactions of four customers in a store.

The *problem of sequence prediction* is defined as follows (adapted from [6], [7] [9]). Let $s = \langle I_1, I_2..., I_{v-1}, I_v, I_{v+1}, I_{w-1}, I_w \rangle$ be a sequence such that $\langle I_1, I_2, I_{v-1} \rangle$ has been observed and that $\langle I_v, ...I_{w-1}, I_w \rangle$ has not yet been observed. Lets *prefix_size* and *suffix_size* be positive integers. The problem of sequence prediction is to predict an item from the subsequence $< I_v, I_{v+1}, ... I_{(v+suffix\_size-1)} >$ based on the subsequence $\langle I_{(v-prefix\_size)}, ...I_{v-2}, I_{v-1} \rangle$.

ID	Sequences
s_1	$\langle\{a, b\}, \{c\}, \{f\}, \{g\}, \{e\}\rangle$
s_2	$\langle\{a, d\}, \{c\}, \{b\}, \{a, b, e, f\}\rangle$
s_3	$\langle\{a\}, \{b\}, \{f\}, \{e\}\rangle$
s_4	$\langle\{b\}, \{f, g, h\}\rangle$

Fig. 1. A sequence database SDB

Consider a sequence of webpages $s = \langle\{c\}, \{a\}, \{b\}, \{e\}, \{g\}, \{c\}\rangle$ that a new user visits. Let $prefix\_size = 2$, $suffix\_size = 2$ and $v = 3$. This represents the problem of predicting an item from $\langle\{e\}, \{g\}\rangle$ based on $\langle\{a\}, \{b\}\rangle$. For this problem a good prediction is e or g. Any other prediction is considered wrong.

3   Two Types of Sequential Rules

To perform prediction with sequential rules, two steps needs to be performed [9], [12]. First, sequential rules are extracted from a training set of sequences (a sequence database). Second, these rules are used to make predictions for new sequences. For example, consider the problem of predicting the next webpages that a user will visit. The approach consists of first extracting sequential rules from logged sequences of webpages visited by previous users. Then, these rules are used to predict the webpages that news users will visit. In this paper, our contribution is to propose to use a new type of sequential rules for improving the quality of predictions. We thereafter define the two types of sequential rules that are compared in this paper.

3.1   Standard Sequential Rules

The first type is *standard sequential rules* [14], [8], [9], [12]. It is the most common type of sequential rules used in the literature. It is defined as follows.

A *sequential pattern* [4], [14] is a sequence that is a subsequence of one or more sequences of a sequence database SDB. Formally, a sequence $s_a = \langle A_1, A_2, ...A_e\rangle$ is said to be a subsequence of a sequence $s_b = \langle B_1, B_2, ...B_f\rangle$ if and only if there exists integers $1 \leq x_1 < x_2 \, ... < x_e \leq f$ such that $A_1 \subseteq B_{x1}$, $A_2 \subseteq B_{x2}$, $...A_e \subseteq B_f$. For instance, consider the sequence database of Figure 1 as SDB. The sequence $\langle\{b\}, \{f\}\rangle$ is a sequential pattern occurring in sequences s_1, s_2, s_3 and s_4. Another example is $\langle\{b\}, \{f\}, \{e\}\rangle$. It is a sequential pattern occurring in sequences s_1 and s_3.

A *standard sequential rule* $s_a \Rightarrow s_b$ is a sequential relationship between two sequential patterns s_a and s_b. The interpretation of a rule $s_a \Rightarrow s_b$ is that if s_a occurs in a sequence, it is likely to be followed by s_b in the same sequence. Formally, a standard sequential rule $\langle A_1, A_2, ... A_e\rangle \Rightarrow \langle B_1, B_2, ...B_f\rangle$ occurs in a sequence $\langle C_1, C_2, ...C_g\rangle$ if and only if there exists integers $1 \leq x_1 < x_2 \, < x_e < y_1 < y_2 < y_e \leq f$ such that $A1 \subseteq C_{x1}$, $A_2 \subseteq C_{x2}$, $...A_e \subseteq C_{xe}$ and $B_1 \subseteq C_{y1}$, $B_2 \subseteq C_{y2}$, $...B_f \subseteq C_{ye}$. For example, the rule $\langle\{b\}, \{f\}\rangle \Rightarrow \langle\{e\}\rangle$ is a sequential rule occurring in sequences s_1 and s_3 of the database depicted in Figure 1. This rule is interpreted as if b is followed by f, it will be followed by e.

Standard sequential rules can be discovered by algorithms such as RuleGen [14]. These algorithms take two thresholds named *minsup* and *minconf* as parameters, which are set by the user. The algorithms return all sequential rules such that their support and confidence are respectively higher than these thresholds. The support and confidence of standard sequential rules are defined as follows.

The support of a standard sequential rule $s_a \Rightarrow s_b$ for a database SDB is denoted as $sup(s_a \Rightarrow s_b)$. It is defined as the number of sequences from SDB where the rule occurs divided by the number of sequences in SDB. For instance, consider the sequence database of Figure 1 as SDB. The rule $\langle\{b\}, \{f\}\rangle \Rightarrow \langle\{e\}\rangle$ has a support of 0.50 because it appears in sequences s_1 and s_3 and there are four sequences in SDB.

The confidence of a standard sequential rule $s_a \Rightarrow s_b$ for a database SDB is denoted as $conf(s_a \Rightarrow s_b)$. It is defined as the number of sequences from SDB, where s_a is followed by s_b, divided by the number of sequence where s_a occurs in SDB. For instance, consider the sequence database of Figure 1 as SDB. The rule $\langle\{b\}, \{f\}\rangle \Rightarrow \langle\{e\}\rangle$ appears in sequences s_1 and s_3 and its antecedent $\langle\{b\}, \{f\}\rangle$ occurs in s_1, s_2, s_3 and s_4. The rule $\langle\{b\}, \{f\}\rangle \Rightarrow \langle\{e\}\rangle$ has therefore a confidence of 2 / 4 = 0.5.

3.2   Partially-Ordered Sequential Rules

The second type of sequential rules that we consider is *partially-ordered sequential rules*, a new type of sequential rules that we have proposed recently [5], [6], [7]. We call these rules partially-ordered because the requirements of a sequential ordering inside the antecedent and inside the consequent of rules are eliminated. But the requirement of a sequential relationship between the antecedent and consequent of a rule is preserved.

A *partially-ordered sequential rule* is a sequential relationship between two unordered itemsets $I_a \Rightarrow I_b$ such that $I_a \cap I_b = \emptyset$ and $I_a, I_b \neq \emptyset$. The interpretation of a partially-ordered sequential rule $I_a \Rightarrow I_b$ is that if the items of I_a occur in a sequence (in any order), the items in I_b will occur afterward in the same sequence (in any order). Formally, we say that a rule $I_a \Rightarrow I_b$ occurs in a sequence $s = \langle I_1, I_2, ...I_n \rangle$ if and only if there exists an integer k such that $1 \leq k < n$, $I_a \subseteq \bigcup_{i=1}^{k} I_i$ and $I_b \subseteq \bigcup_{i=k+1}^{n} I_i$. For instance, consider the sequence database of Figure 1 as SDB. The rule $\{a, b, f\} \Rightarrow \{e\}$ appears in sequence s_1 and s_3. It means that if items a, b and f appears in a sequence in any order, it will be followed by e.

Partially-ordered sequential rules have the interesting property of being more general than standard sequential rules [5], [6], [7]. Several standard sequential rules can be represented by a single partially-ordered sequential rule. For example, the standard sequential rules $\langle\{a, b\}, \{c\}\rangle \Rightarrow \langle f \rangle$ and $\langle\{a\}, \{c\}, \{b\}\rangle \Rightarrow \langle\{f\}\rangle$ are represented by a single partially-ordered sequential rule $\{a, b, c\} \Rightarrow \{f\}$.

Algorithms for mining partially-ordered sequential rules [5] take two thresholds named *minsup* and *minconf* as parameters, which are set by the user. The algorithms return all rules having a support and confidence respectively higher than these thresholds. Support and confidence for partially-ordered sequential rules are defined as follows.

The support of a partially-ordered sequential rule $I_a \Rightarrow I_b$ for a database SDB is denoted as $sup(I_a \Rightarrow I_b)$. It is defined as the number of sequences from SDB where the items in the rule occurs, divided by the number of sequences in SDB. The confidence of a partially-ordered sequential rule for a database SDB is denoted as $conf(I_a \Rightarrow I_b)$. It is defined as the number of sequences in SDB where items the rule occurs divided by the number of sequence where items in I_a appears. For instance, consider the sequence database of Figure 1 as SDB. The rule $\{a, b, c\} \Rightarrow \{e\}$ has a support of 0.5 because it appears in s_1 and s_2. Moreover, its confidence is 1 because its antecedent only appears in s_1 and s_2. Another example is the rule $\{a\} \Rightarrow \{b, c, e\}$, which has a support of 0.5 and a confidence of 0.6.

4   The Comparison Framework

We now present our comparison framework that we have designed to compare the prediction accuracy of standard sequential rules and partially-ordered sequential rules. We named this framework the *Predictor*. It a framework for sequence prediction that can be used with the two types of sequential rules. The predictor works in two phases.

1) Training. The *first phase* consists of mining sequential rules from a sequence database containing a set of training sequences. The user has to provide the sequence database and has to choose the types of sequential rules to be mined. If standard sequential rules are chosen, the RuleGen [14] algorithm is applied. If partially-ordered sequential rules are selected, TRuleGrowth [5] is applied. Note that TRuleGrowth allow specifying a parameter named *window_size*, which is not found in RuleGen. It allows specifying that patterns have to occur within a maximum number of consecutive itemsets. To provide the same functionality for RuleGen, we have modified it to also accept this parameter. As it will be shown in the experimental section, considering this additional constraint can improve the accuracy of predictions. Besides, note that we here only consider sequential rules containing a single item in the consequent because we are interested in predicting one item at a time for the application of web recommendation described in this paper.

2) Prediction. The *second phase* consists of predicting an item that will follow a sequence provided by the user. This phase is accomplished in two substeps. The *first step* is to scan all the rules to identify those that match with the sequence provided by the user. Here, *a rule is said to match with a sequence if* the antecedent appears in the sequence.

The *second step* is to generate a prediction based on the matching rules. This is performed by selecting one of the matching sequential rules. To select a rule, several criteria can be used. We have tested several of them and found that the criterion that gives the best results is to choose the rule with the highest score. We define the score of a rule p as $Score(p) = (c_1 conf(p) + c_2 sup(p)) \times length(p)$, where c_1 and c_2 are constants and $length(p)$ is the number of items contained in p that match with the sequence. In our tests, $c_1 = 0.7$ and $c_2 = 0.3$ generated the best results. Note however that other values of c_1 and c_2 could be used for other datasets and may provide better results. After a sequential rule has been selected based on the score, the *Predictor* makes the prediction by choosing the item in the rule consequent.

5   Experimentation

We have implemented the RuleGen and TRuleGrowth sequential rule mining algorithms, and the Predictor framework in Java.

Experiments were carried with two public click-stream datasets commonly used in the sequential pattern mining literature. The first dataset is **Kosarak** (http://fimi.cs.helsinki.fi/data/). It contains 990,000 sequences of click-stream data from an online news portal. To make the experiment faster, we only used the first 50,000 sequences of Kosarak. Each sequence has an average length of 7.97 items from 21,144 different items. The second dataset is **BM-SWebView1** (BMS). It contains 59,601 sequences of click-stream data from an e-commerce website (http://www.ecn.purdue.edu/KDDCUP/). BMS differs from Kosarak in that sequences are shorter and that the set of different items is much smaller (497 items compared to 21,144 items). The average length of sequences in BMS is short with 2.51 items ($\sigma = 4.85$). But it also contains several long sequences.

We have performed several experiments with the datasets. For each experiment, we have randomly divided each dataset into two sets: a training set and a test set. The division was made according to a parameter *training_ratio* that we have initially set to 50%. This parameter indicates the percentage of sequences from a dataset that is used for training. The training set is used for generating sequential rules. These rules are then used to generate predictions for each sequence of the test set. Statistics are recorded about the number of correct predictions and the total number of predictions generated. This allows computing two measures.

The first measure is the *accuracy*. We define it as the number of good predictions divided by the number of sequences in the test set. The second measure is named *matching rate*. It is defined as the number of sequences where a prediction was generated divided by the number of sequences in the test set. It is important to consider this measure because no prediction is generated for a sequence if there is no matching rule.

The initial parameters for all the experiments are as follows. The parameter *minsup* is set to 0.00055 for BMS and 0.002 for Kosarak. These values were determined has the best values (giving the highest accuracy and matching rate) after executing several preliminary experiments. Similarly, we have found that 0.5 was the best value for *minconf* for both datasets. The parameters *prefix_size* and *suffix_size* (cf. section 2) were set to 3. This means that the problem of sequence prediction is to predict an item from the last three itemsets of a sequence given the three preceding itemsets. Because we used these parameters, we only kept sequence containing at least 6 itemsets in each dataset. Lastly, the *window_size* parameter (cf. section 4) was set to 5.

5.1 Influence of *prefix_size*

The first experiment consists of varying *prefix_size* to assess its influence on the accuracy and matching rate for the two types of sequential rules (*SSR = Standard Sequential Rules, POSR = Partially-Ordered Sequential Rules*). As previously explained, *prefix_size* represents the number of preceding itemsets that are used for making a prediction (cf. section 2). Figure 2 respectively show the impact of varying this parameter from 1 to 10 for BMS and Kosarak. It can be seen that partially-ordered sequential rules can improve accuracy by up to 28% and matching rate by up to 60% compared to standard sequential rules.

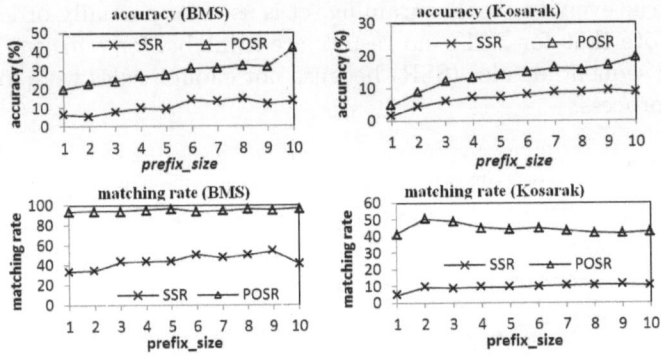

Fig. 2. Influence of *prefix_size* on accuracy and matching rate

5.2 Influence of *suffix_size*

The second experiment consists of varying *suffix_size* to assess its influence on prediction accuracy and matching rate. As explained in section 2, *suffix_size* indicates the number of itemsets that should be considered for making a prediction. Figure 3 respectively show the results obtained of varying *suffix_size* from 1 to 10 for BMS and Kosarak. These results show that partially-ordered sequential rules can improve accuracy by up to 26% and matching rate by up to 60% compared to standard sequential rules.

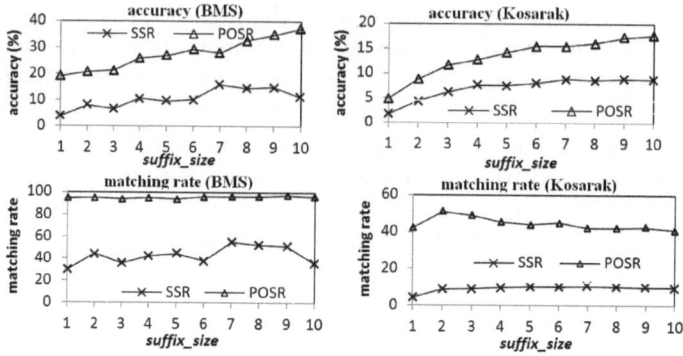

Fig. 3. Influence of $suffix\_size$

5.3   Influence of $training\_ratio$

The third experiment consists of assessing the impact of the size of the training set on the accuracy of predictions by varying $training\_ratio$ from 10% to 90%. Figure 4 respectively show the results obtained with BMS and Kosarak. Predictions based on partially-ordered sequential rules were from 11% to 19% more accurate than predictions based on standard sequential rules. It can be observed that this is true even if a smaller training set is used for partially-ordered sequential rules. Note that for BMS, no results are available for $training\_ratio = 60$ for standard sequential rules (SSR) because not enough rules were found during the mining process.

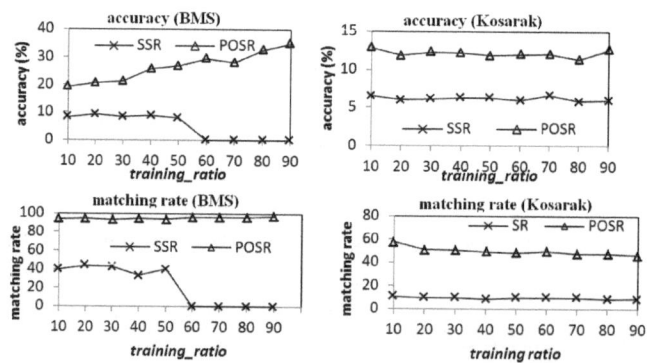

Fig. 4. Influence of the training ratio

5.4   Influence of the Rule Selection Criterion

The fourth experiment consists of assessing the influence of the rule selection criterion to make a prediction when several rules match with a sequence. The criterion previously suggested in section 4 was to select the matching rule with the highest score. An alternative approach that we have tested is to keep the top W matching rules with the highest score, and then to perform a majority vote on the items predicted by the W rules. The item with the most votes is then chosen as the prediction. Note that if W is set to 1, the result is the same as before. Results from his experiment are shown on Figure 5 for $W = 1, 10, 20,$ ...90 for BMS and Kosarak. Results show that using a majority vote improves accuracy by up to 6% for partially-ordered sequential rules, and up to 5% for standard sequential rules.

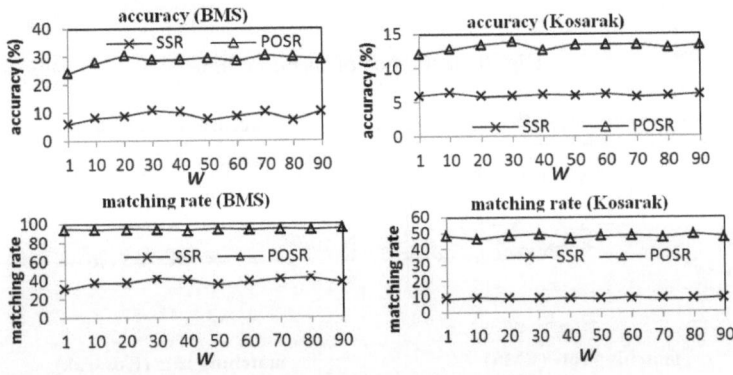

Fig. 5. Influence of the rule selection criterion

5.5   Influence of *window_size*

The fifth experiment consists of assessing the influence of *window_size* on the prediction accuracy by varying *window_size* from 2 to 10. Results of this experiment are shown on Figure 6. Results indicate that setting *window_size* to a value from 3 to 6 provided reasonable accuracy and that using larger values did not provide a major improvement.

5.6   Influence of Making Multiple Predictions

For the sixth experiment, we have tested the possibility of making multiple predictions for each sequence instead of only one. This experiment is motivated by the fact that for some real applications more than one recommendation can be made to the user. For example, for webpage recommendation, more than one webpage recommendation can be made to the user by displaying several links on the same page. To assess the benefits of making multiple predictions, we have added a parameter Q, which is the number of predictions. The definition of accuracy has been adjusted as follows for multiple predictions. The accuracy is calculated as the number of sequences where at least one prediction is correct divided by the total number of sequences in the test set. Results of varying

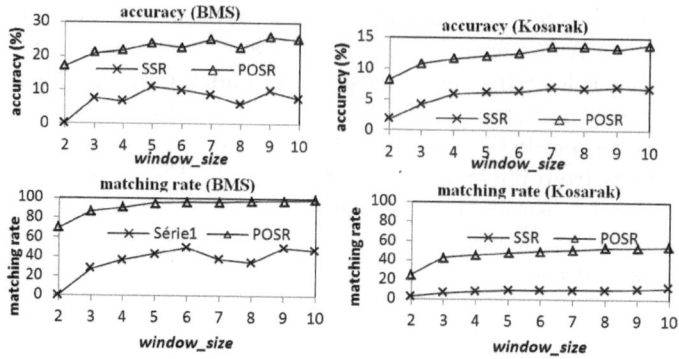

Fig. 6. Influence of *window_size*

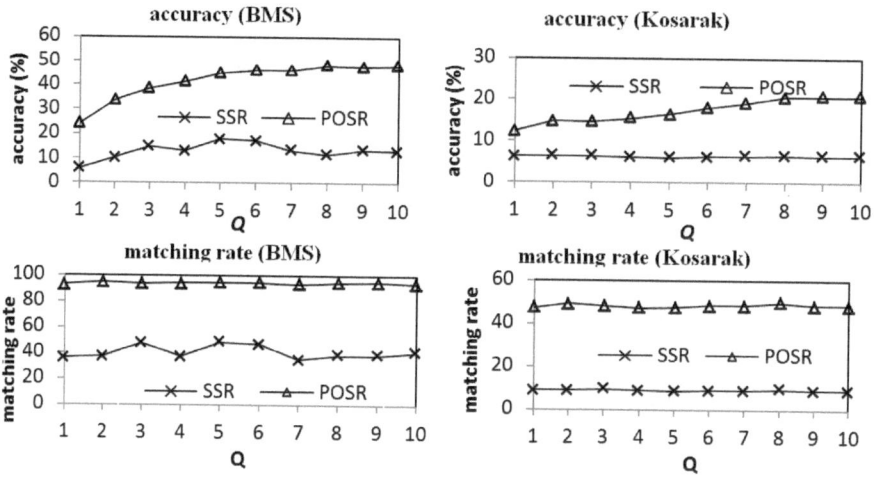

Fig. 7. Influence of making multiple predictions

Q from 1 to 10 are shown on Figure 7. As it was expected, allowing multiple predictions largely increased prediction accuracy (by up to 24% for partially ordered sequential rules and 7% for standard sequential rules).

5.7   Influence of the Number of Rules

The last experiment consisted of varying *minsup* to assess the influence of the sequential rule count on the prediction accuracy. For BMS, we varied *minsup* from 0.0006 to 0.0005. For Kosarak, we varied *minsup* from 0.003 to 0.001. These intervals were chosen because they provide an interesting view of the results. The prediction accuracy and the number of rules generated for BMS and Kosarak are shown in Figure 8. From these results, we can observe that the number of rules for the two types of rules varies differently because their definitions are different. Second, we can see that for the two types of rules, as soon as there are

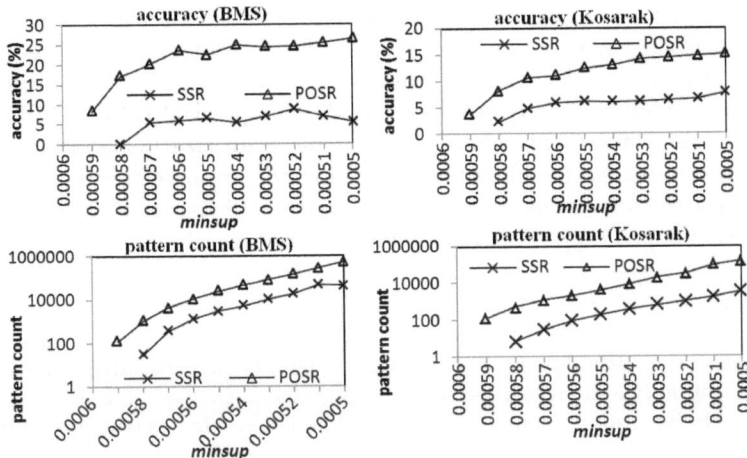

Fig. 8. Influence of the number of rules

approximately 1000 to 10,000 rules, the accuracy remains more or less the same if the number of rules increase. This result is interesting because it provides a solution to the problem of choosing the *minsup* value. The solution is to use an algorithm for mining the top-k sequential rules such as the one that we have designed in previous work [6]. This algorithm can discover the top-k partially-ordered sequential rules where k is set by the user. The advantage of using this algorithm is that the user does not need to find a suitable value for the *minsup* thresholds by hand. For example, the user could set $k = 10,000$ to find the top 10, 000 rules and use them to make predictions.

6 Conclusion

Predicting the next element(s) of a sequence is a research problem with wide applications such as stock market prediction, consumer product recommendation, and web link recommendation. A popular approach to symbolic sequence recommendation is to discover sequential rules in a training set of sequences to then use these rules to make predictions. In this paper, we have explored the possibility of using a new type of sequential rules named partially-ordered sequential rules [5], [6] for sequence prediction. We have compared the prediction accuracy of these rules with standard sequential rules for the task of webpage recommendation under multiple scenarios (different prefix sizes, different suffix sizes, allowing multiple predictions, varying the size of the training set, performing a majority vote, choosing different window sizes and different minsup values). Overall, results show that using partially-ordered sequential rules instead of standard sequential rules can improve the prediction accuracy and matching rate by more than 30%, depending on the scenario.

In this paper, we have focused on improving prediction accuracy and matching rate. For future work, we plan to work on enhancing the performance in terms of execution time. We are working on designing a storage structure that will allow an efficient storage and retrieval of sequential rules to test if they are matching with a sequence. We also plan to consider the problem of sequence prediction with a user profile as it was done for standard sequential rules in [12]. We are also considering integrating ideas such as rule prioritization and rule pruning from works on associative classification to further improve the results (e.g. [8], [1]). The source code of the software presented in this paper is available for free at http://www.philippe-fournier-viger.com/spmf/.

References

1. Antonie, M.-L., Chodos, D., Zaiane, O.: Variations on Associative Classifiers and Classification Results Analyses. In: Zhao, Y., Zhang, C., Cao, L. (eds.) Post-Mining of Association Rules: Techniques for Effective Knowledge Extraction (2008)
2. Begleiter, R., El-Yaniv, R., Yona, G.: On Prediction Using Variable Order Markov Models. Journal of Artificial Intelligence Research 22, 385–421 (2004)
3. Han, J., Kamber, M.: Data Mining: Concepts and Techniques, 3rd edn. Morgan Kaufmann, San Francisco (2011)
4. Pei, J., et al.: Mining Sequential Patterns by Pattern Growth: The PrefixSpan Approach. IEE Trans. on Knowledge and Data Engineering. 16(11), 1420–1440 (2004)
5. Fournier-Viger, P., Wu, C.-W., Tseng, V.S., Nkambou, R.: Mining Sequential Rules Common to Several Sequences with the Window Size Constraint. In: Kosseim, L., Inkpen, D. (eds.) Canadian AI 2012. LNCS, vol. 7310, pp. 299–304. Springer, Heidelberg (2012)
6. Fournier-Viger, P., Tseng, V.S.: Mining Top-K Sequential Rules. In: Tang, J., King, I., Chen, L., Wang, J. (eds.) ADMA 2011, Part II. LNCS, vol. 7121, pp. 180–194. Springer, Heidelberg (2011)
7. Fournier-Viger, P., Nkambou, R., Tseng, V.S.: RuleGrowth: Mining Sequential Rules Common to Several Sequences by Pattern-Growth. In: ACM SAC 2011, pp. 954–959. ACM Press (2011)
8. Liu, B., Hsu, W., Ma, Y.: Integrating Classification and Association Rule Mining. In: Fourth International Conference on Knowledge Discovery and Data Mining (KDD 1998), pp. 80–86. AAAI Press, New York (1998)
9. Liu, D.-R., Lai, C.-H.: A hybrid of sequential rules and collaborative filtering for product recommendation. Information Sciences 179, 3505–3519 (2009)
10. Lo, D., Khoo, S.-C., Wong, L.: Non-redundant sequential rules Theory and algorithm. Information Systems 34(4-5), 438–453 (2009)
11. Pérez-Ortiz, J.A., Calera-Rubio, J., Forcada, M.L.: Online Symbolic-Sequence Prediction with Discrete-Time Recurrent Neural Networks. In: Dorffner, G., Bischof, H., Hornik, K. (eds.) ICANN 2001. LNCS, vol. 2130, pp. 719–724. Springer, Heidelberg (2001)
12. Pitman, A., Zanker, M.: An Empirical Study of Extracting Multidimensional Sequential Rules for Personalization and Recommendation in Online Commerce. In: 10th Intern. Conf. on Wirtschaftsinformatik (2011)
13. Sun, R., Giles, C.L.: Sequence Learning: From Recognition and Prediction to Sequential Decision Making. IEEE Intelligent Systems 16(4) (2001)
14. Zaki, M.J.: SPADE: An Efficient Algorithm for Mining Frequent Sequences. Machine Learning 42(1/2), 31–60 (2001)

Particle Swarm Optimization of Information-Content Weighting of Symbolic Aggregate Approximation*

Muhammad Marwan Muhammad Fuad

Department of Electronics and Telecommunications
Norwegian University of Science and Technology (NTNU)
NO-7491 Trondheim, Norway
marwan.fuad@iet.ntnu.no

Abstract. Bio-inspired optimization algorithms have been gaining more popularity recently. One of the most important of these algorithms is particle swarm optimization (PSO). PSO is based on the collective intelligence of a swam of particles. Each particle explores a part of the search space looking for the optimal position and adjusts its position according to two factors; the first is its own experience and the second is the collective experience of the whole swarm. PSO has been successfully used to solve many optimization problems. In this work we use PSO to improve the performance of a well-known representation method of time series data which is the symbolic aggregate approximation (SAX). As with other time series representation methods, SAX results in loss of information when applied to represent time series. In this paper we use PSO to propose a new minimum distance WMD for SAX to remedy this problem. Unlike the original minimum distance, the new distance sets different weights to different segments of the time series according to their information content. This weighted minimum distance enhances the performance of SAX as we show through experiments using different time series datasets.

Keywords: Particle Swarm Optimization, Bio-inspired Optimization, Time Series Data Mining, Information Loss, Information Content, Symbolic Aggregate Approximation.

1 Introduction

A *time series* is a sequence of real numbers over a period of time. Each of these numbers represents the value of the observed phenomenon at a certain time point. Time series data have been used in many applications such as science, medicine, and engineering. Time series data mining handles several tasks such as similarity search, classification, clustering, and others.

Time series datasets are usually very large so direct search or *sequential scanning* of these datasets is inefficient. In order to overcome this problem transformations can

---

* This work was carried out during the tenure of an ERCIM "Alain Bensoussan" Fellowship Programme. This Programme is supported by the Marie-Curie Co-funding of Regional, National and International Programmes (COFUND) of the European Commission.

S. Zhou, S. Zhang, and G. Karypis (Eds.): ADMA 2012, LNAI 7713, pp. 443–455, 2012.

be applied to reduce the dimensionality of the original time series and to represent them in a space with manageable dimensionality. Such transformations are called *dimensionality reduction techniques*, or *representation methods*. The most widely known methods are *Discrete Fourier Transform* (DFT) [2] and [3], *Discrete Wavelet Transform* (DWT) [5], *Singular Value Decomposition* (SVD) [13], *Adaptive Piecewise Constant Approximation* (APCA) [10], *Piecewise Aggregate Approximation* (PAA) [9] and [30], *Piecewise Linear Approximation* (PLA) [17], and *Chebyshev Polynomials* (CP) [4].

Other methods of time series data mining use multi-resolution approaches. In [16] and [27] a method of multi resolution representation of time series is presented. This symbolic method uses a multi-resolution vector quantized approximation of the time series together with a multi-resolution similarity distance. Using this representation the method keeps both local and global information of the time series data. In [19] and [20] other multi-resolution method are proposed. These methods are based on a fast-and-dirty filtering scheme that iteratively reduces the search space using several resolution levels. The technique presented in [22] couples and fast-and-dirty filter with a multi-resolution representation of the time series.

Indexing time series data usually includes establishing a lower bounding distance on time series in the transformed space to guarantee that the representation method will not cause false dismissals. This is achieved by defining a distance, on the transformed space, that underestimates the distance in the original space. This condition is known as the *lower-bounding lemma.* [2].

Among representation methods of time series data, symbolic representation of time series has several advantages which interested researchers in this field of computer science. One of its main advantages is that symbolic representation permits researchers to benefit from the ample symbolic algorithms known in the text-retrieval and bioinformatics communities [14].

There have been many suggestions to represent time series symbolically. But in general, most of these symbolic representation methods suffered from two main inconveniences [15]; the first is that the dimensionality of the symbolic representation method is the same as that of the original space, so there is no virtual dimensionality reduction. The second drawback is that although distance measures have been defined on the reduced symbolic spaces, these distance measures are poorly correlated with the original distance measures defined on the original spaces.

One of the most widely-known symbolic representation methods of time series is SAX. SAX uses pre-computed distances obtained from lookup tables. This makes SAX fast to compute.

In this work we show how the performance of SAX can be improved by substituting the original similarity measure used with SAX by a new one which assigns different weights to different segments of the time series according to their information content. These weights are set using the particle swarm optimization; a widely used population-based optimization method which has been successful in solving many optimization problems. We show through experiments conducted on different time series dataset how the new similarity measure can give better results than the original one.

The work presented in this paper is organized as follows: in Section 2 we present related background. In Section 3 we introduce the new scheme and we evaluate its performance in Section 4. In Section 5 we give concluding remarks.

2   Background

2.1   The Symbolic Aggregate Approximation (SAX)

The *Symbolic Aggregate approXimation* method (SAX) [14] is one of the most important symbolic representation methods of time series. The main advantage of SAX is that the similarity measure it uses, called MINDIST, uses statistical lookup tables, which makes it is easy to compute with an overall complexity of $O(N)$.

SAX is based on an assumption that normalized time series have Gaussian distribution, so by determining the breakpoints that correspond to a particular alphabet size, one can obtain equal-sized areas under the Gaussian curve. SAX is applied as follows:

1-The time series are normalized.
2-The dimensionality of the time series is reduced using PAA [9], [30]
3-The PAA representation of the time series is discretized by determining the number and location of the breakpoints (The number of the breakpoints is chosen by the user). Their locations are determined, as mentioned above, using Gaussian lookup tables. The interval between two successive breakpoints is assigned to a symbol of the alphabet, and each segment of PAA that lies within that interval is discretized by that symbol.

Table 1. The lookup table of *MINDIST* for alphabet size = 3

	a	b	c
a	0	0	0.86
b	0	0	0
c	0.86	0	0

The last step of SAX is using the following similarity measure:

$$MINDIST(\hat{S}, \hat{R}) \equiv \sqrt{\frac{n}{N}} \sqrt{\sum_{i=1}^{N} (dist(\hat{s}_i, \hat{r}_i))^2} \qquad (1)$$

Where n is the length of the original time series, N is the length of the strings (the number of the segments), $\hat{S}$ and $\hat{R}$ are the symbolic representations of the two time series S and R, respectively, and where the function $dist(\ )$ is implemented by using the appropriate lookup table. For instance, the lookup table of *MINDIST* for an alphabet size of 3 is the one shown in Table 1.

We also need to mention that the similarity measure used in PAA is:

$$d(S,R) = \sqrt{\frac{n}{N}} \sqrt{\sum_{i=1}^{N} (s_i - r_i)^2} \tag{2}$$

Where n is the length of the time series, N is the number of segments.

It is proven in [9], [30] that the above similarity distance is a lower bound of the Euclidean distance applied in the original space of time series. This means that *MINDIST* is also a lower bound of the Euclidean distance, because it is a lower bound of the similarity measure used in PAA. This guarantees no false dismissals.

There are other versions and extensions of SAX [11], [25], [26], [28]. These versions use it for other applications or apply it to index massive datasets, or compute *MINDIST* differently [18]. However, the version of SAX that we presented earlier, which is called classic SAX, is the basis of all these versions and extensions and it is actually the most widely-known one.

2.2   Information Content

Quantifying the content of information a vector carries was first introduced by Shannon in [24]. This is measured by what is known as *entropy* and is defined for a discrete probabilistic system by:

$$H = -\sum_{i} p_i \log p_i \tag{3}$$

where the base of the logarithm is 2.

This concept has many applications in cryptography, data transmission, natural language processing, data compression, and others.

In time series mining, the concept of information content was implicitly or explicitly present in different representation methods of time series data. DFT [2], [3] and DWT [5], for instance are based on the fact that the first coefficients are the most meaningful ones; i.e. they contain most of the information in the time series, so the other coefficients can be truncated without much loss of information. APCA [10] segments the time series into segments of varying lengths such that their individual reconstruction errors are minimal. The intuition behind this idea is that different regions of the time series contain different amounts of information. So while regions of high activity contain high fluctuations, other regions of low activity show a flat behavior, so a representation method with high fidelity should reflect this difference in behavior.

3   Particle Swarm Optimization of Information-Content Weighting of Symbolic Aggregate Approximation (PSOWSAX)

In time series mining the distance that is widely used to compute the similarity between the two time series $S = \{s_1, s_2, ..., s_n\}$ and $R = \{r_1, r_2, ..., r_n\}$ is the Euclidean distance which is defined as follows:

$$L_2(S, R) = \sqrt[2]{\sum_{i=1}^{n} |s_i - r_i|^2} \qquad (4)$$

One of the variations of the Euclidean distance, which is much related to the topic of this paper, is the *Weighted Euclidean Distance*. This distance is defined as:

$$d(S, R, W) = \sqrt{\sum_{i=1}^{n} w_i (s_i - r_i)^2} \qquad (5)$$

where W is the weight vector.

Fig. 1 shows the Euclidean distance and the weighted Euclidean distance between two time series.

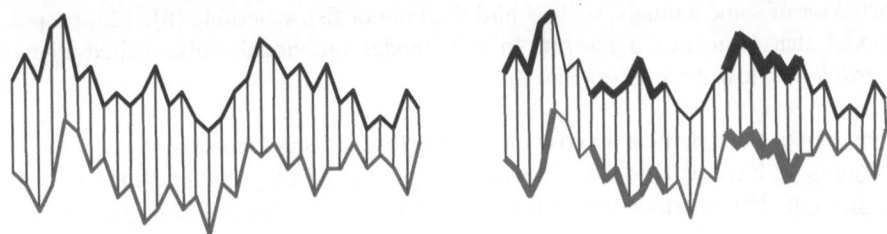

Fig. 1. The Euclidean distance (left) and the weighted Euclidean distance (right)

The intuition behind the weighted Euclidean distance is, again, that some parts of the time series may have more importance than other parts so the distance should reflect these regions differently compared with regions that contain less information.

Although weighing different regions differently seems to be a good solution to distinguish parts with high information content from others with less information, the question remains on how to set the weights. In [29] the authors proposed setting the weights using relevance feedback provided by the user. We can easily see that this solution is highly subjective and also inefficient.

It is important to mention that the weighted distance defined in (5) is applied to the raw time series and not to the reduced, lower-dimensional time series used in representation methods.

In this paper we present a modification of SAX based on the concept of information content.

In Section 2.1, we presented *MINDIST* ; the similarity measure used with SAX. From relation (1) we can derive the following similarity measure which we call the *Weighted Minimum Distance (WMD)*:

$$WMD(\hat{S},\hat{R}) = \sqrt{\frac{n}{N}} \sqrt{\sum_{i=1}^{N} w_i \left(dist(\hat{s}_i, \ \hat{r}_i) \right)^2} ; \qquad w_i \in [0,1] \qquad (6)$$

Notice that if we set $w_i = 1, \ \forall i$ in (6) we obtain *MINDIST* defined in relation (1), so *MINDIST* is in fact a special case of *WMD* . Notice also, which is very important, that from (1) and (6) we can easily see that:

$$WMD(\hat{S},\hat{R}) \le MINDIST(\hat{S},\hat{R}) \qquad (7)$$

Since *MINDIST* is a lower bound of the Euclidean distance in the original space this implies that *WMD* is also a lower bound of the Euclidean distance, so our proposed distance guarantees no false dismissals. .

The weights in (6) will be set for the whole time series in the dataset using the particle swarm optimization.

Particle Swarm Optimization: *Particle Swarm Optimization* (PSO) is a member of a family of naturally-inspired optimization algorithms called *Swarm Intelligence* which are population-based optimization algorithms. PSO was inspired by the social behavior of some animals, such as bird flocking or fish schooling [8]. [23] proposed a model that simulates a swarm. In this model individuals, also called *agents* or *particles*, follow these rules (Fig. 2):

Separation: Each particle avoids getting too close to its neighbors.
Alignment: Each particle steers towards the general heading of its neighbors.
Cohesion: Each particle moves towards the average position of its neighbors.

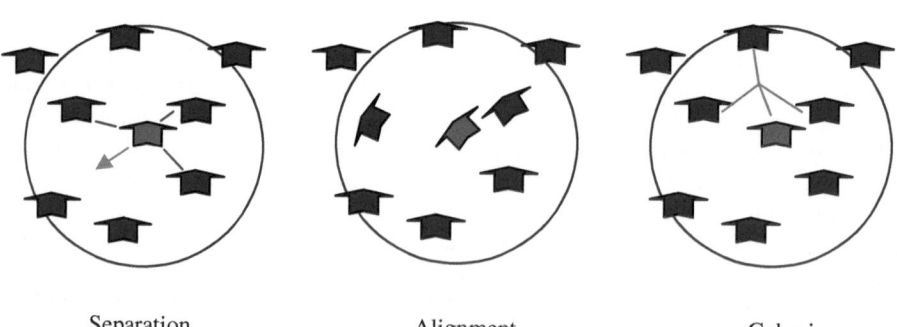

Separation   Alignment   Cohesion

Fig. 2. Simulating swarm's behavior

The above model was elaborated by adding another rule which is *obstacle avoidance* which uses *steer-to-avoid* concept.

As is the case with many other optimization algorithms, there are quite a large number of variations of PSO. In the following we present a standard PSO [7]. PSO starts by initializing a swarm of *sSize* particles at random positions $\vec{X}_i^0$ and velocities $\vec{V}_i^0$ where $i \in \{1,..,sSize\}$.

In the next step each position is evaluated, and for each iteration, using a *fitness function,* also called *objective function* or *cost function.*

The positions $\vec{X}_i^{k+1}$ and velocities $\vec{V}_i^{k+1}$ are updated at time step ($k+1$) according to the following formulae:

$$\vec{V}_i^{k+1} = \omega.\vec{V}_i^k + \varphi_G\left(\vec{G}^k - \vec{X}_i^k\right) + \varphi_L\left(\vec{L}_i^k - \vec{X}_i^k\right) \tag{8}$$

$$\vec{X}_i^{k+1} = \vec{X}_i^k + \vec{V}_i^k \tag{9}$$

where $\varphi_G = r_G.a_G$, $\varphi_L = r_L.a_L$, $r_G, r_L \to U(0,1)$, $\omega, a_L, a_G \in \mathbf{R}$, $\vec{L}_i^k$ is the best position found by particle i , $\vec{G}^k$ is the global best position found by the whole swarm, ω is called the *inertia* , a_L is called the *local acceleration* , and a_G is called the *global acceleration*. These last three parameters are control parameters which are chosen by the algorithm designer.

Algorithm 1. Particle Swarm Optimization (PSO)

Require Number of parameters (*nPar*),swarm size
(*sSize*), number of iterations (*nItr*),local acceleration
(a_L), global acceleration (a_G), inertia ω .

1: Initialize $\vec{X}_i^0$, $\vec{V}_i^0$
2: **for** *itr* =1 to *nItr* **do**
3:    **for all** particles P_i **do**
4:       $r_L, r_G \leftarrow rand$
5:       Update the particle's velocity:
6:       $\vec{V}_i \leftarrow \omega.\vec{V}_i + \varphi_G(\vec{G} - \vec{X}_i) + \varphi_L(\vec{L}_i - \vec{X}_i)$
7:       Move the particle to the new position:
8:       $\vec{X}_i \leftarrow \vec{X}_i + \vec{V}_i$
9:       **if** $f(\vec{X}_i) \le f(\vec{L}_i)$ **then** $\vec{X}_i \leftarrow \vec{L}_i$
10:      **if** $f(\vec{X}_i) \le f(\vec{G}_i)$ **then** $\vec{X}_i \leftarrow \vec{G}_i$
11:   **end for**
12: **end for**

Fig. 3. The particle swarm optimization algorithm

The algorithm continues until a stopping criterion terminates it. Fig. 3 presents a pseudo code of PSO.

As in the case with other evolutionary algorithms, PSO should keep a balance between *exploitation,* and *exploration.* Exploration is defined as the act of searching for the purpose of discovery, and exploitation is defined as the act of utilizing something for any purpose [1].

Diversity in PSO comes from two sources [31]; the first is the difference between the particle's current position and that of its best neighbor, and the other is the difference between the particle's current position and its best historical position. Variation, although provides exploration, can only be sustained for a limited number of generations because convergence of the swarm to the best position is necessary to refine the solution (exploitation).

4   Experimental Validation

We tested our distance on a time series classification task on the datasets available at [12], which is the same repository on which the original SAX was tested. This repository makes up between 90% and 100% of all publicly available, labeled time series data sets in the world, and it represents the interest of the data mining/database community, and not just one group [6].

We tested our method in a classification task based on the first nearest-neighbor (1-NN) rule using leaving-one-out cross validation. This means that every time series is compared to the other time series in the dataset. If the 1-NN does not belong to the same class, the error counter is incremented by 1.

The purpose of the experiments is to compare PSOWSAX (our new method which uses *WMD* as a similarity measure) with the original method (which uses *MINDIST* as a similarity measure) on the classification task and see which one gives the smallest error. This means that for each value of the alphabet size tested we perform the three steps of SAX presented in Section 2.1 to obtain the symbolic representation of the time series, and then we apply *MINDIST* when we test the original method or we apply *WMD* when we want to test ours, which, as indicated earlier, is based on PSO. The weights w_i in relation (6) are obtained during a training phase. This means for each value of the alphabet size tested, we formulate a PSO-based optimization problem where the fitness function is the classification error of the time series (we opt to minimize the classification error) and the outcome of this optimization problem is the weights w_i that yield this minimum value of the classification error, then we use these values on the corresponding testing datasets to obtain the final classification error. As for *MINDIST* , there is no training phase and it is applied directly to the testing datasets.

For PSOWSAX, the swarm size we used in the experiments is 16. Each particle is a vector of *nPar* components representing potential weights of the segments of the time series.

We used a standard PSO, the local acceleration a_L was set to 2, the global acceleration a_G was set to 2. The dimension of the problem $nPar$ is the dimension of the weight vector, which is the number of segments of the times series in the reduced space; i.e. $nPar = \dfrac{length\ of\ the\ original\ time\ series}{compression\ ratio}$. This dimension differs from one dataset to another. The value of inertia ω depends on the current iteration according to the formula $\omega = (nItr - itr)/nItr$ where itr is the current iteration. This means that the influence of ω decreases as the algorithm progresses.

The number of iterations $nItr$ was set to 20. This is a rather small number of iterations and the algorithm could have been left to evolve more. However, the objective of our experiments was to validate our method rather than to show its best performance which could be achieved using more sophisticated PSO algorithms.

We also have to mention that we know from experience that for some datasets the classification error is always high, or always low, for all methods of time series representation, so a threshold of a classification error, related to the dataset tested, could be set as a stopping criterion.

Table 2 summarizes the symbols used in the experiments together with their corresponding values.

Table 2. The symbol table together with the corresponding values used in the experiments

$sSize$	Swarm size	16
$nItr$	Number of iterations	20
a_L	Local acceleration	2
a_G	Global acceleration	2
ω	Inertia	$\omega = (nItr - itr^*)/nItr$

itr^* : The current iteration.

In Fig. 4. we present some of the results we obtained for alphabet size equal to 3, 10, and 20, respectively. We chose to report these values because the first version of SAX used alphabet size that varied between 3 and 10. Then in a later version the alphabet size varied between 3 and 20. So these values are benchmark values.

As we can see from Fig. 4, the classification error of *WMD* is smaller than that of *MINDIST* for all values of the alphabet size and for all the datasets, except for dataset (Beef), alphabet size=20 and dataset (Fish), alphabet size=3 where both *WMD* and *MINDIST* gave the same classification error.

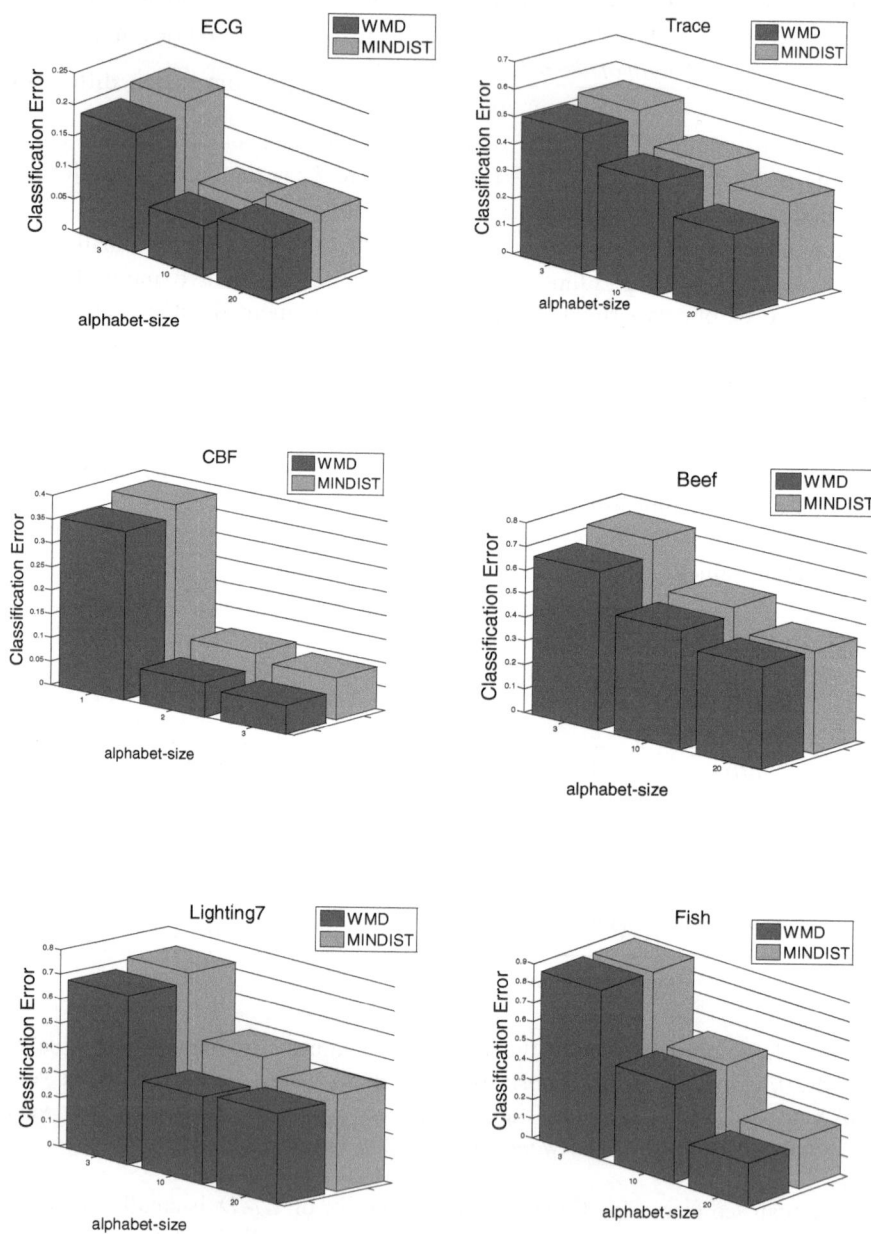

Fig. 4. The classification errors obtained by using *WMD* and *MINDIST*

Finally, we present in Table 3, for reproducibility purposes, the weights of datasets (CBF) and (ECG) (Because of space restrictions, we present these datasets only). As indicated earlier, these weights are obtained by applying PSOWSAX to the training datasets. The final classification errors of WMD (those shown in Fig. 4) are obtained by using those weights, with WMD, on the corresponding testing datasets.

Table 3. The weights of different segments of the time series obtained by using PSOWSAX

Dataset	Alphabet Size	w_i
CBF	3	[0.69 0.256 0.921 0.537 0.105 0.383 0.856 0.581 0.915 0.476 0.426 0.899 0.558 0.565 0.021 0.809 0.381 0.394 0.03 0.556 0.616 0.084 0.955 0.519 0.358 0.143 0.221 0.197 0.086 0.485 0.264 0.159]
	10	[0.722 0.331 0.073 0.662 0.51 0.005 0.107 0.943 0.049 0.869 0.667 0.56 0.396 0.643 0.165 0.573 0.968 0.304 0.492 0.342 0.548 0.372 0.804 0.608 0.874 0.159 0.452 0.514 0.524 0.47 0.488 0.003]
	20	[0.525 0.408 0.311 0.676 0.154 0.215 0.106 0.89 0.575 0.803 0.411 0.543 0.427 0.648 0.639 0.358 0.719 0.03 0.69 0.63 0.547 0.043 0.553 0.313 0.564 0.740 0.798 0.711 0.847 0.173 0.659 0.447]
ECG	3	[0.629 0.538 0.186 0.584 0.772 0.106 0.929 0.143 0.639 0.958 0.409 0.567 0.794 0.2 0.704 0.3 0.351 0.982 0.598 0.337 0.6 0.3 0.662 0.395]
	10	[0.552 0.638 0.967 0.923 0.19 0.027 0.540 0.951 0.966 0.715 0.264 0.857 0.145 0.759 0.901 0.418 0.463 0.358 0.408 0.281 0.918 0.817 0.366 0.0344]
	20	[0.572 0.612 0.804 0.062 0.268 0.685 0.428 0.546 0.3 0.359 0.679 0.408 0.564 0.034 0.797 0.152 0.683 0.758 0.563 0.091 0.931 0.582 0.044 0.006]

5   Conclusion

In this paper we showed through a new scheme PSOWSAX, based on particle swam optimization, how the performance of SAX; one of the most important symbolic representation methods of time series data, can be improved by using a new similarity measure WMD which assigns different weights to different segments of the time series according to their information content. The information loss caused by time series representation methods can be better recovered by setting different weights to different regions of the time series according to their information content.

The optimization process takes place at indexing time so the new scheme has the same low complexity as that of the original SAX.

We validated the new scheme by conducting classification task experiments on different datasets. The experiments showed that our new scheme gives better results than the original one.

A possible future work will be to associate the work presented in this paper with the work presented in [20]. This can be achieved in two ways; the first is to use the optimization scheme presented in [20] to locate the breakpoints then to use the optimization scheme presented in this work to set different weights to different segments according to their information content. The second way is to use a one-step optimization problem to locate the breakpoints together with the corresponding weights of segments.

References

1. Agnes, M.: Webster's New World College Dictionary, Webster's New World (May 2004) ISBN 0764571257
2. Agrawal, R., Faloutsos, C., Swami, A.: Efficient Similarity Search in Sequence Databases. In: Lomet, D.B. (ed.) FODO 1993. LNCS, vol. 730, pp. 69–84. Springer, Heidelberg (1993)
3. Agrawal, R., Lin, K.I., Sawhney, H.S., Shim, K.: Fast Similarity Search in the Presence of Noise, Scaling, and Translation in Time-Series Databases. In: Proceedings of the 21st Int'l Conference on Very Large Databases, Zurich, Switzerland, pp. 490–501 (1995)
4. Cai, Y., Ng, R.: Indexing Spatio-temporal Trajectories with Chebyshev Polynomials. SIGMOD (2004)
5. Chan, K., Fu, A.W.: Efficient Time Series Matching by Wavelets. In: Proc. of the 15th IEEE Int'l Conf. on Data Engineering, Sydney, Australia, March 23-26, pp. 126–133 (1999)
6. Ding, H., Trajcevski, G., Scheuermann, P., Wang, X., Keogh, E.: Querying and Mining of Time Series Data: Experimental Comparison of Representations and Distance Measures. In: Proc. of the 34th VLDB (2008)
7. Fernández-Martínez, J.L., García-Gonzalo, E.: What Makes Particle Swarm Optimization a Very Interesting and Powerful Algorithm? In: Panigrahi, B.K., Shi, Y., Lim, M.-H. (eds.) Handbook of Swarm Intelligence. ALO, vol. 8, pp. 37–65 (2011)
8. Haupt, R.L., Haupt, S.E.: Practical Genetic Algorithms with CD-ROM. Wiley-Interscience (2004)
9. Keogh, E., Chakrabarti, K., Pazzani, M., Mehrotra, S.: Dimensionality Reduction for Fast Similarity Search in Large Time Series Databases. J. of Know. and Inform. Sys. (2000)
10. Keogh, E., Chakrabarti, K., Pazzani, M., Mehrotra, S.: Locally Adaptive Dimensionality Reduction for Similarity Search in Large Time Series Databases. SIGMOD, 151–162 (2001)
11. Keogh, E., Lin, J., Fu, A.: HOT SAX: Efficiently Finding the Most Unusual Time Series Subsequence. In: Proc. of the 5th IEEE International Conference on Data Mining (ICDM 2005), Houston, Texas, November 27-30 (2005)
12. Keogh, E., Zhu, Q., Hu, B., Hao. Y., Xi, X., Wei, L., Ratanamahatana, C.A.: The UCR Time Series Classification/Clustering Homepage. (2011),
    http://www.cs.ucr.edu/~eamonn/time_series_data/
13. Korn, F., Jagadish, H., Faloutsos, C.: Efficiently Supporting Ad Hoc Queries in Large Datasets of Time Sequences. In: Proceedings of SIGMOD 1997, Tucson, AZ, pp. 289–300 (1997)
14. Lin, J., Keogh, E., Lonardi, S., Chiu, B.Y.: A Symbolic Representation of Time Series, with Implications for Streaming Algorithms. DMKD, 2–11 (2003)

15. Lin, J., Keogh, E., Wei, L., Lonardi, S.: Experiencing SAX: a Novel Symbolic Representation of Time Series. DMKD Journal (2007)
16. Megalooikonomou, C.: Multiresolution Symbolic Representation of Time Series. In: Proceedings of the 21st IEEE International Conference on Data Engineering (ICDE), Tokyo, Japan (2005)
17. Morinaka, Y., Yoshikawa, M., Amagasa, T., Uemura, S.: The L-index: An Indexing Structure for Efficient Subsequence Matching in Time Sequence Databases. In: Proc. 5th PacificAisa Conf. on Knowledge Discovery and Data Mining, pp. 51–60 (2001)
18. Muhammad Fuad, M.M., Marteau, P.-F.: Enhancing the Symbolic Aggregate Approximation Method Using Updated Lookup Tables. In: Setchi, R., Jordanov, I., Howlett, R.J., Jain, L.C. (eds.) KES 2010, Part I. LNCS, vol. 6276, pp. 420–431. Springer, Heidelberg (2010)
19. Muhammad Fuad, M.M., Marteau, P.F.: Fast Retrieval of Time Series Using a Multi-resolution Filter with Multiple Reduced Spaces. In: Cao, L., Feng, Y., Zhong, J. (eds.) ADMA 2010, Part I. LNCS, vol. 6440, pp. 137–148. Springer, Heidelberg (2010)
20. Muhammad Fuad, M.M.: Genetic Algorithms-Based Symbolic Aggregate Approximation. In: Cuzzocrea, A., Dayal, U. (eds.) DaWaK 2012. LNCS, vol. 7448, pp. 105–116. Springer, Heidelberg (2012)
21. Muhammad Fuad, M.M., Marteau, P.F.: Multi-resolution Approach to Time Series Retrieval. In: Fourteenth International Database Engineering and Applications Symposium, IDEAS 2010, Montreal, QC, Canada, August 16-18 (2010)
22. Muhammad Fuad, M.M., Marteau, P.F.: Speeding-up the Similarity Search in Time Series Databases by Coupling Dimensionality Reduction Techniques with a Fast-and-dirty Filter. In: Fourth IEEE International Conference on Semantic Computing, ICSC 2010, September 22-24. Carnegie Mellon University, Pittsburgh (2010)
23. Reynolds, C.W.: Flocks, Herds and Schools: A Distributed Behavioral Model. SIGGRAPH Comput. Graph 21(4) (1987)
24. Shannon, C.: A Mathematical Theory of Communication. The Bell Systems Technical Journal 27, 379–423, 623–656 (1948)
25. Shieh, J. and Keogh, E.: iSAX: Disk-Aware Mining and Indexing of Massive Time Series Datasets. Data Mining and Knowledge Discovery (2009)
26. Shieh, J., Keogh, E.: iSAX: Indexing and Mining Terabyte Sized Time Series. In: Proceeding of the 14th ACM SIGKDD International Conference on Knowledge Discovery and Data Mining, Las Vegas, Nevada, USA, August 24-27 (2008)
27. Wang, Q., Megalooikonomou, V., Faloutsos, C.: Time Series Analysis with Multiple Resolutions. Inf. Syst. 35(1) (2010)
28. Wei, L., Keogh, E., Xi, X.: SAXually Explicit Images: Finding Unusual Shapes. In: ICDM (2006)
29. Wu, L., Faloutsos, C., Sycara, K., Payne, T.: FALCON: Feedback Adaptive Loop for Content-Based Retrieval VLDB 2000, pp. 297–306 (2000)
30. Yi, B.K., Faloutsos, C.: Fast Time Sequence Indexing for Arbitrary Lp Norms. In: Proceedings of the 26th International Conference on Very Large Databases, Cairo, Egypt (2000)
31. Zavala, A.M., Aguirre, A.H., Diharce, E.V.: Robust PSO-based Constrained Optimization by Perturbing the Particle's Memory. In: Chan, F.T.S., Tiwari, M.K. (eds.) Swarm Intelligence: Focus on Ant and Particle Swarm Optimization. I-Tech Education and Publishing (2007)

Fast Nyström for Low Rank Matrix Approximation

Huaxiang Zhang*, Zhichao Wang, and Linlin Cao

School of Information Science and Engineering,Shandong Normal University, Jinan
250014, Shandong China,
Shandong Provincial Key Laboratory for Novel Distributed Computer Software
Technology
huaxzhang@hotmail.com

Abstract. Low-rank matrix approximation is a crucial technique for
data analysis and scientific computing, and the Nyström method is one
of the efficient sampling-based low-rank approximation schemes for handling large kernel matrices. The approximation accuracy of Nyström approach highly depends on the number of columns of the subset used, and
it consumes much time on large data sets. This paper presents an accurate and fast Nyström approach to reducing the computational burdens
when handling large kernel matrices, and experimental results show its
competitive performance in both accuracy and efficiency.

Keywords: kernel methods, manifold learning, matrix factorization,
dimension reduction.

1   Introduction

The Nyström method, which was originally designed to solve numerical integration, recently is used to speed up algorithms that require the spectrum of a
kernel matrix and has been successfully used in applications including efficient
learning of kernel-based methods such as spectral clustering [1,2], support vector
machine [3], manifold learning [4] and GP regression [5].

Given a kernel matrix $\mathbf{K}$, the Nyström method first selects a subset from $\mathbf{K}$,
and then forms a low-rank approximation of $\mathbf{K}$ by using the correlations between
the selected subset and the remaining subset. The most straightforward and
popular Nyström scheme is random sampling without replacement [5], and some
sophisticated sampling schemes have also been presented. These schemes include
non-uniform sampling [6], adaptive sampling [7] and deterministic sampling [8].
Although the Nyström approximation achieves better accuracy if the subset has
more sampled columns, but the low-rank approximation becomes very time-
consuming when the selected subset is large.

An ensemble Nyström method [9] is proposed to alleviate this problem by
decomposing the selected columns into p subsets. Each decomposed subset is

---

* Corresponding author.

S. Zhou, S. Zhang, and G. Karypis (Eds.): ADMA 2012, LNAI 7713, pp. 456–464, 2012.

considered as a base learner, and the standard Nyström method is performed on each base learner. The ensemble approximation can be obtained by using the weighted approximation results, and uniform weights are commonly used in the ensemble algorithm to tradeoff the accuracy and the efficiency. However, the eigenvectors of matrix $\mathbf{K}$ cannot be ensembled by the ensemble Nyström algorithm, so it cannot be used in spectral clustering and manifold learning methods. The random Nyström [10] is also an accurate and efficient method. It first selects columns from the kernel matrix $\mathbf{K}$, and then performs randomized singular value decomposition (SVD) on the selected subset. The random Nyström algorithm still requires extra computational cost in the stage of SVD. In this paper, we use a more efficient method to reduce the computational burden when handling SVD on the selected subset. Theoretical analysis and experimental results show that, our algorithm takes less time than the standard Nyström method without sacrificing the accuracy.

The structure of this paper is organized as follows. We give an overview of the Nyström method in section 2, and review the random algorithm for principal component analysis in section 3. We describe and analyze the proposed method in section 4, and experimentally compare our approach with a number of low-rank approximation techniques in section 5. Finally, section 6 concludes the paper.

2 Nyström Method

Let $\mathbf{K}$ be a $n \times n$ symmetric positive semi-definite (SPSD) kernel matrix. The Nyström uses $\mathbf{W}$ and $\mathbf{C}$ in (1) to approximate the kernel matrix $\mathbf{K}$, where

$$\mathbf{K} = \begin{pmatrix} \mathbf{W} & \mathbf{K}_{21}^T \\ \mathbf{K}_{21} & \mathbf{K}_{22} \end{pmatrix} \quad and \quad \mathbf{C} = \begin{pmatrix} \mathbf{W} \\ \mathbf{K}_{21} \end{pmatrix} \tag{1}$$

Note that $\mathbf{W}$ is a $m \times m$ SPSD matrix since $\mathbf{K}$ is SPSD. For an integer $k < m$, the Nyström method calculates a rank-k approximation

$$\widetilde{\mathbf{K}} = \mathbf{C}\mathbf{W}_k^+\mathbf{C}^T \tag{2}$$

where $\mathbf{W}_k = \arg\min_{\mathbf{V}, rank(\mathbf{V})=k} = \|\mathbf{W} - \mathbf{V}\|_F$ is the best k-rank approximation of $\mathbf{W}$, and $\mathbf{W}_k^+$ denotes the pseudo-inverse of $\mathbf{W}$. We assume that the singular value decomposition (SVD) of $\mathbf{W}$ is $\mathbf{U}\mathbf{\Sigma}\mathbf{U}^T$, where $\mathbf{U}$ is orthonormal and $\mathbf{\Sigma} = diag(\sigma_1, \sigma_2, ..., \sigma_n)$ is diagonal matrix with $\sigma_1 \geq \sigma_2 \geq ... \geq \sigma_n$. For $k \leq rank(\mathbf{W})$, $\mathbf{W}_k^+ = \sum_{i=1}^{k} \sigma_i^{-1}\mathbf{u}_i\mathbf{u}_i^t$, where $\mathbf{u}_i$ denotes the ith column of $\mathbf{U}$.

3 Randomized Algorithm for Principal Component Analysis

Recently, a randomized method for principal component analysis (randomized PCA) [11] has been proposed for constructing low rank approximation of matrices, and it obtains approximately the best accuracy. Randomized PCA is efficient for the low-rank approximation of matrices, and produces accuracy very close to the best possible for matrices of arbitrary size. The method has a theoretical bound for the matrix approximation error. For the low rank approximation of $\mathbf{W}$ arisen from the Nyström method, the randomized PCA algorithm makes the following steps:

1. Given a symmetric matrix $\mathbf{W}$ and a target rank k, the randomized PCA forms a real $(k + p) \times m$ matrix $\mathbf{\Omega}$ with entries sampled i.i.d from a domain governed by Gaussian distribution of zero mean and unit variance. Here p is an over-sampling parameter (typically set to 5 or 10);

2. It computes the product matrix $\mathbf{R} = \mathbf{\Omega}(\mathbf{WW}^T)^q\mathbf{W}$, where q is the number of steps of a power iteration (typically set to 1 or 2) which is used to speed up the decay of the singular values of $\mathbf{W}$;

3. It then uses SVD decomposition to form a real $m \times k$ matrix $\mathbf{Q}$ whose columns are orthonormal, such that, there exists a real $k \times l$ ($l = k + p$) matrix $\mathbf{S}$ for which $\|\mathbf{QS} - \mathbf{R}^T\| \leq \rho_{k+1}$, where ρ_{k+1} is the $(k+1)$th greatest singular value of $\mathbf{R}$;

4. After that, the approach computes $\mathbf{T} = \mathbf{Q}^T\mathbf{WQ}$, and forms the SVD of $\mathbf{T} = \mathbf{V\Lambda V}^T$;

5. Finally, it computes $\mathbf{U} = \mathbf{QV}$, and the SVD of $\mathbf{W}$ can be approximated as $\mathbf{W} \approx \mathbf{U\Lambda U}^T$.

In order to construct the approximation to the SVD of $\mathbf{K}$, the randomized PCA algorithm, as analyzed in [11], incurs the following cost: $2qlw_{mul} + O(k^3 + ml^2)$, where w_{mul} denotes the cost of applying $\mathbf{W}$ to a vector.

4 Our Algorithm

The accuracy of Nyström approximation depends on the number of columns selected, and more accurate approximation requires more columns to be sampled. However, computing the eigenvalue decomposition of $\mathbf{W}$, which is constituted by a large number of columns, will be prohibitive on large date sets. Although randomized PCA is more accurate, it is less efficient than Nyström. The main idea behind our proposed algorithm is to make use of the merits of randomized PCA and the Nyström method, and it contains the following steps:

1. Pick m columns of $\mathbf{K}$ uniformly at random without replacement from n columns to form a $n \times m$ matrix $\mathbf{C}$;

2. Let $\mathbf{W}$ be the $m \times m$ matrix consisting of the intersection of these m columns with the corresponding m rows of $\mathbf{K}$, we get the singular vector matrix $\mathbf{U}$ and the singular value matrix $\mathbf{\Lambda}$ of $\mathbf{W}$ using the randomized PCA, and return the approximation of $\mathbf{K} \approx \tilde{\mathbf{U}}\mathbf{\Lambda}\tilde{\mathbf{U}}^T$ based on $\mathbf{K} \approx \mathbf{CW}_k^+\mathbf{C}^T$, where $\tilde{\mathbf{U}} = \mathbf{CU\Lambda}^+$.

We will analyze the error bound for the proposed method based on the conclusions from reference papers [11,12,13] in the following.

Proposition 1 [11]. Suppose that $\mathbf{Q}$ is a matrix produced via the approach described in section 3 and $\mathbf{W}$ is a positive semi-definite matrix. Then, $\|\mathbf{W} - \mathbf{QQ}^T\mathbf{W}\|_2 \leq \lambda\sigma_{k+1}(\mathbf{W})$, where $\lambda = \beta m^{1/4i}$ and β is a constant. Here i is a nonnegative integer specified by the user, and in many applications of PCA, $i = 1$ or $i = 2$.

Proposition 2 [12]. Given $\mathbf{A} \in R^{n \times n}$, then $\|\mathbf{A}(\mathbf{I}-\mathbf{U}\mathbf{U}^T)\|_2^2 = \|\mathbf{A}-\mathbf{A}\mathbf{U}\mathbf{U}^T\|_2^2 = \|\mathbf{A}\mathbf{A}^T - \mathbf{A}\mathbf{U}\mathbf{U}^T\mathbf{A}^T\|_2 \leqslant \|\mathbf{A}\mathbf{A}^T - \mathbf{R}\mathbf{R}^T\|_2$, where $\mathbf{R} \in R^{n \times k}$ and $\mathbf{U}$ is the orthogonal basis of the range of $\mathbf{R}$.

Proposition 3 [12]. Given $\mathbf{A} \in R^{n \times n}$, then, $\mathbf{A}(\mathbf{A}^T\mathbf{A})^+\mathbf{A}^T = \mathbf{U}_A\boldsymbol{\Sigma}\mathbf{V}_A^T(\mathbf{V}\boldsymbol{\Sigma}^2 \mathbf{V}_A^T)\mathbf{V}_A\boldsymbol{\Sigma}\mathbf{U}_A^T$, where $\mathbf{A} = \mathbf{U}_A\boldsymbol{\Sigma}\mathbf{V}_A^T$ and $\mathbf{U}_A\mathbf{U}_A^T$ is the orthogonal projector of $\mathbf{A}^T\mathbf{A}$.

Proposition 4 [13]. Given a matrix $\mathbf{A} \in R^{n \times n}$, $\|\mathbf{A}\mathbf{A}^T - \mathbf{C}\mathbf{C}^T\|_2 \leqslant \varepsilon\|\mathbf{A}\|_F^2 = \varepsilon \max \|G_{ii}\|$, where $\mathbf{C}$ is a $n \times m$ matrix containing the columns sampled from $\mathbf{A}$ and $\varepsilon > 0$ is the corresponding scaling factor.

Theorem 1. Suppose $\mathbf{K}$ is a $n \times n$ SPSD matrix, and $\hat{\mathbf{K}}$ is constructed using our approach. Then, $\|\mathbf{K} - \hat{\mathbf{K}}\|_2 \leq \beta m^{1/4i}\|\mathbf{K} - \mathbf{K}_k\|_2 + (1 + \beta m^{1/4i})\varepsilon \cdot \max \|K_{ii}\|$.

Proof. Let $\mathbf{H}$ be the chosen samples from matrix $\mathbf{X}$ of size $m \times n$, then the matrix $\mathbf{C} = \mathbf{X}^T\mathbf{H}$ consists of the chosen columns of $\mathbf{K}$ (the kernel matrix with size $n \times n$). Similarly, the matrix $\mathbf{W} = \mathbf{H}^T\mathbf{H}$ consists of the intersection between the chosen columns and the corresponding rows of $\mathbf{K} = \mathbf{X}^T\mathbf{X}$. Using proposition 3 and the approach given in section 3, it is easy to obtain

$$\hat{\mathbf{K}} = \mathbf{C}(\mathbf{U}\boldsymbol{\Lambda}\mathbf{U}^T)^+\mathbf{C}^T = \mathbf{C}\mathbf{Q}(\mathbf{Q}^T\mathbf{W}\mathbf{Q})^+\mathbf{Q}^T\mathbf{C}^T$$

$$= \mathbf{X}^T\mathbf{H}\mathbf{Q}(\mathbf{Q}^T\mathbf{H}^T\mathbf{H}\mathbf{Q})^+\mathbf{Q}^T\mathbf{H}^T\mathbf{X} = \mathbf{X}^T\mathbf{U}_R\mathbf{U}_R^T\mathbf{X} \qquad (3)$$

Using proposition 2, we get

$$\|\mathbf{K} - \hat{\mathbf{K}}\|_2 = \|\mathbf{X}^T\mathbf{X} - \mathbf{X}^T\mathbf{U}_R\mathbf{U}_R^T\mathbf{X}\|_2$$

$$= \|(\mathbf{I} - \mathbf{U}_R\mathbf{U}_R^T)\mathbf{X}\|_2^2 = \|\mathbf{X}(\mathbf{I} - \mathbf{U}_R\mathbf{U}_R^T)\|_2^2$$

$$\leq \|\mathbf{X}\mathbf{X}^T - \mathbf{R}\mathbf{R}^T\|_2 \leq \|\mathbf{X}\mathbf{X}^T - \mathbf{H}\mathbf{H}^T\|_2 + \|\mathbf{H}\mathbf{H}^T - \mathbf{R}\mathbf{R}^T\|_2 \qquad (4)$$

The third equation above follows that $\|\mathbf{A}\mathbf{B}\|_2 = \|\mathbf{B}\mathbf{A}\|_2$ [14].

Applying proposition 1 to 4, we have

$$\|\mathbf{H}\mathbf{H}^T - \mathbf{R}\mathbf{R}^T\|_2 = \|\mathbf{H}(\mathbf{H}^T - \mathbf{Q}\mathbf{Q}^T\mathbf{H}^T)\|_2 = \|(\mathbf{H}^T - \mathbf{Q}\mathbf{Q}^T\mathbf{H}^T)\mathbf{H}\|_2$$

$$= \|\mathbf{H}^T\mathbf{H} - \mathbf{Q}\mathbf{Q}^T\mathbf{H}^T\mathbf{H}\|_2 \leq \lambda\sigma_{k+1}(\mathbf{H}^T\mathbf{H})$$

$$\leq \lambda\|\mathbf{X}\mathbf{X}^T - \mathbf{H}\mathbf{H}^T\|_2 + \lambda\sigma_{k+1}(\mathbf{X}^T\mathbf{X}) \qquad (5)$$

The last step of (5) is derived from the perturbation theory of matrices which states that, the difference between the singular value spectrum of two matrices is bounded by the size of the difference between two matrices [15]. Then using proposition 4, we have

$$\|\mathbf{X}\mathbf{X}^T - \mathbf{H}\mathbf{H}^T\|_2 \leq \varepsilon \cdot \max \|K_{ii}\| \qquad (6)$$

Thus,

$$\|\mathbf{K} - \hat{\mathbf{K}}\|_2 \leq \beta m^{1/4i} \sigma_{k+1}(\mathbf{X}\mathbf{X}^T) + (\beta m^{1/4i} + 1)\varepsilon \cdot \max \|K_{ii}\|$$

$$= \beta m^{1/4i}\|\mathbf{K} - \mathbf{K}_k\|_2 + (\beta m^{1/4i} + 1)\varepsilon \cdot \max \|K_{ii}\| \qquad (7)$$

(7) holds since $\|\mathbf{A} - \mathbf{A}_k\|_2 = \sigma_{k+1}(\mathbf{A})$ for matrix $\mathbf{A}$, and it is similar to the standard Nyström bound form given in [12].

Instead of using SVD on $\mathbf{W}$, we use randomized PCA to decompose it. As being analyzed in the next, the proposed approach is as accurate as standard Nyström but more efficient than it. The complexity of the Nyström algorithm is $O(m^3) + O(nmk)$, it takes $O(nmk)$ to calculate $\widetilde{\mathbf{K}}$ defined in (2), and $O(m^3)$ for the SVD of $\mathbf{W}$. Uniform averaging approach with ensemble Nyström algorithm adopted will cost $O(m^3/p^2) + O(nmk)$, its time requirement largely depends on the number p of the base learners. In practice, uniform averaging performs almost as well as the regression method [10]. The total cost of the proposed algorithm is $2qlw_{mul} + O(l^2m + k^3) + O(nmk)$. As discussed in section 3, the first two terms are the cost of decomposing $\mathbf{W}$, and the last one for calculating $\widetilde{\mathbf{K}}$. The time complexity of the randomized Nyström can be obtained similarly, and its total cost is $(2ql + l)w_{mul} + O(l^2m + l^3) + O(nmk)$. Recall that $n > m > l > k$. Thus, comparing with the Nyström algorithm and the randomized Nyström algorithm, the proposed method requires the least time.

5   Experimental Results and Analysis

In this section, we present the experimental results to illustrate the accuracy and efficiency of the proposed method. The performance of different methods is compared for matrix approximation. We also describe the performance of our algorithm for manifold learning. A core(TM)2 Duo PC with 2.26GHz CPU and 4G memory is used. The codes are in matlab.

Table 1. Summary of datasets used in experiments

Dataset	Type of data	# Points(n)	# Features (d)
isomap-face	face images	4096	698
satimage	physical data	4435	36
usps	handwritten text	7291	256
swissroll	manifold data	20000	3
ijcnn1	time series	100000	22

We work with datasets from ISOMAP homepage[1] and LIBSVM archive[2], and describe the datasets in Table 1. These datasets differ from number of features and data points, and both isomap-face and swissroll are manifold datasets.

---

[1] http://isomap.stanford.edu/datasets.html
[2] http://www.csie.ntu.edu.tw/~cjlin/libsvmtools/datasets/

The performance of various methods is compared by measuring their numerical approximation errors on the *RBF* kernel matrix $K(\mathbf{x}, \mathbf{y}) = \exp(-\|\mathbf{x} - \mathbf{y}\|^2/\gamma)$, where γ is calculated by the squared distance between data points and the sample mean. We only calculate the relative error in Frobenius norm $\|\mathbf{K} - \hat{\mathbf{K}}\|_F/\|\mathbf{K}\|_F$, because spectral norm is expensive and prohibitive for even moderately sized datasets, however, Frobenius norm is still a good indication of the accuracy of a low-rank approximation since the inequality $\|\cdot\|_2 \leq \|\cdot\|_F$ holds.

For convenience of comparison, the proposed method is compared with two sets of methods. The first set includes randomized Nyström ('Rnys') and ensemble Nyström ('Ens'). These two approaches are relatively efficient but less accurate, and the results are shown in Fig.1. The second set includes randomized PCA ('RPCA') and Nyström ('Nys'). Both methods are relatively accurate but less efficient than the former two, and the results are shown in Fig.2. To reduce the statistical variability, the experiment with randomness in the sampling process is repeated 15 times, and the average results are obtained.

The over-sampling and power parameter are set to 10 and 3 respectively for both 'Rnys' and the proposed method ('Ours'). We fix rank $k=400$ for all algorithms, and perform them on satimage and usps respectively. The results are shown in Fig.1. 'Ens400' denotes $k=400$ for the ensemble Nyström in Fig. 1. In addition, for any given m, increasing k will decrease the number (m/k) of base learners, thus conduce to good approximation for the ensemble Nyström method, and lead to time consuming. We also choose larger $k=500$ ('Ens500') for the ensemble Nyström method in order to obtain comparable accuracy with the proposed method.

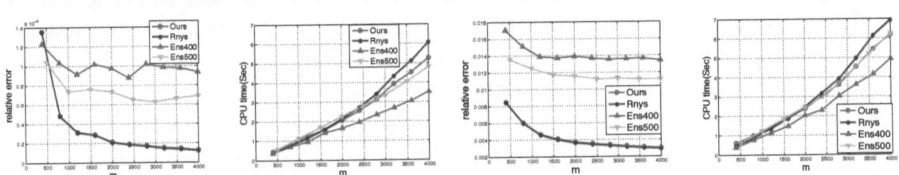

Fig. 1. Relative errors and CPU run times on satimage (left two) and usps (right two)

The results shown in Fig.1 reveal that, on both satimage and usps, 'Ours' and 'Rnys' perform almost the same according to the relative error, but 'Ours' takes less time than 'Rnys'. 'Ens400' is faster than 'Rnys' and 'Ours',but it gets the worst accuracy among all the methods. 'Ens500' performs better than 'Ens400' according to the relative error, but it takes more time than 'Ens400'. This shows that, increasing the rank k can decrease the relative error for the matrix approximation, but increase the running time. The results also show that, 'Ours' outperforms 'Ens500' in terms of relative error with comparable CPU time.

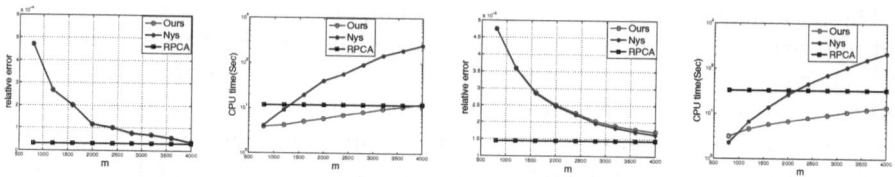

Fig. 2. Relative errors and CPU run times on satimage (left) and usps (right)

Performance comparison of 'Ours', 'Nys' and 'RPCA' is also made on both satimage and usps, and Fig.2 shows the the relative approximation error and the CPU time. We fix $k = 800$ for the three algorithms, and gradually increase the number of sampled columns m. As can be seen, the relative error of 'Ours' and 'Nys' approaches that of 'RPCA' when m is large enough, and on both datasets, 'Ours' is the least expensive one.

In order to show the dependence of the relative error and the cpu time on the sampled columns, we do experiments on ijcnn1 and usps while fixing $k=800$ for the three methods. As can be seen from Fig.3, the CPU time scales with the size of the input matrix. Although 'Ours' obtains slightly lower accuracy than 'RPCA', but it takes less time.

We also implement different low-rank approximation schemes based on Nyström to speed up the spectral method of Laplacian eigenmap. The eigenvectors of the kernel matrix cannot be obtained from the ensemble Nyström method as discussed in section 1. Hence, it cannot be used with Laplacian eigenmap. The Gaussian kernel $\exp(-\|\mathbf{x}\|^2/\gamma)$ is used to construct the similarity. First, we use the isomap-face dataset which has 4096 patterns. The number of the sampled columns and the reduced rank are fixed to 800 and 100 respectively. We also conduct experiments on a large dataset Swissroll (with 20000 patterns), and set $m=600$ and $k=300$. The 2D embeddings given in Fig.4 and Fig.5 reveal that, the performance of our algorithm is very close to 'Nys'. Moreover, to attain the similar embedding structure, our algorithm takes the least amount of computational time on both datasets(3.6 seconds on Isomap-face, and 8 seconds on Swissroll).

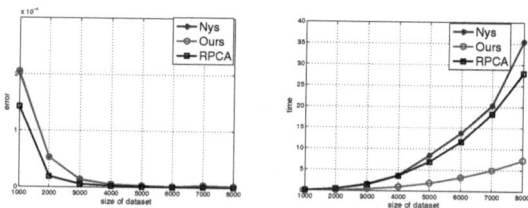

Fig. 3. Relative errors and CPU run times on ijcnn1 (left) and usps (right)

 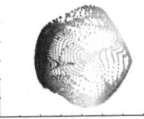 

Fig. 4. Approximated spectral embedding results on Isomap-face (from left to right) (Nys(9.5s), Rnys(4.0s), Ours(3.6s))

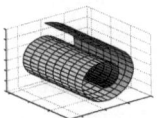 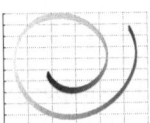 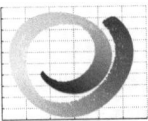 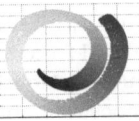

Fig. 5. Approximated spectral embedding results on Swissroll(from left to right) (The original data image, Nys(22s), Rnys(10s), Ours(8s))

6 Conclusions

The Nyström algorithm is an efficient technique for obtaining a low-rank approximation of a large kernel matrix. The quality of the Nyström approximation highly depends on the number of sampled columns. However, the eigenvalue decomposition using SVD on the submatrix will become prohibitive when the number of columns is large. To alleviate the computational burden of the Nyström, this paper presents a novel approach that combines the randomized principal component analysis and the Nyström algorithm. Experimental results on benchmark data sets show that, the proposed approach achieves significant improvement in alleviating computational burdens without sacrificing the approximation accuracy.

Acknowledgments. This research is partially supported by the National Natural Science Foundation of China (No.61170145), the Specialized Research Fund for the Doctoral Program of Higher Education of China(20113704110001), and the Science and Technology Projects of Shandong Province, China (No. ZR2010FM021,2008B0026 and 2010G0020115).

References

1. Ng, A.Y., Jordan, M.I., Weiss, Y.: On spectral clustering: Analysis and an algorithm. In: Advances in Neural Information Processing Systems. MIT, Cambridge (2002)

2. Fowlkes, C., Belongie, S., Chung, F., Malik, J.: Spectral grouping using the nyströom method. IEEE Transactions on Pattern Analysis and Machine Intelligence 26(2), 214–225 (2004)
3. Vapnik, V.: The nature of statistical learning theory. Springer, New York (1995)
4. Belkin, M., Niyogi, P.: Laplacian eigenmaps and spectral techniques for embedding and clustering. In: Advances in Neural Information Processing Systems, pp. 585–591. MIT, Cambridge (2002)
5. Williams, C.K.I., Seeger, M.: Using the nyströom method to speed up kernel machines. In: Advances in Neural Information Processing Systems, pp. 682–688. MIT, Cambridge (2001)
6. Kumar, S., Mohri, M., Talwalkar, A.: Sampling techniques for the nyström method. In: The 12th International Conference on Artificial Intelligence and Statistics, pp. 304–311 (2009)
7. Kumar, S., Mohri, M., Talwalkar, A.: On sampling-based approximate spectral decomposition. In: International Conference on Machine Learning, pp. 553–560 (2009)
8. Zhang, K., Tsang, I., Kwok, J.: Improved nyström low-rank approximation and error analysis. In: International Conference on Machine Learning, pp. 1232–1239 (2008)
9. Kumar, S., Talwalkar, A.: Ensemble nyström method. In: Advances in Neural Information Processing Systems, pp. 1060–1068. MIT, Cambridge (2009)
10. Li, M., Kwok, J.T., Lu, B.-L.: Making large-scale nystrom approximation possible. In: International Conference on Machine Learning, pp. 1–8 (2010)
11. Rokhlin, V., Szlam, A., Tygert, M.: A randomized algorithm for principal component analysis. SIAM Journal on Matrix Analysis and Applications, 1100–1124 (2009)
12. Drineas, P., Mahoney, M.: On the nyström method for approximating a gram matrix for improved kernel-based learning. Journal of Machine Learning Research, 2153–2175 (2005)
13. Drineas, P., Mahoney, M.: Fast monte carlo algorithms for matrices II: Computing a low-rank approximation to a matrix. SIAM Journal on Computing 36(1), 158–183 (2006)
14. Lutkepohl, H.: Handbook of Matrices. John Wiley and Sons (1996)
15. Bhatia, R.: Matrix Analysis. Springer, New York (1997)

An Enhanced Class-Attribute Interdependence Maximization Discretization Algorithm

Kittakorn Sriwanna, Kamthorn Puntumapon, and Kitsana Waiyamai

Computer Engineering Department, Faculty of Engineering, Kasetsart University,
10900 Bangkok, Thailand
{kittakorn.sri,kamthorn}@gmail.com, fengknw@ku.ac.th

Abstract. In this paper, an Enhanced Class-Attribute Interdependence Maximization discretization algorithm (ECAIM) is proposed by 2 extensions to improve a state-of-the-art Class-Attribute Interdependence Maximization discretization algorithm (CAIM). The main drawback that remains unresolved in CAIM is that its stopping criterion depends on the number of target classes. When the number of target classes is large, its performance drops, as CAIM is not a real incremental discretization method. The first extension, ECAIM is extended from CAIM to become a real incremental discretization method by improving the stopping criterion. The stopping criterion is based on the Slope of an *ecaim value* which decreases with an increasing number of intervals. If the slope of *ecaim value* is less than the specified threshold then the discretization terminates. The second extension that we propose is the multi-attribute techniques by simultaneously considering all attributes instead of a single-attribute like CAIM, for accurate and efficient discretizers solution. ECAIM use a feature selection algorithm to select a subset of attributes for reducing the number of attributes, remove irrelevant, redundant attributes and then use multi-attribute techniques only on this subset attributes. Experiment results on 15 real-world datasets show that ECAIM is more efficient than CAIM in terms of accuracy, number of intervals and number of generated rules.

Keywords: Data mining, Classification, Decision tree, Discretization, Class-attribute interdependency maximization.

1 Introduction

Discretization is a process of transforming continuous attribute values into a finite set of intervals [1–5] in order to generate attributes with a smaller number of distinct values. The main process of discretization can be divided into two parts. **The evaluation measure part**, the criterion made for choosing the best cut-point.[1], in order to split or merge the intervals. **The stopping criterion part**, the criterion made for stopping the discretization process in order to get

---

[1] The cut-points are the midpoints of all the adjacent pairs in distinct values of continuous attributes after sorting.

S. Zhou, S. Zhang, and G. Karypis (Eds.): ADMA 2012, LNAI 7713, pp. 465–476, 2012.
© Springer-Verlag Berlin Heidelberg 2012

an appropriate number of intervals. The appropriate number of intervals will increase accuracy and help the users to understand the data.

Discretization methods can be classified into various types [2, 5–8]. **Supervised vs. Unsupervised:** Supervised method discretizes continuous attributes considering class information, but unsupervised method does not consider class information. **Bottom-up vs. Top-down:** Bottom-up method starts with the complete list of cut-points and chooses the best ones to remove by merging intervals. Top-down method start with an empty list of cut-points and chooses the best cut-points to add in the list. **Direct vs. Incremental:** Direct method require users to specify the number of intervals k and then discretize all the continuous attributes into k intervals. Incremental method does not require users to specify the number of intervals. **Single-attribute (univariate) vs. Multi-attribute (multivariate):** single-attribute method discretization only works with a single-attribute but multi-attribute method considers many attributes to determine best cut points [5]. Multi-attribute discretization allows interdependences between attributes that can improve the quality of the overall discretization.

Recently, many discretization algorithms have been proposed. ChiMerge [3] and Chi2 [9] are examples of supervised, bottom-up, incremental and univariate methods. They use statistics to determine the similarity of adjacent intervals. CAIM [1] and CACC [2] are supervised, top-down, incremental and univariate methods. Both algorithms use the same stopping criterion which depends on the minimum number of interval equal to the number of target classes. When the number of target classes is large, their performance drop because they over divide the number of intervals. Further, CAIM usually stops discretization when the number of intervals is equal to the number of target classes. Thus, CAIM algorithm is not a real incremental discretization method.

In this paper, we propose two extensions of the algorithm CAIM. **First,** ECAIM algorithm extends CAIM to become a real incremental discretization method by improving its stopping criterion. The new stopping criterion is based on the slope of *ecam value*. If the slope of *ecam value* is less than the specified threshold then the discretization terminates. **Second,** ECAIM algorithm extends CAIM by using multi-attribute techniques to improve both accuracy and efficiency of the discretization schema.

The rest of the paper is organized as follows. In section 2, we review CAIM discretization algorithm and its drawback. Section 3 explains ECAIM discretization algorithm. Section 4 is the experiment results and analyzed. Finally, we draw conclusion in section 5.

2   CAIM and Its Drawback

CAIM algorithm can be divided into two parts: **The evaluation measure part**, is used to determine the best cut-points that can be computed basd on *caim value*, defined as:

$$caim = \frac{\sum_{r=1}^{n} \frac{max_r^2}{M_{+r}}}{n} \tag{1}$$

where n is the number of intervals, r iterates through all intervals, max_r is the maximum number of class values within the interval r, M_{+r} is the total number of continuous values within the interval r. The largest *caim value* is better generated from the discretization scheme, contrary to smaller *caim value*. **The stopping criterion part**, it is a criterion designed to stop the discretization process. The stopping criterion of CAIM algorithm depends on an evaluating function that is defined by two conditions:

$$s \leq k \quad and \quad caim \leq GlobalCAIM \tag{2}$$

where s is the number of target classes, k is the number of intervals, *caim* is the value that is calculated after dividing the intervals by using Eq. (1) and *GlobalCAIM* is the value that is calculated before dividing the intervals or the last caim value by using Eq. (1). If these two conditions are true, then the discretization process should be stopped.

Age dataset	
Age	**Target classes**
2	care
3	care
4	care
6	care
9	edu
13	edu
18	edu
21	edu
23	work
28	work
39	work
57	work
61	work
71	care
79	care

Sample dataset		
A1	**A2**	**Target classes**
2	0	square
4	0	square
5	0	square
12	0	circlet
30	1	square
45	1	square
50	1	square
52	0	circlet
55	0	circlet
80	1	square
85	0	circlet

Fig. 1. Simple *age* and *sample* datasets

Although CAIM algorithm outperformed the other six state-of-the-art discretization algorithms [1, 2], it still has some drawbacks. Main drawback of CAIM is its simple discretization scheme, in which the number of intervals is very close to the number of target classes [2, 10]. That is to say the number of intervals depends on the number of target classes. For example, if we consider a simple Age dataset that has three classes shown in Fig. 1 (left) , CAIM divides age into three intervals, and chooses the cut-points 7.5 and 22. However, this discretization result is not good; its accuracy is 86.67%. The best discretization consists in four intervals with the following cut-points: 7.5, 22 and 66, it has accuracy 100%. So if a classifier is learnt with such a discretizeation schema, the accuracy will decrease.

The cause of this problem is CAIM stopping criterion. Usually CAIM algorithm stops the discretization process when the number of intervals is equal to the number of target classes. We noticed that its stopping criterion is almost based on only one condition which is $s \leq k$. When discretizations stops, each continuous attribute will always have the number of intervals equal to the number of target classes. For this reason, CAIM is a direct method. Also, CAIM is single attribute discretization algorithm, it does not consider other attributes to determine best cut-points. To improve accuracy and efficiency of the overall discretization, CAIM is extended with an improved stopping criteria and use of multi-attribute techniques. Multi-attribute discretizers have high influence in deductive learning and in complex classification problems where high interaction among multiple attributes exist [5, 11].

3   ECAIM

In this section, we explain the two points that have been improved in ECAIM algorithm. First, we explain how ECAIM improves the stopping criterion, in order to become an incremental discretization method. Then, we present ECAIM algorithm with the use of multi-attribute discretization.

3.1   Improvement of the Stopping Criterion

Stopping criterion of CAIM algoritm is improved in two parts:

Evauation Measure: Evaluation measure of CAIM in Eq. (1) is improved by removing the divider (the number of intervals). The new *ecaim value* becomes:

$$ecaim = \sum_{r=1}^{n} \frac{max_r^2}{M_{+r}} \tag{3}$$

In *ecaim* evaluation measure (Eq. (3)), the divider has been removed in order to stop its decresing value when the number of interval increases. Fig. 2 shows different *ecaim values* when increasing the number of intervals. The *ecaim values* will increase through, their growth decrease. Growth of *ecaim value* (or slope value) varies based on target class distribution of the discretized attribute. For a given attribute, if it has a good distribution of target classes then its growth value will be very high, otherwise its growth value will be very low for bad distribution of target classes. Therefore, the growth or slope of *ecaim value* is used as a criterion for stopping the discretization process.

Stopping Criterion: Stopping criterion of CAIM from Eq. (2), is changed to a new condition as follows:

$$slope < threshold \tag{4}$$

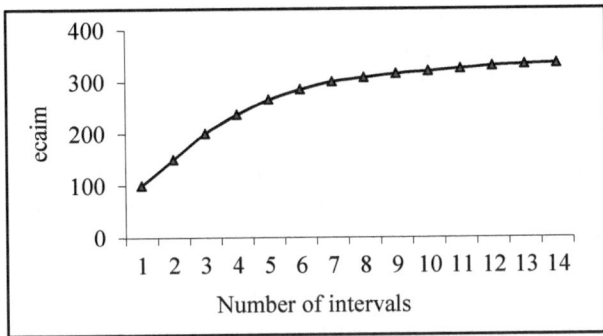

Fig. 2. Growth of *ecaim values* with increasing number of intervals

where *slope* is the increasing of *ecaim values* with increasing number of intervals. And, the *threshold* is defined as follows:

$$threshold = \frac{m(\log s)^3}{s^2} \tag{5}$$

where m is the number of examples, s is the number of target classes. This threshold value is used as a stopping criterion of the discretization process. In a dataset having large number of instances then *ecaim value* will reach its maximum equal to the number of instances. ECAIM algorithm only considers the maximum number of target classes within each interval. If the number of target classes is high then the average maximum number of classes will be low.

In Eq. (4), the condition uses the slope of *ecaim value* as a criterion to terminate the discretization process. If the slope is smaller than threshold then the discretization stops. Usually, when a continuous attribute has a good distribution of the target classes, it will has a fast-growth of *ecaim value*. Therefore, the stopping criterion will be triggered at a slowly rate, and the number of intervals will be large. Inversly, if a continuous attribute has a slow-growth of *ecaim value*, then the stopping criterion will be triggered quickly, and the number of intervals will be small. So each continuous attribute may have different number of intervals. It mostly depends on the data distribution of target classes in each attribute. Therefore, we can say that ECAIM algorithm is an incremental method.

3.2   ECAIM Algorithm

ECAIM is a two-steps discretization algorithm. First step, ECAIM discretizes each of the continuous attributes by condidering its stopping criterion. Second step, multi-attribute discretization technique is applied to the output of discrete data from the first step.

First-step of ECAIM algorithm is presented in Fig. 3. ECAIM discretizes continuous attributes one at a time. In order to explain the process of the first

Input:	Continuous attribute A_i, M examples and S target classes
Output:	The cut-points set of continuous attribute A_i
1:	Sorting the continuous values for an attribute A_i
2:	**for each** intervals I **do**
3:	**if** I_j not yet find the best cut-point **then**
4:	Find the best cut-point by using evaluation measure.
5:	Keep the best cut-point and the increasing of *ecaim value* (slope value).
6:	**end if**
7:	**end for**
8:	Select the interval that has the maximum slope value.
9:	**if** the maximum slope is more than the threshold **then**
10:	Terminate the discretization process.
11:	**else**
12:	Divide selected interval into 2 intervals using the best cut-point.
13:	Add the best cut-point of the selected interval into the set of cut-points.
14:	Go to Line 2.
15:	**end if**

Fig. 3. ECAIM Algorithm (step 1), discretizers by improved the stopping criterion

step clearly, we use the simple age dataset from Fig. 1 as an example again. The process of ECAIM algorithm is shown in Fig. 4. In this figure, the circlet is drawn to represent the continuous values in age dataset. Firstly, all possible continuous values are grouped in one interval A. This interval is sorted by the attribute age. And then ECAIM gets the best cut-point and the slope value of the interval A as shown in Fig. 4. At the beginning, there is only one interval; the selection will select the interval A. When considering the stopping criterion, it is seen to be false because the slope value is greater than the threshold. So interval A is divided into the two intervals, B and C by its best cut-point. Therefore we come again to the beginning of the loop. Secondly, ECAIM finds the best cut-point and the slope value of the intervals B and C. ECAIM selects the interval C because its slope value is higher (the slope value of interval B is 0 because it does not have any cut-point). When calculating the stopping criterion, it appears that the condition is false. So the interval C is divided into the intervals D and E and processing starts again at the beginning of the loop. Thirdly, ECAIM finds the best cut-point and the slope value of the intervals D and E and selects the interval E. The stopping criterion is false again, so the interval E is divided into the interval F and G. Now, the intervals B, D, F and G have a slope that equal to 0. The slope values of those are less than the threshold. Therefore, terminate discretization process.

Second-step; once discretization of all the continuous attributes has been completed from the first-step, the output of discrete data from the first-step will use the multi-attribute techniques. This can be using an example, if we take a *sample* dataset that has two target classes, one continuous attribute A1 and one nominal attribute A2 in Fig. 1 (right). The outputs of the cut-point of discretize

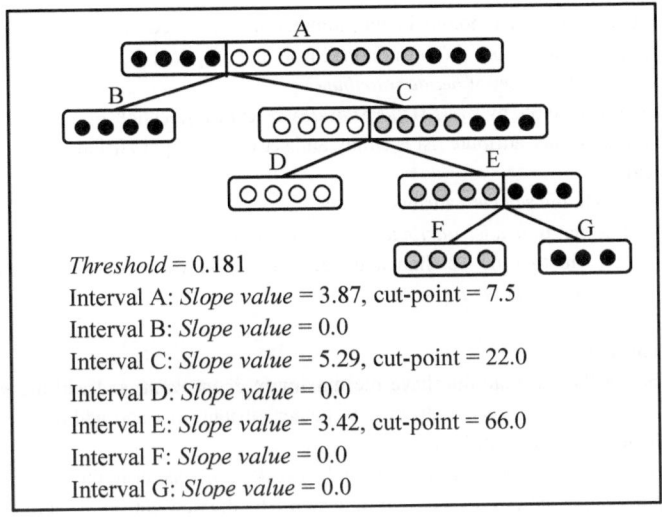

Fig. 4. Process of ECAIM algorithm (step 1) on age dataset

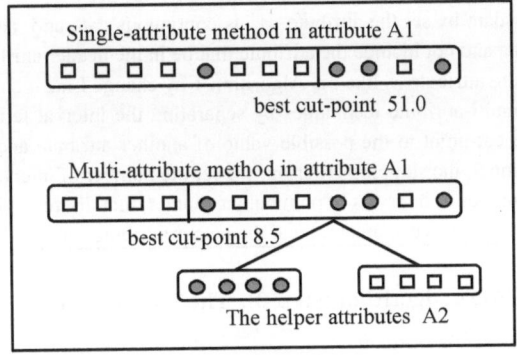

Fig. 5. Single and multi-attribute techniques in sample dataset

the sample dataset are shown in the Fig. 5. In Fig. 5, the circlet and square are drawn to represent the examples in circlet class and square class, respectively. If discretizing the attribute A1 by CAIM algorithm, CAIM found the best cut-point is 51.0 as shown in the upper figure of the Fig. 5. CAIM try to get the best cut-point that has high pure class each interval (the left interval of the cut-point 51.0 is high pure square class and right interval is high pure circlet class). Our idea of multi-attribute techniques not only considers the single attribute A1, it takes other attribute to consider, as shown in the lower figure in the Fig. 5. It found the best cut-point is 8.5. Even if the right interval of the best cut-point is not pure class, but when separate the example of this interval to the possible values of the nominal attribute A2, we can see it have very pure

Input: continuous data and nominal data (nominal data received by transforming continuous to nominal data by use the cut-point set from the output at the step 1)
Output: The cut-points set of *header attribute*
1: Select the attribute by *ConsistencySubsetEval* feature selection from nominal data
2: Set initial header attribute list is empty and set initial helper attribute list is empty
3: **repeat**
4:     **for each** the selected attribute A **do**
5:         **if** A_i not in *header attribute* list and not in *helper attribute* list **then**
6:             Go to Line 12, find and keep the *ecaim* value, cut-points set and the *helper attribute* set of A_i .
7:         **end if**
8:     **end for**
9:     Select the attribute that have the maximum *ecam value* and add the *helper attribute* of this selected attribute to helper attribute list and add this selected attribute to header attribute list.
10: **until** every the selected attribute is in the *header attribute* or in the *helper attribute*
11: Replace some cut-point set of output in step 1 by the new cut-point set of the *header attribute* set

Find the *ecaim value* and the *helper attribute*
Input: Original data, nominal data, attribute A_i and header attribute list
Output: *ecaim value*, cut-points set and the *helper attribute* set
12: Create new data by set the attribute A_i is continuous data and another attribute is nominal data and not include the attribute that be in the header attribute list
13: Discretize the attribute A_i like the Algorithm 1 by change Line 4 and Line 9. In Line 4, we use multi-attribute techniques by separating the interval left and the interval right of the cut-point to the possible value of another attribute and callcuate *ecaim value*. In Line 9, the stopping criterion, we set the number of intervals of attribute A_i equal the number of intervals of nominal attribute A_i that discretize in step 1
14: Return *ecaim value*, cut-points set and the helper attribute set

Fig. 6. ECAIM Algorithm (step 2), The multi-attribute techniques

class. Although the interval that is received by the multi-attribute techniques are less pure than received by CAIM when learning algorithm, the output that is received by multi-attribute techniques will generate an accuracy higher than CAIM. In this example CAIM generate an accuracy of 81.81% but when using the multi-attribute technique, the accuracy is 100%.

In Fig. 5 showing the multi-attribute technique, we call the continuous attribute A1 the header attribute and call the nominal attribute A2 the helper attribute. The header attributes are the attribute that were selected to do multi-attribute discretization. The header attributes must be a continuous attribute. The helper attributes are the attribute that were selected to help the header attributes to separate the example in order to calculate ecame value. The helper attributes must be a nominal attribute. So our multi-attribute method will transform continuous attribute to discrete attribute at a first-step before using the multi-attribute technique.

Some datasets have many features and some of them are redundant or noise. *ConsistencySubsetEval* feature selection algorithm [12] is applied to select features that will be further discretized via multi-attribut discretization technique. It consists in evaluating the worth of a subset of attributes by the level of consistency in the target class values.

The multi-attribute technique is presented in Fig. 6. ECAIM first select the attribute from nominal data by the *ConsistencySubsetEval* algorithm in Line 1. From Line 3 to Line 10, ECAIM add the selected attribute to the header attribute or the helper attributes. In Line 6, ECAIM go to Line 12 to discretize the attribute A_i by use the multi-attribute technique. Finally ECAIM replaces the cut-point set of the output in step 1 by the cut-point set of the header attribute set in Line 11

4   Experiments and Analysis

4.1   Experiment Set-Up

We used fifteen datasets to test CAIM and ECAIM are selected from the UC Irvine machine learning data repository [13] with varying data sizes and varying numbers of target classes. A detailed description of the datasets is shown in Table 1. For each dataset in the experiment, we use 10-fold cross validation of C4.5.

4.2   Result and Analysis

Table 2 shows the comparison of the average number of intervals and the average discretization time of CAIM and ECAIM. Table 3 shows the comparison of the average accuracies and the average number of rules achieved by the C4.5 algorithm. The fifteen datasets were sorted by number of target classes, for an easy comparison, as a large number of target classes affect performance.

Average Number of Intervals: ECAIM generated a discretization scheme with the smallest number of intervals for 10 datasets. For the other 5 datasets, the second smallest number of intervals was generated, but the number of intervals was slightly different. CAIM give a much higher number of intervals than ECAIM algorithm when the number of target classes was higher. For example the average number of intervals of isolet dataset with ECAIM there are 2.49 intervals but CAIM is 25.95 intervals. CAIM usually generates the number of intervals equal the number of target classes. If the number of target classes is high, CAIM will over divide into a large number of intervals.

Average the Execution Time: The execution times depend on the number of interval and number of examples; if those values are high then the execution times high too. ECAIM has highest execution times of all datasets because ECAIM wastes time in the feature selection processes before it use the multi-attribute techniques and also when searching the helper attribute. When using the

Table 1. Major properties of datasets considered in the experimentation

Datasets	#Target classes	#Examples	#Att.	#Continuous Att.
Wisconsin breast cancer database (*breas*)	2	699	9	9
BUPA liver disorders (*bupa*)	2	345	6	6
Statlog project heart disease (*hea*)	2	270	13	6
johns Hopkins university ionosphere (*ion*)	2	351	34	32
iris plants dataset (*iris*)	3	150	4	4
thyroid disease dataset (*thy*)	3	7200	21	6
waveform data set (*wav*)	3	3600	21	21
page blocks classification (*page-blocks*)	5	5473	10	10
glass identification database (*glass*)	6	214	9	9
Statlog project satellite image (*sat*)	6	6435	36	36
image segmentation(*seg*)	7	2310	19	18
Ecoli database (*ecoli*)	8	386	5	5
yeast database (*yeast*)	10	1483	9	7
Deterding vowel recognition (*vowel*)	11	990	13	11
isolated letter speech recognition (*isolet*)	26	1559	617	613

Table 2. Comparison of the average number of intervals and the average discretization time of CAIM and ECAIM

Datasets	Average number of intervals		Average discretization time (s)	
	CAIM	ECAIM	CAIM	ECAIM
breast	**2.00**	2.71	**0.023**	0.121
bupa	**2.00**	2.96	**0.004**	0.048
hea	**2.00**	2.65	**0.006**	0.110
ion	**2.00**	4.93	**0.024**	0.275
iris	**3.00**	3.37	**0.001**	0.006
thy	3.00	**2.16**	**0.162**	3.281
wav	3.00	**2.00**	**0.629**	301.843
page-blocks	5.00	**2.20**	**0.351**	57.409
glass	6.00	**4.54**	**0.005**	0.122
sag	6.00	**4.74**	**1.066**	173.256
seg	6.60	**5.93**	**0.324**	15.996
ecoli	8.00	**3.90**	**0.007**	0.023
yeast	9.00	**2.44**	**0.039**	0.525
vowel	11.00	**3.61**	**0.067**	0.619
isolet	25.95	**2.49**	**12.938**	19.837

multi-attribute techniques, the number of attribute greatly effects to the execution time, if very have the number of attribute then the execution time will very high.

Average C4.5 Accuracy: ECAIM has a high accuracy for 13 datasets. For other 2 datasets, the second highest accuracy was generated. The accuracy of CAIM is dropped when the number of target classes is large because when the

Table 3. Comparison of the average accracies and the average number of rules achieved by the C4.5 algorithm of CAIM and ECAIM

Datasets	Average accuracy (%)		Average number of rules	
	CAIM	ECAIM	CAIM	ECAIM
breast	95.31	**96.19**	**9.90**	10.90
bupa	65.51	**66.67**	**6.50**	18.80
hea	78.52	**81.48**	14.60	**12.50**
ion	89.17	**90.31**	**10.80**	14.50
iris	93.33	**94.67**	3.40	**3.00**
thy	98.53	**99.24**	21.00	**14.00**
wav	76.44	**78.72**	338.00	**160.50**
page-blocks	**96.38**	96.00	81.00	**13.50**
glass	67.76	**72.90**	42.50	**39.90**
sat	**86.09**	85.27	731.50	**620.90**
seg	94.55	**95.50**	163.40	**154.90**
ecoli	77.98	**79.46**	62.20	**30.30**
yeast	32.35	**51.55**	**11.80**	18.30
vowel	70.10	**74.75**	518.40	**200.40**
isolet	45.09	**71.58**	2,749.20	**225.10**

number of target classes is large ECAIM will divide over the number of intervals. But ECAIM still remains highest in accuracy. For example the average accuracy of the isolet dataset with CAIM is 45.09% but with ECAIM it is 71.58%.

Average C4.5 Number of Rules: The experimental results performed with C4.5 algorithm; shows that ECAIM has the smaller number of rules than CAIM. Usually the number of rules depends on the number of intervals. High number of intervals will generate high number of rules. Mostly the number of intervals of ECAIM is smaller than CAIM then the number of rules is smaller too.

5   Conclusion

In this paper, we propose an Enhanced Class-Attribute Interdependence Maximization discretization algorithm (ECAIM). ECAIM is an extended version of CAIM. First, ECAIM is extended from CAIM to become a real incremental discretization method. For each continuous attribute, it determines the appropriate number of discrete intervals by considering the slope of *ecaim value*. Second, the multi-attribute technique is developped. ECAIM algorithm extends CAIM by using multi-attribute technique to improve both accuracy and efficiency of the discretization schema. Experiments demonstrate that ECAIM performs well in both synthetic and real-world datasets.

References

1. Kurgan, L.A., Cios, K.J.: CAIM discretization algorithm. IEEE Transactions on Knowledge and Data Engineering, 145–153 (2004)
2. Tsai, C.-J., Lee, C.-I., Yang, W.-P.: A discretization algorithm based on Class-Attribute Contingency Coefficient. Information Sciences 178, 714–731 (2008)
3. Kerber, R.: ChiMerge: discretization of numeric attributes. In: Proceedings of the Tenth National Conference on Artificial Intelligence, pp. 123–128. AAAI Press, San Jose (1992)
4. Ruiz, F.J., Angulo, C., Agell, N.: IDD: A Supervised Interval Distance-Based Method for Discretization. IEEE Transactions on Knowledge and Data Engineering 20, 1230–1238 (2008)
5. García, S., Luengo, J., Saez, J., Lopez, V., Herrera, F.: A Survey of Discretization Techniques: Taxonomy and Empirical Analysis in Supervised Learning. IEEE Transactions on Knowledge and Data Engineering, 1 (2012)
6. Liu, H., Hussain, F., Tan, C.L., Dash, M.: Discretization: An Enabling Technique. Data Mining and Knowledge Discovery 6, 393–423 (2002)
7. Vyas, O.P., Das, K.: Article: A Suitability Study of Discretization Methods for Associative Classifiers. International Journal of Computer Applications 5, 46–51 (2010)
8. Dougherty, J., Kohavi, R., Sahami, M.: Supervised and Unsupervised Discretization of Continuous Features. In: Prieditis, A., Russell, S. (eds.) Proceedings of the Twelfth International Conference on Machine Learning, Tahoe City, California, USA, pp. 194–202 (1995)
9. Huan, L., Setiono, R.: Feature selection via discretization. IEEE Transactions on Knowledge and Data Engineering 9, 642–645 (1997)
10. Zhu, Q., Lin, L., Shyu, M.-L., Chen, S.-C.: Effective supervised discretization for classification based on correlation maximization. In: 2011 IEEE International Conference on Information Reuse and Integration (IRI), pp. 390–395 (2011)
11. García, M.N.M., Lucas, J.P., Batista, V.F.L., Martín, M.J.P.: Multivariate Discretization for Associative Classification in a Sparse Data Application Domain. In: Graña Romay, M., Corchado, E., Garcia Sebastian, M.T. (eds.) HAIS 2010, Part I. LNCS, vol. 6076, pp. 104–111. Springer, Heidelberg (2010)
12. Huan, L., Setiono, R.: A Probabilistic Approach to Feature Selection - A Filter Solution. In: ICML, pp. 319–327 (1996)
13. Frank, A., Asuncion, A.: Machine Learning Repository (2010)

Towards Normalizing the Edit Distance Using a Genetic Algorithms–Based Scheme[*]

Muhammad Marwan Muhammad Fuad

Department of Electronics and Telecommunications
Norwegian University of Science and Technology (NTNU)
NO-7491 Trondheim, Norway
marwan.fuad@iet.ntnu.no

Abstract. The normalized edit distance is one of the distances derived from the edit distance. It is useful in some applications because it takes into account the lengths of the two strings compared. The normalized edit distance is not defined in terms of edit operations but rather in terms of the edit path. In this paper we propose a new derivative of the edit distance that also takes into consideration the lengths of the two strings, but the new distance is related directly to the edit distance. The particularity of the new distance is that it uses the genetic algorithms to set the values of the parameters it uses. We conduct experiments to test the new distance and we obtain promising results.

Keywords: Edit Distance, Normalized Edit Distance, Genetic Algorithms, Sequential Data.

1   Introduction

Similarity search is an important problem in computer science. This problem has many applications in data mining, computational biology, pattern recognition, and others. In this problem a pattern or a *query* is given and the task is to retrieve the data objects in the database that are "close" to that query according to some semantics that quantify that closeness. This closeness or similarity is depicted using a principal concept which is the *similarity measure* or its more powerful form; the *distance metric*.

Because of its topological properties, the metric model (reflexivity, non-negativity, symmetry, triangle inequality) has been widely used to process similarity queries, but later other models were proposed.

The *edit distance* is the main distance used to measure the similarity between two strings. It is defined as the minimum number of delete, insert, and change operations needed to transform string S into string T.

[*] This work was carried out during the tenure of an ERCIM "Alain Bensoussan" Fellowship Programme. This Programme is supported by the Marie-Curie Co-funding of Regional, National and International Programmes (COFUND) of the European Commission.

S. Zhou, S. Zhang, and G. Karypis (Eds.): ADMA 2012, LNAI 7713, pp. 477–487, 2012.

But the edit distance has its limitations because it considers local similarity only and does not imply any global level of similarity.

In [10], [11], and [12] we presented two new extensions of the edit distance. These new extensions consider a global level of similarity which the edit distance didn't consider. But the parameters used with these two distances were defined using basic heuristics, which substantially limited the search space.

In this paper we propose a new extension of the edit distance. This extension aims to normalize the edit distance using an approach that relates it directly to the edit distance. The new distance uses the genetic algorithms as an optimization method to set the parameters it uses.

Section 2 of this paper presents the related work. Section 3 introduces the new distance. Section 4 validates it through different experiments and Section 5 concludes the paper.

2   Related Work

Strings, also called *sequences* or *words*, are a way of representing data. This data type exists in many fields of computer science such as molecular biology where DNA sequences are represented using four nucleotides which correspond to the four bases: adenine (A), cytosine (C), guanine (G) and thymine (T). This can be expressed as a 4-symbol alphabet. Protein sequences can also be represented using a 20-symbol alphabet which corresponds to the 20 amino acids.

Written languages are also expressed in terms of alphabets with letters (26 in English). Spoken languages are represented using phonemes (40 in English). Texts use alphabets with a very large size (the vocabulary items of a language). These examples show that strings are ubiquitous.

One of the main distances used to handle sequential data is the edit distance (ED) [13], also called the *Levenshtein distance*, which is defined as the minimum number of delete, insert, and substitute operations needed to transform string S into string R.

Formally, ED is defined as follows: Let Σ be a finite alphabet, and let Σ^* be the set of strings on Σ. Given two strings $S = s_1 s_2 .... s_n$ and $R = r_1 r_2 .... r_m$ defined on Σ^*. An *elementary edit operation* is defined as a pair: $(a,b) \neq (\lambda, \lambda)$, where a and b are strings of lengths 0 and 1, respectively. The elementary edit operation is usually denoted $a \rightarrow b$ and the three elementary edit operations are $a \rightarrow \lambda$ (deletion) $\lambda \rightarrow b$ (insertion) and $a \rightarrow b$ (substitution). Those three operations can be weighted by a weighting function γ which assigns a nonnegative value to each of these operations. This function can be extended to edit transformations $T = T_1 T_2 ... T_m$.

The edit distance between S and R can then be defined as:

$$ED\ (S,\ R) = \{ \gamma\ (T) |\ T\ is\ an\ edit\ transformation\ of\ S\ into\ R\ \} \qquad (1)$$

ED is the main distance measure used to compare two strings and it is widely used in many applications. Fig. 1 shows the edit distance between the two strings $S_1 = \{M, A, R, W, A, N\}$ and $S_2 = \{F, U, A, D\}$.

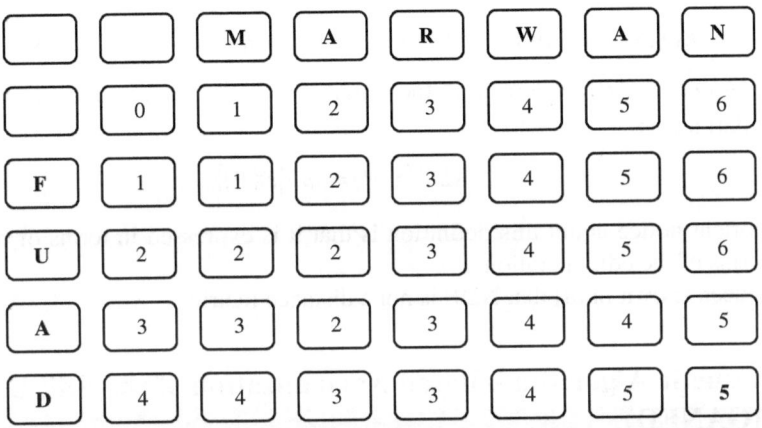

Fig. 1. The edit distance between two strings

ED has a few drawbacks; the first is that it is a measure of local similarities in which matches between substrings are highly dependent on their positions in the strings [5]. In fact, the edit distance is based on local procedures both in the way it is defined and also in the algorithms used to compute it. Another drawback is that ED does not consider the length of the two strings.

Several modifications have been proposed to improve ED. In [10], [11], and [12] two new extensions of ED; the *extended edit distance* (EED) and the *multi-resolution extended edit distance* (MREED) were proposed. These two distances add a global level of similarity to that of ED by including the frequency of characters or bi-grams when computing the distance. The problem with these two distances is that they use parameters which are set using basic heuristics which makes the search process ineffective.

It is worth mentioning that the two distances EED and MREED, as well as ED, are all metric distances.

Another important modification is the *normalized edit distance* (NED) [8]. The rationale behind this distance is that the length of the two strings should be taken into account when computing the distance between them. An *editing path* P between two strings S and R, of lengths n and m, respectively ($n \leq m$), is a sequence of ordered pairs of integers (i_k, j_k), where $0 \leq k \leq m$, that satisfies the following :

i- $0 \leq i_k \leq |S|$, $0 \leq j_k \leq |R|$;

  $(i_0, j_0) = (0,0), (i_m, j_m) = (|S|, |R|)$

ii- $0 \leq i_k - i_{k-1} \leq 1, \ 0 \leq j_k - j_{k-1} \leq 1 \ , \forall k \geq 1$

iii- $i_k - i_{k-1} + j_k - j_{k-1} \geq 1$

The weights can be associated to paths as follows:

$$\omega(P_{S,R}) = \sum_{k=1}^{m} \gamma(S_{i_{k-1}+1...i_k} \rightarrow R_{j_{k-1}+1...j_k})$$

It follows that:

$$ED\ (S,\ R) = min\ \{\omega\ (P)|\ P\ is\ an\ edit\ transformation\ of\ S\ into\ R\ \}$$

Let $\hat{\omega}(P) = \omega(P)/L(P)$, where L is the length of P, the normalized edit distance NDE is defined as:

$$NDE(S,R) = min\{\hat{\omega}(P)\} \tag{2}$$

An important notice about this definition is that it is expressed in terms of paths and not in terms of the edit operations.

It has been shown in [8] that NDE is not a distance metric.

3   Genetic Algorithms-Based Normalization of the Edit Distance (GANED)

ED we presented in Section 2 was mainly introduced to apply to spelling errors. This makes the edit operations a main component of ED. NDE, although takes into consideration the lengths of the two strings, which is an important modification in our opinion, is based on a different principle than that of ED, which, we think, causes it to lose some of the principal characteristics of ED.

In this work we present a new modification of the edit distance that also takes the lengths of the strings into account. However, our proposed distance uses a completely different approach than that of NDE. Our new distance is directly related to the edit distance. In fact, the new distance is a lower bound of the edit distance.

3.1   GANED

Let Σ be a finite alphabet, and let Σ^* be the set of strings on Σ. Let n be an integer, and let $f_{a_n}^{(S)}$ be the frequency of the n-gram a_n in S, and $f_{a_n}^{(T)}$ be the frequency of the n-gram a_n in T, where S, T are two strings in Σ^*.

The GANED distance between S and T is defined as:

$$GANED\ (S,T) = ED(S,T) \times \left[1 - \frac{2 \sum_{n=1}^{max(|S|,|T|)} \lambda_n \left(\sum_{a_n \in A^n} min\left(f_{a_n}^{(S)}, f_{a_n}^{(T)}\right) + (n-1) \right)}{n(|S| + |T|)} \right] \tag{3}$$

where $|S|, |T|$ are the lengths of the two strings S, T respectively, and where $\lambda_n \in [0,1]$. λ_n are called the frequency factors.

Notice that

$$0 \le 2 \sum_{n=1}^{max(|S|,|T|)} \lambda_n \left(\sum_{a_n \in A^n} min\left(f_{a_n}^{(S)}, f_{a_n}^{(T)}\right) + (n-1) \right) \le n(|S| + |T|)$$

So ED is multiplied by a factor whose value varies between 0 and 1, so GANED as presented in (3) includes a form of normalization. In fact:

$0 \le GANED(S,R) \le ED(S,R)$ can be written as $0 \le \dfrac{GANED(S,R)}{ED(S,R)} \le 1$ which is the

common form of normalization, and we could have expressed our new distance in the latter form. However, instead of imposing a condition that the two strings be different (thus ED=0), we preferred to introduce the new distance in the form shown in (3).

Notice also that GANED is a lower bound of ED, so the relation between the two distances is direct.

As mentioned in Section 1, the idea of considering the frequency of characters or bi-grams in computing the distance has previously been proposed in [10], [11], and [12]. However, the definition of the frequency factors remains problematic in these three works. On the one hand, the search space they use is very limited, and on the other hand, generalizing the distances proposed in [10], [11], and [12] using the same basic heuristics to define the frequency factors makes this process inefficient yet limited to very small regions in the search space.

GANED uses one very powerful optimization method; the genetic algorithms, to define the frequency factors λ_n. The use of the genetic algorithms makes the search more effective.

3.2   The Genetic Algorithms

The *Genetic Algorithms* are a member of a large family of stochastic algorithms called *Evolutionary Algorithms* (EAs) which are population-based optimization algorithms inspired by nature, particularly the theory of evolution. In Fig. 2 we show the members of the EAs. These members differ in implementation but they use the same principle.

Of the EAs family, GAs are the most widely known. GAs have the following elements: a population of individuals (also called *chromosomes*), selection according to fitness, crossover to produce new offspring, and random mutation of new offspring [9]. GAs create an environment in which a population of individuals, representing solutions to a particular problem, is allowed to evolve under certain rules towards a state that minimizes, in terms of optimization, the value of a function which is usually called the *fitness function* or the *objective function*.

There are a large number of variations of GAs. In the following we present a description of the simple, classical GAs. GAs start by defining the problem variables

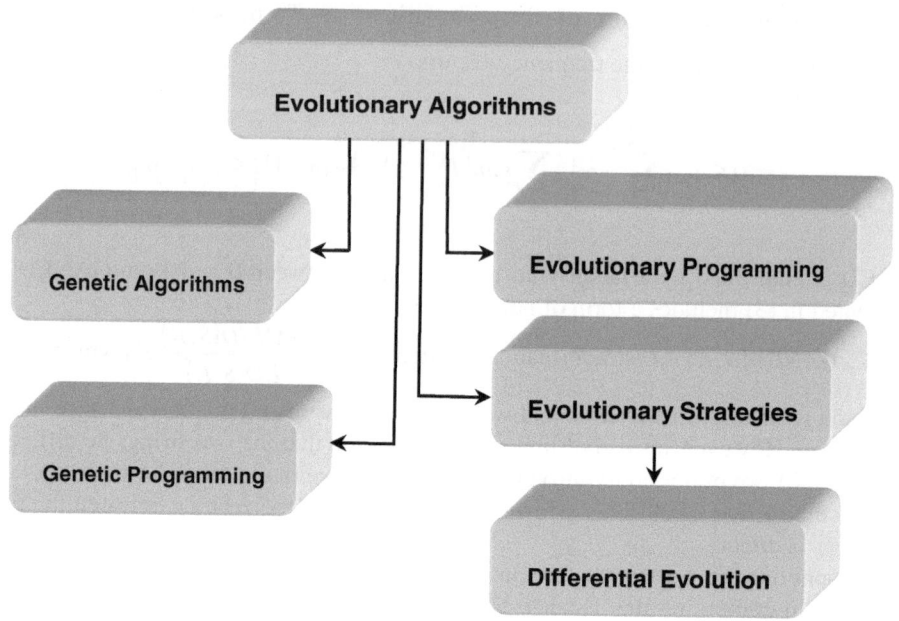

Fig. 2. The family of evolutionary algorithms

(*genes*) and the fitness function. These variables can be bounded or unbounded. A particular combination of variables produces a certain value of the fitness function and the objective of GAs is to find the combination that gives the best value of the fitness function. The terminology "best" implies that there is more than one solution and the solutions are not of equal value [2].

After defining the variables and the fitness function GAs start by randomly generating a number *pSize* of individuals, or *chromosomes*. This step is called *initialization*.

GAs were originally proposed to be binary coded to imitate the genetic encoding of natural organisms [15]. But later other encoding schemes were presented. The most widely used scheme is real-valued encoding. In this scheme a candidate solution is represented as a real-valued vector in which the dimension of the chromosomes is constant and equal to the dimension of the solution vectors [1]. This dimension is denoted by *nPar*. The fitness function of each chromosome is evaluated. The next step is *selection*. The purpose of this procedure is to determine which chromosomes are fit enough to survive and possibly produce offspring. This is decided according to the fitness function of the chromosome in that the higher the fitness function is the more chance it has to be selected for mating. There are several selection methods such as the *roulette wheel selection, random selection, rank selection, tournament selection*, and others [9]. The percentage of chromosomes selected for mating is denoted by *sRate*. *Crossover* is the next step in which offspring of two parents are produced to enrich the population with fitter chromosomes. There are several approaches to perform this process, the most common of which is *single-point* crossover and *multi-point* crossover.

While crossover is the mechanism that enables the GA to communicate and share information about fitter chromosomes, it is not sufficient to efficiently explore the search space. *Mutation*, which is a random alteration of a certain percentage *mRate* of chromosomes, is the other mechanism which enables the GA to examine unexplored regions in the search space. It is important to keep a balance between crossover and mutation. High crossover rate can cause converging to local minima and high mutation rate can cause very slow convergence.

Now that a new generation is formed, the fitting function of the offspring is calculated and the above procedures repeat for a number of generations *nGen* or until a stopping criterion terminates the algorithm.

4   Empirical Evaluation

We tested the new distance GANED on time series because this is our field of expertise, but we believe GANED is highly applicable in bioinformatics and text mining.

Time series data are normally numeric, but there are different methods to transform them to symbolic data. The most important symbolic representation method of time series is the *Symbolic Aggregate approXimation* (SAX) [7]. The first step of SAX is to normalize the time series because SAX is based on the assumption that normalized time series have a Guassian distribution, so SAX can only be applied to normalized time series. The second step is to reduce the dimensionality of the time series by using a time series representation method called *Piecewise Aggregate Approximation* (PAA) [3], [14]. This PAA representation of the time series is then discretized. This is achieved by determining the breakpoints. The number of the breakpoints is related to the desired alphabet size and their locations are obtained using statistical lookup tables.

The distance between the resulting time series after applying the above steps is computed using the following relation:

$$MINDIST(\hat{S}, \hat{R}) \equiv \sqrt{\frac{n}{N}} \sqrt{\sum_{i=1}^{N} (dist(\hat{s}_i, \hat{r}_i))^2} \qquad (4)$$

Where n is the length of the original time series, N is the number of segments, $\hat{S}$ and $\hat{R}$ are the symbolic representations of the two time series S and R, respectively, and where the function $dist(\ )$ is implemented by using the appropriate lookup table.

We tested our new distance GANED on time series classification task based on the first nearest-neighbor (1-NN) rule using leaving-one-out cross validation. This means that every time series is compared to the other time series in the dataset. If the 1-NN does not belong to the same class, the error counter is incremented by 1.

We conducted experiments using datasets of different sizes and dimensions available at UCR [4].

As indicated earlier, we tested GANED on symbolically represented time series This means that the time series were transformed to symbolic sequences using the first three step of SAX presented earlier in this section, but instead of using MINDIST given in relation (4), we use our distance GANED. The parameters λ_n in the definition of GANED (relation (3)) are defined using GAs. This means that for each value of the alphabet size we formulate a GAs optimization problem where the fitness function is the classification error, and the parameters of the optimization problem are λ_n. Practically n can take any value that does not exceed that of the shortest string of the two strings S, T. However, in the experiments we conducted $n \in \{1,2,3\}$ because these are the values of interest for time series. The values of λ_n varied in the interval $[0,1]$.

Notice that GANED can be applied to strings of different lengths, which is one of its advantages since most similarity measures in time series mining are applied only to time series of the same length.

For the GAs we used, the population size $pSize$ was 12, the number of generations $nGen$ was set to 20. The mutation rate $mRate$ was 0.2 and the selection rate $sRate$ was 0.5. The dimension of the problem $nPar$ depends on the number of parameters used in GANED (as mentioned earlier, they were tested for $n \in \{1,2,3\}$). Table 1 summarizes the symbols used in the experiments together with their corresponding values.

For each dataset we use GAs on the training datasets to get the vector λ_n that minimizes the classification error on these training datasets, and then we utilize this optimal λ_n vector on the corresponding testing datasets to get the final classification error for each dataset.

We compared GANED with MINDIST. This means after we represent the time series symbolically as indicated at the beginning of this section we classify them using GANED first then using MINDIST. We chose to compare GANED with MINDIST because we used the same symbolic representation that MINDIST uses. However, GANED can be used with any sequential data.

It is important to mention however that MINDIST has a lower complexity than that of GANED.

In Table 2 we present some of the results we obtained for alphabet size equal to 3, 10, and 20, respectively.

Table 1. The symbol table together with the corresponding values used in the experiments

pSize	Population size	12
nGen	Number of generations	20
mRate	Mutation rate	0.2
sRate	Selection rate	0.5
nPar	Number of parameters	varies

Table 2. Comparison between the classification error of GANED and MINDIST

CBF

α^*	n-gram	GANED		MINDIST
		λ_n	Classification Error	Classification
3	1	[0.77491]	0.026	
	2	[0.81776 0.87965]	0.026	0.382
	3	[0.93285 0.97274 0.75836]	0.023	
10	1	[0.43021]	0.031	
	2	[0.34446 0.32247]	0.031	0.104
	3	[0.13412 0.45606 0.080862]	0.039	
20	1	[0.37819]	0.053	
	2	[0.8962 0.72086]	0.079	0.088
	3	[0.96216 0.091513 0.83706]	0.062	

*:alphabet size

Coffee

A	n-gram	GANED		MINDIST
		λ_n	Classification	Classification
3	1	[0.81472]	0.179	
	2	[0.43021 0.22175]	0.214	0.464
	3	[0.21868 0.19203 0.34771]	0.214	
10	1	[0.89292]	0.179	
	2	[0.97059 0.93399]	0.214	0.464
	3	[0.4899 0.81815 0.08347]	0.143	
20	1	[0.12393]	0.107	
	2	[0.16825 0.9138]	0.107	0.143
	3	[0.81472 0.95717 0.67874]	0.107	

Face Four

α	n-gram	GANED		MINDIST
		λ_n	Classification	Classification
3	1	[0.022414]	0.057	
	2	[0.57462 0.57425]	0.057	0.239
	3	[0.2038 0.60654 0.38334]	0.057	
10	1	[0.015908]	0.045	
	2	[0.16625 0.34168]	0.057	0.182
	3	[0.57997 0.3957 0.21003]	0.057	
20	1	[0.54483]	0.114	
	2	[0.92995 0.18334]	0.102	0.193
	3	[0.57758 0.28758 0.15406]	0.090	

Table 2. (*continued*)

Gun_Point

A	n-gram	GANED λ_n	GANED Classification	MINDIST Classification
3	1	[0.19728]	0.193	
	2	[0.95798 0.58518]	0.193	
	3	[0.40628 0.95213 0.68035]	0.2	0.307
10	1	[0.98445]	0.147	
	2	[0.93927 0.99038]	0.127	
	3	[0.15187 0.40029 0.24364]	0.12	0.233
20	1	[0.16625]	0.06	
	2	[0.5852 0.0038735]	0.06	
	3	[0.32809 0.42736 0.12747]]	0.06	0.12

Olive Oil

A	n-gram	GANED λ_n	GANED Classification	MINDIST Classification
3	1	[0.70608]	0.4	
	2	[0.53732 0.14595]	0.4	
	3	[0.96676 0.15111 0.0015139]	0.4	0.833
10	1	[0.76393]	0.667	
	2	[0.2953 0.41039]	0.667	
	3	[0.97014 0.29259 0.080068]	0.667	0.833
20	1	[0.028529]	0.267	
	2	[0.93581 0.20714]	0.233	
	3	[0.18231 0.09461 0.68031]	0.233	0.833

Trace

α	n-gram	GANED λ_n	GANED Classification	MINDIST Classification
3	1	[0.96149]	0.27	
	2	[0.80699 0.14789]	0.26	
	3	[0.89336 0.01668 0.3959]	0.26	0.54
10	1	[0.76432]	0.08	
	2	[0.42505 0.64252]	0.09	
	3	[0.95513 0.93675 0.99434]	0.04	0.42
20	1	[0.92115]	0.1	
	2	[0.89948 0.8597]	0.12	
	3	[0.69135 0.21079 0.9382]	0.12	0.36

As we can see from the results, the classification errors of GANED are smaller than those of MINDIST for all the datasets and for all values of the alphabet size. The results of other datasets in the archive were similar.

5   Conclusion and Perspectives

In this paper we presented a new normalized edit distance. This new distance, GANED, is related directly to the edit distance and it takes into account the length of the two strings. The particularity of our new distance is that it uses an optimization algorithm; the genetic algorithms, to set the values of its parameters. We tested the new distance by comparing it to another distance applied to strings and we showed how our new distance GANED has a better performance.

The new distance was applied in this work to symbolically represented time series. However, we believe other applications might be more appropriate for our new distance.

References

1. Affenzeller, M., Winkler, S., Wagner, S., Beham, A.: Genetic Algorithms and Genetic Programming Modern Concepts and Practical Applications. Chapman and Hall/CRC (2009)
2. Haupt, R.L., Haupt, S.E.: Practical Genetic Algorithms with CD-ROM. Wiley-Interscience (2004)
3. Keogh, E., Chakrabarti, K., Pazzani, M., Mehrotra, S.: Dimensionality Reduction for Fast Similarity Search in Large Time Series Databases. J. of Know. and Inform. Sys (2000)
4. Keogh, E., Zhu, Q., Hu, B., Hao. Y., Xi, X., Wei, L., Ratanamahatana,C.A.: The UCR Time Series Classification/Clustering Homepage (2011),
   http://www.cs.ucr.edu/~eamonn/time_series_data/
5. Kurtz, S. : Lecture Notes for Foundations of Sequence Analysis (2001)
6. Laguna, M., Marti, R.: Scatter Search: Methodology and Implementations in C. Springer, Heidelberg (2003)
7. Lin, J., Keogh, E., Lonardi, S., Chiu, B.Y.: A Symbolic Representation of Time Series, with Implications for Streaming Algorithms. DMKD, 2–11 (2003)
8. Marzal, A., Vidal, E., Computation, E.: of Normalized Edit Distances and Applications. IEEE Transactions on Pattern Analysis and Machine Intelligence 15(9), 926–932 (1993)
9. Mitchell, M.: An Introduction to Genetic Algorithms. MIT Press, Cambridge (1996)
10. Muhammad Fuad, M.M., Marteau, P.-F.: Extending the Edit Distance Using Frequencies of Common Characters. In: Bhowmick, S.S., Küng, J., Wagner, R. (eds.) DEXA 2008. LNCS, vol. 5181, pp. 150–157. Springer, Heidelberg (2008)
11. Muhammad Fuad, M.M., Marteau, P.F.: The Extended Edit Distance Metric. In: Sixth International Workshop on Content-Based Multimedia Indexing (CBMI 2008), London, UK, June 18-20 (2008)
12. Muhammad Fuad, M.M., Marteau, P.F.: The Multi-resolution Extended Edit Distance. In: Third International ICST Conference on Scalable Information Systems, Infoscale, 2008, ACM Digital Library, Vico Equense, Italy, June 4-6 (2008)
13. Wagner, R.A., Fischer, M.J.: The String-to-String Correction Problem. Journal of the Association for Computing Machinery 21(I), 168–173 (1974)
14. Yi, B.K., Faloutsos, C.: Fast Time Sequence Indexing for Arbitrary Lp norms. In: Proceedings of the 26st International Conference on Very Large Databases, Cairo, Egypt (2000)
15. Yu, X., Gen, M.: Introduction to Evolutionary Algorithms. Springer (2010)

PCG: An Efficient Method for Composite Pattern Matching over Data Streams*

Cheng Ju, Hongyan Li**, and Feifei Li

Key Laboratory of Machine Perception (Peking University), Ministry of Education
School of Electronics Engineering and Computer Science, Peking University
Beijing 100871, P.R. China
{jucheng,lihy,liff}@cis.pku.edu.cn

Abstract. Sequential data segments in data streams are very meaningful in many areas. These data segments usually have complicated appearance and require online processing. But matching these data segments can be time-consuming and there are multiple matching tasks to be proceeded simultaneously. This paper presents a novel data structurepattern combination graph (PCG) and corresponding algorithms to accomplish composite pattern matching over data streams. To make it possible to deal with complicated patterns efficiently, PCG firstly identify similar segments among different segments as basic patterns, and then deal with the composite semantics between basic patterns. In this way, data stream flow into PCG for matching in the form of basic patterns. Later procedures are operated according to the types of nodes in PCG and the final results are returned to users. From the perspective of recall ratio, precision ratio and efficiency, the experimental results on real data sets of medical streams show that PCG is feasible and effective.

Keywords: Pattern Matching, large number, data stream, on-line, optimizing.

1 Introduction

As technology advances, streams of data can be rapidly generated in numerous applications such as financial services, RFID-based tracking and health services. Data Stream Management System (DSMS), such as TelegraphCQ, Stream, Aurora, enables applications to run continuous queries that efficiently process streams of data in real time. However, many applications are becoming so complicated that analysis on simple data points is no longer sufficient to meet our demands. Instead, a continuous segment of data points, which we refer to pattern, can be very meaningful to help us get more information. Pattern matching

---

* This work was supported by Natural Science Foundation of China (No.60973002 and No.61170003), the National High Technology Research and Development Program of China (Grant No. 2012AA011002), National Science and Technology Major Program (Grant No. 2010ZX01042-002-002-02, 2010ZX01042-001-003-05).
** Corresponding author.

S. Zhou, S. Zhang, and G. Karypis (Eds.): ADMA 2012, LNAI 7713, pp. 488–501, 2012.
© Springer-Verlag Berlin Heidelberg 2012

help us find these patterns and it is a significant query technique where data segments are matched against a complex pattern that specifies constraints on them [6, 7, 8].

Modern applications [9, 10] bring several new challenges for pattern matching on data stream, the following example illustrates some of the challenges that we must confront.

Example. Fig.1 shows some typical patterns of diseases in electrocardiograph (ECG) collected in Intensive Care Unit (ICU). The pattern in (a) happens when a patient had cardiac damage, and the pattern in (b) shows the ECG of a patient who gets serious pulmonary infection, and the pattern in (c) represents reperfusion arrhythmia. If someones ECG change into one of the above patterns but doesnt get medical treatment promptly, the patient will probably suffer more serious health problems.

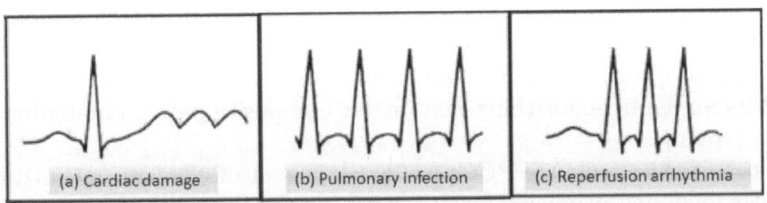

Fig. 1. Different patterns of ECG

The patterns in Fig.1 have the following **characteristics**. All patterns have very complicated appearance. They can be very long and appear in all kinds of shapes. If we consider the entire data segment as a whole and try to accomplish pattern matching task, the complexity will be extraordinarily high and is difficult to be finished online.

Based on the example above, we find these new **challenges** for composite pattern matching:

- **Complex Data Streams:** the arriving data segments vary rapidly and result in complicated forms. This brings troubles to recognize patterns in data stream.
- **Complicated Pattern Matching Tasks:** the composite patterns to match can be very complicated as introduced in the previous example. If we try to match the entire pattern, the complexity could be too very high to be finished online and the accuracy could be low.
- **Multiple Concurrent Tasks:** Take the case in ICU for example, multiple patterns representing different diseases symptoms have to be monitored simultaneously. In a traditional ICU, about 20 to 30 types of symptoms are to be monitored at the same time upon a single patient from nearly 10 medical instruments. The tasks increase proportionally with the number of patients, which call for low-cost matching method.

- **Requirement for On-line Processing:** since data is coming in the form of stream, the result of analysis requires online processing. This is mainly because of two reasons. First, the result may be time-sensitive, an out-of-date diagnose may do tremendous harm to a patient. Second, if the data cannot be handled efficiently, the old data will stay unprocessed and consume the systems memory, which will further affect the later matching task and get into a vicious circle.

In this paper, we provide the following **contributions** to address the challenges:

- A precise description for composite pattern and a formal definition for matching tasks are introduced to lay a solid foundation for composite pattern matching. Firstly, we introduced *basic pattern* to describe the similar *subpatterns* among different data segments. Secondly, we introduced *operators* to describe the composite semantics between basic patterns. Then the *composite pattern* can be expressed as basic pattern with composite semantics. Besides, a precise description for the composite pattern may help make the matching task clearer as well as avoid ambiguity.
- We then propose a novel data structure *Pattern Combination Graph (PCG)* and its matching algorithms that enable composite pattern matching on live data streams.
- On top of this structure PCG, we develop a set of efficient optimizing techniques involving primitive block sharing method, internal block sharing and self-adjust join order which help improve the performance of the algorithm.
- We experimentally demonstrate that the new data structures and shared processing techniques facilitate the pattern matching task, and the scalability issues can be tackled effectively. The performance results of our extensive evaluation show the competitive performance of PCG against other relative approaches.

The rest of the paper is organized as follows: Section 2 analysis the problem and give the framework of the solution. Section 3 discusses the formal definition of composite pattern and examples. Section 4 presents the data structure (PCG) for composite pattern matching and Section 5 describes the shared processing methods using PCG. Section 6 discusses the experimental results for performance. Section 7 discuss the related work and finally Section 8 concludes the paper.

2   Problem Analysis and Framework

2.1   Problem Analysis

To accomplish composite matching, we need to review the example in Fig.1 again and find ways to settle the difficulties. Although all ECG patterns have very complicated appearance, there are still some similarities among them. Take a brief look at the wave peaks in Fig.1 (a) (b) (c), the peaks remain almost the same in different composite patterns. Then take another look at the period of data just before the wave peaks in Fig.1 (a) (c), they also look very much like each other. On the other hand, similarities do not only occur in different patterns, they can also exist in one single pattern. One example is the three peaks with smaller amplitude in Fig.1 (a), they have exactly the same appearance.

Actually, in this example, the similarities among different patterns have reliable reasons. Each movement of a mans heart will be reflected as a segment of data in his ECG. For example, the depolarization action of a heart will be reflected as the subpattern just before the wave peak in Fig.1 (a) (also the similar part in other two composite patterns). We regard these similar patterns as basic patterns (a detailed definition of basic pattern will be given in 3.1), and each of the basic patterns is given a name. The patterns with label of each basic pattern are given as follows in Fig.2, and the rightmost pattern is a healthy mans ECG to be compared with.

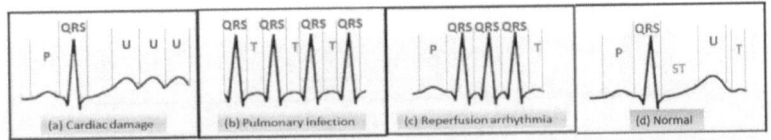

Fig. 2. Composite patterns when each basic pattern is labeled

After each basic pattern is labeled, the composite pattern can be expressed as basic patterns with composite semantics (a detailed definition of composite semantics will be given in 3.2). For example, Fig.2 (b) turns into basic pattern Q-R-S alternate with basic pattern T, and the normal cycle of ECG showed in Fig.2 (d) is a sequence of basic patterns P, Q-R-S, ST, U, T.

The above analysis brings us some exciting ideas to settle the problem. We could firstly find ways to match the basic patterns, and then handle the *composite semantics* between the basic patterns.

2.2 Framework of Composite Pattern Matching

The framework of PCG for composite pattern matching is described clearly in Fig.3. Firstly, data stream is transferred visually in a frame for *basic pattern generator* to define *basic pattern*. Then, *basic pattern dictionary* is generated to avoid overlap and similarity among different basic patterns. By *pattern composite pattern generator*, *target pattern set* is generated, which is to be on matching continuously. *PCG builder* generates the data structure PCG according to the *target pattern set*. In the end, the final composite pattern matching result is generated and show back to user.

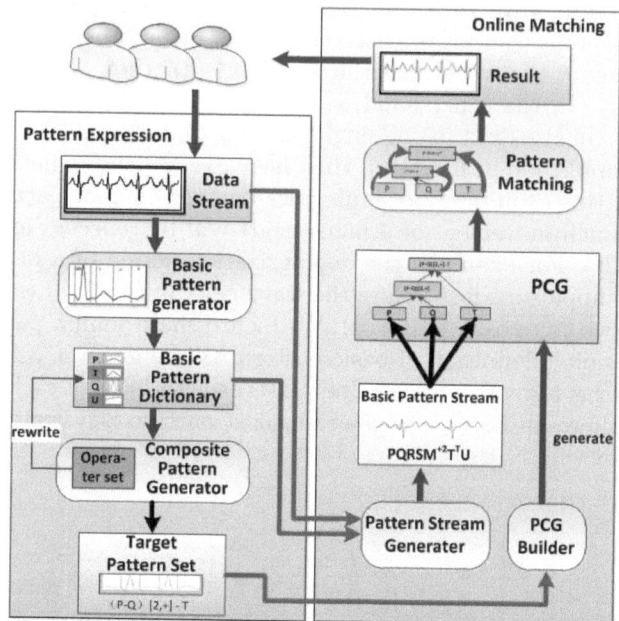

Fig. 3. Framework of PCG for composite pattern matching

3   Definition of Composite Pattern

3.1   Basic Pattern

Definition 1 (Data Stream). A data stream DS is a potentially unbounded sequence of data. Each element in DS is a 2-tuple, (X_i, T_i). T_i is the timestamp and X_i represents the value at time T_i. $DS = \{(X_1,t_1),(X_2,t_2),...,(X_n,t_n)\}$.

DS can be get from the outside world (e.g. from a sensor) that provides detailed information

Definition 2 (Basic Pattern). A basic pattern (bp) is a segment of data points: $bp_t = \{(X_i,t_i),(X_{i+1},t_{i+1}),...,(X_{i+k},t_{i+k})\}$. For any two basic patterns, $bp_t = \{(X_{t1},t_{t1}),(X_{t2},t_{t2}),...,(X_{tk},t_{tk})\}$, $bp_h = \{(X_{h1},t_{h1}),(X_{h2},t_{h2}),...,(X_{hk'},t_{hk'})\}$,Basic patterns have to follow two rules to avoid ambiguity in semantics:

Rule 1 Non-overlaps: $t_{tk} < t_{h1}$ or $t_{hk'} < t_{t1}$

Rule 2 Logical distinguishable:$\nexists t_{h1} \leqslant t_{hp} \leqslant t_{hk'} - t_{tk} - t_{t1} + 1\ subp = X_{hp},t_{hp},...,X_{hp+kh'-1},t_{hp+kh'-1}, s.t.similarity(subp,bp_i) < \varepsilon$, while ε is the threshold to evaluate two base patterns.

Rule 1 and rule 2 are illustrated in Fig.4. Fig.4(a) shows the overlaps between patterni and patternj and it is not allowed. In Fig 4(b1), there are two patterns which cannot be assigned as basic patterns because they share a similar segment (labeled as QRS). In Fig.4 (b2), the two cannot be two basic patterns as well,

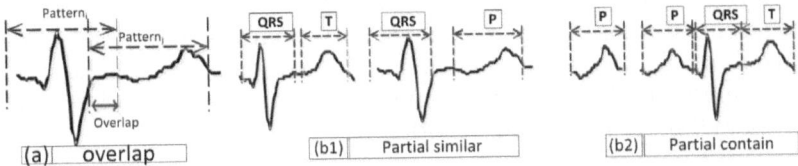

Fig. 4. Basic Pattern Rules

because the left pattern (P) is fully contained in the rightmost pattern. The rules on pattern help avoid ambiguity in pattern matching

3.2   Composite Pattern

Basic pattern is generated from data streams, composite pattern can be expressed based on basic patterns with following considerations. Various combination ways are found between basic patterns, as shown in Fig.1(c) of electrocardiogram, continuous repetition of QRS-wave for more than 3 times represents reperfusion arrhythmia. To support the composite semantics between basic patterns, operators are introduced to express inner-pattern variations and intern-pattern relationships.

Definition 3 (Operators). operators are used to express the relationship between basic patterns. A detailed corresponding list between the symbol and its meaning is listed below.

Symbols	Description
DS	**Data Stream**
Bp	**Basic pattern**: examples of basic pattern, usually expressed by a capital letter, such as X or Y.
Op	**Operator**: show the relationship between patterns
X, Y	**Sequence pattern**: operator finds examples of pattern Y follows pattern X within a specified time range
X & Y	**Conjunction pattern**: both pattern X and Y occur within a specified time range with any order
X | Y	**Disjunction pattern**: either pattern X or pattern Y occurs within a specified time range.
X - Y	**Adjacent pattern**: pattern Y occurs just after pattern X with nothing between them.
X[m,n]	**Loss and Gain Constraints**: [m,n] means pattern X occurs no less than m times and no more than n times. When [change into (, it means the number of X must be more than m and less than n strictly. When there no upper limit, n turn into +.
(X op Y)	**Hierarchical Symbol**: ( )make combined pattern as a group, see Definition 4 for details.

Definition 4 (Hierarchical Semantics). Hierarchical semantics is expressed by ( ), and the basic pattern as well as composite pattern within the parentheses is treated as a basic element for matching. This semantic increase the expressiveness of composite pattern because complicated operators can work on composite pattern as well by hierarchical semantics.

Example: (P-Q-R-T)[3,5],(P-Q-T)[2,+]. Meaning: P, Q, R, T pattern occurs in a row with no extra pattern between them , and repeats 3 to 5 times, then followed by P-Q-R pattern for more than 2 times.

Definition 5 (Composite Pattern). A Composite pattern(cp) consists of basic patterns and operators between basic patterns: cp=(Bp)(Op(Bp))* Usually composite pattern can be very complicated and meaningful. Examples of composite patterns in ECG of Fig.1 is listed as follows (see Fig .5).

Cp Example: Fig1(A)	Cp Example: Fig1(B)	Cp Example: Fig1(C)
cardiac damage	Pulmonary infection	reperfusion arrhythmia
P, (Q-R-S), T, U[3,5]	**((Q-R-S)-T)[3, +]**	**P, (Q-R-S)[3,+], T**

Fig. 5. Examples of composite patterns

4 Composite Pattern Matching with PCG

Composite patterns are highly valuable and difficult to match. This paper presents an online composite pattern matching method over data streams: pattern combination graph (PCG), which enables efficient matching on streams. To support complex forms of target pattern, PCG identify invariant segment as basic pattern. Pattern matching operators are defined over basic patterns. Target pattern is generated based on basic patterns and composite semantics. To accomplish the pattern matching efficiently, PCG firstly recognize data segments as basic patterns and then handle the relationship between them. Later matching procedures are operated according to the types of nodes in PCG.

4.1 Composite Pattern Matching

PCG is created according to the target pattern. And the formal definition of PCG is listed as follows.

Definition 6 (PCG) An PCG is a 5-tuple X= (Q, A, T, F, B), where $Q = q_0, q_1, ..., q_n$ is the set of primitive blocks, responsible for basic pattern matching, $A = a_0, a_1, ..., a_m$ is the set of arcs, $T = T_0, ... T_n$ is the internal blocks, and they handle the composite semantics between patterns, F is the set of root blocks corresponding to each target pattern, B is the sets of buffer for each node.

Each arc a_i is labeled with a pair of transition relationship between primitive blocks and internal blocks. Every root nodes is corresponding to a composite pattern and output results. Blocks are connected by arcs.

For pattern (b) shows the ECG of a patient who get serious pulmonary infection, ECG of this disease is composed of alternative QRS-wave and T-wave. The pattern in (c) represents reperfusion arrhythmia, ECG of which consists of P-wave and QRS-wave(for more than 3 times in a row) and T-wave. The PCG is generated separately as follows.(See Fig .6)

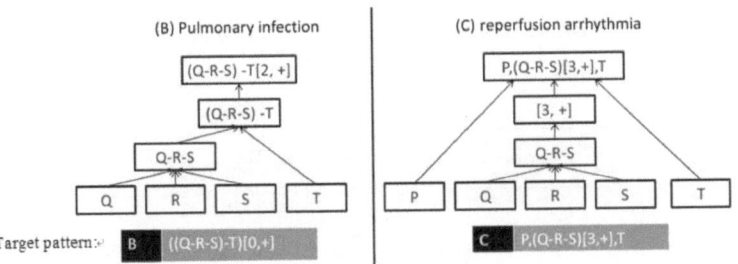

Fig. 6. Examples of composite patterns

4.2   Matching Algorithms

The procedure of processing PCG for pattern matching can be concluded into several steps. First, data streams are recognized as basic patterns which are generated from the selected segments on streams, as the complexity of basic patterns is relatively low and all basic patterns are logically distinguishable. Second, basic patterns flow into PCGs primitive block and send signals to upper internal blocks. If the arriving pattern satisfies the constrains on the block, successful matching signals are sent upward. Internal blocks collect patterns from primitive block if and only if all primitive block deliver successful signals. Finally, composite patterns are generated in root blocks and output results are output. We will explain each of the procedure in detail.

Recognize *basic patterns*: as data stream arrives, the coming sequence is compared with all the basic patterns simultaneously for pattern matching. This procedure ends in the following three cases: 1) when comparing with one pattern, for example pattern P, the accumulated error exceeds the given error bound, then the comparing end. See Fig.7. 2) when ending up comparing with some pattern and the accumulated error is still below the given error bound, such as pattern Q in Fig.7, then the coming sequence is recognized as basic pattern Q. The comparing for this segment of data ends, and the comparing with next segment begins in the same way. 3) when all basic patterns in basic pattern dictionary cannot match the coming sequence (all accumulated error exceeds error bound), the data sequence is not recognized as any pattern and is maybe noise data.

Constrains on different kinds of blocks differ from each other according to the operator and have different algorithms to evaluate. We choose two of the operator algorithms (Alg.1, Alg.2) and Fig.8 give an illustration of matching. The primitive P block gets a P pattern from data stream and sent a successful matching signal upward, then Q,R,S primitive block all get their corresponding

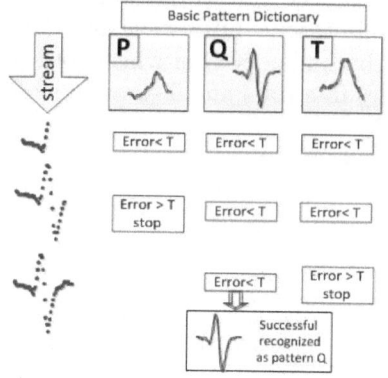

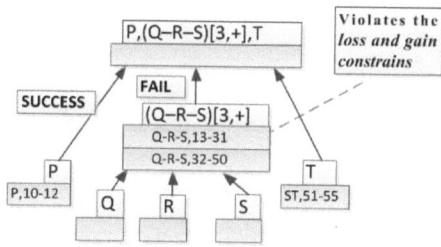

Fig. 7. Recognize Basic Patterns   **Fig. 8.** Pattern Matching in internal nodes

patterns from data stream and combined into their internal nodes buffer. However, when a rightmost primitive block get its pattern (T pattern in Fig.8), it happened that the number of results in the internal nodes ((Q-R-S)[3,+]) just get two patterns, and do not satisfy the nodes operator, which call for more than three patterns in a row. So it sends up a fail signal, and all the buffer need to clean itself for the next match.

5   Optimizing Method

Inefficiencies occur due to duplicated data structures and separate processing of conventional composite patterns when supporting large numbers of queries at the same time.

Algorithm 1: Processing Adjacent pattern in PCG
Input: PCG; Stream   **Output:** Matching Result:
1. Set *pBlock* = select leftmost primitive block
2. *pBasepattern* ← get next base pattern in data stream
3. if *pBasepattern* matches *pBlock*
4.   set pBlock.ismatch = true;
5.   Add *pBasepattern* in *pBlock*'s buffer:
6.   *pBlock* ← get next Block in PCG
7.   *pSubpattern* ← next subpattern in data stream;
8.   if *pBlock* is rightmost childblock
9.   Combine childblocks buffer and insert into
fatherblock Buffer;
10.   Clear childblocks buffer; **goto 1**;
11. Else
12.   Set pBlock.ismatch=false;
13.   For every *buffer* of childblocks;
14.   *buffer*.clear();
15. Goto 2

Algorithm 2: Processing Loss and Gain in PCG
Input: PCG; Stream   **Output:** Matching Result:
1. For every new arriving *pattern* in LGObuffer
2. Update *Bcount*
3. if *Bcount* satisfy *Bconstraint*
4.   set *LGOblock*.ismatch =true;
5.   set pBlock = next block in PCG
6. else
7.   if Bcount < Bconstraint
8.   Set pBlock =LGOblock; get Next pattern;
9.   If Bount > Bconstraint
10.   LGOblock.ismatch = false;
11.   LGObuffer.clear;

5.1   Primitive Block Sharing

Our primitive block sharing approach is based on the idea that a primitive block, specified commonly in multiple queries, can be shared for efficient processing and storage. Moreover, all incoming instances of a given event class, regardless of their sources, can be stored and handled together within a shared storage. Based on the composition patterns of all registered matching tasks, these shared queues can form a single shared graph (PCG) in which processing and storage for each primitive block class are inherently shared by all relevant queries. In Fig.9, primitive block T is shared by both (B) Pulmonary infection and (C) reperfusion arrhythmia.

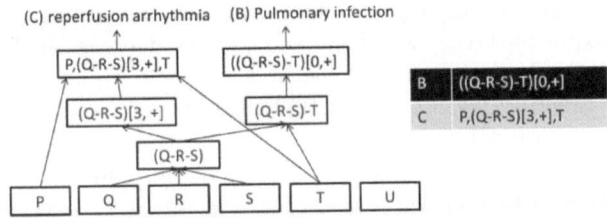

Fig. 9. Examples of composite patterns

5.2   Hierarchical Pattern Sharing

In the PCG discussed so far, individual queries are handled separately during the processing phase. Thus, when different queries share a partial pattern, this pattern is tested multiple times (See Fig.6 for the example of hierarchical pattern Sharing for Fig.1(B) and Fig.1(C), ADJ(Q-R-S) is the sharing part), in our two example queries; not only primitive block are joined together but also the internal blocks. Internal blocks sharing bring great enhancement to improve efficiency, the intermediate result computed by one pattern can be shared to all the others, thus reduplicate calculation for common internal blocks can be avoided.

6   Experiments

The problem solved in this passage is based on real data set in intensive care unit (ICU). The data set consists of three parts, ECG, central venous pressure (CVP), respiratory capacity(RESP). Patterns are more complicated in ECG and RESP, and simple in CVP. The amount of data in each set is up to 500,000.

(1)Memory and Time cost with different optimizing methods on ECG data set. Pattern length indicates the number of basic patterns in all composite patterns.(See Fig.10)

The optimized PCG in Fig.10 means both block sharing and self-adjust join order methods are used to optimize the primitive PCG. We can see the optimized PCG method is superior to all the other 3 methods in both memory cost and

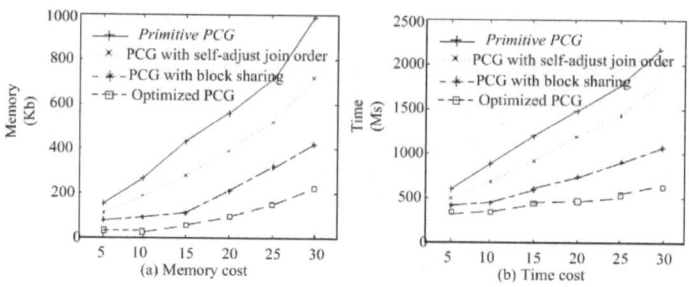

Fig. 10. Efficiency with different optimizing methods

time cost. There are a lot of common internal block and primitive blocks among different composite patterns, which make the cost decrease in both optimized PCG and block sharing method. The method with self-adjust join order has also a lower cost in memory for the reason that unnecessary storage can be reduced when there is no appropriate patterns.

(2) R_{recall} and $R_{precision}$ in different sets of data among PCG and other methods can be seen in Fig.11.

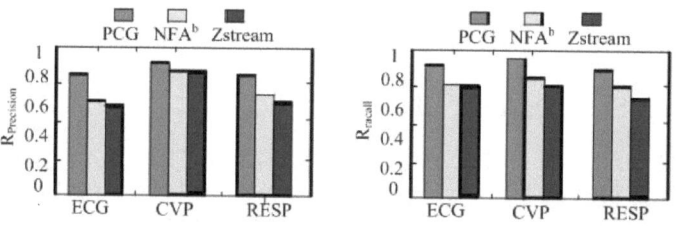

Fig. 11. $R_{precision}$ and R_{recall} between PCG and other method

R_{recall} measures whether the result meet our expectation, N_{real} represent the actual composite patterns that can be get by PCG, and N_{due} represent the number of patterns labeled by experts manually. $R_{recall} = N_{real}/N_{due}$. From Fig.11, we can see that PCG perform much better in ECG and RESP which are more complicated, and the advantages are not so obvious in CVP. The differences come up when recognizing basic patterns from stream. The error bounds used are 0.07(in ECG), 0.02(in CVP) and 0.55(in RESP). The method used in NFAb and ZStream neglect the inter relationship between different basic patterns that they may have overlaps or similarities, which will further result in ambiguity in pattern matching, and the R_{recall} and $R_{precision}$ are affected. $R_{precision}$ measures the correctness of the result. It is the ratio between relevant results and the total number of query results. $R_{precision} = N_{real} / N_{out}$. N_{out} is the number of query results and N_{real} is relevant results.

(3) Performance compared with other methods The performance issue can be classified into two classes, time cost and memory cost. Time cost takes the average response time as an indicator with growing pattern length. As pattern

complexity increases, ZStream spend more increasing time on adjacent pattern calculating while the branches in NFAb grows quickly as well. On the other hand, ZStreams pair comparing results in extra cost in memory management and do not help clean meaningless segment, and singletons in NFAb cost extra memory, shown in Fig.12.

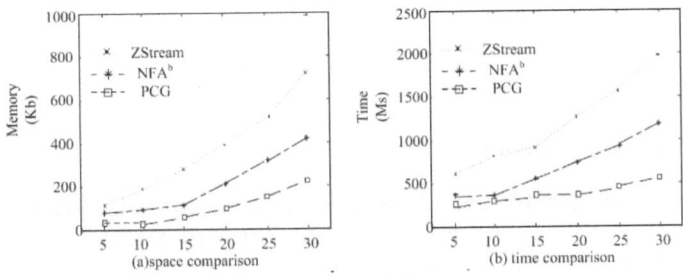

Fig. 12. Performance comparison

7 Related Work

In this section, we discuss the relevant achievements in detail and divided them into the following categories.

7.1 NFA-Based Methods

Most pattern matching algorithms exists in composite (or Complex) event processing (CEP) systems. CEP systems search sequences of incoming events for occurrences of user-specified event patterns[2]. All these method can be divided into two classes as follows. The first type of algorithms is NFA based method, non-deterministic finite automata (NFA) are the most commonly used method for evaluating CEP queries [1,3,4,5,11]. An NFA represents a query pattern as a series of states that must be detected (See Fig.13). A pattern is said to be matched when the NFA transitions into a final state. However, the previously proposed NFA-based approaches have two limitations that we explain in this work: (1) Fixed order of evaluation: NFAs naturally express patterns as a series of state transition. Hence, current NFA-based approaches impose a fixed evaluation order determined by this state transition diagram. If the selectivity of different patterns differs from each other obviously, the fixed order may be very inefficient (2) Negation inefficient: Negation means the items that haven't occurred or processed. If there is a predicate involving B and C (e.g., as in the pattern A followed by C such that there is no interleaving B with B.width ¿ C.width), there is no simple way to evaluate the predicate when a B event arrives, since it requires access to C events that have yet to arrive. Hence, it has difficulty to decide a B event transition. As a result, this NFA cannot be used for negation queries with predicates. For this reason, existing NFA-systems perform negation inefficiently.

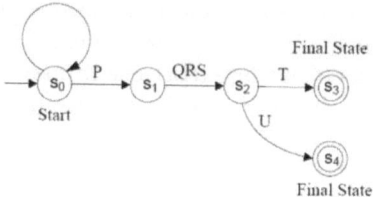

Fig. 13. NFA example

7.2   Tree Based Methods

Mei made a batch-iterator model, using tree-based query Program that models the logical form of the query process[2], but ZStream Method itself is designed for the event stream, the processing unit is a single event. This event cannot change inside within the further processing. The application of the data stream cannot support band inversion, amplitude changes, time span increases and other changes in the internal characteristics. In addition, ZStreams buffer management strategy is highly time-consuming, frequent comparisons are needed to complete certain types of composite pattern, such as adjacent patterns. As the system resources are limited, the data which exceeds the limit threshold will be discarded, which makes matching results unsatisfied.

7.3   Other Methods

The other main kind of matching algorithms is included in DSMS systems to handle queries on live streams with respect to traditional DBMS. DSMSs are used for data processing in a broad range of applications including fraud detection, clickstream analysis, and health services., More famous systems among these are the STREAM system at Stanford University, MIT's Aurora System, Berkeley's TelegraphCQ systems, these systems can be carried out in the online environment and perform the similar functions as ad-hoc queries. However, all systems concerned above only focus on the matching of separate tuples and convert the live matching into a traditional query. In this way, the composite meaning of several data points together which we call pattern is ignored, and the matching between patterns is therefore not able to carry on in these systems.

8   Conclusion

Data streams are highly valuable in many areas and they support abundant semantics. Matching these segments need real-time processing. This paper presents a precise description for composite pattern and a formal definition for matching tasks. To support complex forms of composite pattern, we propose a novel data structure *Pattern Combination Graph (PCG)* and matching algorithm. On top of this structure PCG, we develop a set of efficient optimizing techniques to facilitate online matching. From the perspective of recall ratio, precision ratio and processing efficiency, the experimental results on real datasets of medical streams show that PCG is feasible and effective.

References

1. Jagrati, A., Yanlei, D., Daniel, G., Neil, I.: Efficient pattern matching over event streams. In: Proceedings of the 2008 ACM SIGMOD International Conference on Management of Data, Vancouver Canada, pp. 147–160. ACM, NY (2008)
2. Yuan, M., Samuel, M.: Z Stream a cost-based query processor for adaptively detecting composite events. In: Proceedings of the 35th SIGMOD International Conference on Management of Data, Providence, USA, pp. 193–206. ACM, NY (2009)
3. Cadonna, B., Gamper, J., Böhlen, M.H.: Sequenced Event Set Pattern Matching. In: EDBT (2011)
4. Chandramouli, B., Goldstein, J., Maier, D.: High Performance Dynamic Pattern Matching over Disordered Streams. In: VLDB (2010)
5. Lee, S., Lee, Y., Kim, B., Selçuk Candan, K.: High-Performance Composite Event Monitoring System Supporting Large Numbers of Queries and Sources. In: DEBS (2011)
6. Dindar, N., Fischer, P.M., Soner, M., Tatbul, N.: Efficiently Correlating Complex Events over Live and Archived Data Streams. In: DEBS (2011)
7. Zhang, H., Diao, Y., Immerman, N.: Recognizing Patterns in Streams with Imprecise Timestamps. In: VLDB (2010)
8. Gao, L., Wang, X.S.: Continuous Similarity- Based Queries on Streaming Time Series. IEEE Trans. Knowl. Data Eng. 17(10) (2005)
9. Brenna, L., et al.: Distributed Event Stream Processing with Non-deterministic Finite Automata. In: DEBS (2009)
10. Wu, H., Salzberg, B., Gregory, Jiang, S.B., Shirato, H., Kaeli, D.: Subsequence Matching on Structured Time Series Data. SIGMOD (2005)
11. Wu, E., Diao, Y., Rizvi: High-performance complex event processing over stream. SIGMOD (2006)
12. Li, F., Li, H., Qu, Q., Miao, G.: SPQ: A Scalable Pattern Query Method over Data Streams. Chinese Journal of Computers 33(8), 1481–1491 (2010)
13. Chandramouli, B., Goldstein, J., Maier, D.: High Performance Dynamic Pattern Matching over Disordered Streams. In: VLDB (2010)
14. Lee, S., Lee, Y., Kim, B., Selçuk Candan, K.: High-Performance Composite Event Monitoring System Supporting Large Numbers of Queries and Sources. In: DEBS (2011)

Visual Fingerprinting: A New Visual Mining Approach for Large-Scale Spatio-temporal Evolving Data

Jiansu Pu[1], Siyuan Liu[2], Huamin Qu[1], and Lionel Ni[1]

[1] Department of Computer Science and Technology,
The Hong Kong University of Science and Technology
{jspu,huamin,ni}@cse.ust.hk
[2] iLab, Heinz College, Carnegie Mellon University
syliu@andrew.cmu.edu

Abstract. Spatio-temporal data analysis has important applications in transportation management, urban planning and other fields. However, spatio-temporal data are often highly dimensional, overly large and contain spatial and temporal attributes, which pose special challenges for analysts. In this paper, we propose a new visual aided mining approach, Visual Fingerprinting (VF) for extremely large-scale spatio-temporal feature extraction and analysis. It adopts a visual analytics approach for spatio-temporal data analysis that can generate fingerprints for temporal exploration while preserving the spatial distribution in a region or on a road. Fingerprinting has been proposed to display temporal changes in spatial distributions; for example fingerprints for a region grid can well display temporal changes in the traffic situations of significant spots of a city. These fingerprints integrate important statistical and historical information related to traffic and can be conveniently embedded into urban maps. The sophisticated design of the visualization can better reveal frequent or periodic patterns for temporal attributes. We have tested our approach with real-life vehicle data collected from thousands of taxis and some interesting findings about traffic patterns have been obtained. The experiments validate our methods and demonstrate that our approach can be used for analyzing vehicle trajectories on road networks.

Keywords: Visual analysis, Fingerprinting, Spatiotemporal data.

1 Introduction

Nowadays large-scale movement data are becoming available with the prevalence of mobile devices. Mining or analyzing movement data is important in many different applications such as transportation management, mobility studies, route suggestion, and mobile communication management. There are many existing data mining work on spatial temporal data exploration. However, pure data mining for movement related analysis faces is currently facing new technical challenges as movement data contains both spatial and temporal attributes

S. Zhou, S. Zhang, and G. Karypis (Eds.): ADMA 2012, LNAI 7713, pp. 502–515, 2012.
© Springer-Verlag Berlin Heidelberg 2012

which are often huge in size and high in dimensionality. Firstly places considered 'distinct' may be too numerous. Secondly, time is a complex phenomenon, which is periodical by nature and has hierarchical structures. Therefore a human analyst's sense of space and place is required, which is hard for a machine to achieve. So the demand for a new analysis method for things such as visual analysis is very intense.

Visual displays of movement data shows great potential as they can intuitively present multidimensional spatial-temporal movement data and provide rich interactions, allowing users to explore the data and improve mining processes and results. It can effectively keep humans in the analysis loop to utilize their sense of space and place, their tacit knowledge of inherent properties and relationships, and space-related experiences. To tackle the challenges and assist in the understanding of movement data to improve urban planning in the possible future we have developed a new visual aided mining approach, Visual Fingerprinting (VF) for extreme large-scale spatio-temporal features extraction as illustrated in Fig.1. Our proposed visual mining approach has three advantages: 1) Intuitive anomaly detection; 2) Dynamic and fast exploration for high dimension data analysis; 3) Rich interaction with experts. It can provide more statistical information and transform historical data from numerical knowledge into visual cues like shape, color, size and so on. The approach is designed to keep humans in the analysis loop and take advantage of their analytical abilities, so users can easily analyze any temporal changes in traffic situations over a region segment, or to analyze any changes in the spatial distribution (inside the road network) over time. Users can perceive the correlations among different attributes and filter out noise and irrelevant trajectories for further investigation of interesting cases. Analysts can interactively and progressively refine the settings to improve the results.

We use two case studies to evaluate our method and demonstrate our fingerprint design on real-world taxi GPS data sets of 15,000 taxis running for 92 days

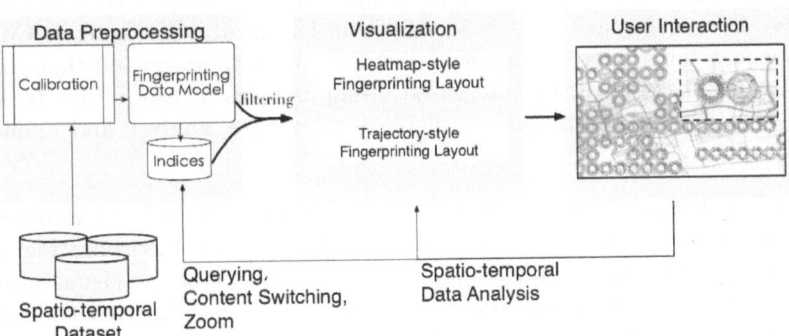

Fig. 1. The framework architecture of our approach. The VF architecture consists of three primary components: (1) a data preprocessing module, (2) a visualization rendering module, and (3) a user interaction module.

in a Chinese city with a population of over 10 million. Experiments show that our approach is capable of effectively finding regular patterns and anomalies in traffic flows. In summary, we have made the following contributions: (1) We have developed a new visually aided mining approach for large scale spatio-temporal data, which can provide temporally related attribute exploration while keeping the spatial distribution layout. It also can be applied to density-based methods or used to evaluate the density/cluster quality. (2) A sophisticated spiral visualization is designed to explore both frequent and periodic patterns in temporally related attribute analysis. (3) This approach guarantees the combination of dynamic dimensions and has good scalability. It can also well explore the temporal evolution in an extreme large-scale spatial level. (4) We utilize large-scale real life data to evaluate our method and provide two case studies with interesting findings.

2   Related Work

Modeling and Querying. Compieta et al. [7] presented a comprehensive study of existing exploratory techniques for spatial temporal data. Alvares et al. [1] proposed a reverse engineering framework for mining and modeling semantic trajectory patterns in a geographic database. Spaccapietra et al. [19] proposed two trajectory-modeling approaches. One was based on a design pattern while the other was based on dedicated data types. They illustrated their differences in the implementation of an extended-relational system. Giannottie et al. [9] extended the sequential pattern-mining paradigm to analyze the trajectories of moving objects. They defined trajectory patterns as frequent behaviors in space and time, and discussed several approaches to mining trajectory patterns. Palma et al. [17] proposed a clustering-based approach to find some interesting but unexpected places in trajectories. Nanni et al. [6] proposed an adaptation of a density-based clustering algorithm for the clustering of trajectories of moving objects. Pelekis et al. [18] classified the similarities of trajectory patterns into two types: similarity in spatial temporal and temporal similarity. Vlachos et al. [20] described a similarity measure algorithm based on LCS (Longest Common Subsequence). This approach allowed stretching in both space and time. In comparison with these work, we integrated visual analysis methods with trajectory density visualization techniques to help users explore, analyze and understand spatio-temporal data.

Visual Analytics. Visual analytics can help users gain insights into massive, heterogeneous, and dynamic volumes of information by incorporating human judgment into the analytical reasoning process with interactive visual interfaces. Crnovrsanin et al. [8] introduced a proximity-based visualization technique to discover human behavior patterns from movement data. Andrienko et al. [5] summarized the approaches in visualizing movement data. Characteristics of movement data and methods to present dynamics, movement, and change are discussed. In [3] they also surveyed existing approaches to the aggregation of movement data and the visual exploration of the aggregates. These authors also defined aggregation methods suitable for movement data and proposed inter-action techniques to represent results of aggregations, enabling comprehensive exploration of the data in [2]. GeoTime [12] displays the 2D path in a 3D space to provide a detailed view of the geographical and temporal changes in move-ment data. Willems et al. [21] visualized vessel movement as well as the vessel density along traces by convolving trajectories with a kernel moving with the speed of the vessel along the path. Guo et al. [10] presented a trajectory visual-ization tool that focuses on visualizing traffic behavior at one road intersection. Microsoft T-drive [22] makes recommendations of the fastest paths taken by taxi drivers. To select a few interesting trajectories from a large number, Hurter et al. [11] proposed a brush-pick-drop interaction scheme to visualize aircraft trajecto-ries. Their methods are focused on 2D trajectory data exploration and provide limited perspectives. We tried to provide more comprehensive perspectives for spatio-temporal data exploration in our approach. In our work, we not only visu-alize large-scale spatial temporal movement data, but also embed traffic analysis results into digital maps. We apply a multidisciplinary approach to develop a framework for the analysis of massive movement data taking advantage of the synergy of a computational, database, and visual techniques. We introduce our method and demonstrate its effectiveness by examples.

3   Methodological Preliminaries

3.1   Dataset

The data used in this study is the taxi trajectory data collected from GPSs in Shanghai, China during a non-continuous eight-month period, totaling 92 days. Each GPS record contains car ID, the latitude and longitude of the taxi, the date, the time of the day in seconds, the taxi's status (loaded/vacant) and the speed and the direction of the taxi. In this work, we adopted a Weighting-based map matching algorithm and an Interpolation algorithm to calibrate the erroneous and low-sampling-rate vehicle GPS trajectory data set. The details are available in our technical report [16] and previous works [13] and [14]. The statistical information for each road segment is computed as preprocessing.

3.2   Visual Fingerprinting

Visual Fingerprinting Definition. We propose a novel visual fingerprint-ing method to discover essential characteristics or "fingerprints", by visually exploiting the multidimensional features through spatial temporal means. Our fingerprinting method has the following benefits: (a) it leads to spatial temporal data feature extraction; (b) it provides a novel visualization to answer queries by flexible and dynamic attributes combination; and (c) it is fast, scalable and easy to compare. The sophisticated visual form is designed to extract spatial tempo-ral features with multi-dimensions and it can well explore the temporal related patterns including periodical and frequent patterns. The "fingerprint" concept used here is to extract the properties of good features by visually/visual analytics to compare trajectories. Good "fingerprints" also can help with anomaly detection and answering similarities queries.

Visual Fingerprinting Approach. Our VF approach has two types of view selection to be displayed on the screen including trajectory-style display and heatmap-style display respectively to study microscopic patterns and discover abnormal behaviors.

(1)Trajectory-style Fingerprinting. We show the trajectory information by connecting the origin and destination locations of the respective geographical places by drawing each sample point in the trajectory with B-Spline curves. The length of the curves encodes the distance of the trip. The color of the curves represents the number of visits to the destination location. The trajectories are colored according to the average speed of each trip from source to destination. The subparts of trajectories are merged together according to the number of visits shared by each trip.

(2)Heatmap-style Fingerprinting. We create heatmap-style fingerprint based on an index matrix that measures the amount of taxi sample points on the road or the average speed of the taxis, even the rank of the number of customers getting in/out of taxis at that location. The higher the numeric values and closer together the sample points, the hotter the map and the higher the density displayed by the fingerprinting.

(3)Design of Fingerprints. In order to provide information with different levels of detail, we design a visual form of a fingerprint that adopts the ringmap-based layout design to display the 24-hour overview distribution for different regions of the city over a selected time period. The circle encodes a daily or monthly distribution which starts from the twelve o'clock position and the clockwise numbers represent increases in time. The time is shown on a circular axis with each major section representing a day, with 24 cells encoding 24 hours. Our fingerprint design can be separated into different ring sectors to display different days' historical information.

4   Visual Analytics Procedure

In this section we present our visual analysis procedure and describe the details of our method. Our approach is designed to help transform the spatial temporal and highly dimensional trajectory data into more intuitive visual cues including size, shape, and color in order to take the full power of human analysts. Hence our visual approach presents the statistics information of multiple regions and provides visual cues to the user about areas possible patterns exist. The whole method can show the spatial temporal changes by presenting the instant values and historical data showing the evolution over a long time period. Fig.3 shows the flowchart of our framework.

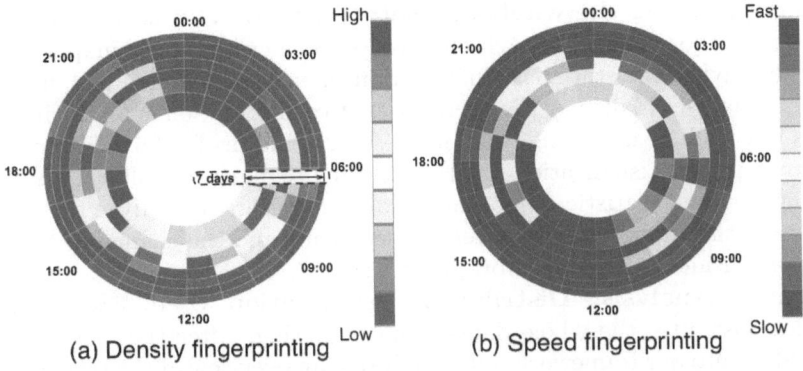

(a) Density fingerprinting   (b) Speed fingerprinting

Fig. 2. The visual fingerprinting design. It uses a ring-map-based radial layout design to help explore historical statistical information. It can use different color encoding scheme to represent density or speed. And the ring sectors around the fingerprint show the value of different time periods of the day. **Ring** Each ring on the fingerprint represents one day. **Sectors** Each sector inside a ring represents one hour, with increases in time by the hour as the angle increases, and time increasing by days as the radius increases. The time of a region's behavior is displayed on each fingerprint's ring circle like a clock to encode a 24-hour distribution. **Color** The color of the bar chart is an intuitive design based on color harmony rule. For e.g. we design the speed scale from green (high) to red (low), which is based on the color coding of traffic light signaling. **Size** The fingerprint's size shows the total number of taxis passing through a selected area. For each fingerprint, we adopt a distortion method to allocate more display space to the inner sectors for better presentation.

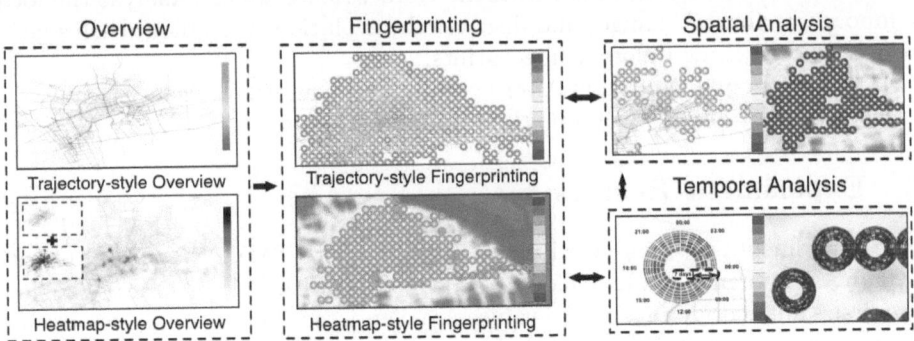

Fig. 3. The analytics workflow. The analytics uses preprocessed vehicle GPS data as input and has three major steps: fingerprinting the whole dataset for overview, spatial analysis, and temporal analysis. The initial fingerprinting displays the whole spatial temporal distribution and then users can look at the ring-map-based radial layout design to discover historical data. There are two different types of display including trajectory style and heatmap style. Users can further analyze interesting locations' temporal features by zooming in to explore visualized fingerprints. After completion, users can query or recheck the raw data to inspect their findings.

(1)Fingerprinting the whole dataset: Overview The most natural way to represent the correlation for each region is to use a geographical map with visual items displaying the statistical information of each region. We first fingerprint the whole dataset to display the overall statistics information of all the important places in the city, suggesting some interesting locations for further exploration. The fingerprints' visualization displays selected segments by filtering both their geographical and statistical information allowing users to analyze the temporal changes to the traffic situation over the segments, and the change to the spatial situation (inside the road network) over time.

(2)Spatial Analysis: Distribution Exploration Users first start from overview just after the whole dataset fingerprinting. After the movement data are loaded onto our framework to be analyzed, an overview of the traffic flow and the statistical information are displayed. Users are free to explore any interesting areas and check any generated fingerprints for details. After that, users can select some interesting segments for further investigation. The aggregation of taxi spatial, temporal, or multidimensional information can be used to compute the hotness of each region and then a map about the features' spatial distribution can be generated. Users can easily compare the evolving data on the map to check the temporal evolution in different regions.

(3)Temporal Analysis: Evolution Exploration When interesting regions are selected, visual fingerprints that represent associated information such as the speed, time, and number of passing taxis can be further explored. All views support user interactions for further exploration. Then we need to display the data distribution of passing taxis over a selected 24 hour time period. The hourly distribution will be colored according to the defined color map and can reveal selected attributer's correlation with the traffic. Hence we can analyze the local temporal attributes' changes and discover the evolution throughout each spatial region by comparing different fingerprints.

(4)Detail Record Analysis After everything is completed, users can query or recheck the raw data to examine their findings.

5   Experiment Results

The experiments are conducted on an Intel(R) Core(TM) 2 2.13GHz PC with 1GB RAM and an NVIDIA Geforce 7900 GS GPU with 256MB RAM. The data are first sanitized and aggregate values are computed during the preprocessing. The preprocessing dealing with the data of thousands of taxis in nine months took about six hours. After that, the VF approach supports interactive query, real-time visual displays and user interactions.

5.1   Case 1: Hotspot Exploration

Validation. For testing our design, we used a dataset where certain spatial and temporal patterns were previously expected. We first used grid computing for hot spot detection by rasterizing the background map into pixels. Then we computed the total number of taxis for each pixel, and after that we used the heat map to reveal the hotness of pixels throughout the 2D map of the city. Fig. 3 also showed the background of the heat map with the red areas representing regions of a high density of customers being picked up/dropped off by taxis while white areas represented regions of relatively low density. Users can use the heat map to choose some interesting regions for further analysis. As shown on the upper side of Fig. 4 (a) we found a rather hot spot of a dark red color, so we checked our visual fingerprints and found that they had revealed this hot spot's major temporal changes over the week. We can clearly see the upper part of the fingerprints is dark red while the rest is blue. It revealed that this region had a group of taxis arriving at midnight and leaving the following morning. We checked the map with our domain knowledge and found this place has a high probability of being a terminus where drivers change over or just go to rest. The results clearly showed that our visual fingerprint design can predict spatial and temporal patterns.

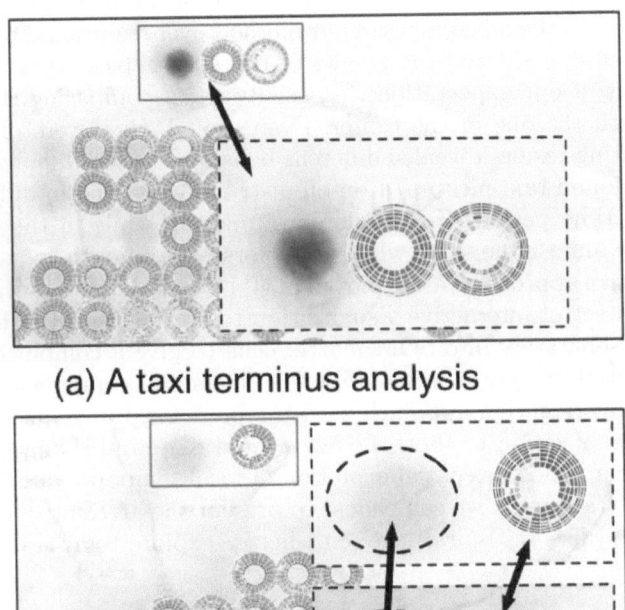

(a) A taxi terminus analysis

(b) Similar colored spot comparison

Fig. 4. Hot spot exploration by visual fingerprint. We selected the heatmap traffic display and applied the region fingerprinting view on taxi density to explore hot spots in the city. We then found that similar-colored hot spots may have different data distributions.

Application. In this part, we used fingerprinting to explore a hot spot identified from heatmap-style fingerprinting results (Fig. 4) (b). We found that similar-colored hot spots may have different data distributions. In this case, we fingerprinted the whole city within one week for all 65,836 roads with all three attributes (vehicle id, average speed, and picking-up/dropping-off id). In order to explore the hot spots that were identified by a dense color we drew the related region-scale level fingerprinting results (see Fig. 2). From Fig. 2, we found that taxis with high mobility can be considered as sensors traveling around the city, which visually back up the assumptions in [22]. The average speed fingerprinting results (Fig. 3) depicted the skeleton of the city's major road network. While this was interesting we can also see that it visually supported the assumptions in [15]], where they employed vehicles as sensors using their instant speed to sense the vicinity of vehicle congestion. They assumed the reported speeds are accurate because they are obtained directly from the speedometers installed in the taxis and sudden changes in speed are rare. Taxi pick-up/drop-off fingerprinting results were a surprise since they were different from traditional heatmap figures. As shown in Fig.4, the advantages of our method over traditional heatmaps were clearly revealed and showed the results of hot spot exploration. We found the results are out of our expectations, especially when comparing the fingerprint at the top with the one in the center. Even though they were of similar heat the fingerprinting results revealed different behaviors. According to the colors of the rings, we found taxi pick-up/drop-off distribution varied in different regions over different time periods if the colors are uneven, while in the center of the city they were almost the same when the colors were smooth.

An integrative approach was employed in [4] by combining self-organizing map (SOM) with a set of interactive visualization tools. They put both feature and index images separately into SOM matrix cells to give a combined representation of the spatial, temporal, and attributive (thematic) components of the data. Another data aggregation approach used the predefined areas in [3]. It applied pixel-based visualization to show the aggregated temporal changes to each grid in the Milan map and the spatial evolution of local temporal variables is clearly visible. In our framework we can choose to present the attributes in a more flexible way and utilize the spiral layout to better explore temporal patterns like periodic events.

5.2 Case 2: Abnormal Detection in Trajectory

Validation. An obvious spatial pattern that can be expected in the distribution of the local temporal variations is that the traffic on the major roads differs from that in the city center (revealed in Fig. 3). To detect such patterns, we had compared the temporal variations in each selected segment with the help of visual fingerprinting on trajectories. As shown in Fig. 5 (a), we visualized the pick-up/drop-off fingerprints for hot spots from the heat map exploration result (dark red spot at the top of (Fig. 4 (a)). The pick-up/drop-off behavior fingerprints were all colored a similar blue around the discovered spot which means this spot had nearly no pick-ups/drop-offs during the week in Fig. 5 (b). This is just as we expected since our domain knowledge told us this place might be a taxi terminal for drivers, so there might not be any customers there. However, we can also see some nearby places are clearly affected by this spot.

Application. We want to detect any abnormal patterns from the traffic data by using our fingerprint design to identify the details. We started with our region-scale fingerprint with density map (Fig. 2). Here we use speed as the variable visualized as a density style map to explore the patterns. Speed, viewed as a dynamic vehicle indicator is very unique, which gives us a multitude of other information, such as spatial information, temporal information or the behaviors of the vehicles. We discovered and checked some abnormal spots from the pick-up/drop-off behavior fingerprints with trajectory map Fig.6 (a). We can easily

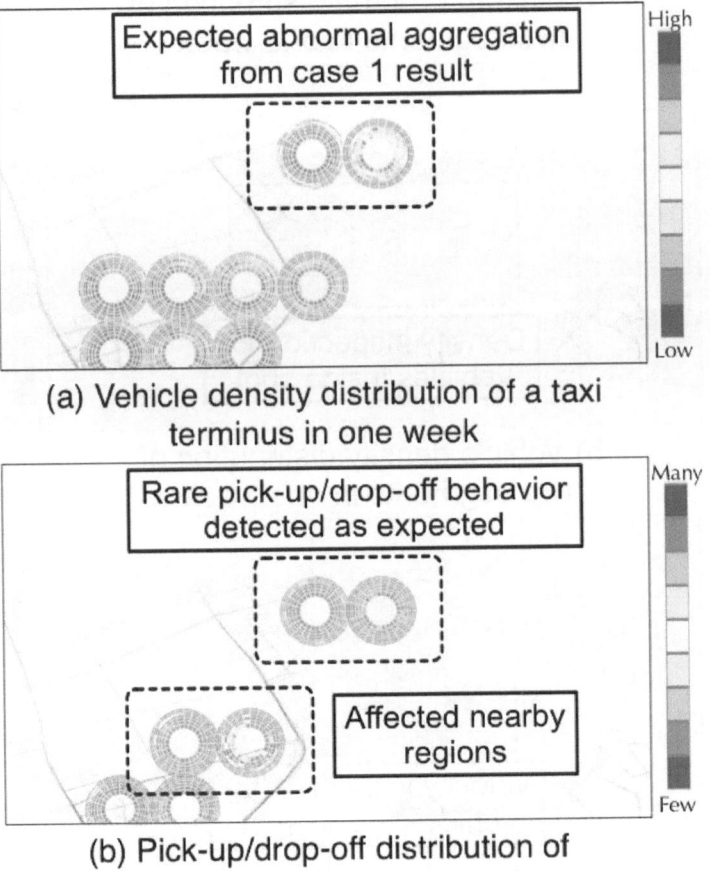

(a) Vehicle density distribution of a taxi terminus in one week

(b) Pick-up/drop-off distribution of a taxi terminus in one week

Fig. 5. Expected abnormal pattern detection. We have evaluated our taxi terminal finding in case one where the spot had many taxis arrive around midnight but no pick-ups/drop-offs in the close vicinity.

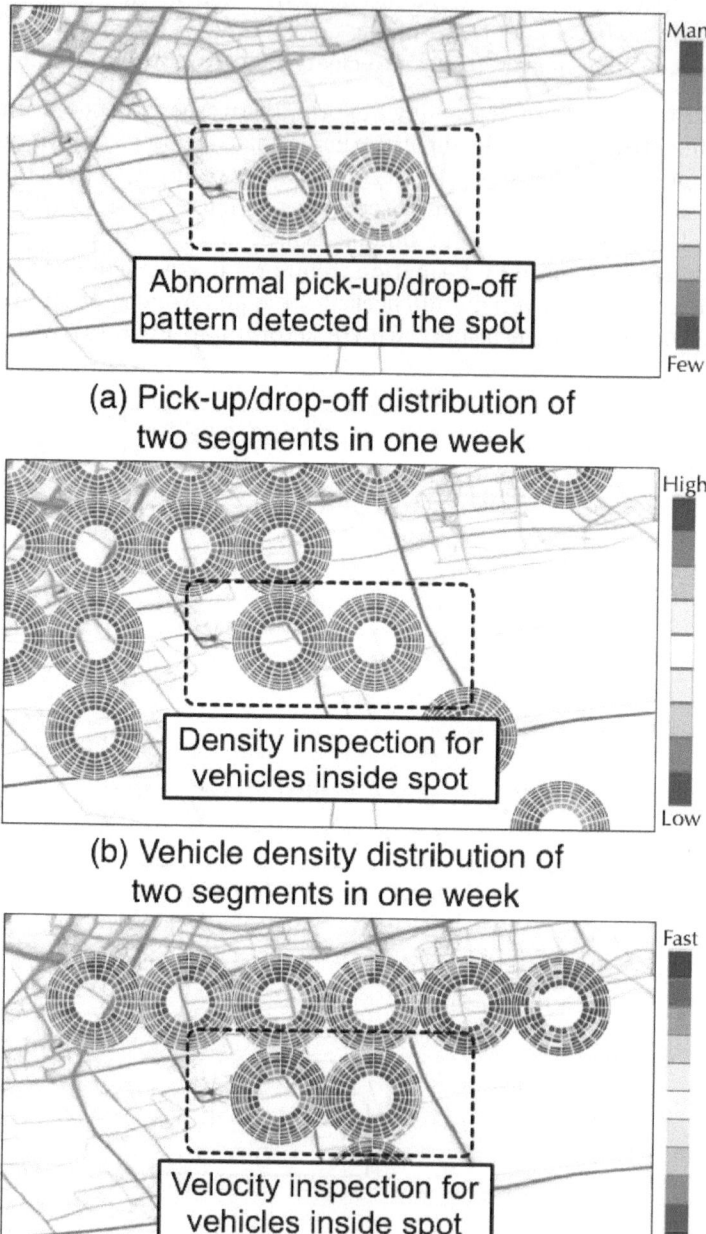

(a) Pick-up/drop-off distribution of
two segments in one week

(b) Vehicle density distribution of
two segments in one week

(c) Average speed distribution of
two segments in one week

Fig. 6. Abnormal Data Detection and Identification. We identified an abnormal pattern where a rather small region had a dramatic number of pick-ups/drop-offs over a week.

see a place with two fingerprints comprising many red dots indicating a heavy pick-up/drop-off area.

So we had found an abnormal spot with a dramatic number of pick-ups/drop-offs in a rather small area, of about two regions,over a week. We thought it interesting and visualized its vehicular density and the average speed by finger-printing and filtering. The density, depicted in blue,told us that very few vehicles frequented the area. The average speed told us these taxis traveled at a very low speed or were static. This did not seem possible so we rechecked the original data, to find the movement could be attributed to one single taxi. The records showed that the driver shifted several times in one hour between the two roads, therefore some peaks formed in the data since the records were affected by the rather stable average speed ($11km/h$). We also checked the driver's history data as his mobility was high and we picked a specific hour to check whether the taxi appeared in two places simultaneously. Here we applied our vehicle fingerprint-ing technique to better reveal the result. We found he appeared in at least three different places, therefore it turned out these peaks were definitely hidden errors. A spiral layout can make periodic data trends easily apparent when the correct period is chosen. In [23] the authors emphasized activity and cyclic time as dom-inant issues in its representation and it was designed to explore patterns from the timeliness of movement, availability and events. Its visual component struc-ture is similar to our road fingerprint design. However, our framework is focused on providing different hierarchical levels of temporal attribute exploration over a single road segment. It can be easily scaled to reveal a one-month pattern of the selected segment, and simultaneously provide visual analytics from multiple aspects of visualizations of spatial, temporal and multi-dimensional perspectives that are linked together.

6   Conclusions

In this paper we have presented a novel visual mining method and a set of visu-alization techniques for large-scale spatio-temporal data analysis. Our methods allow users to explore temporal attributes in the spatio-temporal trajectory data while keeping the layout of their spatial distribution. A sophisticated radial visu-alization is designed to analyze the frequent and periodic patterns in trajectory data. The approach can visualize multiple attributes chosen by users in one dis-play. It has good scalability and can explore temporal evolutions at a large-scale spatial range. It can also be applied to the density-based methods or applications to evaluate the density/cluster quality. We have utilized large scale real life data to evaluate our method and provide two case studies with interesting findings. In future, we will investigate other effective methods to handle the scalability problem. We plan to add a region separation and design a node projection algo-rithm to put nodes in better positions in the map.

Acknowledgment. This work was supported in part by grant HK SRFI11EG15. The authors wish to thank the anonymous reviewers for their valuable comments.

References

1. Alvares, L.O., Fern, V.B.J.A., Macedo, E.D., Moelans, B., Spaccapietra, S.: Dynamic modeling of trajectory patterns using data mining and reverse engineering. In: 26th International Conference on Conceptual Modeling (2007)
2. Andrienko, G., Andrienko, N.: Spatio-temporal aggregation for visual analysis of movements. In: IEEE Symposium on Visual Analytics Science and Technology, VAST 2008, pp. 51–58 (October 2008)
3. Andrienko, G., Andrienko, N.: A general framework for using aggregation in visual exploration of movement data. The Cartographic Journal 47(1), 22–40 (2010)
4. Andrienko, G., Andrienko, N., Bremm, S., Schreck, T., Von Landesberger, T., Bak, P., Keim, D.A.: Space-in-time and time-in-space self-organizing maps for exploring spatiotemporal patterns. Computer Graphics Forum 29(3), 913–922 (2010)
5. Andrienko, G., Andrienko, N., Dykes, J., Fabrikant, S.I., Wachowicz, M.: Geovisualization of dynamics, movement and change: key issues and developing approaches in visualization research. Information Visualization 7(3-4), 173–180 (2008)
6. Auria, M.D., Nanni, M., Pedreschi, D.: Time-focused density-based clustering of trajectories of moving objects. Journal of Intelligent Information Systems 27(3), 267–289 (2006)
7. Compieta, P., Di Martino, S., Bertolotto, M., Ferrucci, F., Kechadi, T.: Exploratory spatio-temporal data mining and visualization. Journal of Visual Languages Computing 18(3), 255–279 (2007)
8. Crnovrsanin, T., Muelder, C., Correa, C., Ma, K.-L.: Proximity-based visualization of movement trace data, pp. 11–18 (2009)
9. Giannotti, F., Nanni, M., Pinelli, F., Pedreschi, D.: Trajectory pattern mining. In: Proceedings of the 13th ACM SIGKDD International Conference on Knowledge Discovery and Data Mining, KDD 2007, pp. 330–339. ACM (2007)
10. Guo, H., Wang, Z.: Tripvista: Triple perspective visual trajectory analytics and its application on microscopic traffic data at a road intersection. In: IEEE Symposium on Pacific Visualization, PacificVis 2010 (2010)
11. Hurter, C., Tissoires, B., Conversy, S.: FromDaDy: spreading aircraft trajectories across views to support iterative queries. IEEE Transactions on Visualization and Computer Graphics 15(6), 1017–1024 (2009)
12. Kapler, T., Wright, W.: GeoTime Information Visualization. Information Visualization, 1–8 (2004)
13. Liu, S., Liu, C., Luo, Q., Ni, L.M., Qu, H.: A visual analytics system for metropolitan transportation. In: Proceedings of the 19th SIGSPATIAL International Conference on Advances in Geographic Information Systems (GIS 2011), pp. 477–480 (2011)
14. Liu, S., Liu, C., Luo, Q., Ni, L.M., Ramayya, K.: Calibrating large scale vehicle trajectory data. In: Proceedings of the 13th IEEE International Conference on Mobile Data Management (IEEE MDM 2012) (July 2012)
15. Liu, S., Liu, Y., Ni, L.M., Fan, J., Li, M.: Towards mobility-based clustering. In: Proceedings of the 16th ACM SIGKDD International Conference on Knowledge Discovery and Data Mining, KDD 2010, pp. 919–928. ACM (2010)
16. Liu, S., Luo, Y., Ni, L.M.: Calibration of vehicle trajectory. Technical Report (2010)
17. Palma, A.T., Bogorny, V., Kuijpers, B., Alvares, L.O.: A clustering-based approach for discovering interesting places in trajectories. In: Proceedings of the 2008 ACM Symposium on Applied Computing, SAC 2008, pp. 863–868. ACM (2008)

18. Pelekis, N., Kopanakis, I., Marketos, G., Ntoutsi, I., Andrienko, G., Theodoridis, Y.: Similarity search in trajectory databases. In: 14th International Symposium on Temporal Representation and Reasoning, TIME 2007, pp. 129–140 (2007)
19. Spaccapietra, S., Parent, C., Damiani, M., Demacedo, J., Porto, F., Vangenot, C.: A conceptual view on trajectories. Data & Knowledge Engineering 65(1), 126–146 (2008)
20. Vlachos, M., Kollios, G., Gunopulos, D.: Discovering similar multidimensional trajectories. In: Proceedings of 18th International Conference on Data Engineering, ICDE 2002, pp. 673–684. IEEE (2002)
21. Willems, N., Van De Wetering, H., Van Wijk, J.J.: Visualization of vessel movements. Computer Graphics Forum 28(3), 959–966 (2009)
22. Yuan, J., Zheng, Y., Zhang, C., Xie, W., Xie, X., Sun, G., Huang, Y.: T-drive: driving directions based on taxi trajectories. In: Proceedings of the 18th SIGSPATIAL International Conference on Advances in Geographic Information Systems (GIS 2010), pp. 99–108. ACM (2010)
23. Zhao, J., Forer, P., Harvey, A.S.: Activities, ringmaps and geovisualization of large human movement fields. Information Visualization 7(3-4), 198–209 (2008)

Stock Trend Extraction via Matrix Factorization

Jie Wang*

Computer Information Systems, Indiana University Northwest, Gary, IN 46408, USA
wangjie@iun.edu
http://www.iun.edu/~cisjw

Abstract. A diversified stock portfolio can reduce investment losses in the stock market. Matrix factorization is applied to extract underlying trends and group stocks into families based on their association with these trends. A variant of nonnegative matrix factorization SSMF is derived after incorporating sum-to-one and smoothness constraints. Two numeric measures are introduced for an evaluation of the trend extraction. Experimental analysis of historical prices of US blue chip stocks shows that SSMF generates more disjointed trends than agglomerative clustering and the sum-to-one constraint influences trend deviation more significantly than the smoothness constraint. The knowledge gained from the factorization can contribute to our understanding of stock properties as well as asset allocations in portfolio construction.

Keywords: portfolio construction, matrix factorization, trend extraction, clustering.

1 Introduction

Building a diversified stock portfolio is one investment strategy for risk management in the stock market. The overall investment risk is controlled by allocating asset portions in a portfolio. One of the most common classifications breaks the stock market into 11 classic sectors according to the primary activities, the products or services of a company (such as basic materials, technology, health care, services, utilities, consumer staples and Financial *etc.*). A trivial way to manage the risks of investment is to distribute assets evenly into groupings/sectors based on their business types. As we know, stock return prices are sensitive to unexpected or expected changes inside and outside of the market and are therefore highly volatile. It has been generally realized that stocks of the same sector may not have similar volatility and exhibit similar behavior in the market. So even though this sector-based strategy is simple enough to perform, the portfolio will not promise an optimal allocation for maximizing the return of investment.

Academics and practitioners of assets management have studied the prediction of stock market volatility in order to avoid huge investment losses. The obstacle in stock modeling and forecasting lies in our inability to discover the

---

* This work has been supported by the Indiana University Northwest Summer Fellowship for Research and Research Project Initiation Grant.

S. Zhou, S. Zhang, and G. Karypis (Eds.): ADMA 2012, LNAI 7713, pp. 516–526, 2012.

true data generating process [1]. One research direction of financial data analysis is an understanding of the underlying dynamics and reconstruction of stock returns. Principle component analysis (PCA) relies on the assumption of linear factors. PCA finds uncorrelated directions and gives projections of the data in the direction of the maximum variance. Independent component analysis (ICA) finds independently nonlinear factors. Andrew D. Back [3] applied joint approximate diagonalization of eigenmatrices algorithm to extract ICs as well as the mixing process. [3] showed ICA reconstructs the stock prices better than PCA for three-year daily returns of 28 Japanese stocks.

Besides ICA and PCA, clustering of stocks attracts a large amount of research of financial analysis. Non-parametric methods represent an optimal strategy when no prior knowledge of clusters is available. These methods make few assumptions about the structure of the data. They employ local criteria for reconstructing the clusters, e.g. by searching for high density regions in the data space. Moreover, as the number of clusters is not selected a priori, they are particularly suited when a hierarchical structure, rather than a fixed partition, of the data should be obtained. The hierarchical structure is the case with stock dynamics and portfolio optimization strategies [2]. Examples of non-parametric methods include the linkage (agglomerative) algorithms, whose output is a dendrogram displaying the full hierarchy of the clustering solutions at different scales either with a cutoff value of inconsistency or a specified number of the clusters. A chaotic map clustering algorithm [2] is used to analyze in pairwise the Dow Jones index companies in order to obtain a hierarchy of classes upon correlation coefficients between time series.

We assume that individual stock prices are actually determined by the fluctuations of several hidden components. The behavior of stocks observed in the complex stock market represents the integrated result of these key but unknown factors which we call latent bases. Then a reasonable assumption is: If stocks are affected in the same way by the same factors, they may act similarly in the market. Hence, if we could extract these latent bases and group stocks into families based on their association with these bases, we can leverage this grouping knowledge to analyze stock properties and assist in portfolio construction. There are a few articles about using nonnegative matrix factorization (NMF) [5,4] in the analysis of stock data. Drakakis examined constrained NMF with real stock data and found a tradeoff exists between noise removal of trends and sparsity of weight matrix [7]. A variant, semi-NMF, was applied in [8] with sparsity constraint to decompose a set of simulated stocks. In these works, there is no numeric measure to evaluate the extracted trends.

In this paper, we focus on trend extraction of stock data. We are concerned with methods based on nonnegative matrix factorization (NMF) for the analysis of historical prices of US blue chip stocks [1]. The problem we are interested in and discuss in this paper can be defined as follows: Suppose there are n stocks $S_1, S_2, \ldots, S_n$; each stock S_i is stored as a row vector whose entries are

---

[1] Blue chip stock is qualified as a high-quality and usually high-priced stock. It has high price because of public confidence in company's long record of steady earnings.

m stock prices at uniformly distributed time intervals, *i.e.*, daily or monthly prices. Assume there are K latent bases, $\mathbf{W}_1, \mathbf{W}_2, \ldots, \mathbf{W}_K$; each basis is a row vector of size m. We attempt to uncover these underlying factors and express each stock S_i as an aggregate of these bases with certain influences.

$$
\begin{aligned}
S_1 &= h_{11}\mathbf{W}_1 + h_{12}\mathbf{W}_2 + \ldots + h_{1k}\mathbf{W}_k + \ldots + h_{1K}\mathbf{W}_K + N_1 \\
S_2 &= h_{21}\mathbf{W}_1 + h_{22}\mathbf{W}_2 + \ldots + h_{2k}\mathbf{W}_k + \ldots + h_{2K}\mathbf{W}_K + N_2 \\
&\vdots \quad \vdots \\
S_i &= h_{i1}\mathbf{W}_1 + h_{i2}\mathbf{W}_2 + \ldots + h_{ik}\mathbf{W}_k + \ldots + h_{iK}\mathbf{W}_K + N_i \\
&\vdots \quad \vdots \\
S_n &= h_{n1}\mathbf{W}_1 + h_{n2}\mathbf{W}_2 + \ldots + h_{nk}\mathbf{W}_k + \ldots + h_{nK}\mathbf{W}_K + N_n
\end{aligned}
\tag{1}
$$

The general formula for the ith stock S_i is

$$
S_i = \sum_{k=1}^{K} h_{ik}\mathbf{W}_k + N_i,
\tag{2}
$$

where h_{ik} is a nonnegative real number and indicates to which degree the stock i is associated with the basis $\mathbf{W}_k$. N_i is an observation noise vector. When using the sum of the product of h_{ik} and $\mathbf{W}_k$ over K bases to approximate S_i, N_i can be seen as the approximation error. When n stocks are considered, stack S_i into a data matrix $S \in \mathbb{R}_+^{n \times m}$ and a matrix notation is appropriate to express all stocks as

$$
S_+ = \mathbf{H}_+ \mathbf{W}_\pm + N,
\tag{3}
$$

where $\mathbf{H} \in \mathbb{R}_+^{n \times K}$, $\mathbf{W} \in \mathbb{R}_\pm^{K \times m}$ and $N \in \mathbb{R}_+^{n \times m}$. S is known in prior. Observably, (3) would be the classical NMF if enforcing a nonnegative constraint in $\mathbf{W}$. In this paper, we introduce a sum-to-one property to $\mathbf{H}$. A constrained NMF is derived to compute two matrices $\mathbf{H}$ (weight matrix) and $\mathbf{W}$ (trend matrix) for a given S and K. $\mathbf{H}$ can be used to partition stocks into K clusters. We specifically investigate the impact of these constraints in real stock return data. Two numeric measures are introduced to evaluate distance and deviation between trends.

2   Constrained Matrix Factorization: SSMF

Classical NMF interprets an object as an additive combination of latent bases. All elements in NMF are nonnegative. For the stock market, the signs of the basis vectors $\mathbf{W}$ is not restricted as shown in (2). Two constraints are introduced into objective functions. Below we describe these constraints and derive update formulas for $\mathbf{H}$ and $\mathbf{W}$.

2.1   Constraints

Smooth Solution. Smoothness constraint is often enforced to regularize the computed solutions in the presence of noise in the data. In order to enforce smoothness in $\mathbf{W}$, the Frobenius norm penalizes the solutions of large Frobenius norm, thus we set

$$\mathcal{J}_1(\mathbf{W}) = \|\mathbf{W}\|_F = \sum_{k=1}^{K} \sum_{j=1}^{m} w_{kj}^2 = \mathbf{tr}(\mathbf{W}\mathbf{W}^T). \tag{4}$$

$\mathbf{tr}$ denotes the trace of a square matrix.

Sum-to-One Constraint. Since the unconstrained NMF gives a linear additive combination of underlying components, rows of $\mathbf{H}$ are used to determine weights of the components in each observed object. For a hard clustering, the most dominant component of an object is indicated by the index of the largest element in its corresponding weight vector in $\mathbf{H}$. When an additive structure is needed to disclose the contributions of all of the components, we can enforce that all of the elements in each row of $\mathbf{H}$ sum to one [6]. This property can be written as a penalty term in the form

$$\mathcal{J}_2(\mathbf{H}) = \|\mathbf{H}e_1 - e_2\|_F^2, \tag{5}$$

where e_1 and e_2 are column vectors with all entries being 1. This penalty term $\mathcal{J}_2$ will compute a $\mathbf{H}$ which approximates to a stochastic vector.

2.2 Updating Algorithm

Now we derive updating rules for a constrained NMF. Multiplicative rules (described in [5]) are used here to form two update rules for $\mathbf{H}$ and $\mathbf{W}$. If using L_1 norm for a sparser $\mathbf{H}$, the optimization problem for (2) becomes

$$\min_{\mathbf{H}\in\mathbb{R}_+^{n\times K}, \mathbf{W}\in\mathbb{R}_\pm^{K\times m}} \mathcal{F} = \{\alpha\mathcal{Q} + \gamma\mathcal{J}_1 + \lambda\mathcal{J}_2\},$$
$$\mathcal{Q} = \|S - \mathbf{H}\mathbf{W}\|_F^2, \tag{6}$$
$$\mathcal{J}_1 = \|\mathbf{W}\|_F,$$
$$\mathcal{J}_2 = \|\mathbf{H}e_1 - e_2\|_F^2.$$

α, γ and λ are regularization parameters to balance the trade-off between the approximation error and the application-dependent restrictions. Now we derive update rules to solve (6), based on an alternating gradient descent mechanism and multiplicative update rules. Take derivatives of $\mathcal{Q}$, $\mathcal{J}_1$ and $\mathcal{J}_2$ with respect to $\mathbf{H}$ and $\mathbf{W}$,

$$\frac{\partial\mathcal{F}}{\partial\mathbf{W}} = -2\alpha\mathbf{H}^T S + 2\alpha\mathbf{H}^T\mathbf{H}\mathbf{W} + 2\gamma\mathbf{W}$$
$$\frac{\partial\mathcal{F}}{\partial\mathbf{H}} = -2\alpha S\mathbf{W}^T + 2\alpha\mathbf{H}\mathbf{W}\mathbf{W}^T + 2\lambda\mathbf{H}e_1 e_1^T - 2\lambda e_2 e_1^T. \tag{7}$$

For some starting matrices $\mathbf{H}^{(0)}$ and $\mathbf{W}^{(0)}$, let $\alpha = 1$, the sequence $\{\mathbf{H}^{(t)}\}$ and $\{\mathbf{W}^{(t)}\}$ are computed by means of the formulas

$$\mathbf{h}_{ik}^{(t)} = \mathbf{h}_{ik}^{(t-1)} \frac{\left[S\mathbf{W}^T\right]_{ik} + \lambda}{\left[\mathbf{H}^{(t-1)}\mathbf{W}\mathbf{W}^T\right]_{ik} + \lambda\left[\mathbf{H}^{(t-1)}e_1 e_1^T\right]_{ik}} \tag{8}$$

$$\mathbf{w}_{kj}^{(t)} = \mathbf{w}_{kj}^{(t-1)} \frac{\left[\mathbf{H}^T S\right]_{kj} - \gamma\mathbf{w}_{kj}^{(t-1)}}{\left[\mathbf{H}^T\mathbf{H}\mathbf{W}^{(t-1)}\right]_{kj}}. \tag{9}$$

At each iteration, $\mathbf{W}$ is normalized. The minus operation in the numerator of (9) may produce negative values. In the classical NMF, a very small value is used to replace any negative results from this minus operation at each iteration. Here, we don't enforce this property because negative entries in $\mathbf{W}$ are allowed in the application of stock data analysis. By adjusting two regularization parameters ranging from 0 to 1, we have flexibility of applying these additional constraints with different weights in the optimization. The matrix factorization defined by (8) and (9) is denoted as SSMF.

2.3   Evaluation Measures

In fact, two stocks with different volatilities may have the same expected return, but the one with higher volatility will have larger swings in values over a given period of time, which represents a higher risk. In order to balance the overall investment risk, a stock portfolio should be diversified so as to be comprised of stocks from most stock families of different volatilities. A good partition should maximize the volatility difference between clusters and minimize the intra cluster difference of volatility. The latent trends should be found deviated from each other as much as possible.

Trend Evaluation. For a quantitative evaluation, we introduce two measures, Frobenius norm (or Euclidean distance) and J-divergence, to compare distance and deviation between trend x and y. Frobenius norm defines an ordinary distance between two vectors as $(\sum(x_i - y_i)^2)^{\frac{1}{2}}$. J-divergence is a symmetric and nonnegative metric based on Kullback-Leiber divergence $D(P||Q)$. J-divergence measures the degree of mutual deviation of two probability distributions P and Q. J-divergence, $J(P||Q)$ is computed as the sum of $D(P||Q)$ and $D(Q||P)$, where $D(P||Q) = \mathbf{E}_{P(x)}[\frac{P(x)}{Q(y)}]$ and $\mathbf{E}_{P(x)}[.]$ denotes the expected value with respect to the probability distribution of x. $J(P||Q)$ becomes 0 if $P(x) = Q(x)$.

Clustering Evaluation. The clustering is evaluated by Silhouette value. The Silhouette validation method, first described by Peter J. Rousseeuw in 1987 [9], provides a succinct graphical representation of how well each data point lies within its cluster. The silhouette value of a data point is a measure of how similar that point is to points in its own cluster compared to points in other clusters. The value ranges from -1 to 1 and is defined in *matlab* code as: $s(i) = \frac{\min\{b(i,:),2\}-a(i)}{\max\{a(i),\min\{b(i,:)\}\}}$ where $a(i)$

is the average distance from the *ith* point to the other points in its cluster, and $b(i, k)$ is the average distance from the *ith* point to points in another cluster k. An $s(i)$ close to one means that the data point i is appropriately clustered. The average $s(i)$ over the entire dataset is a measure of how appropriately the dataset has been clustered.

3   Experimental Results

This group of experiments are carried out on stock data from the Center for Research in Security Prices(CRSP). This real dataset consists of 25 blue chip stocks, each of which contains 120 historic monthly returns over 10 years from 2001 until 2010. The dataset is plotted and the stock tickets are listed in Figure 1. We also use integers 1 through 25 to label these stocks. We take logarithm and normalize the data columnwise as each element value is in the range of $[0, 1]$. Each SSMF run uses random $\mathbf{H}^{(0)}$ and $\mathbf{W}^{(0)}$. Each run iterates 900 times. At each iteration, each row of $\mathbf{W}$ is normalized as $\|\mathbf{W}_k\|_2 = 1$.

The natural number of trends is definitely unknown for this real dataset. In this preliminary work, since we only focus on solutions which give better disjointness of latent trends, we don't investigate estimation of the number of trends K. We simply presume 2 and 3 for our experiments in order to identify straightforward patterns between parameters and trends.

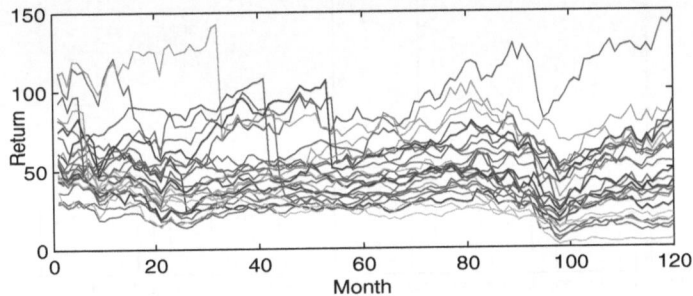

Fig. 1. 10-year monthly returns of 25 US stocks from 2001 until 2010. The stock tickers are MSFT, AA, AXP, BA, BAC, C, CAT, DD, DIS, GE, HD, IBM, INTC, JNJ, JPM, KO, MCD, MMM, MRK, PFE, PG, UTX, VZ, VMT, XOM.

3.1   A Visual Comparison of Trend Extraction

We compare trends from SSMF with centroids from hierarchical clustering using the ward linkage algorithm. A non-parametric agglomerative method, hierarchical clustering, is used to generate dendrograms of stock linkages, shown in Figure 2. The ward linkage appears more appropriate for this stock data than the centroid linkage algorithm since the latter has downward connections and produces a non-monotonic cluster tree.

Four sets of results are plotted in Figure 3. Frobenius norm and J-divergence are recorded in Table 1. Obviously, the trends from SSMF are more volatile than those from agglomerative clustering through pairwise linkages. Even though the agglomerative method produces components that have a larger Frobenius norm, two components, as shown in the top left plot of Figure 3, have very similar time-varying behavior. Not strictly speaking, given this partition, stock returns in these two clusters will react in the similar way to changes in the market. Conversely, two trends from SSMF show opposite oscillations. A portfolio will be well risk-balanced if it consists of stocks from both groups.

Table 1. Measures of four sets of trends in Figure 3

Method	Frobenius norm	J-divergence
Ward linkage	4.3375	3.7760
SSMF($\gamma = 0, \lambda = 0$)	0.9516	3.7166
SSMF($\gamma = 0, \lambda = 0.12$)	0.6981	4.0371
SSMF($\gamma = 0.5, \lambda = 0.12$)	0.6923	4.1218

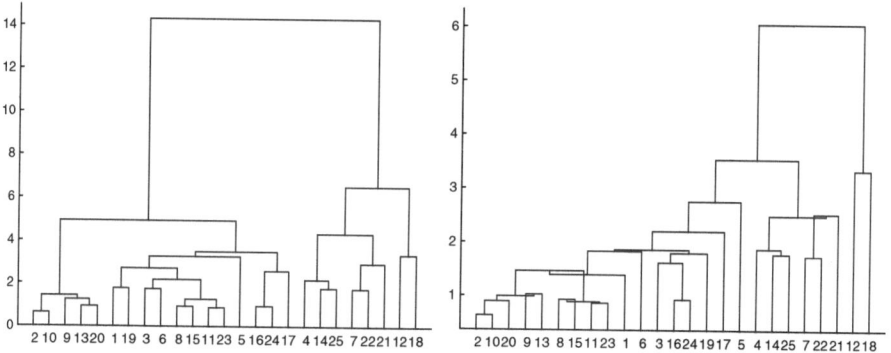

Fig. 2. Two dendrograms of 25 stocks using Euclidean distance. The left dendrogram uses Ward linkage algorithm. The right one uses Centroid algorithm.

3.2 Individual and Combinative Effects of Smoothness and Sum-to-One Constraints

Here we explore the effect of imposing the sum-to-one of **H** by λ and compare results from experiments without and with applying the smoothness of **W** by γ. Consider the case of grouping stocks into two clusters, *i.e.*, $K = 2$, run SSMF with varying smoothness parameter γ and sum-to-one parameter λ. $\gamma \in [0, 1]$ with a step size of 0.1 and $\lambda \in [0, 1]$ that has a step size of 0.01. One inherent feature

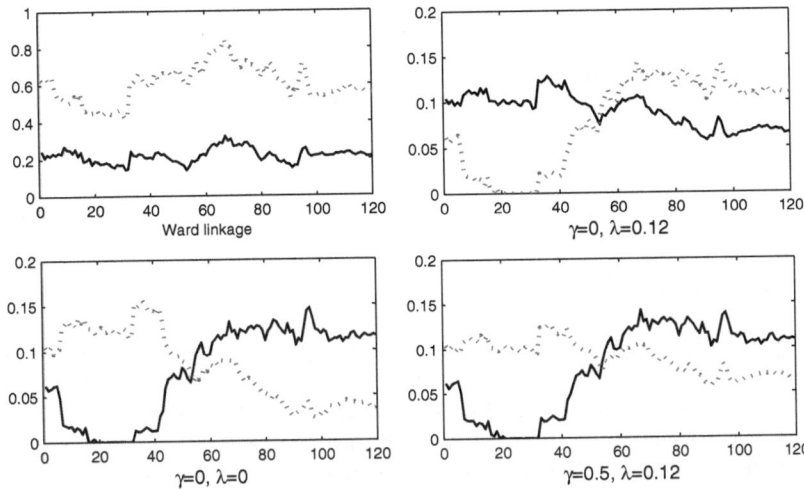

Fig. 3. Four sets of extracted trends. The top left plot shows two centroids using hierarchical clustering. The other three plots show trends from SSMF after imposing a different level of the smoothness and the sum-to-one constraints.

of the SSMF is non-uniqueness of solutions even with exactly the same starting matrices. Therefore, for each value pair of (γ, λ), we run SSMF 20 times and compute the average overall Silhouette value, the average Euclidean distance between two bases, and the average J-divergence. The results are plotted in Figure 4. Figure 5 amplifies the part of the upper three plots in Figure 4 where $\lambda \in [0, 0.1]$.

Figures 4 and 5 reveal the following aspects:

- The sum-to-one constraint of $\mathbf{H}$ (by λ) governs Frobenius norm and J-divergence of two bases. This provides coarse adjustment, while the smoothness of $\mathbf{W}$ brings a slight refinement to the bases.
- The introduction of the sum-to-one constraint decreases Frobenius norm of two bases.
- If $\lambda \in [0, 0.1]$, increment of λ contrarily affects the Frobenius norm and J-divergence. Two bases become more disjointed in the sense of diversity of probability distributions, while Frobenius norm decreases between two bases as a result of less numeric differences in each element pair of two bases.

We also visualize the average overall Silhouette value and the average Frobenius norm of two bases in Figure 6. The left colormap of Figure 6 indicates that concurrently increasing weights of both constraints in the objective function will move two bases closer to each other in the sense of Frobenius norm. This makes sense if assuming more random noise, which is introduced into data due to external market fluctuations, is removed. Moreover, the colormaps illustrate

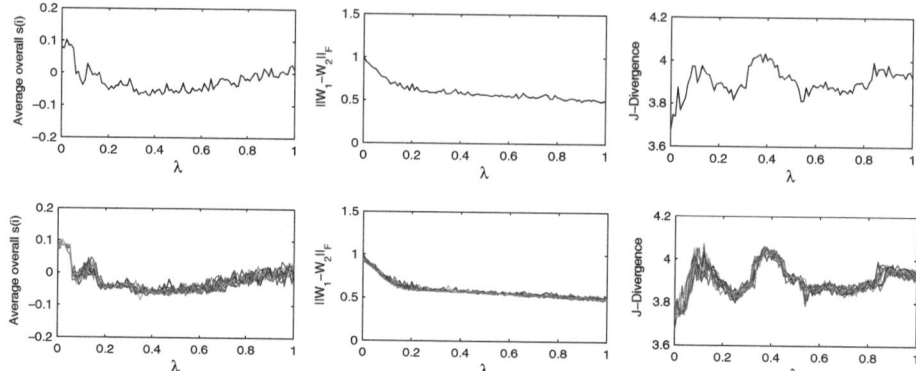

Fig. 4. The effect of the sum-to-one property of **H** by λ. The upper three plots show the change of three measures (from left to right, average overall Silhouette value, the average Euclidean distance between two bases, and the average J-divergence) with respect to the sum-to-one property of **H** by λ, $\lambda \in [0, 1]$ that has a step size of 0.01. The lower three figures show the results after applying two constraints together, the sum-to-one and the smoothness. Each figure on the bottom contains 11 plots, each of which corresponds to a specific value pair (γ, λ). γ varies from 0 to 1. K in SSMF is 2.

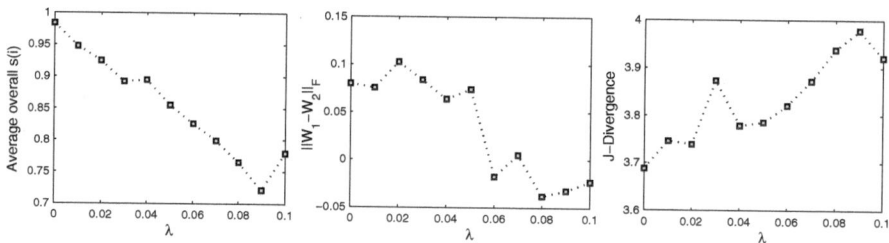

Fig. 5. The plots of three measures (from left to right, average overall Silhouette value, the average Euclidean distance between two bases, and the average J-divergence) with respect to the sum-to-one property of **H** by λ, $\lambda \in [0, 0.1]$ with a step size of 0.01. $\gamma = 0$ and K in SSMF is 2.

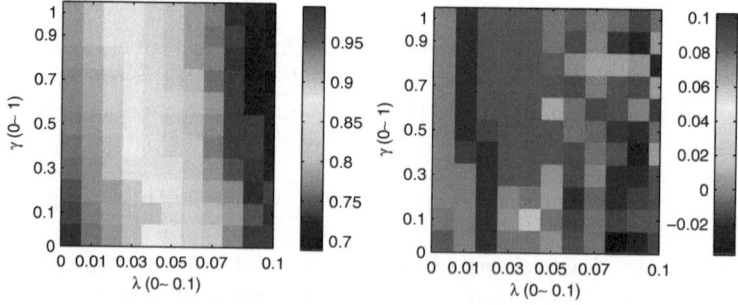

Fig. 6. Two colormaps of Frobenius norm ($\|\mathbf{W}_1 - \mathbf{W}_2\|_F$) (on the left) and average overall Silhouette value $s(i)$ (on the right) with respect to γ and λ. $K = 2$.

that this dataset is more sensitive to the sum-to-one constraint of $\mathbf{H}$ (tuned by λ) than to the smoothness of $\mathbf{W}$(tuned by γ).

The colormap of $s(i)$ is very interesting because it is quite observable that a particular value of $\lambda = 0.05$ straightly splits $s(i)$ into two areas. When $\lambda > 0.05$, regardless of γ, the clustering quality sharply deteriorates. In the area where $\lambda \in [0, 0.05]$, $s(i)$ is positive. For a γ between 0 and 0.4, the best $s(i)$ is achieved at $\lambda = 0.02$. For a γ between 0.5 and 1 (that means $\mathbf{W}$ tends to be smoother than solutions solved using a $\gamma \in [0, 0.4]$), the best $s(i)$ is achieved at $\lambda = 0.01$.

4   Summary

We derived and tested SSMF, a variant of nonnegative matrix factorization after enforcing two properties, the smoothness of trend matrix $\mathbf{W}$ and the rowwise sum-to-one of weight matrix $\mathbf{H}$. Interestingly, the sum-to-one restriction enforced on the weight matrix has stronger influence than the smoothness of the trend matrix. Moreover, the sum-to-one property has some encouraging impact on separating two trends in the sense of J-divergence. Two trends extracted from SSMF in the experiments display opposite oscillations, which implies that the stocks may respond differently to changes in the market if their most associated trends are different. We can apply the clustering results from the factorization to allocate assets so as to construct a well risk-balanced and diversified portfolio. In addition, the smoothness of $\mathbf{W}$ has no evident influence on the overall $s(i)$. Also, the concurrent increase of sum-to-one and smoothness constraints will deteriorate clustering, which means these constraints should not be over weighted beyond a certain degree compared to the basic approximation error between S and $\mathbf{HW}$. Some further work is planned to incorporate the trend disjointness into the objective function of the factorization in order to compute a solution which has a specified level of trend discrepancy.

References

1. Allen, P.G., Morzuch, B.J.: Twenty-five years of progress, problems, and conflicting evidence in econometric forecasting. What about the next 25 years? International Journal of Forecasting 22(3), 475–492 (2006)
2. Basalto, N., Bellotti, R., De Carlo, F., Facchi, P., Pascazio, S.: Clustering stock market companies via chaotic map synchronization. Physica A 345, 196–206 (2005)
3. Back, A.D., Weigend, A.S.: A first application of independent component analysis to extracting structure from stock returns. International Journal of Neural Systems 8(4), 473–484 (1997)
4. Paatero, P., Tapper, U.: Positive matrix factorization: a non-negative factor model with optimal utilization of error estimates of data values. Environmetrics 5(2), 111–126 (1994)
5. Lee, D.D., Seung, H.S.: Learning the parts of objects by non-negative matrix factorization. Nature 401(6755), 788–791 (1999)
6. Berry, M.W., Browne, M., Langville, A.N., Paul Pauca, V., Plemmons, R.J.: Algorithms and applications for approximate nonnegative matrix factorization. Computational Statistics & Data Analysis 52(1), 155–173 (2007)

7. Drakakis, K., Richkard, S., de Frein, R., Cichocki, A.: Analysis of Financial Data Using Non-Negative Matrix Factorization. Internation Mathematical Forum 3(38), 1853–1870 (2008)
8. de Frein, R., Drakakis, K., Rickard, S.: Portfolio diversification using subspace factorizations. In: 42nd Annual Conference on Information Sciences and Systems, pp. 1075–1080 (2008)
9. Rousseeuw, P.J.: Silhouettes: a Graphical Aid to the Interpretation and Validation of Cluster Analysis. Computational and Applied Mathematics 20, 53–65 (1987)

Stock Price Forecasting with Support Vector Machines Based on Web Financial Information Sentiment Analysis

Run Cao, Xun Liang, and Zhihao Ni

School of Information, Renmin University of China, Beijing100872, P.R. China
{caorun,xliang,alexni}@ruc.edu.cn

Abstract. The stock price forecasting has always been considered as a difficult problem in time series prediction. Mass of financial Internet information play an important role in the financial markets,information sentiment is an important indicator reflecting the ideas and emotions of investors and traders. Most of the existing research use the stock's historical price and technical indicators to predict future price trends of the stock without taking the impact of financial information into account. In this paper, we further explore the relationship between the Internet financial information and financial markets,including the relations between the Internet financial information content, Internet financial information sentimental value and stock price. We collect the news of three stocks in the Chinese stock market in the GEMin a few large portals, use the text sentiment analysis algorithm to calculate the sentimental value of the corresponding Internet financial information, combined with the stock price data, implantSupport Vector Machines to analyzes and forecasts on the stock price, the accuracy of the prediction of stock priceshas been improved.

Keywords: Text sentiment analysis, SVM, Stock price prediction.

1 Introduction

The financial market has always been a research hotspot, and stock price forecasting has been considered the most difficult challenges in the time series prediction. Stock price time series are data intensive, noisy, dynamic, unstructured, and have a high degree of uncertainty [1]. Stock price time series is more like a random walk curve [2], using the traditional statistical methods to predict stock price changes has been proven to be very difficult.

With the rapid development of communication technology and the Internet, people can access to the stock news and reviewsthrough the Internet anytime, anywhere. Although the Internet is just one of the channels for an investor to access the financial information, and often overlap with newspapers, television and other media, but Internet information is more quickly andcomprehensive,it could cover other channels, financial Internet information play an important role in the financial markets.

Most of financial information is unstructured text, the main forms are financial news on the major portals, financial analyst's comments, national policies and announcements, and the online community discussion. According to the efficient market hypothesis, the financial information has an important impact on the financial market

S. Zhou, S. Zhang, and G. Karypis (Eds.): ADMA 2012, LNAI 7713, pp. 527–538, 2012.

volatility. The information has two dimensionscharacteristics: information volume and information sentiment, the information sentiment analysis has become an important topic of natural language processing and machine intelligence field.

In the financial markets, information sentiment is an important indicator reflecting the opinions and emotions of investors and traders. Financial information integrates a variety of factors in the market and affects the stock market participants'investment trading behavior to a large extent. The quantify effectiveness of the financial information can be confirmed directly through financial model, combining with existing financial mathematical model. Previous studies have shown that Internet financial information volume and stock market volatility are closely related [3], [4], [5], [6].

In this paper, we further explore the relationship between the Internet financial informationvolume, Internet financial information sentimental value and the stock price. We have three stocks in the Chinese stock market in the GEM as examples, the news of these three stocks in a few large portalshave been collected, then we use the text sentiment analysis algorithm to calculate the sentimental value of the corresponding Internet financial information, combined with the stock price data, implant Support Vector Machines to analyzes and forecasts the stock price, the accuracy of the prediction of stock prices has been improved.

2   Related Work

2.1   Stock Price Forecasting

The stock price forecasting can be divided into two branches, the industry and academia. Traditional industry is mainly based on fundamental analysis and technical analysis. Academics are mainly expanding the stock price forecasts from two aspects: statistical methods and machine learning method.Methods based on statistical are mainly using time series analysis methods, including linear regression prediction, polynomial regression prediction, ARMA modeling, GARCH modeling and so on. Methods based on machine learning often use non-linear prediction and intelligent learning, including the Grey Theory [7], Artificial Neural Networks [8], Support vector machine [9], Markov model, fuzzy network [11]and other.

Methods based on machine learning can solve the noise data problem, learning nonlinear models efficiently, as the stock market is a typical nonlinear complex system. In recent years, the neural network has been successfully applied to financial time series modeling from Stock Price Index [8], [12], to the option price [13]. Unlike the traditional statistical methods, neural networks are data-driven, does not require assumptions and parameters, nonlinear system can be well fitted [14], but the neural network optimization may fall into the Local optimum. The Support Vector Machine developed by Vapnik solve this problem successfully, because of the principle of structural risk optimal, SVM was able to reach the global optimum, and also overcoming the over fitting problem, which making SVM improved a lot compared to ANN in stock price prediction [15], [16] ,[17].

With the development of machine learning methods, integrated forecasting methods have got more and more attention. For example Md.Rafiul and Hassan applied an integration of HMM, ANN, GA model to predict the stock market [18]. Bildirici and Erisn use ANN to restructure on GARCH model, and applied to the Turkish stock market in order to verify the validity of the hybrid model [19]. Wen use Box theory and SVM to build an automatic stock Decision Support System [20].

In machine learning, we focus on the input and output of the data collection. At present, for predicting stock prices, the input often includes a variety of macroeconomic indicators, and stock historical transaction data. In this article, the Internet Financial News as another exogenous input is introduced into the prediction of stock price and thus to improve the prediction accuracy.

2.2   Text Sentiment Analysis

Text sentiment analysis, also known as Opinion Mining, we can acquire the opinion of the textthrough the analyzed information sentiment tendency. For example, by automatically analyzing the text content of the comments of a commodity, we can get the consumer's attitudes and opinions of the goods. At present, the application of sentiment analysis mainly focused on: user comments analysis, decision-making, monitoring public opinion, and the information prediction. Text sentiment analysis is mainly from two aspects, sentimental knowledge-based approach and the method based on feature classification. The former use the existing sentiment dictionary or industry dictionary to calculate and obtain the polarity of the text;while the method based on feature classification use machine learning methods by selecting a large number of significant features to complete the classification task [21].

Present research are mainly use text sentiment analysis in merchandise online reviews, including books, movies, stocks, and electronic products. Mainly use empirical analysis, explore how the sentimental trend of online reviews will affect consumers' purchasing behavior and how to affect the mechanism of the related product sales, and establish the theoretical model [22], [23], [24].Text sentiment analysis has also gradually attracted the attention of many Internet companies. Google has already applied text sentiment analysis into the search engine to provide users with more effective and versatile service. At the same time, the Internet video site YouTube also take text sentiment analysis into application in 2008, each user's comments are divided into "Good Comment" and "Poor Comment", allows users to glance on the previous attitude of the audience on the current video based on this statistics.

Although the Internet text analysis has become the common interest of many researchers, but the analysis specifically on the financial text are still rare. Devitt and other people use the of the Financial Review text sentimental polarity recognition, make predictions to future financial trends [25]. Das and Chen use linear regression method, find that the stock index and online stock analysis sentiment are significant positive correlation, but for a single company, this relationship is not established [6]. Antweiler and Frank discussed the relationship between the number of online stock analysis, stock analysis sentiment and stock trading volume, volatility and other indicators, the significant correlation between some variables show that the influence of online stock analysis on the stock market [3]. Inthis paper, we apply the text sentiment analysis in individual company news and stock analysts, daily information volume and financial information sentimental valueare collected, then a machine learning method is used to predict the ups and downs of the stock price. This study is very innovative and has a strong practical significance.

3   Model and Method

3.1   The Framework of the Method

The framework of the overall project is depicted in Fig.1. with four components. The first component is the collection and computing of experimental information, including the financial text information and stock price information; The second component is the experimental input data preprocessing, including transfer the input data into Libsvm data input format, establish training set and test set, and normalized all input values; The third one is group experiments, all three stocks are divided into four groups separately, conducted SVM training, training the model to predict stock prices, and tested on the test set; The fourth component is calculate the evaluation score of the metrics to measure the performance of the experiment results data .

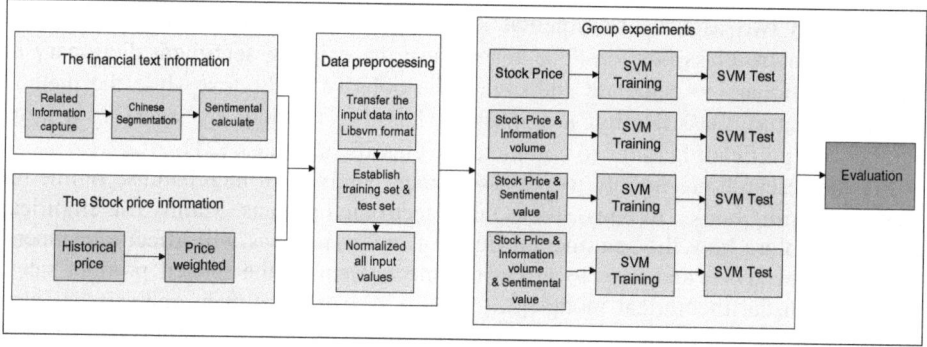

Fig. 1. The experiment framework

3.2   The Calculation of Financial News Sentimental Value

For the Internet text, external factors and internal factorsdetermine their nature.External factors are like the number of text on a particular topic appeared on the Internet within a specified time, and internal factors describe the nature of a single text.For a single text,its nature depends on its content and intensity, content refers to the theme, time, style of the textand intensity refers to the impact factor of the text.

The sentiment of financial text corresponds to the tendency of the attitude embodied in the text, i.e., bullish, bearish or neutral, and the financial text intensity reflects the influence of the text. High intensity of the text has a greater influence on financial markets, the low intensity of the text for the influence of the financial market is relatively weak. A sentimental value that is calculated according to a financial text, then the positive and negative symbols of the value means bullish bearish or neutral, while the absolute value represent the intensity of the text.

In this experiment, we propose the text sentiment algorithm based on Hownetsentiment dictionary. Assume that the current Financial News p, first conduct the segmentation tool to convert it into a sequence constituted by the word, that is, $\{w_1, w_2, w_3, \ldots w_n\}$, the number of total words is n, on each one the $w_i(i=1, 2, 3, \ldots n)$ to calculate an sentimental value of v_i, then the sentimental value of the entire Financial News p is the sum of all the words sentimental value. The FinancialNews sentimental algorithm is as follows:

```
Get the current word Wi, set the sentiment value Vi to 0
if(Wi belongs to Positive Word List)  set Vi=1
else if(Wi belongs to Negative Word List) set Vi=-1
else output Vi=0
if (Wi belongs to Positive Word List|| Wi belongs to
Negative Word List){
get the K words before the current word Wi
if(any word in K belongs to Privative Word List ) set Vi=
-1 *Vi
get M words before Wi and N words after Wi
if(any word in M or N belongs to Degree Word List)
set Vi=Vi* Degree
output Vi
}
```

$$v_p = \sum_{i=1}^{n} V_i$$

```
end
```

In the experiment, we randomly selected three stocks in the Growth Enterprise Market in China Toread(300005) Hanwei Electronics (300007), Huayi Brothers (300027), Their size is moderate, and the company news amount is also more suitable for this experiment. We have a Financial News crawler to get Financial News data, which is supported by Internet Financial Intelligence Laboratory, School of Information, Renmin University of China. Fig.2.is a screenshot of stock financial news text, Fig.3.shows the news number of Huayi Brothers (300027) on the timeline. In the experiment, we use the news number of stocks to represent the relative volume of information that day.

News ID	News Title	Time	Company ID	Newsbody
2010080600000007	（公司）华谊兄弟：《唐山大地震》票房达到 4.8 亿元-新闻频道-和讯网	2010-8-6	300027	全景网 8 月 6 日讯 华谊兄弟 (300027, 股吧) (300027) 周五盘后公告称，截至 8 月 5 日 24 时，上映 15 天，公司 2010 年重点影片《唐山大地震》的票房达到 4.8 亿元……

Fig. 2. The news store in database

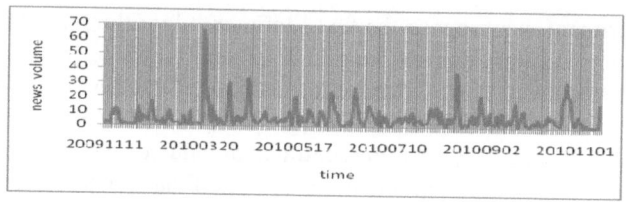

Fig. 3. Thenews number of Huayi Brothers (300027) on the timeline

3.3   Stock Price Information Processing

We obtained the transaction data and historical prices of three stocks from Yahoo Finance (http://finance.yahoo.com.cn),Huayi Brothers (300027)increase in shares in 2010, doubling the total share capital,we need to multiplex the right to calculate the actual price, the amendment shown in Fig.4.

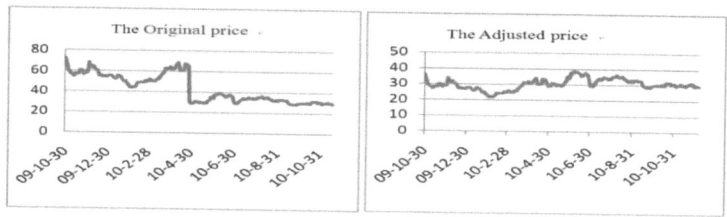

Fig. 4. Theoriginal and adjusted stock price of Huayi Brothers (300027) on the timeline

The experiments selected the closing price, the opening price, the highest price and lowest price of each trading day for nearly a year for the three stocks, take the average of these four price that day the P, P = (openingprice + closing price + highest price + lowest price) / 4).Financial news and stock prices are divided into a time window, we use three-day sliding window to predict the fourth day of the price. Toread (300005) a total of 370 groups (P_{t-2}, P_{t-1}, P_t,P_{t+1}) data, t = 1, 2, 3 ..., 370. Hanwei Electronics (300007) 336 groups (P_{t-2}, P_{t-1}, P_t,P_{t+1}) the data, t = 1, 2, 3 ..., 336, Huayi Brothers (300027) a total of 262 groups (P_{t-2}, P_{t-1}, P_t,P_{t+1}) data, t = 1, 2, 3 ..., 262. Use prediction models for a prediction, that is, input the relevant data during T, T-1, T-2 to forecast the stock priceT+1, and thencomparethe predict valuesto actual values, analysis the forecasting results.

3.4   Data Preprocessing

The format requirements of Libsvm training data collection is : <label><index1>:<value1><index2>:<value2> ... .<label>is the target value of the training data set, for regression it could be any real number, in our experiments,<label>column is the stock price of the next timeperiod. <index>is an integer from 1, may not be continuous. <value>is the value of predictors, or value of explanatory variables.First, we will converted the data to the format required by the above Libsvm, and use the scale Toolbox provided by Libsvmto make each factor data fall into [-1, 1] interval.

Selected training samples in accordance with the training set: test set for the ratio of 4:5,to select 4/5 of all the data as a training set. As a result, Toread (300005) got training set of 296 groups, the test set of 370 groups, Hanwei Electronics (300007) got training set of 268 groups, 336 groups for test sets, Huayi Brothers (300027) got training set of 206 groups, and the test set 262 groups.

3.5   Group Experiments

Here we use the SVM regression function to exploit the relationship between financial news sentiments and stock prices of financial markets, takeLibsvm as the experimental tool. Libsvm is a simple, convenient, fast and efficient SVM pattern recognition and regression package, developed and designed by LinChih-Jen of Taiwan University. There are five different types The SVM, in this experiment, we choose epsilon-SVR to do support vector regression to predict the stock price.

We designed four sets of three-day sliding window SVM prediction comparative experiment. Experiment 1: As shown in Table 1, use the stock price the data of previous three days to forecast the stock price of t+1. Experiment 2: based on Experiment 1, add the corresponding daily number of stock news as the exogenous variables. Experiment 3: based on Experiment 1, add thecorresponding stock news sentimental value of the day. Experiment 4: addingcorresponding daily number of stock news and stock news sentimental value at the same time based on Experiment 1.

Table 1. The input and out variables of the experiment

Indicator	Experiment 1	Experiment 2		Experiment 3		Experiment 4		
Input variables	P_{t-2}	P_{t-2}	V_{t-2}	P_{t-2}	W_{t-2}	P_{t-2}	V_{t-2}	W_{t-2}
	P_{t-1}	P_{t-1}	V_{t-1}	P_{t-1}	W_{t-1}	P_{t-1}	V_{t-1}	W_{t-1}
	P_t	P_t	V_t	P_t	W_t	P_t	V_t	W_t
			V_{t+1}		W_{t+1}		V_{t+1}	W_{t+1}
Output variable	P_{t+1}	P_{t+1}		P_{t+1}		P_{t+1}		

3.6   Calculate the Evaluation Score

We use the following statistical indicators to evaluate the prediction results: mean square error (MSE), the standardized mean square error (NMSE), mean absolute error (MAE), multiple correlation coefficients (SCC), and the direction of symmetry (DS), weighted direction symmetry (WDS), correct upward trend (CP), and correct downward trend (CD).

The definition and calculation formula of the statistical indicators are shown in Table 2. MSE, NMSE and MAE is an indicator which measures the deviation between the true value and the predictive value, MSE, NMSE and MAE is smaller, the predictive value is more close to the true value. The closer the multiple correlation coefficient SCC to 1, the better the forecast performance is, it can be used as an indicator to test the total effect of the forecast. DS, CP and CD are the indicators to evaluate the accuracy of the direction forecast. DS represents the proportion of all correctly predict the direction of the point, totaling 100 for all direction of the points have been correctly predicted, CP indicates the correct upward predict proportion, CD the correct downward predict proportion. So: CP + CD ≤ DS ≤ 100. WDS is an indicator which measures the magnitude and direction of the forecast bias. WDS punish the point of predicting the wrong direction, and reward the point of prediction in the right direction. If WDS is the smaller, the predicted direction accuracy is higher.

Table 2. The definition and calculation formula of the statistical indicators

Metrics	Calculation
MSE	$MSE = \dfrac{1}{n}\sum (y_i - p_i)^2$
SCC	$SCC = \dfrac{(n\sum y_i p_i - \sum y_i \sum p_i)^2}{(n\sum p_i^2 - n(\sum p_i)^2) \times (n\sum y_i^2 - n(\sum y_i)^2)}$
NMSE	$NMSE = \dfrac{1}{n\delta^2}\sum (y_i - p_i)^2 , \delta^2 = \dfrac{1}{n-1}\sum (y_i - \bar{y})^2$
MAE	$MAE = \dfrac{1}{n}\sum |y_i - p_i|$
DS	$DS = \dfrac{100 \times \sum d_i}{n} , d_i = \begin{cases} 1, (y_i - y_{i-1})(p_i - p_{i-1}) \geq 0 \\ 0, \text{otherwise} \end{cases}$
WDS	$WDS = \dfrac{\sum d_i |y_i - p_i|}{\sum d_i' |y_i - p_i|}$ $d_i = \begin{cases} 0, (y_i - y_{i-1})(p_i - p_{i-1}) \geq 0 \\ 1, \text{otherwise} \end{cases}$ $d_i' = \begin{cases} 1, (y_i - y_{i-1})(p_i - p_{i-1}) \geq 0 \\ 0, \text{otherwise} \end{cases}$
CP	$CP = \dfrac{100 \times \sum d_i}{n} ,$ $d_i = \begin{cases} 1, (p_i - p_{i-1}) > 0, (y_i - y_{i-1})(p_i - p_{i-1}) \geq 0 \\ 0, \text{otherwise} \end{cases}$
CD	$CD = \dfrac{100 \times \sum d_i}{n} ,$ $d_i = \begin{cases} 1, (p_i - p_{i-1}) < 0, (y_i - y_{i-1})(p_i - p_{i-1}) \geq 0 \\ 0, \text{otherwise} \end{cases}$

y_i is the actual value, p_i is the predicted value

4   Analysis of the Experimental Results

In the experiment, we used four kernel functions in SVM. With different kernel functions, wealso adjust the parameters respectively to get a better result, as shown in the following tables.

Table 3. Results of Toread (300005)

Experiment	Training data				Test data			
	1	2	3	4	1	2	3	4
MSE	0.0991	0.0759	0.0648	0.0603	0.0919	0.0761	0.2584	0.1261
SCC	0.9871	0.9901	0.9916	0.9921	0.9868	0.9890	0.9638	0.9821
MAE	0.2387	0.2205	0.2004	0.2252	0.2312	0.2184	0.2970	0.2726
NMSE	0.0129	0.0099	0.0084	0.0078	0.0132	0.0109	0.0371	0.0181
DS	65	67	71	72	63	65	68	69
CP	34	34	39	37	31	32	36	34
CD	31	32	32	35	31	33	32	34
WDS	0.4387	0.4697	0.4256	0.3282	0.4455	0.4851	0.5243	0.4292

Table 4. Results of Hanwei Electronics (300007)

Experiment	Training data				Test data			
	1	2	3	4	1	2	3	4
MSE	0.4800	0.3896	0.3873	0.2732	0.4198	0.3588	0.3873	0.2825
SCC	0.9553	0.9638	0.9644	0.9745	0.9614	0.9676	0.9644	0.9744
MAE	0.5224	0.4787	0.4237	0.3850	0.4832	0.4540	0.4237	0.3971
NMSE	0.0446	0.0362	0.0355	0.0254	0.0385	0.0329	0.0355	0.0259
DS	65	68	68	73	64	66	68	70
CP	35	36	36	39	34	34	36	37
CD	29	32	31	33	29	32	31	33
WDS	0.4300	0.4019	0.4210	0.3424	0.4439	0.4069	0.4210	0.4035

Table 5. Results of Huayi Brothers (300027)

Experiment	Training data				Test data			
	1	2	3	4	1	2	3	4
MSE	0.5661	0.5331	0.4345	0.2078	0.5066	0.4434	0.5301	0.4619
SCC	0.9633	0.9652	0.9719	0.9866	0.9591	0.9642	0.9577	0.9634
MAE	0.5596	0.5446	0.4746	0.3821	0.5314	0.5138	0.5049	0.4789
NMSE	0.0367	0.0346	0.0281	0.0135	0.0408	0.0357	0.0427	0.0372
DS	64	64	69	76	65	66	70	73
CP	31	31	34	37	32	32	34	34
CD	33	32	34	39	33	33	35	39
WDS	0.5307	0.5290	0.4588	0.3360	0.4970	0.4926	0.4731	0.4282

Table 6. Results of the average of the three stock

Experiment	Training data				Test data			
	1	2	3	4	1	2	3	4
MSE	0.3817	0.3329	0.2956	0.1804	0.3394	0.2928	0.3920	0.2902
SCC	0.9685	0.9730	0.9760	0.9844	0.9691	0.9736	0.9620	0.9733
MAE	0.4402	0.4146	0.3662	0.3308	0.4153	0.3954	0.4085	0.3829
NMSE	0.0314	0.0269	0.0240	0.0156	0.0308	0.0265	0.0384	0.0271
DS	64.6667	66.3333	69.3333	73.6667	64.0000	65.6667	68.6667	70.6667
CP	33.3333	33.6667	36.3333	37.6667	32.3333	32.6667	35.3333	35.0000
CD	31.0000	32.0000	32.3333	35.6667	31.0000	32.6667	32.6667	35.3333
WDS	0.4664	0.4669	0.4351	0.3355	0.4621	0.4615	0.4728	0.4203

The results data display the performance of the training set and test set in the four groups of experiments. With the increase and change in the input data, the evaluation index has more excellent performance, the accuracy of prediction has been gradually increasing, and this is more evident in the results of the training data set.

Especially from Experiment 2 to Experiment 3, We can see in the experiments of each stock, the input data change from stock price P and information volume V to stock price P and information sentimental value W, the accuracy of the forecast of the price trends DS have been greatly improved, the mean value increased from 65.7% to 68.7%, indicating that to some extent, compared to the financial information volume, financial information sentimental value is more helpful on stock price forecast.

In each stock, experiment 4 are given the best results, indicating the forecast which add information volume and financial information sentimental value based on price is more comprehensive and more accurate.

5   Discussion

In this paper, we conduct experimental research on Support Vector Machines and apply our research to stock price forecasting. Setting historical stock price data, financial information volume and information sentimental value as key factors, we got ideal forecasting result by modeling training and forecasting, analyzed and compared the results. The results indicate that information sentimental value is more influential than information volume on stock price forecasting. Comparing to the experiment with one dimensional data, the result from the experiment with both information volume and information sentimental value is more accurate.

The Internet financial information sentimental value is our major innovation on variable selection. It extends the training and forecasting model from the perspective of information, which not only uses data from the inside transaction, but also show the impact on the stock price of the outside events. The variation of variables will influence the predicted value. In our experiment, we optimized the accuracy of forecasting by minimizing the mean-square error of stock price predicted value via limiting interval.

Our paper improved the existing training and forecasting model. The learning ability of Support Vector Machines is to obtain information from data, by using time window, using more historical data to predict, it can carry more fully information. The result is better than other methods from the perspective of various predictive indicators. Thus, our experiment provides a more accurate stock price forecasting method for investors, which enable them manage investment and risk more efficiently in financial market.

But there are also some limitations in this paper, the sentimental computing is the opinion analysis of online information. The information we use the still is mainly based on news, the trend now is Micro-blogging and other social media has become a generally accepted the opinions platform, use the sentimental value of portal news to represent the overall sentimental is not comprehensive enough. In order to get more effectively and accurately predict and analysis, we need to analyze finance-related government departments, companies, social media, financial experts and commentators, even the opinions of all shareholders in the future work.

Acknowledgment. The work was supported by the Fundamental Research Funds for the Central Universities, and the Research Funds of Renmin University of China (11XNL010), and the Natural Science Foundation of China under grants70871001and 71271211.

References

1. Yaser, S.A.M., Atiya, A.F.: Introduction to Financial Forecasting. J. Appl. Intell. 6, 205–213 (1996)
2. Malkiel, B.G.: A Random Walk Down Wall Street. W.W. Norton and Company Ltd., New York (1973)
3. Werner, A., Murray, Z.F.: Is All That talk Just Noise? The Information Content of Internet Stock Message Boards. J. Fin. 59, 1259–1294 (2004)
4. Liang, X.: Impacts of Internet Stock News on Stock Markets Based on Neural Networks. In: Wang, J., Liao, X.-F., Yi, Z. (eds.) ISNN 2005. LNCS, vol. 3497, pp. 897–903. Springer, Heidelberg (2005)
5. Liang, X.: Mining Associations betweenWeb Stock News Volumes and Stock Prices. J.Int. J. Syst. Sci. 37, 919–930 (2006)
6. Sanjiv, R.D., Mike, Y.C.: Yahoo! For Amazon Sentiment Extraction from Small Talk on the Web. J. Mgt. Sci. 53, 1375–1388 (2007)
7. Wang, Y.F.: Predicting Stock Price Using Fuzzy Grey Prediction System. J.Expert Syst. Appl. 22, 33–39 (2002)
8. Kim, K.J., Han, I.: Genetic Algorithms Approach to Feature Discretization in Artificial Neural Networks for the Prediction of Stock Price Index. J. Expert Syst. Appl. 19, 125–132 (2000)
9. Cao, L.J., Tay, F.E.H.: Financial Forecasting Using Support Vector Machines. J. Neural Comput. Appl. 10, 184–192 (2001)

10. Hassan, M.R., Nath, B.: StockMarket Forecasting Using Hidden Markov Model: A New Approach. In: 5th International Conference on Intelligent Systems Design and Applications, pp. 192–196. IEEE Press, Australia (2005)
11. Chang, P.C., Liu, C.H.: A TSK Type Fuzzy Rule based System for Stock Price Prediction. J. Expert Syst. Appl. 34, 135–144 (2008)
12. Liao, Z., Wang, J.: Forecasting Model of Global Stock Index by Stochastic Time Effective Neural Network. J. Expert Syst. Appl. 37, 834–841 (2010)
13. Liang, X., Zhang, H.S., Xiao, J.G., Chen, Y.: Improving Option Price Forecasts with Neural Networks and SupportVector Regressions. J. Neural Comput. Appl. 72, 3055–3065 (2009)
14. Haykin, S.: Neural Networks: A Comprehensive Foundation. Prentice-Hall International Inc., Englewood Cliffs (1999)
15. Cao, L.J., Tay, F.E.H.: Support Vector Machine with Adaptive Parameters inFinancial Time Series Forecasting. J. IEEE Tran. on Neural Network 14, 1506–1518 (2003)
16. Lee, M.C.: Using Support Vector Machine with a Hybrid Feature Selection Method to the Stock Trend Prediction. Expert Syst. Appl. 36, 10896–10904 (2009)
17. Yeh, C.Y., Huang, C.W., Lee, S.J.: A Multiple-Kernel Support Vector Regression Approach for Stock MarketPrice Forecasting. J. Expert Syst. Appl. 38, 2177–2186 (2011)
18. Hassan, M.R., Nath, B., Kirley, M.: AFusion Model of HMM, ANN and GA for Stock Market Forecasting. J. Expert Syst. Appl. 33, 171–180 (2007)
19. Bildirici, M., Erisn, O.O.: Improving Forecasts of GARCH Family Models with the Artificial Neural Networks: An Application to the Daily Returns in Istanbul Stock Exchange. J. Expert Syst. Appl. 36, 7355–7362 (2009)
20. Wen, Q.H., Yang, Z.H., Song, Y.X., Jia, P.F.: Automatic Stock Decision Support System Based on Box Theory and SVM Algorithm. J. Expert Syst. Appl. 37, 1015–1022 (2010)
21. Zhao, Y.Y., Qin, B., Liu, T.: Sentiment Analysis. J. Softw. 21, 1834–1848 (2010)
22. Chen, P.Y., Wu, S.Y., Yoon, J.S.: The Impact of Online Recommendations and Consumer Feedback on Sales. In: 25th International Conference on Information Systems, pp. 711–724. AIS, Washington (2004)
23. Liu, Y.: Word of Mouth for Movie: Its Dynamics and Impact on Box Office Revenue. J. Marketing 70, 74–89 (2006)
24. Ghose, A., Panagiotis, G.I.: Designing Novel Review Ranking Systems: Predicting the Usefulness and Impact of Reviews. In: 9th International Conference on Electronic Commerce, pp. 303–310. ACM, New York (2007)
25. Devitt, A., Ahmad, K.: Sentiment Polarity Identification InFinancial News: A CohesionbasedApproach. In: 45th Annual Meeting of the Association of Computational Linguistics, pp. 984–991. Association for Computational Linguistics, Prague (2007)

Automated Web Data Mining Using Semantic Analysis

Wenxiang Dou[1] and Jinglu Hu

[1] Graduate School of Information, Product and Systems, Waseda University
2-7 Hibikino, Wakamatsu, Kitakyushu-shi, Fukuoka, 808-0135, Japan
william@ruri.waseda.jp,
jinglu@waseda.jp

Abstract. This paper presents an automated approach to extracting product data from commercial web pages. Our web mining method involves the following two phrases: First, it analyzes the data information located at the leaf node of DOM tree structure of the web page, generates the semantic information vector for other nodes of the DOM tree and find maximum repeat semantic vector pattern. Second, it identifies the product data region and data records, builds a product object template by using semantic tree matching technique and uses it to extract all product data from the web page. The main contribution of this study is in developing a fully automated approach to extract product data from the commercial sites without any user's assistance. Experiment results show that the proposed technique is highly effective.

Keywords: Web data extraction, product data mining, Web mining.

1   Introduction

With the information time coming, more and more companies manage their business and services on the World Wide Web and thus these web sites have an explosive growth. Huge amounts of product have been displayed in respective commercial web sites using fixed templates. It has important meaning to extract these product data for offering valuable services such as comparative shopping and meta-search, etc.

The early approaches use the information extraction technique called wrapper [1], which is a program that extracts data from web pages based on a priori knowledge of their format. The wrapper could either be generated by a human being (called programming wrapper) or learned from labeled data (called wrapper induction). Programming wrapper needs users to find patterns manually from the HTML code to build a wrapper system. This is very labor intensive and time consuming. Systems that use this approach include RAPIER[2], Wargo[5], WICCAP[12], etc. The wrapper induction uses supervised learning to learn data extraction rules from a set of manually labeled examples. Example wrapper induction systems include Stalker[3], WL[6], etc. These wrapper construction systems actually output extraction rules from training examples provided by the designer of the wrapper. But, they have two major drawbacks. Firstly, they require a previous knowledge of the data. Secondly, additional work might be required to adapt the wrapper when the source changes.

S. Zhou, S. Zhang, and G. Karypis (Eds.): ADMA 2012, LNAI 7713, pp. 539–551, 2012.

To overcome these problems, some automatic extraction techniques were developed to mine knowledge or data from web pages [7], [10]. Embley et al. [4] uses a set of heuristics and domain ontologies to automatically identify data record boundaries. But the method requires users to predefine a detailed object-relationship model. Zhai et al. [14] proposes an instance-based learning method, which performs extraction by comparing each new instance to be extracted with labeled instances. This approach also needs users to label pages when a new instance cannot be extracted. In [8], the system IEPAD is proposed to find patterns from the HTML tag string of a page, and then use the patterns to extract data items. However, the algorithm generates many spurious patterns. Users have to manually select the correct one for extraction.

In [9], [11], [13], these automatic extraction techniques require multiple similar pages from the same site to generate a template and extract data. The typical system is RoadRunner [9], which infers union-free regular expressions from multiple pages with the same template. It is a limitation for our opnion: not all web sites can find several pages with the similar structure for every product.

In this paper, we proposed a novel technique to perform automatic data extraction. There are three different features between our method and existing automatic extraction techniques: First, we just need a single page with lists of product data. But most existing methods require multiple pages. Second, our method is a fully automatic extraction technique without any human's labor. Third, existing methods execute tag matching for finding repeated pattern based on whole page. There are two obstacles: 1) The navigator and advertisement bar also contain many repeated structures. It is a problem to accurately estimate which one is a correct repeated pattern for a page with the complex structure. 2) For similar pages of the same site, they have almost the same structure. The automatic extraction methods requiring multiple pages are prone to see a whole page as a data record. So, most existing approaches have a low accuracy. However, our method, firstly, extracts maximal repeated semantic information pattern from a page. Then, it searches the product data region and finds product data in this page by matching the extracted semantic pattern. Finally, it generates the correct repeated pattern from the found product data and uses it to extract all product data from the page. The maximal repeated semantic information pattern is more appropriate to identifying data region than the repeated tag pattern on commercial web pages because the product data with the rich semantic information assure the accuracy of the pattern. So, our method can avoid above two obstacles. Experiment results show that our system has higher accuracy than most existing automatic extraction methods.

The remaining of the paper is organized as follows. Section 2 presents the overall procedure for extracting product data. Section 3 describes the algorithm for finding product data region and extracting product data. We present and analyze experimental results in Section 4. Section 6 concludes our study.

2   Our Approach: Mining Produce Data on the Web Sites

2.1   Product Data Representation Structure

There are three typical representation structures for product data from the view of the page as shown in Fig. 1. However, on the DOM tree, these representations are written in HTML by using following two structures: single data region or multiple data regions structure as shown in Fig. 2. In fact, the data region is one of the sub-trees of DOM tree and all product data are the child sub-trees of the data region. In here, we have to explain two definitions about product data region.

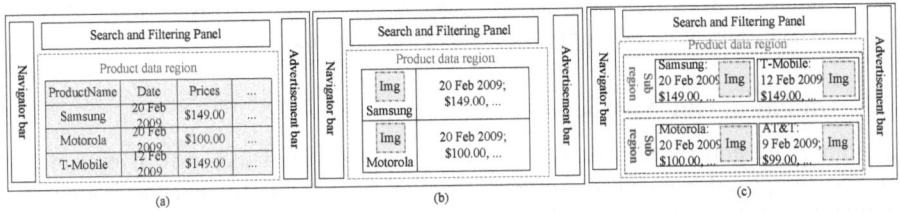

Fig. 1. Typical representation of product data on the commercial web pages. (a) Tabular structure. (b) List structure. (c) Block structure.

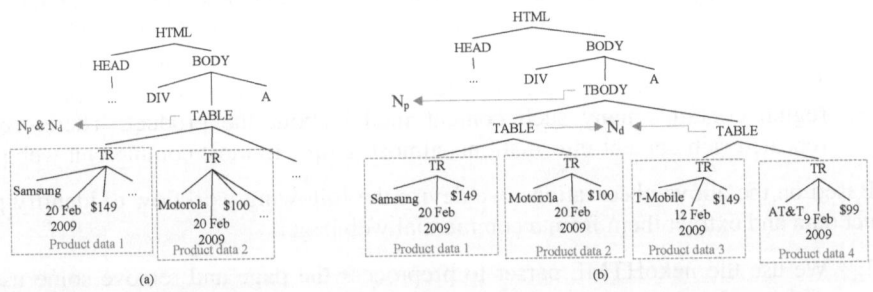

Fig. 2. HTML structure of product data on DOM tree. (a) Single data region structure. (b) Multiple data regions structure.

Definition 1: Product data region is the minimum sub-tree which contains all product data sub-trees in the DOM tree of a pag and N_p is the root node of the sub-tree. In Fig. 2a, the N_p node is the TABLE tag. But the TBODY tag is the N_p node rather than the TABLE tag in Fig. 2b because the TABLE tag does not contain all product data.

Definition 2: Direct product region is the sub-tree that contains product data sub-trees directly in the DOM tree of a pag and N_d is the root of the sub-tree. In Fig. 2a and Fig. 2b, all the TABLE tags are the N_d node. The TBODY tag is not the N_d node because it does not contain product data sub-trees directly.

So, the single data region structure means it just has a N_p node and the N_p node is also a N_d node as shown in Fig. 2a. The multiple data regions structure means it has a N_p node and several N_d nodes, and the N_p node is not a N_d node as shown in Fig. 2b.

From our experiments, we found that the pages containing the tabular structure Fig. 2a must be the single data region structure. Most pages containing the list structure Fig. 2b are the single data region structure. On the contrary, most pages containing the block structure Fig. 2c are the multiple data regions structure. These will be shown in Section 4. Some pages may contain combinations of the above structures, such as a block structure Fig. 2c embedded in an advertisement bar of a page with a list structure Fig. 2b. It is difficult to mine all product data accurately from this page. Fortunately, most pages use only one type of representation.

Therefore, our method will, firstly, find the N_p node and then recognize the N_d node under the N_p node. Finally, we generate the product object template by aligning the product data sub-trees under the N_d node and use the model to extract all product data from the page. The detailed steps will be described in next phase.

2.2   Outline of Our Approach

The important step for structured data extraction is to find the correct repeated pattern. We proposed a novel and effective technique (PDM) to find the boundary of product data. It is based on the following two observations:

1. Product data have the diverse contents such as image, price, title, description, etc. So, product data region has richer semantic information than any other regions which always have the monotone content on the page. For example, the navigator bar consists of links and the advertisement bar basically is dominated by images.
2. Among the product data, their contents are similar. For example as shown in Fig. 1, every phone product are decribed by the similar contents. The product data region contains many such content model about the product. Therefore, the repeated rich content informationis almost unique in most commercial web pages.

Based on the above observations, we devise the following workflow to identify product data and extract them from a commercial web page:

1. We use the nekoHTML parser to preprocess the page and remove some useless nodes and information for our work from the page such as javascript language, etc. Then, we build the DOM tree of the page.
2. The semantic analyzer developed by us identifies the semantic type of the information located at every leaf node of the DOM tree. Next, The semantic information vector is built for every node from down to up on the DOM tree.
3. We align the semantic information vector of nodes in every level of the DOM tree from the down to up and get the maximum repeated semantic vector pattern (MRSV). Then, the pattern is used to find the product data region and N_d node.
4. In final, we build the product object template of the page by matching the product data sub-trees under the N_d node and use this model to extract all product data.

The difference between our approaches with existing methods is that we use the novel semantic analysis and align technique to identify the structured data. It is effective to be applied in the product data extraction because the MRSV always reflects the correct product data region in most commercial sites. Especially, it also can achieve high accuracy when it handles complicated pages from the commercial sites.

3   Product Data Identification

3.1   Semantic Analysis about the Information of the Page

We develop a semantic analyzer to identify the semantic type on the page. In our work, all pair-tags (like <tr></tr>) on DOM tree are nodes and other tags and contents under these pair-tags are the semantic information. We classify the semantic information into seven semantic types: title *(TIT)*, description *(DES)*, price *(PRI)*, number *(NUM)*, image *(IMA)*, single tag *(S_T)* and special *(SPE)*.

The semantic analyzer consists of several if-then rules. These rules are used to match with the semantic information and the result deduced by the best matching rule will be as semantic type of the information. The following show the rules:

Definition of Expressions

X: is variable. $(X)^1$ denotes X has only one appearance, $(X)^?$ denotes X has only one or no appearance, and $(X)^+$ denotes X is larger than or equal one appearance. $\neg X$ denotes it is not X.

W: denotes a word.
C_W: denotes the first letter capitalized word.
T_{ag}:   denotes a single tag such as <br>, <p>, etc.
N_{um}: denotes the number.
Σ: is a alphabet {a, b, c, ...}.
Σv: is a character set {$, ¥, £, ...}. These characters are popular value expressions.
I_{mg}: The information could be an image. So, I_{mg} denotes an image tag <img ...>.
$Num(X)$: the function $Num(X)$ denotes the number of the variable X.
$X \cdot Y$: denotes the union between two variables X and Y.

Semantic Rule Pool

R_{TIT} identifies the title type:

$$If\,(W)^+\ \&\ Num(C_W) >= 1/2 * Num(W)\ then\ TIT. \tag{1}$$

R_{DES} identifies the description type:

$$If\,(W)^+\ \&\ Num(C_W) < 1/2 * Num(W)\ then\ DES. \tag{2}$$

R_{IMG} identifies the image type:

$$If\,(T_{ag})^1\ \&\ I_{mg}\ then\ IMG. \tag{3}$$

R_{S_T} identifies the single tag type:

$$If\,(T_{ag})^1\ \&\ \neg I_{mg}\ then\ S\_T. \tag{4}$$

R_{NUM} identifies the number type including quantity, size, etc:

$$If\,((N_{um})^+\ \&\ (W)^?)\ |\ (N_{um}\cdot(x\in\Sigma))^1\ then\ NUM. \tag{5}$$

R_{PRI} identifies the price type:

$$If\,((x\in\Sigma v)\cdot N_{um})^+\ \&\ (W)^?\ then\ PRI. \tag{6}$$

The information which cannot match any rules will be as special semantic type.

3.2   Building Semantic Information Tree

After obtaining the semantic types of leaf nodes, we start to generate semantic information vectors (SIV) for all non-leaf nodes of the DOM tree. The SIV consists of seven items and every item expresses a semantic type as shown in the follows:

$$SIV = [S\_T, IMG, DES, TIT, NUM, PRI, SPE]$$

The SIV of the node is generated by adding up the value of the SIV of its all child nodes as shown in Fig. 3. For example, the leaf node <br> and <img> in the lowest level of the tree are tagged the semantic type as the S_T and IMG respectively by the semantic analyzer. Then, the SIV of their parent node <td></td> is[1,1,0,0,0,0,0] by adding up [1,0,0,0,0,0,0] and [0,1,0,0,0,0,0]. Final, the SIV [4,6,0,6,0,2,0] of the root node <<body></body> can be generated by iterative computing from down to up. From the DOM tree with semantic information vectors, we can see the distribution of the information of the web page clearly. In the next phases, we will present how to use the semantic information to find product data region and identify product data.

3.3   Maximum Repeated Semantic Vector

As we mentioned in the above, the biggest difference between our work and existing methods is that we identify the data region by finding MRSV (maximum repeated semantic vector) pattern of the page rather than repeated tag pattern. For some complex commercial pages, many repeated tag patterns could be found and it is difficult to find a correct one from them automatically. But, due to many products with rich information contained by most pages, the MRSV always is the correct pattern of product data. For obtaining the MRSV of a page, we need to generate the MRSV of every non-leaf node of the DOM tree from down to up and the MRSV of the root generated in the last will be the MRSV of this page. In the following, we explain how to generate the MRSV of a node. There are two steps:

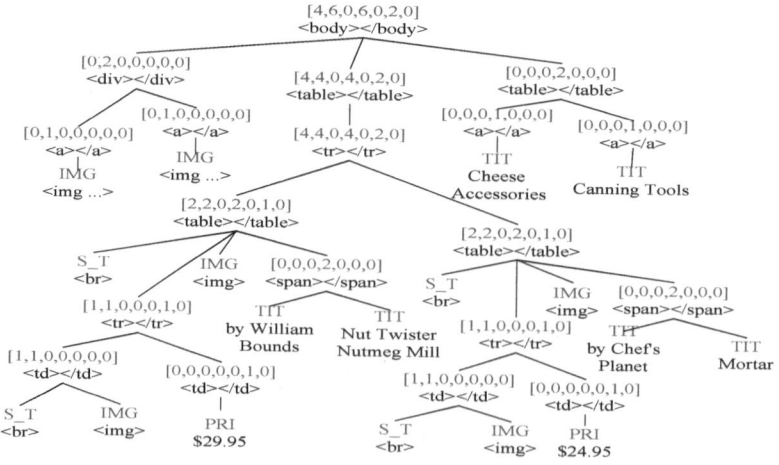

Fig. 3. The DOM tree with semantic information vectors

1. Findind all RSV patterns contained by the node.

Under a node, there could be several RSV patterns. There are two sources for them: 1) the first source is the similar semantic vectors of several child nodes. It can be detected by using vector similarity computation. We designed a new vector similarity formula which can reflect the similarity of semantic types and the information quantity between two SIVs (semantic information vectors). Let us see the definition 3.

Definition 3: Suppose v and w are two semantic information vectors, the similarity of v and w is

$$SIM(v,w) = \frac{\sum_{i=1}^{m}(min(x_i,y_i)/max(x_i,y_i))}{max(n_v,n_w)}, x_i > 0 \text{ or } y_i > 0 \tag{7}$$

Here, x_i, y_i denotes the value of the ith item of vector v and w respectively, and n_v, n_w denotes the number of the items with values larger than zero in vector v and w respectively. We use this formula to detect whether there are similar SIVs among child nodes of the node. If the similarity of two SIVs is larger than a set threshold, then they are similar. After finding similar SIVs, we get their average as the RSV. 2) The second source is the MRSV patterns generated by child nodes of the node.

2. Selecting the maximum one from found RSV patterns.

After finding all RSV patterns contained by the node, we compare them and select the maximum one as the MRSV of the node. So, we need to transform every RSV into a measure value. Definition 4 shows the transforming formula.

Definition 4: Suppose w is a RSV and the information value of w is as follows

$$V(w) = \sum_{i=1}^{m} y_i \cdot n_w \cdot r_w \tag{8}$$

Here, y_i denotes the value of the ith item of vector w, n_w denotes the number of the items with values larger than zero in w and r_w denotes the repeated number of vector w. Definition 4 means more semantic types, information quantity and repeated number the RSV has, larger value it can obtain. Therefore, the RSV with the maximum value will be the MRSV of the node. Similarly, the MRSV of the N_p node (root node of the product data region sub-tree) with rich semantic information has the highest probability to become the MRSV of the whole page.

For example, Fig. 4 shows the detailed searching process of the MRSV with the similarity threshold 0.7. The three nodes <div></div>, <table></table> and <div></div> in the second level of the tree obtain their MRSVs respectively. In the table of the MRSV, n denotes the repeated number and v is the information value of the MRSV. When the algorithm is implemented to the root <body></body>, it firstly detects whether there are similar SIVs among its three child nodes. From the Fig. 4, we can see the SIV [0,2,0,2,0,0,0] of the node <div></div> is similar with the SIV

[0,1,0,2,0,0,0] of another node <div></div> because the similarity is equal to 0.75. So, there is a RSV pattern among child nodes of the root and we extract it by averaging the two SIVs. So, the RSV is [0,1.5,0,2,0,0,0]. Second, the algorithm detects whether the child nodes of the root have their MRSVs. In this example, all three child nodes have their MRSVs. In final, there are four RSV patterns for the root. Through comparing their information values, the RSV [2,2,0,2,0,1,0] with the maximum value 56 is the MRSV of the root. Therefore, it is also the MRSV of the whole page.

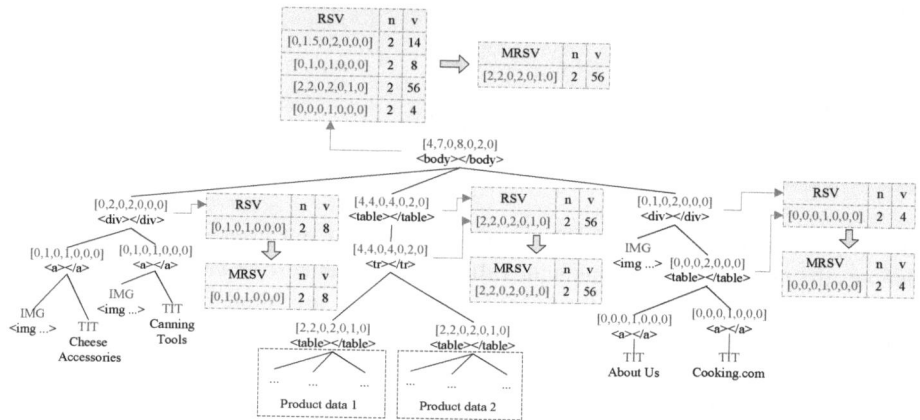

Fig. 4. The searching process of the MRSV pattern with the similarity threshold 0.7

3.4   Identifying the Boundary of the Product Data

In this phase, we will identify the product data using the MRSV. It is easy to identify the product data region (the sub-tree whose root node is the N_p node) because the MRSVs of all nodes are same with the MRSV of the root on the pass from the root to the N_p node if the MRSV of the root is correct pattern of the product data region. But for identifying product data (sub-trees under the N_d node), there are two situations:

1.   The MRSV of the root reflects the semantic information of product data directly. This situation means the N_d node is the lowest level node containing the MRSV of the root node. So, it is simple to identify product data in this situation because the sub-trees of product data are under the N_d node directly. Therefore, we firstly find the N_d node, and then we align the SIV of every child node of the N_d node with the MRSV of the root. If the SIV of some child node is similar with the MRSV, then all sub-trees under the child node is the product data information. While, The N_d node can be found by aligning the MRSV because it is the lowest level node whose MRSV is same with the MRSV of the root. The page with a single data region structure belongs to this situation because the N_p node and the N_d node point to the same node.

2.   The MRSV of the root reflects the semantic information of the sub data region. This situation means the N_p node is the lowest level node containing the MRSV of the root rather than the N_d node. This situation often happens in the page with

multiple data regions because there are a N_p node and several N_d nodes in this page and the N_p node is not the N_d node. When there are several sub data regions sharing most product data averagely in the page, the SIVs of the N_d nodes of these sub data regions will be the MRSV of the whole page. Fig. 5 shows the example about three sub data regions sharing the product data of the page. The <table></table> with the SIV [12,12,0,12,0,6,0] is the N_p node and three <tr></tr> nodes with the SIV [4,4,0,4,0,2,0]   are the N_d nodes. If using the above searching algorithm, the three sub data regions will be mistaken for the product data.

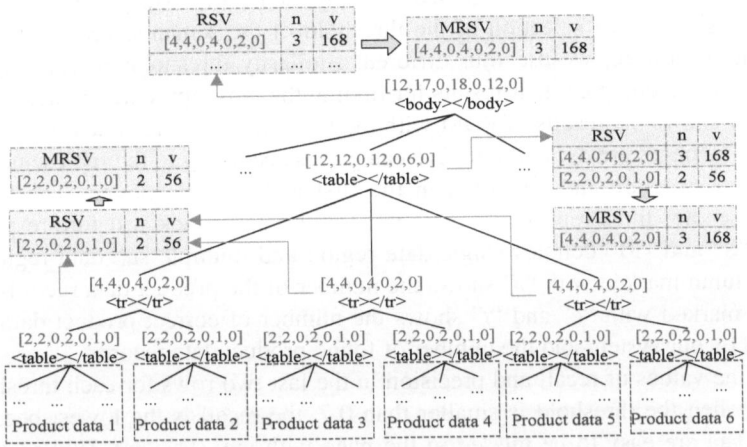

Fig. 5. Identifying the product data on the multiple regions

So, we need to check whether found product data sub-trees are correct. We compare the SIV of the root of the found sub-tree with the product of its MRSV and the repeated number of the MRSV because the SIV of the sub data region basically is the sum of the SIVs of the product data contained by it. If they are similarity, then the found sub-tree is the sub data region rather than product data. So, we continue to extract the real product data from these found sub data regions by aligning the MRSV of them with the SIVs of their child nodes.

For example, let us see Fig.5 again. The three sub data regions with the SIV [4,4,0,4,0,2,0] of the root, firstly, will be extracted and we select one among them to compare its SIV [4,4,0,4,0,2,0] with the product of its MRSV [2,2,0,2,0,1,0] and repeated number 2. Due to the product 2*[2,2,0,2,0,1,0] equal to [4,4,0,4,0,2,0], they are similarity. So, they are not product data and we continue to extract real product data under these sub data regions by comparing the SIVs of child nodes with their MRSV. We can see all <table></table> nodes with the SIV [2,2,0,2,0,1,0] are the root nodes of product data sub-trees.

Therefore, the MRSV information of nodes can help us to find product data region and identify the boundary of the product data. In the last, we generate the product object template from found product data and use it to find all product data.

4   Experiments

In this section, we evaluate our system, PDM (Product Data Mining), which implements the proposed techniques. The system is implemented in Eclipse and runs on a 1.66GHz Double Core with 512MB RAM.

The number of sites in our experiments is 154. The total number of pages is 209 and 85 pages from 30 sites are used to compare our system with RoadRunner system. We use Google engine to select diverse commercial sites randomly from the Web. The selection of pages is also based on the diversity of products and representation structure. These pages are preprocessed by using the nekoHTML parser.

For our system, the selection of the threshold of the semantic vector similarity is important. Therefore, we use four different similarity thresholds to run our system respectively and compare their results for finding the most appropriate one.

Table 1 shows the results of PDM with the four different thresholds 0.65, 0.7, 0.75 and 0.8. The pages are divided into five categories according to the representation and data region structure and are listed in the first and second column. The "L", "T", and "B" denotes three representations List, Tabular and Block structure respectively, and the "S" and "M" denotes single data region and multiple sub data regions. The fourth column marked with "r" shows the number of the product data records and the columns marked with "c" and "f" shows the number of correct product data records extracted by our system and the number of found product data records.

From the values of recall and precision in the last two rows for each threshold, we can see, when the threshold is smaller than 0.7, the recall is the lowest because the product data are easy to be missed in the page with any structure. While, when the threshold is larger than 0.7, the system did not perform better for handling the page with multiple sub data regions. With the threshold be larger, the recall and precision become more and more low. Therefore, the semantic vector similarity threshold with the valueof 0.7 is the most appropriate for our system.

Table 1. Experimental Results

Cate-	gory	No.of sites	Product Data Mining								
				0.65		0.7		0.75		0.8	
			r	c	f	c	f	c	f	c	f
L	S	52	856	792	821	852	854	852	854	852	854
	M	12	151	142	146	134	141	136	142	121	136
T	S	9	151	140	148	151	152	151	152	151	152
B	S	21	447	443	445	443	445	443	445	443	445
	M	60	1105	1043	1057	1061	1073	1044	1113	1028	1101
Total		154	2710	2560	2617	2641	2665	2626	2706	2595	2688
Recall				94.46%		97.45%		96.90%		95.76%	
Precision				97.82%		99.10%		97.04%		96.54%	

In the following experiment, Table 2 shows the comparing results between RoadRunner and our PDM system with 0.7 threshold. The column 1 gives the structure type of pages. The column 2 gives 30 commercial sites selected randomly from the above 154 sites and column 3 shows the product type. The column 3 and 4 give the number of pages and the number of product data records in these pages. The columns marked with "corr." and "found" denote the number of correct product data records extracted by the system and the number of found product data records.

The last two rows of the table give the total number of product data records in each column, the recall and the precision of each system.

see p as a r: It means almost the whole page is extracted as a data record.

see dr as a r: It means sub data regions are extracted as data records and the product data under these regions are viewed as their nest structure items.

Table 2. Comparison of Extraction Results of PDM and RoadRunner

	Source				PDM (0.7)		RoadRunner		
category	site	product	page	records	corr.	found	corr.	found	remark
L / S	Apple Store	iPod	2	28	28	28	28	32	4 wrong
L / S	Marblehead	Socks, Bags	3	20	20	20	20	20	
L / S	Israel Sport Shop	Clothes	2	5	5	5	5	5	
L / S	Wristick	Watches	3	20	20	20	20	20	
L / S	Pcuniverse	Camera Lenses	3	30	20	20	-	-	see p as a r
L / S	Network Camera Store	Network Camera	3	30	30	30	30	30	
L / S	Nobles Camera	Batteries	2	25	25	25	-	-	see p as a r
L / S	AbelCineTech	Cables	3	29	29	29	29	29	
L / S	Ofease	Canon, Scanner,...	3	42	42	42	42	42	
L / S	Hammicks BMA	Business Book	3	60	60	60	60	63	3 wrong
L / S	Biblion	Fiction Book	3	139	139	139	139	141	2 wrong
L / S	IEEE computer Society	Tutorials	3	30	30	30	30	30	
L / S	Amazon	Movies & TV	3	75	75	75	-	-	see p as a r
T / S	The Phone Store	Mobile phone	3	40	40	40	40	40	
T / S	AISC Store	books	2	28	28	28	28	28	
T / S	Net32	Alloys	3	60	60	60	-	-	see p as a r
T / S	Whitakers	Rods, Jackets,...	3	20	20	20	20	20	
T / S	Network Webcams	IP Camera	3	71	71	71	71	73	2 wrong
T / S	Pcmag	Cell Phones	3	30	30	30	30	30	
B / S	Newegg	Books	3	39	39	39	-	-	see p as a r
B / S	Oldnavy.gap	Man's Shirts	2	70	70	70	-	-	see p as a r
B / S	Fashion163	Swiss Watch	3	57	57	57	57	61	4 wrong
B / S	Tracsat.co.uk	Sky	3	13	6	10	-	-	see p as a r
B / M	My Jewelry Box	Diamond Rings	3	72	72	72	39	51	see dr as a r
B / M	Amazon	DVDs	2	28	28	28	28	30	2 wrong
B / M	ShopSunGlassesOnline	Sun Glasses	3	45	45	45	-	-	see p as a r
B / M	Motorola	Batteries & Doors	3	45	45	45	-	-	see p as a r
B / M	Forever Jewelers	Pendants	3	60	60	60	60	60	
B / M	Hartgem	Signet Rings	3	24	24	24	24	24	
B / M	Onhop.ca	Components	3	63	63	63	63	63	
B / M	PerfumeSpace	Men's Perfume	3	48	48	48	48	48	
B / S	Embrace Jewelry	Earrings, Rings,...	3	14	0	6	14	18	4 wrong
	Total		85	1360	1339	1349	935	967	
	Recall/Precision				98.5% / 99.3%		68.9% / 96.7%		

n wrong: It means *n* incorrect data records are found.

The following summarizes the experimental results in Table 2.

1. Our system PDM gives perfect results for every site except for the last one. For three pages of the site, all product data are not found because descriptions of product data in these pages just have an image and a title and surrounding other regions have much richer semantic information than the product data region. From the last two rows, we can see that PDM has a 98.5% recall and99.3% precision. While, RoadRunner just has a recall of 68.9%.

2. When the page has a complex structure, RoadRunner system often happens the *"see p as a r"* error. While, our system will still perform better in these pages.

3. Many repeated patterns will be generated if aligning tag structure by using several pages and the incorrect data records are extracted easily. Therefore, our system has a higher precision than RoadRunner.

The experiment results show that our system has the good effectiveness and high accuracy for mining product data in commercial pages

5   Conclusions

In this paper, we propose a novel and effective approach to extract product data from web sites. Our work is related to the structured data extraction from web pages. Although the problem has been studied by some researchers, existing techniques have respective limitations and most automatic extraction methods have the low accuracy. Our method is a fully automated extraction technique without any human's assistance. We use the novel semantic analysis and align technique to identify the structured data. It is very effective to be applied in the product data extraction. Meanwhile, it also can achieve high accuracy when it handles complicated pages from the commercial sites. Experiments results using a large number of commercial pages showed the effectiveness of the proposed technique.

References

1. Kushmerick, N., Weld, D., Doorenbos, R.: Wrapper induction for information extraction. In: Proc. of the 15th IJCAI (1997)
2. Califf, M.E., Mooney, R.J.: Relational learning of pattern-match rules for information extraction. In: Pro. of the AAAI 1999/IAAI 1999 Conf., pp. 328–334 (1999)
3. Muslea, I., Minton, S., Knoblock, C.: A hierarchical approach to wrapper induction. In: Proc. of the 3th Annual AA Conf., pp. 190–197 (1999)
4. Embley, D.W., Campbell, D.M., et al.: Ontology-Based Extraction and Structuring of Information from Data-Rich Unstructured Documents. In: Proc. CIKM, pp. 52–59 (1998)
5. Raposo, J., Pan, A., Alvarez, M., Hidalgo, J., Vina, A.: The Wargo System: Semi-Automatic Wrapper Generation in Presence of Complex Data Access Modes. In: Proc. of 13th Int'l Workshop Database and Expert Systems Applications, pp. 313–320 (2002)
6. Cohen, W.W., Hurst, M., Jensen, L.S.: A Flexible Learning System for Wrapping Tables and Lists in HTML Documents. In: Proc. 11th Int'l Conf. World Wide Web, pp. 232–241 (2002)

7. Embley, D., Jiang, Y., Ng, Y.-K.: Record-boundary discovery in Web documents. In: Proc. of ACM SIGMOD 1999, pp. 467–478 (1999)
8. Chang, C., Lui, S.: IEPAD: Information Extraction Based on Pattern Discovery. In: Proc. of the 2001 Intl. World Wide Web Conf., pp. 681–688 (2001)
9. Crescenzi, V., Mecca, G., Merialdo, P.: ROAD RUNNER: Towards Automatic Data Extraction from Large Web Sites. In: Proc. of.the 2001 Intl. Conf. on Very Large Data Bases, pp. 108–118 (2001)
10. Bar-Yossef, Z., Rajagopalan, S.: Template Detection via Data Mining and its Applications. In: Proc. WWW, pp. 580–591 (2002)
11. Arasu, A., Garcia-Molina, H.: Extracting Structured Data from Web Pages. SIGMOD (2003)
12. Zhao, L., Wee, N.K.: WICCAP: From Semi-Structured Data to Structured Data. In: Proc. of 11th IEEE Int'l Conf. and Workshop Eng. of Computer-Based Systems (ECBS 2004), p. 86 (2004)
13. Ye, S., Chua, T.S.: Learning Object Models from Semistructured Web Documents. IEEE Transaction on Knowledge and Data Engineering 18(3), 334–349 (2006)
14. Zhai, Y., Liu, B.: Extracting Web Data Using Instance-Based Learning. In: Proc. Sixth Int'l Conf. Web Information Systems Eng. (2005)

Geospatial Data Mining on the Web: Discovering Locations of Emergency Service Facilities

Wenwen Li[1], Michael F. Goodchild[2], Richard L. Church[2], and Bin Zhou[3]

[1] GeoDa Center for Geospatial Analysis and Computation, School of Geographical Sciences and Urban Planning, Arizona State University, Tempe AZ 85287
Wenwen@asu.edu
[2] Department of Geography, University of California, Santa Barbara
Santa Barbara, CA 93106
{good,church}@geog.ucsb.edu
[3] Institute of Oceanographic Instrumentation, Shandong Academy of Sciences
Qingdao, Shandong, China 266001
senosy@gmail.com

Abstract. Identifying location-based information from the WWW, such as street addresses of emergency service facilities, has become increasingly popular. However, current Web-mining tools such as Google's crawler are designed to index webpages on the Internet instead of considering location information with a smaller granularity as an indexable object. This always leads to low recall of the search results. In order to retrieve the location-based information on the ever-expanding Internet with almost-unstructured Web data, there is a need of an effective Web-mining mechanism that is capable of extracting desired spatial data on the right webpages within the right scope. In this paper, we report our efforts towards automated location-information retrieval by developing a knowledge-based Web mining tool, CyberMiner, that adopts (1) a geospatial taxonomy to determine the starting URLs and domains for the spatial Web mining, (2) a rule-based forward and backward screening algorithm for efficient address extraction, and (3) inductive-learning-based semantic analysis to discover patterns of street addresses of interest. The retrieval of locations of all fire stations within Los Angeles County, California is used as a case study.

Keywords: Emergency service facilities, Web data mining, information extraction, information retrieval, ontology, inductive learning, location-based services.

1 Introduction

Although it has only been 22 years since its advent, the World Wide Web (WWW) has significantly changed the way that information is shared, delivered, and discovered. Recently, a new wave of technological innovation - the emergence of Web 2.0 and Web 3.0, such as social networks, government surveillance and citizen sensors [5] - has led to an explosion of Web content, and brought us into the era of Big Data [18]. Statistical reports show that by 2008 the amount of data on the Internet had reached 500bn gigabytes [21], and the indexed Web contained at least 11.5 billion pages [6].

S. Zhou, S. Zhang, and G. Karypis (Eds.): ADMA 2012, LNAI 7713, pp. 552–563, 2012.

This information explosion on the Web poses tremendous challenges for various information-retrieval tasks [12].

Within this massive amount of data, identifying location-based information, such as street address and place name, has become very popular, due to the desire to map this information from cyberspace to the physical world. As a type of spatial data, locations of emergency service facilities are especially important in protecting people's lives and safety, and for government agencies to provide real-time emergency response. Taking fire stations as an example, besides the aforementioned functions, insurance companies need the locations of all fire stations within and near a community to determine the insurance costs to be paid by a household. Decision-makers long for this location information to obtain the urban footprint of each fire station and to plan the optimal placement of fire stations within a region.

Presently, most Internet uses obtain WWW information from search engines [16]. However, the commercial search engines such as Google are designed to index webpages on the Internet instead of considering location information that has smaller granularity as an indexable object. Therefore, these search engines always lead to a low recall rate in search results. Fig.1 shows the search results for fire stations within the city of Santa Barbara, CA, from Google. Red pinpoints are the Googled results and blue pinpoints are the actual locations of fire stations within that city. It can be seen that except for *C* (to the west of the green arrow) overlapping with its actual location (blue pinpoint), all of the results are irrelevant.

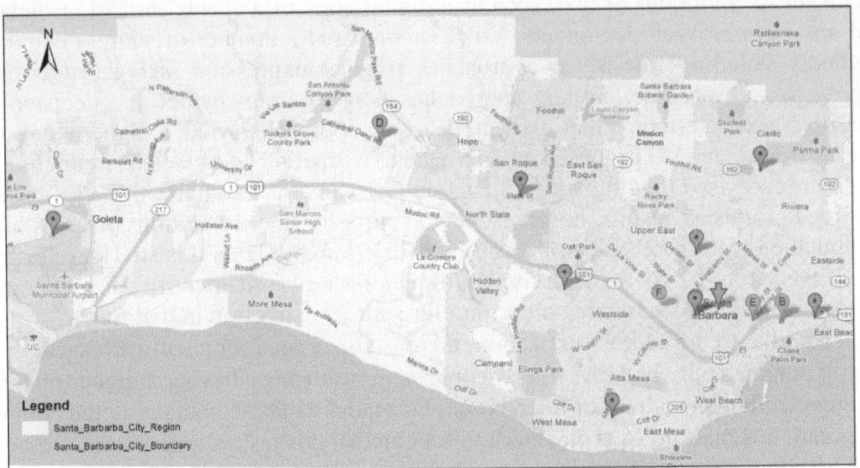

Fig. 1. Results of a Google search for locations of fire stations in the city of Santa Barbara, CA. The pink region shows the geographic extent of Santa Barbara.

In order to retrieve location-based information on the ever-expanding Internet and its almost-unstructured Web data, there is a need of an effective Web-mining mechanism that is capable of extracting desired data on the right webpages within the right scope. In this paper, we report our efforts in developing a knowledge-based Web-mining tool, CyberMiner, that adopts geospatial ontology, a forward- and backward-screening algorithm, and inductive learning for automated location information retrieval. The retrieval of street addresses of all fire stations within Los Angeles County, California is used as a case study because the County's fire department is the one of the largest and most sophisticated fire departments in the US and its fire services are provided through different tiers of local governments [4].

2   Literature

Spatial data mining on the Web, aiming at discovering spatial data or patterns of data, is a part of the Web geographic-information retrieval (GIR) task. In GIR, studies of exploiting geographic aspects of the Web can be categorized by their purposes. One is georeferencing, a way to attach geographical scope to webpages [1]. By identifying the associations between the general content of a webpage and the location information within it, the search engines are believed to better handle location-based queries [9], such as "All Starbucks in Nanjing, China". Another category (also the focus of this paper) is to obtain location information of certain subjects from the Internet. The goal is to build up a global spatial database to support queries and decision-making.

Both of the categories require automatic extraction of location information from the source code of a webpage. Cai et al. [2] present a method combining domain ontology that defines street name, street type, and city name, and a graph matching to extract addresses from the Web. Taghva et al. [19] apply a Hidden Markov Model (HMM) to extract addresses from documents. However, its target is OCR (Optical Character Recognition) of text, such as a digital copy of a check, instead of dynamic and unstructured Web documents. Yu [23] compared a number of address-extraction methods, including rule-based approaches (regular expression and gazetteer-based approach) and machine-learning approaches (word n-gram model and decision-tree classifier) and determined that machine-learning approaches yield a higher recall rate. Loos and Biemann [9] proposed an approach to extracting addresses from the Internet using unsupervised tagging with a focus on German street addresses. These methods provided promising results, however they all suffered from limitations, such as heavy computation load in [2] and low recall rates in [9] and [23]. In this study, we propose an efficient rule-based method combining central keyword identification and a forward- and backward-screening algorithm for address extraction in real time.

In addition to the address extraction, there is also a need for a software agent that is able to automatically pick the right webpages to visit (those having a higher possibility of containing an address of interest) until a spatial database containing the complete information is built up. A typical tool to accomplish this task is a spatial Web crawler, which follows hyperlinks and realizes information extraction from webpages as the process continues. A previous work is [13], which developed a Web crawler to discover the distributed geospatial Web services and their distribution pattern on the Internet. In [13], the determination of whether an extracted URL is a target is straightforward, because those geospatial Web services have a specialized interface. But in our work, the addresses of the desired type have a vague signature, therefore a more intelligent analysis is needed.

In the next sections, we discuss in detail the establishment of the Web crawling framework and use the discovery of all fire stations within Los Angeles County, CA as a case study.

3   Methodological Framework

3.1   Web Crawling Process

Fig. 2 shows the state-transition graph for retrieving location data on the Web. Circles represent states, in which different types of webpages (seed webpages, interlinks, target webpages, and unrelated ones) are being processed. The transition of state is triggered by processing an outgoing link in the webpage being visited. To design an effective Web crawler, it is important to first determine which webpages can be designated as crawling seeds. The proper selection of crawling seeds is important for the whole crawling process because good seeds are always closer to the target webpages in the linked graph of the Web. Second, it is important to identify the target webpage. A target webpage is the one containing the needed location information (in our case, it is the street address for each fire station). This requires the ability to extract all possible existences of street addresses on a target webpage and the ability to filter out those not of interest. It is also important to decide which interlinks will be given higher priority to access during Web crawling. As Fig. 2 shows, a webpage referred to by an interlink may contain hyperlinks to a target webpage or another interlink. An unrelated webpage which comes from an interlink should be filtered out. In the next sections, we will discuss the solutions to these three problems.

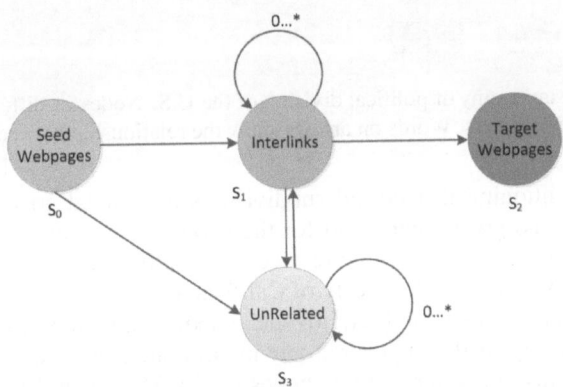

Fig. 2. State transition graph in a Web crawling process

3.2   Geospatial Taxonomy to Aid the Determination of Crawling Seeds

The crawling seeds are the URLs from where the crawling process starts. Seeds selection greatly affects crawling scope, time, and results. Bad seeds may lead to time-consuming crawling without any desired information found. As public-service facilities are mostly operated by local governments (except for a few volunteer service providers) and their locations are publicly listed on the government's website, it is necessary to obtain a knowledge base of the political divisions of the U.S. Fig. 3 is a taxonomy describing such divisions at class level. A class is a subnational entity that forms the U.S. For example, both "County" and "Incorporated Place" are entities (classes) in the taxonomy, and an "Incorporated Place", also known as a "Municipality", is a subdivision (subclass) of a "County". For use in a crawling process, a class needs to be initiated. For example, given the question "How many fire stations exist in Los Angeles (LA) County of California?" one needs to know specifically which subdivisions of LA County have local governments.

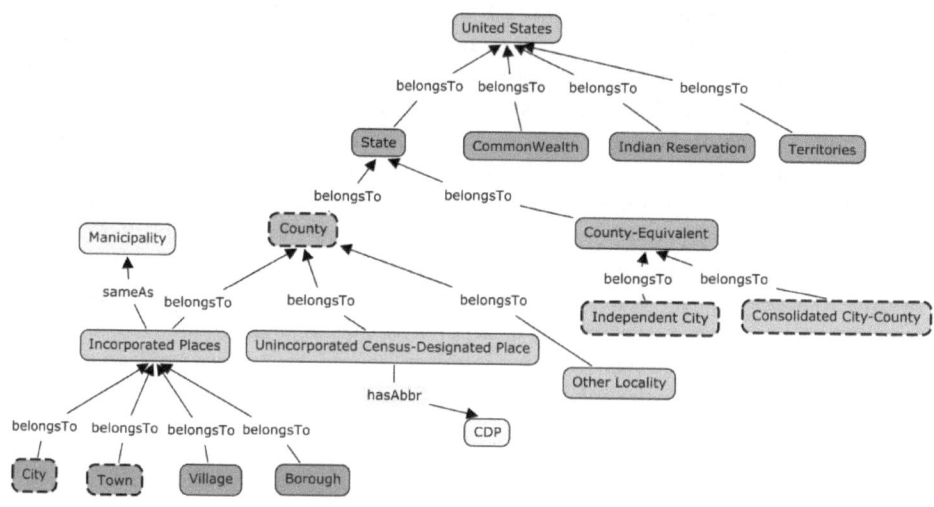

Fig. 3. Geospatial taxonomy of political division of the U.S. Nodes in different colors refer to jurisdictions at different tiers. Words on arrows show the relationship between jurisdictions.

It is worth mentioning that not all subdivisions have local governments, e.g., the CDP always have no government, and for the subdivisions that have a local government, not all of them provide a public service such as fire protection. For example, the fire protection service of Glendora City of California is provided by its parent division Los Angeles County. This would require such subdivisions to be excluded from consideration as crawling seeds. In general, any incorporated area, such as a county, city, town or other county equivalent (dotted boxes in Fig. 3), is more likely to have a local government. These nodes in the geospatial taxonomy are instantiated by the data extracted from the United States Bureau of the Census [20]. The URLs of their official websites are identified from a Google search and populated into the taxonomy. In this way, the starting points of the crawling process can be determined and the scope of search is narrowed to avoid unconcentrated crawling.

3.3   Target Webpage Identification and Street Address Extraction

Target webpages have a prominent characteristic: they contain postal street addresses for public-service facilities. The goal of identifying target webpages is to successfully extract street addresses from them. Though a street address of a public facility can be expressed in multiple ways, a standard form is shown below:

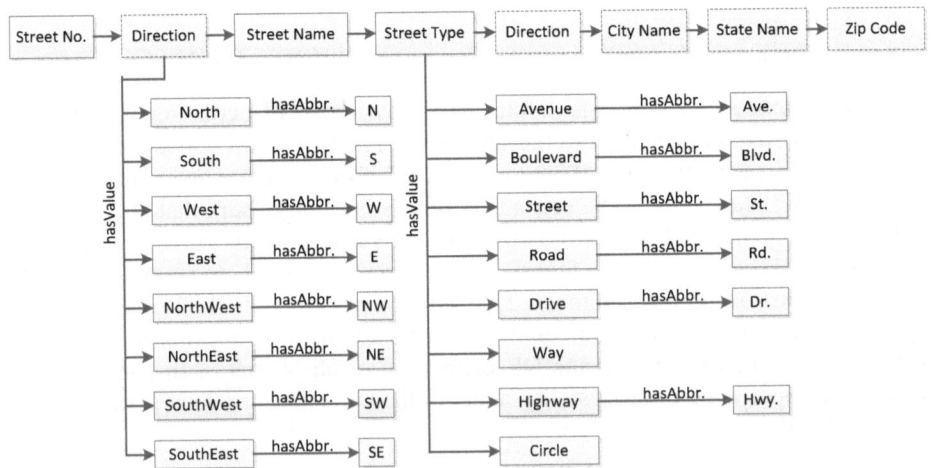

Fig. 4. Common form of street address

In practice, information for direction, city, state name, and ZIP code (boxes with dotted lines in Figure 4) are omitted in a street address. Therefore, "Street Type", such as "Avenue", "Road" and "Highway", becomes a prominent feature (we call it a "central keyword") indicating its existence. Based on a statistical analysis from the 300 known addresses of public facilities, above 90% contain "Avenue", "Boulevard", "Street", "Road", "Drive", "Way", "Highway", or "Circle", and their various abbreviations, such as "Rd" or "Rd." for "Road". Other street types such as "Plaza", "Grove", and "Place" also occur occasionally. Based on this prior knowledge, a street type dictionary was established for quick look-up.

Once the positions of the central keywords on an HTML document are found, backward-screening and forward-screening algorithms are employed to extract the complete address data. Centered on the position of street type p in the text, backward screening inspects the text block before p within a certain radius d_1, aiming at extracting a partial street address including the street number, direction, and street name based on pattern r_1. The forward-screening algorithm inspects the text block after location p within a radius d_2 to find the possible existence of city/county name, state name, and ZIP code as the other part of the address based on pattern r_2. If the extracted text block does not match the given patterns, e.g., the length of the ZIP code is not five or nine digits, even though there is an appearance of a central keyword, it will not be considered as a street address. The definitions of d and r are:

d_1: Distance between p and the location of the foremost digit in the number block closest (before) to location p.

d_2: Distance between p and the location of the last digit of the first number that appears (for detecting 5-digit ZIP code), or the last digit of the second number after p if the token distance of the first and second number block equals 2 (for detecting 9-digit ZIP code, in the format of xxxxx-xxxx).

r_1: regular expression [1-9][0-9]*[\\s\\r\\n\\t]*([a-zA-Z0-9\\.]+[\\s\\r\\n\\t])+

r_2: regular expression "*city-Pattern*"[\\s\\r\\n\\t,]?+("*statePattern*")?+[\\s\\r\\n\\t,]*\\d{5}(-\\d{4})*

Note that a number block in d_1 is defined as a whole word that consists of only numbers and is not a direct neighbor of the street type. The first restriction is to distinguish a street number, e.g., 1034 in "1034 Amelia Ave", from a street name with numbers, e.g., 54 in "W 54th Street". The second restriction is to make sure that street address with name only in numbers can be correctly extracted as well. For example, when "54th Street" is written as "54 Street", the number 54 should be recognized as the street name instead of street number. The "*cityPattern*" in r_1 is the Boolean OR expression of all cities/counties names within our study area (California in this study) and the "*statePattern*" in r_2 is the OR expression of all 50 states. In pattern r_2, we require the appearance of at least city name in the city+state+zipcode pattern for the address identification.

3.4   Semantic Analysis of Addresses

Section 3.3 describes how to extract addresses from an HTML webpage. Apparently, not all addresses are locations of public-service facilities (in our case, fire stations), even though they are extracted from the websites within the domain of city/county governments. Therefore, it is necessary to clarify that an address is truly referring to that of a fire station. We term the set of correct identifications the *positive class* of identifications; the *negative* class is the set of identified addresses that are not fire stations.

To classify an address into a positive class or a negative class, we adopt C4.5 [15], which is a widely used machine-learning algorithm based on decision-tree induction [7]. The basic strategy is to select an attribute that will best separate samples into individual classes by a measurement, 'Information Gain Ratio', based on information-theoretic 'entropy' [11]. By preparing positive and negative examples as a training set, we can produce a model to classify addresses automatically into positive and negative categories. But what semantic information should be used for constructing the training set? In this work, we assume every address on a webpage has navigational information to indicate the semantics of an address, and this navigational information is positioned in a text block right before the position of the address. For example, on the webpage http://www.ci.manhattan-beach.ca.us/Index.aspx?page=124, the fire station address "400 15th Street" has navigational information "fire stations are: Station One" right before it. On the webpage http://www.desertusa.com/nvval/, the address "29450 Valley of Fire Road, Overton, Nevada 89040" has navigational text "Valley of Fire State Park" before its position. Therefore, upon the detection of an address occurrence, we extracted its navigational text block (block size = 40 characters) and used it for semantic analysis.

We prepared four groups of 11 attributes for navigational text T of each address as follows:

Attributes A-D (keyword attributes): the existence of keyword "*fire station*", "*fire*", "*station*", and "*location*" within T. 1 means Yes, 0 means No.

Attribute E (statistical attributes): the ratio of number of keywords that exist and the total number of keywords in T. The value has range of (0,1].

Attribute F (pattern attributes): the existence of "*location*" followed by a number or a pound or hash symbol plus a number. 1 means Yes, 0 means No.

Attributes G-K (keyword attributes): the existence of keyword "fire station", "fire", "station", "location", or "address" in the title of each webpage.

Attributes A-D are based on the keyword frequencies of the navigational information of addresses. Attribute F is also based on the observation that most fire departments list the station number in the navigational text of a fire-station address. Attributes G-K are auxiliary attributes to complement information in the direct navigational text.

Through a semantic analysis based on the C4.5 algorithm, we analyzed the navigational text of 310 addresses, in which 191 are actually of fire stations. These addresses were crawled from the official government websites of Mesa, AZ, Columbus, OH, San Francisco, CA, and some pre-crawled cities in L.A. County. The decision rules shown in Figure 5 were extracted.

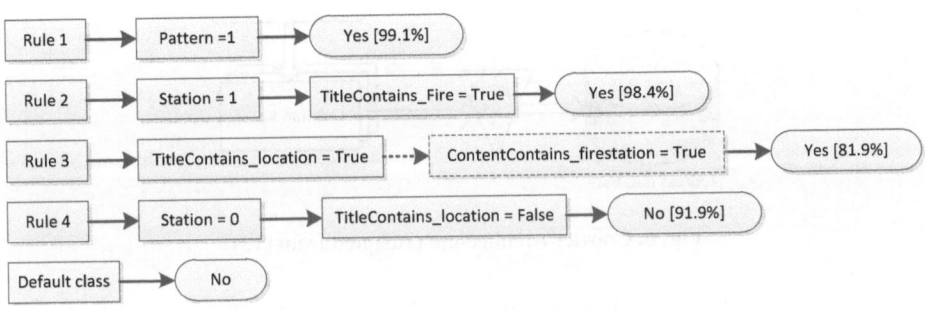

Fig. 5. Decision rules of desired addresses by training data based on semantic information

By default, an address will be placed in the negative class unless the pattern "station+number" exists in the navigational text (Rule 1) or the navigational text contains the keyword "Station" and the title of a webpage contains "fire" (Rule 2). Both Rule 1 and Rule 2 yield very high prediction accuracy, at 99% and 98% separately. Rule 3 determines the address of a fire station by recognizing the keyword "station" in the title. This rule only achieves 81.9% accuracy; therefore, to avoid false positives we added a constraint "if the content contains 'fire station' on a webpage" (in the dotted box). In practice, this rule gives improved prediction accuracy. Using these rules, we made predictions on whether an address is of interest.

4   Implementation

In the previous section we discussed three key techniques for detecting the existence of location information and to predict whether the information is of interest. In this section, we discuss the software architecture of CyberMiner (Fig. 6), which implements the proposed techniques to enable the automatic Web-mining process. *Crawling entry* is where the crawler starts to work. The seeding Web URLs are fed into the entry and a number of initial conditions such as politeness delay and number of threads to start are configured. Politeness delay ensures that the crawler behaves politely to remote servers. The multi-threading strategy is to increase the utilization of a single CPU core, leveraging thread-level parallelism [13]. We currently set the thread number to be 4, considering the relatively small scale of crawling.

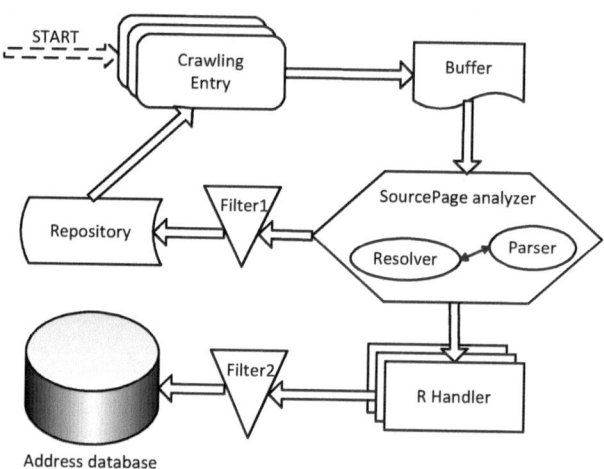

Fig. 6. Crawler Architecture (Adapted from [13])

Buffer selectively caches Web source code linked by URLs for address extraction. *Source page analyzer* is used to analyze the Web source code that has been cached in the buffer, to extract all outgoing links and convert relative URLs to absolute URLs. These URLs go to the *filter1*, which filters out the URLs that have been visited before, within another domain from its parent URL, and the webpages of those having very low possibility to contain the desired addresses, such as a URL of a .css file or an image file. This strategy guarantees the domain purification of the crawling tree inherited from one seed webpage. It also avoids unnecessary cross-domain crawling to reduce search cost.

The *repository* maintains all URLs crawled using a first-in-first-out (FIFO) queue, the head of which is always the URL to be crawled next. Every time a Webpage is being visited, the *R Handler* is initiated to extract all possible addresses from its source code and sends these addresses to *filter 2*, in which the non-desired addresses and duplicated addresses are disregarded based on semantic analysis. The target addresses are inserted into the *Address Database*. This process will continue until the FIFO queue is empty or until it reaches the maximal depth for crawling (the depth is measured by the number of steps between the current URL and its seed URL).

5   Results and Analysis

Fig. 7 shows all the 346 fire stations that have been discovered by CyberMiner within the boundary of Los Angeles County of California, US. These stations are widely distributed in 88 cities within the County, and they show a denser distribution in the cities than that in the rural region (generally the northern part of LA County). The areas without fire station coverage are mountains (the Santa Monica Mountains in the southwest of LA County and the San Gabriel Mountains in northern LA County). To evaluate the performance of CyberMiner, we measured the ratio between the number of retrieved fire locations m to the total number of relevant records on the Web n. This ratio is also known as the recall rate. Here, m equals 346. To compute n ($n=392$), we visited the websites of all city/county governments that have a fire department and manually annotated the street addresses of all fire stations listed on each city's government website. Although 46 stations failed to be detected, our CyberMiner still achieves very satisfying recall rate – at 88%. To share the results of this research, the data and map have been made public through a Web application http://mrpi.geog.ucsb.edu/fire, on which the locations and addresses of fire stations are provided.·

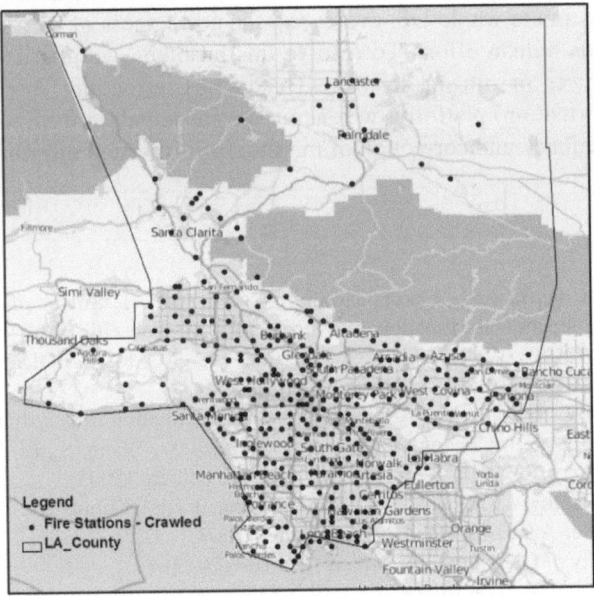

Fig. 7. Locations of all fire stations obtained by CyberMiner

6   Conclusion and Discussion

In this paper, we proposed a method and a software tool CyberMiner for automatic detection and extraction of location information of interest from the WWW. The proposed geospatial taxonomy, containing hierarchical subdivisions of US governments, restrains the search scale to the domains that are most likely to contain the data in need, thereby greatly reducing search cost. The methods of pattern-based address extraction and inductive learning-based prediction contribute to the recall rate. Although locating fire stations is used as a case study in the paper, the proposed work is easily extendable to search for locations of other emergency/public-service facilities, such as police stations and wastewater treatment plants. The proposed algorithms, such as forward and backward screening for automatic address extraction, are beneficial to the general area of GIR. Moreover, this work goes one step further from previous address extraction research [2][9][23], in that it is not only able to extract addresses in general correctly, but it is also able to classify types of address based on the proposed semantic analysis.

In the future, we will continue to improve the performance of the CyberMiner, especially in its tolerance to errors in an address. Since the core of the proposed address detection algorithm is the determination of the central keyword – the street type - the algorithm had a hard time identifying it when it was not contained in the address type dictionary. Addressing this issue requires enriching the dictionary to include a complete list of street types in the US. Another aspect is the quality control of the search results. As our long-term goal is to establish an address database of emergency service facilities in the whole US, evaluating the correctness of these addresses would need tremendous human effort. To resolve this problem, we plan to take the advantage of the power of citizen sensors. That is, by providing a VGI (volunteered geographic information) platform, we encourage participation from the general public in providing feedback and correction of missing or mislocated information.

References

1. Buyukokkten, O., Cho, J., Garcia-Molina, H., Gravano, L., Shivakumar, N.: Exploiting geographical location information of Web pages. In: Proceedings of Workshop on Web Databases (WebDB 1999) held in Conjunction with ACM SIGMOD 1999, Philadephia, Pennsylvania, USA (1999)
2. Cai, W., Wang, S., Jiang, Q.: Address Extraction: Extraction of Location-Based Information from the Web. In: Zhang, Y., Tanaka, K., Yu, J.X., Wang, S., Li, M. (eds.) APWeb 2005. LNCS, vol. 3399, pp. 925–937. Springer, Heidelberg (2005)
3. Chang, G., Healey, M.J., McHugh, J.A.M., Wang, J.T.L.: Mining the World Wide Web, vol. 10. Kluwer Academic Publishers, Norwell (2001)
4. Glendora: City of Glendora Government Website (2012), http://www.ci.glendora.ca.us/index.aspx?page=896 (last Access Date: July 27, 2012)
5. Goodchild, M.F.: Citizens as sensors: the world of volunteered geography. Geo Journal 69, 211–221 (2007)
6. Gulli, A., Signorini, A.: The indexable web is more than 11.5 billion pages. In: Special Interest Tracks and Posters of the 14th International Conference on World Wide Web, pp. 902–903. ACM, Chiba (2005)
7. Han, J., Kamber, M.: Data mining: concepts and techniques. Morgan Kaufmann Publishers, San Francisco (2001)

8. Kofahl, M., Wilde, E.: Location concepts for the web. In: King, I., Baeza-Yates, R. (eds.) Weaving Services and People on the World Wide Web, pp. 147–168. Springer, Heidelberg (2009)

9. Loos, B., Biemann, C.: Supporting web-based address extraction with unsupervised tagging. In: Data Analysis, Machine Learning and Applications 2008, pp. 577–584 (2008)

10. Li, W., Goodchild, M.F., Raskin, R.: Towards geospatial semantic search: exploiting latent semantic analysis among geospatial data. International Journal of Digital Earth (2012), doi:10.1080/17538947.2012.674561

11. Li, W., Yang, C.W., Sun, D.: Mining geophysical parameters through decision-tree analysis to determine correlation with tropical cyclone development. Computers & Geosciences 35, 309–316 (2009)

12. Li, W., Yang, C., Zhou, B.: Internet-Based Spatial Information Retrieval. In: Shekhar, S., Xiong, H. (eds.) Encyclopedia of GIS, pp. 596–599. Springer, NYC (2008)

13. Li, W., Yang, C.W., Yang, C.J.: An active crawler for discovering geospatial Web services and their distribution pattern - A case study of OGC Web Map Service. International Journal of Geographical Information Science 24, 1127–1147 (2010)

14. Ligiane, A.S., Clodoveu Jr., A.D., Karla, A.V.B., Tiago, M.D., Alberto, H.F.L.: The Role of Gazetteers in Geographic Knowledge Discovery on the Web. In: Proceedings of the Third Latin American Web Congress, p. 157. IEEE Computer Society (2005)

15. Quinlan, J.R.: C4.5: Programs for Machine Learning. Morgan Kaufmann Publishers, San Francisco (1993)

16. Rogers, J.D.: GVU 9th WWW User Survey, vol. 2012 (2012), http://www.cc.gatech.edu/gvu/user_surveys/survey-1998-1904/ (last Access Date: July 27, 2012)

17. Sanjay Kumar, M., Sourav, S.B., Wee Keong, N., Ee-Peng, L.: Research Issues in Web Data Mining. In: Proceedings of the First International Conference on Data Warehousing and Knowledge Discovery, pp. 303–312. Springer (1999)

18. Szalay, A., Gray, J.: Science in an exponential world. Nature 440, 413–414 (2006)

19. Taghva, K., Coombs, J., Pereda, R., Nartker, T.: Address extraction using hidden markov models. In: Proceedings of IS&T/SPIE 2005 Int. Symposium on Electronic Imaging Science and Technology, San Jose, California, pp. 119–126 (2005)

20. USCB: GCT-PH1 - Population, Housing Units, Area, and Density: 2010 - State - Place and (in selected states) County Subdivision (2012), http://factfinder2.census.gov/faces/tableservices/jsf/pages/productview.xhtml?pid=DEC_10_SF11_GCTPH11.ST10 (last Access Date: July 27,2012)

21. Wray, R.: Internet data heads for 500bn gigabytes. The guardian, Vol. 2012. Guardian News and Media, London (2009), http://www.guardian.co.uk/business/2009/may/2018/digital-content-expansion (last Access Date: July 27,2012)

22. Yasuhiko, M., Masaki, A., Michael, E.H., Kevin, S.M.: Extracting Spatial Knowledge from the Web. In: Proceedings of the 2003 Symposium on Applications, p. 326. IEEE Computer Society (2003)

23. Yu, Z.: High accuracy postal address extraction from web pages. Thesis for Master of Computer Science. 61p. Dalhousie University, Halifax, Nova Scotia (2007)

Summarizing Semantic Associations Based on Focused Association Graph

Xiaowei Jiang[1], Xiang Zhang[2], Wei Gui[1], Feifei Gao[1],
Peng Wang[2], and Fengbo Zhou[3]

[1] College of Software Engineering, Southeast University, Nanjing, China
{xiaowei,guiwei,ffgao}@seu.edu.cn
[2] School of Computer Science and Engineering, Southeast University, Nanjing, China
{x.zhang,pwang}@seu.edu.cn
[3] Focus Technology Co., Ltd, Nanjing, China
zhoufengbo@made-in-china.com

Abstract. As the explosive growth of online linked data, there is an urgent need for an efficient approach to discovering and understanding various semantic associations. Research has been done on discovering semantic associations as link paths in linked data. However, few discussions have been given on how we can understand complex and large-scale semantic associations. Generating human understandable summaries for semantic associations is a good choice. In this paper, we first give a novel definition of semantic association, and then we describe how we discover semantic associations by mining link patterns. Next, a notion of Focused Association Graph is proposed to characterize merged associations among a set of focused objects. Then we focus on summarizing of Focused Association Graph. Concise summaries are generated with the help of Steiner Tree problem. Experiments show that our approach is feasible and efficient in generating summaries for semantic associations.

Keywords: linked data, semantic association, link pattern, summarization.

1 Introduction

As the rapid growth of semantic web in this decade, there is an exponential growth in the scale of online linked data. Every day, enormous amount of linked data are produced by social communities, companies, and even by end-users. Linked data provide a good practice for connecting and sharing semantic objects by URI and RDF.

An important knowledge we can discovered in linked data is the explicit or hidden relationship between or among semantic objects, which is terminologically named as semantic associations. An early statement of semantic association can be found in [1], in which semantic associations are connections between two objects, and are represented as semantic paths in RDF graph. Discussions of mining semantic associations in linked data have lasted for near ten years. Most

S. Zhou, S. Zhang, and G. Karypis (Eds.): ADMA 2012, LNAI 7713, pp. 564–576, 2012.
© Springer-Verlag Berlin Heidelberg 2012

relevant works adopt a path-based definition of semantic association, and a series of efficient path discovery algorithms have been proposed.

The problem of current study of semantic associations lies in two aspects: first, path-based definition of semantic associations has limitations. It can only characterize pairwise relations between two objects, but is unable to represent group-links among multiple objects. Besides, current definition does not consider frequency or typicality to measure whether discovered semantic associations are meaningful. Second, there are few studies on how human readers can understand semantic associations with ease. Linked data is essentially a complex network as stated in [2]. Given a set of objects, there may be a great amount of semantic associations among them in linked data, each of which may be complex in structure. This will bring a huge barrier for human understanding. A concise and comprehensible representation is needed.

In this paper, we first propose a novel definition of semantic associations as a sub-graph structure connecting multiple objects. This graph model can characterize more complex relationship among objects than the simple path-based model. A notion of link pattern is used to ensure that the pattern of discovered semantic associations should be frequent and thus typical in linked data. Inspired by the approach of text summarization [3], we propose a summarization approach on semantic associations. Given a set of focused objects, a Focused Association Graph is built as graph model and an association tree is generated as a summary by a Steiner Tree algorithm.

2   Architecture

We give an overview of the architecture of our summarization approach in this section. As shown in Figure 1, the input of the system is a linked data, as well as a set of specified focused objects. Our goal is eventually producing a concise and comprehensible summary of the semantic associations among focused objects, and users can understand the summary with a low time-cost.

Given a linked data, the derived RDF graph is transformed into a Typed Object Graph (TOG in short) in **TOG Builder**. Statistics of the occurring frequency of each URI in linked data will be analyzed in **Frequency Analyzer** to provide useful information for **Weight Evaluator**. **Association Extractor** first discovers link patterns from TOGs by applying frequent pattern mining algorithm, and then semantic associations are extracted in an instantiation process. Given a set of focused objects, **Association Merger** first extracts all related semantic associations, and then merges them into a single large graph named Focused Association Graph (FAG in short). There are two association extraction policies: intersection-extraction and union-extraction, which generate iFAG and uFAG respectively. In **Weight Evaluator**, a weighting scheme is applied to assign a value to each vetex (object) and edge (relational term) in the FAG. Each assigned value represents a potential time cost that a user has to pay for understanding the semantic association. At last, a concise association tree will be generated as summary in Summary Generator by an implementation of Steiner-tree problem [4].

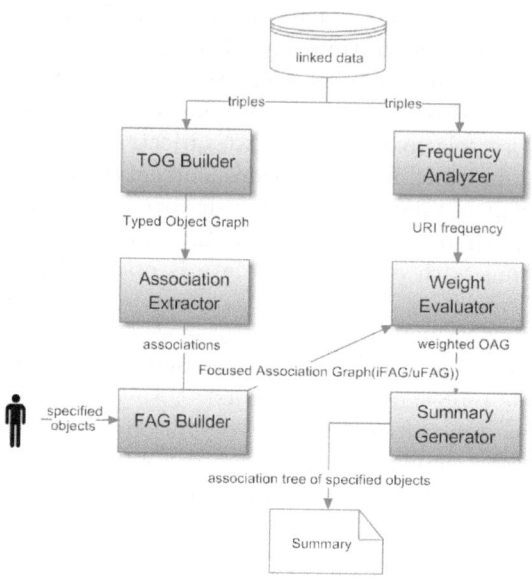

Fig. 1. The architecture of the summarization system of semantic associations

3   Discovering Semantic Associations

The notion of semantic association is traditionally defined as paths between two objects. Declared in [1], two objects are semantically associated if they are semantically connected by a semantic path in RDF graph, or they are semantically similar by lying on two similar semantic paths. Anyanwu proposed a notion of property sequence in [5] to define the semantic association between two objects. In [6], Kochut used Defined Directionality Path to characterize semantic associations.

From real linked data, a complex association among six objects is shown in Figure 2(a): Tim Berners-Lee together with his five friends are all members of W3C, which is a typical group of working partners association. However, path-based association model is difficult to represent this complex graph structure. We also made an observation that the corresponding pattern behind this graph structure is frequently occurred in Falcons, which is shown in Figure 2(b). That indicates this type of associations is typical and should be familiar to users. We name it as Link Pattern.

3.1   Mining Link Patterns

Link patterns are frequent styles of how different types of object are interlinked in linked data. Each link pattern is a graph connecting object types, and its number of support should exceed a specified threshold frequency. In [7], we have given a full discussion of the definition of link pattern, as well as our mining approach.

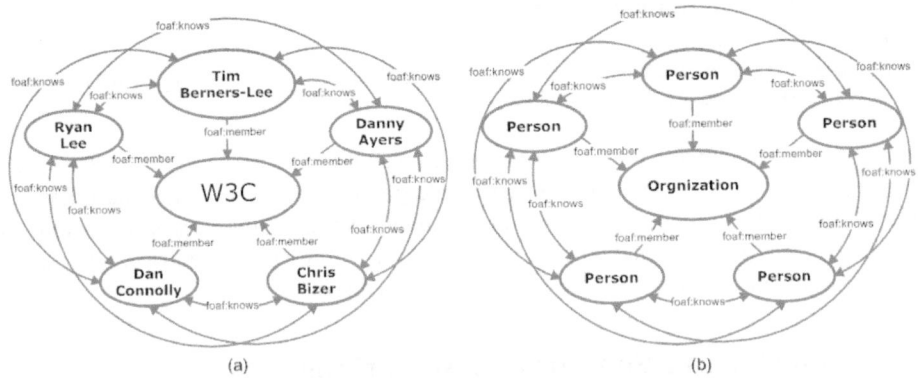

Fig. 2. (a)An example of semantic associations (b)corresponding link pattern

Link patterns cannot be directly mined from the RDF graph of linked data, because object types are core elements in link patterns. However, in RDF graphs, object types are implicit and can only be determined by reasoning according to RDF semantics. We have proposed Typed Object Graph (TOG in short) in [7] as the graph model for mining link patterns. A TOG is derived from an RDF graph in linked data, in which each triple is extended to a link quintuple, additionally containing the types of the subject and the object in the original triple.

Our mining approach of link patterns follows the idea of pattern-growth-based frequent pattern mining. We adopt gSpan [8] as the mining algorithm. Its kernel ideas are the minimum DFS code and the rightmost extension. The minimum DFS code is introduced to canonically identify a pattern by a DFS traverse path; the rightmost extension is used to produce candidates based on mined patterns. Both ideas can reduce the generation of duplicated candidates. Original gSpan algorithm is designed for undirected and simple graphs, while TOG is directed and non-simple graphs. Self-loops and multiple edges should be taken into consideration. We modified gSpan algorithm, especially the DFS coding, to make it adaptable to TOG.

3.2   Discovering Semantic Associations

Link patterns provide a schema-level template for mining semantic associations. They also ensure that each discovered semantic association must be frequent and typical in linked data. A simple and stepwise mining approach is: we first discover all link patterns in linked data, and then we traverse the instance-level of linked data and check whether there are instantiations of patterns.

Definition 1. (Instantiation of Link Pattern): *Given linked data d and RDF graph $g = \langle V(g), E(g) \rangle$ derived from d.a discovered link pattern $p = \langle V(p), E(p) \rangle$ can be instantiated by g, iff: (1)$\forall u \in V(g)$, u is an object, (2)$\forall v \in V(p), \exists v\prime \in V(g)$ and $type(v\prime, d) = v$, (3)$\forall(u, v) \in E(p), \exists(u\prime, v\prime) \in E(g)andtype(u\prime, d) = u$ and $type(v\prime, d) = v$. (4) g is minimal. $type(v)$ is the type of vertex v.*

Definition 2. (Semantic Association):   *Given a set of objects O={ $o_1, o_2, \ldots, o_i$} in a linked data d, a semantic association sa(O,E) is an RDF graph derived from d, iff sa(O,E) is an instantiation of a discovered link pattern in d. O is the vertex set of sa(O,E) , and E is the edge set.*

Instead of the stepwise mining, we use a more efficient mixed approach to mining semantic associations in practice: link patterns and semantic associations are mined simultaneously in each iteration of gSpan. That means all semantic associations will be mined and the number of semantic associations equals to the summation of total supports of all link patterns.

4   Summarizing Semantic Associations

For a large-scale linked data, an extremely large volume of semantic associations can be discovered in it. Supposing a user intends to find out whether there are associations among specified objects, hundreds or even thousands of related semantic associations will bring a great barrier to the human understanding.

Given a set of specified objects, we first merge related semantic associations into a Focused Association Graph, and then a cost value is assigned to each vertex and edge in the graph, reflecting the level of difficulty for understanding. The final summary is an association tree. Comparing to a large set of related semantic associations, we believe a concise association tree is intuitive and easy to understand.

4.1   Building Focused Association Graph

To understand associations among specified objects, a na?ve approach is generating summaries directly on original RDF graph. But this approach is inefficient. The original RDF graph can be extremely large. Famous linked data, for example DBpedia, has more than 3 hundred million triples, and some giant data sources contain even billion triples. Any graph summarization algorithm will be inapplicable on this scale. Pruning the original RDF graph to get a related sub-graph is a better idea. But without the knowledge of discovered semantic associations, the pruning process is prone to loss of key information. Seeming unrelated neighboring objects will be probably discarded in summarization, but in fact, they may play a key role of intermediary of specified objects. This case is gaining a growing concern especially in the area of national security [9]. In our approach, related semantic associations of specified objects will be merged to build Focused Association Graph, as a source graph of summarization. This will greatly reduce the loss of information.

Definition 3. (Focused Association Graph, FAG in short): *Focused Association Graph $G=\langle V, \Psi, E, W \rangle$ is a weighed and directed graph, in which V is the vertex set. $\Psi \subseteq V$ is called Focused Set, which is a subset of V, and each element in is called a Focused Object. W is a weighting scheme for edges. For each edge as an RDF triple $e=\langle s, p, o \rangle \in E$, W(e) is a value between (0,1).*

From the definition, we can see FAG is inherently an RDF graph characterizing a complex group-links among a set of objects. Especially, there is a focused set of objects in FAG, which comprises objects interested by users. Building FAG is a process of merging related semantic associations of all focused objects. There are two policies for association merging: merging by intersection or union operation. It is obvious that Union-merge will produce a much larger FAG than Intersection-merge. It will further reduce the loss of information, but will meanwhile bring burden to the summarization process.

Intersection-Merge Policy: Given a focused set of objects O={ $o_1, o_2, \ldots, o_i$ }, and a set of related semantic association set $\Phi = \{SA(o_1), SA(o_2), \ldots, SA(o_i)\}$, in which $SA(o_i)$ is a set of all semantic associations concerning object o_i, we can build an FAG containing semantic associations from $SA(o_1) \cap SA(o_2) \cap \ldots \cap SA(o_i)$, and O is the focus set of the FAG. Using this policy, we denote the resulted FAG as iFAG.

Union-Merge Policy: The only difference to intersection-merge policy is the union operation on related semantic associations, instead of the intersection operation. Using this policy, we denote the resulted FAG as uFAG.

4.2   Weighting Focused Association Graph

FAG plays a role as a representation of a meaningful and comprehensive association among focused objects. Each part of a FAG will be comprehensible for human readers, because it is guaranteed that the each link pattern is frequent and thus typical. However, FAG can still be large-scale, and we have to evaluate which part of it is easy to understand. A weighing scheme is needed to quantify the effort that a human has to pay to understand. We made the observation that if the subject, predicate and object of a triple are popular, human readers can understand it with less effort.

Thus, a weighting scheme can be defined in equation (1) to (5). Given an edge in FAG represented as a triple $e = \langle s, p, o \rangle$, freq(s) / freq(p) / freq(o) are their frequency of occurrence in linked data respectively, and an understanding cost of each part of a triple can be computed using a inverse number of log frequency. The weighting scheme $W(e)$ is calculated using a normalization of a total cost of each edge.

$$cost(s) = \frac{1}{\log_2(freq(s) + 1)}; \tag{1}$$

$$cost(p) = \frac{1}{\log_2(freq(p) + 1)}; \tag{2}$$

$$cost(o) = \frac{1}{\log_2(freq(o) + 1)}; \tag{3}$$

$$totalCost(e) = cost(s) + cost(p) + cost(o); \tag{4}$$

$$W(e) = \frac{totalCost(e)}{max\{totalCost(\varepsilon)|\varepsilon \in E\}} \tag{5}$$

4.3   Generating an Association Tree as Summary

An uFAG can be very large. In our test cases, the largest uFAG with a small focus set of only three objects contains even more than one thousands of objects and two thousands of links. On the contrary, intersection-merge policy usually results in very small or even empty iFAG. Over 90 percents of randomly selected focus set have no iFAG in our test cases. Graph model of uFAG reserves to the most extent the useful associations among focused objects. Comparing to iFAG, uFAG is more suitable for characterizing semantic associations among focused objects, but its large scale makes it difficult for human understanding. We select a tree structure for modeling summaries, and we call each summary an association tree of focused objects.

Definition 4. (Association Tree): *Give a FAG $G = \langle V, \Psi, E, W \rangle$, an association tree $\tau(G)$ is a spanning tree of G, in which the vertex set of $\tau(G)$ is a superset of focused set Ψ in G, and the total cost of all edges in $\tau(G)$ should be minimum.*

Association tree is a special minimum spanning tree of FAG. It is required that all focused objects together with none or several intermediary objects are included in the summary for a complete understanding of focused objects, and meanwhile the cost of the summary should be minimum for a best human understanding experience. Traditional minimum-spanning-tree algorithms do not guarantee that focused objects are all included in the summary. But it can be perfectly resolved by a classic variation of minimum-spanning-tree, which is called Steiner Tree problem.

The Steiner Tree problem, named after Jakob Steiner, is originally a problem in combinatorial optimization. It can be briefly stated as following: Given an undirected graph, a distance function on its edges and a subset of its vertex as terminals (focused objects in our scenario), the Steiner Tree problem is to find a tree that span the terminals with minimum total distance. Steiner Tree is now widely applied in circuit layout, network design, and even in semantic web research [10].

It is recognized that Steiner Tree is a classic NP-complete problem. The accurate solutions of Steiner Tree are time-consuming. However, with the help of KMB algorithm [4], we can utilize an approximate implementation that can be solved in polynomial time. KMB algorithm is heuristic, which considers the similarity of minimal Steiner Tree and minimal spanning tree to approach the final solution. It is approved that the worst case upper bound of approximation ratio is $2(1-\frac{1}{\Psi})$, where Ψ is the focused set. Its worst case time complexity is $O(|\Psi||V|^2)$.

An example of summarization is given in Figure 3. There are three focused objects: Jack, Mary and Steve. The uFAG of these objects are constructed by four semantic associations: the family association among Jack, Mary and their family members; a business association among Jack, Mary, Steve and other partners and companies; an alumni association among Jack, Steve and other graduates of a college; a reader-writer association among Steve, Marys father and a book. Each edge in uFAG is weighted according to the weighting scheme. With the help of KMB algorithm, a final association tree is generated as the summary of these four associations, which is highlighted in green and bold.

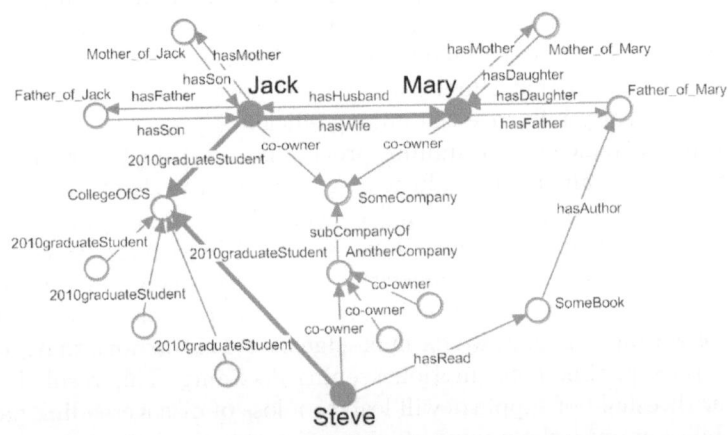

Fig. 3. An example of FAG and a summarized association tree

5   Experiments

We evaluate our approach on a subset of DBpedia[1]. In our evaluation, we mainly discuss the time and space performance of both mining and summarization of semantic associations. Experiments of mining are performed on a 3GHZ Intel Core2 Duo PC with 4G memory.

The entire DBpedia dataset contains a completely extracted data from Wikipedia. As of September 2011, the DBpedia dataset describes more than 3.64 million objects using over 1 billion triples. Majority type of objects in DB-pedia includes persons, places, music albums, films, video games, organizations, species and diseases and so on. DBpedia has a broad scope of objects covering different areas of human knowledge, and is widely used for the research of semantic knowledge management or semantic search. However, the massive volume of the entire DBpedia dataset is a great challenge for semantic web mining. In [7], we have proposed a clustering algorithm for mining link patterns in large-scale linked data. However, for the simplicity and clarity of the problem, clustering is not used in this experiment, and a subset of DBpedia is extracted, which can fit into the memory.

Our dataset is composed of 4 RDF documents, including more than 10,048 RDF triples, in which there are 2,532 object links between 1,306 objects.

---

[1] DBpedia: http://dbpedia.org/

5.1   Evaluation on Mining Semantic Associations

From our dataset, we find 104 link patterns, and 11,303 semantic associations. Each link pattern corresponds to an average number of 108 semantic associations. We use a mixed approach, instead of a stepwise approach, to mining link patterns together with semantic associations. That makes the mining of semantic association very fast. In our mining process, time consumed for generating semantic associations only covers less than 5% of all the mining time.

There are two parameters in the mining process: max-edge, which limits the maximal edges of discovered semantic associations, and min-sup, which is a threshold of the support that a link pattern must have. The setting of these two parameters will greatly influence the mining efficiency.

The time performance of the mining process is shown in Figure 4. We first fix min-sup to 5 to evaluate the influence of various max-edge to the time performance. As max-edge increases, time consumed in mining process increase dramatically in an exponential way. This is caused by a huge lexicographical search tree produced in the gSpan algorithm. Mining frequent patterns in large-scale dataset is still challenging and remains an open question. When evaluating the impact of various min-sup, we fix max-edge to 4. It is obvious that, with the increase of min-sup, time consumption keeping declining. This result indicates that a higher threshold of supports will lead to a loss of discovered link patterns, but meanwhile improves the mining efficiency.

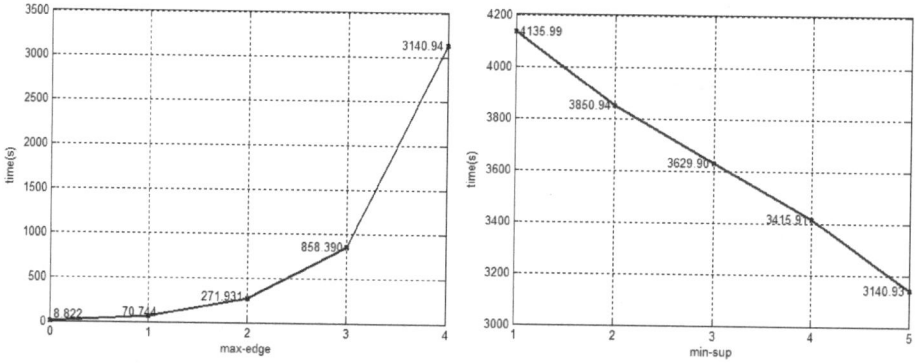

Fig. 4. Time Performance with various max-edge and min-sup

5.2   Evaluation on Summarizing Semantic Associations

In our dataset, there are 1,306 objects, and 11,303 discovered semantic associations. We randomly select k objects ($3 \leq k \leq 10$) as focused objects and extract related semantic associations. Associations are further merged into uFAG by union-merge policy. We have performed experiments on the correlation of the number of focused objects and the number of total objects and links in uFAG, the average time and space consumption with various numbers of focused objects, and a complete evaluation on time and space consumption when uFAG grows.

Shown in Figure 5, when we randomly select a set of focused objects, the resulted uFAG can be a huge graph. If we only select three focused objects, the uFAG can have an average size of more than two hundreds of total objects and four hundreds of object links. However, there is no rapid increase of uFAG when we randomly select more focused objects. When ten focused objects are selected, the uFAG contains an average size of six hundreds of objects and more than one thousands of object links.

Figure 6 presents the averaged time and space consumption in summarization with various numbers of focused objects. We also randomly select 3 to 10 objects as focused objects. When we select more and more focused objects, we can notice that time and space consumptions in summarization are near linear. Summarizing semantic associations among three focused objects will cost only 1 seconds and 4.5M memory in average. Summarization on ten focused objects will cost 7 seconds and 16M memory in average. It indicates that our approach is scalable for summarizing a large-scale set of focused objects.

The scatter diagram in Figure 7 presents a complete evaluation on time and space consumption on our approximate implementation of Steiner Tree. The horizontal ordinate $|V|$ represents the total number of vertex in uFAG. With the increase of $|V|$, time and space consumptions for producing association trees grow in a polynomial manner, as stated in [4]. It indicates that although Steiner Tree problem is NP-complete, our implementation is still feasible for summarizing semantic associations in real applications.

6   Related Works

Summarization of semantic associations is a novel research topic. Our approach is motivated by the traditional research of text summarization, which is widely used in understanding unstructured documents. Defined by Mani [11], text summarization is the process of distilling the most important information from sources to produce an abridged version for user understanding. Typical applications of text summarization include producing indicative summaries of web pages by search engines to help users quickly understand search results.

Research of ontology summarization and semantic snippet generation are both closely related to our topic. Ontology summarization aims at improving the scalability of reasoning or the efficiency of human understanding. Fokouel proposed in [12] an approach to summarizing Abox in secondary storage by reducing redundancy to make reasoning scalable for very large Aboxes. It is an alternative approach with KAON2 [13], which reduces an SHIQ(D) ontology to a disjunctive datalog program and makes it naturally applicable to Aboxes stored in deductive databases. In [14], Zhang proposed an ontology summarization approach based on salience ranking of RDF sentences in ontology. Snippet generation is usually performed by semantic search engine, and aims at helping users understand resulted ontology or linked data. In [15], Cheng gave a discussion on how to generate query-relevant snippet for ontology search. In [16], Penin used a hierarchical clustering algorithm to group RDF sentence into topics in

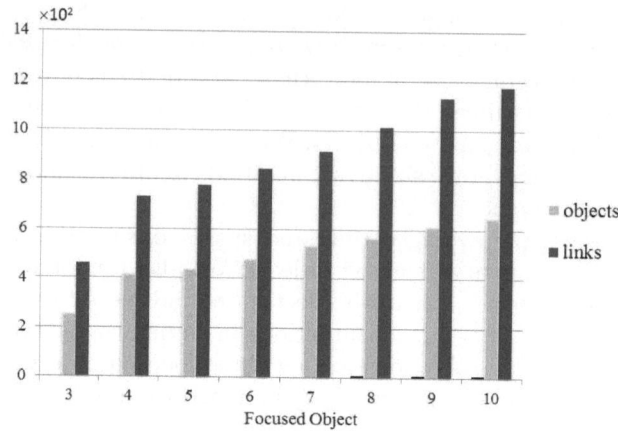

Fig. 5. Correlation of the number of focused objects and the volume of correspoding uFAG

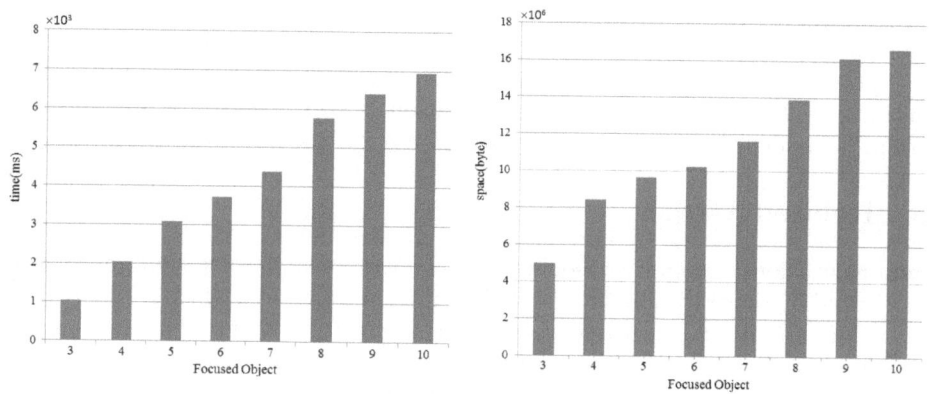

Fig. 6. Averaged time and space consumptions with various number of focused objects

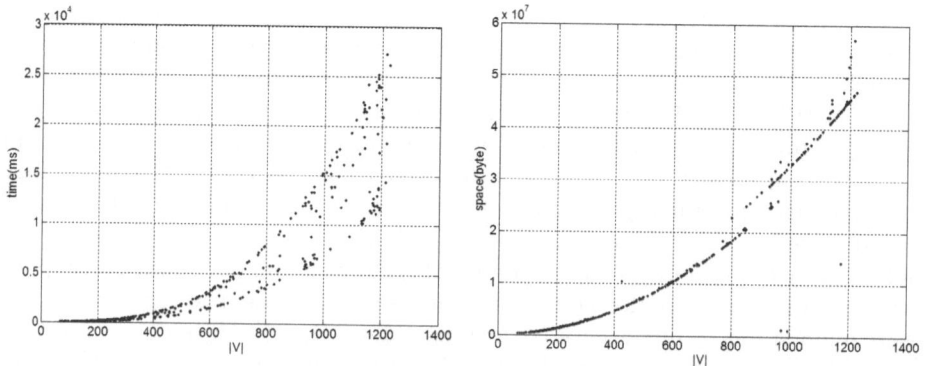

Fig. 7. Complete evaluation on time and space consumptions with various $|V|$

snippet generation. Similarly, Bai proposed in [17] a topic-related and query-related snippet generation approach.

7 Conclusion and Future Works

There is an urgent need from users that they require an efficient approach to discovering and understanding semantic associations in linked data. In this paper, we present a summarization approach based a novel graph model. We first introduce the mining of semantic associations with the help of link patterns, and then we fully explain how concise and comprehensible summaries are generated from Focused Association Graph with the help of Steiner Tree problem. Experiments on real linked data show that our approach is feasible on both time and space efficiency.

In our future work, we will design more experiments on the quality of the summaries. A search engine of semantic associations is under construction. We will explore in our search engine that how summaries of semantic associations can help users fulfill their understanding tasks. Another study will be given on how to generate query-relevant summaries.

Acknowledgements. The work is supported by the NSFC under Grant 61003055, 61003165, and by NSF of Jiangsu Province under Grant BK2009136, BK2011335. We would like to thank Cuifang Zhao for her valuable suggestions and work on related experiments.

References

1. Aleman-Meza, B., Halaschek-Wiener, C., Arpinar, I.B., Sheth, A.P.: Context-aware Semantic Association Ranking. In: Proceedings of the 1st International Workshop on Semantic Web and Databases, pp. 33–50 (2003)
2. Ge, W., Chen, J., Hu, W., Qu, Y.: Object Link Structure in the Semantic Web. In: Aroyo, L., Antoniou, G., Hyvönen, E., ten Teije, A., Stuckenschmidt, H., Cabral, L., Tudorache, T. (eds.) ESWC 2010, Part II. LNCS, vol. 6089, pp. 257–271. Springer, Heidelberg (2010)
3. Hovy, E., Lin, C.Y.: Automated Text Summarization and the SUMMARIST System. In: Proceedings of TIPSTER Workshop, pp. 197–214 (1998)
4. Kou, L., Markowsky, G., Berman, L.: A Fast Algorithm for Steiner Trees. Acta Informatica 15(2), 141–145 (1981)
5. Anyanwu, K., Sheth, A.: p-Queries: Enabling Querying for Semantic Associations on the Semantic Web. In: Proceedings of the 12th International World Wide Web Conference, pp. 690–699 (2003)
6. Kochut, K.J., Janik, M.: SPARQLeR: Extended Sparql for Semantic Association Discovery. In: Franconi, E., Kifer, M., May, W. (eds.) ESWC 2007. LNCS, vol. 4519, pp. 145–159. Springer, Heidelberg (2007)
7. Zhang, X., Zhao, C., Wang, P., Zhou, F.: Mining Link Patterns in Linked Data. In: Gao, H., Lim, L., Wang, W., Li, C., Chen, L. (eds.) WAIM 2012. LNCS, vol. 7418, pp. 83–94. Springer, Heidelberg (2012)

8. Yan, X., Han, J.W.: gSpan: Graph-based Substructure Pattern Mining. In: Proceedings of the IEEE International Conference on Data Mining, pp. 721–724 (2002)

9. Sheth, A., Aleman-Meza, B., Arpina, I.B., et al.: Semantic Association Identification and Knowledge Discovery for National Security Applications. Journal of Database Management 16(1), 33–53 (2005)

10. Li, H.Y., Qu, Y.Z.: KREAG: Keyword Query Approach over RDF Data Based on Entity-Triple Association Graph. Chinese Journal of Computers 34(5), 825–835 (2011)

11. Mani, I.: Automatic Summarization. John Benjamins Publishing Company (2001)

12. Fokoue, A., Kershenbaum, A., Ma, L., Schonberg, E., Srinivas, K.: The Summary Abox: Cutting Ontologies Down to Size. In: Cruz, I., Decker, S., Allemang, D., Preist, C., Schwabe, D., Mika, P., Uschold, M., Aroyo, L.M. (eds.) ISWC 2006. LNCS, vol. 4273, pp. 343–356. Springer, Heidelberg (2006)

13. Hustadt, U., Motik, B., Sattler, U.: Reducing SHIQ Description Logic to Disjunctive Datalog Programs. In: Proceedings of the 9th International Conference on Knowledge Representation and Reasoning, pp. 152–162 (2004)

14. Zhang, X., Cheng, G., Qu, Y.Z.: Ontology Summarization Based on RDF Sentence Graph. In: Proceedings of the 16th International Conference on World Wide Web, pp. 707–716 (2007)

15. Cheng, G., Ge, W.Y., Qu, Y.Z.: Generating Summaries for Ontology Search. In: Proceedings of the 20th International Conference on World Wide Web, pp. 27–28 (2011)

16. Penin, T., Wang, H., Tran, T., Yu, Y.: Snippet Generation for Semantic Web Search Engines. The Semantic Web, 493–507 (2008)

17. Bai, X., Delbru, R., Tummarello, G.: RDF Snippets for Semantic Web Search Engines. In: Meersman, R., Tari, Z. (eds.) OTM 2008, Part II. LNCS, vol. 5332, pp. 1304–1318. Springer, Heidelberg (2008)

News Sentiment Analysis
Based on Cross-Domain Sentiment Word Lists
and Content Classifiers*

Lun Yan and Yan Zhang**

Department of Machine Intelligence, Peking University, Beijing 100871, China
pkualan@gmail.com, zhy@cis.pku.edu.cn

Abstract. The main task of sentiment classification is to automatically judge sentiment polarity (positive or negative) of published sentiment data (e.g. news or reviews). Some researches have shown that supervised methods can achieve good performance for blogs or reviews. However, the polarity of a news report is hard to judge. Web news reports are different from other web documents. The sentiment features in news are less than the features in other Web documents. Besides, the same words in different domains have different polarity. So we propose a self-growth algorithm to generate a cross-domain sentiment word list, which is used in sentiment classification of Web news. This paper considers some previously undescribed features for automatically classifying Web news, examines the effectiveness of these techniques in isolation and when aggregated using classification algorithms, and also validates the self-growth algorithm for the cross-domain word list.

Keywords: Sentiment classification, Seed words, Domain adaption, Opinion mining.

1 Introduction

Opinion mining or sentiment classification aims at classifying sentiment data into polarity categories (positive or negative). As a result, opinion mining has attracted much attention recently, for example, opinion summarization, opinion integration and review spam identification, etc. Many researches are mining the opinions of reviews or blogs [13,17,4]. However, few researches are about sentiment classification of news. With the explosion of Web 2.0 services, more and more news are published on the Web. So the analysis of large scale of news data is meaningful. Sentiment classification of web news is very different from other sentiment classification methods and very important. For example, the analysis of news reports about a new product will give the opinions of different people and different media sources, and also the analysis will show the potential problem of the product. With this analysis, the company of this product will have

---

* Supported by NSFC under Grant No. 61073081.
** Corresponding author.

S. Zhou, S. Zhang, and G. Karypis (Eds.): ADMA 2012, LNAI 7713, pp. 577–588, 2012.

enough time to upgrade this product to solve this problem, or to do the crisis public relations to get better opinions. Further, the same event can have different report versions from different sources. It is meaningful to analyze the difference to get the bias of different sources.

News are the objective reports of the latest events, so there are less sentiment features in news. The effective method does not perform well in the news sentiment classification. We need to find some new methods to detect the opinion or the polarity of the news. In previous work, supervised machine learning algorithms have been proved promising and widely used in sentiment classification. So we choose a supervised method to detect the opinion of news. However, because of the characteristic of the news, we should find more features to achieve a high performance. In the selection of the characteristics, we not only consider the score of the sentiment words, but also consider some other features such as the polarity of the title, the length of the news. We consider some content and structure features of the news and make some statistics analysis of these features. We choose the good features which perform differently in the positive news and the negative news so that it is useful in the sentiment classification and integrate them to make an effective classifier.

Of all the features, the score of the sentiment units in one piece of news needs a word list to get the score. The key of this feature is the quality of the sentiment word list. Also, the same word may play different roles in different domains. So we need to generate a domain-independent word list. In this paper, we present a self-growth sentiment word list algorithm to make an accurate sentiment dictionary based on the data and seed words. That is to say, the word list will change according to the news data. We assume that an entity will keep the same polarity in same news. With this process, the sentiment word list can grow up and become more and more accurate for the classification.

The rest of this paper is organized as follows. The overview of our approach and the data set is presented in Section 2. Section 3 shows the details of the self-growth algorithm for the sentiment word list. The details of the statistics analysis of features are presented in Section 4. The decision tree and the experimental results are introduced in Section 5. Section 6 surveys related work. The final section gives the conclusion and proposes future work.

2   Experimental Framework and Data Set

The news data used in our work is crawled from some famous news web sites in China, such as sina (http://www.sina.com) and sohu (http://www.sohu.com). We manually label 1208 pieces news, among whom 1000 are chosen as training data and the rest are left as testing data. All of the news are written in Chinese.

Since same words may have different polarity in different domains, we use a domain-independent sentiment word list to start our analysis. However, such a list is small initially and thus we need a self-growth algorithm to bring up it.

Considering the characteristics of the news, we analyze some features of the news documents such as the length of the article, the polarity score of the sentiment words, the amount of the negation in one document, the polarity of the title, the first and the last sentence and so on. The features are about the structure or the content. In these features, the polarity of the title, first sentence, last sentence, the topic sentence and the whole document are all labeled by volunteers.

We use the decision-tree method and SVM method to combine the features to generate sentiment classifiers. We evaluate the classifiers and compare the results. We also validate the self-growth algorithm for the sentiment word list. The experimental results are shown in Section 5.

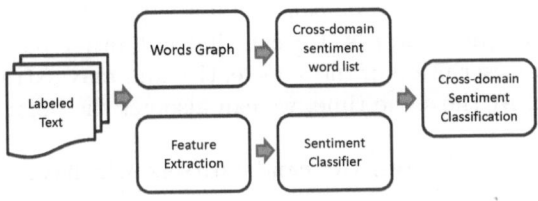

Fig. 1. The Experimental Framework

3   Cross-Domain Sentiment Word List

The performance of the sentiment word list affects the accuracy of the sentiment score of a sentence. As mentioned above, same words may have different polarity in different domains. If using a domain-dependent sentiment word list in different domains, it will either make mistakes or misunderstand the real polarity of a news report. Therefore, we need to construct a cross-domain sentiment word list and employ such a list in sentiment analysis. Obviously, such a list can grow up in applications.

In this section, we will discuss the construction method and the self-growth process of this domain-independent sentiment word list.

3.1   Initialization

The sentiment word list, represented by V_{sen}, contains weighted sentiment units which are words or phrases. Usually there are two methods for the initialization. In the first method, the sentiment word list is generated with the existed Chinese sentiment dictionary, which contains a large amount of positive and negative words. Then a training set is used to choose the words and produce the weights. In the second method, people can calculate the entropy based on the probability distribution of the words in the training set, and use a feature-selected algorithm to generate the seed list. Unfortunately, the feature-selected algorithms are not suitable for our news data due to insignificant features in news reports. Therefore, we choose the first method to give birth to a seed word list.

This method uses the existed sentiment dictionary and there are some ambiguous words in the seed list. Thus we filter the list based on the training set. We also calculate the inverse document frequency (idf) values of these words and regard the values as their weights.

3.2   The Candidate Set

After the initialization, we can begin our self-growth algorithm to produce the domain-independent word list. First, we need to choose a candidate set. Based on our observation, we have a hypothesis here.

Hypothesis 1: In one news report, the polarity descriptions for the same entity are interrelated.

We notice that the author will keep the same opinion on the same object or entity in the same sentence of his/her report. Suppose we have a sentence which describe an entity. Once we can decide the polarity of the sentence based on the existing sentiment words, we can find out other part of this sentence which contain the description of the same entity and may extract new sentiment words from them. At the same time, we can also get the polarities of these new sentiment words.

After this step, we will obtain two candidate sets, which contain positive words and negative words, respectively.

3.3   Graph Method

Graph-based approaches are widely used in sentiment detection applications. For example, Brody and Elhadad employ it for detecting sentiment in reviews [5], and Velikovich et al find out sentiment terms in a giga-scale web corpus [16]. In our work, we also make use of a graph method and regard all candidate words as nodes. We choose top 100 words in terms of frequency in each category as seeds. The weight of the edge between two words are decided by the distance of the two words in a same sentence.

3.4   Growth Algorithm

After the construction of the graph, we need a propagation algorithm to choose the words from the candidate set. We consider two methods which we will introduce in the next two paragraphs.

The first formula is from TrustRank [9]. The words in the positive seed set are assigned a polarity score of 1, while the words in the negative seed set are assigned -1. All the words in the candidate set except for the seeds start with a score of 0. We use Formula 1 to compute the score of the words in the candidate set. For each word x, $p(x)$ is the polarity score.

$$p(x) = \sum_{y \in N(x)} \frac{p(y)}{r(y)} \tag{1}$$

$N(x)$ is the word set of x's neighbors. $p(y)$ is the polarity score of the word y. $r(y)$ is the number of neighbors of word y.

The second method is based on probability model. The words in the positive seed set are assigned a polarity score of 1, while the words in the negative seed set are assigned 0. All the words in the candidate set except for the seeds start with a score of 0.5. Once the graph is constructed, we use Formula 2 to update the sentiment scores of the words. The following update-rule is applied:

$$p(x_i) = MAX_{j \in N_i \& j \neq i}(|p(x_i|x_j) * p(x_j)|) * S_j \qquad (2)$$

$$p(x_i|x_j) = 1 - \frac{l_{ij}}{L} \qquad (3)$$

$p(x_i)$ is the probability of word x to be the polarity words. N_i is the set of neighbors of x, and $p(x_i|x_j)$ is the weight of the edge between x_i and x_j as in Formula 3. The l_{ij} is the distance of word x_i and word x_j in the same sentence while L is the length of this sentence. S_j is the symbol of the word x_j of which the word x_j make the $p(x_i)$ get the largest absolute value.

The framework of the algorithm is shown as Algorithm 1.

Algorithm 1. Self-Growth Algorithm of the Cross-Domain Word List

---

Require: V_{sen}: a set of general word list; D: a set of documents; p and n: two thresholds;

Ensure: V_c: a cross-domain word list;

1: Get the candidate set C of the words with the help of V_sen and D using Hypothesis 1;
2: Put the top 100 most frequent words of V_{sen} into the seed set S;
3: Build the graph with C and S, take the words as nodes, and compute weights of the edges according to the co-occurrence relations of the two nodes;
4: Use Formula 1 or Formula 2 to compute the score of the words in C.
5: Choose the words whose scores are higher than threshold p as new positive words and whose scores are less than threshold n as new negative words. Add these words to the set V_c
6: **return** V_c;

---

4   Feature Analysis and Discussion

News are generally objective, which makes them different from other Web documents. Usually there are few explicit sentiment factors in a typical news report. Thus we need to dig out some features to strengthen the sentiment factors so that it is more practical to extract the polarity of the news. In this section, we will present some features and analyze their contribution for judging the polarity of a news report.

All the features are content-based and they are independent from each other. Some are related to the language while some others not.

The language of the data in this paper is Chinese.

4.1   The Polarity Score of the Whole Document

Though there are few sentiment words in news, the scores of the sentiment words are still an important feature. The sentiment words in one document usually reflect the opinion of the author and thus the sentiment of the news report. Considering this point, we calculate the score of the news with the formula shown in Formula 4[13]. We compare the scores for positive and negative news, and the result is shown in Figure 1(a).

$$S_i = \frac{L_d^2}{L_{phrase}} S_d N_d \tag{4}$$

In Formula 4, S_i stands for the score of the sentiment words in a sentence, S_d is the weight of the sentiment words, and N_d stands for the negation of this sentence. If there is a negation, N_d is -1, otherwise it is 1. L_d means the length of the words while L_{phrase} means the length of the sentence. This factor is added when we consider that if the sentence is long, the sentiment of the words is less important in the whole sentence than that occurs in a short sentence. Using this formula and the word list described in Section 3, we can have some results, as shown in Figure 2(a).

The horizontal axis depicts the range of the score. The vertical axis depicts the number of the reports that fall into a particular range.

The result in Figure 2(a) shows that, most reports (67.98%) fall into the range of 0-10. It validates the rule that the news reports uses sentiment words very carefully.

4.2   The Polarity of the Title

The title of a news report usually summarizes the main content of the report. We observe that the polarity of the title is consistent with the polarity of the whole report.

Let us see the statistic result shown in Figure 2(b). The polarities of the titles are manually labeled by volunteers.

The horizontal axis depicts the polarity of the title. The vertical axis depicts the number of the reports whose title falls into a particular polarity class.

The figure shows that all the news reports with a positive title are positive, and all the reports with a negative title are negative. However, there are too many reports whose titles are neutral.

4.3   The Polarity of the First Sentence

The situation of the first sentence is similar to that of the title. The first sentence usually implies the main opinion. According to [3], some researches have discussed the position of the sentence in a document. It turns out that the first sentence has something to do with the subject of the document. We also label the polarities of the first sentences manually, and the result is shown in Figure 2(c).

The horizontal axis depicts the polarity of the first sentence. The vertical axis depicts the number of the documents whose first sentence falls into a particular polarity class.

The figure shows that the polarity of the first sentence accords with the polarity of the whole report perfectly. All the news reports with a positive first sentence are positive. 96.41% reports which have a negative first sentence is either negative. However, the pity is there are still many reports whose first sentences are neutral.

4.4   The Polarity of the Last Sentence

The last sentence sometimes summarizes the whole report or expresses the opinion of the author. Some researches have been done in [3] which shows that people like to present their opinions in the last sentence. We also analyze the polarity relations between the last sentence and the whole document. The result is shown in Figure 2(d).

The horizontal axis depicts the polarity of the last sentence. The vertical axis depicts the number of the reports whose last sentence falls into a particular polarity class.

We can find that the last sentences of most reports have no polarities. For those whose last sentence has polarity, the polarity relation between the last sentence and the whole document is a nice match. If the last sentence is negative, the probability of the report to be negative is 95.29%. If it is positive, the probability of the report to be positive is 93.15%.

4.5   The Polarity of the Topic Sentence

The topic sentence is obviously concerned with the article. It is the summarization of the whole article. It will reflect all the characteristics of the whole article. However, it is not easy to find the sentence just by position. In this paper, we use the Topic Sentence Extraction Algorithm [6] to find the topic sentence. We skip the details of the algorithm and just use the sentences it extracts. After the extraction, we label these sentences. Same analysis is done and the result is shown in Figure 2(e).

The horizontal axis depicts the polarity of the topic sentence. The vertical axis depicts the number of the texts whose topic sentence falls into a particular polarity class.

The polarity of the topic sentence admirably matches that of the whole report. If the topic sentence is negative, the probability of the report to be negative is 97.29%. If it is positive, the positive probability of the report is 96.77%.

There are similar features which seem useful but not play an important role in sentiment analysis according to the statistics in practice, such as the inversion of the polarity showed in Figure 2(f). The inversion of the polarity is that if a sentiment word co-occurs with a negation word, the polarity of this sentiment word will be inverted. The statistic result of this feature shows that the inversion of polarity is not important. Therefore, we only consider the above features except the inversion of the polarity.

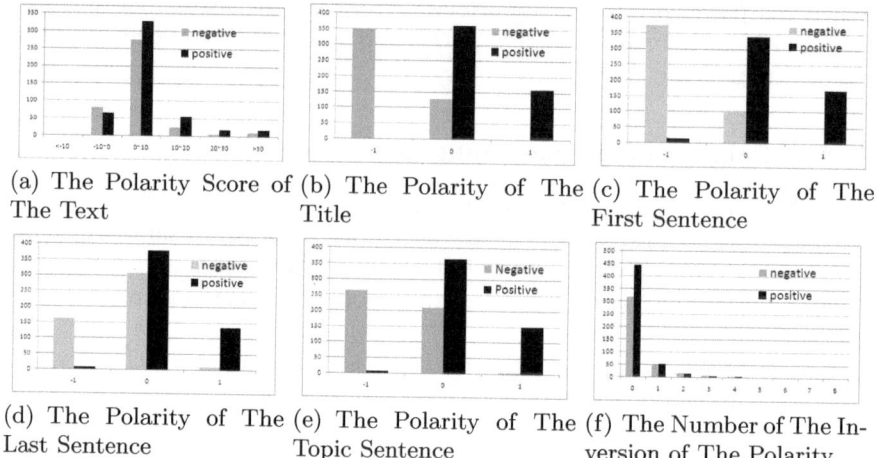

(a) The Polarity Score of The Text

(b) The Polarity of The Title

(c) The Polarity of The First Sentence

(d) The Polarity of The Last Sentence

(e) The Polarity of The Topic Sentence

(f) The Number of The Inversion of The Polarity

Fig. 2. The Statistics of The Features

5   Using Classifiers to Combine the Features

In the previous section, we present a number of features for sentiment classification. That is, we measure several characteristics of news, and find out ranges of these characteristics which are correlated with the polarity of news. Nevertheless, when used individually, no features can perform well in the sentiment classification. Some may make big mistakes. In this section, we study whether we can classify the news more efficiently by combining these features. Our goal is that if we apply multiple features and then combine the outcome of each feature, we will be able to classify more news with greater accuracy. One way of combining our features is to regard the sentiment extraction problem as a classification problem. In this case, there are many ways to combine the features.

We use 1000 pieces of news as the training set, 208 pieces of news as the testing set. According to the result of the previous section, the features we choose are the score of the sentiment sentences, the polarity of the title, first sentence, last sentence, and topic sentence, and the number of inversion of the polarity.

We use some machine learning methods to combine the polarity features of news reports. From the experimental results with different classification techniques, we find that the decision-tree-based techniques and *SVM* perform well. We use *weka* [1] to realize these two algorithms.

One of the decision-tree is shown in Figure 3. The nodes in rectangle are classifying features. The PN-Score node is the score of the sentiment sentences. The nodes in oval-shaped are the classification results.

We also use *SVM*(Support Vector Machine) [1] to combine the features and estimate the result of the classification. The result is shown in Table 1.

---

[1] http://www.cs.waikato.ac.nz/ml/weka/

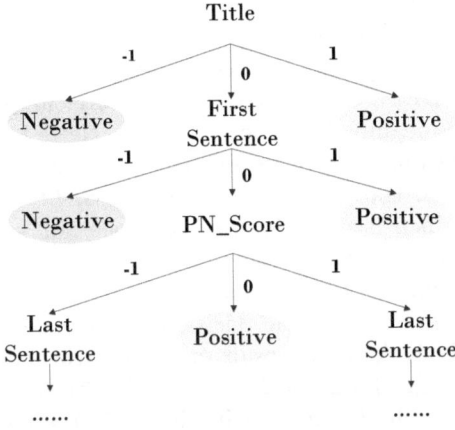

Fig. 3. The example of Decision Tree

5.1 Experiment

In order to evaluate our methods, we make some contrastive experiments. The first is the different rule of the decision tree. The second contrastive experiment is the *SELC* model [13].

The *SELC* model is a classification model of integrating the Lexicon-based method and the Corpus-based method. The main idea of the *SELC* model is based on the self-growth sentiment word list and the ratio control to make a first classification without a training set. Then use the result of this phase as the training set and use the machine learning method to generate a classifier.

5.2 Result Analysis

The evaluation criterions are precision, recall and the F1 value. The number of news reports in the data set is 1208. We use 1000 news reports as the training set and the rest as the testing set. The result is shown in Table 1.

Table 1. The average precision, recall, F-measure values for the sentiment classification with different methods

	Precision	Recall	F1
SVM+Growth*	96.18%	96.08%	96.13%
SVM+Growth	95.74%	95.60%	95.67%
C4.5+Growth*	95.30%	95.10%	95.20%
C4.5+Growth	95.63%	95.18%	95.40%
C4.5	91.21%	90.21%	90.70%
ID3	90.61%	89.23%	89.91%
SELC	64.26%	53.88%	58.61%

The result of the *SELC* model is worse than that in [13] because that in [13], the data set contains reviews which have much more sentiment features than that in news. Besides, the sentence in reviews is usually very short. The growth algorithm in *SELC* is comparing the words in positive and negative reviews, which is not meaningful in news reports. News is longer than reviews, which means that news have more words. The most words in news do not have sentiment meanings. So the growth algorithm in *SELC* is not appropriate in news. The method which is good for reviews does not work for news reports either. This also shows that the sentiment features we selected from news are effective for news sentiment classification.

The result of *C4.5+Growth* is the decision-tree method with self-growth algorithm with Formula 1. And the result of *C4.5+Growth** is the method with self-growth algorithm and Formula 2. The *SVM+Growth* is the result of SVM combined with self-growth algorithm using Formula 1 while the *SVM+Growth** is the result of SVM combined with self-growth algorithm using Formula 1. The decision-tree with different rules have different results, though the difference is not large. In a word, *C4.5* is better than *ID3*. The *C4.5* rule decision-tree method with self-growth algorithm is much better. The reason is that the self-growth algorithm can make the polarity score of the document more accurate. We can also see the performance of Formula 1 is similar to that of Formula 2.

6   Related Work

6.1   Corpus-Based Method

Most corpus-based methods are supervised. The sentiment classifiers use standard machine learning techniques such as SVM and NBm[1]. It has been investigated that different factors affect the machine learning process. For example, linguistic, statistical and n-gram features are used in [7]. [17] uses semantically oriented methods to identify the polarity at the sentence level. [12] uses graph-based techniques to identify and analyze only subjective parts of documents. Selected words and negation phrases are investigated in [10]. These approaches need large labeled corpora which are available for training to work well. But in a different domain [2], topic or time period [14] of the training data may decrease the performance of corpus-based methods.

To solve this problem, a cross-domain sentiment classification via SFA is used [11]. SFA discovers a robust representation for cross-domain data by fully exploiting the relationship between the domain-specific and domain-independent words via simultaneously co-clustering them in common latent space.

6.2   Lexicon-Based Method

Some of the lexicon-based methods use general sentiment word lists acquired from dictionaries or the Internet. [12] shows that lexicon-based methods perform worse than statistical models built on sufficiently large training sets in the movie review domain. It is shown in [8] that the performance of systems using general word lists is comparable to that of supervised machine learning methods on some domains such as product reviews.

Other approaches just need some seed words, and then get more sentiment words and phrases with some growth algorithm. For example, [15] uses two human-selected seed words (poor and excellent) in combination with a very large text corpus. Then using the association (measured by point-wise mutual information) of the phases and the seed words to get the semantic orientation of phases. Calculate the sentiment of a document with the average semantic orientation of all such phases. In [18] uses linguistic pattern to generate seed words automatically. The experiment shows that the performance of this method is similar to that of supervised methods.

6.3 Machine Learning Methods

Machine learning techniques have been successfully used in opinion mining. It always needs a large amount of labeled training data. In this paper, we use the decision tree method as the classifying method for news sentiment classification to combine the selected features.

7 Conclusion and Future Work

This paper considers the characteristics of news and analyzes different features to build a model of sentiment classification for news. The contributions of this paper are: (1)Proposing a self-growth algorithm to generate a cross-domain sentiment word list; (2)Analyzing the sentiment features of news reports, finding out some useful features for the news sentiment classification; (3)Generating a model of sentiment classification for news reports, integrating the lexicon-based methods and the supervised machine learning methods. The experiment validates the effectiveness of the self-growth algorithm, and evaluates the proposed model by comparing with different methods.

There is still a lot work needed to accomplish. First, the semantic information of the words which is not used in this paper. The semantic information of the words is strongly correlated with the sentiment information. Second, the self-growth of the word list is not accomplished perfectly. We only use the information of the training set. It is considerable to use the extended information from the Internet such as *Wikipedia* to improve the accuracy of the classification.

References

1. Alpaydin, E.: Introduction to machine learning. The MIT Press (2004)
2. Aue, A., Gamon, M.: Customizing sentiment classifiers to new domains: A case study. In: ICRA in NLP (2005)
3. Becker, I., Aharonson, V.: Last but definitely not least: On the role of the last sentence in automatic polarity-classification. In: ACL (2010)

4. Brody, S., Diakopoulos, N.: Cooooooooooooooolllllllllllllll!!!!!!!!!!!!!! using word lengthening to detect sentiment in microblogs. In: EMNLP (2011)
5. Brody, S., Elhadad, N.: An unsupervised aspect-sentiment model for online reviews. In: ACL (2010)
6. Bun, K.K., Ishizuka, M.: Topic extraction from news archive using tf*pdf algorithm. In: WISE (2002)
7. Dave, K., Lawrence, S., Pennock, D.M.: Mining the peanut gallery: Opinion extraction and semantic classification of product reviews. In: WWW (2003)
8. Gamon, M., Aue, A.: Automatic identification of sentiment vocabulary: Exploiting low association with known sentiment terms. In: ACL (2005)
9. Gyongyi, Z., Garcia-Molina, H., Pedersen, J.: Combating web spam with trustrank. In: VLDB (2004)
10. Na, J., Sui, H., Khoo, C., Chan, S., Zhou, Y.: Effectiveness of simple linguistic processing in automatic sentiment classification of product reviews. In: ISKO (2004)
11. Pan, S.J., Ni, X., Sun, J.-T., Yang, Q., Chen, Z.: Cross-domain sentiment classification via spectral feature alignment. In: WWW (2010)
12. Pang, B., Lee, L.: A sentiment education: Sentiment analysis using subjectivity summarization based on minimum cuts. In: ACL (2004)
13. Qiu, L., Zhang, W., Hu, C., Zhao, K.: Selc: A self-supervised model for sentiment classification. In: IKM (2009)
14. Read, J.: Using emoticons to reduce dependency in machine learning techniques for sentiment classification. In: ACL (2005)
15. Turney, P.D.: Thumbs up or thumbs down? semantic orientation applied to unsupervised classification of reviews. In: ACL (2002)
16. Velikovich, L., Blair-Goldensohn, S., Hannan, K., McDonald, R.: The viability of web-derived polarity lexicons. In: ACL (2010)
17. Yu, H., Hatzivassiloglou, V.: Towards answering opinion questions: Separating facts from opinions and identifying the polarity of opinion sentences. In: EMNLP (2003)
18. Zagibalov, T., Carroll, J.: Automatic seed word selection for unsupervised sentiment classification of chinese text. In: COLING (2008)

Integrating Data Mining and Optimization Techniques on Surgery Scheduling

Carlos Gomes[1,2], Bernardo Almada-Lobo[1,3], José Borges[1,3],
and Carlos Soares[1,2]

[1] INESC-TEC
[2] Faculdade de Economia/Universidade do Porto, Portugal
[3] Faculdade de Engenharia/Universidade do Porto, Portugal

Abstract. This paper presents a combination of optimization and data mining techniques to address the surgery scheduling problem. In this approach, we first develop a model to predict the duration of the surgeries using a data mining algorithm. The prediction model outcomes are then used by a mathematical optimization model to schedule surgeries in an optimal way. In this paper, we present the results of using three different data mining algorithms to predict the duration of surgeries and compare them with the estimates made by surgeons. The results obtained by the data mining models show an improvement in estimation accuracy of 36%. We also compare the schedules generated by the optimization model based on the estimates made by the prediction models against reality. Our approach enables an increase in the number of surgeries performed in the operating theater, thus allowing a reduction on the average waiting time for surgery and a reduction in the overtime and undertime per surgery performed. These results indicate that the proposed approach can help the hospital improve significantly the efficiency of resource usage and increase the service levels.

Keywords: Surgery Scheduling, Data Mining, Optimization.

1 Introduction

Technological advances, medical breakthroughs, better and more efficient services are constantly changing the world, improving the quality of life and, in the long run, life expectancy. Hospitals, which can now provide more and better care to patients who once were unable to receive treatment, face pressures due to the rise on demand and the elevated waiting time for treatments.

Looking at the operating theater, which is considered by many the largest budget consumer in hospitals [1], the waiting time problem is worsened due to the nature of the treatments involved (surgeries) and the immediate attention they naturally require. In fact, reducing waiting lists for surgery has always been a priority of sovereign governments and therefore of many researchers. The Portuguese government has successfully reduced the median waiting time for surgery from 8.6 to 3.3 months between 2004 and 2009, as a result of the introduction

S. Zhou, S. Zhang, and G. Karypis (Eds.): ADMA 2012, LNAI 7713, pp. 589–602, 2012.

of an incentive program to perform additional surgeries. Still, despite the efforts and success in tackling the long waiting lists for surgery, Portugal along with the United Kingdom ranked last among other 31 European countries regarding waiting times for treatment [2]. Moreover, incentive strategies imply great costs, and in the current economic context can easily be ceased. An OECD (Organisation for Economic Co-operation and Development) report also lists other countries with waiting time for surgery issues, such as Italy and Norway which have more than 25% of their patients waiting for more than 12 weeks [3].

In this work, we introduce a combination of data mining and optimization techniques applied to operating theater planning. In the process of scheduling surgeries, surgeons have to estimate empirically how long the combination of surgical procedures will take in order to book the operating room. The accuracy of these estimations will define the quality of the operating theater schedule, since every deviation from the estimates leads either to schedule disruptions (surgeries exceeding their allocated time) or to unoccupied time (surgeries finishing earlier than estimated). This wasted time is valuable, not only for the hospital but to the patients who see their health conditions and overall satisfaction quickly deteriorating throughout time. On the other hand, since scheduling is a complex combinatorial problem, subject to different rules and constraints, it represents an opportunity to use optimization techniques to improve its quality. Herein, we make use of data mining algorithms to estimate the duration of surgeries and of an optimization model to optimize the surgeries' schedule. We achieve good results with both techniques separately, but it is their combination that enhances the final scheduling solution and enables the automation of the entire surgery scheduling process. In summary, the contributions in this work are:

- Development of an automatic and effective mechanism to estimate surgery duration, based on historical surgery records, patient and surgeon information;
- An approach that optimizes operating theater resource utilization based on the predicted durations of surgeries.

Next we will provide further details about the problem and a review of related work in Sections 2 and 3, respectively. In Section 4 the estimation problem will be presented and the surgery scheduling optimization model will be described in Section 5. Experimental results are shown in Section 6 and final remarks are given in Section 7.

2 Problem

Our work was conducted at a large Portuguese hospital, enabling the analysis and gathering of the necessary inputs to develop and validate the approach presented. The scope of the experiments is restricted to the outpatient department, also know as ambulatory surgery department, which deals with patients who are not hospitalized after the surgical intervention. On the other hand, inpatient surgeries require hospitalization for the patients' recovery, and this recovery is

the true bottleneck of the operating theater and hospital resources [4–6]. Therefore, we have decided to start by focusing on outpatients, for which the efficient
time management of the operating theater becomes crucial to make the most of
the existing resources. To illustrate how the predictions made by surgeons, which
are currently used as basis for operating room scheduling, are deviated from reality , Figure 1 plots those estimates against the real surgery times recorded on
this department between 2010 and 2011. It is possible to observe a ladder effect
created by the coarse time granularity used by surgeons on their estimates (15,
30 or 60 min.). This prevents fine tuning of the surgery schedule and therefore
reduces its quality. Furthermore, surgeons tend to make predictions according
to their own interests and undervalue the objectives of the hospital. In fact,
Macario states that surgeons intentionally overbook operating rooms to block
them to other surgeons or underestimate the duration of surgeries in order to
squeeze more surgeries in their available time [7]. Both scenarios lead to wasted
time, but the latter actually produces schedule disruptions due to the delays
caused, often leading to the postponing of surgeries.

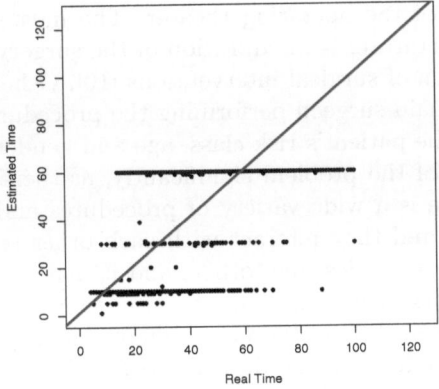

Fig. 1. Comparison between the estimated duration of surgeries made by surgeons and
the real values. The diagonal line shows the desirable scenario where the estimated
duration is always correct.

The economic and social effects of an improvement in efficiency through accurate estimates can be quantified in the increase of the number of surgeries carried
out and the revenue obtained by hospitals. Concerning public health, this enables
the possibility to reduce waiting lists for surgery with marginal costs. However,
the challenge is not only methodological but also of change management. The
introduction of decision support systems in this environment leads to a reduction in the decision making power of doctors and is prone to face resistance.
Nonetheless, this work is only concerned with the former challenge and refrains
from discussing the implementation issues.

3   Background

Operating theater planning is a major challenge throughout the scientific literature. The reviews by Cardoen et al., and Guerriero and Guido are good indicators of the vast amount of research that is being done in this field recently [1, 8]. Although there are many publications on modeling and predicting surgical durations, the most prevalent research field in the operating room is operational research, which typically addresses scheduling and planning problems. The operating theater planning is normally divided into three decision levels: (i) operational, (ii) tactical and (iii) strategic. Our work focuses on the first planning level. We combine methods from the operations research and data mining fields for the periodic (weekly) scheduling of patients to the available operating rooms. The tactical and strategic decision levels concern longer term decisions on capacity definition and allocation of resources to the different surgical specialties, also known as the master surgery schedule and case-mix planning problems.

The surgical process is naturally characterized by deep uncertainty [9], due to different sources of variability emerging since the patient arrival to his postoperative recovery. These factors affect the total duration of the patient's stay in the different areas of the operating theater. The most important, and the most studied in the literature, is the duration of the surgery, including anesthesia and the combination of surgical interventions [10]. Other significant sources of variability are the main surgeon performing the procedure and his team, the anesthesia type, and the patient's risk class, age and gender [11]. These factors increase the difficulty of the problem significantly, as there are many different possibilities (e.g., there is a wide variety of procedures and a large number of surgeons in hospitals) and they interact with each other (e.g., each surgeon is more effective on some surgeries than others and it may be easier to perform a given surgery on patients with some characteristics than others) [7]. Researchers have been modeling surgical times targeting different management decisions but most studies aim to predict surgery duration before it starts (off-line scheduling) [12, 13, 9, 10, 14]. Other tasks include predicting the time remaining during surgery execution (on-line scheduling) [15] and the duration of a series of surgeries, aiming to reduce the total overtime incurred into [16]. However, it is also recognized that, due to the uncertain nature of surgical procedures, it is often better to know the upper and lower bounds of the duration than a point estimate [14].

We found that most of the research performed in this field determines the most important factors of variation between surgeries but does not come up with a generalized estimation model that could be applied transversely in the operating theater [17, 18]. As stated by Combes et al., the statistical models represent an average phenomenon and not the variation in the subsets of observations that that average represents [12]. Nonetheless, prediction models have been developed to successfully improve the accuracy of predictions of surgery duration. Wright et al., in the mid 1990s reduced the mean absolute error relative to surgeon estimates in almost 20% using regression-based methods [19]. Marinus et al. also show significant reductions in average overtime and undertime per surgery on a vast set of surgical procedures using a regression model [13]. Stepaniak et al., estimate the effect of several medical variables that affect a surgery in two different hospitals in the Netherlands by means of ANOVA models and use those models to estimate the duration of surgical cases, obtaining a 15 % improvement [20]. An alternative approach, in which surgeons are given tools to improve their estimates, rather then being replaced by prediction models, was proposed by Combes et al. Motivated by a real case scenario, they developed a data exploration framework to help surgeons estimate the duration of surgeries based on their past performance. However, this study limited itself by applying the methodology to a small subset of gastric procedures [12].

4   Surgery Duration Estimation

The main goal of our work is to reduce the uncertainty of the estimated duration of surgeries. This can be addressed as a regression problem, where the surgery duration is the dependent variable y and the known environment settings and patient characteristics constitute the vector of independent variables $\mathbf{x}$ that influence the duration of a surgery: $y_i = f(\mathbf{x}_i)$. The data used to carry out this work was extracted from different databases, allowing us to enrich the dataset with characteristics of the patient, surgeon and surgical procedures.

4.1   Dataset

The data used in this work describes the outpatient surgeries performed in one of the largest Portuguese public hospitals from 2006 to 2011, containing approximately 9.500 completed surgical cases.

The most relevant attributes, selected using a attribute subset evaluator algorithm, are: Gender, Priority, Week Day, Shift, ICD Disease, ICD Procedure 1 and 2, Number of Interventions to Date, the existence of circulatory problems and the average total duration of the procedure.

In order to give an overview of the surgery duration distribution, Figure 2 plots the density histogram of surgery duration on the test set. It fits a log-normal distribution ($\mu = 26.123$, $\sigma = 1.862$ found by maximum likelihood estimation) for a significance level of 0.05, verifying the distribution type fit supported by Strum et al. and Spangler et al. [18, 17].

4.2   Evaluation

The accuracy of the methods was evaluated using standard error measures. The Mean Absolute Error (MAE):

$$MAE = \frac{\sum_{i=1}^{n} |y_i - \hat{y}_i|}{n} \tag{1}$$

where y_i and $\hat{y}_i$ are the true and predicted duration values and n is the number of predictions (i.e., the number of surgeries for which a prediction was made). MAE

Table 1. Independent variables used for predictive modeling

#	Type	Description
1	Nominal	Patient gender
2	Numeric	Patient age
3	Ordinal	Patient priority
4	Numeric	Patient waiting time for surgery
5	Nominal	Surgery month
6	Nominal	Surgery weekday
7	Nominal	Surgery shift
8	Nominal	Patient diagnosed disease
9	Numeric	Number of interventions to be performed
10	Nominal	Intervention code 1
11	Nominal	Intervention code 2
12	Nominal	Intervention code 3
13	Numeric	Number of surgeries performed on the patient to date
14	Numeric	Number of interventions performed on the patient to date
15	Binary	If the patient has undergone surgery on other specialties
16	Nominal	Surgeon identification
17	Nominal	Surgeon gender
18	Numeric	Number of times the surgeon has dealt with this disease
19	Numeric	Number of times the surgeon has performed the main intervention
20	Binary	If the patient has other diagnosis
21	Binary	If the patient has any circulatory problem
22	Binary	If the patient has diabetes or renal problems
23	Binary	If the current diagnosis is recidivist
24	Numeric	Average total surgery duration
25	Numeric	Surgery real duration

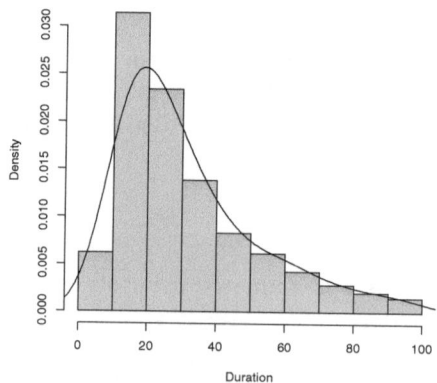

Fig. 2. Distribution of surgery durations (duration in minutes)

values are in the same scale as the dependent variable (i.e., surgery duration in this case). Alternatively, we can use Mean Absolute Percentage Error (MAPE), which rescales the values of MAE in the interval $[0, 1]$:

$$MAPE = \frac{1}{n} \sum_{i=1}^{n} \frac{|y_i - \hat{y}_i|}{y_i} \tag{2}$$

In many applications, such as this one, the impact of a small deviation is often irrelevant. So more importance should be given to larger ones. In such cases, the Mean Squared Error (MSE) measure may be more appropriate than MAE and MAPE:

$$MSE = \frac{\sum_{i=1}^{n}(y_i - \hat{y}_i)^2}{n} \tag{3}$$

An additional measure is the correlation between the predictions and the real values:

$$\tau = \frac{\sum_{i=1}^{n}(y_i \hat{y}_i) - n\bar{y}\bar{\hat{y}}}{(n-1)(s_y s_{\hat{y}})} \tag{4}$$

where $\bar{y}$ and s_y is the mean and variance of y.

These metrics do not take into account if the deviations are originated by over-or under-estimation. This is significant, due to the different effects they have on the operating theater schedule. Overestimation is when a surgery lasts less than predicted ($y_i < \hat{y}_i$), leading to idle time in the operating room time or under-utilization. Underestimation occurs when a surgery lasts longer than predicted ($y_i > \hat{y}_i$), having a greater impact on the operating room since it may disrupt and delay subsequent surgeries. Dexter proposes an asymmetric absolute error measure involving under-utilization and over-utilization costs [9]. As there are no estimates of those costs available in our application, we do not use this measure. However, in Section 6, we provide some analysis of the results in terms of over-and under-estimated time.

4.3   Experimental Setup

We compare three different data mining techniques: Linear Regression (LR), Random Forests (RF) and the M5 Rules (M5) algorithm [21]. However, rather than individual models, we have used ensembles of each of those models using the bootstrap aggregation algorithm (Bagging) [22], to reduce the variance and increase the estimation accuracy.

To develop and evaluate our models we separated the data into two sets. The first includes the surgeries from 2006 to 2009 and is used for training the models, while the second (surgeries in 2010 and 2011) is used for testing those models. The dataset consists of 25 variables, of which one is ordinal, five are binary, seven are numeric and the remaining 12 are nominal (Table 1). The identification of the surgical interventions and diseases are coded using the International Classification of Diseases norm (ICD-9) published by the World Health Organization.

5   Surgery Scheduling

The models obtained in the previous section are used to estimate the duration of a set of surgeries. The next step is to find the optimal schedule for those surgeries assuming that their duration is the predicted one. The capacity available (time) might be spread by different operating rooms, in different days and also different parts of the day (i.e., morning and afternoon shifts).

Our optimization model is also able to take into account constraints concerning the availability of surgeons. It is based on the 1-0 Multiple Knapsack Problem, proved to be NP-Hard [23]. In our case a knapsack is an operating room shift with a given time capacity. We want to fill the knapsacks with surgeries, which consist of pairs of patients and surgeons. Table 2 summarizes the notations used throughout this section.

Table 2. Variables used in the optimization model

Symbol	Description
N	Set of patients
R	Set of operating rooms
S	Set of surgeons
D	Set of scheduling days
T	Set of shifts (morning/afternoon)
s_i	Surgeon assigned to patient i
d_i	Duration estimated for the surgery of patient i (minutes)
A_{rdt}	Availability of operating room r on day d and shift t (binary)
S_{jdt}	Availability of surgeon j on day d and shift t (binary)
u	Operating room clean up time (constant)
c	Shift capacity (constant)

5.1   Decision Variables and Objective Functions

As mentioned above, the mathematical model devised resembles the binary multiple knapsack model. Thus, the binary decision variables x_{irdt} define the assignment of a patient i to an operating room r on a given day d and shift t, and the binary decision variable y_{jrdt} assigns the surgeon j to an operating room r on day d and shift t. Note that the model presented assigns patients to a period of time and local (knapsack) but does not sequence patients inside a knapsack. The sequence can be obtained using a simple method such as random ordering, since we constrain the problem to fix a surgeon to one operating room per shift.

The main goal of our work is to increase the efficiency of the operating room, which can be translated to the maximization of surgeries performed in a given period:

$$Max_{f_1} = \sum_{i \in N} \sum_{r \in R} \sum_{d \in D} \sum_{t \in T} x_{irdt} \tag{5}$$

Yet, increasing surgeries performed decreases the operating room utilization, due to the setup (clean-up of rooms) times between surgeries. The following expression represents the maximization of the mean utilization of all operating rooms in R over the set of days in D.

$$Max_{f_2} = \frac{\sum_{i \in N} \sum_{r \in R} \sum_{d \in D} \sum_{t \in T} x_{irdt} d_i}{c \sum_{r \in R} \sum_{d \in D} \sum_{t \in T} A_{rdt}} \tag{6}$$

5.2   Constraints

First of all, given that there is a surgeon responsible for each patient, the binary variables y_{jrdt} are linked to x_{irdt} (Eq 7). This enables us to ensures that the surgeons cannot be assigned to multiple procedures simultaneously. Additionally, the following constraints are also included in our model:

$$x_{irdt} \leq y_{jrdt}, \forall i \in N, r \in R, d \in D, t \in T, j \in S : j = s_i \tag{7}$$

$$\sum_{i \in N} x_{irdt}(d_i + u) \leq A_{rdt} c, \forall d \in D, r \in R, t \in T \tag{8}$$

$$y_{jrdt} \leq S_{jdt}, \forall j \in S, d \in D, r \in R, t \in T \tag{9}$$

$$\sum_{r \in R} \sum_{t \in T} y_{jrdt} \leq 1, \forall j \in S, d \in D \tag{10}$$

$$x_{irdt}, y_{jrdt} \in \{0,1\} \tag{11}$$

Equation 8 represents the constraint for the available capacity per operating room. It prevents overtime, i.e., planning surgeries on a shift that take longer than the available time, and restricts patients from being assigned to operating rooms if the room is unavailable on a given day and shift. Equation 9 limits the allocation of surgeons according to their availability. Equation 10 prevents the surgeons from being allocated to different operating rooms in the same shift and day. Finally, Equation 11 states that the decision variables are binary.

5.3   Experimental Conditions

The experiments are limited to a single week and we compared the results obtained with the optimization model based on the predicted duration of surgeries with the observed schedule. We used a patient waiting list composed by 394 real patients, the total duration of the surgery is known, as well as the estimates made by the surgeons. These figures will be used to compare the performance in terms of planned and real utilization rate, number of surgeries performed, overtime and undertime. We assume there is only one operating room with one 4 hour shift available per day, on a Monday to Saturday week (Friday off). It is also assumed a constant clean-up time of 17 minutes between surgeries, and that every surgeon (8 in total) and patient is immediately available for surgery at the scheduled time. The experiments were performed using CPLEX 12.2.

6   Results

Initially we will focus on the quality of the predictions (Section 6.1) and then we analyze the results of the combination of the predictions with the optimization model (Section 6.2).

6.1   Surgery Duration Estimation

Table 3 compares the accuracy of the three data mining algorithms tested: Linear Regression (LR), Random Forests (RF) and M5 Rules (M5) with two baselines. The first is the surgeon estimates and the second is the median duration, that is, the median of the surgeries of the same type.

Table 3. Estimation errors

Metric	Surgeon	Median	LR	RF	M5
MAE	13.84	11.19	9.55	9.63	8.89
MSE	312	224	171	158	144
MAPE	47%	54%	39%	44%	37%
Correlation	0.69	0.75	0.78	0.79	0.81

The results show there is clear advantage of using the data mining algorithms for predicting surgery duration, when compared to the estimates made by surgeons, which is the method currently used in the hospital, and also to the median estimates. The M5 algorithm shows an average absolute improvement of 4.95 minutes per surgical case, which represents an improvement of almost 36% in estimation accuracy compared to the surgeon estimates. Additionally, the mean squared error decreased to half, meaning that the largest deviations were greatly reduced. The estimates are plotted against the real durations in Figure 3, showing that the ladder effect seen in Figure 1 disappeared.

Focusing just on the results obtained by the M5 algorithm and the surgeon predictions and separating them according to whether they over- or underestimate the true duration, it is noticeable that the greatest improvement comes from surgeries whose prediction was overestimated by the surgeons (Table 4). This observation shows that surgeons are mostly prone to over-estimate the duration of surgeries, thus apparently preferring to block their colleagues [7]. On the other hand, as expected, the data mining algorithms distribute their errors evenly.

Table 4. Total surgery overtime and undertime relative to the predictions (minutes)

	Surgeon	M5	Difference	Improvement
Overtime	11605	9894	1711	-15%
Undertime	18714	9588	9126	-49%

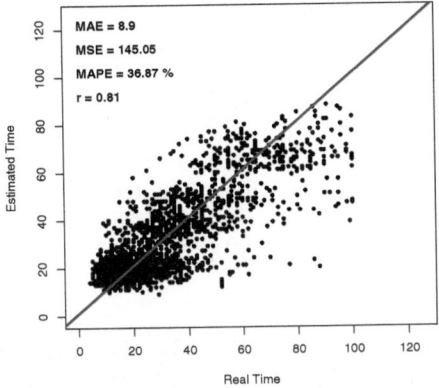

Fig. 3. Comparison between the bagged M5-Rules algorithm estimates and the real duration

Although the absolute gain per surgery may seem small (4.95 minutes) the relative gain of 36% is significant, where one could assume that the M5 algorithm provides 36% more time to perform surgeries. Considering that the average duration of surgeries is 31 minutes and that the approach proposed here could reduce the wasted time by 10837 minutes in 2010 and 2011, then this system would have enabled up to an extra 343 surgeries to be performed. Despite being a rough estimation, this still indicates that a significant improvement in efficiency could be achieved with the approach proposed here.

6.2   Schedule Optimization

Two experiments were executed to schedule an entire week of surgeries. In the first, we compare the schedule obtained by maximizing the number of surgeries performed (Eq. 5) based on the duration estimates made by the surgeons with the true schedule. This isolates the effect of the optimization component in the results. The second experiment compares the schedules generated by the optimization model using each objective functions, number of surgeries (Eq. 5) and operation room utilization (Eq. 6). In both cases the optimization is based on the durations estimated by the data mining models. This experiment aims to compare the two objective functions.

The results for the first experiment are presented in Table 5. As expected, the optimization model increased the number of surgeries scheduled but the magnitude is surprising: almost three-fold. In fact, the optimization model performed abnormally due to the selection of surgeries with very low estimated duration. This resulted in a schedule consisting solely on surgeries with underestimated duration, which naturally leads to overtime in the use of the operating theater. The surgeries scheduled would require 124% of the available time.

Table 5. Comparison of the real schedule with the one optimizing the number of surgeries (Eq. 5) based on the estimates made by the surgeons

	True Schedule						Maximize Surgeries					
	Mon	Tue	Wed	Thu	Sat	Total	Mon	Tue	Wed	Thu	Sat	Total
Number of Surgeries	4	2	3	4	5	18	10	10	10	10	10	50
Planned Utilization	45%	37%	32%	58%	69%	48%	21%	23%	21%	21%	21%	21%
Real Utilization	35%	29%	10%	58%	63%	39%	126%	131%	135%	118%	111%	124%
Overtime	40	18	55	18	54	185	252	259	273	233	216	1233
Undertime	17	0	4	20	39	80	0	0	0	0	0	0

Table 6 compares the schedules obtained by the optimization of each of the two objective functions, using data mining estimations. As stated, the first one maximizes the number of surgeries performed while the second the utilization of the operating rooms. By maximizing the utilization, less surgeries are carried out than what happened in fact, due to the clean-up time (setup) between surgeries. Therefore, the results illustrate that we may obtain results that are better in terms of the use of resources but worse in terms of the reduction of surgery waiting lists.

Table 6. Comparison of the optimized schedules using the two objective functions (Eq. 5 and Eq. 6) based on the predictions by the data mining models

	Maximize Surgeries						Maximize Utilization					
	Mon	Tue	Wed	Thu	Sat	Total	Mon	Tue	Wed	Thu	Sat	Total
Number of Surgeries	7	7	7	7	8	36	3	3	3	3	3	15
Planned Utilization	29%	29%	29%	29%	42%	31%	82%	90%	69%	74%	88%	80%
Real Utilization	47%	60%	48%	48%	65%	53%	75%	75%	75%	54%	75%	71%
Overtime	43	73	49	44	56	265	40	54	15	47	36	192
Undertime	0	0	4	0	0	4	24	19	29	0	5	77

Additionally, when comparing the schedule obtained by maximizing the number of surgeries using each of the two methods to estimate duration, surgeon estimates (right-hand side of Table 5) and data mining models (left-hand side of Table 6), we observe that the latter not only provides more accurate predictions (as observed in Section 6.1) but also generates a more reasonable schedule: the real utilization of the operating rooms is under 100%. Finally, when compared to the real schedule (left-hand side of Table 5), the schedule obtained using the method to optimize the number of surgeries based on the predictions made by the data mining models (left-hand side of Table 6), we observe a very significant improvement in both evaluation measures: twice as many surgeries are performed, resulting in an increase in real utilization time from 39% to 53%. These results clearly show the advantage of using data mining to predict the duration of surgeries.

7   Conclusions

Although the stimulus initiatives introduced by the Portuguese government to reduce waiting lists for surgery have shown positive results, they imply significant costs and are now being reduced due to the current financial crisis. We address the problem of optimizing the schedules of operating rooms with a combination of data mining and optimization techniques. To the best of our knowledge this is a novel approach. It was tested on outpatient surgeries, a subset of all the surgeries for which the accurate prediction of the duration is particularly important because it is the most important factor for the optimization of operating room usage. Our results show that it is possible to significantly increase the utilization of the rooms and the number of surgeries performed.

The performance achieved by the data mining and optimization models is sufficient to encourage the development of this work and its experimentation on different surgical departments. However, the optimization model will have to be adjusted to take into consideration downstream resources following the operating room (e.g., recovery wards).

Acknowledgements. This work was partially supported by FCT projects "An integrated framework for operating room capacity planning and scheduling" (PTDC/EGE-GES/102681/2008) and "Evolutionary algorithms for Decision Problems in Management Science" (PTDC/EGE-GES/099741/2008).

References

1. Cardoen, B., Demeulemeester, E., Beliën, J.: Operating room planning and scheduling: A literature review. European Journal of Operational Research 201(3), 921–932 (2010)
2. Bjornberg, A., Garrofe, B.C., Lindblad, S.: Euro health consumer index. Technical report, Health Consumer Power House (2009)
3. Siciliani, L., Hurs, J.: Explaining waiting-time variations for elective surgery across OECD countries. Economic Studies, vol. (38). OECD (2004)
4. Chang, J.K., Calligaro, K.D., Lombardi, J.P., Dougherty, M.J.: Factors that predict prolonged length of stay after aortic surgery. Journal of Vascular Surgery 38(12), 335–339 (2003)
5. Lazar, H.L., Fitzgerald, C., Gross, S., Heeren, T., Aldea, G.S., Shemin, R.J.: Determinants of length of stay after coronary artery bypass graft surgery. Circulation 92(9), 20–24 (1995)
6. Tu, J., Mazer, C.: Can clinicians predict ICU length of stay following cardiac surgery? Canadian Journal of Anesthesia 43(8), 789–794 (1996)
7. Macario, A.: Truth in scheduling: Is it possible to accurately predict how long a surgical case will last? Anesthesia & Analgesia 108(3), 681–685 (2009)
8. Guerriero, F., Guido, R.: Operational research in the management of the operating theatre: a survey. Health Care Management Science 14(1), 89–114 (2011)
9. Dexter, F., Traub, R.D., Qian, F.: Comparison of statistical methods to predict the time to complete a series of surgical cases. Journal of Clinical Monitoring and Computing 15(1), 45–51 (1999)

10. Li, Y., Zhang, S., Baugh, R.F., Huang, J.Z.: Predicting surgical case durations using ill-conditioned CPT code matrix. IIE Transactions 42(2), 121–135 (2009)
11. Strum, D.P., May, J.H., Vargas, L.G.: Modeling the uncertainty of surgical procedure times: comparison of log-normal and normal models. Anesthesiology 92(4), 1160–1167 (2000)
12. Combes, C., Meskens, N., Rivat, C., Vandamme, J.P.: Using a KDD process to forecast the duration of surgery. International Journal of Production Economics 112(1), 279–293 (2008)
13. Eijkemans, M.J.C., van Houdenhoven, M., Nguyen, T., Boersma, E., Steyerberg, E.W., Kazemier, G.: Predicting the unpredictable: A new prediction model for operating room times using individual characteristics and the surgeon's estimate. Anesthesiology 112(1), 41–49 (2010)
14. Stepaniak, P.S., Heij, C., Mannaerts, G.H.H., Quelerij, M.D., Vries, G.D.: Modeling procedure and surgical times for current procedural terminology-anesthesia-surgeon combinations and evaluation in terms of case-duration prediction and operating room efficiency: a multicenter study. Anesthesia & Analgesia 109(4), 1232–1245 (2009)
15. Dexter, F., Epstein, R.H., Lee, J.D., Ledolter, J.: Automatic updating of times remaining in surgical cases using bayesian analysis of historical case duration data and instant messaging updates from anesthesia providers. Anesthesia & Analgesia 108(3), 929–940 (2009)
16. Alvarez, R., Bowry, R., Carter, M.: Prediction of the time to complete a series of surgical cases to avoid cardiac operating room overutilization. Canadian Journal of Anesthesia 57(11), 973–979 (2010)
17. Spangler, W., Strum, D., Vargas, L., May, J.: Estimating procedure times for surgeries by determining location parameters for the lognormal model. Health Care Management Science 7(2), 97–104 (2004)
18. Strum, D.P., May, J.H., Vargas, L.G.: Modeling the uncertainty of surgical procedure times: comparison of log-normal and normal models. Anesthesiology 92(4), 1160–1167 (2000)
19. Wright, I.H., Kooperberg, C., Bonar, B.A., Bashein, G.: Statistical modeling to predict elective surgery time. comparison with a computer scheduling system and surgeon-provided estimates. Anesthesiology 85(6), 1235–1245 (1996)
20. Stepaniak, P.S., Heij, C., De Vries, G.: Modeling and prediction of surgical procedure times. Statistica Neerlandica 64(1), 1–18 (2010)
21. Holmes, G., Hall, M., Frank, E.: Generating Rule Sets from Model Trees. In: Foo, N.Y. (ed.) AI 1999. LNCS, vol. 1747, pp. 1–12. Springer, Heidelberg (1999)
22. Breiman, L.: Bagging predictors. Machine Learning 24(2), 123–140 (1996)
23. Khuri, S., Bäck, T., Heitkötter, J.: The zero/one multiple knapsack problem and genetic algorithms. In: Proceedings of the 1994 ACM Symposium on Applied Computing, SAC 1994, pp. 188–193. ACM (1994)

Using Data Mining for Static Code Analysis of C

Hannes Tribus, Irene Morrigl, and Stefan Axelsson

Blekinge Institute of Technology

Abstract. Static analysis of source code is one way to find bugs and problems in large software projects. Many approaches to static analysis have been proposed. We proposed a novel way of performing static analysis. Instead of methods based on semantic/logic analysis we apply machine learning directly to the problem. This has many benefits. Learning by example means trivial programmer adaptability (a problem with many other approaches), learning systems also has the advantage to be able to generalise and find problematic source code constructs that are not exactly as the programmer initially thought, to name a few. Due to the general interest in code quality and the availability of large open source code bases as test and development data, we believe this problem should be of interest to the larger data mining community. In this work we extend our previous approach and investigate a new way of doing feature selection and test the suitability of many different learning algorithms. This on a selection of problems we adapted from large publicly available open source projects. Many algorithms were *much* more successful than our previous proof-of-concept, and deliver practical levels of performance. This is clearly an interesting and minable problem.

Keywords: software engineering, static analysis, application.

1 Introduction

The finding and fixing of bugs, and other problems, is important to the writing of quality software. It is an important part of software development. Several different approaches have been proposed. These range from the widely used; *coding rules*, which focus on avoiding the introduction of faults, via *manual inspection* (*code review*), and *testing*, to more sophisticated methods like tool supported *error prediction* or *detection*. Today the latter is becoming popular as they can help directly by pointing to the places in the source code where an actual fault is presumed to exist. Depending on how these tools operate, they are able to find more or less complicated faults. One class of approaches is *static analysis*, by which is meant that the source code of the program is analysed for flaws that can be found without having to execute the program. This approach is especially useful in finding code that hides flaws that are more difficult to find with some form of dynamic analysis, for example those relating to non-functional requirements such as security flaws etc. This is due to the problem of the *coverage* of testing tools. Unusual flows of execution, such as those that involve error conditions, may receive insufficient testing unless *special* care is

S. Zhou, S. Zhang, and G. Karypis (Eds.): ADMA 2012, LNAI 7713, pp. 603–614, 2012.

taken. Static analysis has many other advantages; it can be applied early in the development process to provide early fault detection, the code does not have to be fully functional, or even executable, and test cases do not have to have be developed.

Several tools of this nature already exist. Some are even meant to be programmer adaptable (such as *Coverity* [6]), but in actual practise this feature is seldom taken advantage of, due to its perceived difficulty [6]. In order to address esp. the problem of programmer adaptability, we have previously turned to supervised machine learning, in order to provide trivially *programmer extensible* static code analysis. In this approach the programmer first trains the analyser by providing it with examples of correct, and problematic, code. The trained analyser is then run on code with potential problems and (hopefully) points out problematic situations, without too many false alarms, or missed problems. Another advantage of machine learning, is its capacity to generalise from a set of given examples. Correctly applied, machine learning has the capacity to *surprise*, by producing results that were previously unanticipated, but still relevant, and correct.

In order to utilise a machine learning framework, example and evaluation data has to be prepared. Hence, the raw source code has to be converted into suitable input for a machine learning algorithm. Thus, one of the goals of this project was the development of a feature selection model for a representative procedural language (the C programming language in this case) by using an appropriate parser. We will used the data to evaluate possible machine learning algorithms, and then compared and evaluated them in terms of accuracy and false positive/negative rate. We used source code from various open-source projects for the experiment. We sought to answer: What are the relevant source code features needed to classify an instance of code as faulty or correct? How can those features be transformed and represented as input to the machine learner? How accurate is the resulting static analyser/machine learning algorithm used?

The rest of this paper is organised as follows: We start by describing related work in section 2. Then, in section 3, the types of software flaws we focused on are described. The actual approach to converting *C language* source code to machine learning feature vectors is described in section 4. The experiments and results, to validate the approach are explained in section 5, with discussion, conclusions and future work finishing the paper in section 6.

2   Related Work

Both machine learning and static code analysis have been widely studied. However, when it comes to the application of machine learning *to static* analysis we know of only our own previous work in the area: Sidlauskas et.al. [1]. The work demonstrated that this approach was possible and that the static analyser even managed to generalise from e.g. a faulty *strcpy* example to detecting a similarly incorrect application of *strcat*. However, the work also had several limitations that we try to address here: only one machine learning algorithm based on the normalised compression distance (NCD) was tried, the experimental data was limited in scope, and the results in terms of accuracy etc. were several percentage points (even tens of percent) worse than what we achieve here.

While there are no direct analogs, there is work in the related fields; *fault prediction* and *fault detection*. Most of the work done so far in the area of *fault prediction* deals with the prediction of faults in source code. The basic idea here is to extract properties, or attributes, of the code which allows the drawing of conclusions about the likelihood of the presence of faults. Properties in this case could be metrics such as *lines-of-code* (LOC), or *cyclomatic complexity* (CC), or in the case of object oriented programming even *number-of-children* (NOC) etc. These are also known collectively as "CK-Metrics". A study by Fenton et.al. [7] criticised the models presented so far, called *single-issue models*, and suggested instead the use of machine learning techniques in order to arrive at a more general model for predicting faults in software. This approach was confirmed by the studies of Turhan and Kutlubay [16]. Most of the work in the prediction area deal with the code metrics above, but there are a few different approaches. Challagulla [4] showed that similar results using code metrics can be obtained by using design or change metrics, such as the number of times a file has been changed, the expertise of the person affecting the change etc. This was supported by Heckman and Williams [9] and Moser et.al. [13]. Another contribution to the area by Jiang et.al. [11] who compared the performance of design and code metrics in fault prediction models. They came to the conclusion that models using a combination of these metric outperform models that use either code or design metrics alone.

Despite the approaches above being more or less successful, they all try to draw conclusions about the presence of errors in the source code by studying meta data *about* the project instead of using the actual source to detect the faults. Burn et.al. showed in a case study [3] that support vector machines (SVM) and decision trees (DT), previously trained on faulty code execution vectors and corrected versions, can be used to identify errors during program execution. Despite their approach using dynamic analysis instead of static, it demonstrated that machine learning could be used for fault detection. A similar approach by Kreimer [12] is a tool that is able to detect design flaws during execution. Similarly, Song et al. [14] used the *association rule mining* (ARM) machine learner to find dynamic execution patterns that are similar or related to previously found errors. Finally Jiang et.al. [10] the authors successfully applied neural networks to the same problem.

Approaches to static analysis ranges from the very simple (searching for pure textual patterns of known problems) to the very complex. Most advanced methods depend on some form of formal method, based on e.g. denotational/axiomatic/operational semantics, or abstract interpretation. Practical implementations of these theories include:

Model-checking. Given a model of the system (e.g. a FSM), and a specification (e.g. a temporal specification) determine if the model meets the specification.

Data-flow analysis. At control points in the program information about the possible set of values for e.g. variables is tracked and this information propagated to later stages of the analysis.

Abstract interpretation. The partial abstract execution of a program that preserves and gathers information about data-flow and control-flow such that important information about properties of the programme can be studied.

The area has seen extensive study in the past decades, and we can scarcely do it justice here.

3   Types of Software Faults

In order to develop a method of detecting faults in source code, we need to define what we mean by fault in this context. A multitude of different kinds of software flaws have been categorised. We have decided to limit our research to the ones described below, as they are important in that they have shown to be the cause of many problems, security and otherwise. Most of these flaws have been adapted from [5].

Buffer Overflow. A buffer overflow or buffer overrun, results when a program inadvertently stores data outside an allocated buffer, most commonly by continuing to write data past the end of the allocated storage. If the buffer is allocated on the stack, this allows an attacker to overwrite active parts of the stack, including the return address of the current function. This leads trivially to a severe security flaw. Even the inadvertent triggering of a buffer overrun leads to data corruption, and most often a program crash. This flaw was popularised in the nineties and exploited heavily which lead to some calling it "the vulnerability of the nineties."

Memory Handling. This second kind of error regards dynamic memory allocation and de-allocation. The two main errors in this group are the use of the memory after it has been freed, and the freeing of memory that has already been freed before (*double-free*). Both can lead to a buffer overflow attack. A third problem, which is usually not that serious, but a waste of memory is whenever memory is allocated, but not subsequently used.

Dereferencing a Null pointer. A very common fault is the dereferencing of a null pointer. This fault occurs whenever a pointer to NULL is used, meaning that a pointer variable is used before e.g. it has been assigned a memory address. Dereferencing such a pointer will cause a program crash.

Control Flow. By this we mean failing to check a return value and, failure to release a resource. Whenever a functions return status is not checked, for example, assigned variables could be left with undefined values. This often leads to a null pointer reference.

Signed to Unsigned Conversion. The problem here is the automatic conversion between signed to unsigned values in the C-language. If a programmer checks a signed value and finds that it is smaller than a e.g. a buffer size (due to it being a negative value of large magnitude) this value could then be automatically converted to a large positive value when it is passed as a parameter to a function that take an unsigned value. This often leads to a situation where much more data is allocated/read than there is room for, and hence a buffer overflow.

4   Static Analysis by Data Mining

We now arrive at the problems of how to transform static source code to a suitable format for data mining, which data to use to train the classifier, and which classifier/learning algorithm that bests fits the problem.[1]

In order to perform our experiments we have chosen to us the WEKA data mining framework [8]. The main reasons were its popularity, and suitability, for the task, including its offering a wide variety of different learners. Furthermore all these learners can be used with the same kind of input and the same configuration. WEKA also provides a good framework for performing evaluation of the experiments.

In order to address the problem of feature extraction we wrote a prototype tool (*CMore*), that analyses C source code by extracting information about basic functions and procedures. The basic idea is that this should enable us to follow the execution flow of a program and capture the state of all the variables involved. Having all the state, it is possible to detect several kind of possible memory access faults such as are null-pointer-references and misuse of memory (see above).

CMore first divides the input into function definitions (with parameters and types), definitions, variable declarations, types, structure types etc. Information about these are stored in a data base for future reference. When the general structural information about the program have been identified, CMore goes into a more detailed processing stage where each line of the source code is analysed to see if it contains a mathematical expression, an assignment, a declaration, function call, control flow instruction etc.

```
@RELATION cmore

@ATTRIBUTE operation      {Functioncall , Assignment , Usage}

@ATTRIBUTE left_type      {FILE,  Function ,  header ,  char ,  int ,  Reference ,  Simple ,  void ,
    ...}
@ATTRIBUTE left_name      {open ,  alloc ,  close ,  free ,  gets ,  printf ,  seek ,  strdup ,strcpy ,
    strncpy ,  strcat}
@ATTRIBUTE left_isRef     {true , false}
@ATTRIBUTE left_isNull    {true , false}
@ATTRIBUTE left_isAss     {true , false}
@ATTRIBUTE left_isUse     {true , false}
@ATTRIBUTE left_isChk     {true , false}

[...]

@ATTRIBUTE orig_stmt      STRING
@ATTRIBUTE err_free       {true , false}

@DATA
Assignment , int ,? , true , true , false , false , false , Reference ,? , true , false , false , true , false
    ,? ,? ,? ,? ,? ,? ,? ,? ,? ,? ,? ,? ,? ,? ,? ,? ,? ,? ,? ,? ,? ,? ,? ,? ,? ," co_=_&cp" , false
Assignment , int ,? , true , true , false , false , false , Reference ,? , true , false , false , true , false
    ,? ,? ,? ,? ,? ,? ,? ,? ,? ,? ,? ,? ,? ,? ,? ,? ,? ,? ,? ,? ,? ,? ,? ,? ,? ," com_=_&ptr1" , false
...
```

Listing 1.1. Snapshot of the final ARFF

---

[1] The treatment here is necessarily short, due to space constraints. Much more details can be found in [15].

After this analysis, an attribute file is written that details the information we have collected about the program. Listing 1.1 shows an example of the output from this stage. The structure is quite simple and emulates the structure of a typical statement. The first attribute states what kind of operation the statement is; *assignment, usage* or *function call*. Depending on this attribute one or more of the groups (*left, right, par1–parX*) are emitted. A group contains seven attributes. The last two attributes contain the real statement needed to identify the faulty line in the code and the *error-free* flag.

The following examples should clarify the behaviour of the single groups and why they are emitted:

a = b; $-->$ In this case *a* is used for the *left* attribute and *b* to fill the *right* group

aProcedure(a); $-->$ In this case, *aProcedure* is the *right* attribute and *a* is used for *par1*

a = aFunction(b,c); $-->$ In this case *a* is *left*; *aFunction* is *right* and *b*, and *c* are the two parameters

Note that one statement in the source code can produce multiple instances in the attribute file.

In order to help in training it is useful to have help in determining the differences between two versions of a program. This is because we need labelled instances of faulty and correct versions. Our input data will most often be in the form of two different versions of a source code file. One with an error ("bug") and one with the error fixed. In order to identify these differences we developed a tool, *Holmes*, to be able to pick out and flag these differences when given two ARFF files as input.

5   Experiment

In order to test this approach we performed an experiment using a number of publicly available software projects. Suitable projects for such an experiment have to be available (obviously), have a bug data base (that enables us to identify interesting software problems), and a version control repository (that enables us to access different versions of the software, one with bugs fixed).[2] However, in some cases we were unable to identify the precise flaw, and in those cases we resorted to artificially injecting a bug in the previous version. As we were uncertain of our initial feature selection, we reduced the number of features as the experiment progressed. We display the results of both the full and the reduced set. As a last step the best learners were tested on the complete input from one of the open source projects previously used to generate the training set for the learners, to evaluate its performance on a real data set. This to show that the classifiers are working in a more real world scenario.

---

[2] The connection between bug database and source code control is important, as it enables us to connect a bug report, which gives us a clue about the type of fault, with the actual change in the source code that fixed that error.

After extensive study we selected six candidates for the experiment: *Algoview*; a small program to visualise a program as a graph (*Sourceforge*), *Boxc*; a vector graphics language to simplify their creation (*Sourceforge*), *ClamAV*; an anti-virus program for Linux (*Sourceforge*), *PocketBook*; an open source firmware for e-book readers (*Sourceforge*), *FalconHTTP*; a HTTP server (*Google Code*), and *Siemens Programs*; a collection of programs with several versions, where faults have been introduced for educational purposes (*Other*[3]).

From the programs selected above it was possible to extract correct and faulty instances using the tools described above. In total we developed 496 instances of faulty and correct versions, and also some correct instances reflecting the most important standard situations (to give contrasting examples of correct usage). The latter were added to reduce false alarms on correct standard situations in source code (important for the second experiment).

We used the most common measures of success; *accuracy, false positive* and *false negative* rate. All experiments were run using ten-fold cross validation.

5.1 Accuracy

The resulting ranking in terms of accuracy is shown in figure 1.[4] In the figure, of the seven classifiers performing better than 95%, three are based on nearest neighbour algorithms (*Ib1, Ibk, NNge*), three are tree based (*LMT, RotationForest* and *ADABoost* performed on a *BFTree*) and the the last based on artificial neural network; *MultiLayerPerceptron*.

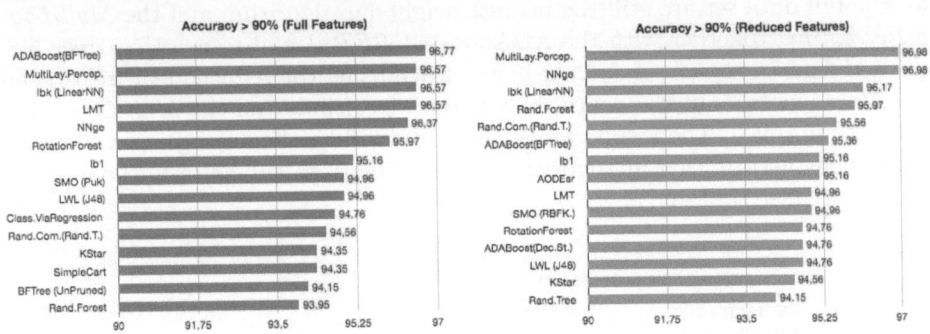

Fig. 1. Accuracy (Full)   **Fig. 2.** Accuracy (Reduced)

Despite the fact that trees perform quite well on the given data set, study of the attributes used, showed that most of the time the reasoning was based on the real data type of the involved variables. To see what effect the removal of this data had, these features were removed in the second step of the experiment. The results in terms of accuracy of this change is shown in figure 2. As expected the performance

---

[3] http://pleuma.cc.gatech.edu/aristotele/Tools/subject/index.html

[4] Due to space constraints, some of the graphs are truncated to focus on the interesting parts for the conclusions drawn, the complete data set is available on request.

Table 1. Average Accuracy per Category

Category	Accuracy		False positive		False negative	
	Full features	Reduced feature	Full features	Reduced feature	Full features	Reduced feature
Total average	82.0	82.4	0.24	0.23	0.14	0.14
Lazy	90.7	91.2	0.08	0.06	0.10	0.11
Tree	85.9	85.6	0.19	0.20	0.10	0.10
Functions	82.2	81.7	0.23	0.20	0.14	0.17
Bayes	81.4	84.0	0.17	0.13	0.20	0.18
Meta	80.4	80.8	0.30	0.30	0.11	0.11
Rule	79.4	79.7	0.22	0.17	0.14	0.10
Misc	61.8	65.0	0.14	0.09	0.56	0.55

of the tree algorithms was reduced and *ADABoost* could not sufficiently improve matters. However, the performance of the *MultiLayerPerceptron* increased to be the highest overall, together with the *NNge* nearest neighbour learner.

This result is supported by the left table (*accuracy*) in 1 which shows the average accuracy per category (These categories are taken from the categories used in WEKA to describe the various classifiers). It shows that almost all classifiers improve by not taking the real name of the involved types under consideration, except the tree learners.

5.2   False Positive Rate

The false positive analysis paints a similar picture. Figure 3 shows the results for the full feature set and figure 4 for the reduced one. Among the best classifiers for the full data set are still the nearest neighbour algorithm and the *MultiLay-erPerceptron*, together with the ADAboosted *BFTree* and some other trees like the *LMT* and a version of *J48*. As before when using the reduced feature set, the trees fared worse, but the nearest neighbour algorithms and the *MultiLayerPerceptron* improved. The results are again supported by the averages per category which demonstrate that all algorithms except the trees gained from the reduced feature set (see table 1).

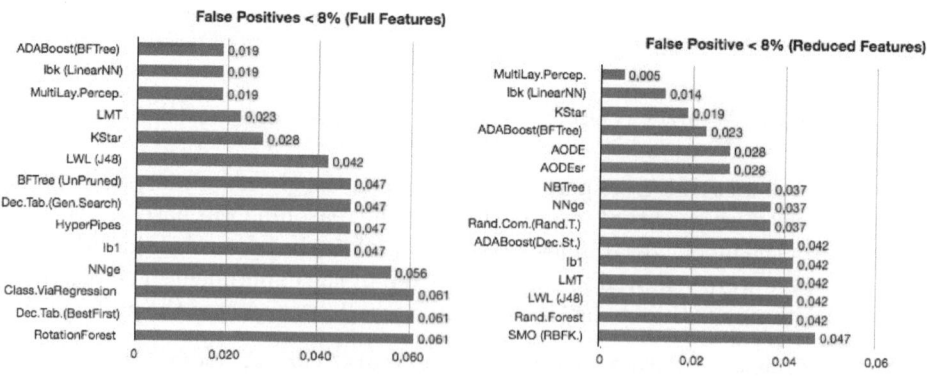

Fig. 3. False Positive (Full)   **Fig. 4.** False Positive (Reduced)

It is interesting to note that the *NNge* nearest neighbour algorithm was not as good when it came to false positive rate and also the feature reduction did not significantly increase its performance. This in combination with its good accuracy in fact meant that this classifier should have a low value when it comes to false negative rate

5.3   False Negative Rate

According to the results by Baca [2] it is very important for static analysis tools to have a low false positive rate. Otherwise, the users will quickly tend to distrust the results of the tool, even when they are correct, or even worse, introduces new faults at the location the tool indicates.

On the other hand, it is also important to have a low false negative rate, that is, that the tool finds most of the flaws present in the code. Thus, we need to find an acceptable trade off between false positives and false negatives.

The results in terms of false positive rate are shown in figure 5 for the complete set, and figure 6 for the reduced one.

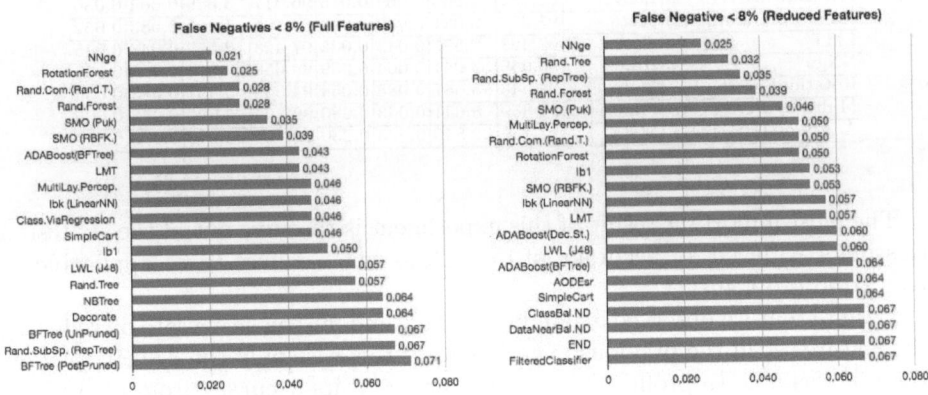

Fig. 5. False Negative (Full)   **Fig. 6.** False Negative (Reduced)

As expected the best performing algorithm according to this measure is *NNge* due to its bad performance in the false positive measure. However it is worth noticing that the tree algorithms perform quite well, and most of them are able to gain from the reduced feature set. Unfortunately as they are not as good in terms of accuracy. This means that their average false positive rate is also high. Generally this measure suffers from the reduction of the feature set, as the number of learners below 8% decreases. The average values of the various categories do not change much due to the decrease shown in table 1. The only real notable difference is the rule based learners, but they were not among the top learners before the reduction, and they did not gain sufficiently to rank at the top afterwards.

5.4   Selected Project Evaluation Results

In the previous experiment there were only well sorted examples that had pairs of two or more instances showing the difference between correct and faulty source code. In order to test more real world performance of the method, we ran an experiment with the best performing machine learning algorithms against a complete data set obtained from *FalconHTTP*. *CMore* was able to extract over 4000 feature instances from this software. Using fault injection, 44 feature instances were marked as faulty, the remaining were considered fault free, even though we of course cannot prove this correct. For this experiment the whole set of instances used for the first experiment was used to train the six previously best performing classifiers. The new generated set of instances was used to test the performance. The experiment was performed for both the full and the reduced feature set. The results are listed in table 2, sorted by accuracy for the full feature set.

Table 2. Results from WEKA for the test experiment

		Full Features				Reduced Features			
		corr	incorr	FN	FP	corr	incorr	FN	FP
Ibk (LinearNN)	Lazy	98,434	1,566	0,064	0,015	96,311	3,689	0,064	0,036
Rand.Com.(RandT.)	Meta	97,979	2,021	0,083	0,019	94,795	5,205	0,083	0,052
LMT	Tree	97,423	2,577	0,043	0,025	97,423	2,577	0,043	0,025
NNge	Rule	96,059	3,941	0,064	0,039	96,059	3,941	0,064	0,039
RotationForest	Meta	94,543	5,457	0,083	0,054	94,543	5,457	0,083	0,054
MultiLay.Perc.	Functions	91,359	8,641	0,043	0,086	99,394	0,606	0,043	0,005
AdaBoost(BFTree)	Meta	77,160	22,840	0,083	0,230	73,421	26,579	0,083	0,268

The most important result of this experiment is that five out of the six tested classifiers achieved an accuracy of over 90%, which makes them applicable to real world problems.

Despite that we observed a gain in accuracy and false positive rate by reducing the feature set in the previous experiment, in this case only the *MultiLayerPerceptron* was able to profit. This learner increased its accuracy from ca. 91% to over 99%, reducing the false positive rate from over 8%, down to 0.5%.

Probably the most surprising result from this experiment, is the poor performance of the *ADABoost* classifier using both feature sets. The major difference between the data set used for this experiment and the one used in the previous (which is the training set for this experiment) is the relation between faulty and correct instances. Lines of correct code should (hopefully) greatly exceed faulty in a real world project. So the test set contained over 4000 feature instances of which only 44 were considered faulty, which is a realistic number, on the high side even.

6   Discussion, Conclusions and Future Work

Despite that our results show that this method could be a possible approach for future static analysis tools, there are some limitations of the current work. The most important limitation come from the feature extraction process and what ability it has to track sufficient information. Even though the proposed feature extraction method produces a good result, and is not completely *ad-hoc* like our previous approach, much research remains before this area has been sufficiently studied. The approach here should be seen as a first step.

Also, even though we have performed an experiment on several substantial open source projects, there is of course much to be done, to expand the data set, both in terms of identifying projects of different kinds (commercial etc.), and identifying important and relevant flaws and other software problems.

The results of our experiment show that the best performing machine learning algorithms perform well enough in terms of accuracy, false positive and false negative rate, to be useful in practise. It was possible to train seven classifiers to demonstrate an accuracy of over 95%, five to have a false positive rate below 4% and six to have a false negative rate below 4% for the full feature set and eight with accuracy over 95%, nine with false positive below 4% and four below 4% for false negative on the reduced feature set. To validate the results a second experiment was performed by running the best six machine learning algorithms on a more realistic problem. This showed that five out of six classifiers could meet the requirements, while the one using *ADABoost* on a *BFTree* suffered from reduced accuracy, from over 96.77% in the first experiment down to below 77.15% for the full feature set, and from 95.36% down to 73.43% for the reduced feature set (Similarly for false positive/negative rates).

In summary; the timely identification of software bugs is an industrially important problem. We feel we have demonstrated that this is clearly a data minable problem, with readily and publicly available data from which to build data sets. We would like to bring this data mining problem to the attention of the larger community.

References

1. Axelsson, S., Baca, D., Feldt, R., Sidlauskas, D., Kacan, D.: Detecting defects with an interactive code review tool based on visualisation and machine learning. In: The 21st International Conference on Software Engineering and Knowledge Engineering, SEKE 2009 (2009)
2. Baca, D.: Automated static code analysis: a tool for early vulnerability detection. Department of Systems and Software Engineering, School of Engineering, Blekinge Institute of Technology, Karlskrona (2009), licentiatavhandling Ronneby: Blekinge tekniska högskola (2009)
3. Brun, Y., Ernst, M.D.: Finding latent code errors via machine learning over program executions. In: Proceedings of the 26th International Conference on Software Engineering, ICSE 2004, pp. 480–490. IEEE Computer Society, Washington, DC (2004)
4. Challagulla, V.U.B., Bastani, F.B., Yen, I.L., Paul, R.A.: Empirical assessment of machine learning based software defect prediction techniques. In: Proceedings of the 10th IEEE International Workshop on Object-Oriented Real-Time Dependable Systems, WORDS 2005, pp. 263–270. IEEE Computer Society, Washington, DC (2005)

5. Chess, B., West, J.: Secure Programming with Static Analysis, 1st edn. Addison Wesley Professional, Erewhon (2007)
6. Engler, D., Chelf, B., Chou, A., Hallem, S.: Checking system rules using system-specific, programmer-written compiler extensions. In: Proceedings of the 4th Symposium on Operating System Design and Implementation (OSDI 2000). USENIX, San Diego (2000)
7. Fenton, N.E., Neil, M.: A critique of software defect prediction models. IEEE Trans. Softw. Eng. 25(5), 675–689 (1999)
8. Hall, M., Frank, E., Holmes, G., Pfahringer, B., Reutemann, P., Witten, I.H.: The weka data mining software: An update. SIGKDD Explorations 11(1) (2009)
9. Heckman, S., Williams, L.: A model building process for identifying actionable static analysis alerts. In: Proceedings of the 2009 International Conference on Software Testing Verification and Validation, ICST 2009, pp. 161–170. IEEE Computer Society, Washington, DC (2009)
10. Jiang, M., Munawar, M.A., Reidemeister, T., Ward, P.A.S.: Detection and diagnosis of recurrent faults in software systems by invariant analysis. In: Proceedings of the 2008 11th IEEE High Assurance Systems Engineering Symposium, HASE 2008, pp. 323–332. IEEE Computer Society, Washington, DC (2008)
11. Jiang, Y., Cuki, B., Menzies, T., Bartlow, N.: Comparing design and code metrics for software quality prediction. In: Proceedings of the 4th International Workshop on Predictor Models in Software Engineering, PROMISE 2008, pp. 11–18. ACM, New York (2008)
12. Kreimer, J.: Adaptive detection of design flaws. Electron. Notes Theor. Comput. Sci. 141(4), 117–136 (2005)
13. Moser, R., Pedrycz, W., Succi, G.: A comparative analysis of the efficiency of change metrics and static code attributes for defect prediction. In: Proceedings of the 30th International Conference on Software Engineering, ICSE 2008, pp. 181–190. ACM, New York (2008)
14. Song, Q., Shepperd, M., Cartwright, M., Mair, C.: Software defect association mining and defect correction effort prediction. IEEE Trans. Softw. Eng. 32(2), 69–82 (2006)
15. Tribus, H.: Static Code Features for a Machine Learning based Inspection An approach for C. Master's thesis, School of Engineering, Blekinge Institute of Technology, SE371 79 Karlskrona, Sweden (June 2010), Computer Science Thesis no: MSE-2010-16
16. Turhan, B., Kutlubay, O.: Mining software data. In: Proceedings of the 2007 IEEE 23rd International Conference on Data Engineering Workshop, ICDEW 2007, pp. 912–916. IEEE Computer Society, Washington, DC (2007)

Fraud Detection in B2B Platforms Using Data Mining Techniques

Qiaona Jiang, Chunxiang Hu, and Liping Xu

Focus Technology Co., LTD, Nanjing 210061
{jiangqiaona,huchunxiang,xuliping}@made-in-china.com

Abstract. In order to predict the potential fraud users of B2B platform, this paper constructed an anti-fraud system by employing the Decision tree and Association analysis. Firstly, based on the research of the platform users' operation behavior a predictive model was built by using the Decision tree and in the model each user was given a fraud warning score. Secondly, to improve the accuracy of the predictive model, the FP-growth algorithm was adopted. The similarity identification of the correlation analysis was applied to further amend the warning score of the model. The members were divided into high-risk fraud group and low-risk one according to a certain threshold limit. In the end, whether the users had high-risk fraud rating was identified through two iteration of the Association analysis. It turned out that the effect of the model application can meet demand well in B2B platform and the anti-fraud system we built is more targeted and convinced.

Keywords: Fraud, Decision Tree, FP-growth, Association Analysis.

1 Introduction

With the rapid development of the Internet, there are more and more people becoming the Internet users, and a variety of online trading patterns dominated the market, there is no doubt that fraud behavior comes accompany with the Internet transactions. Fraud has the most outstanding performance in banking, insurance, securities, telecommunications, and electronic commerce industry. B2B e-commerce platform as a type of electronic commerce industry is different from other e-commerce platforms, its users are enterprises. Once the fraud becomes true on the platform, it will bring the victims direct or indirect economic loss, usually the loss is bigger than the others in Internet transactions, it will also damage the reputation of B2B e-commerce platform. Up to now, many B2B e-commerce also in a later, passive and rely on human stage when facing fraud users. So it is very significant to effectively prevent the fraud on B2B e-commerce platform.

For both common and diverse fraud behavior, lots of researchers have been constructed kinds of credit evaluation models to reduce the risk of fraud. Currently, well-known credit evaluation systems are Beta Reputation System[1], iCLUB[2], TRAVOS[3], Personalized[4] and so on in abroad. Wee Keong Ng [5] and his partners evaluated these models and constructed a combined model for credit evaluation which received a visible effect in electronic market. Based on credit card users' basic information, credit rating and the record of consumption details, C.

S. Zhou, S. Zhang, and G. Karypis (Eds.): ADMA 2012, LNAI 7713, pp. 615–624, 2012.

Whitrow[1] and his partners constructed a credit card anti-fraud system adopted clustering algorithm and integrated score methods. Michael Kwan[6] used social network analysis and Decision tree algorithm to analysis the relationship between transaction data and behavior and set a predictive system for auction fraud which could efficiently identify the online auction fraud. In China, there are many people do researches for anti-fraud system, research methods containing Decision tree algorithm, outlier data mining, case-based reasoning, rough set and so on[7]. Anti-fraud models or systems above acquired effective results which depended on detail transaction data, detail capital flow information, that is, the establishment of the model could not do without accurate financial data or transaction data.

In view of the characteristic of B2B industry, the platform is easy to record users' basic information and behavior. Through the analysis of users' information we find that fraudulent users always have more than one account and abnormal behavior. This paper made full use of users' basic information and behavior characteristics, Decision tree, FP-growth algorithm and similarity identification analysis, constructed a combine model for B2B e-commerce anti-fraud system. The B2B e-commerce anti-fraud system had received a good practical effect when this model was applied for a long time.

2   General Design of the Method

2.1   Fraud in B2B Platform

The main means of fraud in B2B industry are as follows :Not shipped to the buyers; Providing inferior products; Shorting freight cycle; Releasing information for virtual goods with low price; Intentionally supplying misleading description for the product, especially for the foreigners, taking advantage of the flaw on different language, habits and customs for fraud. In view of these fraud behaviors, it is extremely important to take an effective measure to prevent. In the B2B industries up to now, the platform system can only capture its online operation of behavior traces and basic information after a series of operation of users through website platform, and it is almost impossible to obtain its offline trade information detail. So in B2B industry the starting points of fraud analysis are only behavior information and basic material information shown in Fig.1.

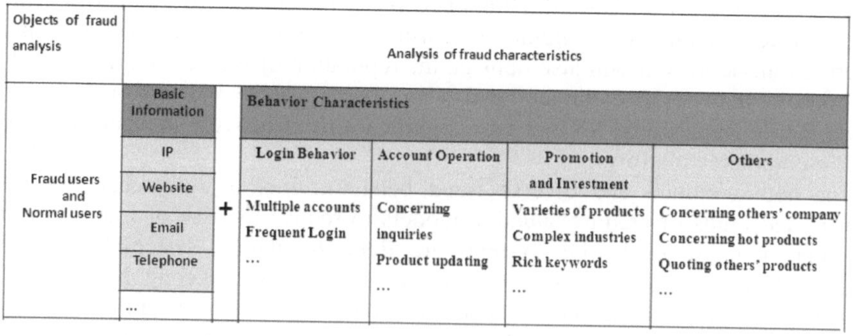

Objects of fraud analysis	Analysis of fraud characteristics				
Fraud users and Normal users	Basic Information	Behavior Characteristics			
	IP	Login Behavior	Account Operation	Promotion and Investment	Others
	Website				
	Email	Multiple accounts Frequent Login ...	Concerning inquiries Product updating ...	Varieties of products Complex industries Rich keywords ...	Concerning others' company Concerning hot products Quoting others' products ...
	Telephone				
	...				

Fig. 1. Objects and behavior characteristics of fraud analysis

2.2   System Design

In this paper, through the users' behavior information and basic information mainly, we constructed a combined anti-fraud system to analyze and predict the fraud risk for website users, using Decision tree and FP-growth algorithm. The overall design for the anti-fraud system was shown in Fig.2.

The system included two modules, the first one described in 3.1 mainly applying Decision tree algorithm based on the users' behavior characteristics to predict the fraud suspected members; Based on the first module, the second one described in 3.2 adopting the FP-growth algorithm which combined the users'(predictive users, fraud users) basic information to further strengthen the intensity of the predictive model.

Following, the iteration steps about association were adopted to amend the results of Decision tree. The first step was to compare the history fraud uses' basic information with the predictive users'. If their basic information is in accordance, the predictive users may be fraud users. The second step was to divide the predictive users into high-risk users and low-risk ones according to predetermined threshold, and then the Association analysis was applied once more. Finally the two groups of different risk level obtained more accurate warning scores through which we could judge the high-risk users.

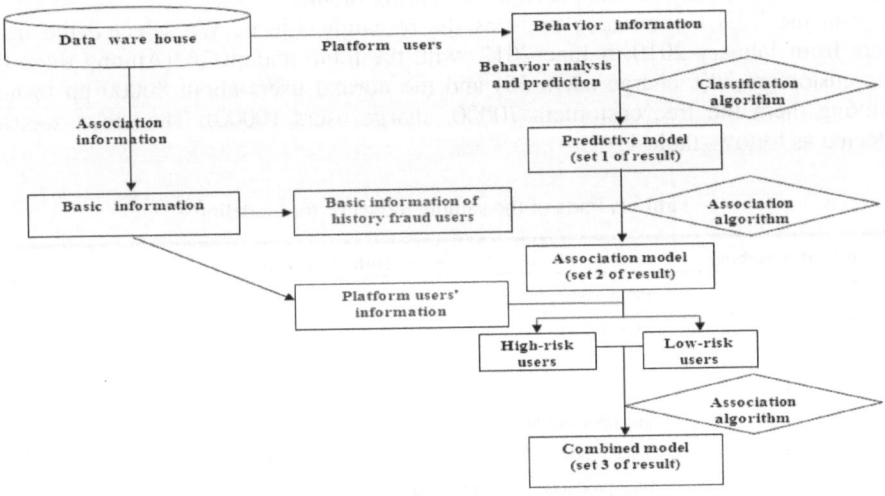

Fig. 2. System modeling ideas and research methods

3   Core Algorithms of Anti-fraud

3.1   Fraud Detection Using Decision Tree

In the classification algorithm, compared to neural networks, Bayesian algorithm, Decision tree algorithm has some advantages: 1) The construction process does not require any domain knowledge or parameter settings, so in practical applications for the discovery of exploratory knowledge, Decision tree is more applicable; 2) The Decision tree is good at handling non-numerical data to avoid lots of data preprocessing steps; 3)

Simple Decision tree is relatively easy to explain, its accuracy is also guaranteed; 4) Decision tree algorithm has very good robustness facing the noise interference, and the redundancy attributes have little adverse effects to its accuracy.

B2B anti-fraud system built in this paper is now in the exploratory knowledge discovery phase, so an explanatory conclusion with high accuracy is critical. In addition, through the experimental comparison of several classification algorithms, we found the model of Decision tree in the anti-fraud prediction is better. After comprehensive consideration of all factors, the Decision tree algorithm was adopted to build the anti-fraud predictive model.

3.1.1   Preparation and Description for Experimental Data

The platform users' specific performance behavior was monitored and analyzed for some time. Ultimately the users' related information such as login information, account management, promotion and information dissemination were taken into account, IP amount, logging on days, logging on times, the number of inquiries read, the number of products crossing industries, the number of companies visited or searched and their derived indicators were picked up for the follow-up modeling. Data sample included the fraud users and the normal users, the fraud users' data was intercepted from the previous month before its account closed and the normal users' data was intercepted from the previous or nearing month.

With the B2B e-commerce users as the research objects, We selected the fraud users from January 2010 to June 2012, with the total amount 764(Among them the free customers 709, charge users 55) and the normal users about 80000 up to now (Among them the free customers 70000, charge users 10000). The index sets are selected as follows table 1:

Table 1. Parts of the sample indicator for modeling

Indicator number	Indicator name
1	login days
2	login times
3	ip amount
4	inquiries read
5	inquiries replied
6	Total products(From the registration date to now)
7	total products (in recent 30 days)
8	products crossing categories
9	products crossing categories(in recent 30 days)
10	per product key words (in recent 30 days)
11	per product key words (in recent 30 days)
12	update product times
13	companies searched
14	companies visited
15	the total number of search
16	he total number of visits
17	the quantity of buying the service
18	the total cost of service

Each company corresponds to a specific behavior records, and the representation of the data was shown in Table 2.

Table 2. Parts of the sample data for modeling

company id	login days	login times	companies searched	companies visited	inquiries read	products crossing categories	IP amount	...
644970022	1	2	1	0	0	0	1	...
645310712	1	1	0	0	0	0	1	...
654538745	15	20	0	2	0	2	13	...
607105734	19	27	0	2	2	1	12	...
...	...	...	...	...	...	...	...	...

3.1.2 Predictive Model

The prepared data was divided into the training set and the predictive set, on which the neural network and C5.0 were tentatively used for modeling. Because of the obvious difference on behavior between buyers and suppliers, members for fees and free members, the indicators chosen for modeling should be flexibly adjusted according to the researched objects. It is important to emphasize that many charge members' behavior is consistent to fraud users' due to frequent management account. To avoid high-quality users being misjudged as fraud users, it is necessary to remove the high quality users when modeling. By comparing we find that the 17th (the quantity of   buying the service) and the 18th (the total cost of service) index mentioned above are the main difference between the normal users and the fraud users. It is show that the 17th index of fraud users is less than four as well as the 18th index is no more than ten thousand. Obviously, the decision criteria of high-quality users are the quantity of   buying the service and the total cost of service but the free members will not be considered. We will model the anti-fraud system with the indexes and the data format above. Take the charge suppliers for example, by   using the classification models and the formula as (1) and (2),   the effect of several classification models is demonstrated in Table 3.

R: Classification rule
S: prediction set
D:fraud users
A: predicted fraud users

$$Coverage\ (r) = \frac{|A \cap D|}{|D|} \tag{1}$$

$$Accuracy\ (r) = \frac{|A \cap D|}{|A|} \tag{2}$$

Table 3. The effect comparison among the classification and predictive models

$|S|$	$|D|$	$|A|$		$|A \cap D|$		Accuracy		Coverage	
		Decision tree	59	Decision tree	39	Decision tree	66.1%	Decision tree	70.9%
5000	55	Neural networks	69	Neural networks	28	Neural networks	40.6%	Neural networks	50.5%
		Bayesian	72	Bayesian	22	Bayesian	30.5%	Bayesian	40.0%

The effect of the Decision tree was better as it is illuminated in table3, the specific process of the Decision tree modeling is shown in Fig.3.

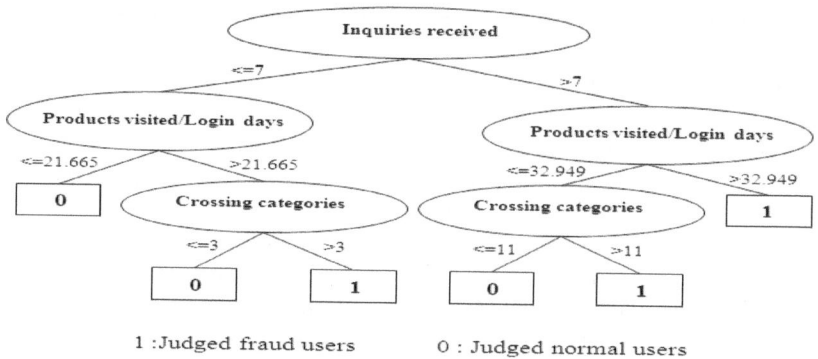

1 :Judged fraud users   0 : Judged normal users

Fig. 3. Partial results of the Decision tree model

From the right branch of the tree in Fig.3, we obtained the rules that 1) when the amount of inquiries received by the user greater than 7, the average products visited per day greater than 33, we could judge the use as a fraud user. 2) when the amount of inquiries received by the user greater than 7, the average products visited per day less than 33, and the number of crossing categories greater than11,we could also judge the use as a fraud user. The other branch of the tree could also be explained like this. The rules from the Decision tree model were in line with the actual business logic which had good explanation.

From the analysis above, the Decision tree was selected as the final predictive model, each user by the confidence level had a warning score $S_1 (com_i)(i = 1,2,...n)$, com_i was one of the total company.

3.2   Improving Accuracy Using the FP-Growth Algorithm

In section 3.1, we relied on users' behavior for making predictions, the advantage of this method is to monitor abnormal behavior of individual timely but research result turns out that most of fraud users will register more than one account, while some accounts are frequently used, the others are not. The method mentioned in 3.1 can only monitor the frequently used ones, so how to monitor the same users' other accounts? In fact, there is some connection of different accounts of the same user such as the IP, email, telephone, website, and so on. For the reasons given above, the correlation analysis will be used in the following paragraph.

The main role of the correlation analysis is to find strong correlation events efficiently and accurately from the massive data. The correlation analysis gives a great help to find the intrinsic hidden relationship between the data, which can be used to monitor the trade behavior of multiple users. These provide a valid path for the commercial fraud detection referring to cross-account users. Although the fraudster's behavior looks like normal superficially, we can still use the correlation analysis to carry out cross-account collaborative detection to find the special conditions, that is, if a user is identified as the fraud user, then the users who are more closely associated with him are considered as fraud users. Therefore, to improve the effect of prediction the correlation analysis was employed to excavate potential fraud users which were not recognized from the behavior analysis.

In this paper, the FP-growth algorithm was adopted for Association analysis. The new mining algorithm FP-growth for frequent item sets completely detached from the Apriority algorithm which should generate candidate item sets traditionally, having the idea to generate candidate item sets based on FP-tree structure, opened up new ideas for mining association rules. FP-growth algorithm is significantly better than the Apriority algorithm in the efficiency of mining, especially in dense databases, if the length of frequent item sets is too long, the advantages of FP-growth algorithm are more obvious.

Associated predictive analysis made full use of the predictors' basic information (mail, telephone, website, IP, etc.) to optimize and improve the accuracy of predictive model. Here we constructed a similarity identification algorithm fitting for a specific environment, the detailed steps are shown as follows:

1) Based on the user's basic information (email, phone, etc.), respectively calculates the number of frequent one item set $m_t(t = 1,2,\cdots,n)$ for com_i and the number of frequent two item set $p_t(t = 1,2,3...n)$ for $com_i, com_j(i \neq j)$ on the tth basic information using the FP-growth correlation analysis. Set the distance $d_t(t = 1,2,3...n)$ between $com_i, com_j(i \neq j)$ on the basic information.

2) When the basic information is accurately matched , one of the e-mail, website, telephone information is shared by com_i, com_j, then the relationship between the company is determined absolutely association with the correlation degree 1, otherwise with the correlation degree 0 ; when the basic information is not accurately matched, just as IP information sharing, the relationship between the company is determined relatively association with a correlation degree between 0 and 1 measured according to their shared proportion.

   a.   Condition of accurately matched, if $p_t > 0$, then $d_t = 1$, if $p_t = 0$, then $d_t = 0$;

   b.   Condition of not accurately matched, the distance of information between the companies is $d_t = p_t / m_t (0 < d_t < 1)$.

3) The formula for similarity identification between $com_i, com_j(i \neq j)$ is

$$Dis(com_i, com_j) = \prod_{t=1}^{n} d_t \tag{3}$$

4) Assuming com_i as the reference company, com_j as the company to predict, conjunction with the predictive warning values $S_1(com_i), S_1(com_j)$ calculated as (1) and (2), we acquire the new predictive warning scores $S_2^i(com_j)$ for com_j is

$$S_2^i(com_j) = \begin{cases} S_1(com_i) \times Dis(com_i, com_j), Dis(com_i, com_j) < 1 (i \neq j) \\ S_1(com_i), Dis(com_i, com_j) >= 1 (i \neq j) \end{cases} \qquad (4)$$

5) Compare $S_2^i(com_j)$ acquired from the previous step with $S_1(com_j)$, and choose the maximum value as the final predictive score $S_3(com_j)$ for com_j is

$$S_3(com_j) = \max(\max(S_2^i(com_j)), S_1(com_j)) \qquad (5)$$

4   Experimental Evaluation

4.1   The Application of the Two Iteration of Association Analysis

Following is an example of company id called 608076534(shown in Table 4) for whom further illustration below is given for the model's accuracy using associated algorithms. The predict time of the model is May 2011.

1) The basic information between the predictive user and fraud users were not associated, combined with the predictive warning score S_1, we obtained the predictive warning score $S_2 = S_1 = 0.75$.

2) Here setting the critical value of 0.8 to distinguish whether the predictive company was high-risk fraud user or low-risk fraud one, as you seen in 1) $S_2 = 0.75$ <0.8, so we had the conclusion that the company belong to the low-risk fraud groups.

3) Made correlation analysis for the company with high-risk fraud groups, on one hand, we found the company presenting accurately match with high-risk fraud groups in mailbox, telephone, website, so the similarity value was 1; on the other, the company presenting not accurately match in IP, with the similarity value 0.8 after calculation.

4) With the similarity value Dis and secondary predictive score S_2 of correlative companies, the final conclusion $S_3 = \max(\max(1*0.8, 0.95*1, 0.88*1, 0.91*1), 0.75) = 0.95$ get which indicated that the user belong to the fraud groups of high-risk.

Table 4. Example of forecasting the fraud association

predictive companies' info			correlative companies' info		associated info	companies' similarity	predictive results	
company id	S_1	S_2	company id	S_2		Dis	associated score	S_3
608076534	0.75	0.75	635947052	1.00	IP	0.8	0.80	
			646463475	0.95	website	1	0.95	0.95
			628457565	0.88	E-mail	1	0.88	
			608076534	0.91	telephone	1	0.91	

From the data in table 3, the user's final warning score S_3 is bigger than initial S_1 of Decision tree, in other words, the fraud suspicion of the predictive user company id called 608076534 is strengthened after two iteration of Association analysis. In fact, in March 2012 the company 608076534 has been complaint as a fraud user. This shows that the model we improved using FP-growth will play a role in predicting the platform fraud users in the future.

4.2   The Effect of the Model for High-Risk Fraud Users

To verify rationality and practical application effect of the anti-fraud system proposed in this paper, the Decision tree model and the final anti-fraud system models to be predicted as follows, the results shown in Fig.4, the model predicted score 1 is the result of Decision tree model for suspected fraud users prediction score, the model predicted score 2 is the result of final anti-fraud model for suspected fraud users predictive score.

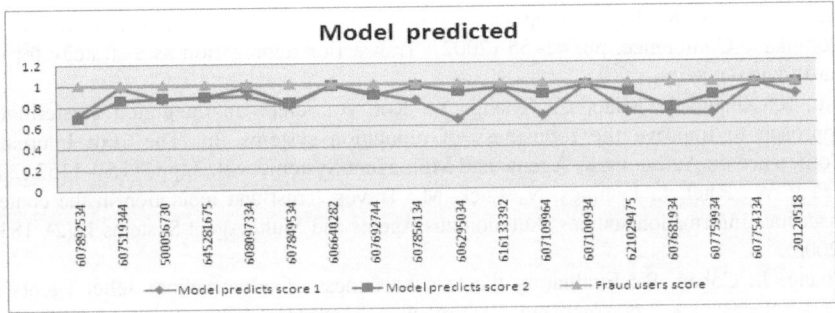

Fig. 4. Simulation and prediction of anti-fraud system for fraudulent user

From the results above, comparing the final anti-fraud model with the one before, the predictive accuracy has been improved obviously and the error has been controlled in the reasonable range which showed that our model was effective and feasible. The anti-fraud system would be conducive to business person's professional judge which could be believed to be able to enhance the overall efficiency of anti-fraud platform.

Although our model is effective and feasible, there is still much more to do next, such as refining users' behavior on the website to find out more convincing indicators for model and using the combined method of several algorithm instead of Decision tree model, We may apply Decision tree to classify and Bayesian to fore cast, by this mean, the accuracy and the coverage of model will be improved.

5   Conclusion

The conflict of accuracy and coverage for fraud predictive model in the actual operation process is inevitable; generally, it is difficult to keep the balance. In this paper, the anti-fraud system is highly effective which used Decision tree algorithm according to the characteristics of users' behavior in the Website. We gave the initial warning score to each member in the system. Compared the basic information between history fraud users and the non-fraud ones, their similarity can be identified by using FP-growth algorithm. We obtained more accurate modified warning score. Finally, the model accuracy and coverage respectively reached up to 80% and 75% after two iteration of the Association analysis. It has been proved that the system we constructed achieved the balance of accuracy and coverage, and could satisfy the meet in practical application.

Acknowledgement. This research is supported by Focus Technology Co., Ltd.. The views and conclusions contained herein are those of the authors and should not be interpreted as necessarily representing the official policies or endorsements, either expressed or implied, of the above organizations or any person connected with them.

References

1. Jang, A., Ismail, R.: The beta reputation system. In: Proceedings of the 15th Bled Electronic Commerce Conference, pp. 41–55 (2002); Transaction aggregation as a strategy for credit card fraud detection
2. Liu, S., Zhang, J., Miao, C., Theng, Y., Kot, A.: iclub: an integrated clusteringbased approach to improve the robustness of reputation systems. In: The 10th International Conference on Autonomous Agents and Multiagent Systems, vol. 3, pp. 1151–1152 (2011)
3. Teacy, W., Patel, J., Jennings, N., Luck, M.: Travos: Trust and reputation in the context of inaccurate information sources. Autonomous Agents and Multi-Agent Systems 12(2), 183–198 (2006)
4. Zhang, J., Cohen, R.: Evaluating the trustworthiness of advice about seller agents in e-marketplaces: A personalized approach. Electronic Commerce Research and Applications 7(3), 330–340 (2008)
5. Zhang, L., Jiang, S., Zhang, J., Ng, W.K.: Robustness of Trust Models and Combinations for Handling Unfair Ratings. In: Dimitrakos, T., Moona, R., Patel, D., McKnight, D.H. (eds.) Trust Management VI. IFIP AICT, vol. 374, pp. 36–51. Springer, Heidelberg (2012)
6. Kwan, M., Overill, R., Chow, K.-P., Silomon, J., Tse, H., Law, F., La, P.: Evaluation of Evidence in Internet Auction Fraud Investigations. In: Chow, K.-P., Shenoi, S. (eds.) Advances in Digital Forensics VI. IFIP AICT, vol. 337, pp. 121–132. Springer, Heidelberg (2010)
7. Ye, X.: Data mining in fraud risk analysis of application,Commercial science and technology (2006)

Efficiently Identifying Duplicated Chinese Company Names in Large-Scale Registration Database

Shaowu Liu[1], Jiyong Wei[2], and Shouwei Wang[1]

[1] Focus Technology Co., LTD, Nanjing, China
{liushaowu,wangshouwei}@made-in-china.com
[2] Department of E-Business School, Nanjing University
mg1102192@smail.nju.edu.cn

Abstract. It is always a challenge for large E-commerce platforms to audit mass information in real time manner, especially to identify multi-registrations efficiently. In this paper, we design a novel method for detecting multi-registrations in Chinese E-commerce platforms. In the proposed method, company names in Chinese are first divided into regional attribute, template attribute and the key attribute according to most companies' naming rules, by utilizing the Chinese word segmentation technology. This greatly narrows down the searching range with the extracted key attribute. Then, the similarity between the company names are computed by a dynamic threshold-based string matching algorithm. Finally, the company names with high similarity are detected. This method is evaluated by using the dataset from a real E-commerce company, and the results show this method has better accuracy, efficiency and scalability, compared with other methods. The proposed method is more precision and more time-saving than artificial means, therefore, it can save a lot of human cost for B2B industry.

Keywords: Multi-Registration Detection, Information Audition, Data/Text Mining.

1 Introduction

1.1 Background

With the development of E-commerce, more and more companies are executing business activities through the network platforms, especially in the foreign trade area. As the network resource is limited, some opportunistic companies began to use improper means, such as duplicate registrations, getting extra accounts for free in order to increase exposure in the net world. For many B2B platforms can online compare the company : names of the register with its database to judge if the account has exsited before, those opportunistic companies will use a different but similar company name to avoid being detected. However, most B2B platforms adopt the manual method.

1.2 The Related Method

Though there is almost no the same problem settled down in current related references, especially in multi-registration field, there still exist some similar

S. Zhou, S. Zhang, and G. Karypis (Eds.): ADMA 2012, LNAI 7713, pp. 625–634, 2012.

problems like string similarity problem's methods, such as sentence similarity and string similarity.

Many researches on Chinese sentence, text or paper similarity computing are made, building methods like method based on semantic dictionary, TF-IDF (short for term frequency–inverse document frequency)method, method combining grammar and sequence of the words, dependency trees method , method based on semantic representation and etc[1- 5]. However, most of those above methods are based on the model in vector space or consider semantic which will decrease the efficiency in the large-scale data background and are not very suitable for the solution of our problem as some part of the names do not have semantics.

The methods of computing string similarity[6] can be classified into three main categories: Edit Distance[7-8] , Jaro-Winkler algorithm[9] and Cosine theorem[10]; The mainly methods solving the mass duplicate data detection are MPN(short for Muti-Pass sorted-Neighborhood) method based on Edit Distance and N-grams based on the clustering.

Edit distance is a common measure method for the string distance, which has a wide range of applications in determining the similarity of the two strings. The improved MPN [11] method is done by sorting the records of the keyword, reducing the edit distance calculation of string similarity computation. But it is suitable for static database, performing generally in incremental dynamic database .

Jaro–Winkler distance is a measure of similarity between two strings. The Jaro–Winkler distance metric is designed to be best suited for short strings such as person names, but the performance in dealing with high-volume database is not so satisfied.

Cosine theorem converts the computing of two strings to the judgment of the size of the angle between corresponding vectors. So it requires the words in the vectors space must be orthogonal. Before getting the eigenvectors of the text, disambiguation and standardization must be done first, which is cumbersome and inefficiency.

According to the N-gram subspace sequence inherent similarity level, similar records(fields) were merged by hierarchical cluster[12-14]. The advantages are that the complexity of time is O(n), and it spends less I / O and requires less memory space.

The results show that different fields similarity calculation algorithm is particularly effective to a specific string type[15].

1.3   The Organizational Structure

The remainder of this discussion is organized as follows. The related work in the string similarity field is introduced in Section 2 . The general process of the proposed method is introduced in Section 3.A new effective string similarity computing algorithm and one example are given in Section 4. Finally, the experiment and conclusion are given in Section 5.

2   Posing the Method of Judging Duplicate

2.1   Patterns of Company Names

Statistical analysis of more than 300 million companies' names in MIC BI warehouse shows that,without loss of generality, the company name structure can be expressed as follows:

Company Name=region+core words+ Industry (Main product) +nature+Company (Factory)

The core words are composed of at least two-word. For   example:"江苏省南京市雨 花台区张三电器有限公司", the word "江苏省南京市雨花台区" as a geographical information, and "电器"as category information,"有限公司"as the nature of the company,"张三" is the core words of the company name.

The company name can be decomposed three properties according to the company named rules: geographical attribute, key attribute and template attribute, which are defined as follows:

Geographical attribute: the ingredients of company name which represent the Region of company.

Key attribute: Neither all the ingredients will be useful, nor each the ingredients have the same contribution to the method, such as the core word and its role were significantly greater than other ingredients. Therefore, the ingredient which can play a greater role for repeated testing was defined as key attribute, which was regarded as the main reference to duplicate detection.

Template attribute: the ingredient which has a high frequency in the company name was referred to as template attribute, such as [the nature of the company + company (factory)]or [industries (Main Product) + nature + (Factory)].

2.2   Underlying Assumptions

According to the analysis above, under normal circumstances, the company's core word, industry and nature were always unchanged, and only its geographical information was enlarged or narrowed mostly, compared to original company name. The general duplicate registration strategy is to abandon the structure of outside the core word, or change geographic information. Therefore, if the names of the two companies are suspected to be duplicate, the core word will be the same.

2.3   General Process

The basic data which was cleaned by its validity, deletion, and abnormalcy was extracted and loaded into data warehouse. On this basis, the judging duplicate strategy is as follows:

Firstly, the company field was split and its key attribute was extracted, together with the high frequency lexicon and the Chinese word segmentation; secondly, the smaller company set of the same core word were searched by the core word which narrowed down the range of judging duplicate; finally, the similarity computing was taken within the smaller similar company set obtained by the previous step. The core processes were shown in figure1.

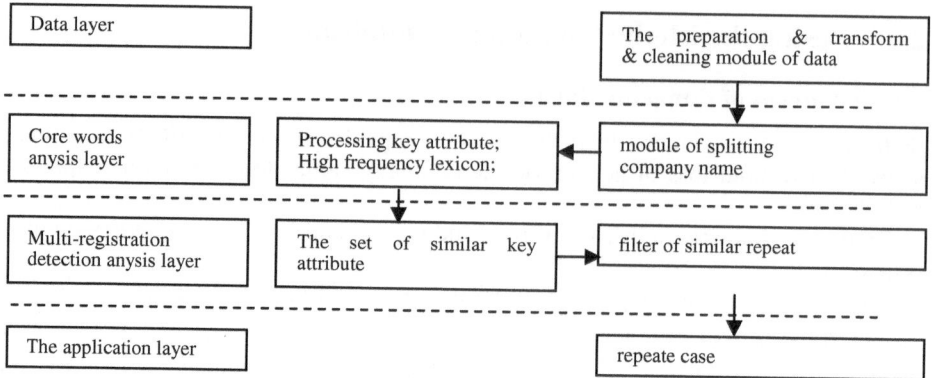

Fig. 1. The duplicate detecting strategy of the company name

2.3.1   Self-learning Thesaurus Building (High-Frequency Thesaurus)

Historical data was split one word by one word and so on, until five by five regularly. For example, the split result of two words by two words of the name "江苏省南京市雨花台区张三电器有限公司" was "江苏", "苏省", "省南", "南京", "有限", "限公"and "公司". Then the split data was stored in the database by column. By ordering the above data by quantity, deleting words which were under some setting threshold and comparing the remaining ones with Baidu Thesaurus, we can get the high-frequency thesaurus. Most vocabularies of the template attribute of the company name can be obtained from the high-frequency thesaurus whose regional attribute was removed.

2.3.2   Matching Range Reduction

By previous reasoning, we can reach the conclusion that two words are similar or duplicate only when their core words are similar and at the same time at least two consecutive words of the core words are the same.

First, the core words of the historical company name were split two words by two words and stored in the warehouse by column. Secondly, the same operation was done to the new ones. Then via database technique we can quickly obtain all the historical records which have the same two consecutive words with the other one.

Then the matching range was narrowed significantly which ensured both the precision and the efficiency of the similarity computing.

2.3.3   Filtering of Similar Repeat

The first step of filtering repeat companies is computing the similarity between each of the searched company and the ad referendum company. Then we remove the companies with lower similarity, and this is a question of text(string) similarity, and traditional algorithm has problems of filtering and inaccurate sorting . the main reason is that the company names' key attribute is usually short, even small inconsistency will still make great effect on the result of the similarity computing. Therefore, in the consideration of the interval and order's weight among strings, the imperfection of the above three algorithms will also significantly effect the computing result. We propose a new algorithm in Section 3 and compare it with traditional algorithms.

3   String Similarity Computing

3.1   The Description of the Algorithm

Besides the disadvantages discussed in the introduction, Above all, a new effective string similarity computing algorithm is proposed in this paper, which considers the interval, weight and matching numbers of the string. Following is the brief description:

Set string $A = \text{'} a_1 a_2 a_3 ...... a_{n-1} a_n\text{'}$, string $B = \text{'} b_1 b_2 b_3 ... b_{m-1} b_m\text{'}$, ($n <= m$);

$|A \cap B|$ represents the numbers of the same Chinese word;

$score(A,B)$ is the accumulated value in similarity computing;

int (A,B) is the sum of the interval between two words and their matched two words;

sim (A, B) is the similarity of the tow strings, and the two strings are thought to be duplicate when their $sim(A,B)$ is greater than the threshold T. The formula is shown as follows:

$$sim(A,B) \quad = \ f\left(\|AB\|,I(AB)\right) = \frac{\sum_{i=1}^{n}\sum_{j=1}^{m} score(A_i, B_j)}{n} \qquad (1)$$

In the above formula, n is the numbers of the iterations.
    The threshold function is defined as:

$$F(\text{ int } (A,B)) \ = \begin{cases} \text{int } (A,B), & 0 <= \text{ int } (A,B) \ <= \ 3; \\ 4, & \text{int } (A,B) \ >= \ 4; \end{cases} \qquad (2)$$

Thus the dynamic threshold value T is:

$$T = 1 - 0.1 \times \left(|A \cap B|\right)/n \times \left(F\left(\text{int}(A,B)\right)+1\right) \qquad (3)$$

The accumulated value in similarity computing is defined as:

$$f(\text{ int } (A,B)) \ = \begin{cases} (1 - 0.1 \times \text{int } (A,B)) \ , & \text{int } (A,B) \ <= \ 3; \\ 0.7, & \text{int } (A,B) \ >= \ 4; \end{cases} \qquad (4)$$

3.2   The Flow Char of the Algorithm

Following is the flow char of the algorithm described above:

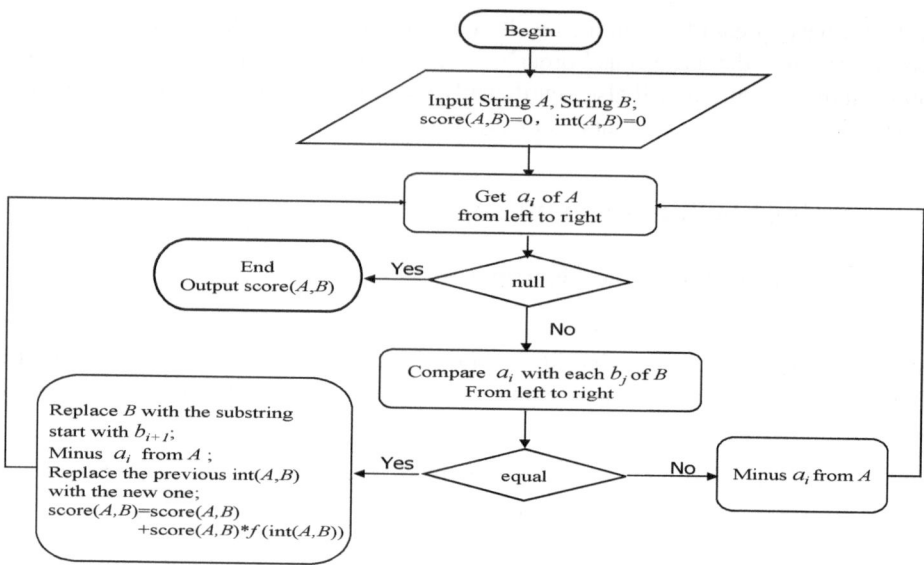

Fig. 2. The flow char of the proposed algorithm

4   Evaluation

4.1   Experiments Environment

To verify the performance of the algorithm in this paper, we compare this method with two famous methods in detecting the similarity of strings, and one is edit distance and the other is Jaro-Winkler distance. Ten to one hundred thousand records were selected from the warehouse of MIC BI department.   The number of duplicated company names records which has been confirmed by the information Department is 6900, 15700, 21200, 28100, 34700, 40800, 47000, 52900, 60100 and 66300. The experimental environment is shown as follows:

Operating system: Windows 7 Professional edition 64 bit
CPU : Intel Pentium（R）3.2GHz ;
Memory : 4G ;
Database : Oracle 10g ;

4.2   Results and Analysis of Proposed Algorithm

According to the above scheme, the experimental results are shown in Figures 3 to 5.The horizontal axis represents corresponding data scale, i.e., the numbers of records from the warehouse, and the vertical axis represents the corresponding performance metrics.

Figure 3 shows the precision of each algorithm. It can be seen from Figure 3 that the accuracy of the Edit Distance algorithm and the proposed algorithm are much higher than the Jaro-Winkler distance algorithm during the experiment. When the scale of the data is small, for example, less than 40000, the accuracy of the proposed algorithm and Edit Distance algorithm differs in small range. But within the growth of the scale of the data, the accuracy of the proposed algorithm is obviously superior to the Edit Distance algorithm, which reflects the superiority of the proposed algorithm in processing massive data.

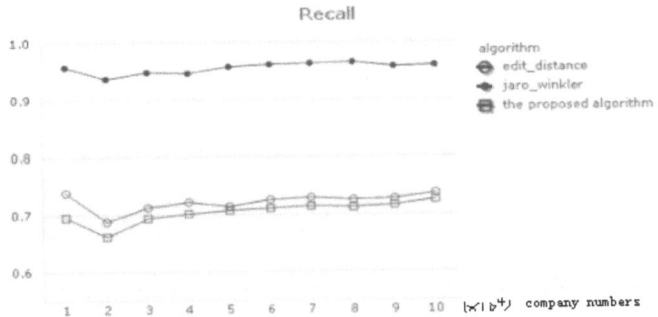

Fig. 3. The precision distribution of the three algorithms

Figure 4 is the recall distribution of the three algorithms, from which we can see that none algorithms' recall is lower than 60% and are all increasing. It represents that the coverage is acceptable.

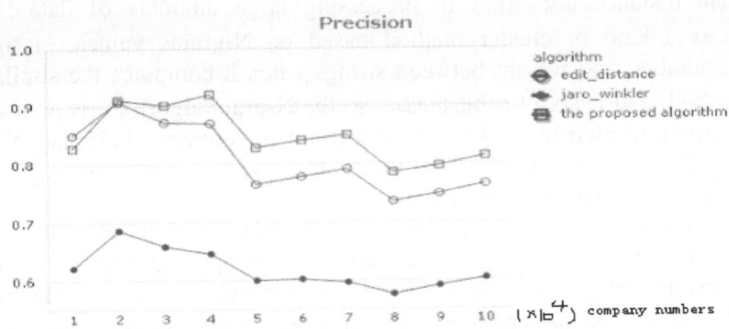

Fig. 4. The recall distribution of the three algorithms

Figure 5 is the distribution of the *F-measure*. For this indicator balancing recall and precision, it can reflect the performance more comprehensive. It can be seen from Figure 5 that when the scale of the data is small, the comprehensive performance of the proposed algorithm is poorer. But with the growth of the data scale, it performs better and better, keeping first until the end.

The conclusion above confirms the reliability and superiority of the proposed algorithm dealing with massive data more directly.

As duplicate detecting is usually working in big scale data environment, runtime is also an important indicator when assessing the performance. Thus, following is the analysis about runtime besides recall and precision.

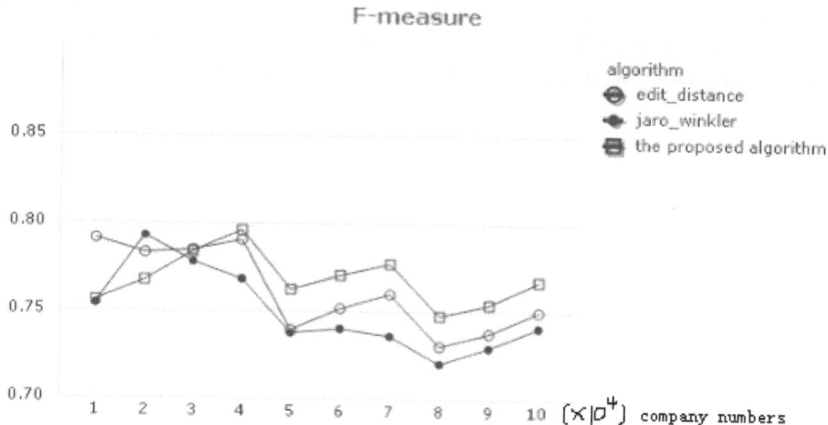

Fig. 5. The distribution of the F-measure of the three algorithms

4.3   Comparison with Other Duplicated-Detection Methods

Other common ways dealing with duplicated-detection problem are MPN method and N-grams method, they compute the similarity between strings of Levenshtein distance algorithm andstrings of N-grams algorithm independently. We can get the result from the previous chapters that the proposed algorithm is better than the Levenshtein distance algorithm in processing large amounts of data ; N-grams algorithm is a kind of cluster method based on N-grams values, it ignores the sequence, number, and weight between strings when it computes the similarity with cluster method, and lacks of rubustness, so the N-gram algorithm is poor compared with the proposed algorithm. The MPN method in reference [11] and N-grams in reference [12] are chosen to compare the runtime with the proposed algorithm, and details are shown in the following table 1.

Table 1 shows that, because we narrow the record range first before computing the similarity, dividing one big record set into small groups and computing within the smaller ones, the compare numbers between records and computing scale are reduced greatly. And therefore, the runtime of the proposed algorithm is always the smallest and the performance is always the best during the entire experiment.

Table 1. Runtime comparison

Rows of records (Ten thousand)	Runtime /s		
	MPN method	N-grams method	The proposed method
1	32.987	20.004	0.997
10	319.863	207.268	10.157
100	3345.589	2234.241	102.589
200	6637.486	4450.689	205.478
300	9967.783	6988.254	309.258

4.4   Result

Above all, we can conclude that in the massive data environment, the overall performance of the proposed method in recall, precision and runtime is better than the other typical methods. Thus the proposed method is feasible and effective in solving the company name duplicate detecting problem in massive data environment.

5   Conclusion

Detecting duplicated company names is a key task of information audit in B2B platforms, which requires high accuracy and high efficiency. Given the existing methods cannot identify the Chinese company names very well, this paper develops a real-time multi-registration detection method. The core of this method is to utilize Chinese word segmentation technology to narrow down the matching range of data. And also, a new string similarity computing method is adopted in this study. The proposed approach is evaluated by using the real data from a B2B platform, and demonstrates the excellent performance.

Through the proposed method, we test the 2000 new companies increased every day, more than 1000 of them are duplicated, and these companies do not need to be checked, This will probably save about three quarters of the human cost. Obviously, the judgment, which takes only one factor—name consideration , will effect the precision in the real world to a certain extent, so taking other factors like IP, city, leader name and etc into account will make the result more comprehensive in future. In addtion, given the different pattern of different companies, to solve the problem, more duplicated-detection methods will be considered, in the future.

Acknowledgements. This research is supported by Focus Technology Co., Ltd.. The views and conclusions contained herein are those of the authors and should not be interpreted as necessarily representing the official policies or endorsements, either expressed or implied, of the above organizations or any person connected with them.

References

1. Sui, Z., Yu, S.: A model of computing sentences similarity based on dependency tree. In: Proceedings of ICCIP 1998, pp. 458–465. Tsinghua University Press, Beijing (1998)
2. Li, S.: Research of relevancy between sentences based on semantic computation. In: Computer Engineering and Applications, pp. 75–76 (2002)
3. Lv, X., Ren, F., Huang, Z., et al.: Sentence similarity model and the most similar sentence search algorithm. Journal of Northeastern University: Natural Science ed., 531–534 (2003)
4. Qin, B., Liu, T., Wang, Y., et al.: Question answering system based on frequently asked questions. Journal of Harbin Institute of Technology, 1179–1182 (2003)
5. Chen, K., Fan, X.-Z., Liu, J., Jia, K.-L.: Calculation Method of Chinese Question Semantic Similarity Based on Question Semantic Representation. Transactions of Beijing Institute of Technology, 1073–1076 (2007)

6. Minton, S.N., Nanjo, C., Knoblock, C.A.: A heterogeneous field matching method for record linkage. In: Proceedings of the 5th International Conference on Data Mining, pp. 314–321. IEEE Computer Society, Washington (2005)

7. Cohen, W.W.: Integration of Heterogeneous Databases without Common Domains Using Queries Based Textual Similarity. In: Proc. ACM SIGMOD Conf. on Management of Data, pp. 201–212 (1998)

8. Hu, D.-B., Ding, J.: Learning String-edit Distance. Study on Similar Engineering Decision Problem Identification Based on Combination of Improved Edit-Distance and Skeletal Dependency Tree with POS. Systems Engineering Procedia, 406–413 (2011)

9. William, E.W.: Overview of Record Linkage and Current Research Directions. Tech. Rep. US Census Bureau, Washington, USA (2006)

10. Masek, W., Paterson, M.A.: A Faster Algorithm for Computing String Edit Distance. Computer System Science, 18–31 (1980)

11. Liu, W., Cao, X.-B.: Improvement for the Algorithm of Detecting Approximately Duplicate Database Records Based on MPN. Control & Automation, 152–154 (2005)

12. Kukich, K.: Techniques for automatically correcting words in text. ACM Computing Surveys, 377–439 (1992)

13. Jin, L., Li, C., Mehrotra, S.: Efficient record linkage in large data sets. In: The 8th Int'l Conf. on Database Systems for Advanced Applications, Kyoto, Japan (2003)

14. Ananthakrishna, R., Chaudhuri, S., Ganti, V.: Eliminating fuzzy duplicates in data warehouses. In: Proc 28th VLDB, pp. 586–597. Morgan Kaufmann, San Francisco (2002)

15. Ukkonen, E.: Approximate string-matching with q-grams and maximal matches. Theoretical Computer Science, 191–212 (1992)

16. Ananthakrishna, R., Chaudhuri, S., Ganti, V.: Eliminating Fuzzy Duplicates in Data Warehouses. In: Proceedings of the 28th VLDB Conference, Hong Kong, China (2002)

Keyword Graph: Answering Keyword Search over Large Graphs

Dong Wang, Lei Zou*, Wanqiong Pan, and Dongyan Zhao

Peking University, Beijing, China
zoulei@pku.edu.cn

Abstract. In this paper, we focus on IR style queries, *keyword search*, over large disk-resident graphs. Since most existing approaches cache the whole graph and indexing structure in memory, these approaches cannot be applied into large graphs, such as RDF graphs and social networks. In this paper, we design a novel indexing structure,*(kernel) keyword graph* to summarize the structure of original graph. Based on (kernel) keyword graph, we propose an efficient keyword search algorithm. Extensive experiments confirm that our method can scale up to large graphs with millions of nodes and edges. The performance of our approach outperforms state-of-the-art algorithms by at least one order of magnitude.

Keywords: Keyword Search, Graph Data, Keyword Graph.

1 Introduction

Due to increasing interests over sematic web and social network data, how to manage graph-structured data efficiently proposes new challenges to web data management. The graph size in these web applications can be very large. To query graph data, some structural query languages have been proposed, such as XQuery[3], SPARQL[1] and so on. The DB-style queries require users to master the complex syntax and the underlying schema. Besides, some IR-style queries like keyword search [2,9,8,6,13,12] over graph data have been proposed.

In this paper, we focus on keyword search over a large disk-resident graph. Different from most existing methods, our approach is disk-based, which can scale up to millions of nodes and edges in graph data. We construct an index called keyword graph, which summarizes the original graph, and takes keywords as distinct nodes. Besides, we extended an disk-based algorithm for approximate steiner tree(ST) problem called STAR[11] to answer keyword queries.

The main contribution of this paper lies in the novel data structure, *(kernel-based) keyword graph*, which has two following advantages:

1) (less number of random accesses) In a disk-based scenario, one traversal corresponds to one random access to disk. Let k_1 and k_2 denote two different keywords. k_1 and k_2 may appear in many adjacent node pairs in G. Therefore,

---

* Corresponding author.

[1] http://www.w3.org/TR/rdf-sparql-query/

S. Zhou, S. Zhang, and G. Karypis (Eds.): ADMA 2012, LNAI 7713, pp. 635–649, 2012.

one traversal step on keyword graph covers several traversals on the original graph. Thus, the number of random accesses to disk can be reduced.

2) (smaller graph size) In (kernel-based) keyword graph, each node corresponds to one distinct keyword. In many real large graphs, the number of keywords is much less than that of nodes. It also means that the (kernel-based) keyword graph has much fewer nodes. Obviously, smaller graph size in a (kernel-based) keyword graph leads to smaller search space during online traversal.

2   Keyword Graph – A Basic Solution

In this section, we present a conceptually simple scheme to help illustrate the main idea of our method. Some optimization issues are addressed in Section 3.

Definition 1. *Given a graph G and a subtree T in G, the cost of T is defined as follows: $C(T) = |T|$, where $|T|$ denotes the edge number of T.*

Definition 2. *Given a graph G and n keywords $k_1, ..., k_n$, each keyword k_i ($i = 1, ..., n$) corresponds to one cluster C_i of nodes in G, where each node in C_i contains keyword k_i. The minimal result of keyword search is a subtree T if and only if 1) T contains at least one node in each C_i; and 2) $C(T)$ is minimal.*

Top-k keyword search *is defined as to find the top-k minimal subtrees that contain at least one node in each C_i.*

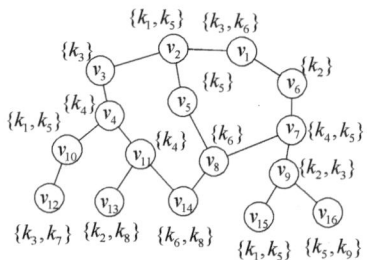

Fig. 1. Data Graph G

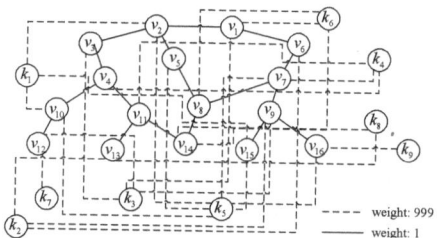

Fig. 2. Augmented Data Graph $G^{\triangledown}$

2.1   Index Construction-Keyword Graph

Let us consider a data graph $G(V, E)$ in Figure 2, where each node $v(\in V)$ contains a set of keywords $\{k\}$. $k \in v$ means that keyword k appears in v. The corresponding keyword graph is denoted as $G'(V', E')$. Each node $v' \in V'$ is called *keyword node*, which corresponds to one keyword k. To distinguish, nodes and edges in G are called as *data nodes* and *data edges*. Analogously, nodes and edges in keyword graph are called *keyword nodes* and *keyword edges*. For the consistency of notations, we use v and e to denote data nodes/edges, respectively; and $v'(k)$ and e' to denote keyword nodes/edges, respectively.

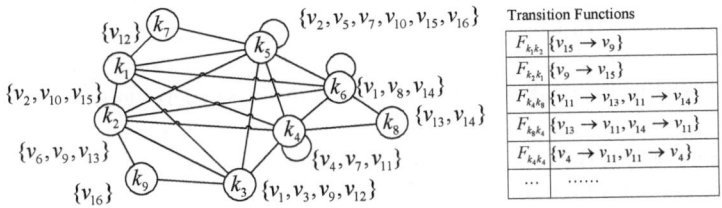

Fig. 3. Keyword Graph G'

Figure 2 and Figure 3 show a data graph G and the corresponding keyword graph G'. For each keyword node k, it is associated with a set $k.S = \{v_i | v_i \in V \wedge k \in v_i\}$, meaning that keyword k appears in these data nodes.

Then, we discuss how to introduce keyword edges in G'. Let us consider two keyword nodes k_{i_1} and k_{i_2}. The principle is that there is a keyword edge between k_{i_1} and k_{i_2} in G' *if and only if* there exist at least two nodes v_{i_1} and v_{i_2} in data graph G, where $k_{i_1} \in v_{i_1}$ and $k_{i_2} \in v_{i_2}$ and v_{i_1} is adjacent to v_{i_2} in G. We also store the link information between k_{i_1} and k_{i_2}. Specifically, we need to record how two keywords k_{i_1} and k_{i_2} are connected in the original data graph G. For example, for keyword edge $\overline{k_1 k_3}$ in G', its edge label is a *transition function* $F_{\overline{k_1 k_3}} = \{v_2 \to v_3\}$, which means that there is one edge $\overline{v_2 v_3}$ in data graph G, where $k_1 \in v_2 \wedge k_3 \in v_3$, as shown in Figure 3(b).

2.2 Query Algorithm Based on Keyword Graph

Overview of Query Algorithm

Example 1. Given a keyword search $Q = \{k_7, k_8, k_9\}$ over graph G (in Figure 2), we want to find the top-1 minimal result subtree (according to Definition 2) in graph G. Besides, the GST problem can be seen as a Steiner Tree problem by introducing virtual keyword nodes in graph G^∇ shown in Figure 2.

Algorithm 1. Framework of Search Over Keyword Graph

Require: Data graph G, keyword graph G', keywords $k_1, ..., k_n$ and size of result set k.

Ensure: Top-k results RS.
1: Set result set $RS = \phi$.
2: **while** $|RS| < k$ **do**
3:    Find the next keyword search result T' and T''s image subtrees $I(T')$ over G'
4:    **for** each image subtree T in $I(T')$ **do**
5:       **if** $T \notin RS$ **then**
6:          Put T into RS
7: Output all result subtrees in RS.

Algorithm 1 outlines the framework. For answering query Q, we first locate each keyword in G' (in Figure 3). Then, we find the minimal tree T' interconnecting the three keywords in G'. Finally, we find T''s *image subtrees*. By iteratively calling Algorithm 2, top-k result trees are computed incrementally.

Operations over Keyword Graphs. Some basic operations over keyword graphs should be redefined, since the graph structure is greatly different. We first discuss an important operation over G': *traversal*. When we traverse from keyword k_i to k_j, the traversal function should be obeyed. For example, the traversal from keyword k_1 to k_3 is illustrated as follows: $k_3.S = k_3.S \cap F_{\overline{k_1 k_3}}(k_1.S) = \{v_3, v_9\} \cap \{v_3\} = \{v_3\}$. When the traversal from keyword k_i to k_j is done, if $k_j.S = \phi$, the traversal is *invalid*.

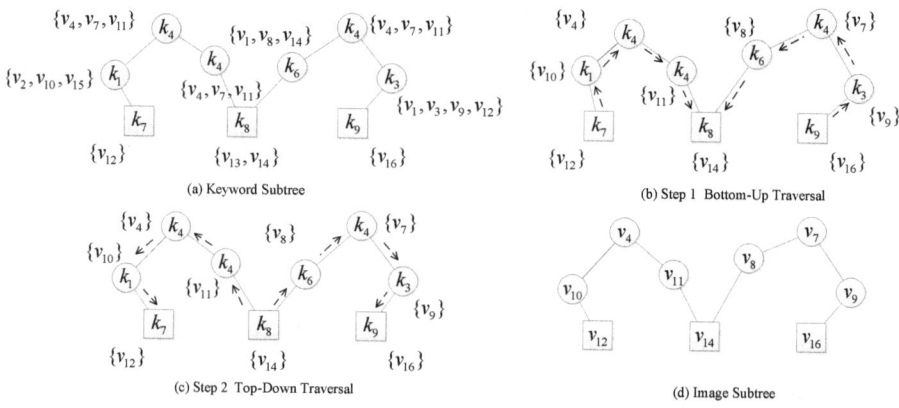

Fig. 4. Subtree Stabilization

A path p' in G' can be regarded as a series of traversal steps over G'. Note that, k_i may occur multiple times in a valid path since each of them corresponds to different data nodes, i.e., $k_i.S$ are different.

Definition 3. *Given a subtree $T'(k_1, ..., k_n)$ in G', a subtree $T(v_1, ..., v_n)$ in data graph G is called an* image subtree *of T' iff. $k_i \in v_i, i = 1, ..., n$.*

Given $T'(k_1, ..., k_n)$, if vertex v in $k_i.S$ does not appear in any image subtree of T', v is redundant for T'. The process to remove redundant vertices for all keyword nodes along the subtree T' is called stabilization. *After stabilization, if $\exists i \in [1, n]$, $k_i.S = \phi$, T' is invalid. Note that path is a special subtree.*

Figure 4 illustrates the stabilization process. Given a subtree $T'(k_1, ..., k_n)$ in a keyword graph G', the process to find all image subtrees of T' is called *instantiated*. The instantiated process generate all possible image subtrees and remove the invalid trees.

Theorem 1. *Given a keyword set K and a keyword subtree T_1', if T_1' is the top-1 subtree in G', its image subtree(s) is also the top-1(or minimum) subtree(s) in G^2.*

Given two keyword subtrees T_1', T_2' in graph G', T_1 and T_2 are their corresponding images in graph G, respectively. The following equation holds: $C(T_1') < C(T_2') \Leftrightarrow C(T_1) < C(T_2)$, where $C(T)$ is the size of T defined in Definition 1.

Proof. (Sketch) According to definition 3, if $\exists k \in T'$, there's no data node can be derived from k, there's no image tree for T', and we can delete k from T'. As a result, if we obtain an image tree T from T', there's $N(T') = N(T)$. Thus, $C(T_1') < C(T_2') \Leftrightarrow C(T_1) < C(T_2)$, where $C(T_1) = C(T_1')$ and $C(T_2) = C(T_2')$.

Query Algorithm. In keyword graph G', each keyword corresponds to one node in G'. We modify STAR algorithm (denoted as KGraph) to find approximate results for keyword search problem. Specifically, the algorithm has two phases:

First Phase: Given n keywords k_i, $i = 1, ..., n$, we can first find the corresponding n keyword nodes v_i'. Then, we adopt BFS (breath-first-search) strategy to obtain an initial answer tree. Then, we *stabilize* (see Definition 3) the initial answer tree. Note that, the traversal over G' needs to follow the transition function, as mentioned in Section 2.1. An running example of the first phase is shown in Figure 5.

Algorithm 2. Get Next Subtree

---

get_next_tree(G, G', K)

1: Locate keyword nodes k_i in G', $i = 1, ..., n$
2: Find the first tree T' over G' to interconnect k_i, $i = 1, ..., n$ by BFS search
3: Find all loose paths of T' and insert them into priority queue PH
4: **for** *longest path* $p \in |PH|$ **do**
5:     Delete p from T' and T' is separated into T_1' and T_2'
6:     Find the shortest path p' between T_1' and T_2' in G'
7:     **if** $|p'| < |p|$ **then**
8:         Re-connect T_1' and T_2' by p' to form the updated T'
9:         Update PH using T'
10: Return T'

Second Phase: In the second phase, KGraph algorithm needs to upgrade the answer tree iteratively to find an approximate result. While upgrading an answer tree, KGraph deletes the longest loose path (Definition 4), and then connects the two generated subtrees up with the shortest path between the two subtrees. Figure 5 and Figure 6 illustrate the upgrading steps based on Example 1. The detail information can be seen in Algorithm 2.

---

[2] If T_1' has more than one image subtree, these image subtrees have the same cost and they are all top-1 subtrees in G.

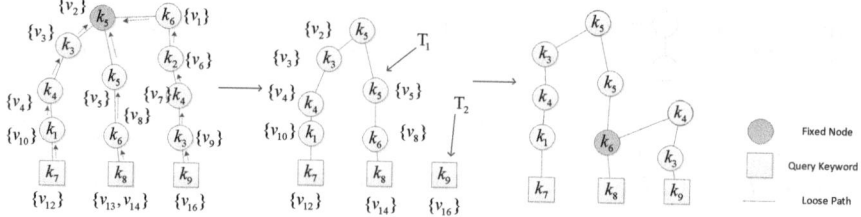

Fig. 5. After The First Iteration

Definition 4. *A node is a* fixed node *in G′ if and only if it either contains at least a query keyword or its degree is no less than 3. A path is a* loose path *if it interconnects two fixed nodes, i.e., the path has two fixed nodes as its endpoints.*

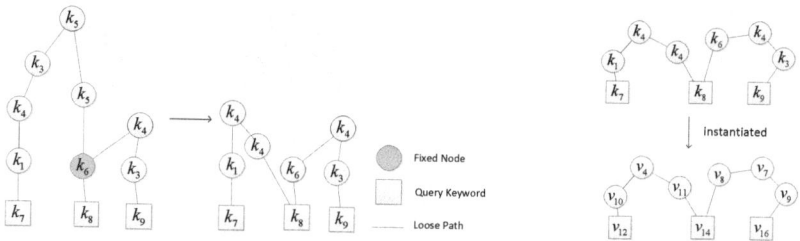

Fig. 6. After Second Iteration   **Fig. 7.** Final Result

At the end of Algorithm 2, the current optimized subtree is found. Algorithm 3 describes how to compute top-k trees. To compute the top-2 result tree in the running example, all edge weights in the top-1 result tree are increased by δ, where δ is a system parameter to control the quality of top-k results[11]. Larger δ leads to larger approximate rate, and smaller δ causes longer running time.

Algorithm 3. Keyword Search Algorithm (KGraph)

Require: Input: Data graph G, keyword graph $G′$ and keywords $k_1, ..., k_n$.
    Output: Top-K keyword search results.
1: Set result set RS=ϕ
2: $T′$=get_next_tree($G, G′, K$) // Call Algorithm 2
3: Instantiate all image subtrees of $T′$, and put them into RS
4: **for** each edge e of $T′$ **do**
5:    $e.weight = e.weight \times (1 + \varepsilon)$
6: **while** $|RS| < K$ **do**
7:    $T′$=get_next_tree($G, G′, K$)
8:    Instantiate all image subtrees of $T′$, and put them into RS

Theorem 2. *The STAR algorithm over keyword graph G′ is a $(4\lceil logN \rceil + 4)$-approximation algorithm for the optimal top-1 subtree for keyword search (according to Definition 2).*

Proof. (Sketch) It can be proved according to Theorem 1 in [11].

3   Optimizations

3.1   Kernel Keyword Graph

Since each neighboring keyword pair have an edge, space cost of G' is very expensive. In order to address the problem, we propose "virtual keywords" to cover all nodes in original graph G. Algorithm 4 outlines how to generate kernel keyword graph G^*. First, we generate a virtual keyword set N randomly distributed in $V \in G$, where each data node v_j contains just one virtual keyword n_i (denoted as $n_i \in v_j$). Next, the *link* edge is introduce between virtual keyword nodes. We also record the transition function between n_{i_1} and n_{i_2}, as discussed in Section 2.1. Then, the *co-occurrence* edge is introduced to indicate the co-occurrence relation between virtual keywords and real keywords. The kernel graph $G^* = \{e_{link}\} \cup \{e_{co-occurrence}\} \cup \{n\} \cup \{k\}$, where the subgraph induced by all virtual keyword nodes is called *kernel*. Figure 8 shows an example of the kernel keyword graph corresponding to graph G in Figure 2.

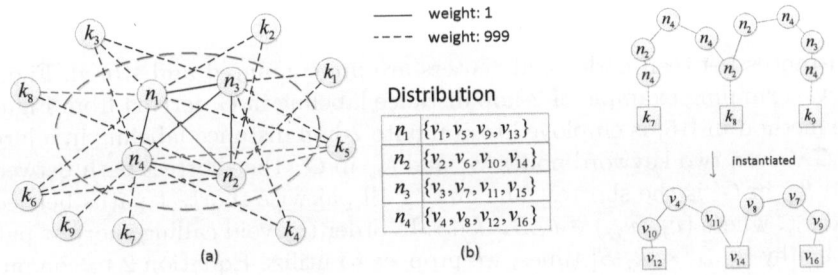

Fig. 8. Kernel of Virtual Keywords

Fig. 9. Instantiated Procedure of Kernel

Algorithm 4. Building kernel keyword graphs

Require: Input: Data graph G and the number of virtual keywords m.
    Output: Virtual kernel graph G^*.
1: Generate a virtual keyword set $N = \{n_1, n_2, \dots, n_m\}$
2: Distribute N into $V \in G$
3: $\exists link$ edge $e^*_l = \overline{n_{i_1} n_{i_2}}$ iff. $\exists e = \overline{v_{j_1} v_{j_2}}$, where $n_{i_1} \in v_{j_1}$ and $n_{i_2} \in v_{j_2}$
4: The transition function F between n_{i_1} and n_{i_2} is recorded in G^*
5: $\exists co-occurrence$ edge $e^*_c = \overline{k_i n_j}$ iff. $\exists v \in G, k \in v, n \in v$
6: $G^* = E^* = \{\{e^*_l\} \cup \{e^*_c\}\} \cup N = \{n\} \cup K = \{k\}$
7: Return G^*

We have some guides to determine m. Obviously, large m leads to large kernel size. On the other hand, smaller m means that the frequency of each kernel is larger. Consequently, the kernel in G^* is very dense, which leads to more I/O cost during online traversal over the kernel.

This optimization just impact the *instantiated* procudure. We instantiate the keyword subtree and delete all data nodes v corresponding to the keyword nodes. Figure 9 gives a running example of the modified image tree generating procedure. Since the keyword nodes are connected with the *co-occurrence edges*, the delete operation causes no information lost.

3.2   Distance Labeling

In order to speed up shortest path computation, we propose to utilize distance labeling technique in Line 6 in Algorithm 2.

A 2-hop distance labeling method [4] over a large graph G assigns to each node $u \in V(G)$ a label $L(u) = (L_{in}(u), L_{out}(u))$, where $L_{in}(u), L_{out}(u) \subseteq V(G)$. Nodes in $L_{in}(u)$ and $L_{out}(u)$ are called *centers*. We can compute $Dist_{sp}(u_1, u_2)$ using the following equation, where $Dist_{sp}(u_1, u_2)$ is the shortest path distance between nodes u_1 and u_2.

$$
\begin{aligned}
Dist_{sp}(u_1, u_2) = \min\{&Dist_{sp}(u_1, w) + Dist_{sp}(u_2, w)| \\
&w \in (L_{out}(u_1) \cap L_{in}(u_2))\}
\end{aligned} \tag{1}
$$

The distances between nodes and centers are pre-computed and stored. Figure 10 shows a running example of 2-hop distance labeling in G derived from Figure 2. The method in [15] is employed to compute 2-hop distance labeling in a large graph G. Given two keyword nodes k_{i_1} and k_{i_2} in G', the shortest path between k_{i_1} and k_{i_2} in G' is the shortest one among all pairwise shortest paths between v_{j_1} and v_{j_2}, where $(v_{j_1}, v_{j_2}) \in k_i.S \times k_j.S$. In order to avoid calling shortest path algorithm by $|k_i.S| \times |k_j.S|$ times, we propose to utilize Equation 2 to compute shortest path between k_i and k_j in keyword graph G'.

$$
\begin{aligned}
L_{out}(k_i) &= \{(w, Dist(k_i, w))\} = \\
&\{(w, \min(Dist_{sp}(w, v)))|w \in \bigcup_{v \in k_i.S} L_{out}(v), v \in k_i.S\} \\
L_{in}(k_j) &= \{(w, Dist(w, k_j))\} = \\
&\{(w, \min(Dist_{sp}(v, w)))|w \in \bigcup_{v \in k_j.S} L_{out}(v), v \in k_j.S\} \\
Dist_{sp}(k_i, k_j) &= \min\{Dist_{sp}(k_i, w) + Dist_{sp}(w, k_j)| \\
&w \in (L_{out}(k_i) \cap L_{out}(k_j))\}
\end{aligned} \tag{2}
$$

According to Equation 2, the shortest path computing cost between two keyword nodes in G' is greatly reduced, where $Dist_{sp}(x, y)$ means the distance of the shortest path between x and y. See Figure 10, given two keyword k_1 and k_2 in G', supposing $k_1.S = \{v_2, v_{10}\}$ and $k_2.S = \{v_6, v_9\}$, we first get $L_{out}(k_1)$ and $L_{in}(k_2)$ in Figure 10. According to 2-hop labeling and Equation 2, $Dist_{sp}(k_1, v_1) = Dist_{sp}(v_2, v_1)$ and $Dist_{sp}(v_1, k_2) = Dist_{sp}(v_1, v_6)$. Then, we find the path $P = sp(k_1, v_1) + sp(v_1, k_2)$, which is an image shortest path between k_1 and k_2 in keyword graph G', where sp means the shortest path. We have the analogue method to compute the shortest path in G^* by Equation 2. Due to space limitation, we omit the details.

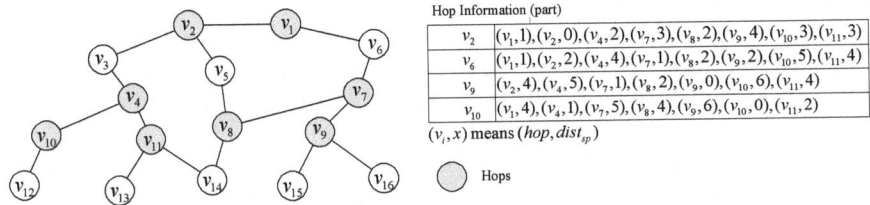

	Hop Information (part)
v_2	$(v_1,1),(v_2,0),(v_4,2),(v_7,3),(v_8,2),(v_9,4),(v_{10},3),(v_{11},3)$
v_6	$(v_1,1),(v_2,2),(v_4,4),(v_7,1),(v_8,2),(v_9,2),(v_{10},5),(v_{11},4)$
v_9	$(v_2,4),(v_4,5),(v_7,1),(v_8,2),(v_9,0),(v_{10},6),(v_{11},4)$
v_{10}	$(v_1,4),(v_4,1),(v_7,5),(v_8,4),(v_9,6),(v_{10},0),(v_{11},2)$

(v_i, x) means $(hop, dist_{sp})$

⬤ Hops

Fig. 10. Distance Labeling

3.3   Put It All Together

Off-Line Process. First, we employ Algorithm 4 to compute a kernel graph G^*. Then, we compute 2-hop distance labeling over G by the method in [15].

On-Line Process. We have presented the keyword query algorithm in Algorithm 3. In order to speed up query processing, we utilize 2-hop distance labeling to replace Line 6 in Algorithm 2. Furthermore, instantiated procudure is modified, as discussed in Section 3.1, i.e., deleting the node $n \in T'$ instead of data node replacing.

4   Experimental Results

In this section, we compare KGraph algorithm with the state-of-the-art algorithms for keyword search over graph data on external storage, such as BANKS[2], STAR[11] and EMKS(External Memory Keyword Search) [5][3]. Since STAR algorithm cannot work for keyword search on graphs directly, thus, we generate an augmented graph $G^\triangledown$, as shown in Figure 2. For comparison, we run BANKS and EMKS algorithm in the same graph.

We choose two real datasets with different sizes for performance study. One is extracted from Freebase[4], which is smaller and denser. The other one is part of YAGO[5], which is larger and sparser.

All experiments were performed on a 2.19GHz Inter Core2 Duo CPU E4500 with 3GB of main memory and an MySQL (version 5.1) database to persist the underlying data structure. All implementations are in Java.

4.1   Data Sets

1) Freebase is a collaborative knowledge base describing relations between entities such as people, organizations, and so on. We extract one data set from Freebase about football. This data set (denoted as Soccer Set) consists of 34,011 nodes, 54,240 edges, and 30,620 keywords are extracted as shown in Table 1.

---

[3] The java implementations of BANKS and STAR are kindly provided by Gjergji Kasneci, author of [11]. We implement EMKS by ourselves.

[4] http://download.freebase.com/datadumps/

[5] http://www.mpi-inf.mpg.de/yago-naga/yago/downloads.html

2) Yago automatically extracts RDF triples from Wikipedia, WordNet and GeoNames. The latest version of YAGO has knowledge more than 10 million entities and contains more than 460 million facts (statements, or RDF triples). We extract part of YAGO as our test dataset (denoted as YAGO Set). The part of YAGO contains more than 3.3 million nodes and 3.6 million edges. We scan the whole set and treat all English alphabetics and numbers split by symbols or blank spaces as keywords. Thus, we have got 304,032 keywords.

Table 1. Statistics of Data Sets

	Soccer Set	YAGO Set
Nodes	34,011	3,380,359
Edges	54,240	3,663,610
Keywords	30,620	304,032

Table 2. Time Evaluation of Kernels (Soccer Set)

Kernel Size	3 keywords	6 keywords
500	615ms	1,412ms
1000	571ms	565ms
3000	509ms	540ms
10000	456ms	512ms

The statistics of the Soccer Set and the YAGO Set are given in Table 1. Furthermore, we evaluate the ratio of (keyword number)/(data node number) in other data sets, for verifying the assumption that the size of keyword graph is stable with the growing of data graph's size. Table 2 confirms our analysis.

Table 3. Edge count: Kernel-based vs Keyword-based

Data Set	Kernel-based Keyword Graph					Original Keyword Graph	
	Kernel Size	Link Edges	Co-occurrence Edges	Totle Edges	# of Records	Edges	# of Records
Soccer	500	42,733	79,541	122,274	133,781	264,173	386,091
YAGO	10,000	3,445,998	6,411,368	9,857,366	10,074,978	5,117,304	30,020,521

4.2   Experiments of Optimizations

Experiment 1(Evaluating Distance Labeling). In Sec.3.2, we introduce distance labeling to boost searching performance. In this experiment, we make a comparison of STAR with distance labeling, STAR with breadth-first-search, and Backward Search like BANKS over keyword graph (without kernel).

Due to space limit, we only use the Soccer Set in this experiment. We randomly select ten keyword sets, each with 3 keywords as a query, to be test set. We denote the test set as $S_t = S_1, S_2, \ldots, S_{10}$, where $S_i = \{k_{i1}, k_{i2}, k_{i3}\}$.

We measured the time cost of each algorithm for computing top-6 and the result is shown in Figure 11. The x axis represents different queries from 1 to 10, and the y axis represents time cost (truncated at 25,000ms) of each query with different algorithm. STAR with distance labeling outperforms the competitors across most queries. With the increase of graph scale, the exploring cost would increase rapidly, which means the gap would be further expanded.

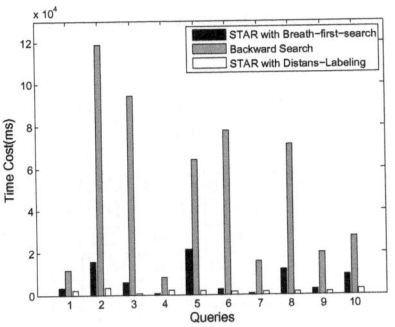

Fig. 11. Distance Labeling Evaluating

Table 4. Space Evaluation of Kernels (Soccer Set)

Kernel Size	Link Edges	Total Edges	# of Records
500	42,733	122,274	133,781
1000	50,665	132,184	135,759
3000	53,616	137,079	137,703
10000	54,144	138,224	138,320
G^∇	54,240	138,696	138,696

Experiment 2 (Evaluating Kernel Keyword Graph). In this experiment, we will verify the effectiveness of kernel keyword graph. We generate both kernel-based graph and keyword-based graph for the Soccer Set and the YAGO Set, and list the edge count and record count. Since the optimized kernel size is hard to determine, we also design a experiment to show the trade-off. We randomly choose 30 keywords and divide them into ten queries.

Table 5. Keywords and Nodes in Different Data Sets

	Keywords	Nodes	Keywords /Nodes
Soccer Set	30,620	34,011	0.90
IMDB	38,769	68,139	0.57
Freebase (English Film)	78,146	166,181	0.47
Freebase (Film)	157,193	355,599	0.44
DBLP	583,929	3,103,613	0.19
YAGO Set	304,032	3,380,359	0.09

Table 6. Time Evaluation of Kernel Graph

Data Set	Kernel Size	3 keywords	6 keywords
Soccer	500	615ms	1,412ms
	keyword	530ms	2,086ms
YAGO	10,000	3,802ms	12,097ms
	keyword	19,784ms	N/A

N/A: Out of memory or Out of time limit

Table 3 shows the statistics information. The column "# of Records" indicate the total records count in databases, which means the real space cost. Clearly kernel keyword graph requires less space cost than keyword graph. Besides, smaller kernel leads to fewer edges. Table 4 shows the storage cost of the four kernel keyword graphs. In extreme cases, suppose there is only one virtual keyword in kernel, the edge count would be reduced to keyword scale (i.e., about 300,000 in YAGO Set, or about 30,000 in Soccer Set). Table 5 shows the run time cost with different kernel size. Noticeably, the run time decreases with the increasing of kernel size, and the storage cost grows. In most cases, the memory limit is tighter, and we tend to choose the kernel with smaller size.

Furthermore, we designed an experiment to make a comparison between the searching efficiencies over kernel-based graph and keyword-based graph, and the kernel size is set to 500 for Soccer Set and 10,000 for YAGO Set. Table 6 shows the running result. Notice that kernel-based graph has low efficiency with Q_1

Table 7. Soccer test: modified STAR on keyword graph, STAR and BANKS

		3 keywords				6 keywords			
	top-1	KGraph	STAR	BANKS	EMKS	KGraph	STAR	BANKS	EMKS
	avg. weight	4.60	4.93	3.67	**3.47**	10.85	**10.10**	10.15	11.50
	avg. run time(s)	**0.58**	0.82	0.87	0.71	1.53	1.92	1.97	**1.21**
High	top-3	KGraph	STAR	BANKS	EMKS	KGraph	STAR	BANKS	EMKS
	avg. weight	5.99	5.67	3.94	**3.79**	11.65	10.88	**10.20**	11.86
Frequency	avg. run time(s)	1.10	**0.96**	1.07	0.97	2.99	1.95	2.14	**1.47**
	top-6	KGraph	STAR	BANKS	EMKS	KGraph	STAR	BANKS	EMKS
	avg. weight	6.59	6.05	**4.14**	4.27	14.97	11.70	**10.24**	12.00
	avg. run time(s)	2.78	**1.07**	1.16	1.23	3.64	2.00	2.35	**1.57**
	top-1	KGraph	STAR	BANKS	EMKS	KGraph	STAR	BANKS	EMKS
	avg. weight	11.32	12.16	11.56	**8.03**	23.93	25.53	27.60	**20.93**
	avg. run time(s)	**0.12**	1.65	1.43	0.62	**0.49**	3.94	3.80	1.31
Low	top-3	KGraph	STAR	BANKS	EMKS	KGraph	STAR	BANKS	EMKS
	avg. weight	12.02	12.77	11.84	**8.43**	24.24	26.42	27.76	**21.28**
Frequency	avg. run time(s)	**0.27**	1.74	1.62	0.91	**0.87**	3.97	3.97	1.45
	top-6	KGraph	STAR	BANKS	EMKS	KGraph	STAR	BANKS	EMKS
	avg. weight	12.66	13.13	12.07	**8.58**	25.78	26.98	27.94	**21.68**
	avg. run time(s)	**0.52**	1.86	1.74	1.08	2.11	4.03	4.56	**1.53**

of the Soccer Set, while in the other tests it has better performance. This phenomenon is due to that most of keywords have very low frequency among the data set, and a few keywords appears very frequently such as "the", "of" [7]. If the searching process in keyword graph involves the keywords with high frequency, the query efficiency may below the acceptable range. Otherwise, i.e., if the searching process in keyword graph just involves the keywords with low frequency, it may performs better than in kernel keyword graph.

4.3 Experiments of Performance

In these experiments, we compare our method (KGraph) with the state-of-the-art algorithms, that are STAR, BANKS and EMKS. As in [5], the block size of EMKS is set to 100. In this experiment, we use both data sets.

Queries: For comprehensiveness, we construct 4 sets of queries with 3/6 and high/low frequency keywords, each include 20 queries respectively. The low-frequency keyword sets are randomly selected (mostly between 1-10), and the high-frequency keyword sets consist of the keywords with maximum frequency.

Metrics: There are two measures to be used, the average number of edges in answer subtrees (i.e., avg. weight) and the query response time (i.e., avg. run time). Since we use 20 queries for each query set, we report the average values in Tables 7 and 8. Furthermore, we report top-1, top-3 and top-6 results in Tables 7 and 8.

Results: We compare KGraph with other methods in the quality of answer subtrees and the average query response time. Though quality of answer subtrees has no significant difference, KGraph is much faster than other methods when data graph is large or keywords have lower frequency. Table 8 shows that KGraph

Table 8. YAGO test: modified STAR on keyword graph, STAR and BANKS

		3 keywords				6 keywords			
		KGraph	STAR	BANKS	EMKS	KGraph	STAR	BANKS	EMKS
High Frequency	top-1	KGraph	STAR	BANKS	EMKS	KGraph	STAR	BANKS	EMKS
	avg. weight	3.33	3.17	**3.00**	**3.00**	7.00	**6.00**	**6.00**	N/A
	avg. run time(s)	**24.68**	127.08	45.28	1858.71	**67.06**	219.80	276.48	N/A
	top-3	KGraph	STAR	BANKS	EMKS	KGraph	STAR	BANKS	EMKS
	avg. weight	3.63	3.17	**3.00**	**3.00**	7.25	**6.00**	6.25	N/A
	avg. run time(s)	**37.09**	179.65	46.81	2292.13	**91.43**	267.08	306.61	N/A
	top-6	KGraph	STAR	BANKS	EMKS	KGraph	STAR	BANKS	EMKS
	avg. weight	4.53	3.28	**3.00**	3.04	8.04	**6.00**	6.25	N/A
	avg. run time(s)	**46.02**	248.21	48.17	2785.45	**133.50**	339.64	310.08	N/A
Low Frequency	top-1	KGraph	STAR	BANKS	EMKS	KGraph	STAR	BANKS	EMKS
	avg. weight	**9.20**	10.5	10.00	11.12	**19.86**	N/A	N/A	N/A
	avg. run time(s)	**0.29**	57.57	46.88	2295.90	**1.30**	N/A	N/A	N/A
	top-3	KGraph	STAR	BANKS	EMKS	KGraph	STAR	BANKS	EMKS
	avg. weight	**9.91**	11.08	10.42	11.75	**20.19**	N/A	N/A	N/A
	avg. run time(s)	**0.64**	62.62	64.92	3338.47	**2.28**	N/A	N/A	N/A
	top-6	KGraph	STAR	BANKS	EMKS	KGraph	STAR	BANKS	EMKS
	avg. weight	**9.94**	11.33	10.5	12.08	**21.83**	N/A	N/A	N/A
	avg. run time(s)	**1.14**	70.67	70.94	4195.55	**3.53**	N/A	N/A	N/A

N/A: can not get result in reasonable time

outperforms other methods at least one or two orders of magnitudes. The key reason is that large graph leads to more I/O cost in other methods, but, I/O cost is reduced by keyword graphs in our method. We find that the query response time in EMKS is very large in Yago dataset and it cannot answer six keyword queries, since EMKS needs to expand many blocks in large graph, which leads to frequent I/O swap in/out. However, when keywords have high frequency in Table 7, the advantage of our method is not clear, since the average number of edges in answer subtree is small, which means that other algorithms need a small number of exploring steps.

5 Related Work

Due to the wide application of graph data, keyword search over large graphs has attracted much attention in research community, such as [2,9,8,6,13,12]. Usually, keyword search over graph is modeled as "minimal group Steiner Tree (GST)" problem, which is a NP-complete problem. In order to avoid the inherent hardness, most keyword search algorithms re-define the query semantics as "distinct core" to reduce the complexity except for DPBF [6]. BANKS[2] adopt the traversal-based approach to find the root. BLinks[8] pre-compute some indexing structures to speed up the traversal. DPBF is a dynamic programming algorithm to find the optimal result for GST problem [6]. Although keyword search algorithms have been well studied, most existing methods are memory-based, i.e, assuming the whole graph is cached in memory and no I/O cost during the traversal. Obviously, these methods cannot scale to massive disk-resident graphs.

EMKS[5] is a disk-based keyword search algorithm. EMKS partitions a large graph into several blocks to construct a summary graph, and expands the blocks incrementally during backward search until all answer trees are made up of real nodes [5]. The query performance of EMKS suffers two main problems, 1)how to find a good partition to speed up query processing, and 2), if there are many blocks need to be expanded, the performance will degrade greatly due to frequent swap in/out. Recently, STAR algorithm is proposed to interconnect n nodes v_i in a large graph efficiently [11]. STAR algorithm is an approximate solution with $(4\lceil logN \rceil +4)$-approximation to address *steiner tree* (ST) problem, a special case of GST problem. Note that, finding ST is also a NP-complete problem.

Besides, some recent work attempts to define other query semantics. Kargar et al introduce a new query semantic called "r-clique"[10]. "r-clique" means the longest distance between each keyword pair is less than r. [1] and [14] both treat keyword search task as a translating problem, and they both translate keyword queries to meaningful structured queries instead of GST, which are different from our problem definition.

6   Conclusions

In order to answer keyword search over a large disk-resident graph, we propose a novel index structure, keyword graph. To decrease storage cost and to improve query efficiency, we introduce some optimizations over keyword graph. Extensive experiments confirm that our method outperforms state-of-the-art algorithms by at least one order of magnitude in large graphs.

Acknowledgments. This work was supported by NSFC under Grant No.61003009, 61272344, the National High Technology Research and Development Program of China under Grant No. 2012AA011101 and RFDP under Grant No. 20100001120029.

References

1. Bergamaschi, S., Domnori, E., Guerra, F., Lado, R.T., Velegrakis, Y.: Keyword search over relational databases: a metadata approach. In: SIGMOD Conference 2011, pp. 565–576 (2011)
2. Bhalotia, G., Hulgeri, A., Nakhe, C., Chakrabarti, S., Sudarshan, S.: Keyword searching and browsing in databases using banks. In: ICDE, pp. 431–440 (2002)
3. Chamberlin, D.D.: Xquery: A query language for xml. In: SIGMOD Conference, pp. 1–1 (2003)
4. Cohen, E., Halperin, E., Kaplan, H., Zwick, U.: Reachability and distance queries via 2-hop labels. SIAM J. Comput. 32(5), 937–946 (2003)
5. Dalvi, B.B., Kshirsagar, M., Sudarshan, S.: Keyword search on external memory data graphs. PVLDB 1(1), 1189–1204 (2008)
6. Ding, B., Yu, J.X., Wang, S., Qin, L., Zhang, X., Lin, X.: Finding top-k min-cost connected trees in databases. In: ICDE, pp. 836–845 (2007)

7. Ha, L.Q., Sicilia-Garcia, E.I., Ming, J., Smith, F.J.: Extension of zipf's law to words and phrases. In: COLING, p. 1 (2002)
8. He, H., Wang, H., Yang, J., Yu, P.S.: Blinks: ranked keyword searches on graphs. In: SIGMOD Conference (2007)
9. Hristidis, V., Papakonstantinou, Y.: Discover: Keyword search in relational databases. In: VLDB (2002)
10. Kargar, M., An, A.: Keyword search in graphs: Finding r-cliques. PVLDB, pp. 681–692 (2011)
11. Kasneci, G., Ramanath, M., Sozio, M., Suchanek, F.M., Weikum, G.: Star: Steiner-tree approximation in relationship graphs. In: ICDE, pp. 868–879 (2009)
12. Ning, X., Jin, H., Jia, W., Yuan, P.: Practical and effective ir-style keyword search over semantic web. Inf. Process. Manage. 45(2) (2009)
13. Tran, T., Wang, H., Rudolph, S., Cimiano, P.: Top-k exploration of query candidates for efficient keyword search on graph-shaped (rdf) data. In: ICDE (2009)
14. Xin, D., He, Y., Ganti, V.: Keyword++: A framework to improve keyword search over entity databases. PVLDB, 711–722 (2010)
15. Zou, L., Chen, L., Özsu, M.T., Zhao, D.: Answering pattern match queries in large graph databases via graph embedding. VLDB J. 21(1), 97–120 (2012)

Medical Image Retrieval Method Based on Relevance Feedback

Rui Wang, Haiwei Pan*, Qilong Han, Jingzi Gu, and Pengyuan Li

College of Computer Science and Technology, Harbin Engineering University, Harbin
Heaven_007cn@yahoo.com.cn

Abstract. The current image retrieval systems are almost based on content, and facing the main problem of semantic gap between low level features and high level semantic. So the relevance feedback technology is used to solve this problem. In this paper, we propose a medical image retrieval system based on relevance feedback framework. In the framework, Region of Interest (ROI) is extracted in the preprocessing as the semantic information of medical images, and then the Genetic Algorithm is designed for ROI clustering. According to user's feedback information, the Diverse Density algorithm proposed in the Multiple Instance Learning Framework is adopted to capture user's real intention and realize effectively medical image relevance. Experimental results show that our algorithm has higher precision and recall ratio.

Keywords: Medical Image, Relevance Feedback, Multiple Instance Learning, Diverse Density.

1 Introduction

With the development of modern medical imaging equipments and technology, hospitals produce large amounts of complex information-rich digital images (such as CT images, ECT images, magnetic resonance images, etc.) everyday. However, doctors' diagnosis always depend on their eyes and subjective judgment which result in some shortages [1]: (1) The resolution of eyes is low, and it is easy to miss some small changes in the characteristics; (2) Doctors observations are subjective so the diagnostic reliability varies from person to person; (3) Different doctors have different experience which may result in misdiagnosis or misdiagnosed. It can be an urgent problems that how those medical images are used to help doctors diagnose correctly.

Text-Based Image Retrieval is a traditional image retrieval technology, in which images are labeled by manual analysis and the textual information is taken advantage to index and retrieve images. Clearly, there are many limitations, such as time-consuming, subjective, etc. To solve the above shortcomings, the Content-based Image Retrieval (CBIR) is used.

CBIR is the process of processing and understanding images by the visual features of images and its purpose is to search images with similar features or con-tents

---

* Corresponding author.

S. Zhou, S. Zhang, and G. Karypis (Eds.): ADMA 2012, LNAI 7713, pp. 650–662, 2012.

with your query image. However, the CBIR is facing two main problems: (1) the semantic gap between low level features and high level concept; (2) curse of dimensionality. This paper aims to build a framework to alleviate those two problems.

The semantic gap is that the existing color, shape, texture and other features in image cannot intuitively express the true meaning of image. And it is not the characters of color, shape, texture and so on, but the true meaning of image that people really understand on image. Therefore, the role of human is particularly important in the images retrieval process, and currently the most widely used technology is relevance feedback (RF). The existing search feedback algorithms are divided into three categories:

1) Feature weight optimization [2]: The basic idea is to analyze user feedback and dynamically adjust the weight of image features to achieve the purpose that improve retrieval accuracy;

2) Query point movement [3]: The basic principle is that each image is represented as a vector(point) in feature vector space, , and the system adjusts the query vector based on user feedback to make the query center close to the positive examples, away from the negative examples;

3) Probabilistic model-based method [3]: The basic idea is to make statistical inferences based on user feedback information, then estimate association prob-ability among images and then query and return the images with high probability.

Most of the existing relevance feedback techniques consider each image as a whole which is represented by an N dimensional feature vector. However, in the same parts of the medical images such as brain CT, users' query interest is part of the diseased area in the image (such as brain tumors) rather than the whole image. Therefore, each image in this paper is viewed as a semantic package which has a number of instances (semantic regions). We adopt Multiple Instance Learning (MIL) [4] to learn users' interested area from users' feedback information.

MIL was first proposed by Dietterich et al in 1997. In MIL, the label of an individual instance is unknown. Only the label of a set of instances is available, which is called the label of a bag. The goal of MIL is to estimate the labels of the test image instances based on the learned information from the labeled images in the training set. Neural Network based Multiple Instance Learning algorithm with relevance feedback [5] is a typical example of MIL, and its main idea is to mine knowledge of interest in a similar group to build new neural network based users' feedback information.

In Section 2, we present an overview of the basic system framework of medical image retrieval based on relevance feedback (MIRRF). The preprocessing of medical image is presented in Section 3. In Section 4, the cluster method of medical image regions is discussed. The detailed learning and retrieval approach is discussed in Section 5. In Section 6, system performance evaluation with experimental results is presented. Section 7 is the conclude part.

2   The Framework of Medical Image Retrieval Based on Relevance Feedback

Response to above problems, we proposed a framework of medical image retrieval based on relevance feedback. This framework is divided into two phases: the preprocessing stage and medical image retrieval based on relevance feedback.

The preprocessing phase includes three parts of the region of interest in medical image extraction, the region feature extraction and region clustering. A medical image is represented by a number of regions after segmentation. Feature extraction of the region is to extract characteristics which can express regions from the region to facilitate comparison. Region clustering stage is to put regions into clusters based on some similarity criteria to make the query space narrowed to a few clusters when retrieval.

In medical image retrieval based on relevance feedback phase, when the system first receives a query image that user submits, the system knows nothing about users' interests. Therefore, the system uses Euclidean distance to calculate the similarity between images, to return an initial query results. The query results are marked positive and negative examples, and the system learns from the user feedback information to get users' interests. Then view the user's interest as the query point to query again. The system framework is shown in Fig.1.

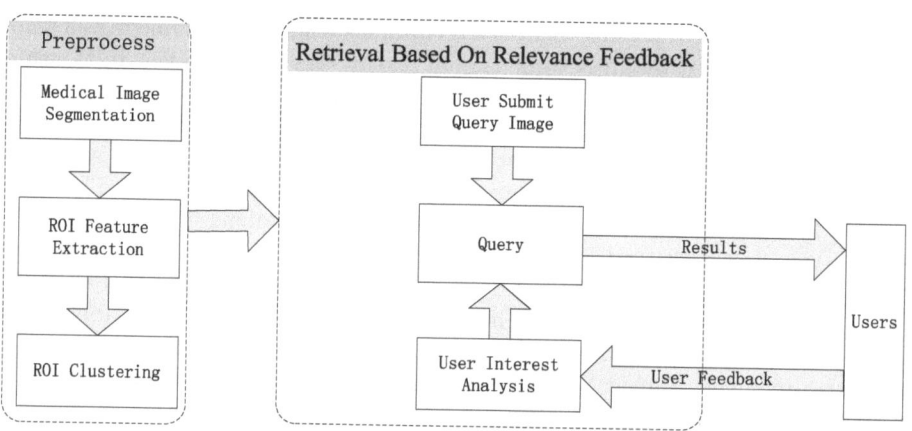

Fig. 1. The system framework for MIRRF

There are a variety of image segmentation methods but most for natural images. However, medical images and natural images are very different, such as the medical images are gray and the number of colors is less than natural images. So the segment methods for natural images is not good for medical images and need to find a specific method for medical images which is the focus of this paper.

After segmentation, an image is expressed by a few areas to enlarge the search space by many times and slowdown the search speed. Therefore, putting ROIs which are similar into a cluster can make the query space reduce. K-means [6] is a traditional clustering method and has been widely used in image clustering.

However, it is incapable of finding non-convex clusters and tends to fall into local optimum especially when the number of data objects is large. So it is necessary to find a new cluster method to overcome the shortcomings of the k-means.

In addition, at present, the RF methods just only modify the query vector and cannot accurately capture user query interest. In this paper, we propose a new algorithm to capture users' query intention, which is the focus of this paper.

3   Segmentation and Feature Extraction of Medical Image Retrieval

Existing computer-aided medical diagnosis systems are almost knowledge-based expert ones, and often have the shortages of knowledge acquisition method, knowledge vulnerability, reasoning monotony and so on. These systems have the difficulties in the acquisition of knowledge and expression, and with a certain degree of subjectivity, intelligence and robustness poorer. View of the complexity of the excavation object, the medical image data mining is no longer the simple application and extension of traditional data mining ideas, but a combination of methods of data mining and medical knowledge to study.

3.1   Segmentation of Medical Image Region of Interest

It is found in the same parts of the medical image, the image most of the information contained in the specific area, we call the region of interest (Region of Interest, ROI). ROI contains the main information of medical images which play an important role in the study of medical images, so the segmentation of medical images, we extract only the ROIs.

We use self-Adaptive Water Immersion algorithm [7] to segment images into ROI and scan image to find gray-scale maximum and minimum values, then calculate the average value, based on which to find the ROI pixels, the region outside the ROI pixel assignment 255. Specific is shown in Fig.2.

3.2   Feature Extraction of ROI

We need to extract part of the characteristics to represent each ROI. Domain knowledge of the human brain images shows that the brain is symmetrical, that is the density distributions of left brain and right brain are the same. The areas of the lesion are represented as the irregular gray level distribution in CT images, which may destroy the symmetry. Because of that, we have adopted the following characteristics to properly show ROI:

1)Gray level (Gr): ROIs extracted from CT images of the human brain using the grayscale threshold are either darker areas or brighter areas. So, we define the gray level (GL), GL $= 0$(brighter), and GL$=1$(darker);

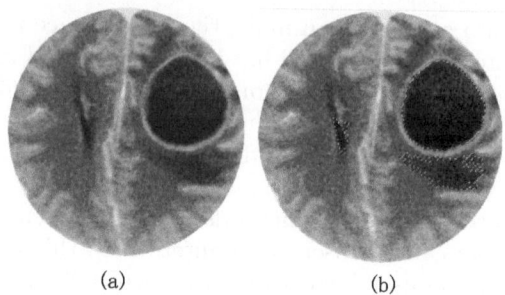

(a)   (b)

Fig. 2. Medical image segmentation: (a) initial image, (b) it is the key pixel region in the dotted lines

2)Area (Ar): The ROI is calculated by the number of pixels in the region, and it is a relative ratio for standardization that is the ratio of the number of pixels contained in the region with the image contains the number of pixels;

3)Location (L): It is also defined as a relative ratio, which is the ratio of the absolute horizontal and vertical coordinates of the ROI's centroid with vertical and horizontal coordinates of the image;

4)Circle-Like (CL): It is the value of a [0, 1], as specified in Equation (1) shown in the where ROI. Area is the ROI area that the number of pixels within the ROI, ROI.Perimeter is the ROI perimeter, namely the number of ROI boundary pixels;

$$CL = 4\pi \times ROI.Area \div ROI.Perimerter^2 \tag{1}$$

5)Direction (Di): It is defined as angle between the ROI long axis and the x-axis, but also to standardize [0,1].

Table 1 shows the image formula after feature extraction: part1 is the image number, part2 ROI label, part3 is the ROI characteristics.

Table 1. The image expression after feature extraction

Part1	Part2	Part3
ImageID	ROIID	Gr Ar L Cl Di
Im1	R11	... ...
Im1	R12	... ...
Im1	R13	... ...
Im2	R21	... ...
Im2	R22	... ...
...	...	... ...
Imn	Rnm	... ...

4   ROI Cluster Algorithm

After preprocess, the medical images are concluded with many areas which make the searched volume expand by many times and the search speed slow. The studies in [8] have shown that after classification we query image in some classes and does not affect the accuracy of the query but the query would be very efficient.

In contrast, Genetic algorithm is known for its robustness and ability to approximate global optimum. In this study, we adapted it to suit our needs of clustering image regions. The genetic algorithm was first proposed by John Holland in 1970s [9] , and the flow chart of GA is shown in Fig.3. The goal of clustering is to minimize Equation (2).

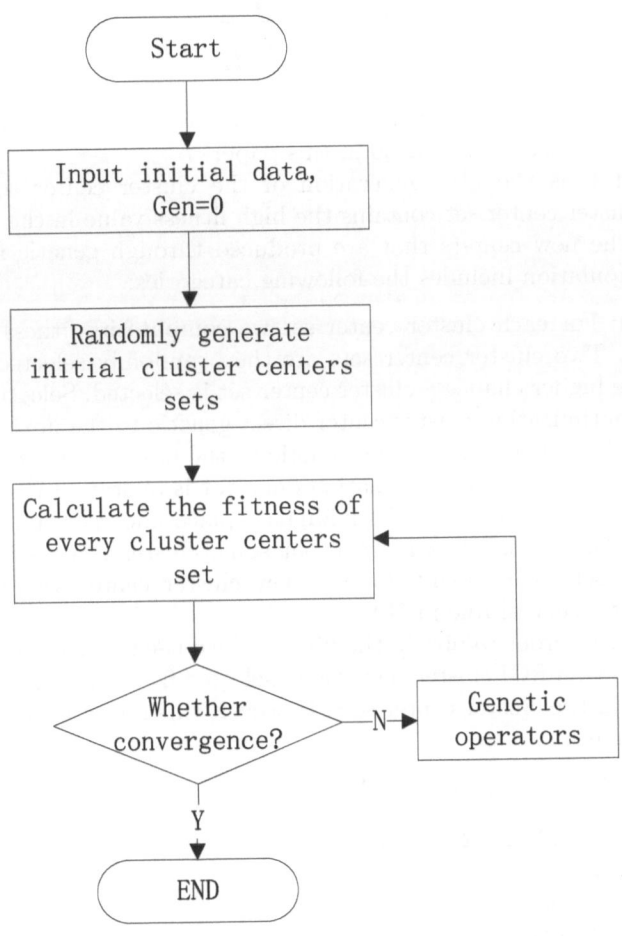

Fig. 3. The GA flow chart

$$F(R) = \sum_{j=0}^{k} \sum_{i=0}^{n} d(p_i, rep[p_i, R_j]) \tag{2}$$

pi is a ROI in the cluster R_j which is represented by a representative image region $rep[p_i, R_j]$. n is the total number of ROI and k is the number of clusters. The value of k is determined experimentally as there is no prior knowledge about how many clusters are there. A too large k value would result in over-clustering and increase the number of false negatives, while a too small k value would not help much in reducing the search space.

GA on ROI clustering algorithm pseudo code is shown in Algorithm 1.

The fitness Fit (i) mentioned in Algorithm 4.1 is calculated as Equation (3).

$$Fit(i) = \frac{f_i}{\sum_{i=1}^{l} f_i} \tag{3}$$

f_i is calculate as the inverse values of the objective function. l is the number of the cluster center sets, which is 50 in this paper.

The initial P is the first generation of the cluster center set. The next-generation cluster center set contains the high fitness value in the previous generation and the new centers that are produced through genetic manipulation. Genetic manipulation includes the following categories:

1)Selection: For each cluster center set we compute its fitness according to Equation 4.1. Two cluster center sets are then randomly selected. The higher the fitness the higher chance a cluster center set is selected. Selection purpose is to make the optimization cluster center direct genetic to the next generation or new individuals by match and cross genetic to the next generation.

2)Recombination: The recombination operator is operating that the parts of the structure in the two parent individuals replace and reorganize to generate new individuals. In other words, take out some cluster centers from two ROI cluster center sets to exchange for the new cluster center set. The crossover operator plays a central role in GA.

3)Mutation: In order to obtain the diversity, we need newly-born cluster center, that is a part of ROI cluster center is re-selected from the remain ROIs which is not appear in the cluster center sets. However, this mutation is operated at a very low frequency.

Algorithm 1. Genetic Algorithm(GA)

```
Input: k: the number of clusters
        R:ROI sets
Threshold: threshold of fitness
Output: k clusters
BEGIN
(1)Initialization P: random generate some cluster center sets
   from R;
```

```
(2)calculate the fitness Fit(i) of every cluster center set in
   P;
(3)do{
(4)The cluster centers with high fitness selected from group P
   and its later generations produced by genetic operators com-
   posite new groups of P';
(5) P'instead of P;
(6)calculate the fitness Fit(i) of every cluster center set in
   P;
(7)} While(Fit(i)<threshold)
(8)choose the cluster centers with best fitness;
(9)Each ROI is assigned to closest cluster;
END
```

4.1   User Interest Analysis

In general, it is the part of the diseased area in the image, such as brain tumors, rather than the whole image in the medical images (such as brain CT) retrieval that user more interested in. In this paper, we assume that points of interest of users search in a ROI in the same parts of the medical image, such as brain tumors, and are expected to search out the medical images that contain similar semantic region ROI.

In the initial search process, user submits a medical image as a query image, but users' interests in this image are not known. The distance between the query image and the image in image database is expressed by the minimum distance be-tween ROIs in the two images. The distances between these images are stored in ascending order, and the top 50 images are returned to the users.

Users identify a returned image as positive if it is what they want; otherwise users label it negative. With this information at hand, we estimate the users interests which specific ROI of the query image users interested in. In this paper, the Diverse Density algorithm (DD)[10] is used to achieve this goal.

DD is firstly proposed by Maron et al (Multiple Instance Learning MIL) in the multi-instance learning framework. In DD, every bag (image) is considered as a set of instances (ROI) and the task of DD is to find the instance with the greatest diversity density within the instance space. The diversity density is a measure refers to the more positive examples around at that point, and the less negative examples, the greater the diversity density of this point.

With Diverse Density approach, an objective function called DD function is de-fined to measure the co-occurrence of similar ROI from different images. The target of DD is to find a ROI which is closest to all the positive images and farthest from all the negative images.

We denote the positive bags as $B_1^+, B_2^+,, B_n^+$ and negative bags as $B_1^-, B_2^-,,$ B_m^-. The jth instance of bag B_i is represented as B_{ij}, and every instance is represented by a k dimensional vector where k is a constant.

For any ROI $r = \{r_1, r_2, , r_k\}$, DD is defined by the probability of it being our target point, given all the positive and negative bags. So the point we are looking for is the one that maximize the probability below.

$$argmaxPr\left(r|B_1^+, B_2^+, ..., B_n^+, B_1^-, B_2^-, ..., B_m^-\right) . \tag{4}$$

Assuming a uniform prior over the concept location $Pr(r)$ and conditional independence of the bags given the target concept p, the above formula equals to

$$argmax\prod_{i=1}^{n} Pr\left(r|B_i^+\right) \prod_{j=1}^{m} Pr\left(r|B_j^-\right) . \tag{5}$$

The probability that ROI r appear in the image B_{ij} is calculated by the Equation (6).

$$Pr\left(B_{ij} = r\right) = exp\left(-\sum_k \left(\boldsymbol{B_{ijk}} - \boldsymbol{r_k}\right)^2\right) . \tag{6}$$

The goal is to find such a ROI r which can make the above function reach its maximum. We combine Expectation-Maximization (EM) algorithm as proposed by Zhang et al. [11] and DD above mentioned. The detail is shown in Algorithm 2.

Algorithm 2. EM-DD Algorithm

```
Input:positive bags:B1+,B2+,,Bn+,
      negative bags:B1-,B2-,,Bm-,
Output:ROI which user is interested in
BEGIN
(1)select a ROI in the positive images, r={r1,r2,,rk};
(2)calculate the probability that r in positive and
   negative;
(3)If the probability is largest then quit, otherwise
   the Execution (4);
(4)Maximize the probability function to find a r';
(5)r' instead of r, repeat (2) - (5);
END
```

We then use the final result r, the point of the users' interests, to find the cluster this ROI r belongs to. Hence all the other ROIs in this cluster can be located. However, we cannot simply reduce the search space to this cluster alone because it is very common that a particular region is closer to some region in another cluster than some regions within the same cluster. Therefore, in our system, we choose three clusters whose centroids are the closest to the query region. In addition, as an image is composed of several ROI, but the ROIs may fall into different clusters. We then group all the images that have at least one ROI fall into the three clusters mentioned above, and take these images as the reduced search space.

4.2   Analysis of Experimental Results

The medical images we used in this research are all real brain CT image from hospitals. Brain image position is very important in medical image. Brain tissue is the senior nerve center of the human body, whose function is particularly important. Diseases that occur in the brain have always attached great importance to by the medical profession. So we use brain CT.

According to the field of medicine, we know that the role of small size of the ROI region is not large, so we went in addition to ROI that area is less than the threshold, as shown in Fig.4.

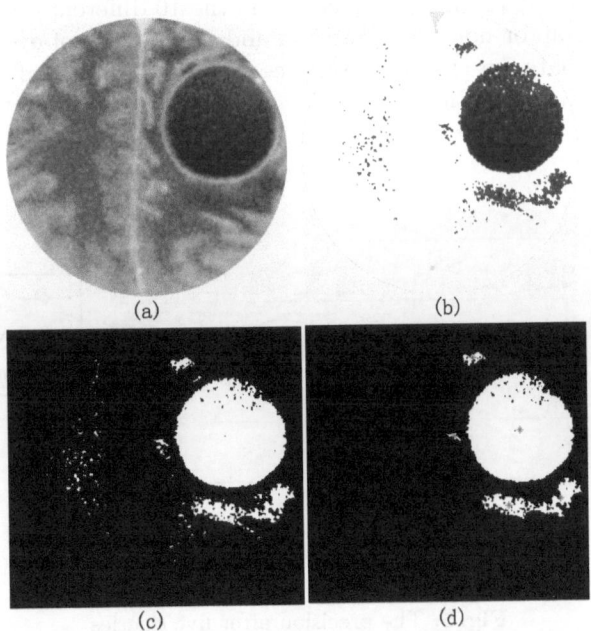

Fig. 4. The extraction process of medical image ROI:(a) initial image, (b) ROI extracted, (c) image anti-significant, (d) remove the small size ROI and the red * mark is the center of each ROI.

In this experimental, we use Euclidean distance to calculate the similarity be-tween images. After user submits a query image, we return an initial query results to users, and then according to the hypothesis generated by EM-DD, we pull out the three closest clusters as the reduced search space.

The key in GA clustering is to find the balance between reduction of search space and accuracy. In MIRRF, every image is segmented into 5 9 ROIs, so the 1000 original image in the search space into the 7136 regions. In the experiments, we tested the system performance under different clustering schemes by dividing the entire set of ROIs into 50 to 100 clusters. Each time we increase the number of clusters by 5 and find that when the number of clusters k=75, the above balance is reasonable.

In addition, the average accuracy has been more than 75% after user mark four times. Therefore, taking into account the efficiency of user retrieval, each retrieval process contains four rounds of relevance feedback. It is worth mentioned that the number of positive images increases steadily through each round.

We compare the performance of MIRRF proposed system with the ones that are showed in two other relevance feedback algorithms: Neural Network based Multiple Instance Learning (NNMIL) algorithm with relevance feedback and Query Vector Modification (QVM).

In this experiment, we conducted five groups experiments, and each group experiment included 10 different queries and the 10 different queries from every group may overlap or not. The precision and recall radio of every group which are used to evaluate our system are expressed by the average of each 10 queries, as is shown in Fig.5and Fig.6.

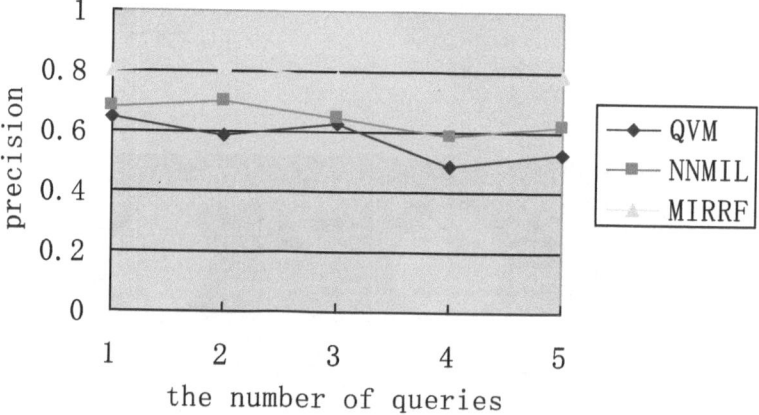

Fig. 5. The precision after five queries

It can be gleaned from Fig.5 and Fig.6 that when the number of images is small, the precision and recall radio of three systems have no significant differences and all close to 1. But with the number of images increasing, the precision and recall radio of our system is 15% 20% higher than the other two systems. It also can be seen that the stability of our system is better than the other two systems.

The Fig.7 shows the consuming time of three systems after five queries. It can be seen, for a database with 1000 brain CT images, the consuming time of MIRRF is about 22.17 seconds less than QVM and 40.18 seconds than NNMIL. In addition, the most time-consuming stage in each query is the initial query without feedback, about 12s in MIRRF and 8s in QVM and NNMIL. The above time is all exclusive of time that users mark positive examples and negative examples.

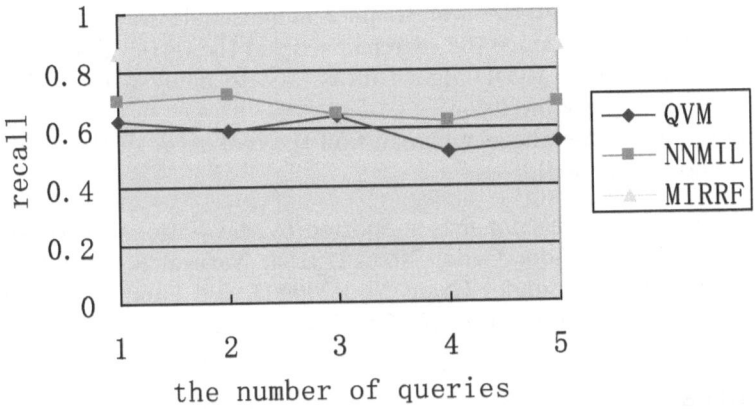

Fig. 6. The recall radio after five queries

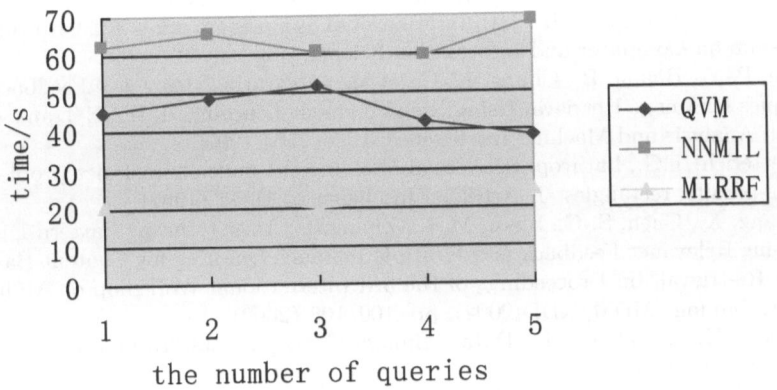

Fig. 7. The consuming time of three systems

Clearly, that we cluster ROIs with GA algorithm to reduce the search space can effectively reduce the search time and does not make the precision and recall radio lower. Besides, using the EM-DD algorithm to analyze the users' query intent can significantly improve system performance.

4.3 Conclusions

In this paper, we propose a medical images retrieval based user feedback system. This system mainly solves the semantic gap between low level feature and high level concept. And in image segmentation, our segmentation algorithm which applies the medical domain knowledge is better in ROI feature extraction than other algorithms in medical images. After segmentation an image is expressed by several regions so that the search space increases. However, we apply GA algorithm clustering to reduce the search space and the retrieval accuracy is not lower. In addition, the EM-DD algorithm is used to analysis users' interest to capture the user's query intention.

The above shows that the system can find the real needs of the user and can remove the noise very well.

Acknowledgments. The paper is supported by the National Natural Science Foundation of China under Grant No.61272184, Natural Science Foundation of Heilongjiang Province under Grant No.F200903, and Fundamental Research Funds for the Central Universities No.GK2060260118.

References

1. Wang, X., Erdelez, S.: Medical Image Describing Behavior: A comparison between an expert and novice. In: Proceedings of the 2011 iConference, pp. 792–793. ACM, New York (2011)
2. Guldogan, E., Gabbouj, M.: Dynamic weights with relevance feedback in content-based im-age retrieval. In: 24th International Symposium on: IEEE 24th Int'l Symposium on Computer and Information Sciences, pp. 56–59 (2009)
3. Yin, P.-Y., Bhanu, B., Chang, K.-C., et al.: Integrating Relevance Feedback Techniques for Image Retrieval Using Reinforcement Learning. J. IEEE Trans. on Pattern Analysis and Machine Intelligence, 1536–1551 (2005)
4. Dietterich, T.G., Lathrop, R.H., et al.: Solving the multiple-instance problem with axis-parallel rectangles. J. Artificial Intelligence, 31–71 (1997)
5. Huang, X., Chen, S.-C., Shyu, M.-L., Zhang, C.: User Concept Pattern Discovery Us-ing Relevance Feedback and Multiple Instance Learning for Content-Based Image Re-trieval. In: Proceedings of the 3rd International Workshop on Multimedia Data Mining (MDM/KDD 2002), pp. 100–108 (2002)
6. Jiawei, H., Kamber, M.: Data Mining: Concepts and Techniques, 2nd edn., pp. 402–407 (2008)
7. Pan, H., Li, J., Zhang, W.: Incorporating domain knowledge into medical image clustering. Applied Mathematics and Computation 185(2), 844–856 (2007)
8. Kalpathy-Cramer, J., Hersh, W.: Multimodal Medical Image Retrieval Image Categorization to Improve Search Precision. In: Proceedings of the International Conference on Multimedia Information Retrieval, MIR 2010, pp. 165–173. ACM, New York (2010)
9. Holland, J.H.: Adaptation in Natural and Artificial Systems. University of Michigan Press, Michigan (1975)
10. Maron, O., Lozano-Perez, T.: A Framework for Multiple Instance Learning. In: Advances in Netural Information Processing System 10. MIT Press, Cambridge (1998)
11. Zhang, Q., Goldman, S.A.: EM-DD: An Improved Multiple-Instance Learning Technique. In: Advances in Neural Information Processing Systems, NIPS, Denver (2002)
12. May, P., Ehrlich, H.-C., Steinke, T.: ZIB Structure Prediction Pipeline: Composing a Complex Biological Workflow Through Web Services. In: Nagel, W.E., Walter, W.V., Lehner, W. (eds.) Euro-Par 2006. LNCS, vol. 4128, pp. 1148–1158. Springer, Heidelberg (2006)

Personalized Diversity Search Based on User's Social Relationships

Ming Li[1,2], Juanzi Li[1], Lei Hou[1], and Hai-Tao Zheng[2]

[1] Department of Computer Science and Technology
Tsinghua University, Beijing 100084, P.R. China
[2] Tsinghua-Southampton Web Science Laboratory at Shenzhen
Graduate School at Shenzhen, Tsinghua University, Shenzhen 518055, P.R. China
liming09@mails.sz.tsinghua.edu.cn,
{ljz,houlei}@keg.cs.tsinghua.edu.cn,
zheng.haitao@sz.tsinghua.edu.cn

Abstract. Keyword-based web search is nowadays the most popular and convenient means for people to access information, whereas traditional search methods usually disappoint people with inaccurate, insufficient or redundant results, because those methods often fail to understand user's search intents and interest preference. Diversity search is an effective approach to present various kinds of results so that average people may satisfy with as least one result, however, most existing diversity search methods are uniformly applied to all users and queries, the returned results generally reflect the masses' needs, individual users' requirements are not fully considered. To deal with this issue, we present a systematic method named personalized diversity search based on user's social relationships (PDSSR), this method is a combination of personalization and diversification, which enables computer better understand user's search intents and interests, consequently returns a personalized and reduced diversified result set. Besides, we introduce social relationships into the personalization, which helps to avoid "cold start" and "data sparsity" problems. Empirical experiments conducted show that the proposed method outperforms the baseline in terms of nDCG, which proves the effectiveness of our method.

Keywords: personalized search, diversity search, social relationships.

1 Introduction

Web search has become a predominant and convenient way for people to access information, whereas with the issue of information explosion and ineffective search methods, user's fulfillment is usually not assured because of the inaccurate, insufficient or redundant results. Some personalized or diversity search methods have been proposed to deal with these issues[1] [2], however, these methods are uniformly applied to all users and queries, the returned results generally reflect the masses' needs, which sometimes are different from individual users' real needs. Considering two scenarios:

S. Zhou, S. Zhang, and G. Karypis (Eds.): ADMA 2012, LNAI 7713, pp. 663–674, 2012.

(1) the research areas of data mining are composed of classification, clustering, association rule, regression etc. Suppose most research work focuses on the first two areas, then when one researcher wants to review the literature of data mining by providing keywords "data mining", the works of the first two areas are returned preferentially and may be predominant according to the probability. However, if we know that some close people of this researcher (such as coauthor, advisor, student) have most of their work on "association rule" and "regression", then should the original result set still be returned then?

(2) when talk about "clothing fashions", people may think about the brand, price, designer, fashion show, styles, fashion element etc. General people may concern style, brand or price much more, and such information would be returned predominantly with keywords "clothing fashions", while if we know the searcher have several friends working in fashion, then could we assume that he/she may also concern or concern more about the fashion show, fashion element etc.?

From the above scenarios, we would find there exist three problems:

First, users have various information needs, some of them have interests that are different from the masses', thus they may be disappointed with search results provided by general search engines which aim to meet the masses' needs.

Second, sometimes search results are redundant, certain kinds of results which are similar to each other may dominate the returned result set when this kind of information overwhelms other kinds, accordingly if one result fails to satisfy the user, so do the others.

Last, the search intents are usually ambiguous, it is quite difficult for a search engine to understand the exact user search intent just based on a few keywords, extra knowledge should be introduced to help search.

Personalized search is adopted as a solution to address the first problem, it can incorporate information about the individual users so that provide specific results to them. Diversity search is an effective way to solve the second problem, since the general target of diversity search is to ensure the result be both relevant to the query (similarity or relevance) and contains minimal similarity to previous selected results (novelty or diversity). Both diversity search and personalized search could help to deal with the search intents problem. If we can combine the personalized search and diversity search together, not only user's search intents could be clarified to a great extent (diversity search provides as many probable results as possible, the personalization further narrows down the results), but also the various needs of different users and result redundancy could be solved. However, personalization means enough information should be collected to build effective user model, and diversification means the categories of results should be recognized correctly. And to our knowledge there are few studies[3] [4] combining them together. In this paper we propose a personalized diversity search approach based user's social relationships – PDSSR. The contributions are summarized as follows:

(1) we introduce personalization into diversity search on which most of the current methods are uniformly applied to all queries and users, the results of those methods reflect the masses' needs, while the proposed method provides diversified results specific to individual users, it is especially useful to satisfy users who have different interests with the masses.

(2) we assign both queries and documents to several topics, which helps computer cover multiple search intents. In the process of personalization, we exploit user's social relationships to build a comprehensive user model, which overcomes the "cold start" and "data sparsity" problems.

(3) we build a data set to validate PDSSR, and empirical experiments based on this data set proves that the proposed method improve user satisfaction.

The rest of this paper is organized as follows: in Section 2 we discuss the related work of diversity search, personalized search, personalized diversity search and their measurements, Section 3 proposes a user model building method, Section 4 elaborates the search method and algorithm, Section 5 presents the empirical experiment and shows the comparison of the results with a baseline, the last section gives the conclusion and the future work.

2 Related Work

2.1 Diversity Search

J. Carbonell and J. Goldstein regarded "relevant novelty" as the potentially superior criterion in search, measuring relevant novelty is to measure relevance and novelty independently and provide a linear combination as the metric, they called the linear combination "marginal relevance", the target of diversity search to maximize marginal relevance, they proposed a method – Maximal Marginal Relevance (MMR), which uses two similarity functions to balance the similarity and diversity, one measures the similarity between result to query, the other measures the similarity among results[5]. B. Smyth and P. McClave argued the importance of retrieved case's diversity in case-based recommender system, while increasing diversity of recommendations generally means reducing their similarity to the target query, so they proposed a number of techniques for improving the diversity with considering the trade-off between similarity and diversity [1]. The diversity strategy in this work is adopted in [6] [7].

To cover the multiple potential search intents is a major step to diversify the search result, Zhu et al. proposed a model based on the intrinsic query manifold, this model makes full use of the relationships among queries to find relevant and salient queries [8]. While topics were applied on search intent in the retrieval process [9] [10].

2.2 Personalized Search

User profile modeling is an important component of any personalization system. The information collection for user profile construction could be explicit or implicit. Due to the flaws of the explicit way, such as the changing of user interests, incorrect information provided by user or unwilling to feedback, many efforts are put on modeling the user profile implicitly[11]. Such implicit modeling methods can be found in [12].

Z. Dou et al. described that the profile modeling can be based on content analysis and user group [13]. In the content analysis approach, the contents may be composed of user's click history, tagging etc. User interests are represented by topical categories [14] [15] or lists of keywords [16]. In user group approach, user profile is created by incorporating information provided through user's social network [17].

2.3   Personalized Diversity Search

To our knowledge, there are few studies bringing personalization into diversity search. F. Radlinski and S. Dumais presented a strategy for diversify the personalized reranking of search results by means of query-query reformulations [3]. R. Agrawal et al. classified both the queries and documents into categories, with this information they developed an objective function to maximized the probability that the average user satisfies with at least one desired result in the top k results. For that objective function was proved to be NP-hard, the authors defined a Greedy Algorithm (IA-SELECT) for computing a solution to the objective function [18]. Based on R. Agrawal's work P. Lubell-Doughtie employed search logs to build the user model by extracting latent topic distribution from clicked documents and summing over these latent topic distributions. And then the author introduced the user model into the objective function of diversity [4]. R.L. Santos et al. presented an intent-aware approach for search result diversification [19]. However these studies did not consider the social network in the user model.

3   User Profile Modeling

User profiles convey approximations of user interests, it can be built with the information that provided by user, which is the explicit way; or it can be built by collecting relevant information, e.g. behavior logs from user or user group, and this is the implicit way. In PDSSR, we exploit LDA [20] to generate a list of topics from relevant documents from user u and u's friends, then the topics are used to represent the interests of u, and preference for topics indicate the degree of u's interests. For example, if u represents a researcher, the documents that we collected to train model are the papers, the topics reflects his/her research interests. Therefore the user model could be defined as an interest preference vector which is shown as follows:

$$\boldsymbol{u} = (t_1, t_2, ..., t_n) = \frac{\sum_{m \ documents \ of \ u}(\theta_{m1}, \theta_{m2}, ..., \theta_{mn})}{m} \tag{1}$$

where θ is the topic distribution over document, and t_i is weight assigned to topic i.

Inspirited by the idea of social network types and social relationships in [21] and [22], we present an $infl(u_1, u_2)$ function to measure the social influence that u_2 exerts on u_1:

$$infl(u_1, u_2) = (1 + sim(u_1, u_2))(1 + fam(u_1, u_2)) \qquad (2)$$

$infl(u_1, u_2)$ reflects the information of similarity between users' interests as well as familiarity between users.

$sim(u_1, u_2)$ is the *similarity* function measured by cosine distance between user interest vector.

$$sim(u_1, u_2) = \cos \frac{\boldsymbol{u_1} \cdot \boldsymbol{u_2}}{||\boldsymbol{u_1}|| \, ||\boldsymbol{u_2}||} \qquad (3)$$

This function is symmetric, i.e. $sim(u_1, u_2) = sim(u_2, u_1)$, it describes the relationship of interests, it provides a quantitative measurement on how related or how similar of interests between two users, however, it can not describe the relationship of users themselves, two users with quite similar interests may not know each other. Therefore, we take advantage of another measurement $fam(u_1, u_2)$ to indicate user's *familiarity* in their social network. The computation of $fam(u_1, u_2)$ depends on the data set, for social relationships can be expressed and measured in various ways. For example, assume u_1 and u_2 are two researchers, then we define the *familiarity* of u_2 to u_1 as the number of papers that u_1 and u_2 co-authored dividing the number of all u_1's papers:

$$fam(u_1, u_2) = \frac{|papers\ co\text{-}authored\ by\ u_1\ and\ u_2|}{|papers\ of\ u_1|} \qquad (4)$$

Note that, in this definition, $fam(u_1, u_2)$ is asymmetric.

After getting the user model and the social influence between users, we could apply them to search process, which enables computer find results that involve user preference.

4   Methodology

In this section, we present the formulation of PDSSR's objective and give the detailed algorithm.

4.1   Objective Function

IA-SELECT [18] is mentioned as the one of state-of-the-art approaches to diversification by some researchers [19] [23]. In this work we take it as the starting point, IA-SELECT assumed the queries and documents belong to more than one categories according to existing taxonomy information, the objective of result diversification is stated to maximize the probability that average user finds at least one useful result within the top k results, given that user only considers the top k returned results. The objective are formalized as follows:

DIVERSIFY(K): Given a query q, a set of documents D, a probability distribution of categories for the query $P(c|q)$, the quality values of documents $V(d|q, c)$, and an integer k, find a set of documents $S \subseteq D$ with $|S| = k$ that maximizes

$$P(S|q) = \sum_c P(c|q)(1 - \prod_{d \in S}(1 - V(d|q, c)))$$ (5)

The authors proved DIVERSIFY(K) is NP-hard and presented a greedy algorithm for it, i.e. IA-SELECT.

One main drawback of this method is that it is uniformly applied to all users, any user will get the same result set S with the same order, individual's interest preference is not taken into account. Even though average user is able to find as least one useful result in S, the user would get unequal degree of satisfaction with this result ranked the first and ranked the last.

To address this problem, P. Lubell-Doughtie introduced user's preference vector t_u into Eq. (5) [4]:

$$P(S|q, t_u) = \sum_i P(t_i|q, t_{u,i})(1 - \prod_{d \in S}(1 - V(d|q, t_i, t_{u,i})))$$ (6)

With Eq. (6), user's interest would be reflected in the result. Whereas, the information used for user modeling in that work are the search logs issued by the user, accompanying which is the "cold start" problem (no search log) or "data sparsity" problem (few search logs), the user model may be inaccurate or incorrect.

To settle these problems, we further improve the DIVERSIFY(K) by introducing social network in user profile modeling, not only the information owned by user u but also the information owned by other users who have social relationships with u would be exploited to extract the user model. We restate the DIVERSIFY(K) as follows:

Given a query q issued by user u, a set of documents D, a probability distribution of topics $P(t_i|q, u_j)$ for the query and all users related to u , the quality values of documents under certain query, topic and user $V(d|q, t_i, u_j)$, the social influence $infl(u, u_j)$ of u_j to u, and an integer k, find a set of documents $S \subseteq D$ with $|S| = k$ that maximizes:

$$P(S|q, u) = \sum_i \sum_j P(t_i|q, u_j)(1 - \prod_{d \in S}(1 - V(d|q, t_i, u_j)))infl(u, u_j)$$ (7)

4.2   Parameter Estimation

In Eq. (7) there exist three parameters: $P(t_i|q, u_j)$, $V(d|q, t_i, u_j)$ and $infl(u, u_j)$. We have defined $infl(u, u_j)$ in Section 3, in the following we give the estimation of the other two parameters.

We know a query q consists of several terms $w_1, w_2, ..., w_x$, search engines generally assume these terms are connected with OR relationship, i.e. the returned result does not necessarily match all the terms, but the more terms it matches, the more satisfying it usually is. Made this clear, we can get the probability of a given query q belonging to given topics in an matched document d (d contains at least one of the terms $w_1, w_2, ..., w_x$):

$$P(t_i|q) = P(t_i|w_1, w_2, ..., w_x) = \frac{P(t_i)P(w_1, w_2, ..., w_x|t_i)}{P(w_1, w_2, ..., w_x)} \tag{8}$$

In the rightmost of Eq. (8), $P(w_1, w_2, ..., w_x)$ is independent of t_i, and all the topics t_1, t_2, ... $t_i...t_n$ have equal probabilities, i.e. $P(t_1) = P(t_2) = ...P(t_i) = ...P(t_n)$, so the equation could be deduced as:

$$P(t_i|q) \propto P(w_1, w_2, ..., w_x|t_i) \tag{9}$$

Considering LDA is based on bag-of-words assumption, and as we have mentioned above, the keywords $w_1, w_2, ..., w_x$ have the OR relationships, we get:

$$\begin{aligned} P(w_1, w_2, ..., w_x|t_i) \\ = P(w_1|t_i) + P(w_2|t_i) + ... + P(w_x|t_i) \\ = \phi_{i1} + \phi_{i2} + ... + \phi_{ix} \end{aligned} \tag{10}$$

i.e.

$$P(t_i|q) \propto \phi_{i1} + \phi_{i2} + ... + \phi_{ix} \tag{11}$$

where ϕ denotes the term distribution over topics, it is a parameter defined in LDA.

We further define the probability of a given document d belonging to given topics when the user is u_j:

$$P(t_i|q, u_j) = t_{u_j,i}P(t_i|q) \tag{12}$$

where $t_{u_j,i}$ is weight of topic i for user u_j, it is calculated by Eq. (1).

The quality values of documents under certain query, topic and user $V(d|q, t_i, u_j)$ is defined as follows:

$$V(d|q, t_i, u_j) = t_{u_j,i}V(d|q, t_i) \tag{13}$$

where $V(d|q, t_i)$ denotes the quality of documents given a query q and a topic i, it is calculated with the method proposed in [4]:

$$V(d|q, t_i) = \lambda t_{d,i} + (1 - \lambda)t_{d,i}^{1+\log(rank(d))} = \lambda\theta_{di} + (1 - \lambda)\theta_{di}^{1+\log(rank(d))} \tag{14}$$

In this equation, $t_{d,i}$ is the probability of topic i for document d, in fact it is θ, the topic distribution over document in LDA. $rank(d)$ denotes the original rank of document d. λ is used to control the amount of diversity with $\lambda = 1$ supplying the maximum diversity and $\lambda = 0$ the minimum.

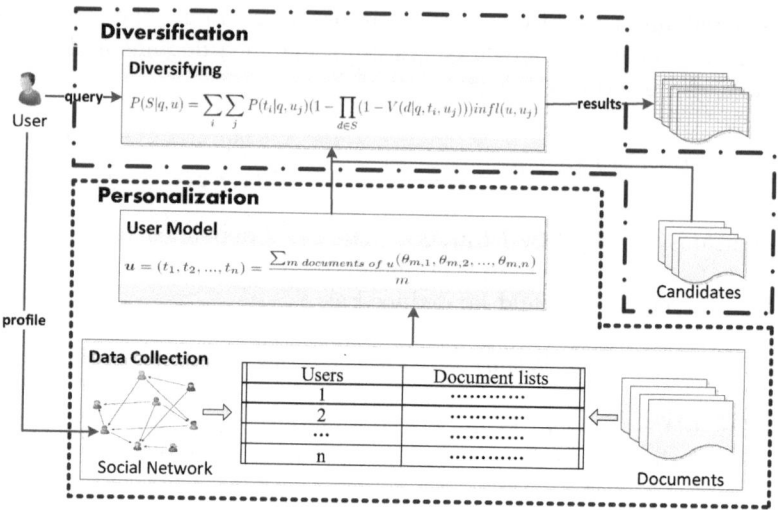

Fig. 1. Framework of PDSSR

4.3 *SIA-SELECT* Algorithm

With the user model and diversification objective function, the framework of proposed method PDSSR is shown in Fig. 1, from which the two procedures of the method are clearly illustrated. Namely, the personalization procedure based on social information, and the diversification procedure.

According to the knowledge mentioned above, we revise the IA-SELECT algorithm and named it as *SIA-SELECT*, the main changes are the definitions of $U(c|q, S)$ and $g(d|q, c, S)$.

Originally, $U(c|q, S)$ denotes the conditional probability that query q belongs to category c, given that all documents in set S fail to satisfy the user, at the beginning, before any document is selected, $U(c|q, \emptyset) = P(c|q)$. In *SIA-SELECT*, this parameter has been changed to $U(t_i|q, u_j, S)$, which adds the impact of u_j. Initially, when $S = \emptyset$, $U(t_i|q, u_j, S) = P(t_i|q, u_j)$.

In IA-SELECT, a value $g(d|q, c, S)$ is defined as the *highest marginal utility*, the marginal utility is interpreted as the probability that selected documents satisfies the user given that all documents that come before it fails to do so. $g(d|q, c, S)$ is computed as the product of $U(c|q, S)$ and $V(d|q, c)$ in IA-SELECT, while in *SIA-SELECT*, $g(d|q, c, S)$ is redefined as $g(d|q, t_i, S)$, this value is computed as the product of $U(t_i|q, u_j, S)$, $V(d|q, t_i, u_j)$ and $infl(u, u_j)$.

The *SIA-SELECT* algorithm is described in Algorithm 1: (1) give the document set size k, query q, result set of documents $R(q)$ when issued q, topics contained in document $T(d)$, topic distribution $P(t_i|q, u_j)$ and quality of document $V(d|q, t_i, u_j)$; (2) initialize the $U(t_i|q, u_j, S)$ when $S = \emptyset$; (3) compute the $g(d|q, t_i, S)$ of each document d in $R(q)$, and find the document d^* with the highest $g(d|q, c, S)$; (4) add d^* in S, and delete it from $R(q)$; (5) update $U(t_i|q, u_j, S)$; (6) stop while the size of S equals k and returned S.

Algorithm 1. SIA-SELECT

---

1 **INPUT**: $k, q, R(q), T(d), P(t_i|q, u_j), V(d|q, t_i, u_j)$
2 $S = \emptyset$
3 $\forall j, i, U(t_i|q, u_j, S) = P(t_i|q, u_j)$
4 **while** $|S| < k$ **do**
5   **for** $d \in R(q)$ **do**
6    $g(d|q, t_i, S) \leftarrow \sum_j \sum_i U(t_i|q, u_j, S)V(d|q, t_i, u_j)infl(u, u_j)$
7   **end for**
8   $d^* \leftarrow argmax\ g(d|q, t_i, S)$[ties broken arbitrarily]
9   $S \leftarrow S \cup d^*$
10   $R(q) = R(q) \setminus \{d^*\}$
11   $\forall t_i \in T(d^*), U(t_i|q, u_j, S) = (1 - V(d^*|q, t_i, u_j))U(t_i|q, u_j, S \setminus \{d^*\})$
12 **end while**
13 **OUTPUT**: S

---

5 Evaluation

5.1 Experiment Setup

In order to evaluate the effectiveness of PDSSR, we exploit the data crawling from Arnetminer [1] to carry out the experiments. We utilize the paper titles and abstracts of the given researchers and his/her coauthors to build user model. Then we predefine several queries and issued them on Arnetminer, the returned documents of each query compose its result set $R(q)$. Next we invite the researcher who is the object user of our model to give the rank of documents in each result set, and the rank is regarded as the ideal order, which is used to measure the performance of our method.

In the following, we show the experiment process and result of the researcher with id "83180" and "1516105" in Arnetminer. There are 2299 papers related to "83180", among which 86 are directly related (published by "83180" with her 79 coauthors) and 2213 are indirectly related (other papers published by her coauthors). Similarly, "1516105" have 328 related papers in total, among which five are directly related (with nine coauthors) and 323 papers indirectly related. On the related papers, we utilized LDA with Gibbs Sampling technique to generate topic models with predefined topic number K, the hyperparameters are fixed to the empirical value i.e. $\alpha = \frac{50}{K}$ and $\beta = 0.1$. Then we build the user model with Eq. (1) for "83180" and "1516105" and their coauthors. And next, we predefined 6 queries for "83180" and "1516105" respectively, the queries are relevant to their research interests.

We apply *normalized discounted cumulative gain*(nDCG) to measure the performance of PDSSR, and the search result of the method proposed in [4] is used as the baseline.

---

[1] http://arnetminer.org/

5.2   Experiment Result and Discussion

We conduct the experiments with two steps:

First, we evaluate the impact of λ in Eq. (14) to performance. Fig. 2 shows the the average nDCG of six queries in terms of λ given a topic number. From these figures we find the performance becomes better with λ increasing, however some curves in Fig.2(b) appears not so strict in keeping with this trend, which we think is influenced by the data set size. Anyway when the $\lambda = 0.9$ or $\lambda = 1.0$ the performance is better than that with λ being other values in those curves. Therefore, we fix $\lambda = 1.0$ in the following experiment.

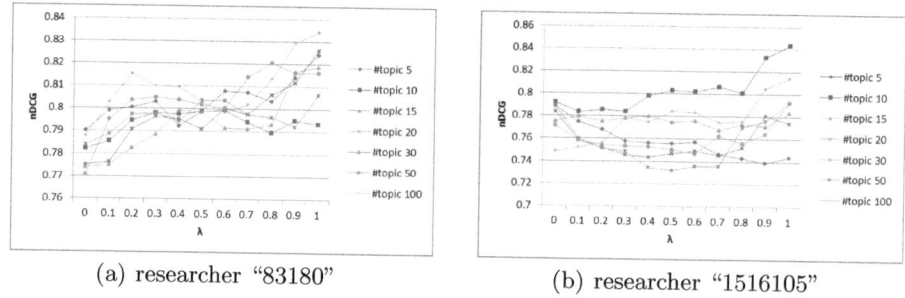

(a) researcher "83180"   (b) researcher "1516105"

Fig. 2. The average nDCG in terms of λ given certain topic number

Table 1. The nDCG improved by PDSSR in comparison with baseline($\lambda = 1.0$)

(1st row is topic number)		05	10	15	20	30	50	100
83180	baseline	0.8105	0.7824	0.8334	0.8165	0.8117	0.8085	0.8052
	PDSSR	0.8242	0.7936	0.8188	0.8260	0.8063	0.8161	0.8343
	improved(%)	1.6925	1.4331	-1.7509	1.1680	-0.6599	0.9430	3.6194
1516105	baseline	0.3797	0.3541	0.3942	0.4173	0.4252	0.4184	0.3587
	PDSSR	0.7443	0.8432	0.7835	0.7741	0.7923	0.7926	0.8142
	improved(%)	95.9966	138.1265	98.7733	85.4976	86.3533	89.4216	126.9928

Next, we continue our experiments with comparing the performance between PDSSR and the baseline. Table 1 shows the results. From the table we could find the improvement by PDSSR is conspicuous to researcher "1516105", while it improves little even performs worse to researcher "83180". This is because that PDSSR builds the user model with all related papers, directly and indirectly, while the baseline only with directly related ones. When the directly related papers are few (data sparsity) or none (cold start), the user model is inaccurate, so the personalized results based on the model are far from perfect. Consequently, the performance of baseline is worse than the PDSSR, which is proved by the dataset of "1516105". However, when directly related paper data set are large enough (as that of "83180"), the performance of PDSSR is not so notable.

6   Conclusion

In this paper, we propose a personalized diversity search method based on user's social relationships – PDSSR, in which we introduce personalization into an state-of-the-art diversity search approach, this step improves the uniformity problem of search results to all users. Besides in the process of personalization, user's social relationships are taken into account to build user model, thus the "cold start" and "data sparsity" problems can be avoided. We conduct empirical experiments to comparing PDSSR and the baseline method, the results show that PDSSR outperforms the baseline method in terms of nDCG, it validates the effectiveness of the proposed method.

In the future we would like research the influence of changing user interests to the search results.

Acknowledgments. The work is supported by the Natural Science Foundation of China (No. 61035004, No. 60973102), 863 High Technology Program (2011AA01A207), European Union 7th framework project FP7-288342, and THU-NUS NExT Co-Lab.

References

1. Smyth, B., McClave, P.: Similarity vs. Diversity. In: Aha, D.W., Watson, I. (eds.) ICCBR 2001. LNCS (LNAI), vol. 2080, pp. 347–361. Springer, Heidelberg (2001)
2. Shen, X., Tan, B., Zhai, C.: Implicit user modeling for personalized search. In: Proceedings of the 14th ACM International Conference on Information and Knowledge Management, CIKM 2005, pp. 824–831. ACM, New York (2005)
3. Radlinski, F., Dumais, S.: Improving personalized web search using result diversification. In: Proceedings of the 29th Annual International ACM SIGIR Conference on Research and Development in Information Retrieval, SIGIR 2006, pp. 691–692. ACM, New York (2006)
4. Lubell-Doughtie, P.: Applying diversity and novelty to personalized search (2010), http://peet.ldee.org/wp-content/uploads/2011/01/ls_project_p_lubell-doughtie.pdf
5. Carbonell, J., Goldstein, J.: The use of mmr, diversity-based reranking for reordering documents and producing summaries. In: Proceedings of the 21st Annual International ACM SIGIR Conference on Research and Development in Information Retrieval, SIGIR 1998, pp. 335–336. ACM, New York (1998)
6. McSherry, D.: Diversity-Conscious Retrieval. In: Craw, S., Preece, A.D. (eds.) ECCBR 2002. LNCS (LNAI), vol. 2416, pp. 27–53. Springer, Heidelberg (2002)
7. Halvey, M., Punitha, P., Hannah, D., Villa, R., Hopfgartner, F., Goyal, A., Jose, J.M.: Diversity, Assortment, Dissimilarity, Variety: A Study of Diversity Measures Using Low Level Features for Video Retrieval. In: Boughanem, M., Berrut, C., Mothe, J., Soule-Dupuy, C. (eds.) ECIR 2009. LNCS, vol. 5478, pp. 126–137. Springer, Heidelberg (2009)
8. Zhu, X., Guo, J., Cheng, X., Du, P., Shen, H.W.: A unified framework for recommending diverse and relevant queries. In: Proceedings of the 20th International Conference on World Wide Web, WWW 2011, pp. 37–46. ACM, New York (2011)

9. Ziegler, C.N., McNee, S.M., Konstan, J.A., Lausen, G.: Improving recommendation lists through topic diversification. In: Proceedings of the 14th International Conference on World Wide Web, WWW 2005, pp. 22–32. ACM, New York (2005)
10. Yin, D., Xue, Z., Qi, X., Davison, B.D.: Diversifying search results with popular subtopics. In: Proceedings of the 18th Text REtrieval Conference, TREC 2009 (2009)
11. Speretta, M., Gauch, S.: Personalized search based on user search histories. In: Proceedings of the 2005 IEEE/WIC/ACM International Conference on Web Intelligence, WI 2005, pp. 622–628. IEEE Computer Society, Washington, DC (2005)
12. Lu, D., Li, Q.: Personalized search on flickr based on searcher's preference prediction. In: Proceedings of the 20th International Conference Companion on World Wide Web, WWW 2011, pp. 81–82. ACM, New York (2011)
13. Dou, Z., Song, R., Wen, J.R., Yuan, X.: Evaluating the effectiveness of personalized web search. IEEE Trans. on Knowl. and Data Eng. 21, 1178–1190 (2009)
14. Pretschner, A., Gauch, S.: Ontology based personalized search. In: Proceedings of the 11th IEEE International Conference on Tools with Artificial Intelligence, ICTAI 1999, pp. 391–398. IEEE Computer Society, Washington, DC (1999)
15. Ma, Z., Pant, G., Sheng, O.R.L.: Interest-based personalized search. ACM Trans. Inf. Syst. 25 (February 2007)
16. Matthijs, N., Radlinski, F.: Personalizing web search using long term browsing history. In: Proceedings of the Fourth ACM International Conference on Web Search and Data Mining, WSDM 2011, pp. 25–34. ACM, New York (2011)
17. Shapira, B., Zabar, B.: Personalized search: Integrating collaboration and social networks. J. Am. Soc. Inf. Sci. Technol. 62, 146–160 (2011)
18. Agrawal, R., Gollapudi, S., Halverson, A., Ieong, S.: Diversifying search results. In: Proceedings of the Second ACM International Conference on Web Search and Data Mining, WSDM 2009, pp. 5–14. ACM, New York (2009)
19. Santos, R.L., Macdonald, C., Ounis, I.: Intent-aware search result diversification. In: Proceedings of the 34th International ACM SIGIR Conference on Research and Development in Information Retrieval, SIGIR 2011, pp. 595–604. ACM, New York (2011)
20. Blei, D.M., Ng, A.Y., Jordan, M.I.: Latent dirichlet allocation. J. Mach. Learn. Res. 3, 993–1022 (2003)
21. Carmel, D., Zwerdling, N., Guy, I., Ofek-Koifman, S., Har'el, N., Ronen, I., Uziel, E., Yogev, S., Chernov, S.: Personalized social search based on the user's social network. In: Proceedings of the 18th ACM Conference on Information and Knowledge Management, CIKM 2009, pp. 1227–1236. ACM, New York (2009)
22. Zanardi, V., Capra, L.: Social ranking: uncovering relevant content using tag-based recommender systems. In: Proceedings of the 2008 ACM Conference on Recommender Systems, RecSys 2008, pp. 51–58. ACM, New York (2008)
23. Capannini, G., Nardini, F.M., Perego, R., Silvestri, F.: Efficient diversification of web search results. Proc. VLDB Endow. 4(7), 451–459 (2011)

Towards a Tricksy Group Shilling Attack Model against Recommender Systems

Youquan Wang[1], Zhiang Wu[2,*], Jie Cao[2,1], and Changjian Fang[2]

[1] College of Computer Science and Engineering,
Nanjing University of Science and Technology, Nanjing, China
[2] Jiangsu Provincial Key Laboratory of E-Business,
Nanjing University of Finance and Economics, Nanjing, China
{youq.wang,zawuster}@gmail.com,
{caojie690929,jselab1999}@163.com

Abstract. The robustness of recommender systems has drawn recently more and more attention of both industry and academia. Although a multitude of studies have been devoted to shilling attack modeling and detection, few of them focus on *group* shilling attack. The attackers in a shilling group work together to manipulate the output of the recommender system. Meanwhile, since the rating profiles in a shilling group are carefully designed, it is hard to detect them by traditional methods. This paper presents a generative model to create shilling group in which every pair of attackers has high diversity. In particular, both strict and loose versions of group shilling attack generation algorithm are proposed. Experimental results on MovieLens data set demonstrate that the shilling group generated by the our model can not only exert large negative effect to recommender systems, but also avoid the detection by the traditional methods.

Keywords: shilling attack, group shilling attack, recommender systems.

1 Introduction

The security and robustness of recommender systems has become a hot topic in recent years, in the context of *profile injection* or *shilling attacks* [1, 2]. As the prevalence of online shopping, shilling attackers have a natural profit incentive to promote their own products (or suppress their competitors' products) by creating biased online comments and/or extreme ratings. Therefore, shilling attacks are seriously threatening the sound development of e-commerce.

A large body of work has been devoted to discuss the shilling attack generative models [3, 4, 5], define classification metrics for genuine users shilling attackers [3, 6, 7, 8], and design the shilling attack detectors [6, 7, 9, 10, 11, 12]. The existing detectors are fairly effective to the shilling attackers working separately. However, as the development of the attack tricks, shilling attackers are not working separately yet today, but working together to commit swarm and

---

* Corresponding author.

S. Zhou, S. Zhang, and G. Karypis (Eds.): ADMA 2012, LNAI 7713, pp. 675–688, 2012.
© Springer-Verlag Berlin Heidelberg 2012

Table 1. Generative models of shilling attacks

Attack models	Push	Nuke
Random attack	$I_s = \emptyset$; $I_F = r_{ran}$; $i_t = r_{max}$	$I_s = \emptyset$; $I_F = r_{ran}$; $i_t = r_{min}$
Average attack	$I_s = \emptyset$; $I_F = r_{avg}$; $i_t = r_{max}$	$I_s = \emptyset$; $I_F = r_{avg}$; $i_t = r_{min}$
Segmented attack	$I_s = r_{max}$; $I_F = r_{min}$; $i_t = r_{max}$	$I_s = r_{min}$; $I_F = r_{max}$; $i_t = r_{min}$
Bandwagon attack	$I_s = r_{max}$; $I_F = r_{max}$; $i_t = r_{max}$	$I_s = r_{min}$; $I_F = r_{ran}$; $i_t = r_{min}$

Note: (1) r_{ran}: random ratings; (2) r_{avg}: average ratings;
(3) I_s in segmented attack: a set of similar items of the target item;
(4) I_s in bandwagon attack: a set of frequently rated items.

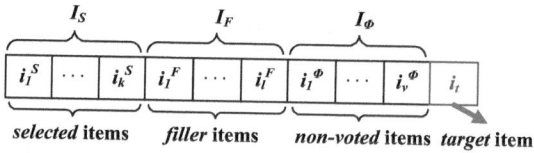

Fig. 1. The illustration of a shilling profile

massive attacks after a premeditated planning. This new type of shilling attack was coined as "group shilling attacks" [13]. Every attacker in the shilling group has carefully designed its ratings to conceal the malicious intention. An individual attacker may not successfully promote (or suppress) the target products, but when they work together, they will attain the attack goal.

This paper focuses on a generation method to construct the effective group shilling attacks which can not only avoid the detection of the existing detectors but also successfully modify the recommendation output of the system. Based on the observation that the *low diversity* attackers can be easily recognized by the traditional detectors [14], we propose two versions group shilling generative models to create *highest diversity* and *high diversity* profiles, respectively.

The remainder of this paper is organized as follows. In Section 2, we briefly introduce background knowledge about shilling attack, define the problem and summarize the characteristics of group shilling attack. In Section 3, we present two versions of group shilling attack generation algorithm. Section 4 shows the experimental results. We finally present the related work and conclude our work in Section 5 and Section 6, respectively.

2   Preliminaries

Shilling attackers can be classified as *push* and *nuke* attacks according to their intent–making a target item more likely (push) or less likely (nuke) to be recommended [6]. The rating records of a user on various items construct the profile of that user. The profile of a shilling attacker (or a *shilling profile* for short) usually consists of the ratings on four types of items: target item, filler items, selected items, and non-voted items, as show in Fig. 1. Target item often has the highest rating in a push attack, or lowest rating in a nuke attack. Filler items can make a shilling profile look normal and yield profound impact against a recommender system. Selected items are often used to make friends with as many genuine users as possible. Finally, non-voted items are the remaining unrated items.

Four generative models, i.e. random, average, segment, and bandwagon, are often used to generate shilling profiles by carefully rating the four types of items, as illustrated in Table 1. In the literature [1, 3, 4, 7, 8, 10, 11], the random rating r_{ran} is often generated by Gaussian distribution. For instance, in the 5-grade marking data set `MoiveLens`, r_{ran} is assumed to obey a Gaussian distribution with $\mu = 3.6$ and $\sigma = 1.1$ [1].

2.1 Problem Definition

We consider a rating matrix of which the row vectors represent a set of user profiles and every element is the rating that a user gave for a item. Let $\mathbb{B} = \{\mathbf{B}_1, \cdots, \mathbf{B}_l\}$ denote l groups of shilling attackers. Actually, $\mathbf{B}_i (1 \leq i \leq l)$ is a $n_i \times m$ rating matrix where n_i is size of $\mathbf{B}_i$ and m is the number of items. If let b_j denote the j-th row vector of $\mathbf{B}_i$, we have $\mathbf{B}_i = \{b_1, \cdots, b_{n_i}\}$.

From the perspective of the attacker, the best attack against a system is one that yields the biggest impact for the least amount of effort [3]. The descriptive dimensions of single shilling attacks have been discussed in [3], while we are primarily interested in the following dimensions for group shilling attacks:

- Attack intent: There are usually two intents: *push* and *nuke*. The group shilling attackers aim to make the target items more likely (push) or less likely (nuke) to be recommended.
- The number of target items T_i: All shillers in a group work together to promote or demote a set of target items rather than one item. For example, a shilling group may works for a brand consisting of many products.
- The number of groups l: Many shilling groups may co-exist in a recommender system.
- The size of group shilling attacks n_i: Big size of shilling group leads to the big cost of the creation of the group, because on-line registrations require human intervention.
- Profile size $|b_j|$: The number of ratings assigned by an attacker in a shilling group is equal to *filler size* in the single shilling attack.

It is interesting to note that as the increase of T_i, l and $|b_j|$, the group shilling attacks may threaten the recommender systems more seriously, however, the cost for generating the attack profiles is also increasing. Therefore, the tricksy attackers should balance the power for damaging the recommender systems and the cost required for generating the attack profiles.

2.2 Characteristics of Group Shilling Attack

Before modeling the group shilling attack, we summarize two important characteristics to support the proposed group shilling attack model in Section 2. The two fundamental properties are presented as following:

Property 1. Effective group shilling attacks can push (or nuke) target items successfully and affect the recommender system as much as possible.

Property 2. Every individual shiller in group attacks looks like normal users and should not be discovered by detector designed for single shilling attack.

The property 1 is the original target of shilling attackers. Group shilling attack is not the combination of many single shilling profiles generated by attack models, since the group shilling attacks can be easily filtered by the traditional detectors. Profiles in group shilling attacks are quite different from traditional attack profiles, for they have well designed filling ratings to conceal their malicious intentions. So, we present the property 2. The generation model of group shilling attacks should not only make every profiles in the group looks like normal users to avoid the detection by traditional algorithms, but also keep a grave threat to recommender systems.

3   The Generation Model of Group Shilling Attacks

Shilling profiles created by above-mentioned four models, i.e. random, average, segment, and bandwagon, have high similarity. These low-diversity attackers can be easily detected by the existing detection algorithm such as C4.5-, PCA- and clustering-based detectors [3, 9, 10]. In this section, we target at proposing a method to generate high-diversity shilling profiles that are used to create group shilling attacks. The idea behind the method is that every pair of shilling attackers in the group keeps low similarity to avoid the detection by the existing methods designed for single shilling attack. Specially, we propose two versions for the generative model of group shilling attacks. The first is a strict version denoted as $GSAGen_s$, which can guarantee the PCC similarity of each pair of profiles in a group be -1. However, the second version $GSAGen_l$ employs a loose condition which guarantee the PCC similarity of each pair of profiles in a group be -1 or 0.

3.1 $GSAGen_s$: A Strict Version

Our group shilling attack generative model takes a set of shilling profiles created by attack models, i.e. random or average, as the input. Then, we aims to generate a group of shilling attackers $\mathbf{B}_i$ with high diversity satisfying the following:

Definition 1 (The Strict Condition of Group Shilling Attack). *Let* $\mathbf{B}_i = \{b_1, \cdots, b_{n_i}\}$ *be a group of shilling attackers. We say* $\mathbf{B}_i$ *satisfies the strict condition iff the following three conditions are satisfied simultaneously: (1)* $\forall u \neq v, b_u \cap b_v \neq \emptyset$; *(2)* $\forall u \neq v \neq t, b_u \cap b_v \cap b_t = \emptyset$; *(3)* $\forall u \neq v, PCC_{uv} = -1$.

In Definition 1, PCC_{uv} is the Pearson Correlation Coefficient (PCC) between b_u and b_v. As is known, PCC is in [-1,1] and $PCC_{uv} \leq 0$ indicates that b_u and b_v are not similar, and thus $PCC_{uv} = -1$ implies that b_u and b_v have the highest diversity. The procedure of $GSAGen_s$ as shown in Algorithm 1 generates a group of shilling attackers satisfying Definition 1 based on a set of shilling profiles. In line 2, $\mathbf{B}_i$ is initialized with a shilling profile a_1. In lines 3-17, every profile $a_u \in \mathbf{A}_i$ is examined, and b_u is constructed in terms of a_u and current rating records in $\mathbf{B}_i$. In lines 4-15, every rated item of a_u is traversed and if the current item is not be rated by all attackers in $\mathbf{B}_i$, the rating value of j-th item by a_u is assigned to b_u (see lines 12-14). Lines 5-11 describe the case that the j-th item is rated by a_u and at least one attacker in $\mathbf{B}_i$. If the j-th item is rated by only one attacker in $\mathbf{B}_i$, b_{uj} is set to $2\overline{b_{v,\kappa}} - b_{vj}$. Note that $\overline{b_{v,\kappa}}$ is the average value on the intersection of a_u and b_v. If there exists one item rated by a_u that has been rated by more than one attacker in $\mathbf{B}_i$, a_u cannot be utilized to generate b_u and thus $GSAGen_s$ jumps to examine the next shilling profile a_{u+1}(see lines 9-11).

As can be seen from the procedure of $GSAGen_s$, we can easily observe that $\forall u \neq v, b_u \cap b_v \neq \emptyset$ and $\forall u \neq v \neq t, b_u \cap b_v \cap b_t = \emptyset$. In the following theorem, we will prove that $\forall u \neq v, PCC_{uv} = -1$.

Theorem 1. $\forall b_u, b_v \in \mathbf{B}_i,\ PCC_{uv} = -1$.

PROOF:Let $\kappa = b_u \cap b_v$ and $PCC_{uv} = \dfrac{\sum_{j\in\kappa}(b_{uj}-\overline{b_{u,\kappa}})(b_{vj}-\overline{b_{v,\kappa}})}{\sqrt{\sum_{j\in\kappa}(b_{uj}-\overline{b_{u,\kappa}})^2}\sqrt{\sum_{j\in\kappa}(b_{vj}-\overline{b_{v,\kappa}})^2}}.\because b_{uj} =$

$2\overline{b_{v,\kappa}} - b_{vj}$, and $\overline{b_{u,\kappa}} = \overline{b_{v,\kappa}}$,
$\therefore b_{uj} - \overline{b_{u,\kappa}} = -(b_{vj} - \overline{b_{v,\kappa}})$,
$\therefore PCC_{uv} = -1$, which completes the proof.   □

From Theorem 1, the group of shilling attack generated by $GSAGen_s$ possesses the strict condition as shown in Definition 1. So, every pair of shilling profiles in the group has the *highest diversity*, i.e.$PCC = -1$, which is also the reason that we call $GSAGen_s$ the strict version.

Algorithm 1. The strict version of Group Shilling Attack Generation Algorithm

---

1: **procedure** $GSAGen_s(\mathbf{A}_i = \{a_1, \cdots, a_{n_i}\})$
2:     $b_1 \leftarrow a_1, \mathbf{B}_i \leftarrow \{b_1\}$;
3:     **for** $u \leftarrow 2 : n$ **do**
4:         **for** $j \leftarrow 1 : m$ **do**
5:             **if** $\exists b_v \in \mathbf{B}_i, a_{uj} \neq 0\ \&\&\ b_{vj} \neq 0$ **then**
6:                 **if** $\forall b_t \in \mathbf{B}_i, b_{tj} = 0$ **then**
7:                     $b_{uj} \leftarrow 2\overline{b_{v,\kappa}} - b_{vj}, \kappa = a_u \cap b_v$;
8:                 **end if**
9:                 **if** $\exists b_t \in \mathbf{B}_i, b_{tj} \neq 0$ **then**
10:                    **Goto** line 3;
11:                **end if**
12:            **else if** $\forall b_v \in \mathbf{B}_i, a_{uj} \neq 0\ \&\&\ b_{vj} = 0$ **then**
13:                $b_{uj} \leftarrow a_{uj}$;
14:            **end if**
15:        **end for**
16:        $\mathbf{B}_i \leftarrow \mathbf{B}_i \cup \{b_u\}$;
17:    **end for**
18:    **return** $\mathbf{B}_i$;
19: **end procedure**

3.2 $GSAGen_l$: A Loose Version

Although the shilling attack group generated by $GSAGen_s$ bids fair to escape the detection of the existing algorithms, the group size may be limited due to the rigorous condition for the shilling group. Meanwhile, $PCC \leq 0$ indicates the dissimilarity of two attackers. Therefore, to find a loose condition of the group shilling attack is a natural idea, and thus we have the following definition:

Definition 2 (The Loose Condition of Group Shilling Attack). *Let* $\mathbf{B}_i = \{b_1, \cdots, b_{n_i}\}$ *be a group of shilling attackers. We say* $\mathbf{B}_i$ *satisfies the loose condition iff* $\forall u \neq v, \kappa = b_u \cap b_v, \kappa = \emptyset$ *or* $\kappa \neq \emptyset, \exists j \in \kappa, \forall t \neq u \neq v, b_{tj} = 0$, *and* $\forall u \neq v, PCC_{uv} \leq 0$.

In Definition 2, we allow the intersection of any pair of attackers to be empty, however, if the intersection is not empty, there exists at least one item that is rated only by these two attackers. Furthermore, we can guarantee the PCC between any pair of attackers be smaller or equal to 0.

Algorithm 2. The loose version of Group Shilling Attack Generation Algorithm

---

1: **procedure** $GSAGen_l(\mathbf{A}_i = \{a_1, \cdots, a_{n_i}\})$
2:     $b_1 \leftarrow a_1, \mathbf{B}_i \leftarrow \{b_1\}$;
3:     **for** $u \leftarrow 2 : n$ **do**
4:         **bool** $flag \leftarrow$ TRUE;
5:         **for** $t \leftarrow 1 : |\mathbf{B}_i|$ **do**
6:             let $\kappa = a_u \cap b_t$;
7:             **if** $\forall j \in \kappa, |\mathbf{B}_i[j]| > 1$ **then**
8:                 $flag \leftarrow$ FALSE;
9:                 break;
10:            **else**
11:                $\forall j \in \kappa, |\mathbf{B}_i[j]| = 1, b_{uj} \leftarrow 2\overline{b_t} - b_{tj}$;
12:            **end if**
13:        **end for**
14:        **if** $flag =$ TRUE **then**
15:            $\forall j \in [1, m], a_{uj} \neq 0$ && $|\mathbf{B}_i[j]| = 0, b_{uj} = a_{uj}$;
16:            $\forall j \in [1, m], a_{uj} \neq 0$ && $|\mathbf{B}_i[j]| > 1, b_{uj} = \overline{b_u}$;
17:            $\mathbf{B}_i \leftarrow \mathbf{B}_i \cup \{b_u\}$;
18:        **end if**
19:    **end for**
20:    **return** $\mathbf{B}_i$;
21: **end procedure**

The pseudocode of $GSAGen_l$ is shown in Algorithm 2. The input of $GSAGen_l$ is also a set of shilling profiles created by attack models, i.e. random or average. In line 2, $\mathbf{B}_i$ is initialized with a shilling profile a_1. In lines 3-19, every profile $a_u \in \mathbf{A}_i$ is examined, and a boolean variable $flag$ is defined to indicate the current a_u whether can be used to generate the b_u. For any b_t contained in the current $\mathbf{B}_i$, if all items in the intersection between a_u and b_t is rated by more than one user in

$\mathbf{B}_i$, $flag$ is set to FALSE and thus the current a_u is skipped(see lines 7-9). Note that $|\mathbf{B}_i[j]|$ is the number of ratings to the j-th item by all profiles $\{b_1, \cdots, b_{u-1}\}$ in $\mathbf{B}_i$. There are three different cases for the assignments to the nonzero items in a_u: (1) if none attacker in $\mathbf{B}_i$ has rated the item, just keep the rating value of a_u (see line 15); (2) if only one attacker in $\mathbf{B}_i$ has rated the item, select this profile $b_t \in \mathbf{B}_i$ and update the rating b_{uj} (see line 11); (3) if more than one attackers in $\mathbf{B}_i$ have rated the item, fill the ratings of these items with the average value of b_u (see line 16). In lines 15-17, the boolean variable $flag$ =TRUE indicates the cases "$\kappa \neq \emptyset, \exists j \in \kappa, \forall t \neq u \neq v, b_{tj} = 0$" or "$a_u$ has no intersection with the current $\mathbf{B}_i$" happened. However, Algorithm 2 does not tell us $\forall u \neq v, PCC_{uv} \leq 0$ directly, which will be shown in the following theorem.

Theorem 2. $\forall b_u, b_v \in \mathbf{B}_i$, $PCC_{uv} = \begin{cases} 0 & if \ \ b_u \cap b_v = \emptyset \\ -1 & otherwise \end{cases}$

PROOF: Let $\kappa = b_u \cap b_v$ and $PCC_{uv} = \dfrac{\sum_{j \in \kappa}(b_{uj} - \overline{b_u})(b_{vj} - \overline{b_v})}{\sqrt{\sum_{j \in \kappa}(b_{uj} - \overline{b_u})^2}\sqrt{\sum_{j \in \kappa}(b_{vj} - \overline{b_v})^2}}$. Obviously, $PCC_{uv} = 0$ when $\kappa = \emptyset$. When $\kappa \neq \emptyset$, $\forall j \in \kappa, b_{uj} = \overline{b_u}$, the j-th item does not contribute to numerator and denominator of PCC_{uv}, since $b_{uj} - \overline{b_u} = 0$. However, $flag$ =TRUE in line 15 of Algorithm 2 guarantee that there is at least one item satisfying $a_{uj} \neq 0$ and $|\mathbf{B}_i[j]| = 1$.
$\because \exists j, b_{uj} = 2\overline{b_v} - b_{vj}$, and $\overline{b_u} = \overline{b_v}$.
$\therefore \exists j, b_{uj} - \overline{b_u} = -(b_{vj} - \overline{b_v})$, and thus $PCC_{uv} = -1$,which completes the proof. $\square$

It is interesting to note that the computation of PCC in Theorem 2 utilizes the average on the whole rating, while the average on the intersection set κ is used in Theorem 1. In the literature, these two kinds of PCC computation methods are often exchanged equivalently. For instance, the average on the whole rating is adopted in [9, 15], while the average on the intersection set is employed in [14].

Discussion: Indeed, $GSAGen_s$ is similar to the method proposed in [14] for generating *high diversity* single attack profiles. In this paper, we borrow ideas from [14] to design the strict version of generative model for shilling group. However, the size of the output profiles of $GSAGen_s$ is rather scarce due to the strict condition for the generated attackers, which will be shown in Section 4.2. The new $GSAGen_l$ extends the generative condition in order to create more profiles in a shilling group that can maintain the big negative effect on recommender systems and have the good anti-detection performance.

4   Experimental Study

In this section, we conduct sets of experiments on MovieLens[1] data set to illustrate the effectiveness of the proposed group shilling attack model. And the experimental results show: (1) $GSAGen_l$ is more suitable to generate the shilling

---

[1] http://movielens.umn.edu/

group including more attack profiles than $GSAGen_s$. (2) The attackers in the shilling group works together to exert big negative effect to recommender systems. (3) The group shilling attacks generated by both $GSAGen_s$ and $GSAGen_l$ can avoid the detection by the existing methods.

4.1   Experiment Setup

Data Sets. MovieLens data set is published by GroupLens and consists of 100,000 ratings on 1682 movies by 943 users. All ratings are integer values ranged from 1 to 5 where 1 indicates disliked and 5 indicates most liked. MovieLens data set is widely used in the realm of shilling attack detection [1, 3, 4, 7, 8, 10, 11].

Vulnerability Measures. For the purpose of evaluating the vulnerability of the recommender system against shilling attacks, two kinds of measures are employed, i.e., Average Prediction Shift ($\overline{\Delta}$) for rating prediction and Hit Ratio (HR) for ranking prediction.

$$\overline{\Delta} = \sum_{(u,i) \in N} \frac{|p'_{ui} - p_{ui}|}{|N|}. \tag{1}$$

where p'_{ui} represents the prediction rating after the attack and p_{ui} before, and N is the set of missing ratings of normal users.

$$HR = \frac{\#hits}{K \cdot T_i}. \tag{2}$$

where $\#hits$ indicates the number of target items in the top K recommendation list, and T_i is the number of target items. Since whether target items enter the recommendation list is the key issue that attackers considered, HR can better reflect the power of shilling attacks.

Detection Measures. The widely-used recall(R), precision(P), and F-measure(F) are adopted for the detection performance evaluation.

$$P = \frac{TP}{TP + FP}, R = \frac{TP}{TP + FN}, F = \frac{2PR}{P + R}. \tag{3}$$

with TP being the number of truly identified attackers, TN the number of truly identified normal users, FP the number of wrongly identified attackers, and FN the number of missed attackers. In general, R and P highlight the completeness and accuracy of a detector, respectively, and F provides a global view.

Generators and Detectors. We have implemented $GSAGen_s$ and $GSAGen_l$ as two generators in C++. Meanwhile, two traditional shilling detectors, i.e. C4.5 [3] and PCA (PCASelectUsers) [9] are employed for anti-detection performance evaluation. PCA was coded in MATLAB to facilitate the principal-component computation. C4.5 was the J48 version provided by WEKA[2] with the default settings.

---

[2] http://www.cs.waikato.ac.nz/ml/weka/

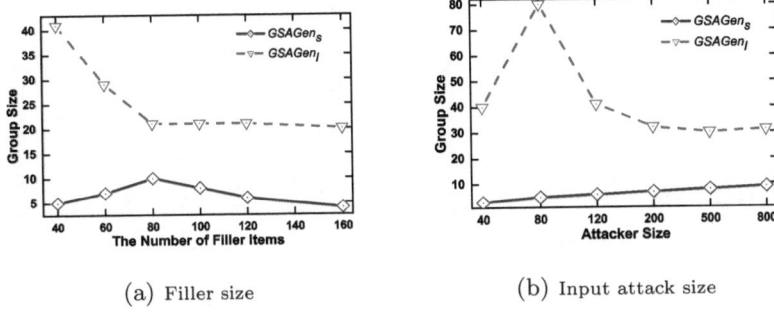

(a) Filler size   (b) Input attack size

Fig. 2. A comparison on the generated shilling group size

Procedure. We assumed the users in the MovieLens data set were normal. Then, we generated attacker profiles according to the attack models mentioned in Table 1, and utilized two generators to produce group shilling attacker profiles which were injected into MovieLens data set. Finally, various detectors were invoked to classify the normal users and shilling attackers.

4.2 $GSAGen_s$ vs. $GSAGen_l$ on Group Size

In Section 3.2, we have mentioned that the group size generated by $GSAGen_s$ is much smaller than $GSAGen_l$. Here, we demonstrate it through experiments. Two factors, i.e. the filler size b_j and the input attack size $|\mathbf{A}_i|$, affect the size of output shilling group. Obviously, the smaller filler size b_j, the larger group size, and the larger attack size $|\mathbf{A}_i|$, the larger group size. Fig. 2(a) shows the effect of filler size on the size of shilling group under $|\mathbf{A}_i| = 100$, and Fig. 2(b) depicts the effect of input attack size under $b_j = 40$. Note that every movie is selected as the filler item with equal probability. As can be seen from Fig. 2, $GSAGen_l$ can indeed generate more profiles in the shilling group than $GSAGen_s$. Specially, when b_j and $|\mathbf{A}_i|$ are small, $GSAGen_l$ can transform all profiles in $\mathbf{A}_i$ to high diversity profiles, since the overlapping between any pair of profiles are small (see two points $|\mathbf{A}_i| = 40$ and 80 in Fig. 2(b)). However, $GSAGen_s$ still generates less than 10 profiles in the shilling group.

4.3   Vulnerability Analysis

Here, we investigate the vulnerability of the recommender system against the group shilling attack. The User-based Collaborative Filtering (UCF) method [16] is employed as the kernel algorithm of the recommender system. We set the number of nearest neighbors $k = 20$ and the length of recommendation list $K = 10$. A shilling group with 3 target items is inserted into user-item database. We utilize both $GSAGen_s$ and $GSAGen_l$ based on random and average attack model to generate 4 kinds shilling groups, respectively.

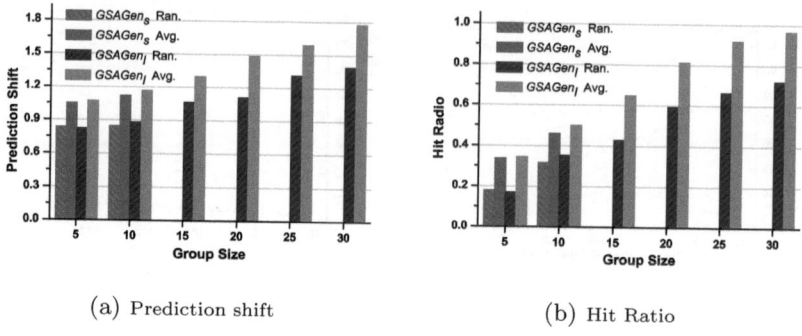

(a) Prediction shift   (b) Hit Ratio

Fig. 3. Effect of shilling group to UCF based Recommender Systems

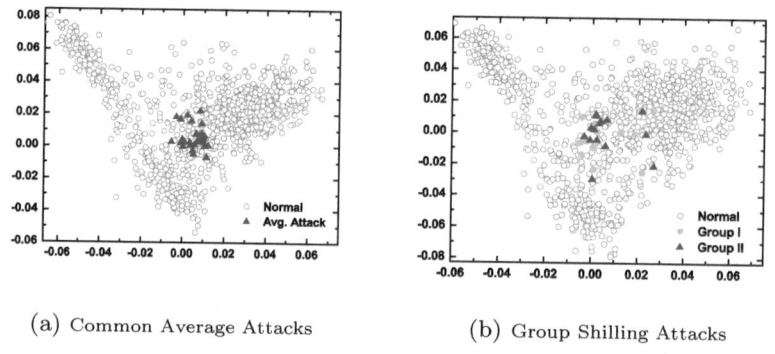

(a) Common Average Attacks   (b) Group Shilling Attacks

Fig. 4. Dispersion shapes of common average attacks and group shilling attacks

As can be seen from Fig. 3, the shilling group generated by $GSAGen_l$ exerts bigger negative effect on the recommender system than $GSAGen_s$, and the shilling group created based on average attack model affects more than the random attack model. $GSAGen_s$ often generated less than 10 profiles in the shilling group, which led to the missing bars for the cases of "$GSAGen_s$ Ran." and "$GSAGen_s$ Avg." in Fig. 3. When we take a deep look at the hit ratio shown in Fig. 3(b), as the increase of the group size, the hit ratio becomes fairly high. For instance, the hit ratio a shilling group containing 20 attackers is larger than 0.8, which indicates the target items have entered the recommendation list of over 80% normal users!

4.4   The Anti-detection Performance

In this subsection, we ascertain the anti-detection performance of group shilling attack. Two kinds of detectors are chosen as the representatives of the existing methods, i.e., C4.5 for supervised learning based method and PCA for unsupervised learning based method. Since C4.5 is a feature-based detector, we use a feature selection algorithm *MC-Relief* proposed in our previous work [11] to select effective features on MovieLens data set. Finally, we obtain 5 features among 10 features for C4.5, i.e., *Entroy, LengthVar, MeanVar, RDMA* and *DegSim*. PCA is a rating-based detector. So, we can run PCA on rating records directly. In addition, we assume that PCA know the number of group size $|\mathbf{B}_i|$, and return the top-$|\mathbf{B}_i|$ users as attackers. The number of filler items is set to 40 and the input attack size is 80.

Table 2. The anti-detection performance comparison in terms of *FMeasure*

Gen. Method	Attack Detector	Group Size					
		5	10	15	20	25	30
None	*Ran.* C4.5	0.774	0.882	0.912	0.0.936	0.953	0.979
	Ran. PCA	1	1	1	1	1	1
	Avg. C4.5	0.756	0.877	0.903	0.93	0.944	0.953
	Avg. PCA	0.8	0.9	0.933	0.95	1	1
GSAGen_s	*Ran.* C4.5	0.375	0.346	–	–	–	–
	Ran. PCA	0.6	0.7	–	–	–	–
	Avg. C4.5	0.23	0.232	–	–	–	–
	Avg. PCA	0.6	0.7	–	–	–	–
GSAGen_l	*Ran.* C4.5	0.368	0.324	0.308	0.291	0.286	0.283
	Ran. PCA	0.4	0.6	0.6	0.65	0.56	0.567
	Avg. C4.5	0.277	0.268	0.262	0.259	0.255	0.252
	Avg. PCA	0.4	0.6	0.533	0.5	0.52	0.533

Table 2 shows the results in terms of *FMeasure*. As can be seen, for common average and random attackers, PCA performs nearly perfect and C4.5 also performs fairly good. However, when the average and random attacks are transformed to group shilling attacks, the detection performance of two detectors decreases remarkably. Although PCA can recognize about half attackers, PCA is assumed to know the number of injected attackers. Therefore, the performance of PCA will be degraded in practise, which in turn, justify that the shilling groups generated by the proposed methods have good anti-detection performance.

Furthermore, we take a deeper look at the detection performance of PCA according to the dispersion shapes of genuine users and several shilling groups. Fig. 4 shows the dispersion shapes of common average attacks and group shilling attacks. As can be seen, the common average attackers tend to be gathered at the center of normal users. After we apply $GSAGen_l$ on these common average attackers to generate two shilling groups, the attackers in a group become scattered. However, attackers in different shilling groups may have overlapping, since different shilling groups often lack for premeditation. Fig. 4 implies that the proposed $GSAGen_l$ can successfully make the high similar attackers be more diverse.

5   Related Work

The pioneer work on shilling attack detection can be traced to [1, 2]. Since then, several generative models of shilling attacks as shown in Table 1 have been proposed [3, 4]. From machine learning perspective, three kinds of shilling attack detection methods were presented in the literature, i.e., supervised learning based method [3, 7, 8, 17], unsupervised learning based method [9, 10, 18], and semi-supervised learning based method [19, 12, 11]. Different from the work on defending the shilling attack, Cheng and Hurley [14] studied the problem from the attackers' perspective, and presented an effective diverse and obfuscated attacks which has inspired our work on the strict version of group shilling attack generation algorithm.

To date, few of research has been carried out on the modeling and detecting of group shilling attacks. Only preliminary discussion on group shilling was presented in [13]. However, many research has been devoted to *group review spam* detection [20, 21, 22]. Our paper indeed fills this crucial void by proposing the $GSAGen_s$ and $GSAGen_l$ to model the group shilling attacks.

6   Conclusion and Future Work

In this paper, we proposed a group shilling attack generation approach under the observation that the attackers with low similarity are difficult to be discovered by the traditional detectors. Specifically, a strict version of generative method named $GSAGen_s$ was presented to create shilling profiles with highest diversity, i.e. $PCC = -1$. However, $GSAGen_s$ suffered from the small output group size. Therefore, we extended the strict version to obtain the loose version $GSAGen_l$ which guaranteed the output shilling profiles have high diversity, i.e. $PCC \leq 0$. Experimental results on MovieLens data set demonstrate that the shilling groups generated by the proposed model have great negative influence on recommender systems and can effectively avoid the detection by the traditional methods. Future study will include additional examinations on more tricksy shilling groups as well as their obfuscated techniques. The design of a new detector for group shilling attacks combing the HySAD detector [11] toward developing a complete shilling detectors is planned for future direction.

Acknowledgments. This research is supported by National Natural Science Foundation of China under Grants No.61103229, 71072172 and 41201486, Industry Projects in the Jiangsu S&T Pillar Program under Grants No. BE2011198, Jiangsu Provicial Colleges and Universities Outstanding S&T Innovation Team Fund under Grants No. 2011013, Key Project of Natural Science Research in Jiangsu Provincial Colleges and Universities under Grants No. 12KJA520001,

National Key Technologies R&D sub Program in 12th five-year-plan under Grants No. SQ2011GX07E03990,the Natural Science Foundation of Jiangsu Province of China under Grant BK2012863, and International S&T Cooperation Program of China under Grants No. 2011DFA12910.

References

[1] Lam, S., Riedl, J.: Shilling recommender systems for fun and profit. In: Proceedings of the 13th International Conference on World Wide Web (WWW 2004), pp. 393–402 (2004)

[2] O'Mahony, M., Hurley, N., Kushmerick, N., Silvestre, G.: Collaborative recommendation: A robustness analysis. Transactions on Internet Technology (TOIT) 4(4), 344–377 (2004)

[3] Williams, C.: Profile injection attack detection for securing collaborative recommender systems. Technical report, DePaul University (2006)

[4] Mobasher, B., Burke, R., Bhaumik, R., Williams, C.: Toward trustworthy recommender systems: An analysis of attack models and algorithm robustness. Transactions on Internet Technology (TOIT) 7(4), 1–40 (2007)

[5] Chiang, M., Peng, W., Yu, P.: Exploring latent browsing graph for question answering recommendation. World Wide Web: Internet and Web Information Systems, WWWJ (2011)

[6] Zhang, S., Chakrabarti, A., Ford, J., Makedon, F.: Attack detection in time series for recommender systems. In: Proceedings of the 12th ACM SIGKDD International Conference on Knowledge Discovery and Data Mining, KDD 2006 (2006)

[7] Chirita, P., Nejdl, W., Zamfir, C.: Preventing shilling attacks in online recommender systems. In: Proceedings of the 7th Annual ACM International Workshop on Web Information and Data Management (WIDM 2005), pp. 67–74 (2005)

[8] Burke, R., Mobasher, B., et al.: Classification features for attack detection in collaborative recommendation systems. In: Proceedings of the 12th ACM SIGKDD International Conference on Knowledge Discovery and Data Mining, KDD 2006 (2006)

[9] Mehta, B., Nejdl, W.: Unsupervised strategies for shilling detection and robust collaborative filtering. User Modeling and User-Adapted Interaction 19(1-2), 65–97 (2009)

[10] Lee, J., Zhu, D.: Shilling attack detection—a new approach for a trustworthy recommender system. INFORMS Journal on Computing 24(1) (2011)

[11] Wu, Z., Wu, J., Cao, J., Tao, D.: HySAD: A semi-supervised hybrid shilling attack detector for trustworthy product recommendation. In: Proceedings of the 18th ACM SIGKDD International Conference on Knowledge Discovery and Data mining (KDD 2012), pp. 985–993 (2012)

[12] Cao, J., Wu, Z., Mao, B., Zhang, Y.: Shilling attack detection utilizing semi-supervised learning method for collaborative recommender system. World Wide Web: Internet and Web Information Systems (WWWJ) (2012)

[13] Su, X., Zeng, H.J., Chen, Z.: Finding group shilling in recommendation system. In: Proceedings of the 14th International Conference on World Wide Web, WWW 2005 (2005)

[14] Cheng, Z., Hurley, N.: Effective diverse and obfuscated attacks on model-based recommender systems. In: Proceedings of ACM Conference on Recommender Systems (RecSys 2009), pp. 141–148 (2009)

[15] Rashid, A.M., Karypis, G., Riedl, J.: Influence in ratings-based recommender systems: An algorithm-independent approach. In: Proceedings of SIAM International Conference on Data Mining, SIAM 2005 (2005)

[16] Herlocker, J., Konstan, J., Terveen, L., Riedl, J.: Evaluating collaborative filtering recommender systems. ACM Transactions on Information Systems (TOIS) 22(1), 5–53 (2004)

[17] Mobasher, B., Burke, R., Williams, C., Bhaumik, R.: Analysis and detection of segment-focused attacks against collaborative recommendation. In: WebKDD Workshop (2006)

[18] Hurley, N., Cheng, Z., Zhang, M.: Statistical attack detection. In: Proceedings of ACM Conference on Recommender Systems, RecSys 2009 (2009)

[19] Wu, Z., Cao, J., Mao, B., Wang, Y.: SemiSAD: Applying semi-supervised learning to shilling attack detection. In: Proceedings of ACM Conference on Recommender Systems (RecSys 2011), Chicago, IL, USA, pp. 289–292 (2011)

[20] Mukherjee, A., Liu, B., Glance, N.S.: Spotting fake reviewer groups in consumer reviews. In: Proceedings of the 21th International Conference on World Wide Web (WWW 2012), pp. 191–200 (2012)

[21] Mukherjee, A., Liu, B., Wang, J., Glance, N.S., Jindal, N.: Detecting group review spam. In: Proceedings of the 20th International Conference on World Wide Web WWW 2011 (Companion Volume), pp. 93–94 (2011)

[22] Leung, C., Chan, S., Chung, F., Ngai, G.: A probabilistic rating inference framework for mining user preferences from reviews. World Wide Web: Internet and Web Information Systems (WWWJ) 14(2), 187–215 (2011)

Topic-Centric Recommender Systems
for Bibliographic Datasets

Aditya Pratap Singh, Kumar Shubhankar, and Vikram Pudi

IIIT Hyderabad, India
{aditya.pratap,kumar.shubhankar}@research.iiit.ac.in,
vikram@iiit.ac.in

Abstract. In this paper, we introduce a novel and efficient approach for Recommender Systems in the academic world. With the world of academia growing at a tremendous rate, we have an enormous number of researchers working on hosts of research topics. Providing personalized recommendations to a researcher that could assist him in expanding his research base is an important and challenging task. We present a unique approach that exploits the latent author-topic and author-author relationships inherently present in the bibliographic datasets. The objective of our approach is to provide a set of latent yet relevant authors and topics to a researcher based on his research interests. The recommended researchers and topics are ranked on the basis of authoritative scores devised in our algorithms. We test our algorithms on the *DBLP* dataset and experimentally show that our recommender systems are fairly effective, fast and highly scalable.

Keywords: Recommender Systems, Topic Mining, Author-topic Graph, Topic Evolution, Authoritative Score.

1 Introduction

The world of academia is growing at a tremendous rate with thousands of research papers being published every year. For a researcher looking for new dimensions of research, it is becoming an increasingly difficult task to identify the relevant yet novel domains of research. There is a definite need of a system that could help the researchers in their decision-making by interacting with large academic information space. Recommender systems are aimed at performing this very task. In this paper, we propose a novel and efficient approach for author and topic recommendation to an author in a bibliographic dataset that exploits the hidden relationships in such datasets as well as addresses the issues of scalability and sparsity. Our author recommendation algorithm aims at suggesting latent yet relevant authors to a researcher based on his research topics. For a pair of authors, their shared research interests as well as the relative diversity between their research interests forms the basis of the similarity measure in our algorithm. The objective of our topic recommendation algorithm is to suggest a set of relevant topics to a researcher based on the research interests of similar

S. Zhou, S. Zhang, and G. Karypis (Eds.): ADMA 2012, LNAI 7713, pp. 689–700, 2012.

authors. The recommendation algorithm considers the importance of a topic as well as the importance of the suggesting authors to assign authoritative scores to the recommended topics. We also exploit the hierarchy of topics and their temporal property to prune the irrelevant topics. We tested our algorithms on the *DBLP* dataset to produce a ranked set of recommended authors and topics. We study our algorithms on various parameters and validate our experiments by dividing the dataset into training and test portions. Our algorithms have a host of applications like author and topic recommendation, retrieval of authors for a given topic query and vice-versa, building communities of similar authors etc.

2   Related Work

Recommender Systems [1] form a specific type of information filtering system that attempts to reduce the information overload and provide customized information access for targeted domains that is likely to be of their interest. A recommender system usually compares a user profile to some reference characteristic and seeks to predict the interestingness value of an item on behalf of the users. These characteristics may be from the information item (the content-based approach [2] or the knowledge-based approach [3]) or the users social environment (the collaborative filtering approach [4][5]). The collaborative filtering based algorithms are mainly designed for transactional datasets. Our problem belongs to the bibliographic domain, which has a lot of latent information contrary to the transactional domain where the user-item relationship is quite straightforward, with the users rating the items they like, either explicitly or implicitly. The bibliographic datasets are also highly sparse in nature. For example, our *DBLP* dataset contains $1,035,532$ authors who have studied a total of $12,057$ topics. On an average an author works on 20 topics, which is 0.16 percent of the total number of topics. The probability of finding the k-nearest neighbors [4] for an active author using collaborative filtering is thus very low. The issue of sparsity is addressed through matrix factorization in [6]. We address the issue of sparsity in our algorithm by using only the relevant information and pruning the unnecessary and irrelevant information at the onset for a given author. The item-based collaborative filtering [5] and the content-based approaches consider only item-item similarity, not user to user similarity thus are not suitable for author recommendation for a given author.

To tackle the aforementioned problems and to exploit the latent information present in a bibliographic dataset, we represent the author-topic relationships as a graph of author and topic nodes. For a given author node, we can directly prune the author and topic nodes present in the graph that cannot be reached from the given node. Also, not all the nodes that are reachable from a node are equally relevant. The relevance goes on decreasing as we move further from the node being considered. This maps to our graphical structure as traveling only to a certain distance from a node and pruning the rest of nodes as having little relevance. [7] applies graph-based approach on a bibliographic dataset for co-author relationship prediction. We have used a topic-centric graph-based model for author and topic recommendations based on the relative shared research interests between authors as well as the importance of authors and topics in the dataset. Temporal recommendation is discussed in [8] where the users preferences are weighted over time to incorporate the temporal effects. We use the evolution of topics for temporal aspects of topic recommendation.

3   Problem Definition

The goal of our algorithms is to suggest relevant authors and topics to a given author A_i based on his topic list τ_{A_i}. Let us consider a list of authors $A = \{A_1, A_2, A_3,...\}$ and a list of topics $T = \{T_1, T_2, T_3,...\}$. Each author A_i has a list of topics $\tau_{A_i} \subseteq T$, that contains all the topics that A_i has worked on. It is to be noted that for any author A_i, $\tau_{A_i} \neq \phi$. For A_i, we also have a list of co-authors $\chi_{A_i} \subseteq A$. There exists a distinguished author $A_i \in A$ called the active author for whom the task is to find the *Top-K* authors and the *Top-N* topics to be recommended. For an active author A_i, we determine:

- **Author Recommendation:** The set of recommended authors $RA_{A_i} = \{A_1, A_2, A_3,...\}$, that contains the authors who according to our recommender algorithms share similar research interest with the active author A_i, but are not the co-authors χ_{A_i} of A_i. Thus,

$$RA_{A_i} = \{A_j \mid A_j \notin \chi_{A_i}, A_j \in A\} \tag{1}$$

- **Topic Recommendation:** The set of recommended topics $RT_{A_i} = \{T_1, T_2, T_3,...\}$, that contains the topics that according to our recommender algorithms are relevant to the active author A_i, but do not belong to the topic list τ_{A_i} of A_i. Thus,

$$RT_{A_i} = \{T_j \mid T_j \notin \tau_{A_i}, T_j \in T\} \tag{2}$$

4   Our Approach

In this section, we present our recommender algorithms. For a given author A_i, our algorithms first determine a list of authors RA_{A_i} whose research interest is *similar* to that of A_i. Using these authors, we determine a ranked list of topics RT_{A_i} that our algorithms detect to be the most interesting and relevant to the author A_i. Before proceeding to our recommender algorithm, we give some background regarding the results of our previous works which we use in assigning recommendation scores. The recommender algorithms proposed in this paper use these concepts and derivations as input. In our previous works, we developed algorithms for author ranking [9] and topic detection and ranking [10].

- **Author Ranking.** For a given author A_i, the author score is represented by AS_{A_i}.
- **Topic Ranking.** For a given author T_j, the author score is represented by TS_{T_j}.

4.1   Author Recommendation

The intuition behind our author recommendation is that for a given author A_i, a new author A_k is a candidate recommendation author if A_k shares considerable research interest with A_i and yet is relevant, novel and latent to A_i. The topic list τ_{A_i} of an active author forms the basis of our recommender algorithm for A_i. For a topic $T_j \in \tau_{A_i}$, the author list α_{T_j} gives the authors that share a research interest with A_i. It is to be noted that an author $A_k \in \chi_{A_i}$ cannot be a candidate recommendation author for A_i because A_k is not latent to A_i. If an author A_k shares considerable research interest with A_i, A_k becomes a candidate recommendation author for A_i. To quantify the notion of considerable shared research interest, we define λ as the threshold for the number of shared topics between the active and the candidate authors. For a given author A_i, we first determine the candidate author A_k, for whom $| \tau_{A_i} \cap \tau_{A_k} |$ is maximum. We call this value γ, which represents the maximum topic match possible for A_i. At this step, we prune the authors who fail to suffice at least a certain fraction of γ. Thus, we define the threshold $\lambda = C * \gamma$, where C is a control parameter having value in the range $[0, 1]$. Having fixed λ, we determine the set ξ_{A_i} of candidate recommendation authors having number of matching topics not less than λ.

Now, to filter the *Top-K* recommended authors from ξ_{A_i}, we assign a similarity score to each author $A_k \in \xi_{A_i}$ that ranks the authors in decreasing order of similarity. For computing similarity between the active author and an author belonging to ξ_{A_i} we apply the *Jaccard Index* on their topic lists. We also incorporate the overall *importance* of the candidate author A_k in the similarity score by including the author score AS_{A_k} in the similarity computation. Thus, the similarity score for an author $A_k \in \xi_{A_i}$ is given by:

$$RS_{A_k} = AS_{A_k} * \frac{| \tau_{A_i} \cap \tau_{A_k} |}{| \tau_{A_i} \cup \tau_{A_k} |} \tag{3}$$

For *Top-K* similarity values, we select the recommended K authors RA_{A_i} Since our algorithm is topic-centric, the set of recommended authors and the active author form a *community* of authors who have worked/are working on similar research areas. C is an important parameter in identifying the number of similar authors. For high values of C, the number of identified similar authors is less and vice-versa. Thus, C can be used to control the *compactness* of the *community*. For high values of C, the similarity of the authors in the community is high, forming a *close community*. For low values of C, the community consists of authors with varied research interest, forming a *diverse community*. We study our algorithm for various values of C to be discussed in Section 5.

4.2   Topic Recommendation

At this stage, for a given author A_i, we have the list of authors RA_{A_i} to be recommended to him. Intuitively, the research interests of the recommended authors are *relevant* to the active author. Thus, the topics that the authors belonging to RA_{A_i} *have worked on* become the candidate recommendation topics. Experimentally, we found that the candidate recommendation topics contain topics that are *redundant* and *dead* as far as the task of recommendation is concerned. The *redundant* topics are the ones that do not contain any extra information whereas the *dead* topics are the ones on which no significant work is being done. We prune such topics in the following two steps:

- **Hierarchical Pruning:** We use the hierarchical structure of the topics to prune the *redundant* topics for the task of recommendation. For any two candidate topics T_j and T_l, we prune T_j if it lies above T_l in the topic hierarchy.
- **Temporal Pruning:** For topic recommendation, temporal pruning is required to remove the *dead* topics. For a candidate topic T_j, we prune T_j if its *best year* was before a user-defined year. For our experiments, we set this user-defined year to 2000.

Let ψ_{A_i} be the set of pruned topics. The *importance* of a topic T_j for recommendation depends on the number of recommended authors suggesting T_j, the overall *importance* of T_j in the corpus and the *importance* of the recommending author. If an *important* author suggests a topic, the suggestion should have more influence on the active author. The *importance* of an author A_k and a topic T_j are given by the author score AS_{A_k} and the topic score TS_{T_j} respectively. It is to be noted that a candidate topic T_j should not belong to the topic list τ_{A_i} of the active author A_i. For an active author A_i, we define recommendation score RS_{T_j} for a candidate topic T_j as:

$$RS_{T_j} = TS_{T_j} * \sum_{A_k \in RA_{A_i}, T_j \in \tau_{A_k}} AS_{A_k}, \text{ where } T_j \notin \psi_{A_i}, T_j \notin \tau_{A_i} \qquad (4)$$

This recommendation score captures the three properties of a recommended topic as mentioned above. Having computed the authoritative recommendation scores for all the candidate topics, we produce the recommendation topic list RT_{A_i} as the *Top-N* topics from amongst the candidate topics.

5   Experimental Study

5.1   Dataset Description and Preprocessing

We use the *DBLP XML Records* available at *http://dblp.unitrier.de/xml/* to show the results of our algorithms. The *DBLP* dataset contains information about $1,632,442$ research papers from various fields published over the years. The total number of topics derived by the topic detection algorithm [15] is $12,057$ and there are $1,035,532$ authors who have worked on one or more topics. All the keywords present in the titles of the research papers were stemmed using the Porter Stemming Algorithm in [9]. For our recommender algorithms, we preprocess the *DBLP* dataset to construct the look-up tables mentioned in Section 4.1. It is to be noted that the *DBLP* dataset used by us contains research papers with citation information till *July* 2010 only.

5.2   Results

A purely objective and quantitative evaluation of the results obtained is difficult due to the lack of standard formal measures for recommendation task in bibliographic dataset. But the recommended author and topic list produced by our recommender algorithms when examined by field experts were found to be relevant. We experimentally validate our results in the next section by dividing the dataset into *training* and *test* sets. In this section, we discuss the results produced by our algorithms. The results for a few authors are given in the following tables. Table 1 shows the top 3 authors recommended by our author recommendation algorithm for the control parameter value $C = 0.7$.

Table 1. Top 3 Recommended Authors for some Authors along with their Jaccard Coefficient and the Number of Matching Topics

Active Author (Number of Topics)	Recommended Author	Jaccard Coefficient	Number of Matching Topics
Divyakant Agarwal (365)	Beng Chin Ooi	0.0237	105
	Hector Garcia-Molina	0.0230	101
	Kian-Lee Tan	0.0225	104
Robin D. Burke (52)	Joemon M. Jose	0.0139	21
	Barry Smyth	0.0116	20
	Jian Pei	0.0089	19
Won Kim (219)	Pierangela Samarati	0.0181	49
	W. Bruce Croft	0.0166	44
	Hector Garcia-Molina	0.0143	52

From Table 1, the following observations were obtained:

- Our author recommendation suggests *Beng Chin Ooi* and *Kian-Lee Tan* to *Divyakant Agrawal*. Our dataset does not contain research papers that were published after *July* 2010. *Divyakant Agrawal* collaborated with *Beng Chin Ooi* for publications in 2011 and 2012 and with *Kian-Lee Tan* in *late* 2010 and 2011. Thus for an active author, our recommendation algorithm is able to suggest authors who can be possible collaborators in future.
- Overlapping recommendations lead to the formation of *communities*. An example of this kind of overlapping recommendation can be seen from the recommendation of *Hector Garcia-Molina* to both *Divyakant Agrawal* and *Won Kim*. In this case, the intersection of topic lists of the three authors is significant and thus *Hector Garcia-Molina* is recommended to both *Divyakant Agrawal* and *Won Kim*.

Using the recommended authors from the author recommendation algorithm, the topic recommendation algorithm recommends the *important* topics to a given author. Table 2 show the top 3 topics recommended for some authors by our algorithm.

Table 2. Top 3 Recommended Topics for some Authors along with the Topic Authoritative Scores

Active Author	Recommended Topic	Score
Sergey Brin	*recommend*	0.1162
	answer	0.1057
	social network	0.1051
Robin D. Burke	*search engin*	0.0920
	summar	0.0812
	data mine	0.0807
Jiawei Han	*multi agent*	0.1692
	relev	0.1593
	watermark techniqu	0.1016

From Table 2, the following observations were obtained:

– Our topic recommendation suggests *relev* and *watermark techniqu* to *Jiawei Han*. As mentioned earlier, our dataset does not contain research papers published after *July 2010*. *Jiawei Han* has publications on the recommended topics *relev* and *watermark techniqu* in 2010 and 2011 respectively.
– The research interest of *Sergey Brin* was concentrated on *web search* with his last publication being in the year 2005. Over the years, the research interest of the people working on *web search* has shifted to recommendations and social networks. This notion is captured in our topic recommendation result for *Sergey Brin*.

5.3   Validation

We validate our experiments by dividing the dataset into *training* and *test* portions. Also, to determine the optimum value of the control parameter C, we test the results on different values of C.

Author Recommendation. For validation of our author recommendation algorithm, we removed the check on co-author occurrence. Thus for a given active author, a co-author can appear in his recommendation list. One measure to test the accuracy of our results is *recall* of the co-authors in the *Top-K* recommended authors for different values of C.

We define *recall* for the whole dataset as the mean of the *recall* of all the active authors present in it. Thus:

$$recall = \mu(\sum_{A_i} \frac{Number\ of\ CoAuthors\ Retrieved}{Total\ Number\ of\ CoAuthors}) \tag{5}$$

From Figure 1, we observe that the *closeness* of the recommended authors depends on the values of C. Higher the value of C, higher is the value of λ, and consequently more similar are the recommended authors though less in number. Now, for high values of C, we only get very similar co-authors in the recommended set. Since the number of such co-authors is limited, the *recall* remains nearly the same on increasing the value of K. As we lower the value of C, more

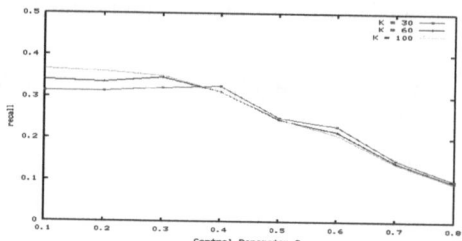

Fig. 1. Variation of *Recall* with Control Parameter C, $K = \{30, 60, 100\}$ for Author Recommendation

co-authors overcome the threshold λ to become candidate for recommendation. In Information Retrieval, *(1 - recall)* value represents the ratio of misclassified and irrelevant instances. But in our algorithm, the set of authors which are not the co-authors but still appear in the *Top-K* retrieved authors, are not misclassified instances. They are similar to the active authors in terms of their relative shared research interests. Thus, even a moderate *recall* value can be used to infer quality results.

Topic Recommendation. For the validation of our topic recommendation algorithm, we divided the topic list of each author into *training* and *test* portions such that the topics in *training* portion precede the topics in *test* portion in the active author's topic study timeline. We kept the topics studied by the author in the initial ρ percentage of his *timeline* as *training* set and the rest as *test* set. Thus if Y is the year when the author first studied a topic and if Y lies in the first ρ percentage of his active years, then it belongs to *training* topics, otherwise it belongs to *test* topics. We test the accuracy of our results by the *recall* of the topics for different values of C. For our experiments we keep $\rho = 70$. We define *recall* for the whole dataset as the mean of the *recall* of all the active authors present in it. Thus:

$$recall = \mu(\sum_{A_i} \frac{Number\ of\ Test\ Topics\ Retrieved}{Total\ Number\ of\ Test\ Topics}) \tag{6}$$

In Figure 2, the *recall* graph shows that for higher values of N, *recall* value is higher. This is because as the value of N increases, more and more *test* topics come in the recommended topic set. It is to be noted that we do not use *precision* because *precision* is not an ideal measure for result evaluation in recommendation systems [11] as the recommended topics though not in the *test* set are not irrelevant or misclassified instances.

5.4 Performance Evaluation

To evaluate the performance of our author and topic recommendation algorithm, we show the runtime of our algorithm with respect to varying number of authors in Figure 3.

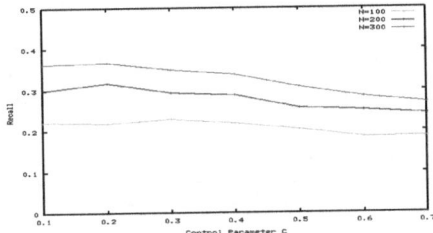

Fig. 2. Variation of *Recall* with Control Parameter C, $N = \{100, 200, 300\}$ for Topic Recommendation

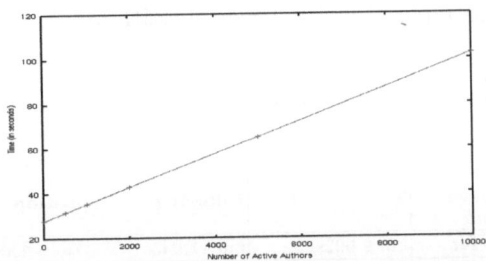

Fig. 3. Graph showing the Variation of Runtime of our Recommendation Algorithm with the Number of Active Authors

The preprocessing time of our algorithm, *i.e.* the time taken to construct the information tables *ACI*, *TAI* and *ATI* is 28.03 *seconds*, shown corresponding to number of authors = 0 in the figure. As we increase the number of active authors for which the author and topic recommendation is generated, the runtime increases linearly. The linear runtime with respect to the number of authors shows that our algorithm is highly scalable. The look-up tables ensure constant *amortized* runtime for each active author and hence the linearity of runtime.

5.5   Comparison

In this subsection, we compare our author and topic recommendation algorithms with a few popular recommendation algorithms: *Memory-based Neighborhood Collaborative Filtering* [13], *k-Nearest Neighbor* [12], *Item-based Collaborative Filtering* [5] and *Probabilistic Matrix Factorization* [14].

Our dataset does not contain explicit ratings for the author-topic pairs. We generate the implicit ratings from the topic list of the authors. For a given author A_i, we first determine $|P_{A_i, T_j}|$, the number of research papers in which A_i has studied the topic $T_j \in \tau_{A_i}$. Now, we scale down/up the value of $|P_{A_i, T_j}|$ for each topic $T_j \in \tau_{A_i}$ to give a rating of 10.0 to A_i's most studied topic. Once we have the ratings for the author-topic pairs, we run each of the four algorithms mentioned above in the experimental settings show in Table 3.

Table 3. Experimental Settings for Comparison

Experiment Number	Paper Count	Number of Authors	Number of Topics	Number of Active Authors	K	N
1	1, 000	1, 385	1, 404	1, 385	40	100
2	10, 000	17, 224	7, 019	17, 224	50	100
3	100, 000	127, 606	11, 384	2, 000	100	200
4	1, 632, 442	1, 036, 950	12, 057	10, 000	200	300

Table 4 shows the runtime of the algorithms under comparative study. We call our algorithm topic-centric recommendation. It is clear from the table that as the size of the dataset grows, our recommender algorithms become more and more time-efficient compared to the other algorithms.

Table 4. Comparison of the Time Taken by the Recommendation Algorithms

Experiment Number	Memory Based Collaborative Filtering	k-Nearest Neighbor	Item-based Collaborative Filtering	Probabilistic Matrix Factorization	Topic-centric Recommendation
1	0m 4.544s	0m 9.992s	6m 11.943s	1m 0.990s	0m 5.276s
2	13m 47.408s	35m 15.049s	1080m 18.828s	61m 10.096s	10m 39.497s
3	11m 23.875s	62m 16.372s	734m 39.832s	141m 27.070s	6m 54.562s
4	159m 56.395s	4402m 40.829s	1636m 13.773s	2402m 43.941s	128m 28.836s

For comparing the quality of results of our author recommendation algorithm, we draw the histograms for *precision, recall* and *F1-Score* obtained on running the three algorithms, viz. *Memory-based Collaborative Filtering, k-Nearest Neighbor* and *Topic-centric Recommendation*. The *test* set for a given active author in case of author recommendation is the list of his co-authors. We run all the three algorithms with the setting that co-authors are allowed to be recommended to a given author. The histograms for *precision, recall* and *F1-Score* are shown in figures 4.

We see that the *precision* value of *Memory-based Collaborative Filtering* algorithm is very high compared to the other algorithms in the experimental setting 1. Thus, even though the *recall* value of our algorithm is quite high in experimental setting 1, the *F1-Score* is more for *Memory-based Collaborative Filtering*. This is because memory-based collaborative filtering performs well on dense datasets. As the dataset grows and the sparsity increases, the performance of memory-based collaborative filtering starts to drop as seen in the experimental settings 2, 3 and 4. As the dataset grows, our algorithm outperforms the other two algorithms on all the three parameters. It is to be noted that our algorithm consistently fairs better on *recall* as compared to the other two algorithms.

To compare our topic recommendation algorithm, we draw the histograms for *precision, recall* and *F1-Score* for the five topic recommendation algorithms. The histograms for *precision, recall* and *F1-Score* are shown in figures 5. Similar

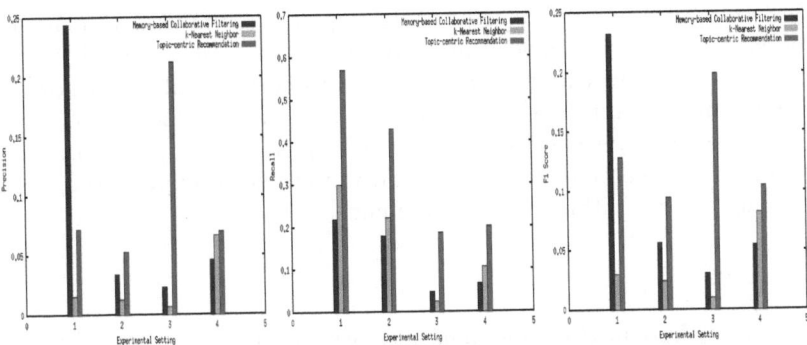

Fig. 4. Bar Graph showing *Precision*, *Recall* and *F1-Score* respectively for the Author Recommendation Algorithms

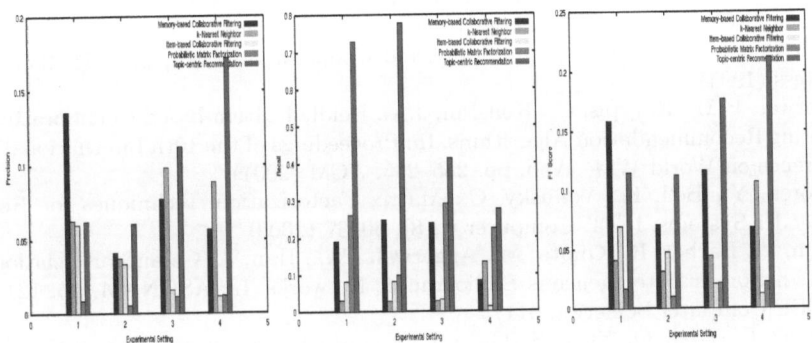

Fig. 5. Bar Graph showing *Precision*, *Recall* and *F1-Score* respectively for the Topic Recommendation Algorithms

to the case of author recommendation, *Memory-based Collaborative Filtering* performs well on *precision* and hence *F1-Score* on a small dataset. But as the dataset grows, our algorithm outperforms other algorithms.

6   Conclusions and Future Work

In this paper, we have proposed a novel approach for author and topic recommendation in a bibliographic dataset by exploiting the latent author-topic and author-author relationships. For a given active author, we derive a ranked list of recommended authors based on the similarity scores designed in our algorithm. Based on the importance of a topic and the recommending author, we provide a ranked list of recommended topics. The various pruning steps make our algorithm fast and scalable. We analyzed the recommendation results on various parameters and validated them by dividing the dataset into training and test portions and also compared to some of the popular recommendation algorithms. As mentioned earlier, our algorithms have a host of applications. In future, we would like to implement active learning by considering explicit author opinion, community mining, author social network analysis, etc. One future work is to test our algorithms beyond accuracy metrics [15]. Also, having proposed the framework for a recommender system, we would like to explore the possibility of extending our algorithms to other domains.

References

1. Resnick, P., Varian, H.R.: Recommender Systems. Commun. ACM 40(3), 56–58 (1997)
2. Pazzani, M.J., Billsus, D.: Content-Based Recommendation Systems. The Adaptive Web 2, 325–341 (2007)
3. Burke, R.D.: Knowledge-based Recommender Systems. Encyclopedia of Library and Information Systems 69(suppl.32), 175–186 (2000)
4. Resnick, P., Iacovou, N., Suchak, M., Bergstrom, P., Riedl, J.: GroupLens: An Open Architecture for Collaborative Filtering of Netnews. In: Proceedings of the 4th Conference on Computer Supported Cooperative Work, pp. 175–186. ACM Press (1994)
5. Sarwar, B.M., Karypis, G., Konstan, J.A., Reidl, J.: Item-based Collaborative Filtering Recommendation Algorithms. In: Proceedings of the 10th International Conference on World Wide Web, pp. 285–295. ACM (2001)
6. Koren, Y., Bell, R., Volinsky, C.: Matrix Factorization Techniques for Recommender Systems. IEEE Computer 42(8), 30–37 (2009)
7. Sun, Y., Barber, R., Gupta, M., Aggarwal, C.C., Han, J.: Co-author Relationship Prediction in Heterogeneous Bibliographic Networks. In: ASONAM, pp. 121–128. IEEE Computer Society (2011)
8. Xiang, L., Yuan, Q., Zhao, S., Chen, L.: Temporal Recommendation on Graphs via Long- and Short-term Preference Fusion. Journal of KDD 29(2), 723–731 (2010)
9. Singh, A.P., Shubhankar, K., Pudi, V.: An Efficient Algorithm for Ranking Research Papers based on Citation Network. In: Proceedings of the 3rd Conference on Data Mining and Optimization, DMO 2011, Bangi, Malaysia, pp. 88–95 (2011)
10. Shubhankar, K., Singh, A.P., Pudi, V.: An Efficient Algorithm for Topic Ranking and Modeling Topic Evolution. In: Hameurlain, A., Liddle, S.W., Schewe, K.-D., Zhou, X. (eds.) DEXA 2011, Part I. LNCS, vol. 6860, pp. 320–330. Springer, Heidelberg (2011)
11. Herlocker, J.L., Konstan, J.A., Terveen, L.G., Riedl, J.T.: Evaluating Collaborative Filtering Recommender Systems. ACM Transactions on Information Systems 22(1), 5–53 (2004)
12. Rashid, A.M., Lam, S.K., LaPitz, A., Karypis, G., Riedl, J.T.: Towards a Scalable k-NN CF Algorithm: Exploring Effective Applications of Clustering. WEBKDD, 147–166 (2006)
13. Su, X., Khoshgoftaar, T.M.: A Survey of Collaborative Filtering Techniques. In: Adv. in Artif. Intell., pp. 4:2–4:2. Hindawi Publishing Corp. (2009)
14. Salakhutdinov, R., Mnih, A.: Probabilistic Matrix Factorization. In: NIPS (2007)
15. McNee, S.M., Riedl, J.T., Konstan, J.A.: Being Accurate is Not Enough: How Accuracy Metrics have hurt Recommender Systems. In: CHI Extended Abstracts, pp. 1097–1101 (2006)

Combining Spatial Cloaking and Dummy Generation for Location Privacy Preserving

Nana Xu[1,2], Dan Zhu[1,2], Hongyan Liu[3,*], Jun He[1,2,*], Xiaoyong Du[1,2],
and Tao Liu[1,2]

[1] Key Labs of Data Engineering and Knowledge Engineering, Ministry of Education, China
[2] School of Information, Renmin University of China
{xunana,zhudan,hejun,duyong,tliu}@ruc.edu.cn
[3] Department of Management Science and Engineering, Tsinghua University
hyliu@tsinghua.edu.cn

Abstract. Location privacy preserving is attracting more and more attentions with the wide use of accurate positioning devices. Two kinds of methods based on k-anonymity have been proposed for location privacy preserving. One is spatial cloaking and the other is dummy generation. However, simple cloaking may result in large cloaking region, especially when user distribution is sparse. Large cloaking region means low efficiency and high time cost. Dummy generation will also arouse great extra system cost. In this paper, we propose an approach that combines the two techniques to achieve an ideal complement. On one hand, cloaking algorithm is used to blur precise location into a region so that location privacy can be well protected. On the other hand, to those users whose cloaking region is large, generating limited number of dummies can help decrease the size of region. The effectiveness of our methods has been demonstrated by comprehensive experiments.

Keywords: Location Privacy, k-Anonymity, Spatial Cloaking, Dummy Generation.

1 Introduction

In recent years, location based services (LBSs) have become widely used in all aspects of our lives. For example, a user wants to find an ideal path with the GPS device, or a user wants to find the nearest restaurant. There is no doubt that they have provided great convenience for us. However, privacy issues consequently happen. For example, mobile phone is one of the most common service terminals nowadays; the adversary can link a phone number to a certain user with a high probability, which discloses the real identity of the user and lead to the disclosure of some sensitive information, such as medical condition, home address, and so on. In conclusion, location privacy preserving is quite a necessary and urgent task.

A lot of works have been done on this problem. For all of them, k-anonymity [4] serves as the fundamental idea. They can be categorized into two types: dummy based

---

* Corresponding authors.

S. Zhou, S. Zhang, and G. Karypis (Eds.): ADMA 2012, LNAI 7713, pp. 701–712, 2012.
© Springer-Verlag Berlin Heidelberg 2012

and cloaking based. For the first type, the Anonymizer generates k-1 false positions along with the true position to LBS server. For the adversary, it cannot tell which the true position is. The main idea of the second type is blurring the true position of a user into a region by a trusted middle-server. At the same time, the server must make sure that there are at least k different users in this region.

Although existing works contribute much to location privacy preserving, there are still some limitations. For cloaking based methods, unlimited cloaking will result in large cloaking area, which has two negative effects on user services. Firstly, large cloaking region usually means less accurate search results. For example, given a user who is located in Renmin University and wants to find the nearest restaurant, his position was blurred to Zhongguancun Street after cloaking. If the service provider wants to find the target restaurant, it has to search all the restaurants on and near the long Zhongguancun Street. Besides, if the user is not familiar with this region, it is difficult for him to pick up the best solution further. This is not a good experience for the user. Also, if the size of the cloaking region is limited to a small size, k-anonymity may not be able to meet. As to dummy generation, the server has to process a large number of false queries, which are meaningless. This no doubt increases the system cost, especially the responding time of users' queries. Busy in handling false queries, true queries may be delayed to answer.

To cope with these problems, it is reasonable for us to combine these two kinds of methods. To satisfy users' anonymity needs, cloaking includes two steps. Firstly, enlarge the cloaking region to a certain extent. Then if k-anonymity is not satisfied, generate a certain number of dummies to meet the k-anonymity as much as possible. This procedure can achieve complement between simple cloaking and dummy generation. Cloaking region can be controlled in an acceptable range, while the system does not pay much extra cost. The contributions of the current study are summarized as follows.

• We propose a strategy to combine cloaking based and dummy based methods. In addition, we do some improvement to make generated dummies be used more effectively.

• We define several evaluation metrics to measure the effectiveness of the privacy preserving method, and conduct a series of experiments to prove the effectiveness of our proposed approaches.

The rest of the paper is organized as follows: Section 2 describes some related works. Section 3 introduces our solutions to location privacy preserving. Experimental studies are presented in Section 4. Finally, the paper is concluded in Section 5.

2   Related Work

There is a basic concept of k-anonymity in lots of anonymity methods when publishing data in relational databases. Besides explicit identifier (e.g. name), there are other attributes (e.g. zip code), called quasi identifiers [4], that can potentially identify users. The notion of k-anonymity is that if one user has some quasi identifier values, there are at least k-1 users have the same quasi identifier values in a table, then the probability that each user is identified is less than $1/k$. Furthermore, l-diversity [4] is introduced to strengthen the privacy protection. Within each group of users with the same quasi identifier value, it asks l different sensitive attribute values. The privacy strategy of most location-based privacy preserving measures is also k-anonymity, that is, each user cannot be distinguished from k-1 other users. Existing methods can be classified into two broad categories: spatial cloaking [1, 2, 5, 6, 10, 11, 13, 14] and dummy generation [3, 7, 8, 9, 15].

Spatial cloaking is the most popular mechanism. It assumes that there is a trusted middle-server between the clients and the service provider. The middle-server receives users' locations, and for each user, it tries to find a region that includes the current user and k-1 other users. Then the regions are sent to the server instead of the exact locations of users. Since there are at least k users in the cloaking region, the probability that each user is recognized is less than $1/k$. Apart from the parameter k, other parameters are used by different measures, including the minimum cloaking region, the maximum cloaking region, the maximum time tolerance, and so on. Casper [11] and PrivacyGrid [1] divide the whole space into a lot of basic square grids. The regions started from one grid are expanded until the parameter k is satisfied. Xstar [13] introduces l-diversity. According to it, most users are moving on the road networks, and if a cloaking region contains just one road segment, users' locations can be identified easily. Consequently, Xstar hides a user's location with a set of road segments instead of a square cloaking region. There is a method [10] focuses on delivering ads. Besides cloaking regions, users' profile information (including age, gender, and so on) are also sent to the server. The middle-server is required to find a region containing k users that have similar profile information.

Dummy generation methods believe that when the density of users is sparse, the cloaking region generated by spatial cloaking methods is too large. Large cloaking area will lead to high processing cost of service provider and large result sets sent to users. In dummy generation methods the clients generate a number of fake users and send them to the server directly. Kido et al. [7, 8] first proposed the dummy generation mechanism, and they focus on snapshot scenarios. You et al. [15] attempt to generate human-like trajectories which imply users' long term movements to protect users' location privacy. PAD [9] generates dummies which distribute over an area of certain size.

However, dummy generation methods also have drawbacks. Since dummies are considered to be real users, the service provider has to process a large number of false queries, which leads to high system cost. Therefore, we adopt the trusted middle-server architecture and combine the two popular approaches to prevent their drawbacks.

3 Anonymous Methods

3.1 System Architecture

In this part, we introduce the system architecture and its working mechanism. Middle service server was first proposed in [5]. In this paper, our system is also based on the trusted middleware, which will be called Location Anonymizer in the rest of the paper. When a user wants to request service from a service provider, he needs to send the following information: k, A_{max}, position, and query. k defines that the user wants to be k-anonymity. A_{max} indicates that the size of cloaking region of the user should be smaller than or equal to A_{max}. Position represents the user's exact position and query describes what the user wants the LBS to do. Figure 1 shows a simple description of how the system works.

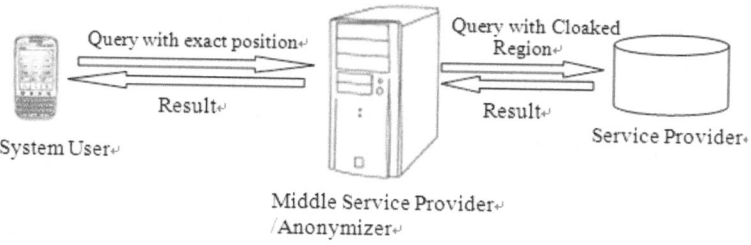

Fig. 1. Illustration of System Architecture

3.2   Problem Definition

In this section, we introduce the concepts and the proposed algorithms: the *Basic* and the *Adaptive*. Our algorithms are gird-based. The basic data structure is similar to PrivacyGrid [1].

Definition 1. (Initial Query) For a user U, his current position is p, his query requirement is q, his anonymity need is k and the region tolerance is A_{max}, i.e., this user cannot be distinguished from at least $k - 1$ other users within A_{max} grids.

Then a user's initial query is defined as:

$$Q = (p, q, k, A_{max})$$

Different users have different demands for k and A_{max}.

Definition 2. (Cloaked Query) after users exact position has been processed by the middleware, the cloaked query can be described as:

$$CQ = (R, q)$$

R describes the cloaking region of user, $R = \{x, y, l, w\}$, where x and y are the coordinates of left-top gird of R, l and w represents the length and width of the region respectively. For generated dummies, they are processed as true users so their queries have the same form as mentioned above. When we call users in the following section, we mean both true users and generated dummies.

3.3   The *Basic* Algorithm

The previous cloaking algorithms focus on providing more efficient strategy so as to form the minimum region while k-anonymity is satisfied. In our algorithm, we introduce dummy generation to cloaking method. Thus, for each step of the algorithm, there exist two choices, enlarging the size of region or generating dummies. For example, in Figure 2 (a), there is a user U_1, whose anonymity requirement is 10 and region tolerance is 6. According to our strategy, the system can generate 5 other dummies and keep the cloaking region size 3. Or the system may enlarge the cloaking region to 6 and generate 3 dummy. As we have mentioned before, the ideal solution should have smaller cloaking region and fewer dummies. Then we need a measurement to evaluate different solutions. The measurement is defined by the following equation.

$$V = (\mid R_s \mid \times \rho^{\Delta_k})^{-1} \tag{1}$$

R_s represents the cloaking region and $|R_s|$ is its size. For the same size, the shape of region can be different, which means different total number of users. In this paper, $|R_s|$ is measured by the number of grids. ρ is an system defined argument that controls the weight of dummy generation. Δ_k is the number of dummies that are needed to fulfill k-anonymity within s. V is in inverse proportion to the size of the region and value related to the number of dummies generated, which is consistent with the common sense. Then the problem converts to finding R_s and Δ_k to make V the largest, which can be described as a triple (R_s, Δ_k, V_L).

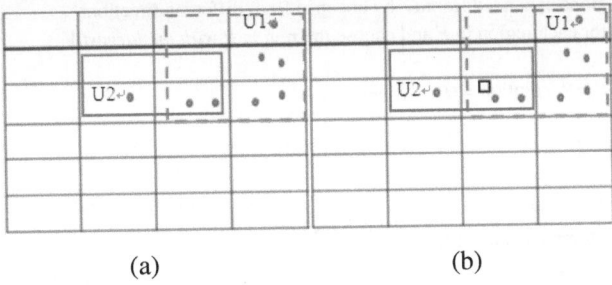

(a)   (b)

Fig. 2. An example to show that changing processing sequence is reasonable

The initial cloaking region is the grid in which the user is located and initial V_L is calculated according to grid condition. The algorithm performs by following steps. It first checks if k-anonymity and spatial constraints are satisfied within the current region. If yes, the system will take this region as the cloaking results for the user. Otherwise, the algorithm will expand the cloaking region and calculate V. The algorithm will calculate the value V corresponding to the region R_s that has the most users inside to compare with the previous V_L. If the new V is larger, then the best solution is replaced by (R_s, Δ_k, V). All the solutions with the same size *region* will be saved for the next round. This process will continue until k-anonymity is reached within A_{max} or the size of cloaking region is larger than A_{max}. In order to control the total number of dummies, we define that the number of dummies that need to generate for a user should be no more than 20% of the user defined k.

As the number of dummies is limited, there still exist some users who are located in an extremely sparse region that anonymity cannot be satisfied within A_{max}. We call these users unsuccessful cloaking users and the system will return a region which has the maximum number of users within size A_{max}. For successful cloaking users, the system will take R_s as the cloaking region and generate Δ_k dummies for users at randomly chosen positions. The region information is then updated so that the later users can reuse those generated dummies.

In order to prevent endless loop, dummies are processed in a simple way that we do not consider their degree of anonymity, .i.e., there is no need to generate dummies for dummies. For each dummy, choose a value for s according to distribution of real users' value of s and find the cloaking region that has the largest number of users inside. The major steps of the algorithm, which we call Basic is shown as follows.

Algorithm 1. Basic (User position p, anonymity degree k, region tolerance A_{max})

1:   $K = k, Max = A_{max}$
2:   R_s is the gird where the user is located
3:   $V = (|R_s| \times \rho^{\Delta_k})^{-1}, solution = (R_s, \Delta_k, V)$
4:   $V_T = V; s = 1; G = R_s$
5:   while $s < Max$ and R doesn't meet k
6:     F =set of regions expanded from G by one grid from different directions
7:     calculate V for each region in F and choose the *region R with the largest V*
8:     if $V > V_T$
9:        $V_T = V; R_s = R; solution = (R_s, \Delta_k, V_T)$
10:    end if
11:    $s = s + 1; G = F$
12:   end while
13: if the user is successful
14:    generate Δ_k dummies randomly
15:    update the region
16:    return solution
17: else
18:    return a region of size of A_{max} with the most users inside
19: end if

3.4   *Adaptive* Algorithm

Dummies are useful for improving success rate (the ratio of the number of successful users over the number of total users), while the number of dummies has to be controlled to decrease system cost. As a result, making a full use of each dummy is an important way to improve the previous method. To achieve this goal, we put forward *Adaptive* algorithm that changes the way dummies are placed and the sequence users are processed. The reasons why we make these two changes are discussed below.

Commonly, users are processed according to their arriving time, which is not an efficient way for dummy reusing. Taking the simple situation in Figure 2 as an example, suppose user U_1's query is $(6, 10, p_1, q_1)$ and it comes to the system earlier than U_2's query $(4, 4, p_2, q_2)$. Traditionally, the algorithm will process U_1 first and compute the optimum solution. The dotted line box in Figure 2 is cloaking region of U_1 with 7 users (represented by dots) inside and three dummies are needed to fulfill 10-anonymiy. However, as the number of dummies of U_1 should be no more than 2 so that U_1 is regarded as unsuccessful users by algorithm *Basic*. Figure 2 (*b*) shows the result that U_2 is processing earlier than U_1. From this figure, we can see that one dummy (square box) is generated in U_2's cloaking region and coincidently this dummy is within U_1's cloaking region so that U_1 can be changed from unsuccessful to successful user.

Motivated by this situation, we propose to change the processing sequence of users who arrive at the system within a short time interval according to value of k/A_{max}. We intend to process those users whose anonymity requirements are easy to fulfill earlier regardless of their arriving time, which means that the smaller k/A_{max} is, the easier to meet the user's requirement, the earlier this user will be processed. Generally, users who have a large value of k but small A_{max} are more likely to be regarded as unsuccessful by the algorithm. Because the time interval is quite small, it can be omitted compared to the time taken for cloaking. Therefore, users will not feel any difference when using this service.

Another factor that has an influence on reuse rate of dummies is how dummies are placed. In the *Basic* method, the algorithm places Δ_k dummies randomly in the cloaking region for a user, which is not an effective way to use dummies and not practical in real situation. To solve this problem we propose to put all Δ_k dummies in the grid which has the most users within the current cloaking region and the grid is placed dummies for the first time for a user. The reasons why we choose this strategy are as follows. Firstly, the grid which has the most users usually corresponds to prosperous area that is visited more in real life. Dummies in this grid may be reused more by later users so that they can be successful users with a higher probability. What is more, placing all the dummies of a user in one grid can help decrease the size of cloaking region compared to separating Δ_k dummies to the whole cloaking region. The later user can get Δ_k more users with one grid other than $|R_s|$ grids as long as his maximum cloaking region includes this grid. But if one gird has been placed dummies before, there is a good chance that the total number of dummies is the most among the grids in the whole region. Therefore, the next round of placement should exclude this grid.

4   Experimental Evaluation

4.1   Evaluation Metrics and Dataset

In order to verify the effectiveness of our approaches, we introduce the following three evaluation metrics: the size of cloaking region, the success rate, and the number of dummies. In the following part, we will show the performances on these metrics of three algorithms, i.e., the *Basic*, the *Adaptive*, and the *Baseline*, which means that simple cloaking is implemented for comparison with various arguments.

Decreasing the size of cloaking region is the main purpose of our study. From this perspective, it is an important metric of measuring the effectiveness of the algorithm. Since the distribution of users is sparse, users' degree of anonymity may not all be satisfied. The anonymous process is considered to be successful if a user's degree of anonymity can be satisfied within A_{max} cloaking region which is defined by the user self. The success rate is defined as the percentage of successful users. The higher success rate is a sign of improvement. With the introduction of dummies, the success rate should be improved theoretically. On one hand, dummies can help decrease the size of cloaking region and increase success rate; on the other hand, they can cause extra system cost. To make a balance between the two sides is critical and that is why we put forward the algorithm *Adaptive*. Thus, we make a comparison to examine whether the *Adaptive* has a higher reuse rate of dummies than the algorithm *Basic*, which means higher success rate with less cost.

For dataset, we use the well known Network-based Generator of Moving Objects [12] to simulate moving objects on road network and we make some adjustments to make it more suitable for our experiments. It takes a real map of Oldenburg, Germany as input. This city covers about $15 \times 15\ km^2$. The output is a certain number of users. The default parameters of our experiments are listed in table 1.

Table 1. Default parameters

Name	Default Value
The number of users in whole region	6000
The number of queries	1000
Degree of anonymity, k	20-25
Range of A_{max}	15-25

4.2   Experimental Results

4.2.1   The Size of Cloaking Region

In this part, we study the performance of three algorithms on the metric of the size of cloaking region. Because different users have different k values, the value is represented by an interval. The comparison result of three algorithms is plotted in Figure 3(a).

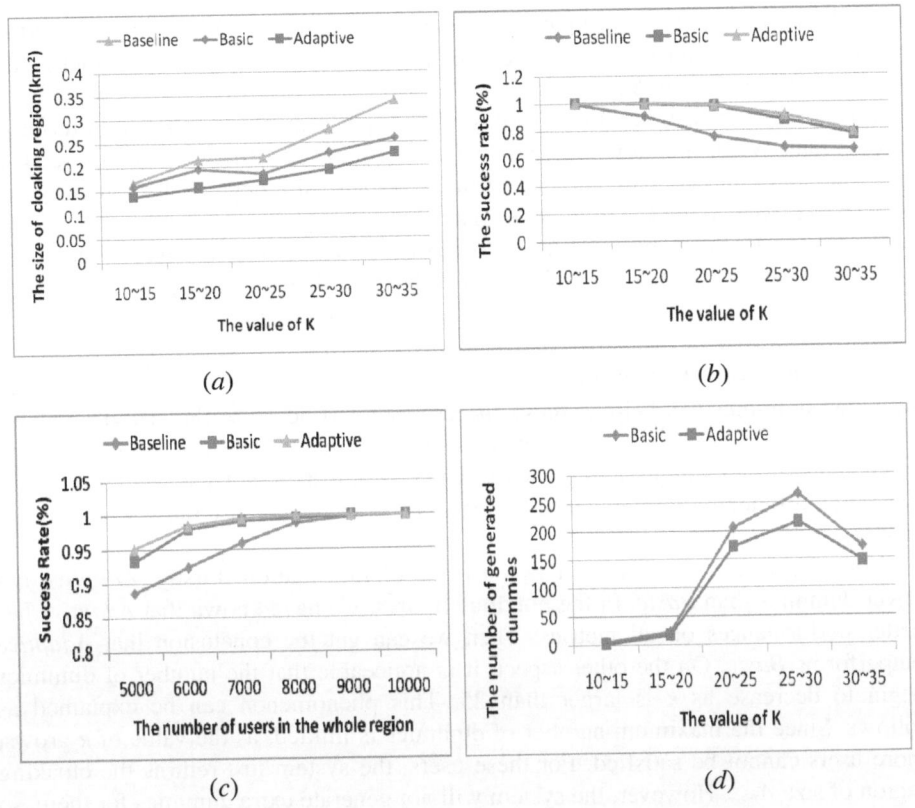

Fig. 3. The comparison of three methods on different metrics

From Figure 3(*a*), we can see that the size of cloaking region using *Basic* is smaller than *Baseline*. For *Baseline*, if the anonymity degree of a user can not be satisfied, then the system will return the A_{max}. For *Basic*, with the use of dummies, more users can achieve *k*-anonymity within A_{max}. Besides, we can see that *Adaptive* has a smaller cloaking region than *Basic*. *Adaptive* has a higher reuse ratio of dummies than *Basic*.

4.2.2   The Success Rate

Introducing dummies can not only help decrease the size of cloaking region, but also increase the percentage of successful cloaking users. Successful users are those whose anonymity requirement can be fulfilled within A_{max}. In this part, we will compare the success rate of three methods. Figure 3(*b*) shows performance result of the three algorithms.

From Figure 3(*b*), we can see that compared with *Baseline*, success rate of the proposed methods increases greatly. The reason is that introducing dummies results in transferring a part of unsuccessful users to successful users. Two aspects ensure this kind of effect. Firstly, a user can reuse dummies that are generated for other users earlier. Secondly, if the user's anonymity degree cannot be satisfied, the maximum number of generated dummies is limited. However, for some users who have a high demand of k and small A_{max}, or who are located in extremely sparse region, adding dummies may still not be able to reach k-anonymity. For these reasons, there exist unsuccessful users using *Basic*.

Besides, we can see that *Adaptive* has a better performance than *Basic* in terms of success rate. When the number of k is between 10 and 25, the two methods seem to have the same unsuccessful number. That is because the degree of anonymity can easily be satisfied when k is relatively small.

We also vary the number of users to see how success rate changes. Figure 3(c) presents the results. That is not difficult to understand that anonymity of a user can be more easily to satisfy with more users located nearby within the same A_{max}.

4.2.3   The Number of Dummies

Although dummies can help increase success rate and decrease the size of cloaking region, they cause extra cost. From this perspective, making full use of dummies can make dummy generation strategy work better. *Basic* does not pay attention to the utilization efficiency. *Adaptive* takes some steps to increase reuse ratio, which means less dummies, higher efficiency.

From Figure 3(d), we can get two aspects of information. Firstly, *Adaptive* has fewer dummies than *Basic*. In the previous figures, we have known that *Adaptive* has better performances on all metrics. Then we can get the conclusion that *Adaptive* outperforms *Basic*. On the other aspect, it is noticeable that the number of dummies begin to decrease as k is larger than 25. This phenomenon can be explained as follows. Since the maximum number of dummies is limited, as the value of k grows, more users cannot be satisfied. For these users, the system just returns the cloaking region of size A_{max}. However, the system will not generate extra dummies for them, so the total number of dummies decreases.

4.2.4   Other Results

In order to show *Adaptive* has a higher reuse ratio of dummies than *Basic*, we calculate the average reuse ratio of them. Figure 4 (a) shows the result.

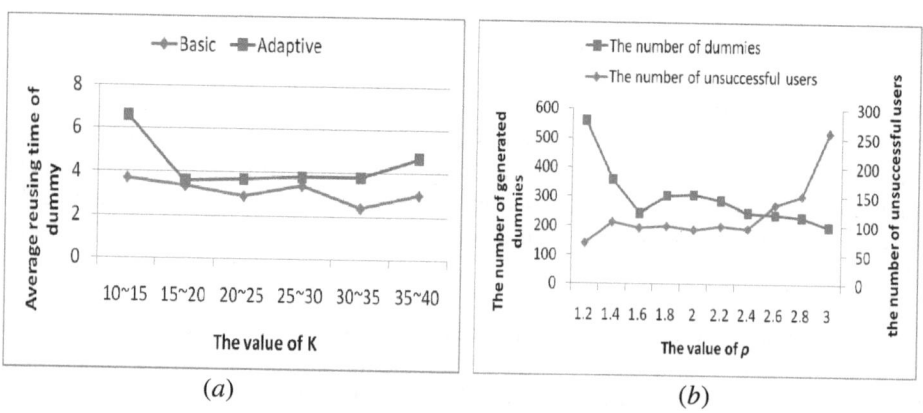

Fig. 4. Explanation of experiment phenomenon

From Figure 4(a), we can see that for *Adaptive*, the average reuse times are larger than that of *Basic*. This explains why fewer dummies can result in better performance.

In our algorithm, there is an argument ρ, which controls the weight of enlarging region and generating dummies. The value of ρ has an obvious effect on the result. When ρ becomes smaller, it means that generating dummies is prior to enlarging region when choosing the best solution. In order to study how ρ influences the results, we conduct a series of experiments with the *Basic*. The results are shown in Figure 4(b).

From Figure 4(b), we can see that when the value of ρ is between 1.6 and 2.6, there are fewer dummies and unsuccessful cloaking users. When ρ is small, dummy generation is preferable for choosing the best solution. Thus more dummies are generated. Of course these dummies can help improve success rate. But large number of dummies increase system cost. Therefore, it is not an ideal choice. When ρ is larger than 2.6, the algorithm prefers to enlarge the size of cloaking region. Thus fewer dummies are generated and reused, which contributes to lower success rate. According to this experiment, we can come to the conclusion that setting the value of ρ between 1.6 and 2.6 is a preferable choice. In the previous experiments, the value of ρ is 2.0.

5   Conclusion

In this paper, we propose two algorithms, the Basic and Adaptive, which can take advantage of spatial cloaking and dummy generation methods to protect location privacy. Comparison with previous works, our methods have the following advantages. First, cloaking method can help blur precise location into a region so that location privacy will not be sacrificed. Second, to users who are located in sparse region, our methods can generate limited number of dummies to decrease the size of region; to users that k-anonymity cannot be satisfied within the maximum cloaking region, generating dummies can help meet their requirements with more chances. The experiment results demonstrate that our methods are effective.

Acknowledgement. This work was supported in part by the National Natural Science Foundation of China under Grant No. 70871068, 70890083, 71110107027, 61033010 and the Fundamental Research Funds for the Central Universities, and the Research Funds of Renmin University of China (10XNI018).

References

1. Bamba, B., Liu, L., Pesti, P., Wang, T.: Supporting anonymous location queries in mobile environments with PrivacyGrid. In: Proceedings of the International Conference on World Wide Web (2008)
2. Chow, C.Y., Mokbel, M.F., Liu, X.: Spatial cloaking for anonymous location-based services in mobile peer-to-peer environments. GeoInformatica 15(2), 351–380 (2011)

3. Duckham, M., Kulik, L.: Simulation of Obfuscation and Negotiation for Location Privacy. In: Cohn, A.G., Mark, D.M. (eds.) COSIT 2005. LNCS, vol. 3693, pp. 31–48. Springer, Heidelberg (2005)
4. Fung, B.C.M., Wang, K., Chen, R., Yu, P.S.: Privacy-preserving data publishing: A survey of recent developments. ACM Computing Surveys 42(4), 14:1–14:53 (2010)
5. Gruteser, M., Grunwald, D.: Anonymous usage of location-based services through spatial and temporal cloaking. In: Proceedings of the International Conference on Mobile Systems, Applications, and Services (2003)
6. Gedik, B., Liu, L.: A customizable k-anonymity model for protecting location privacy. In: ICDCS (2005)
7. Kido, H., Yanagisawa, Y., Satoh, T.: An anonymous communication technique using dummies for location-based services. In: Proceedings of IEEE International Conference on Pervasive Services (2005)
8. Kido, H., Yanagisawa, Y., Satoh, T., Satoh, T.: Protection of Location Privacy using Dummies for Location-based Services. In: ICDE (2005)
9. Lu, H., Jensen, C.S., Yiu, M.L.: PAD: Privacy-Area Aware, Dummy-Based Location Privacy in Mobile Services. In: MobiDE (2008)
10. Mano, M., Ishikawa, Y.: Anonymizing User Location and Profile Information for Privacy-aware Mobile Services. In: LBSN (2010)
11. Mokbel, M.F., Chow, C.Y., Aref, W.G.: The new casper: Query procesing for location services without compromising privacy. In: Proceedings of the International Conference on Very Large Data Bases (2006)
12. Brinkhoff, T.: Network-based Generator of Moving Objects, http://iapg.jade-hs.de/personen/brinkhoff/generator/
13. Wang, T., Liu, L.: Privacy-aware mobile services over road networks. In: Proceedings of the International Conference on Very Large Data Bases (2009)
14. Yiu, M.L., Jensen, C., Huang, X., Lu, H.: Spacetwist: Managing the trade-offs among location privacy, query performance, and query accuracy in mobile services. In: Proceedings of the IEEE International Conference on Data Engineering (2008)
15. You, T.H., Peng, W.C., Lee, W.C.: Protecting moving trajectories with dummies. In: Proceedings of the International Workshop on Privacy-Aware Location-Based Mobile Services (2007)

Modeling Outlier Score Distributions

Mohamed Bouguessa

Département d'informatique
Université du Québec à Montréal
Montreal, Quebec, Canada
bouguessa.mohamed@uqam.ca

Abstract. A common approach to outlier detection is to provide a ranked list of objects based on an estimated outlier score for each object. A major problem of such an approach is determining how many objects should be chosen as outlier from a ranked list. Other outlier detection methods, transform the outlier scores into probability values and then use a user-predefined threshold to identify outliers. *Ad hoc* threshold values, which are hard to justify, are often used. Outlier detection accuracy can be seriously reduced if an incorrect threshold value is used. To address these problems, we propose a formal approach to analyse the outlier scores in order to automatically discriminate between outliers and inliers. Specifically, we devise a probabilistic approach to model the score distributions of outlier scoring algorithms. The probability density function of the outlier scores is therefore estimated and the outlier objects are automatically identified.

Keywords: Outlier score distributions, beta mixtures, outlier detection.

1 Introduction

An outlier could be broadly defined as an observation which appears to be inconsistent with the remainder of that set of data [1]. Several data mining algorithms consider outliers as noise that must be eliminated because it degrades their accuracy. For example, most of the clustering algorithms, consider outliers as points that are not located in clusters and should be captured and eliminated because they hinder the clustering process. Finding outliers in a data set is also a task is of practical relevance in several real-life applications such as fraud detection, intrusion detection, medical diagnosis, and many others.

Outlier detection has been well studied, for which appropriate approaches have been proposed. Such approaches can be classified in supervised-learning-based methods, where each example must be labeled as exceptional or not, and the unsupervised-learning-based ones, where the label is not required [2]. The latter approach is more general because in real situations, we do not have such information. In this paper, we deal with unsupervised methods only. Further details and surveys on outlier detection can be found in [3].

Many existing unsupervised outlier detection algorithms calculate some kind of score per data object which serves as a measure of the degree of outlier.

S. Zhou, S. Zhang, and G. Karypis (Eds.): ADMA 2012, LNAI 7713, pp. 713–725, 2012.

Scores are used in ranking data points such that the top n points are considered as outliers. For example, the statistical-based approach proposed in [4], uses a Gaussian mixture model to represent normal behaviours and each datum is given a score on the basis of changes in the model. A high score indicates a high possibility of being an outlier.

Generally, distance-based approaches exploit the distance from a data point to its neighbourhood to determine whether it is an outlier or not. The DB-outlier algorithm [5] was the pioneering distance-based approach to outlier detection. Scoring variants of the DB-outlier are proposed in [6], [7], basically using the distance to the k-nearest neighbours (kNN). For example, the kNN algorithm [6] and the weighted kNN algorithm [7], assign an outlying score to each data object o based on the k nearest neighbourhood of o in such a way that inliers are characterized by low scores while outliers are characterized by high scores. After ranking data points based on the estimated scores, the top n points are identified as outliers.

The density-based approach presented in [8], introduces a new notion of local outlier which measures the degree of an object to be an outlier with respect to the density of the local neighbourhood. This degree is called Local Outlier Factor (LOF) and is assigned to each object. The higher the LOF value of an object o is, the more distinctly it is o considered to be an outlier. In [9], the notion of the outlier factor has been merged with the distance-based notion of outliers, resulting in the local distance-based outlier detection approach (LDOF). LDOF uses the relative distance from an object to its neighbours to measure how much the object deviates from its scattered neighbourhood. The higher the violation degree of an object, the more likely the object is an outlier. In [10], the authors introduce the OPTICS-OF algorithm which also exploits the notion of local outlier in the sense that the outlier degree of an object is determined by taking into account the clustering structure in a bounded neighbourhood of the object. INFLO [11] is another density-based outlier scoring algorithm, which mine outliers based on a symmetric neighborhood relationship. In [12], a reference based outlier detection algorithm is proposed. This algorithm uses the relative degree of density with respect to a fixed set of reference points to calculate the neighborhood density of a data point. Outliers are those objects with the highest scores.

Virtually, outlier scoring methods have so far been only used to obtain an object ranking, expecting the outliers to come first. On the other hand, less effort has been invested on how to automatically discriminate between outliers and inliers. In fact, most of existing methods aim to make outlier detection more effective in retrieving the top n outliers only. The weakness of such an approach resides in the unprincipled selection of the value of n. In general, the value of n is often chosen in *ad hoc* manner. With such an informal approach, it is impossible to be objective or consistent. Furthermore, setting the value of n manually causes practical difficulties in applying outlier scoring approaches to real applications, in which prior knowledge about the data under investigation is not always available. In this setting, thresholding has turned out to be important in detecting outlier objects, since a ranked list has a particular disadvantage: there no clear cut-off point where to stop consulting results.

In [13], calibration methods (sigmoid functions and mixture modeling) have been proposed to translate scores of different outlier detection algorithms into probability values. In order to identify outlier, the approach developed in [13] requires the user to specify a threshold, so that data points whose estimated probability value exceeds the threshold will be declared as outlier. In order to identify the appropriate threshold value, the authors in [13] use the Bayesian risk model, which minimizes the overall risk associated with some cost function. For example, in the case of a zero-one loss function any observation whose estimated posterior probability exceeds 0.5 is declared as an outlier.

In [1], several normalization techniques have been proposed in order to the convert outlier scores into probability estimates. The main goal of [1] is to convert the outlier scores, produced by any outlier detection algorithm, to values in the range [0,1] interpretable as values describing the probability of a data object of being an outlier. However, the authors in [1] have not demonstrated how to formally distinguish between outliers and inliers. In general, the identification of outliers is related to a user-predefined threshold. *Ad hoc* threshold values (e.g., 0.5), which are hard to justify, are often used. Outlier detection accuracy can be seriously reduced if an incorrect threshold value is used. Furthermore, in any case, the optimal threshold depends on the outlier detection algorithm being used and there is no single threshold suitable for all purposes.

In this paper, we develop a systematic approach to automatically discriminate between outliers and inliers. In our approach, we propose to model the outlier scores, produced by an outlier scoring algorithm, as a finite mixture distribution. Specifically, the outlier scores can be considered as coming from several under-lying probability distributions. Each distribution is a component of the mixture representing scores with close values, and all the components are combined into a comprehensive model by mixture form. To this end, we use a beta mixture model approach to divide the outlier scores, produced by an outlier detection algorithm, into several populations so that the large scores that characterize outliers[1] can be identified. Finally, note that the main motivation of using the beta distribution to model the outlier scores is due to its great shape flexibility. In fact, the beta distribution is very versatile and it is therefore capable to model a variety of uncertainties [14], [15]. The shape flexibility of the beta distribution encourages its empirical use in a wide range of applications.

2 Proposed Approach

A first step toward modeling outlier score distributions is to transform these scores into some form which exhibits better distributional properties. In fact,

---

[1] In this paper, we assume that outliers are characterized by high score values in comparison to inliers. In fact, most of the exiting outlier scoring approaches assign high score values to outliers, while inliers receive low score values. In the case where an outlier detection algorithm initially assigns small score values to outliers, we simply perform linear inversion so that high scores will correspond to outliers.

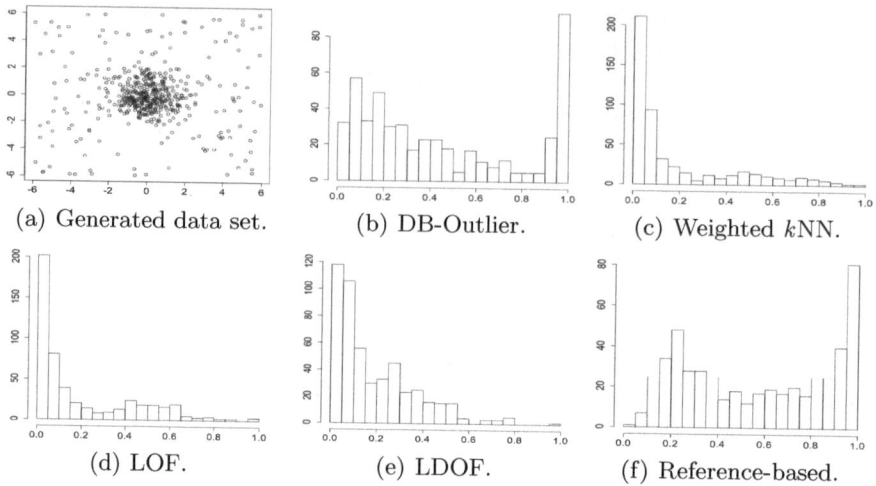

(a) Generated data set.   (b) DB-Outlier.   (c) Weighted kNN.

(d) LOF.   (e) LDOF.   (f) Reference-based.

Fig. 1. Histograms of the outlier scores

outlier scores often vary a lot in their value range, and it makes sense to perform some transformation on these scores [1]. In principle, any monotonic transformation of the scores produced by an outlier detection algorithm (which necessarily leaves the ranking unchanged) would be a candidate for this process. In our approach, since the outlier scores produced by existing algorithms are positives, we first perform a log-transformation to all the outlier scores provided by a specific outlier detection algorithm. Such a log-transformation aims to squeeze together the large score values that characterize outliers and stretch out the smallest values, which correspond to inliers. This squeezing and stretching yields comparable score values and also contributes to enhance the contrast between outlier and inlier scores. Then, we normalize the log-transformed scores of each algorithm in the interval [0,1]. As a result, without loss of generality, outlier scores will have comparable normalized values. In the remainder of this paper, we use only the normalized values of the outlier scores.

Statistical properties of the outlier scores, displayed by the shape of their distributions, can be used to determine the appropriate statistical model that best fit these scores. In this setting, estimating the histogram is a flexible tool to describe some statistical properties of the outlier scores. For the purpose of clarification, consider the data set depicted in Fig. 1(a). This is a simple 2-d data set containing a large bivariate normal distribution (inliers) and number of outlier points which are sparsely distributed. Figures 1(b)-1(f) illustrate the histograms of the normalized values of the outlier scores produced by several state-of-the-art outlier detection methods when they were applied to the data depicted by Fig. 1(a).

The histograms presented in Fig. 1 suggest the existence of components with different shapes. It is therefore necessary to use a statistical model which is suitable for dealing with such shape variation. In this paper, we propose to use the beta distribution in order to model the outlier scores. The beta distribution has been known for its flexible shapes and is therefore widely used to describe data from various experiments [14]. In fact, as mentioned in [15], in contrast to other univariate distributions, the beta distribution may be L-shaped, U-shaped, J-shaped, skewed to the left, skewed to the right or symmetric. Such great shape flexibility enables the beta distribution to provide an accurate fit of the outlier scores.

It is important to note that we do not draw any assumptions on the distribution of the original data, but on the distribution of the outlier scores produced by an outlier detection method. In our approach, we assume that the data under investigation contains a sufficient amount of outliers, so that the estimated scores can be effectively modeled. A general perception in the field is that the number of outliers in a data set is likely to be very small. This is true for some applications. However, as mentioned in [16], the data available nowadays are usually noisy and rich of outliers.

2.1   The Probabilistic Model Framework

Let x_i $(i = 1, \ldots, N)$ denote the normalized outlier scores and N denotes the number of objects in the data set under investigation. Formally, we expect that x_i follow a mixture density of the form

$$f(x) = \sum_{l=1}^{c} p_l Beta_l(x, a_l, b_l),\ 0 \leq x \leq 1 \tag{1}$$

where $Beta_l(.)$ is the lth beta distribution; c denotes the number of components in the mixture; a_l and b_l $(a_l, b_l > 0)$ are the shape parameters of the lth component; and $p_l(l = 1, \ldots, c)$ are the mixing coefficients, with the restriction that $p_l > 0$ for $l = 1, \ldots, c$ and $\sum_{l=1}^{c} p_l = 1$.
The density function of the lth component is given by

$$Beta_l(x, a_l, b_l) = \frac{\Gamma(a_l + b_l)}{\Gamma(a_l)\Gamma(b_l)} x^{a_l - 1}(1 - x)^{b_l - 1} \tag{2}$$

where $\Gamma(.)$ is the gamma function given by $\Gamma(\lambda) = \int_0^\infty y^{\lambda - 1} \exp(-y)dy;\ y > 0$.
A common approach for estimating the parameters a_l and b_l of the beta component is the maximum likelihood technique [14]. The likelihood function of the lth component is defined as

$$L_{Beta_l}(a_l, b_l) = \prod_{x \in Beta_l} Beta_l(x, a_l, b_l)$$

$$= \left(\frac{\Gamma(a_l + b_l)}{\Gamma(a_l)\Gamma(b_l)} \right)^{N_l} \prod_{i=1}^{N_l} (x_i)^{a_l - 1} \prod_{i=1}^{N_l} (1 - x_i)^{b_l - 1} \tag{3}$$

where N_l is the size of the lth component. Usually, it is more convenient to work with the logarithm of the likelihood function which is equivalent to maximizing the original likelihood function.The logarithm of the likelihood function is given by

$$\log(L_{Beta_l}(a_l, b_l)) = N_l log(\Gamma(a_l + b_l)) - N_l log(\Gamma(a_l)) - N_l log(\Gamma(b_l))$$
$$+ (a_l - 1) \sum_{i=1}^{N_l} log(x_i) + (b_l - 1) \sum_{i=1}^{N_l} log(1 - x_i) \quad (4)$$

To find the values of a_l and b_l that maximize the likelihood function, we differentiate $\log(L_{Beta_l}(a_l, b_l))$ with respect to each of these two parameters and set the result equal to zero:

$$\frac{\partial log(L_{Beta_l}(a_l, b_l))}{\partial a_l} = N_l \big[\psi(a_l + b_l) - \psi(a_l)\big] + \sum_{i=1}^{N_l} log(x_i) = 0 \quad (5)$$

and

$$\frac{\partial log(L_{Beta_l}(a_l, b_l))}{\partial b_l} = N_l \big[\psi(a_l + b_l) - \psi(b_l)\big] + \sum_{i=1}^{N_l} log(1 - x_i) = 0 \quad (6)$$

where $\psi(.)$ is the digamma function given by $\psi(y) = \frac{\Gamma'(y)}{\Gamma(y)}$.

There is no closed-form solution to equations (5) and (6), so the parameters $\hat{a}_l$ and $\hat{b}_l$ can be estimated iteratively using the Newton-Raphson method, a tangent method for root finding. Specifically, we estimate the vector of parameters $\hat{V}_l = (\hat{a}_l, \hat{b}_l)^t$ iteratively:

$$\hat{V}_l^{(I+1)} = \hat{V}_l^{(I)} - \nu_i^t \cdot M_l^{-1} \quad (7)$$

where I is the iteration index, ν_l and M_l are respectively the vector of the first derivatives and the matrix of the second derivatives of the log likelihood function of the lth component.

The Newton-Raphson algorithm for the update of (7) converges, as our estimates of a_l and b_l change by less than a small positive value ϵ with each successive iteration, to $\hat{a}_l$ and $\hat{b}_l$. Note that in our implementation we have used the method of moments estimators of the beta distribution [17] to define starting values for $\hat{V}_l^{(0)}$ in (7).

Usually, the maximum likelihood of the parameters of the distribution is estimated using the Expectation-Maximization algorithm [18]. With the EM algorithm, the mixture model is expressed in terms of missing data. Accordingly, we augment the data by introducing the latent indicator variable z_{il}, $(i = 1, \ldots, N)$, $(l = 1, \ldots, c)$ for each x_i. Here, z_{il} indicates to which component x_i belongs such that: If $z_{il} = 1$ then x_i belongs to component l, otherwise $z_{il} = 0$. The complete data is thus defined by the sets of values $\{z_{il}\}$ and $\{x_i\}$. The likelihood function for the complete data is:

$$L_f(\theta, p, z) = \prod_{i=1}^{N} \prod_{l=1}^{c} \big(p_l Beta_l(x_i, a_l, b_l)\big)^{z_{il}} \quad (8)$$

and the complete log-likelihood is:

$$log(L_f(\theta, p, z)) = \sum_{i=1}^{N} \sum_{l=1}^{c} z_{il} log\left(p_l Beta_l(x_i, a_l, b_l)\right) \tag{9}$$

where $\theta = \{a_1, b_1, \ldots, a_c, b_c\}$ denotes the set of unknown parameters of the mixture and $z = \{z_1, \ldots, z_N\}$ such that $z_i = (z_{i1}, \ldots, z_{ic})^t$ denotes the vector of indicator variables z_{il}.

The EM algorithm produces a sequence estimate $\{\hat{\theta}\}^{(I)}, (I = 0, 1, 2, \ldots)$ by alternatingly applying two steps (until the change in the value of the log-likelihood in (9) is negligible):

E-step:

- Compute $\hat{z}_{il}^{(I)} = \frac{\hat{p}_l^{(I)} Beta_l(x_i, \hat{a}_l^{(I)}, \hat{b}_l^{(I)})}{\sum_{j=1}^{c} \hat{p}_j^{(I)} Beta_l(x_i, \hat{a}_j^{(I)}, \hat{b}_j^{(I)})};$

M-step:

- Compute the mixing coefficients $\hat{p}_l^{(I+1)} = \frac{\sum_{i=1}^{N} \hat{z}_{il}^{(I)}}{N};$
- Estimate the vector of parameters $\hat{V}_l^{(I+1)}$ using (7);

Note that, the EM algorithm requires the initial parameters of each component. Since EM is highly dependent on initialization, it will be helpful to perform initialization by means of clustering algorithms [19]. For this purpose, we implement the Fuzzy C-Means (FCM) algorithm [20] in order to partition the set $\{x_i\}$ into c components. Based on such partition we can estimate the parameters of each component and set them as initial parameters to the EM algorithm. Once the EM converges, we can derive a classification decision about the membership of x_i to each component in the mixture. Since we assume that outlier detection algorithms assign a high score values to outliers, we are therefore interested by the beta component containing the highest values of x_i. Accordingly, data points associated with the set values of x_i that belong to such a component correspond to outliers.

Lets now focus on how to estimate c. One popular approach to estimating the number of components c is to increase c from 1 to c_max (the maximal number of components in the mixture) and to compute some particular performance measures in each run, until partition into an optimal number of components is obtained. For this purpose, we implement a standard two-steps process. In the first step, we calculate the maximum likelihood of the parameters of the mixture for a range of values of c (from 1 to c_max) using the aforementioned EM algorithm. The second step involves calculating an associated criterion and selecting the value of c which optimizes the criterion. A variety of approach has been proposed to estimate the number of components in the data. In our method, we use a penalized likelihood criteria, called the Bayesian Information Criterion (BIC) [21]. BIC is given by

$$BIC(c) = -2L_c + d \log(N) \tag{10}$$

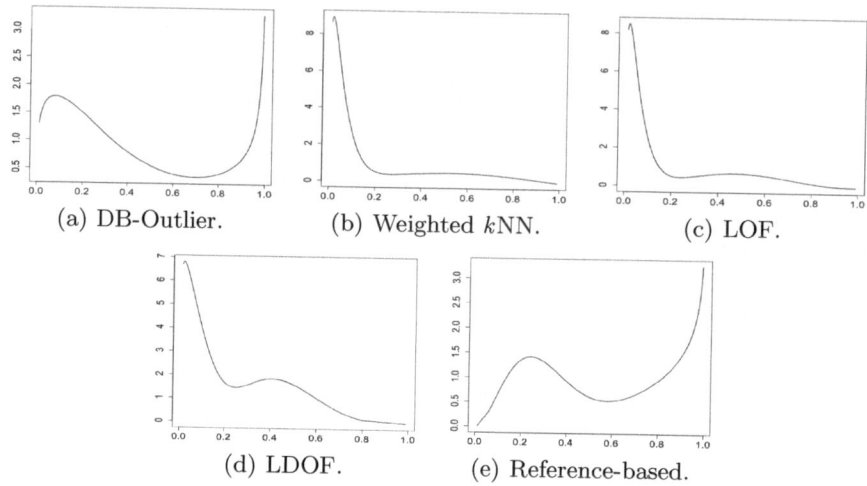

(a) DB-Outlier.   (b) Weighted kNN.   (c) LOF.

(d) LDOF.   (e) Reference-based.

Fig. 2. The synthetic 2-d data set: Density curves of the outlier scores

where L is the logarithm of the likelihood at the maximum likelihood solution for the investigated mixture model, and d is the number of parameters estimated. The number of components that minimize $BIC(c)$ is considered to be the optimal value for c.

2.2 Summary of Our Approach

The steps that follow can be implemented to model the outlier scores produced by an outlier detection algorithm. The probability density function is therefore estimated and the outlier points are automatically identified.

1. For a given data set, estimate the outlier scores using an outlier scoring algorithm;
2. Perform a log-transformation to the estimated outlier scores;
3. Normalize the transformed scores in [0,1];
4. Estimate the probability density function of the normalized scores with different values of c where $c = 1, \ldots, c\_max$;
5. Select the optimal number of components $\hat{c}$, such that $\hat{c} = \arg_{min} BIC(c)$;
6. Select the beta component that corresponds to the highest score values;

We used our approach to model the outlier scores produced by the each of the outlier detection algorithm represented in Fig. 1. We found that the outlier scores are well fitted by two beta components. For the purpose of illustration, the probability density function of the outlier scores of each algorithm considered in Fig. 1 is illustrated in Fig. 2. The knowledgeable reader can observe in this rendering that, based on the distribution of the outlier scores shown in the histograms in Fig. 1, the two beta components suggested by BIC (see Fig. 2) provide a relevant partitioning of the outlier scores. The second component of the mixture represents outliers with the highest score values.

3   Experiments

In this section, we put our approach to work using both synthetic and real data sets, in order to model the outlier scores produced by eight different outlier detection algorithms: DB-outlier [5], kNN [6], weigthed kNN ($wkNN$) [7], LOF [8], LDOF [9], OPTICS-OF [10], Reference-based [12], and INFLO [11]. In the following, we describe the data sets that we have used, and then we report the results of our experiments. Note that the implementation of all the outlier detection algorithms considered in our experiments is available in the framework ELKI[2] [22].

3.1   Synthetic Data Sets

We used the generator implemented in ELKI to generate data sets that contain normally clustered data with uniform background noise. Specifically, we generated three different data sets in 8-d space with clusters of different sizes and densities. In each data set, we fixed the number of objects N to 1000, while the percentage of outliers PO varies from 5 to 25 percent of N. Each generated data is denoted by $Synthetic\_PO$.

3.2   Real Data Sets

It worth noting that there is a shortage of standard benchmark data which can be used for the purpose of outlier detection. Most of publicly available labeled data are primarily designed for classification and machine learning applications. This makes the evaluation of the proposed method a challenging task. In view of this, like other outlier detection studies [1], [9], we have adopted a principled way of evaluating the approach presented in this paper. In this section, we saliently illustrate the suitability of our method on real data sets taken from the UCI Machine Learning Repository[3]. A brief description of the data sets that we have used is given below.

Wisconsin Breast Cancer (WBC): This set originally contains 458 objects labeled as Benign and 241 objects labeled as Malignant. The data has 10 integer-valued dimensions. To produce a data set for use in outlier detection, the malignant class was randomly down sampled to 100. Note that the WBC data set contains some objects with missing values. In our experiments, we have simply ignored those objects. Thus, the final set contains 544 total points with 444 non-outlier points (Benign) and 100 outlier points (Malignant). Such an approach to produce data for use in outlier detection is adopted from previous studies [1], [9]. The same principle is also applied on the remaining real data sets used in our experiments.

Pen-Based Recognition of Handwritten Digits (Pen Digits): This data set contains 10 classes such that each class corresponds to a single digit between

---

[2] http://elki.dbs.ifi.lmu.de/
[3] http://archive.ics.uci.edu/ml/

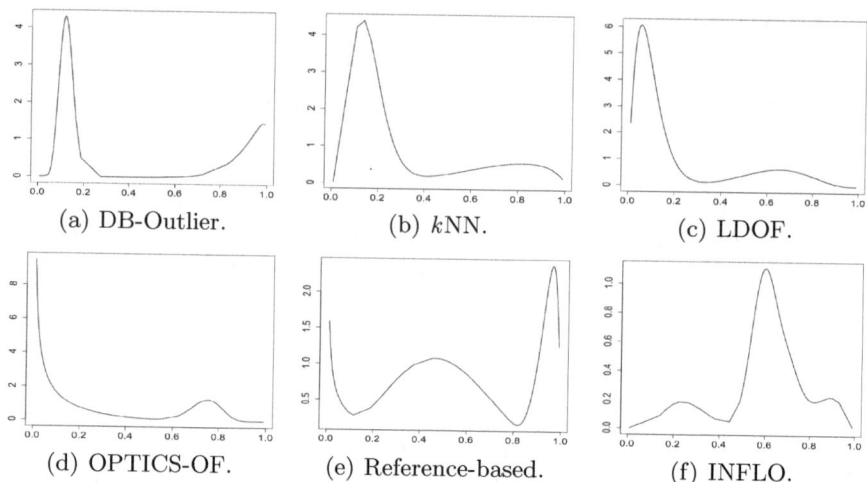

(a) DB-Outlier.   (b) kNN.   (c) LDOF.

(d) OPTICS-OF.   (e) Reference-based.   (f) INFLO.

Fig. 3. WBC: Density curves of the outlier scores

0 and 9. The data has 16 integer-valued dimensions. We regarded the class corresponding to digit 0 as normal data. In our experiments, we used all 780 objects of the class corresponding to digit 0 as normal objects and we added 120 objects randomly selected from the remaining nine classes (corresponding to digit 1 to 9) as outliers.

Image Segmentation (Image): This data set contains 7 classes namely Sky, Brickface, Foliage, Cement, Window, Path and Grass. The data has 19 real-valued dimensions. To produce data sets for use in outlier detection, we considered all 330 objects of the class Sky as normal objects and we randomly selected 80 objects from the remaining 6 classes and considered them as outliers.

3.3   Results

We applied DB-Outlier, kNN, wkNN, LOF, LDOF, OPTICS-OF, Reference-based, and INFLO to each of the six data sets considered in our experiments. We then used our approach to model the scores produced by each algorithm for each data set. To this end, we set c_max to 5 and then we selected the optimal number of components that minimize BIC. We found that the optimal number of components in a mixture varies from 2 to 3. For the purpose of illustration, and due to space limits, we show in Fig. 3 the probability density function of the outlier scores produced by DB-Outlier, kNN, LDOF, OPTICS-OF, Reference-based and INFLO when they were applied to WBC only. In each subfigure of Fig. 3, the beta component which represents large values of the outlier scores corresponds to outliers.

In all the six data sets used in our experiments, data points have labels to designate whether an object is an outlier or an inlier. We ignored these labels when we have used our approach to model the outlier scores, but we have used

		DB-Outlier	kNN	wkNN	LOF	LDOF	OPTICS-OF	Ref.-based	INFLO
Synthetic_5	TP(%)	100.00	92.00	100.00	92.00	88.00	92.00	98.00	88.00
	FP(%)	4.74	0.32	0.21	0.32	0.11	0.32	1.05	0.00
Synthetic_15	TP(%)	92.67	90.00	96.00	89.33	94.67	89.33	99.00	94.67
	FP(%)	1.53	0.47	0.35	0.47	1.06	0.47	4.82	0.24
Synthetic_25	TP(%)	97.20	94.00	90.00	94.00	89.20	94.00	96.80	95.60
	FP(%)	3.20	0.67	0.13	0.27	0.13	0.27	0.13	2.93
Average	TP(%)	96.62	92.00	95.33	91.78	90.62	91.78	97.93	92.76
	FP(%)	3.16	0.48	0.23	0.35	0.43	0.35	2.00	1.06

(a) Synthetic data sets.

		DB-Outlier	kNN	wkNN	LOF	LDOF	OPTICS-OF	Ref.-based	INFLO
WBC	TP(%)	91.00	97.00	97.00	96.00	98.00	96.00	98.00	90.00
	FP(%)	1.58	4.05	4.50	4.28	6.08	4.28	6.53	4.73
Pen Digits	TP(%)	99.00	99.00	99.00	99.00	100.00	99.00	100.00	92.00
	FP(%)	8.33	6.15	5.00	6.79	10.51	4.36	12.05	2.95
Image	TP(%)	100.00	93.75	98.75	96.25	91.25	96.25	95.00	78.75
	FP(%)	9.39	0.00	0.00	0.00	0.00	0.00	0.00	0.61
Average	TP(%)	96.67	96.58	98.25	97.08	96.42	97.08	97.67	86.92
	FP(%)	6.43	3.40	3.17	3.69	5.53	2.88	6.19	2.76

(b) Real data sets.

Fig. 4. Quality of results

them as ground truth to measure the accuracy of results. To this end, we used the
following two standard metrics: (1) True Positive rate (TP) which corresponds
to the ratio of the number of outliers correctly identified as outliers over the total
number of outliers and (2) False Positive rate (FP), the ratio of the number of
normal objects erroneously flagged as outliers over the total number of normal
objects. The results are summarized in Fig. 4. Here, it is important to note
that, in our experiments, we are more interested in illustrating the increase of
usability of different outlier detection methods that can be gained by applying
our approach rather than evaluating such methods for outlier detection. This
would be out of the scope of this study.

Results in Fig. 4(a) suggest that, after using our approach to model the scores
produced by the different algorithms, we are able to accurately identify most of
the outlier objects in both real and synthetic data sets. On synthetic data sets,
the average TP rate varies from 90.62% to 97.93%, while on real data sets the
average TP rate varies from 86.92% to 98.25%. From Fig. 4, we also note that a
certain number of normal objects were misclassified as outliers. In fact, as can be
seen from Fig. 4(a), the average FP rate achieved on synthetic data sets varies
from 0.23% to 3.16%. On real data sets, as depicted by Fig. 4(b), the average
FP rate varies from 2.76% to 6.43%.

Overall, the results presented in Fig. 4, suggest that, using the proposed approach, we are able to identify most the outliers on both synthetic and real data
sets, with the expense of also selecting a small number of non-outlier objects.
This demonstrates that the application of the proposed approach can clearly
increase the usability of outlier scores. In fact, our approach enhance the use of
existing outlier scoring methods by avoiding the use of *ad hoc* threshold values
or manual selection of outliers form a ranked list.

4   Conclusion

In this paper, we devised a probabilistic approach to model outlier score distributions. Specifically, we proposed a beta mixture model approach, and derived a corresponding EM algorithm for parameter estimation. Our goal is to divide the outlier scores into several components, so that the beta component that corresponds to outliers is identified. To the best of our knowledge, no formal method for analyzing the outlier scores, in order to automatically discriminate between outliers and inliers, has yet been published. In our continuing research, we will explore various extensions of the approach proposed in this paper. One interesting possibility is to extend the proposed model in order to combine the results obtained from different outlier scoring algorithms. Combining the outputs of different algorithms could further discriminate between outliers and inliers and therefore enhance the outlier detection accuracy.

Acknowledgments. The author would like to thank the reviewers for their valuable comments and suggestions. This work is supported by research grants from the Natural Sciences and Engineering Research Council of Canada (NSERC).

References

1. Kriegel, H.-P., Kroger, P., Schubert, E., Zimek, A.: Interpreting and Unifying Outlier Scores. In: 11th SIAM International Conference on Data Mining (SDM 2011), pp. 13–24 (2011)
2. Tan, P.-N., Steinbach, M., Kumar, V.: Introduction to Data Mining. Addison Wesley (2006)
3. Chandola, V., Banerjee, A., Kumar, V.: Anomaly Detection: A Survey. ACM Computing Surveys 41(3) (2009)
4. Yamanishi, K., Takeuchi, J.-I., Williams, G., Milne, P.: On-line Unsupervised Learning Outlier Detection Using Finite Mixtures with Discounting Learning Algorithms. In: 6th ACM SIGKDD International Conference on Knowledge Discovery and Data Mining (KDD 2000), pp. 320–324 (2000)
5. Knorr, E.M., Ng, R.T.: Algorithms for Mining Distance-Based Outliers in Large Datasets. In: 24th International Conference on Very Large Data Bases (VLDB 1998), pp. 392–403 (1998)
6. Ramaswamy, S., Rastogi, R., Shim, K.: Efficient Algorithms for Mining Outliers from Large Data Sets. In: ACM SIGMOD International Conference on Management of Data (SIGMOD 2000), pp. 427–438 (2000)
7. Angiulli, F., Pizzuti, C.: Fast Outlier Detection in High Dimensional Spaces. In: Elomaa, T., Mannila, H., Toivonen, H. (eds.) PKDD 2002. LNCS (LNAI), vol. 2431, pp. 15–26. Springer, Heidelberg (2002)
8. Breunig, S., Kriegel, H.-P., Ng, R., Sander, J.: LOF: Identifying Density-Based Local Outliers. In: ACM SIGMOD International Conference on Management of Data (SIGMOD 2000), pp. 93–104 (2000)
9. Zhang, K., Hutter, M., Jin, H.: A New Local Distance-Based Outlier Detection Approach for Scattered Real-World Data. In: Theeramunkong, T., Kijsirikul, B., Cercone, N., Ho, T.-B. (eds.) PAKDD 2009. LNCS, vol. 5476, pp. 813–822. Springer, Heidelberg (2009)

10. Breunig, M.M., Kriegel, H.-P., Ng, R., Sander, J.: OPTICS-OF: Identifying Local Outliers. In: Żytkow, J.M., Rauch, J. (eds.) PKDD 1999. LNCS (LNAI), vol. 1704, pp. 262–270. Springer, Heidelberg (1999)

11. Jin, W., Tung, A., Han, J., Wang, W.: Ranking Outliers Using Symmetric Neighborhood Relationship. In: Ng, W.-K., Kitsuregawa, M., Li, J., Chang, K. (eds.) PAKDD 2006. LNCS (LNAI), vol. 3918, pp. 577–593. Springer, Heidelberg (2006)

12. Pei, Y., Zaiane, O.R., Gao, Y.: An Efficient Reference-based Approach to Outlier Detection in Large Datasets. In: 6th IEEE International Conference on Data Mining (ICDM 2006), pp. 478–487 (2006)

13. Gao, J., Tan, P.-N.: Converting Output Scores from Outlier Detection Algorithms into Probability Estimates. In: 6th IEEE International Conference on Data Mining (ICDM 2006), pp. 1–10 (2006)

14. Ma, Z., Leijon, A.: Beta Mixture Models and the Application to Image Classification. In: 16th IEEE International Conference on Image Processing (ICIP 2009), pp. 2045–2048 (2009)

15. Bouguila, N., Ziou, D., Monga, E.: Practical Bayesian Estimation of a Finite Beta Mixture Through Gibbs Sampling and its Applications. Statistics and Computing 16(2), 215–225 (2006)

16. Zuliani, M., Kenny, C.S., Manjunath, B.S.: The Multiransac Algorithm and its Application to Detect Planar Homographies. In: 12th IEEE International Conference on Image Processing, ICIP 2005 (2005)

17. Bain, L.J., Engelhardt, M.: Introduction to Probability and Mathematical Statistics, 2nd edn. Duxbury Press (2000)

18. Dempster, A., Laird, N., Rubin, D.: Maximum Likelihood from Incomplete Data via the EM Algorithm. Journal of Royal Statistical Society (Series B) 39, 1–37 (1977)

19. Figueiredo, M.A.T., Jain, A.K.: Unsupervised Learning of Finite Mixture Models. IEEE Transactions on Pattern Analysis and Machine Intelligence 24(3), 381–396 (2002)

20. Bezdek, J.C.: Pattern Recognition with Fuzzy Objective Function Algorithms. Plenum, New York (1981)

21. Schwarz, G.: Estimating the Dimension of a Model. Annals of Statistics 6(2), 461–464 (1978)

22. Achtert, E., Goldhofer, S., Kriegel, H.-P., Schubert, E., Zimek, A.: Evaluation of Clusterings - Metrics and Visual Support. In: 28th IEEE International Conference on Data Engineering (ICDE 2012), pp. 1285–1288 (2012)

A Hybrid Anomaly Detection Framework in Cloud Computing Using One-Class and Two-Class Support Vector Machines

Song Fu[1], Jianguo Liu[2], and Husanbir Pannu[2]

[1] Department of Computer Science and Engineering, University of North Texas,
Denton, TX 76203, USA
[2] Department of Mathematics, University of North Texas,
Denton, TX 76203, USA
{Song.Fu,jgliu}@unt.edu, HusanbirPannu@my.unt.edu

Abstract. Modern production utility clouds contain thousands of computing and storage servers. Such a scale combined with ever-growing system complexity of their components and interactions, introduces a key challenge for anomaly detection and resource management for highly dependable cloud computing. Autonomic anomaly detection is a crucial technique for understanding emergent, cloud-wide phenomena and self-managing cloud resources for system level dependability assurance. We propose a new hybrid self-evolving anomaly detection framework using one-class and two-class support vector machines. Experimental results in an institute wide cloud computing system show that the detection accuracy of the algorithm improves as it evolves and it can achieve 92.1% detection sensitivity and 83.8% detection specificity, which makes it well suitable for building highly dependable clouds.

Keywords: cloud computing, dependable computing, anomaly detection, classification, clustering.

1 Introduction and Related Work

Anomaly detection is an important problem that has been researched within diverse research areas and application domains. Anomalies arise due to various reasons such as mechanical faults, changes in system behavior, fraudulent behavior, human error and instrument error. Detection of anomalies can lead to identification of system faults so that administrators can take preventive measures before they escalate and also may enable detection of new attacks. With ever-growing complexity and dynamics of cloud computing systems, anomaly detection is an effective approach to enhance system dependability [30]. Failure predictions are the key to such techniques. It forecasts future failure occurrences in cloud systems using runtime execution states of the system and the history information of observed failures. It provides valuable information for resource allocation, computation reconfiguration and system maintenance [22].

S. Zhou, S. Zhang, and G. Karypis (Eds.): ADMA 2012, LNAI 7713, pp. 726–738, 2012.
© Springer-Verlag Berlin Heidelberg 2012

The online detection of anomalous system behavior caused by operator errors [23], hardware/software failures [28], resource over-/under-provisioning [18,19] and similar causes is a vital element of operations in large-scale data centers and cloud computing systems. Failure detection based on analysis of performance logs has been the topic of numerous studies. Reference [16] provided an extensive survey of anomaly detection techniques developed in machine learning and statistical domains. Most often the task of failure detection can be solved by estimating a probability density of the normal data. For example, in [4] the density is estimated by a Parzen density estimator, whereas in [25] a Gaussian distribution is used. In [39], fault detection was modeled as a classification problem, which was solved by convex programming.

Recently, data mining and statistical learning theories have received growing attention for failure detection and failure management. These methods extract failure patterns from systems' normal behaviors, and detect abnormal observations based on the learned knowledge [26]. In [40] the authors presented several methods to forecast failure events in IBM clusters. Reference [21] examined several statistical methods for failure prediction in IBM Blue Gene/L systems and [20] investigated meta-learning based method for improving failure prediction. In [11], the authors developed a proactive failure management framework for networked computing systems.

Other related work include [35] in which the authors presented Kahuna, an approach that aims to diagnose performance problems in MapReduce systems. Kahuna uses peer-similarity and the observation that a node behaving differently is likely culprit of a performance problem. Mantri, a system that monitors tasks and culls outliers using cause- and resource-aware techniques was introduced in [1]. Tiresias [41] is a system that makes black-box failure-prediction possible by transparently gathering, and then identifying escalating anomalous behavior in various node-level and system-level performance metrics. ALERT [36] is an adaptive runtime anomaly prediction system which aims at raising advance anomaly alerts to achieve just-in-time anomaly prevention. A stream-based mining algorithm for online anomaly prediction was presented in [12]. The scheme combines Markov models and Bayesian classification methods to predict when a system anomaly will appear in the foreseeable future and what are the possible anomaly causes. The authors in [13] explored a new predictive failure management approach that employs online failure prediction to achieve more efficient failure management than previous reactive or proactive failure management approaches. The RACH algorithm was introduced in [15] which explores the compactness property of the rare categories. It is based on an optimization framework which encloses the rare examples by a minimum-radius hyper ball. The SPIRIT system presented in [24] can incrementally find correlations and hidden variables, summarize the key trends in the entire stream collection, and be used to spot potential anomalies. A comprehensive list of references on anomaly detection algorithms can be found in [7].

In contrast to classical reliability methods, our anomaly detection method is based on runtime monitoring, current state of a system and the past experience as well. It is self-evolving in nature and it integrates classification and clustering as a hybrid approach. The algorithm includes two components. One is detector determination and the other is detector retraining and working data set selection.

There are several approaches for classification and clustering. Our preliminary experiments show that support vector machines (SVM) and one-class support vector machines (or support vector data description (SVDD)) work very well. Reference for SVM and one-class SVM can be seen in [3,32,37,38]. We will focus on these two methods and show how to use them together. An ensemble incorporated with other methods, such as k-means, neural networks, Bayesian networks, and rule based, can easily be designed using the same framework.

The paper is organized as follows. Section 2 provides an overview for a cloud system and cloud metric (feature) extraction. Section 3 presents anomaly detection mechanism. Experimental results are included in Section 4. Conclusion and remarks on future work are presented in Section 5.

2   Cloud Metric (Feature) Extraction

A metric (feature) in the runtime performance dataset refers to any individual measurable variable of a cloud server or network being monitored. It can be a statistic of usage of hardware, virtual machines, or cloud applications. In production cloud computing systems, hundreds of performance metrics are usually monitored and measured. The large metric dimension and the overwhelming volume of cloud performance data make the data model extremely complex. Moreover, the existence of interacting metrics and external environmental factors introduce measurement noises in the collected cloud performance data.

To achieve efficient and accurate failure detection, the first step is to extract the most relevant performance metrics to characterize a cloud's behavior and health. This step transforms the cloud performance data to a new metric space with only the most important attributes preserved. Given the input cloud performance dataset D including L records of N metrics $M = \{m_i, i = 1, \ldots, N\}$, and the classification variable c, metric extraction is to find from the N-dimensional measurement space, $\mathbb{R}^N$, a subspace of n metrics (subset S), $\mathbb{R}^n$, that optimally characterizes c. For a two-class failure detection, the value of variable c can be either 0 or 1 representing the "normal" or "failure" state. In a multi-class failure detection, each failure type corresponds to a positive number that variable c can take.

Anomaly Detector first extracts those metrics, which jointly have the highest dependency on the class c. To achieve this goal, Anomaly Detector quantifies the mutual dependence of a pair of metrics, say m_i and m_j. Their mutual information (MI) [8] is defined as $I(m_i; m_j) = H(m_i) + H(m_j) - H(m_i m_j)$, where $H(\cdot)$ refers to the Shannon entropy [33]. Metrics of the cloud performance data usually take discrete values. The marginal probability $p(m_i)$ and the probability mass function $p(m_i, m_j)$ can be calculated using the collected dataset. The MI of m_i and m_j is computed as $I(m_i; m_j) = \sum_{m_i \in M} \sum_{m_j \in M} p(m_i, m_j) \, log(\frac{p(m_i, m_j)}{p(m_i) p(m_j)})$. We choose the mutual information for metric extraction because of its capability of measuring any type of relationship between variables and its invariance under space transformation.

Anomaly Detector applies two criteria to extract cloud metrics: finding the metrics that have high relevance with the class c (*maximal relevance criterion*) and have low mutual redundancy between each other (*minimal redundancy criterion*). The metric relevance and redundancy are quantified as follows.

$$relevance = \frac{1}{|S|} \sum_{m_i \in S} I(m_i; c), \quad redundancy = \frac{1}{|S|^2} \sum_{m_i, m_j \in S} I(m_i; m_j), \tag{1}$$

where $|S|$ is the cardinality of the extracted subset of cloud metrics S. The N metrics in the metric set M defines a 2^N search space. Finding the optimal metric subset is NP-hard. To extract the near-optimal metrics satisfying Criteria (1), we apply the incremental metric search algorithm [10].

From our experiments, we find the resulting subset S may still contains too many cloud metrics. Therefore, we extract the cloud metrics further by applying metric space separation to reduce the dimension. This is done by the independent component analysis (ICA) method. ICA is particularly suitable for separating a multivariate signal of the non-Gaussian source. Principal component analysis (PCA) could be used for dimension reduction. But for this application, ICA works better than PCA.

3   A Hybrid Anomaly Detection Mechanism

Our proposed hybrid anomaly detection framework includes two components. The first one is detector determination. The detector is self-evolving and constantly learning. For a newly collected data record, the detector will calculate an abnormality score. If the score is below a threshold, a warning will be triggered, possibly with the type of abnormality which may help a system administrator to pin point the anomaly. The second component is detector retraining and working data set selection. The detector needs to be retrained when certain new data records are included in the working data set. In addition, working data set selection is imperative since the size of available health-related data from large-scale production systems may easily reach hundreds and even thousands giga-bytes. The detector can not blindly use all available data. Metric selection and extraction described in Section 2 work in a horizontal fashion while working data selection is vertical or sequential. Clearly, all these components are important and they will be orchestrated to achieve accurate and efficient real time anomaly detection.

Without loss of generality, we assume the given cloud system is newly deployed or managed. Health-related system status data, such as system logs, will be gradually collected. The size of the data set will quickly grow from zero to something very large. Initially, all the data records are normal. As time goes by, a small percentage of abnormal records will appear. Those abnormal records can be labeled according to their anomaly types.

At an early stage of the deployment of the anomaly detection mechanism, the working data set includes only normal data. The detector will be a function generated by the one-class SVM. To be more specific, let D be the working data set including N records $x_i \in R^m$ $(i = 1, 2, ..., N)$. Let ϕ be a mapping from R^m to a higher dimensional feature space where dot products can be evaluated by some simple kernel functions: $k(x, y) = \langle \phi(x) \cdot \phi(y) \rangle$ where $\langle \cdot \rangle$ denotes the inner product in the corresponding vector space. A common kernel function is the Gaussian kernel $k(x, y) = exp(-\frac{\|x-y\|^2}{2\sigma^2})$. The idea of one-class SVM is to separate the data set from the origin by solving a minimization problem:

$$\min_{w,b,\xi} \frac{1}{2}\|w\|^2 - b + \frac{1}{\nu N} \sum_i \xi_i \quad \text{s.t. } \langle w \cdot \phi(x_i) \rangle \geq b - \xi_i \text{ and } \xi_i \geq 0 \; \forall \; i$$

where w is a vector perpendicular to the hyperplane in the feature space, b is the distance from the hyperplane to the origin, and ξ_i are soft-margin slack variables to handle outliers. The parameter $\nu \in (0, 1)$ controls the trade-off between the number of records in the data set mapped as positive by the decision function $f(x) = sgn(\langle w \cdot \phi(x) \rangle - b)$ and having a small value of $\|w\|$ to control model complexity. In practice, the dual form is often solved. Let α_i $(i = 1, 2, ..., N)$ be the dual variables. Then the decision function can be written in the dual variables as $f(x) = sgn(\sum_i \alpha_i k(x_i, x) - b)$. A newly collected data record x is predicted to be normal if $f(x) = 1$ and abnormal if $f(x) = -1$. One of the advantages of the dual form is that the decision function can be evaluated by using the simple kernel function instead of the expensive inner product in the feature space. As the working data set grows, it will eventually contain some abnormal records. In other words, two classes or multiple classes of data records will be available. Therefore, SVM will become a natural choice for anomaly detection since SVM is a powerful classification tool and has been successfully applied to many applications. A soft-margin binary SVM can be formulated using the slack variables ξ_i :

$$\min_{w,b,\xi} \frac{1}{2}\|w\|^2 - b + C \sum_i \xi_i \quad \text{s.t. } y_i(\langle w \cdot \phi(x_i) \rangle + b) \geq 1 - \xi_i, \; \xi_i \geq 0 \; \forall \; i$$

where $C > 0$ is a parameter to deal with misclassification and $y_i \in \{+1, -1\}$ are given class labels. A data record x_i is normal if the corresponding class label $y_i = 1$ and abnormal if $y_i = -1$. Once again, a dual form is solved and the decision function is $f(x) = sgn(\sum_i \alpha_i k(x_i, x) + b)$. A newly collected data record x could be predicted to be normal if $f(x) = 1$ and abnormal if $f(x) = -1$. Multi-class classification can be done using binary classification.

3.1   Detector Determination

A challenge to SVM is that the working data set is often highly unbalanced: normal data records outnumber abnormal data records by big margin. Classification accuracy of SVM is often degraded when applied to unbalanced data sets. However, as the percentage of abnormal data records increases, the performance of SVM will improve. Our numerical experiments show that SVM starts to perform reasonably well for this particular unbalanced problem once the percentage of abnormal data reaches 10%. Our detector is determined by combining one-class SVM and SVM with a sliding scale weighting strategy. This strategy can easily be extended to including other classification methods.

The weighting is based on two factors. One is credibility score and the other is the percentage of abnormal data records in the working data set. The method with a higher credibility score will weigh more and more weight will be given to SVM as the percentage of abnormal data records increases. For a given method, let $a(t)$ denote the numbers of attempted predictions and $c(t)$ denote the number of correct predictions where t is any given time. The credibility score is defined to be

$$s(t) = \begin{cases} \frac{c(t)}{a(t)} & \text{if } a(t) > 0 \text{ and } \frac{c(t)}{a(t)} > \lambda \\ 0 & \text{if } a(t) = 0 \text{ or } \frac{c(t)}{a(t)} \leq \lambda \end{cases}$$

where $\lambda \in (0, 1)$ is a parameter of zero trust. A good choice is $\lambda = 0.5$. Let $s_1(t)$ and $s_2(t)$ be the credibility scores of one-class SVM and SVM, respectively. Let $p(t)$ denote the percentage of abnormal data records in the working data set. Suppose $f_1(x)$ is the decision function generated by one-class SVM and $f_2(x)$ is generated by SVM where x is a newly collected data record at time t. Then the combined decision function is given by

$$f(x) = \begin{cases} f_1(x)s_1(t) & \text{if } p(t) = 0 \\ \frac{1}{2}(f_1(x)s_1(t) + f_2(x)s_2(t)) & \text{if } p(t) \geq \theta \\ f_1(x)s_1(t)(1 - \frac{p(t)}{2\theta}) + f_2(x)s_2(t)(\frac{p(t)}{2\theta}) & \text{if } 0 < p(t) < \theta \end{cases}$$

where $\theta \in (0, 1)$ is a parameter of trust on SVM related to the percentage of abnormal data records. A reasonable choice is $\theta = 0.1$. An anomaly warning is triggered if $f(x)$ is smaller than a threshold τ, say, $\tau = 0$. When multiple labels are available for abnormal data records, a multi-class SVM can be trained to predict the type of anomaly if a new data record is abnormal.

3.2   Detector Retraining and Working Data Set Selection

Detector retraining and working data set selection are part of a learning process. The basic idea is to learn and improve from mistakes and maintain a reasonable size of the data set for efficient retraining. Initially, all data records are included in the working data set to build up a good base to train the detector. Once the data set reaches a certain size and the detection accuracy is stabilized, the inclusion will be selective. A new data record x is included in the working data set only if one or more of the following is true:

- The data record corresponds to an anomaly and $p(t) < 0.5$. It is ideal to include more abnormal data records in the working data set but not too many.
- One of the predictions by $f_1(x)$, $f_2(x)$, or $f(x)$ is incorrect. The detector will be retrained to learn from the mistake.

Table 1. Definition of USV, BSV and NSV

Points	Condition	Details
USV	$0 < \alpha_i < C$	unbounded (or margin) support vectors
BSV	$\alpha_i = C$	bounded support vectors (errors)
NSV	$\alpha_i = 0$	non support vectors (within the sphere)

- The data record may change the support vectors for SVM. This happens when the absolute value of $(\sum_i \alpha_i k(x_i, x) + b)$ is less than 1, where we assume $f_2(x) = \text{sgn}(\sum_i \alpha_i k(x_i, x) + b)$. The detector will be adjusted to have better detection accuracy.

The decision functions $f_1(x)$ and $f_2(x)$ will be retrained whenever one of the predictions by $f_1(x)$ or $f_2(x)$ is incorrect. The retraining can be done quickly since the size of the data set is well maintained. In addition, the solutions of the old one-class SVM and SVM can be used as the initial guesses for the solutions of the new problems. Solving one-class SVM and SVM is an iterative process. Having good initial guesses will make the iterations converge fast to the new solutions.

To update the working dataset, the trained data are partitioned into three categories based on the KKT conditions which are explained in *Table 1*.

The computational complexity of an anomaly detection method can be proportional to the size of dataset so the increment of data size may cause scale problems in detector retraining. The spatial complexity can be even more serious if all trained data have to be preserved.

To make detector retraining more scalable in a real large problem like utility clouds, we need to remove useless data. In our approach, we exploit complexity reduction method by removing useless data based on the sample margin [17].

Detector retraining of our anomaly detection algorithm is to find a new decision boundary considering only data trained up to present. Because all data are not trained, the current data description is not optimal for whole dataset but it can be considered as an optimal data description for trained data up to now. We can eliminate every NSVs classified by the current hyperplane. However it is risky because important data which have a chance to become USVs might be removed as learning proceeds incrementally. In that case, the current hyperplane may not converge to the optimal hyperplane.

Therefore we need to cautiously define removable NSVs using sample margin. To handle the problem of removing data which become USVs, we choose data whose sample margin is in the specific range as removable NSVs. As shown in Figure 1, we select data in the region above the gray zone as removable NSVs. The gray region is called the epsilon region. It is defined to preserve data which may become USVs. A removable NSV is a data point x that satisfies the following condition:

$$\gamma(x) - \gamma(SV) \geq \epsilon(\gamma_{max} - \gamma(SV))$$

where $\epsilon \in (0, 1]$ is the user defined parameter, $\gamma(SV)$ is the sample margin of a support vector which is on the boundary and $\gamma_{max} = max_{i \in R} \gamma(x)$. As in

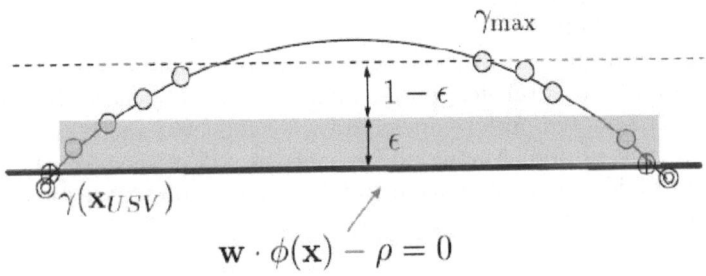

Fig. 1. The candidates of removable NSV and ϵ region

Figure 1, by preserving data in the ϵ region, an incremental detector retraining using sample margin information can obtain the same data description as original incremental anomaly detector with less computational and spatial load. The parameter ϵ should be chosen between 0 and 1. In the extreme case when $\epsilon = 0$, all data lying on the upper side of the hyperplane will be candidates for removable NSVs. That will make learning unstable. On the other extreme when $\epsilon = 1$, we can hardly select any removable NSVs. The effect of speeding up and storage reduction will be meager.

4   Performance Evaluation

We have implemented a proof-of-concept prototype of our hybrid anomaly detection system and tested it in a cloud computing environment on campus. In this section, we present the experimental results.

4.1   Experiment Settings

The cloud computing system consists of 362 servers, which are connected by gigabit Ethernet. The cloud servers are equipped with two to four Intel Xeon or AMD Opteron cores and 2.5 to 8 GB of RAM. We have installed Xen 3.1.2 hypervisors on the cloud servers. The operating system on a virtual machine is Linux 2.6.18 as distributed with Xen 3.1.2. Each cloud server hosts up to eight VMs. A VM is assigned up to two VCPUs, among which the number of active ones depends on applications. The amount of memory allocated to a VM is set to 512 MB. We run the RUBiS [6] distributed online service benchmark and MapReduce [9] jobs as cloud applications on VMs. The applications are submitted to the cloud computing system through a web based interface. We have also developed a fault injection program, which is able to randomly inject four major types with 17 sub-types of faults to cloud servers. They mimic faults from CPU, memory, disk, and network. We exploit the third-party monitoring tools, such as SYSSTAT [34] to collect runtime performance data in Dom0 and a modified PERF [27] to obtain the values of performance counters from the Xen hypervisor on each cloud server. In total, 518 metrics are profiled 10 times per hour for one month (in summer 2011). They cover the statistics of every component of a cloud server, including CPU usage, process creation, task switching activity, memory and swap space utilization, paging and page faults, interrupts, network activity, I/O and data transfer, power management, and more. In total, about 601.4 GB health-related performance data were collected and recorded from the cloud in that time period. Among all the metrics, 112 of them display zero variance, which provides no contribution to failure detection. After removing them, we have 406 non-constant metrics left.

4.2   Performance of Hybrid Anomaly Detection

We evaluate the performance of our anomaly detection framework by measuring the detection sensitivity and specificity. Sensitivity = (detected failures)/(total failures), Specificity = (detected normals)/(total normals).

Our anomaly detector identifies possible failures in the cloud performance data. It adapts itself by learning the verified detection results and observed but undetected failure events reported from the cloud operators. Figures 2 and 3 show anomaly detector's adaptation of the data description contours for failure detection (green and red stand for normal and anomaly class data respectively).

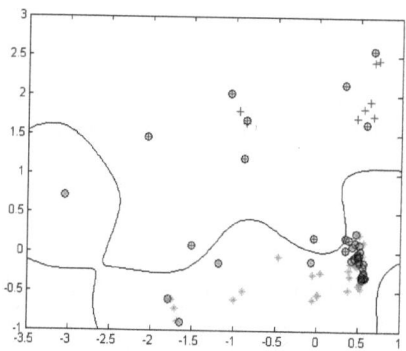 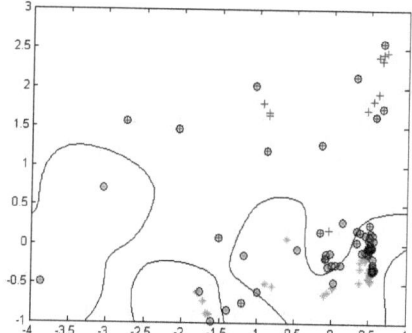

Fig. 2. Contours of decision function during Adaptation-I

Fig. 3. Contours of decision function during Adaptation-II

Less than 1% of the training data points are plotted in the figure for better readability. From the figures, we can see as more verified detection results are exploited, the contours become tighter, which improves the accuracy of failure detection. Figure 4 shows that credibility of one and two class SVM gets stabilize over period of time. Figure 5 depicts the detection performance after three selected rounds of adaptation by hybrid anomaly detector I, II, and III, respectively. Each adaptation phase is a result of multiple retraining sessions, because although the learning performance improves with each retraining but it should be reported only when it is significant. Only three adaptations are selected for better readability of the figure. After adaption III, detection performance gets stabilizes and does not increase further. These adaptations are recorded at 2 sec, 20 sec, and 1.5 min of the algorithm's execution. The figures show that both the detection sensitivity and the detection specificity improve as the detector adapts. After adaptation III is applied, the anomaly detector achieves 92.1% detection sensitivity and 83.8% detection specificity. These results indicate that our anomaly detector is well suitable for building dependable cloud computing systems. We have also compared our algorithm with failure detectors using some advanced learning algorithms. Subspace Regularization [42] performs better than other smoothness-based methods, including Gaussian Random Field [43] and Manifold Regularization [2]. In our experiments, the failure detector using Subspace Regularization achieves 67.8% sensitivity. The ensemble of Bayesian sub-models and decision tree classifiers that we proposed in[14] could only have 72.5% detection sensitivity. Our anomaly detector can achieve 92.1% detection sensitivity and 83.8% detection specificity, which are much higher than the other detectors.

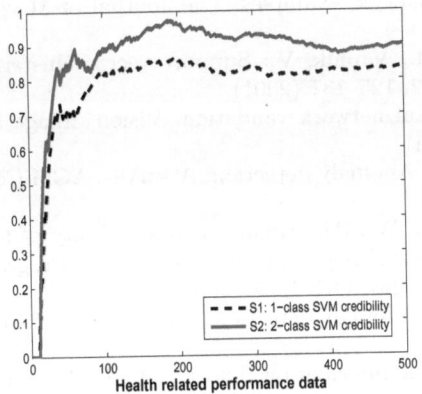 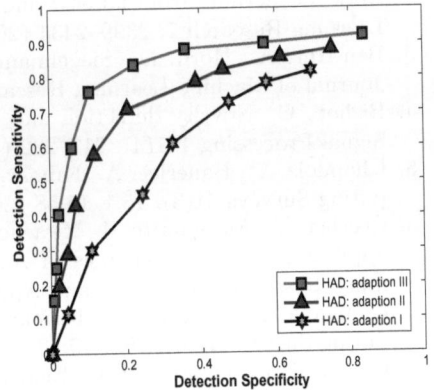

Fig. 4. Credibility scores improvement and stabilization of 1-class and 2-class SVM during hybridization

Fig. 5. Sensitivity and Specificity of Hybrid Anomaly Detection Algorithm (HAD) with adaptation

5   Conclusions

These days large-scale and complex cloud computing systems are susceptible to software and hardware failures and human mistakes, which significantly affect the cloud dependability and performance. In this paper, we employ *Hybrid Anomaly Detection* mechanism based on 1-class and 2-class SVM for adaptive failure detection. Different from other failure detection approaches, it does not require a prior failure history and it can self-adapt by learning from observed failure events at runtime. Thereby, it is capable of finding failures not yet seen in the past. Based on the cloud performance data, it detects possible failures, which are verified by the cloud operators. We note that even with the most advanced learning methods, the accuracy of failure detection cannot reach 100%. Reactive approaches, such as checkpointing and redundant execution, should be included to handle mis-detections. We plan to integrate the proactive and reactive failure management approaches to achieve even higher cloud dependability.

Acknowledgments. We would like to thank the anonymous reviewers for their constructive comments and suggestions. This research was supported in part by U.S. NSF grant CNS-0915396 and LANL grant IAS-1103.

References

1. Ananthanarayanan, G., Kandula, S., Greenberg, A., Stoica, I., Lu, Y., Saha, B., Harris, E.: Reining in the Outliers in Map-Reduce Clusters using Mantri. In: Proc. of the 9th USENIX Conf. on Operating Systems Design and Implementation (OSDI), pp. 1–16. USENIX Association (2010)
2. Belkin, M., Niyogi, P., Sindhwani, V.: Manifold Regularization: A geometric framework for learning from labeled and unlabeled examples. The Journal of Machine Learning Research 7, 2399–2434 (2006)
3. Ben-Hur, A., Horn, D., Siegelmann, H., Vapnik, V.: Support vector clustering. Journal of Machine Learning Research 2, 125–137 (2001)
4. Bishop, C.: Novelty detection and neural network validation. Vision, Image and Signal Processing 141(4), 217–222 (1994)
5. Chandola, V., Banerjee, A., Kumar, V.: Anomaly detection: A survey. ACM Computing Surveys 41(3), 15:1–15:58 (2009)
6. Cecchet, E., Marguerite, J., Zwaenepoel, W.: Performance and scalability of EJB applications. In: Proceedings of ACM Conference on Object-Oriented Programming, Systems, Languages, and Applications, OOPSLA (2002)
7. Chandola, V., Banerjee, A., Kumar, V.: Anomaly detection: A survey. ACM Computing Surveys 41(3), 15:1–15:58 (2009)
8. Cover, T.M., Thomas, J.A.: Elements of Information Theory. Wiley, New York (1991)
9. Dean, J., Ghemawat, S.: Mapreduce: simplified data processing on large clusters. Communications of the ACM 51(1), 107–113 (2008)
10. Fu, S.: Performance metric selection for autonomic anomaly detection on cloud computing systems. In: Proc. of IEEE Global Communications Conference, GLOBECOM (2011)
11. Fu, S., Xu, C.: Quantifying event correlations for proactive failure management in networked computing systems. Journal of Parallel and Distributed Computing 70(11), 1100–1109 (2010)
12. Gu, X., Wang, H.: Online Anomaly Prediction for Robust Cluster Systems. In: Proc. of the 2009 IEEE Intl. Conf. on Data Engineering (ICDE), pp. 1000–1011. IEEE Computer Society (2009)
13. Gu, X., Papadimitriou, S., Yu, P.S., Chang, S.-P.: Toward Predictive Failure Management for Distributed Stream Processing Systems. In: Proc. of the 28th Intl. Conf. on Distributed Computing Systems (ICDCS), pp. 825–832. IEEE Computer Society (2008)

14. Guan, Q., Zhang, Z., Fu, S.: Ensemble of bayesian predictors and decision trees for proactive failure management in cloud computing systems. Journal of Communications 7(1), 52–61 (2012)
15. He, J., Tong, H., Carbonell, J.: Rare Category Characterization. In: Proc. of the IEEE International Conference on Data Mining (2010)
16. Hodge, V., Austin, J.: A survey of outlier detection methodologies. Artificial Intelligence Review 22(2), 85–126 (2004)
17. Pyo Jae Kim, H.J.C., Choi, J.Y.: Fast incremental learning for one-class support vector classifier using sample margin information (2008) IEEE, 978-1-4244-2175-6/08
18. Kumar, V., Schwan, K., Iyer, S., Chen, Y., Sahai, A.: A state-space approach to SLA based management. In: Proc. of IEEE Network Operations and Management Symposium, NOMS (2008)
19. Kumar, V., Cooper, B.F., Eisenhauer, G., Schwan, K.: iManage: Policy-Driven Self-management for Enterprise-Scale Systems. In: Cerqueira, R., Campbell, R.H. (eds.) Middleware 2007. LNCS, vol. 4834, pp. 287–307. Springer, Heidelberg (2007)
20. Lan, Z., Gu, J., Zheng, Z., Thakur, R., Coghlan, S.: A study of dynamic meta-learning for failure prediction in largescale systems. Journal of Parallel and Distributed Computing 70(6), 630–643 (2010)
21. Liang, Y., Zhang, Y., Sivasubramaniam, A., Jette, M., Sahoo, R.K.: BlueGene/L failure analysis and prediction models. In: Proc. of IEEE/IFIP International Conference on Dependable Systems and Networks, DSN (2006)
22. Oliner, A.J., Sahoo, R.K., Moreira, J.E., et al.: Fault-aware job scheduling for BlueGene/L systems. In: Proc. of the 18th International Parallel and Distributed Processing Symposium, IPDPS (2004)
23. Oppenheimer, D., Ganapathi, A., Patterson, D.: Why do Internet services fail, and what can be done about it. In: Proc. of USENIX Symposium on Internet Technologies and Systems, USITS (2003)
24. Papadimitriou, S., Sun, J., Faloutsos, C.: Streaming Pattern Discovery in Multiple Time-Series. In: VLDB 2005, pp. 697–708 (2005)
25. Parra, L., Deco, G., Miesbach, S.: Statistical independence and novelty detection with information preserving nonlinear maps. Neural Computation 8(2), 260–269 (1996)
26. Peng, W., Li, T.: Mining logs files for computing system management. In: Proc. of IEEE International Conference on Autonomic Computing, ICAC (2005)
27. PERF. Performance counters on linux, http://linuxplumbersconf.org/2009/slides/Arnaldo-Carvalho-de-Melo-perf.pdf
28. Pertet, S., Narasimhan, P.: Causes of failure in web applications. Technical Report Technical report, CMU-PDL- 05-109, MIT Laboratory for Computer Science (2005)
29. Rosenblum, M., Garfinkel, T.: Virtual machine monitors: current technology and future trends. IEEE Computer 35(5), 39–47 (2005)
30. Sahoo, R.K., Oliner, A.J., Rish, I., Gupta, M., Moreira, J.E., Ma, S., Vilalta, R., Sivasubramaniam, A.: Critical event prediction for proactive management in large-scale computer clusters. In: Proc. of ACM International Conference on Knowledge Discovery and Data Dining, KDD (2003)
31. Salfner, F., Lenk, M., Malek, M.: A survey of online failure prediction methods,42:10:110:42. In: Proc. of ACM Computing Surveys (2010)
32. Scholkopf, B., Williamson, R.C., Smola, A.J., Shawe-Taylor, J., Platt, J.C.: Support vector method for novelty detection. In: Proc. of Conference on Advances in Neural Information Processing Systems (2000)

33. Shannon, C.E.: A Mathematical Theory of Communication. Bell System Technical Journal 27(3), 379–423 (1948)
34. SYSSTAT, http://sebastien.godard.pagespersoorange.fr/
35. Tan, J., Pan, X., Marinelli, E., Kavulya, S., Gandhi, R., Narasimhan, P.: Kahuna: Problem Diagnosis for Mapreduce-based Cloud Computing Environments. In: Network Operations and Management Symposium (NOMS), pp. 112–119. IEEE (2010)
36. Tan, Y., Gu, X., Wang, H.: Adaptive System Anomaly Prediction for Large-Scale Hosting Infrastructures. In: Proc. of the 29th ACM SIGACT-SIGOPS Symposium on Principles of Distributed Computing (PODC), pp. 173–182. ACM (2010)
37. Tax, D.M.J., Duin, R.P.W.: Support vector data description. Machine Learning 54(1), 45–66 (2004)
38. Vapnik, V.N.: The nature of statistical learning theory, 2nd edn. Springer, New York (2000)
39. Vapnik, V.: Pattern seperation by convex programming. Journal of Mathematical Analysis and Applications 10(1), 123–134 (1998)
40. Vilalta, R., Ma, S.: Predicting rare events in temporal domains. In: Proc. of IEEE International Conference on Data Mining, ICDM (2002)
41. Williams, A.W., Pertet, S.M., Narasimhan, P.: Tiresias: Blackbox Failure Prediction in Distributed Systems. In: 21st Intl. Parallel and Distributed Processing Symposium (IPDPS), pp. 1–8 (2007)
42. Zhang, Y.-M., Hou, X., Xiang, S., Liu, C.-L.: Subspace Regularization: A New Semi-supervised Learning Method. In: Buntine, W., Grobelnik, M., Mladenić, D., Shawe-Taylor, J. (eds.) ECML PKDD 2009, Part II. LNCS (LNAI), vol. 5782, pp. 586–601. Springer, Heidelberg (2009)
43. Zhu, X., Ghahramani, Z., Laerty, J.: Semi-supervised learning using Gaussian elds and Harmonic functions. In: Proceedings of the International Conference on Machine Learning, pp. 912–919 (2003)

Residual Belief Propagation for Topic Modeling

Jia Zeng[1,2], Xiao-Qin Cao[3], and Zhi-Qiang Liu[3]

[1] School of Computer Science and Technology,
Soochow University, Suzhou 215006, China
[2] Shanghai Key Laboratory of Intelligent Information Processing, China
j.zeng@ieee.org
[3] School of Creative Media,
City University of Hong Kong, Tat Chee Ave 83, Hong Kong
caoxiaoqin@gmail.com, zq.liu@cityu.edu.hk

Abstract. Fast convergence speed is a desired property for training topic models such as latent Dirichlet allocation (LDA), especially in online and parallel topic modeling algorithms for big data sets. In this paper, we develop a novel and easy-to-implement residual belief propagation (RBP) algorithm to accelerate the convergence speed for training LDA. The proposed RBP uses an informed scheduling scheme for asynchronous message passing, which passes fast convergent messages with a higher priority to influence those slow convergent messages at each learning iteration. Extensive empirical studies confirm that RBP significantly reduces the training time until convergence while achieves a much lower predictive perplexity than several state-of-the-art training algorithms for LDA, including variational Bayes (VB), collapsed Gibbs sampling (GS), loopy belief propagation (BP), and residual VB (RVB).

Keywords: Topic modeling, residual belief propagation, latent Dirichlet allocation, fast convergence speed.

1 Introduction

Probabilistic topic modeling [1] is an important problem in machine learning and data mining. As one of the simplest topic models, latent Dirichlet allocation (LDA) [2] requires multiple iterations of training until convergence. Recent studies confirm that the convergence speed determines the efficiency of topic modeling for big data sets. For example, online topic modeling algorithms [3] partition the entire data set into many small mini-batches, and optimize sequentially each mini-batch until convergence. Another example lies in parallel topic modeling algorithms [4], which optimize the distributed data sets until convergence and then communicate/synchronize the global topic distributions. Therefore, the faster convergence speed leads to the faster online and parallel topic modeling algorithms for big data sets.

Recent training algorithms for LDA can be broadly categorized into variational Bayes (VB) [2], collapsed Gibbs sampling (GS) [5] and loopy belief propagation (BP) [6]. We can interpret VB, GS and BP within a unified message

S. Zhou, S. Zhang, and G. Karypis (Eds.): ADMA 2012, LNAI 7713, pp. 739–752, 2012.

passing framework, which infers the posterior distribution of topic label for the word referred to as *message*, and estimates parameters by the iterative expectation-maximization (EM) algorithm according to the maximum-likelihood criterion [7]. They mainly differ in the E-step of EM algorithm for message update equations. For example, VB is a synchronous variational message passing algorithm [8], which updates variational messages by complicated digamma functions, slowing down the overall training speed [9,6]. In contrast, GS updates messages by discrete topic labels randomly sampled from the message in the previous iteration. Obviously, the sampling operation does not keep all uncertainties encoded in the previous messages. In addition, such a Markov chain Monte Carlo (MCMC) sampling process often requires more iterations until convergence. To avoid sampling, BP directly uses the previous messages to update the current messages. Such a deterministic process often takes the less number of iterations than GS to achieve convergence. According to a recent comparison [6], VB requires around 100 iterations, GS takes around 300 iterations and synchronous BP (sBP) needs around 170 iterations to achieve convergence in terms of training perplexity [2], which is a widely-used performance metric to compare different training algorithms of LDA [3,9].

In this paper, we develop a residual belief propagation (RBP) [10] algorithm to accelerate the convergence speed of topic modeling. Compared with sBP, RBP uses an informed scheduling strategy for asynchronous message passing, in which it efficiently influences those slow-convergent messages by passing fast-convergent messages with a higher priority. Through dynamically scheduling the order of message passing based on the residuals of two messages resulted from successive iterations, RBP in theory converges significantly faster and more often than general synchronous BP on cluster/factor graphs [10]. This paper studies RBP for probabilistic topic modeling, and demonstrates RBP's fast convergence speed for training LDA. Although jumping from synchronous BP to RBP is a simple idea on inference for general cluster/factor graphs [10], the RBP for training specific hierarchical Bayesian models such as LDA remains largely unexplored, especially the RBP's convergence property on LDA. Extensive experimental results demonstrate that RBP in most cases converges fastest while reaches the lowest predictive perplexity when compared with other state-of-the-art training algorithms, including VB [2], GS [5], sBP [6], and residual VB (RVB) [11,12].

Similar to the proposed RBP, residual VB (RVB) algorithms for LDA [11,12] have also been proposed from a matrix factorization perspective. Because VB is in nature a synchronous message passing algorithm, it does not have the direct asynchronous residual-based message passing counterpart. So, RVB is derived from online VB (OVB) algorithms [3], which divide the entire documents into mini-batches. Through dynamically scheduling the order of mini-batches based on residuals, RVB is often faster than OVB to achieve the same training perplexity. Indeed, there are several major differences between RVB and the proposed RBP. First, it is obvious that they are derived from different OVB and sBP algorithms, respectively. While OVB can converge to the VB's objective

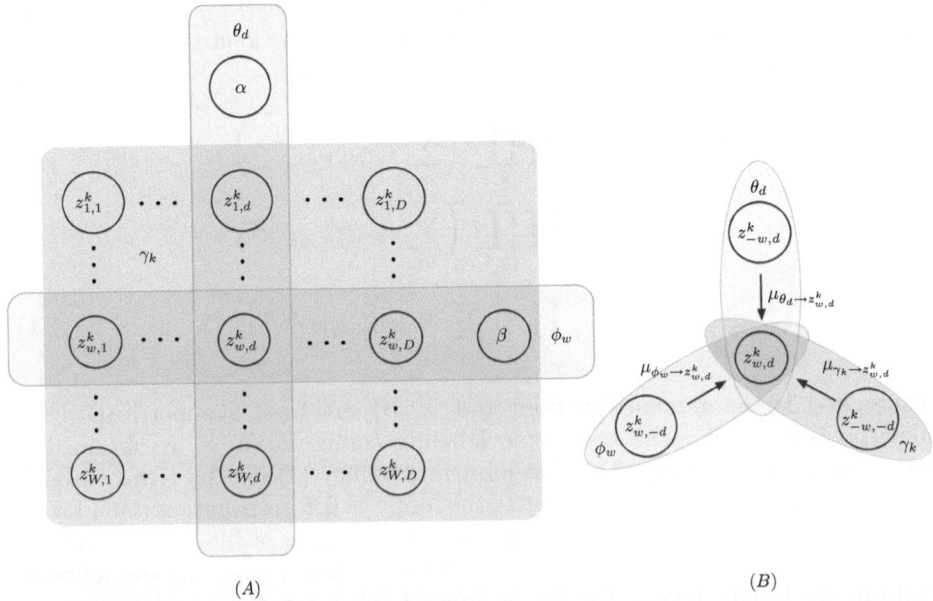

Fig. 1. (A) Hypergraph representation for the collapsed LDA and (B) message passing on hypergraph

function, it practically involves complicated digamma functions for biases and the slowness [9,6]. Second, RVB randomly generates a subset of mini-batches from a complicated residual distribution for training, while the proposed RBP simply sorts residuals in a descending order for either documents or vocabulary words. Notice that the random sampling process often misses those important mini-batches with largest residuals, but the sorting technique ensures to locate those top documents or vocabulary words with largest residuals. Because larger residuals correspond to more efficiency [11,12], our simple sorting technique in RBP is more efficient than the random sampling strategy in RVB.

2 Synchronous Belief Propagation (sBP) for LDA

We begin by briefly reviewing the sBP algorithm for training the collapsed LDA [6]. The probabilistic topic modeling task can be interpreted as a labeling problem, in which the objective is to assign a set of thematic topic labels, $\mathbf{z}_{W \times D} = \{z_{w,d}^k\}$, to explain the observed elements in document-word matrix, $\mathbf{x}_{W \times D} = \{x_{w,d}\}$. The notations $1 \le w \le W$ and $1 \le d \le D$ are the word index in vocabulary and the document index in corpus. The notation $1 \le k \le K$ is the topic index. The nonzero element $x_{w,d} \ne 0$ denotes the number of word counts at the index $\{w,d\}$. For each word token, there is a topic label $z_{w,d,i}^k = \{0,1\}$, $\sum_{k=1}^K z_{w,d,i}^k = 1, 1 \le i \le x_{w,d}$, so that the topic label for the word index $\{w,d\}$ is $z_{w,d}^k = \sum_{i=1}^{x_{w,d}} z_{w,d,i}^k / x_{w,d}$. After integrating out the document-specific topic proportions $\theta_d(k)$ and topic distribution over vocabulary words $\phi_w(k)$ in LDA, we obtain the joint probability of the collapsed LDA [13],

$$
p(\mathbf{x}, \mathbf{z}; \alpha, \beta) \propto \prod_d \prod_k \Gamma\left(\sum_w x_{w,d} z_{w,d}^k + \alpha\right) \times
$$

$$
\prod_w \prod_k \Gamma\left(\sum_d x_{w,d} z_{w,d}^k + \beta\right) \times
$$

$$
\prod_k \Gamma\left(\sum_{w,d} x_{w,d} z_{w,d}^k + W\beta\right)^{-1}, \tag{1}
$$

where $\Gamma(\cdot)$ is the gamma function, and $\{\alpha, \beta\}$ are fixed symmetric Dirichlet hyperparameters [5]. The best topic labeling configuration $\mathbf{z}^*$ is obtained by maximizing (1) in terms of $\mathbf{z}$. The joint probability (1) can be represented by the factor graph [14], which facilitates the loopy belief propagation (BP) for approximate inference [6]. The factor graph is an example of the hypergraph [15], because the factor nodes can be replaced by the hyperedges. In the following section, we briefly review BP for training LDA using the novel hypergraph representation.

2.1 Hypergraph Representation

Fig. 1A shows the hypergraph for the joint probability (1). There are three types of hyperedges $\{\theta_d, \phi_w, \gamma_k\}$ denoted by the yellow, green and red rectangles, respectively. For each column of $\mathbf{z}_{W \times D}$, the hyperedge θ_d, which corresponds to the first term in (1), connects the variable $z_{w,d}^k$ with the subset of variables $\mathbf{z}_{-w,d}^k$ and the hyperparameter α within the document d. For each row of $\mathbf{z}_{W \times D}$, the hyperedge ϕ_w, which corresponds to the second term in (1), connects the variable $z_{w,d}^k$ with the subset of variables $\mathbf{z}_{w,-d}^k$ and the hyperparameter β at the same word w in the vocabulary. Finally, the hyperedge γ_k, which corresponds to the third term in (1), connects the variables $z_{w,d}^k$ with all other variables $\mathbf{z}_{-w,-d}^k$ and the hyperparameter β on the same topic k. The notations $-w$ and $-d$ denote all word indices except w and all document indices except d, and the notations $\mathbf{z}_{-w,d}^k$, $\mathbf{z}_{w,-d}^k$ and $\mathbf{z}_{-w,-d}^k$ represent all neighboring labeling configurations through three types of hyperedges $\{\theta_d, \phi_w, \gamma_k\}$, respectively. Therefore, the hypergraph can completely describe the local dependencies of the topic configuration $\mathbf{z}$ in (1).

2.2 Message Passing

Fig. 1B shows that the variable $z_{w,d}^k$ is influenced by the subsets of neighboring variables $\{\mathbf{z}_{-w,d}^k, \mathbf{z}_{w,-d}^k, \mathbf{z}_{-w,-d}^k\}$ through three types of hyperedges $\{\theta_d, \phi_w, \gamma_k\}$. For a better illustration, we do not show the hyperparameters $\{\alpha, \beta\}$ in Fig. 1B.

BP [6] is an approximate inference method that calculates the posterior probability, $\mu_{w,d}(k) = p(z_{w,d}^k, x_{w,d}|\mathbf{z}_{-w,-d}^k, \mathbf{x}_{-w,-d})$, referred to as *message*. The message can be normalized efficiently using a local computation, i.e., $\sum_{k=1}^K \mu_{w,d}(k) = 1, 0 \leq \mu_{w,d}(k) \leq 1$. According to the Bayes' rule and the joint probability (1), we obtain

$$p(z_{w,d}^k, x_{w,d}|\mathbf{z}_{-w,-d}^k, \mathbf{x}_{-w,-d}) = \frac{p(\mathbf{z}_{w,d}^k, \mathbf{x}_{w,d})}{p(\mathbf{z}_{-w,d}^k, \mathbf{x}_{-w,d})}$$

$$\propto \frac{\Gamma(\sum_w x_{w,d} z_{w,d}^k + \alpha)}{\Gamma(\sum_w x_{-w,d} z_{-w,d}^k + \alpha)} \times \frac{\Gamma(\sum_d x_{w,d} z_{w,d}^k + \beta)}{\Gamma(\sum_d x_{w,-d} z_{w,-d}^k + \beta)} \times$$

$$\frac{\Gamma(\sum_{w,d} x_{-w,-d} z_{-w,-d}^k + W\beta)}{\Gamma(\sum_{w,d} x_{w,d} z_{w,d}^k + W\beta)}. \tag{2}$$

According to Fig. 1B, the first term of (2) is the message $\mu_{\theta_d \to z_{w,d}^k}$, the second term is the message $\mu_{\phi_w \to z_{w,d}^k}$, and the third term is the message $\mu_{\gamma_k \to z_{w,d}^k}$ through three types of hyperedges $\{\theta_d, \phi_w, \gamma_k\}$, respectively. Thus, the message $\mu_{w,d}(k)$ is proportional to the product of three incoming messages [16],

$$\mu_{w,d}(k) \propto \mu_{\theta_d \to z_{w,d}^k} \times \mu_{\phi_w \to z_{w,d}^k} \times \mu_{\gamma_k \to z_{w,d}^k}. \tag{3}$$

From (3) and the property $\Gamma(x+1) = x\Gamma(x)$, we obtain the approximate message update equation [6],

$$\mu_{w,d}(k) \propto \frac{[\boldsymbol{\mu}_{-w,d}(k) + \alpha] \times [\boldsymbol{\mu}_{w,-d}(k) + \beta]}{\boldsymbol{\mu}_{-w,-d}(k) + W\beta}, \tag{4}$$

where

$$\boldsymbol{\mu}_{-w,d}(k) = \sum_{-w} x_{-w,d}\mu_{-w,d}(k), \tag{5}$$

$$\boldsymbol{\mu}_{w,-d}(k) = \sum_{-d} x_{w,-d}\mu_{w,-d}(k), \tag{6}$$

$$\boldsymbol{\mu}_{-w,-d}(k) = \sum_{-w,-d} x_{-w,-d}\mu_{-w,-d}(k). \tag{7}$$

After updating (4), we normalize messages locally by the normalization factor $Z = \sum_{k=1}^K \mu_{w,d}(k)$. The normalization of each message requires K iterations. Messages are passed until convergence or the maximum number of iterations is reached. Using the converged messages, we can estimate the document-specific topic proportion θ_d and the topic distribution over vocabulary ϕ_w by the expectation-maximization (EM) algorithm [13],

$$\theta_d(k) = \frac{\boldsymbol{\mu}_{\cdot,d}(k) + \alpha}{\sum_k [\boldsymbol{\mu}_{\cdot,d}(k) + \alpha]}, \tag{8}$$

$$\phi_w(k) = \frac{\boldsymbol{\mu}_{w,\cdot}(k) + \beta}{\sum_w [\boldsymbol{\mu}_{w,\cdot}(k) + \beta]}. \tag{9}$$

The synchronous schedule f^s updates all messages (4) in parallel simultaneously at iteration t based on the messages at previous iteration $t-1$:

$$f^s(\mu_{1,1}^{t-1}, \ldots, \mu_{W,D}^{t-1}) = \{f(\mu_{-1,-1}^{t-1}), \ldots, f(\mu_{-w,-d}^{t-1}) \ldots, f(\mu_{-W,-D}^{t-1})\}, \qquad (10)$$

where f^s is the message update function (4) and $\mu_{-w,-d}$ is all set of messages excluding $\mu_{w,d}$.

3   Residual Belief Propagation (RBP) for LDA

The asynchronous schedule f^a updates the message of each variable in a certain order, which is in turn used to update other neighboring messages immediately at each iteration t:

$$f^a(\mu_{1,1}^{t-1}, \ldots, \mu_{W,D}^{t-1}) = \{\mu_{1,1}^t, \ldots, f(\mu_{-w,-d}^{t,t-1}), \ldots, \mu_{W,D}^{t-1}\}, \qquad (11)$$

where the message update equation f is applied to each message one at a time in some order. The basic idea of RBP for LDA is to select the best updating order based on the messages' residuals $r_{w,d}$, which are defined as the p-norm of difference between two message vectors at successive iterations,

$$r_{w,d} = x_{w,d} \| \mu_{w,d}^t - \mu_{w,d}^{t-1} \|_p, \qquad (12)$$

where $x_{w,d}$ is the number of word counts. For simplicity, we choose the L_1 norm with $p = 1$. In practice, the computational cost of sorting (12) is very high because we need to sort all non-zero residuals $r_{w,d}$ in the document-word matrix at each learning iteration. Obviously, this scheduling cost is expensive in case of large-scale data sets. Alternatively, we may accumulate residuals based on either document or vocabulary indices,

$$r_d = \sum_w r_{w,d}, \qquad (13)$$

$$r_w = \sum_d r_{w,d}. \qquad (14)$$

These residuals can be computed during message passing process at a negligible computational cost. For large-scale data sets, we advocate (14) because the vocabulary size is often a fixed number W independent of the number of documents D. So, initially sorting r_w requires at most a computational complexity of $\mathcal{O}(W \log W)$ using the standard quick sort algorithm. If the successive residuals are in almost sorted order, only a few swaps will restore the sorted order by the standard insertion sort algorithm, thereby saving time. In our experiments (not shown in this paper due to the page limit), RBP based on (14) uses little computational cost to sort r_w while retains almost the same convergence rate as that of sorting (12). We see that Eq. (13) is also useful for small-scale data sets, because in this case $D < W$ as shown in Table 1.

```
input    : x, K, T, α, β.
output   : θ_d, φ_w.
```
1 $\mu^1_{w,d}(k) \leftarrow$ random initialization and normalization;
2 $\mathcal{W}^1 \leftarrow$ random order;
3 **for** $t \leftarrow 1$ **to** T **do**
4   **for** $w \in \mathcal{W}^t$ **do**
5   **for** $d \leftarrow 1$ **to** D, $k \leftarrow 1$ **to** K, $x_{w,d} \neq 0$ **do**
6           $\mu^{t+1}_{w,d}(k) \propto$
            $\frac{\mu^t_{-w,d}(k)+\alpha}{\sum_k[\mu^t_{-w,d}(k)+\alpha]} \times \frac{\mu^t_{w,-d}(k)+\beta}{\sum_w[\mu^t_{w,-d}(k)+\beta]}$;
7           $\mu^{t+1}_{w,d}(k) \leftarrow$ normalize$(\mu^{t+1}_{w,d}(k))$;
8   **end**
9        $r^{t+1}_w \leftarrow \sum_d \sum_k x_{w,d}|\mu^{t+1}_{w,d}(k) - \mu^t_{w,d}(k)|$;
10   **end**
11   **if** $t = 1$ **then**
12       $\mathcal{W}^{t+1} \leftarrow$ quick sort$(r^{t+1}_w, \text{'descending'})$;
13   **else**
14       $\mathcal{W}^{t+1} \leftarrow$ insertion sort$(r^{t+1}_w, \text{'descending'})$;
15   **end**
16 **end**
17 $\theta_d(k) \leftarrow [\mu^T_{\cdot,d}(k) + \alpha]/\sum_k[\mu^T_{\cdot,d}(k) + \alpha]$;
18 $\phi_w(k) \leftarrow [\mu^T_{w,\cdot}(k) + \beta]/\sum_w[\mu^T_{w,\cdot}(k) + \beta]$;

Fig. 2. The RBP algorithm for LDA

Fig. 2 summarizes the proposed RBP algorithm based on (14), which will be used in the following experiments. First, we initialize messages randomly and normalize them locally (Line 1). Second, we start a random order of $w \in \mathcal{W}^1$ (Line 2) and accumulate residuals r^{t+1}_w during message updating (Line 9). At the end of each learning iteration t, we sort r^{t+1}_w in the descending order to refine the updating order $w \in \mathcal{W}^{t+1}$ (Line 12 or 14). Finally, after $1 \leq t \leq T$ iterations, RBP stops and estimates two multinomial parameters $\{\theta, \phi\}$ (Line 17 to 18). Intuitively, the message residuals reflect the convergence speed of message updating. The larger message residuals correspond to the faster-convergent messages. In the successive learning iterations, RBP always start passing fast-convergent messages with a higher priority in the order $\mathcal{W}^{t+1}$. Because the asynchronous message passing influences the current message updating by the previous message updating, passing the fast-convergent messages will speed up the convergence of those slow-convergent messages. In conclusion, it is more efficient to pass a message whose current value is quite different from its previous value, while passing a message whose current value is very similar to its value in the previous iteration is almost redundant.

We demonstrate that the RBP algorithm in Fig. 2 has a faster convergence rate than sBP [6] for training LDA. We assume that the message update equation f in (4) a contraction function under some norm [10], so that

$$\|\boldsymbol{\mu}^t - \boldsymbol{\mu}^*\| \leq \gamma\|\boldsymbol{\mu}^{t-1} - \boldsymbol{\mu}^*\|, \tag{15}$$

for some global contraction factor $0 \leq \gamma \leq 1$. Eq. (15) guarantees that the messages $\boldsymbol{\mu}^t = \{\mu^t_{1,1}, \ldots, \mu^t_{W,D}\}$ will converge to a fixed point $\boldsymbol{\mu}^* = \{\mu^*_{1,1}, \ldots, \mu^*_{W,D}\}$

in the synchronous schedule (10). This assumption often holds true for sBP in training the collapsed LDA [6] based on the cluster/factor graphical representation [17]. According to [10], the asynchronous schedule (11) will also converge to a fixed point μ^* if f is a contraction mapping and for each message $\mu_{w,d}(k)$, there is a finite time interval, so that the message update equation f is executed at least once in this time interval. As a result, the RBP algorithm in Fig. 2 will converge to a fixed point μ^* as sBP.

To speed up convergence in the asynchronous schedule (11), we choose to update the message $\mu_{w,d}$ so as to minimize the largest distance $\|\mu_{w,d}^t - \mu_{w,d}^*\|$ first. However, we cannot directly measure the distance between a current message and its unknown fixed point value. Alternatively, we can derive a bound on this distance that can be calculated easily. Using the triangle inequality, we obtain

$$
\begin{aligned}
\|\mu_{w,d}^t - \mu_{w,d}^{t-1}\| &= \|\mu_{w,d}^t - \mu_{w,d}^* + \mu_{w,d}^* - \mu_{w,d}^{t-1}\| \\
&\le \|\mu_{w,d}^t - \mu_{w,d}^*\| + \|\mu_{w,d}^{t-1} - \mu_{w,d}^*\| \\
&\le \gamma\|\mu_{w,d}^{t-1} - \mu_{w,d}^*\| + \|\mu_{w,d}^{t-1} - \mu_{w,d}^*\| \\
&= (1+\gamma)\|\mu_{w,d}^{t-1} - \mu_{w,d}^*\|.
\end{aligned}
\tag{16}
$$

According to (15) and (16), we derive the bound (17) as follows

$$
\begin{aligned}
\|\mu_{w,d}^t - \mu_{w,d}^*\| &\le \gamma\|\mu_{w,d}^{t-1} - \mu_{w,d}^*\| \\
&= \|\mu_{w,d}^{t-1} - \mu_{w,d}^*\| - (1-\gamma)\|\mu_{w,d}^{t-1} - \mu_{w,d}^*\| \\
&\le \|\mu_{w,d}^{t-1} - \mu_{w,d}^*\| - \frac{1-\gamma}{1+\gamma}\|\mu_{w,d}^t - \mu_{w,d}^{t-1}\|.
\end{aligned}
\tag{17}
$$

which is bounded by some fraction (less than 1) of the difference between the message before and after the update $\|\mu_{w,d}^t - \mu_{w,d}^{t-1}\|$. Because we do not know the fixed point $\mu_{w,d}^*$ in (17), alternatively, we can maximize the corresponding difference $\|\mu_{w,d}^t - \mu_{w,d}^{t-1}\|$ in order to minimize $\|\mu_{w,d}^t - \mu_{w,d}^*\|$. Notice that the difference is the definition of the message residual (12). Therefore, if we always update and pass messages in the descending order of residuals (12) dynamically, the RBP algorithm in Fig. 2 will converge faster to the fixed point μ^* than sBP [6] for training LDA.

4   Experimental Results

We carry out experiments on six publicly available data sets: 1) 20 newsgroups (NG20)[1], 2) BLOG [17], 3) CORA [18], 4) MEDLINE [19], 5) NIPS [20], and 6) WEBKB[2]. Table 1 summarizes the statistics of six data sets, where D is the total number of documents in the corpus, W is the number of words in the vocabulary, N_d is the average number of word tokens per document, and W_d is

---

[1] people.csail.mit.edu/jrennie/20Newsgroups
[2] csmining.org/index.php/webkb.html

Table 1. Statistics of six document data sets

Data set	D	W	N_d	W_d
NG20	7505	61188	239	129
BLOG	5177	33574	217	149
CORA	2410	2961	57	43
MEDLINE	2317	8918	104	66
NIPS	1740	13649	1323	536
WEBKB	7061	2785	50	29

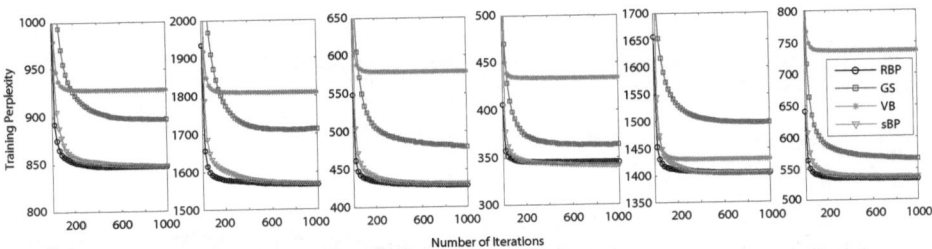

Fig. 3. Training perplexity as a function of the number of iterations when $K = 10$

the average number of word indices per document. All subsequent figures show results on six data sets in the above order. We compare RBP with three state-of-the-art approximate inference methods for LDA including VB [2], GS [5], and sBP [6] under the same fixed hyperparameters $\alpha = \beta = 0.01$. We use MATLAB C/C++ MEX-implementations for all these algorithms [21], and carry out the experiments on a desktop computer with CPU 2.4GHz and RAM 4G.

Fig. 3 shows the training perplexity [9] at every 10 iterations in 1000 iterations when $K = 10$ for each data set. All algorithms converge to a fixed point of training perplexity within 1000 iterations. Except the NIPS set, VB always converges at the highest training perplexity. In addition, GS converges at a higher perplexity than both sBP and RBP. While RBP converge at almost the same training perplexity as sBP, it always reaches the same perplexity value faster than sBP. Generally, the training algorithm converges when the training perplexity difference at two consecutive iterations is below a threshold. In this paper, we set the convergence threshold to 1 because the training perplexity decreases very little after this threshold is satisfied in Fig. 3.

Fig. 4 illustrates the number training iterations until convergence on each data set for different topics $K \in \{10, 20, 30, 40, 50\}$. The number of iterations until convergence seems insensitive to the number of topics. On the BLOG, CORA and WEBKB sets, VB uses the minimum number iterations until convergence, consistent with the previous results in [6]. For all data sets, GS consumes the maximum number of iterations until convergence. Unlike the deterministic message updating in VB, sBP and RBP, GS uses the stochastic message updating scheme accounting for the largest number of iterations until convergence.

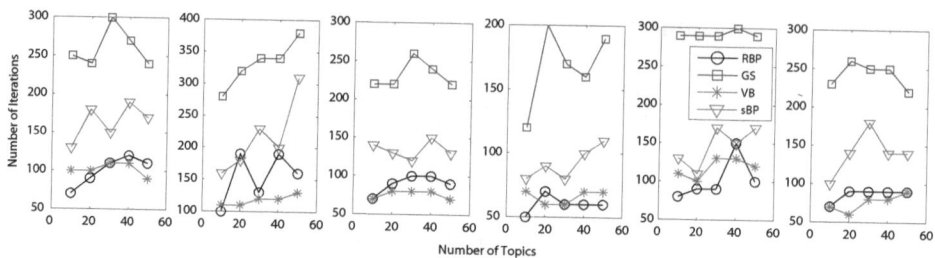

Fig. 4. The number of training iterations until convergence as a function of number of topics

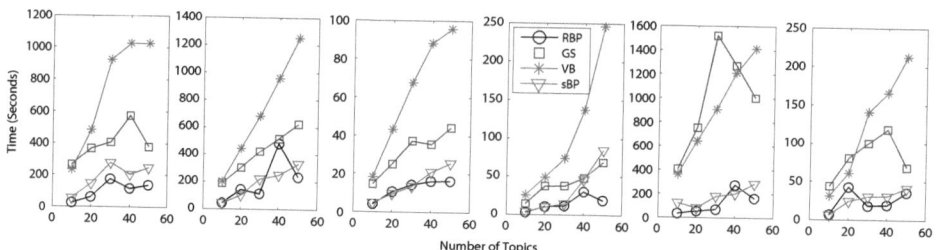

Fig. 5. Training time until convergence as a function of number of topics

Although sBP costs the less number of iterations until convergence than GS, it still uses the much more number of iterations than VB. By contrast, through the informed dynamic scheduling for asynchronous message passing, RBP on average converges more rapidly than sBP for all data sets. In particular, on the NG20, MEDLINE and NIPS sets, RBP on average uses a comparable or even less number of iterations than VB until convergence.

Fig. 5 shows the training time in seconds until convergence on each data set for different topics $K \in \{10, 20, 30, 40, 50\}$. Surprisingly, while VB usually uses the minimum number iterations until convergence, it often consumes the longest training time for these iterations. The major reason may be attributed to the time-consuming digamma functions in VB, which takes at least triple more time for each iteration than GS and sBP. If VB removes the digamma functions, it runs as fast as sBP. Because RBP uses a significantly less number of iterations until convergence than GS and sBP, it consumes the least training time until convergence for all data sets in Fig. 5.

We also examine the predictive perplexity of all algorithms until convergence based on a ten-fold cross-validation. The predictive perplexity for the unseen test set is computed as that in [9]. Fig. 6 shows the box plot of predictive perplexity for ten-fold cross-validation when $K = 50$. The plot produces a separate box for ten predictive perplexity values of each algorithm. On each box, the central mark is the median, the edges of the box are the 25th and 75th percentiles, the whiskers extend to the most extreme data points not considered outliers, and outliers

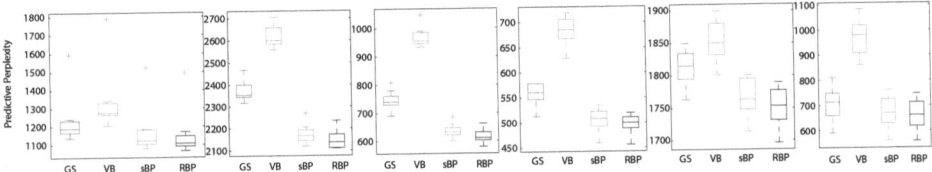

Fig. 6. Predictive perplexity for ten-fold cross-validation when $K = 50$

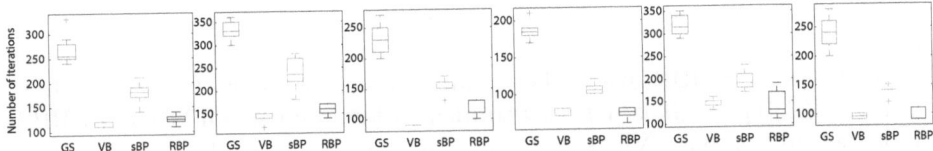

Fig. 7. The number of training iterations until convergence for ten-fold cross-validation when $K = 50$

are plotted individually by the red plus sign. Obviously, VB yields the highest predictive perplexity, corresponding to the worst generalization ability. GS has a much lower predictive perplexity than VB, but it has a much higher perplexity than both sBP and RBP. The underlying reason is that GS samples a topic label from the messages without retaining all possible uncertainties. The residual-based scheduling scheme of RBP not only speeds up the convergence rate of sBP, but also slightly lowers the predictive perplexity. The reason is that RBP updates fast-convergent messages to efficiently influence those slow-convergent messages, reaching fast to the local minimum of the predictive perplexity.

Figs. 7 and 8 illustrate the box plots for the number of iterations and the training time until convergence for ten-fold cross-validation when $K = 50$. Consistent with Figs. 4 and 5, VB consumes the minimum number of iterations, but has the longest training time until convergence. GS has the maximum number of iterations, but has the second longest training time until convergence. Because RBP improves the convergence rate over sBP, it consumes the least training time until convergence.

To measure the interpretability of inferred topics, Fig. 9 shows the top ten words of each topic when $K = 10$ on CORA set using 500 training iterations. We observe that both sBP and RBP can infer almost the same topics as other algorithms except the topic one, where sBP identifies the "pattern recognition" topic but RBP infers the "parallel system" topic. It seems that both sBP and RBP obtain slightly more interpretable topics than GS and VB especially in topic four, where "reinforcement learning" is closely related to "control systems". For other topics, we find that they often share the similar top ten words but with different ranking orders. More details on subjective evaluation for interpretability of topics can be found in [22]. However, even if GS and VB yield comparably interpretable topics as RBP, we still advocate RBP because it consumes less

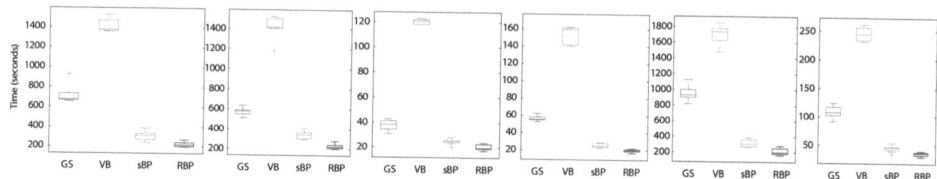

Fig. 8. Training time until convergence for ten-fold cross-validation when $K = 50$

training time until convergence while reaches a much lower predictive perplexity value.

We also compare RBP with other residual-based techniques for training LDA such as RVB [11,12]. It is not easy to make a fair comparison because RBP is an offline learning but RVB is an online learning algorithm. However, using the same data sets WEBKB and NG20 [11], we can approximately compare RBP with RVB using the training time when the predictive perplexity converges. When $K = 100$, RVB converges at the predictive perplexity 600 using 60 seconds training time on WEBKB, while it converges at the predictive perplexity 1050 using 600 seconds training time on NG20 [12]. With the same experimental settings as RVB (hyperparameters $\alpha = \beta = 0.01$), RBP achieves the predictive perplexity 540 using 35 seconds for training on WEBKB, while it achieves the predictive perplexity 1004 using 420 seconds for training on NG20. The significant speedup is because RVB involves relatively slower digamma function computations, and adopts a more complicated sampling method based on residual distributions for dynamic scheduling.

5 Conclusions

This paper presents a simple but effective RBP algorithm for training LDA with fast convergence speed. Using the dynamic residual-based scheduling scheme, we demonstrate that RBP significantly improves the convergence rate of sBP but adding only an affordable scheduling cost for large-scale data sets. On six publicly available data sets, RBP reduces around $50 \sim 100$ training iterations until convergence, while achieves a relatively lower predictive perplexity than sBP [6]. For the ten-fold cross-validation on six publicly available document sets when $K = 50$, RBP on average reduces 63.7% and 85.1% training time until convergence than two widely-used GS [5] and VB [2] algorithms, respectively. Meanwhile, it on average achieves 8.9% and 22.1% lower predictive perplexity than GS and VB, respectively. Compared with other residual techniques like RVB [11,12], RBP reduces around 30% $\sim$ 50% training time to achieve a much lower predictive perplexity. Although RBP is a simple extension of sBP [6] by introducing the dynamic scheduling for message passing, its theoretical basis and strong experimental results in this paper will support its promising role in the probabilistic topic modeling field. In our future work, we shall develop online [23] and parallel topic modeling algorithms [24] based on the proposed RBP.

Topic 1	model visual network recognition data neural learning object patterns input learning network model networks feature neural features selection models paper model visual recognition system patterns object structure sequences network human parallel performance paper system implementation efficient execution memory machine data
Topic 2	learning algorithm model show algorithms number class general queries time algorithm learning model algorithms results class number show error problem algorithm learning number function model algorithms results show class bounds algorithm learning function model show number results algorithms functions class
Topic 3	system paper learning knowledge design case reasoning problem approach theory system paper design case reasoning knowledge systems memory performance planning design system reasoning case knowledge paper cases systems approach planning design system reasoning knowledge case paper problem planning approach cases
Topic 4	learning paper system control approach systems robot results reinforcement environment learning algorithm problem paper method results problems reinforcement system function learning control reinforcement paper state agent environment robot problem dynamic learning model reinforcement control system environment visual robot agent behavior
Topic 5	bayesian belief probability network networks show model problem probabilistic revision paper algorithm data bayesian learning show results algorithms methods genetic learning paper theory knowledge examples rules system problem approach concept learning paper examples rules theory knowledge system problem approach concept
Topic 6	genetic problem search algorithms paper algorithm problems results optimization programming problem genetic learning paper system search approach design knowledge problems genetic problem search algorithms programming paper results problems optimization evolutionary genetic problem search algorithms programming problems paper results optimization algorithm
Topic 7	model models algorithm data distribution markov state analysis paper methods method algorithm network networks model neural data results markov distribution model models bayesian data markov probability distribution methods analysis belief model models bayesian data markov probability distribution methods analysis method
Topic 8	learning algorithm training decision method paper data algorithms classification results learning paper decision problem algorithms system algorithm results methods model data decision training algorithm classification methods method performance learning algorithms decision training data classification algorithm learning methods method algorithms performance
Topic 9	neural network networks learning input training function algorithms recurrent algorithm neural networks network learning paper model time data algorithm method network neural networks learning input time paper training algorithm recurrent network neural networks learning input training time algorithm recurrent hidden
Topic 10	research report grant university technical science supported department part national research learning grant paper models supported model university science part research report technical grant university science supported department part paper research report technical grant university supported science department part paper

Fig. 9. Top ten words of $K = 10$ topics for GS (blue), VB (red), sBP (green) and RBP (black) on CORA set

Acknowledgements. This work is substantially supported by NSFC (Grant No. 61003154), Natural Science Foundation of the Jiangsu Higher Education Institutions of China (Grant No. 12KJA520004), the Shanghai Key Laboratory of Intelligent Information Processing, China (Grant No. IIPL-2010-009), a grant from Baidu to JZ, and a GRF grant from RGC UGC Hong Kong (GRF Project No.9041574) and a grant from City University of Hong Kong (Project No. 7008026) to ZQL.

References

1. Blei, D.M.: Introduction to probabilistic topic models. Communications of the ACM, 77–84 (2012)
2. Blei, D.M., Ng, A.Y., Jordan, M.I.: Latent Dirichlet allocation. J. Mach. Learn. Res. 3, 993–1022 (2003)

3. Hoffman, M., Blei, D., Bach, F.: Online learning for latent Dirichlet allocation. In: NIPS, pp. 856–864 (2010)
4. Newman, D., Asuncion, A., Smyth, P., Welling, M.: Distributed algorithms for topic models. J. Mach. Learn. Res. 10, 1801–1828 (2009)
5. Griffiths, T.L., Steyvers, M.: Finding scientific topics. Proc. Natl. Acad. Sci. 101, 5228–5235 (2004)
6. Zeng, J., Cheung, W.K., Liu, J.: Learning topic models by belief propagation. IEEE Trans. Pattern Anal. Mach. Intell. (2012); arXiv:1109.3437v4 [cs.LG]
7. Dempster, A.P., Laird, N.M., Rubin, D.B.: Maximum likelihood from incomplete data via the EM algorithm. Journal of the Royal Statistical Society, Series B 39, 1–38 (1977)
8. Winn, J., Bishop, C.M.: Variational message passing. J. Mach. Learn. Res. 6, 661–694 (2005)
9. Asuncion, A., Welling, M., Smyth, P., Teh, Y.W.: On smoothing and inference for topic models. In: UAI, pp. 27–34 (2009)
10. Elidan, G., McGraw, I., Koller, D.: Residual belief propagation: Informed scheduling for asynchronous message passing. In: UAI, pp. 165–173 (2006)
11. Wahabzada, M., Kersting, K.: Larger Residuals, Less Work: Active Document Scheduling for Latent Dirichlet Allocation. In: Gunopulos, D., Hofmann, T., Malerba, D., Vazirgiannis, M. (eds.) ECML PKDD 2011, Part III. LNCS, vol. 6913, pp. 475–490. Springer, Heidelberg (2011)
12. Wahabzada, M., Kersting, K., Pilz, A., Bauckhage, C.: More influence means less work: fast latent Dirichlet allocation by influence scheduling. In: CIKM, pp. 2273–2276 (2011)
13. Heinrich, G.: Parameter estimation for text analysis. Technical report, University of Leipzig (2008)
14. Kschischang, F.R., Frey, B.J., Loeliger, H.A.: Factor graphs and the sum-product algorithm. IEEE Transactions on Inform. Theory 47(2), 498–519 (2001)
15. Berge, C.: Hypergraphs. North-Holland, Amsterdam (1989)
16. Bishop, C.M.: Pattern recognition and machine learning. Springer (2006)
17. Eisenstein, J., Xing, E.: The CMU 2008 political blog corpus. Technical report, Carnegie Mellon University (2010)
18. McCallum, A.K., Nigam, K., Rennie, J., Seymore, K.: Automating the construction of internet portals with machine learning. Information Retrieval 3(2), 127–163 (2000)
19. Zhu, S., Takigawa, I., Zeng, J., Mamitsuka, H.: Field independent probabilistic model for clustering multi-field documents. Information Processing & Management 45(5), 555–570 (2009)
20. Globerson, A., Chechik, G., Pereira, F., Tishby, N.: Euclidean embedding of co-occurrence data. J. Mach. Learn. Res. 8, 2265–2295 (2007)
21. Zeng, J.: A topic modeling toolbox using belief propagation. J. Mach. Learn. Res. 13, 2233–2236 (2012)
22. Chang, J., Boyd-Graber, J., Gerris, S., Wang, C., Blei, D.: Reading tea leaves: How humans interpret topic models. In: NIPS, pp. 288–296 (2009)
23. Zeng, J., Liu, Z.Q., Cao, X.Q.: Online belief propagation for topic modeling (2012) arXiv:1210.2179 [cs.LG]
24. Yan, J., Liu, Z.Q., Gao, Y., Zeng, J.: Communication-efficient parallel belief propagation for latent Dirichlet allocation (2012) arXiv:1206.2190v1 [cs.LG]

The Author-Topic-Community Model: A Generative Model Relating Authors' Interests and Their Community Structure

Chunshan Li[1,2,*], William K. Cheung[3], Yunming Ye[1,2], and Xiaofeng Zhang[1,2]

[1] Shenzhen Graduate School, Harbin Institute of Technology, China
[2] Shenzhen Key Laboratory of Internet Information Collaboration, China
[3] Department of Computer Science, Hong Kong Baptist University, Hong Kong SAR
lichunshan.hit@gmail.com, william@comp.hkbu.edu.hk,
yeyunming@hit.edu.cn, zhangxiaofeng@hitsz.edu.cn

Abstract. In this paper, we introduce a generative model named Author-Topic-Community (ATC) model which can infer authors' interests and their community structure at the same time based on the contents and citation information of a document corpus. Via the mutual promotion between the author topics and the author community structure introduced in the ATC model, the robustness of the model towards cases with spare citation information can be enhanced. Variational inference is adopted to estimate the model parameters of ATC. We performed evaluation using both synthetic data as well as a real dataset which contains SIGKDD and SIGMOD papers published in 10 years. By contrasting the performance of ATC with some state-of-the-art methods which model authors' interests and their community structure separately, our experimental results show that 1) the ATC model with the inference of the authors' interests and the community structure integrated can improve the accuracy of author topic modeling and that of author community discovery; and 2) more in-depth analysis of the authors' influence can be readily supported.

Keywords: graphical model, community discovery, user modeling.

1 Introduction

People relation information has been made widely available due to the recent development of online social media sites like twitter, digg, and digital libraries of scientific literature. This makes user modeling using data mining techniques highly viable. Compared with the number of individuals considered, people relation information is usually sparse. it is common that like-minded individuals or people with shared expertises are only partially marked. Open research issues include (1) how to infer individuals' interest or expertise, and (2) how to discover the hidden community structure.

While user expertise modeling and community discovery have been widely studied in recent years, the two issues are usually treated independently. For instance, the Author Topic (AT) model is a representative example which adopts the generative model

---

[*] This work was completed during the exchange visit of Chunshan Li at CS Department, Hong Kong Baptist University.

S. Zhou, S. Zhang, and G. Karypis (Eds.): ADMA 2012, LNAI 7713, pp. 753–765, 2012.
© Springer-Verlag Berlin Heidelberg 2012

approach to model author interests as the inferred topics embedded in documents [1]. It does not consider the community structure of the authors, which usually is useful and important. Take Twitter as an example. Suppose user A posts just a few sports related messages but retweets a lot of messages from user B who is a big fan of sports. Without considering the social relationships as reflecting by the retweeting, user A's interest in sport cannot be derived. Meanwhile, many efforts have been devoted to community discovery based on the social links among users [2]. Also, the LDA based topic modeling method have been extended to model both the documents' topics and their links defined based on co-authorship and/or citation relations [3,4]. While these methods are effective for analysing topics and community for documents, and sometimes do consider authors interests as well [4], author community structure is seldom explicitly considered and modeled.

In this paper, we propose the Author-Topic-Community (ATC) model to address the two aforementioned research issues. The ATC model is designed to uncover author interests and at the same time the author community structure via a unified mutual promotion process. In particular, we hypothesize that the linkage between two authors can be inferred from not only the fact that they author documents with shared topics but also the community structure tying them close. The design of the model is inspired from the Author Topic (AT) model [1] and the Topic-Link LDA (TLLDA) model [3]. Each author is associated with two multinomial distributions over topics and communities respectively and each word is assigned to a multinomial distribution over author interests. The words in documents and links among authors are modeled as a joint probability distribution. We estimate the model parameters using the variational Expectation-Maximization (EM) approach. For perfomance evaluation, we apply the ATC model to both synthetic datasets and a real dataset of DBLP research papers and compare its performance with the AT model and the TLLDA model which consider author interests and author communities separately. The experimental results obtained show that the ATC model outperforms the AT model and the TLLDA model. At the same time, ATC can discover community structure and themes, as well as the authors' individual interests.

The remaining of the paper is organized as follows. Section 2 reviews some related works. The ATC model is presented in Section 3. Section 4 describes in the model inference and parameter estimation steps. Experiments and evaluation results are given in Section 5. Section 6 concludes the paper.

2   Related Work on Topic Modeling

In the literature, a number of latent variable models have been proposed for analyzing documents. Latent Dirichlet Allocation (LDA) [5] models a document as a mixture of topics and each word in a document obeys a multinomial distribution over the topics. Nallapati *et al.* [6] proposed Linked-LDA model which governs link generation on the basis of LDA. Mei *et al.* [7] presented a TMN model which can uncover topics in the text and maps a topic onto a community. Topic-Link LDA model [3] represents a generative process for both documents and links. It assumes that both topics for documents and community relations for authors determine the generation for a link between two documents. However, the interests of the authors are not inferred. Author topic model

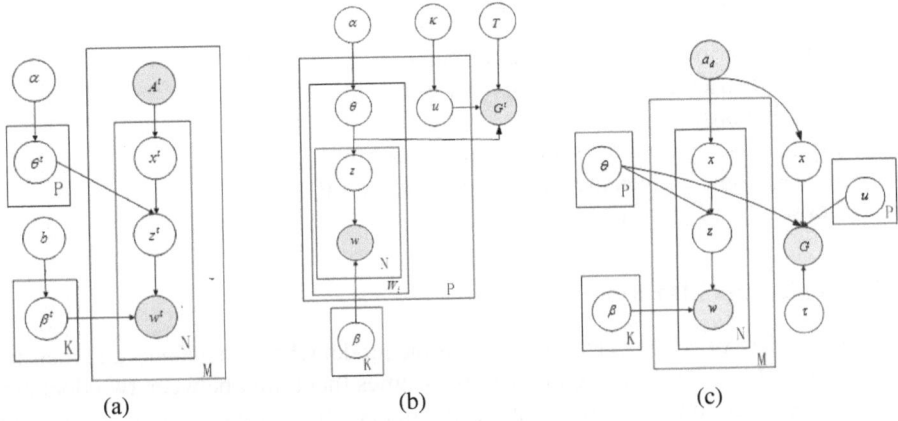

Fig. 1. (a) AT model (b) TLLDA model (c) ATC model

(ATM) [1] is one of the early attempts to model author interests as a distribution of topics. It assumes that each word obeys a multinomial distribution defined over topics which are associated with specific authors. Bhattacharya *et al.* [8] extended the AT model to recommend references to authors on the basis of author interests. By considering the citation ties of authors, Shiozaki et al. [9] proposed an entity-entity relationship model to represent the generative process for topic and named-entities in articles. Tu *et al.* [10] and Kataria *et al.* [4] proposed Citation Author Topic (CAT) model and Author Link Topic (ALT) model to represent author interests and topic influences among cited authors at the same time. These models assume that the cited network indicates how the authors of the cited document have influenced the contributions of the citing authors. However, none of them considered the author community.

3   Generative Models Relating Authors' Interests and Their Community Structure

In this section, we first present two generative models mostly related to the ATC model, namely the Author-Topic model and Topic-Link LDA model. Then, we describe in detail the proposed ATC model and the related parameter estimation methods.

3.1   Author-Topic Model

The Author Topic (AT) model [1] extends the Latent Dirichlet Allocation (LDA) model to infer authors' interests as mixtures of probability distributions over topics based on the documents they authored, as shown in Figure 1(a). It is designed so that the process of generating a document is based on topics of words which is in turn affected by the interests of the authors, thus also called authors' topics. Suppose a set of authors, a_d, want to write a document d. The generative process of the AT model is as follow:

For each document d in a corpus D:

1. Choose distribution θ over topics of author $\sim Dirichlet(\alpha)$.
2. Choose distribution β over topics of word $\sim Dirichlet(b)$.
3. for each word w in d do
   choose a user x from the set of authors uniformly
   choose a topic z given the author $x \sim Multi(x, \theta)$
   choose a word w from the chosen topic $z \sim Multi(z, \beta)$

3.2   Topic-Link LDA Model

Topic-Link LDA (TLLDA) model [3] is a topic model where documents and their links are generated, as shown in Figure 1(b). It assumes that a link between two documents is not purely due to their content similarity but also the community information of the authors. The generative process of the TLLDA model is as follow:

1. For each document d in a corpus D:
   (a) Choose distribution θ over topics of document $\sim Dirichlet(\alpha)$.
   (b) Choose distribution β over topics of word
   (c) for each word w in d do
       choose a topic z_n given the author $\sim Multi(\theta)$
       choose a word from the chosen topic $\sim Multi(z_n, \beta)$
2. For the Graph G:
   (a) Choose distribution u over community of author $\sim Dirichlet(\kappa)$.
   (b) for each edge e in G do
       choose an edge $\sim Bernolli(\sigma(\rho(u, \theta)))$.

where σ is the sigmoid function, and ρ represents the probability between two authors.

3.3   Author-Topic-Community Model

The Author-Topic-Community (ATC) model is designed to generate documents and links based on both the interests and community ties of authors. It assumes that a link between two authors is due to not only the interest similarity of authors, but also the relation or links of the authors (e.g., their social activities). Thus, if there exists a connection between two authors, it is assumed that they should be active in the same community and at the same time have similar interests. The ATC model combines the virtues from the AT model and the TLLDA model and its detailed graphical representation is shown in Figure 1(c). Its generative process is as follow:

1. For each document d in a corpus D:
   (a) for each word w in d do
       choose an users x from the set of authors uniformly
       choose a topic z given the author $x \sim Multi(x, \theta)$
       choose a word w from the chosen topic $z \sim Multi(z, \beta)$

2. For the Graph G:
    (a) for each edge e in G do
        choose a pair of users x_i, x_j from the overall set of authors uniformly
        introduce an edge $e \sim Bernolli(\sigma(\rho(u, \theta)))$

where σ is the sigmoid function and ρ represents the probability between two users.

One key difference between the ATC model and the TLLDA model is how content information of the documents contribute to the formation of the author communities. The community distribution for each author in the TLLDA model is related to the words of the documents he/she authors in the following way: For two documents with a citation link between them; the probability of generating the link will be strengthened according to the *similarity* of the two documents' topic distributions. So the inference for the author community structure is not directly related to some characterization of the authors, but indirectly via the documents they author where citation links exist. This will cause problems when the citation links are sparse or even missing. The suggested ATC model estimates the distributions of author interests so that their *similarity* can be directly considered for the generation of author community links.

4   Inference and Parameter Estimation

By estimating the parameters θ, β and u of the ATC model, we can readily derive the information about the author interested topics, the community the authors are typically in, as well as the theme of the community. In this section, we describe how to infer the model parameters. Among the several approximate inference methods proposed for topic model inference, e.g. variational inference [5], expectation propagation [11], and Gibbs sampling [12]. we adopt the variational inference approach (see Table 1 for the adopted notations).

We first define a similarity measure between two authors and thus the probability of forming a link between them [13], given as

$$\rho(u, \theta)_{i,j} = \tau_1 u^T{}_i u_j + \tau_2 \theta_i^T \theta_j + \tau_3 \tag{1}$$

where the sigmoid function $\sigma(x) = 1/(1 + exp(-\rho(x)))$ is for converting the similarity measure to a probability value [14].

The joint distribution of the model is thus defined as

$$P(G, W) = P(G, W, z, \beta, \theta, u, x) = P(G|u, \theta)P(z|\theta, x)P(W|z, \beta)$$

and the log likelihood function is thus given as

Table 1. Notations used in the ATC model

Parameter	description	Parameter	description
V	size of the word vocabulary	τ	impact coefficient for topic and community
M	the number of documents	G_{ij}	link between u_i and u_j
P	the number of authors	$\theta_{(P \times T)}$	topic probability for given author
a_d	the set of authors	$u_{(P \times H)}$	community probability for given author
x	an author	$\beta_{(T \times V)}$	word probability for given topic
N	the number of words in document d	H	the number of communities
W	the set of documents	K	the number of topics
w_d	the word in document d	G	the set of links

$$L(P(G, W, z | \theta, \beta, u, x))$$

$$= \log(\prod_{i=1}^{P} \prod_{j=1}^{P} p(G_{ij} | u, \theta) \prod_{m=1}^{M} \prod_{n=1}^{N} p(z_{m,n} | \theta_{m,n}, x) p(w_{m,n} | z_{m,n}, \beta_{m,n})$$

$$= \log(\sum_{i=1}^{M} \sigma(\rho_{ij})^{G_{ij}} (1 - \sigma(\rho_{ij}))^{(1-G_{ij})} \qquad (2)$$

$$+ \sum_{m=1}^{M} \sum_{n=1}^{N} p(z_{m,n} | \theta_{m,n}) + \sum_{m=1}^{M} \sum_{n=1}^{N} p(w_{m,n} | z_{m,n}, \beta_{m,n})$$

Note that the second term in (2) shows the likelihood of the observed links, and the third shows the likelihood of the observed words and that of the hidden topics from author which hints another important difference between our ATC model and the TLLDA model.

4.1   Variational Inference

The fundamental principle of variational inference is to make use of Jensen's inequality to find the tightest possible lower bound to original expression. This method can obtain a group of simple computable variational parameters, which can be easily estimated easily by an approximation of the objective distribution. We will consider the following variational distribution:

$$q(u, \theta, z, \beta) = q(\beta)q(u)q(\theta) \prod_{n=1}^{N} p(z_n | \phi_n) \qquad (3)$$

where $z_n \sim Multi(\phi_n)$, and ϕ_n are the free variational parameters. Having specified a variational distribution, the next step of the inference is to determine the free variational parameter ϕ_n, so that the KL-divergence between the approximate and the true distributions is minimized.

To estimate the free variational parameter ϕ_n, the expectation of $\sigma(x)$ function is needed and yet is also intractable. The variational methods is used again with the following variational bound [3]

$$\sigma(x) \geq \sigma(\varepsilon) \exp(\frac{x - \varepsilon}{2} + g(\varepsilon)(x^2 - \varepsilon^2)) \quad g(\varepsilon) = (\frac{1}{2} - \sigma(\varepsilon))/2\varepsilon$$

where ε is the free variational parameter.

Therefore the update equations of variational parameters are found by minimizing the Kullback-Leibler (KL) divergence between the variational distribution and the true posterior distribution, such that $\phi_{nk} \propto \beta_{nk} \exp(\frac{\sum_{i=1}^{M} w_{in} \log \theta_{pnk}}{m_n})$ and $\varepsilon_{ij} = (E_q[\rho_{ij}^2])^{\frac{1}{2}}$.

4.2   Parameter Estimation

As the variational inference provides us with a computable lower bound on the log likelihood, we can thus fix the values of the variational parameters ϕ, ε, and maximizes the lower bound for the model parameters β, θ, u and τ. Because of space limit, we

define $\Delta = \tau_2 \sum\limits_{j=1}^{M_i} \theta_{jk}((G_{ip} - \frac{1}{2}) + g(\varepsilon_{ij})(\tau_2 \sum\limits_{l=1, l\neq k}^{K} \theta_{il}\theta_{jl} + 2\tau_3 + 2\tau_1 u^T{}_i u_j)).$

$$\beta_{nk} \propto \sum_{d=1}^{M} \sum_{j=1}^{N_d} \phi_{djk} w_{djk} \tag{4}$$

$$u_{ik} = \frac{\tau_1 \sum\limits_{j=1}^{m_j} u_{jk}((G_{ip} - \frac{1}{2}) + g(\varepsilon_{ij})(\tau_1 \sum\limits_{l=1, l\neq k}^{H} u_{il}u_{jl} + 2\tau_3 + 2\tau_2 \theta^T{}_i \theta_j))}{2\tau_1{}^2 \sum\limits_{j=1}^{m_j} g(\varepsilon_{ij}) u_{jk}{}^2} \tag{5}$$

$$\theta_{ik} = \frac{-\Delta \pm \sqrt{\Delta^2 - 8\tau_2{}^2 \sum\limits_{j=1}^{M_i} g(\varepsilon_{ij})\theta_{jk}{}^2 m_i \sum\limits_{v=1}^{V} w_{np}\phi_{nk}}}{4\tau_2{}^2 \sum\limits_{j=1}^{M_i} g(\varepsilon_{ij})\theta_{jk}{}^2} \tag{6}$$

$$\tau_1 = \frac{\sum\limits_{i=1}^{M} \sum\limits_{j=1}^{M} (G_{ij} - \frac{1}{2})(u^T{}_i u_j) + g(\varepsilon_{ij})(2\tau_3 u^T{}_i u_j + 2\tau_2 u^T{}_i u_j \theta_i{}^T \theta_j))}{-2 \sum\limits_{i=1}^{M} \sum\limits_{j=1}^{M} g(\varepsilon_{ij}) \sum\limits_{h=1}^{H} \sum\limits_{l=1}^{H} u_{ih}u_{il}u_{jh}u_{jl}} \tag{7}$$

where τ_2, τ_3 are omitted, due to space limit and they have similar forms as τ_1.

5   Experimental Results

We implemented the ATC model as well as the two algorithms —the AT model and the TLLDA model for baseline comparison. Extensive experiments were performed based on both synthetic and real datasets to evaluate (1) to what extent the authors interest modeling accuracy is improved with the author community factor considered, 2) whether the community discovered using the ATC model is of higher quality when compared with those derived based on the TLLDA model, and 3) whether the ATC model can be used to explain the reasons why authors interact with others in the same as well as different communities.

5.1   Synthetic Datasets

We first created synthetic datasets for evaluation. Each dataset contains an author set A, a vocabulary W with 100 words, a topic set T which contains 2 topics (T_1 and T_2) and a community set C which contains 2 communities (C_1 and C_2). The reference word-topic distribution is set so that $P(w_i|T_1) = 0.9$ and $P(w_i|T_2) = 0.1$ for $w_{i\in[1,50]} \in W$, and $P(w_j|T_1) = 0.1$ and $P(w_j|T_2) = 0.9$, for $w_{j\in[51,100]} \in W$. The community-author distribution is set so that $P(C_1|x_i) = 0.9$ and $P(C_2|x_i) = 0.1$ for $x_{i\in[1,30]} \in a_d$, and $P(C_1|x_j) = 0.1$ and $P(C_2|x_j) = 0.9$, for $x_{j\in[31,60]} \in A$. Topic T obeys the topic-author distribution on which $P(T_1|x_i) = 0.9$ and $P(T_2|x_i) = 0.1$ for $x_{i\in[1,30]}$, and $P(T_2|x_j) = 0.9$, $P(T_1|x_j) = 0.1$ for $x_{j\in[31,60]}$. Each document is assumed to have one author and can have a random number of words sampled from $[1, 200]$. For the generation, for each word in a document, a topic will first be picked according to the topic-author distribution. Then, the word will be picked according to the word-topic distribution. A link, denoted as $< x_i, d_m, x_j, d_n >$, indicates that author x_i wrote a document d_m which cited another document d_n written by x_j. The within-community citation links are generated by (1) randomly picking pairs of authors from the same community according to the community-author distribution, and (2) randomly picking a document for each of them to establish a citation link. We also generate cross-community citation links by (1) selecting an author in a specific community and another one in the other, and then (2) randomly picking a document for each of them to establish a citation link. For our evaluation, we generated three synthetic datasets which consist of 200 within-community citations with 0%, 5%, 10% cross-community citations as described.

5.2 Real Datasets

We use a research paper corpus downloaded from the DBLP web service as the real dataset. The corpus contains research papers published over ten years in two international conferences KDD (1999-2009) and SIGMOD (2000-2010). For each paper, the title and abstract were extracted as the document content and the references were extracted to establish the links among authors. For instance, if paper B is cited by paper A, all the authors of A will have links to those authors of paper B. Authors who didn't publish any paper in the two conferences during the period of time considered (i.e., those just being cited) will be ignored. All the words in the vocabulary were preprocessed by stemming and stop words removal before the model inference step. Also, the words that occurred less than 3 times were also deleted from the word vocabulary. This results in a vocabulary with 3199 distinct words and a text corpus of 2341 documents and a link set involving 4205 authors.

For correctness checking, as each paper in the corpus contains several subject tags (provided at the ACM Digital Library), we estimate the "true" research interests of the authors by aggregating the subject tags of their authored documents so as to compare with those being inferred. For instance, an author who wrote two documents with one tagged as *clustering*, and another *search*. Then, the author will be tagged with *clustering* and *search*. In following experiments, we utilize those tags to estimate the performance of the ATC model.

5.3 Performance Evaluation on Author Topic Modeling

To evaluate the effectiveness of the ATC model for author topic modeling, we compared its performance empirically with that of the AT model. The performance measure is *least squared error* $= \sum_i \sum_k (\theta_{ik} - \theta'_{ik})^2$ where θ' is the original topic-author distribution used to generate synthetic dataset and θ is the parameter estimated using the ATC/AT model. Figure 2(a) shows the comparison results for author topic modeling on the synthetic data where the number of topics K and the number of communities H are both set to be 2. As shown in the figure, the ATC model outperforms the AT model for all the synthetic datasets with 0%, 5%, 10% cross-community links. The ATC model can extract author interests not only from those documents that he/she wrote, but also from the interests of the neighbors who were in similar communities, and thus is more robust against the more random part of the citations.

Table 2. Performance comparison in terms of top words, topic tags and representative authors for 3 chosen topics inferred from the research paper corpus

AT			ATC		
				topic-1	
top words	topic tags	authors	top words	topic tags	authors
flat	Relational databases	W.Lee	databas	Relational databases	J. Doumlrre
truth	Information filtering	K. Mok	provid	Parallel databases	P. Gerstl
mention	Classifier design and evaluation	R. Seiffert	support	Cache memories	R. Seiffert
systemat	Transaction processing	W. Potts	list	Transaction processing	H. Kauderer
oracl	Distributed databases	A. UzTansel	describ	Graphs and networks	F. Artiles
spot	Retrieval models	J. Zhang	integr	Classifier design and evaluation	H. Jeromin
embryo		M. Kelly	comput		D. Chatziantoniou
transit		J. Doumlrre	approach		S. Harp
				topic-2	
top words	topic tags	authors	top words	topic tags	authors
hypothesi	Search process	T.Fawcett	mine	Statistical	K. Lin
grn	Statistical	S. Sahar	perform	Classifier design and evaluation	N. Lesh
exactli	Distributed databases	C. Elsen	applic	Transaction processing	M. Ogihara
dual	Relational databases	M. Wasson	list	Distributed databases	M. Kelly
style	Query languages	J. Q. Louie	pattern	Clustering	B. Petzsche
fanout	Indexing methods	B. Petzsche	class	Parallel databases	H. Spiegelberger
robot		D. Dunn	combin		S. Chen
output		S. Zhou	oper		T. Fountain
				topic-3	
top words	topic tags	authors	top words	topic tags	authors
develop	Search process	R. Hackathorn	featur	Distributed databases	R. Hackathorn
phase	integrity	B. Larsen	compar	Search process	G. Mainetto
band	Security	B. Masand	statist	integrity	Y. Aumann
fail	Web-based services	J. Wolf	select	Security	Y. Lindell
physic	Transaction processing	D. Geiger	associ	Relational databases	D. Pelleg
teradata	Distributed databases	J. Cerquides	level	Algorithms	W. Buntine
pda		K. Lin	best		B. Fischer
drug		C. Davatzikos	discoveri		T. Pressburger

Table 2 illustrates the experimental results obtained based on the research paper corpus using the AT model and the ATC model ($K=5$). The topic tags refer to the subjects of authors with top interest in the topic as explained in the previous section. In particular, the top 6 subjects in terms of tag occurrence frequency are considered as the topic tags. We trained both models, each with five topics. For the sake of comparison, we manually aligned the five topics of the two models which are semantically closer to each other. From the table, we can conclude that the ATC model can infer more cohesive topics than the AT model. For instance, the ATC model obtained several similar tags such as "Relational databases, Parallel databases, Cache memories, Transaction processing" and ranked them high, but the AT model can find only "Relational databases, Parallel databases, Transaction processing". Also, for Topic 2 inferred by ATC, the top words, "mine, perform, applic, pattern, class, combin, oper" is semantically very close to the corresponding topic tags "Statistical, Classifier design and evaluation, Clustering". Meanwhile, only two words, "hypothesi and exactli", can be matched to the tags from the AT model.

5.4   Performance Evaluation on Community Discovery

To evaluate the effectiveness of the ATC model for community discovery, we compared the quality of those discovered by the ATC model with those discovered based on the

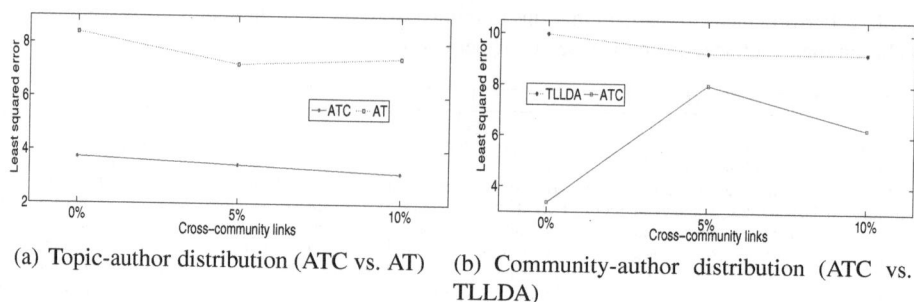

(a) Topic-author distribution (ATC vs. AT)   (b) Community-author distribution (ATC vs. TLLDA)

Fig. 2. Performance comparison on estimation accuracy based on synthetic datasets

TLLDA model. The least squared error is used again, defined as $\sum_i \sum_k (u_{ik} - u'_{ik})^2$ where u' is the original community-author distribution, and u is the estimated parameters of the ATC/TLLDA model. Figure 2(b) shows the comparison results on community discovery based on the synthetic data. The number of topics and the number of communities were both set to be 2 for the ease of illustration. As shown in Figure 2(b), the ATC model outperforms the TLLDA model for community discovery based on the synthetic data with 0%, 5%, 10% cross-community links.

In addition, we have applied the TLLDA and ATC models to the research paper corpus. To evaluate the quality of the community discovered, we first compute the set of tags for a community which is defined as the union of the subjects of the authors with the highest probability belonging to the community defined based on u. Then, the specificity of the community structure can be measured using the entropy score:

$$entropy(C) = -\frac{|a_{C_i}|}{\sum_j^H |a_{C_j}|} \sum_{j=1}^{|T_{C_i}|} p(t_j) \log(p(t_j)), \qquad (8)$$

where T_{C_i} denotes the number of tags of C_i, $p(t_j) = n_j / \sum_j n_j$, t_j represents the tag j, and n_j is the number of times t_j appearing in C_i, and $|a_{C_i}|$ denotes the number of authors belonging to the community C_i.

Figure 3 shows the experimental results, where the community number is set to be 10. The ATC model can end up with more communities with much smaller entropy value, that is each community is more focused on some topics when compared with those derived based on the TLLDA model.

5.5   Ranking Authors Based on Cross-Community Linkage

To gain further insight on the interactions among authors, the ATC model could be used to analyze the authors based on the within-community and cross-community links. To illustrate that, we made use of the ATC model learned based on the research paper corpus and extracted the authors with the most cited (authoritative) papers and the most citing (resourceful) papers in general. At the same time, we extracted those only based on the cross-community links. Authors with more cross-community links may indicate making a higher influence and/or being more resourceful in multiple research areas. The extracted results are shown in Figure 4, where top 5 authors of four groups are illustrated.

The upper-left group shows the authors with the most citing papers in general. Authors in this group hold a wide range of research interests. For example, based on the dataset we used, Anthony K. H. Tung has the most citing links. He published 85 papers, cited 18 times by other authors, and had 9 interests, such as "data mining , performance attributes , query languages , distributed databases, data warehouse and repository , experimental design, network management, logging and recovery, selection process".

community	ATC model	TLLDA model
1	11.2514832	0.367343187
2	**12.47653497**	74.51602885
3	**63.40152376**	84.36579125
4	**9.503438203**	16.10029447
5	**25.24059093**	41.27500116
6	65.18868635	3.288517137
7	**5.321687956**	8.554737674
8	**5.321687956**	16.10029447
9	**7.614424551**	17.45304784
10	**23.66485519**	40.49181987
total	**55.52414575**	66.13413158

Fig. 3. Entropy comparison

Most citing	Most cited
Anthony K. H. Tung	Samuel Madden
Jeffrey Xu Yu	Jiawei Han
Huan Liu	Christos Faloutsos
Chris Jermaine	Philip S. Yu
Jianyong Wang	Ravi Kumar
Most citing across communities	**Most cited across communities**
Zachary G. Ives	Pedro Domingos
Anthony K. H. Tung	Philip S. Yu
Naren Ramakrishnan	Jiawei Han
Johannes Gehrke	Jian Pei
Michael J. Carey	Samuel Madden

Fig. 4. Comparing Top 5 authors computed by the ATC model by different considerations of cross-community links

The upper-right group shows the authors with the most cited papers. We found two types of authors in it. One can be considered focused experts. E.g., Samuel Madden, based on the ATC model, is computed to be one of the most influential authors and his papers hold mainly two topics — "Query processing" and "Distributed database". Another type is multiple-domain experts. They not only published a lot of papers, but also hold a wide range of research interests. For example, Jiawei Han is a multiple-domain expert, and his research scope covers up to 12 topics as computed by the ATC model.

The lower-left group have a conspicuous feature. Their citing papers covers different communities. Authors in this group have few research topics of interest (more focused) and their interests are different from those authors being cited by them. Their research works are usually cross domains. For example, Naren Ramakrishnan is computed to be an author of this type and held 2 subjects — "induction" and "computing equipment management". According to Naren Ramakrishnan's home page, his interests cover "mining scientific datasets in domains such as systems biology, neuroscience, sustainability, and intelligence analysis". It seems that the computed results do make some sense.

The lower-right group includes the set of authors quite close to the upper-right group. Researchers like Jianwi Han and Philip Yu remain to be mostly cited even based only on cross-domain links. But in addition, some who are relatively speaking ranked not so high for citations in general become the top 5 candidates as the works are cited more by authors of other disciplines as grouped under different communities. Pedro Domingos and Jian Pei are computed to be the authors of this type.

While the validity of the above analysis is needed to be further verified, the empirical results show that the ATC model is an effective tool for achieving accurate author topic and community discovery.

6   Conclusion

In this paper, a novel generative model which allows both author interests and the author community structure to be inferred at the same time was proposed. Empirical comparison based on both synthetic and a real dataset of research paper corpus demonstrated the superiority of the ATC model over the AT and TLLDA models for author topic and community discovery. For future work, we are interested in studying the evolution of the author interests and the dynamic formation of the corresponding communities.

Acknowledgement. This work is supported in part by NSFC under Grant No. 61073195, National Key Technology R&D Program of MOST China under Grant No. 2012BAK17B08, Shen-zhen Strategic Emerging Industries Program under Grant No. ZDSY20120613125016389, Natural Scientific Research Innovation Foundation in HIT under Grant No. HIT.NSFIR.2010128, HIT Innovation Fund No. HIT.NSRIF.2013105 and General Research Fund (HKBU210410) from the Research Grant Council of the Hong Kong Special Administrative Region, China.

References

1. Rosen-Zvi, M., Griffiths, T., Steyvers, M., Smyth, P.: The author-topic model for authors and documents. In: Proceedings of the 20th Conference on Uncertainty in Artificial Intelligence, pp. 487–494. AUAI Press (2004)
2. Zhou, D., Manavoglu, E., Li, J., Giles, C., Zha, H.: Probabilistic models for discovering e-communities. In: Proceedings of the 15th International World Wide Web Conference, pp. 173–182. ACM (2006)
3. Liu, Y., Niculescu-Mizil, A., Gryc, W.: Topic-link LDA: Joint models of topic and author community. In: Proceedings of the 26th Annual International Conference on Machine Learning, pp. 665–672. ACM (2009)
4. Kataria, S., Mitra, P., Caragea, C., Giles, C.: Context sensitive topic models for author influence in document networks. In: Twenty-Second International Joint Conference on Artificial Intelligence (2011)
5. Blei, D., Ng, A., Jordan, M.: Latent dirichlet allocation. The Journal of Machine Learning Research 3, 993–1022 (2003)
6. Nallapati, R., Ahmed, A., Xing, E., Cohen, W.: Joint latent topic models for text and citations. In: Proceeding of the 14th ACM SIGKDD International Conference on Knowledge Discovery and Data Mining, pp. 542–550. ACM (2008)
7. Mei, Q., Cai, D., Zhang, D., Zhai, C.: Topic modeling with network regularization. In: Proceedings of the 17th International World Wide Web Conference, pp. 101–110. ACM (2008)
8. Bhattacharya, I., Getoor, L.: A latent dirichlet model for unsupervised entity resolution. Technical Reports of the Computer Science Department (2005)

9. Shiozaki, H., Eguchi, K., Ohkawa, T.: Entity Network Prediction Using Multitype Topic Models. In: Washio, T., Suzuki, E., Ting, K.M., Inokuchi, A. (eds.) PAKDD 2008. LNCS (LNAI), vol. 5012, pp. 705–714. Springer, Heidelberg (2008)
10. Tu, Y., Johri, N., Roth, D., Hockenmaier, J.: Citation author topic model in expert search. In: Proceedings of the 23rd International Conference on Computational Linguistics: Posters, pp. 1265–1273. Association for Computational Linguistics (2010)
11. Minka, T.: Expectation propagation for approximate Bayesian inference. In: Proceedings of the Seventeenth Conference on Uncertainty in Artificial Intelligence, pp. 362–369. Morgan Kaufmann Publishers Inc. (2001)
12. Griffiths, T., Steyvers, M.: Finding scientific topics. Proceedings of the National Academy of Sciences of the United States of America 101(suppl. 1), 5228 (2004)
13. Lin, Y., Chi, Y., Zhu, S., Sundaram, H., Tseng, B.: Analyzing communities and their evolutions in dynamic social networks. ACM Transactions on Knowledge Discovery from Data (TKDD) 3(2), 8 (2009)
14. Chang, J., Blei, D.: Relational topic models for document networks. In: Artificial Intelligence and Statistics, pp. 81–88 (2009)

Constrained Closed Non Derivable Data Cubes

Hanen Brahmi[1] and Sadok Ben Yahia[2]

[1] Faculty of Sciences of Tunis, Computer Science Department,
Campus University, 1060 Tunis, Tunisia
hanenbrahmi@gmail.com
[2] Institut Mines-TELECOM, TELECOM SudParis,
UMR CNRS Samovar, 91011 Evry Cedex, France
sadok.benyahia@fst.rnu.tn

Abstract. It is well recognized that data cubes often produce huge outputs. Several efforts were devoted to this problem through Constrained Cubes allowing the user to focus on a particular set of interesting tuples. In this paper, we investigate reduced representations for the Constrained Cube (*e.g.*, Constrained Closed Cube and Constrained Quotient Cube). That is why we introduce a new and concise representation of data cubes: the Constrained Closed Non Derivable Data Cube (*CCND-Cube*). The latter captures all the tuples of a data cube fulfilling a combination of monotone/anti-monotone constraints. It can be represented in a very compact way in order to optimize both computation time and required storage space. The results of our experiments confirm the relevance of our proposal.

Keywords: Data warehouses, Data cubes, Constrained cubes, Closed patterns, Non derivable patterns, Minimal generators.

1 Introduction

A data warehouse is a collection of technologies aimed at enabling the knowledge worker (executive, manager, analyst, etc.) to make better and faster decisions. In a data warehouse architecture, data coming from multiple external sources are extracted, filtered, merged, and stored in a central repository. The content of a data warehouse is analyzed by On Line Analytical Processing (OLAP) applications in order to discover trends, behaviors, and anomalies as well as to find hidden dependencies between data [7].

Analysts pose complex OLAP queries that extensively use aggregation in order to group together "similarly behaving tuples". The response time of such queries over extremely large fact tables in modern data warehouses can be prohibitive. This issue inspired Gray *et al.* to propose a new operator called "CUBE" [9]. It is a multidimensional extension of the relational operator "GROUP BY". The "CUBE" operator computes "GROUP BY" operations corresponding to all possible combinations of the dimension attributes over a fact table in a data warehouse. Then, we obtain the so called *Data Cube*.

Given a base relation R with n attributes, the number of tuples in a cuboid (GROUP BY) with k attributes, such as ($0 < k \leq n$), is the number of tuples in R that have distinct attribute values on the k attributes. The size of a cuboid is possibly close to the size of R. Since the complete cube of R consists of 2^n cuboids, the size of the union of 2^n cuboids

S. Zhou, S. Zhang, and G. Karypis (Eds.): ADMA 2012, LNAI 7713, pp. 766–778, 2012.
© Springer-Verlag Berlin Heidelberg 2012

is much larger than the size of *R*. Consequently, the *Input/Output* cost for storing the cube result tuples becomes dominative as indicated in [2].

It is obvious that computing data cubes is a combinatory problem. In fact, the size of a cube exponentially increases according to the number of dimensions. Furthermore, the problem worsens since we deal with large datasets. For instance, Ross and Srivastava exemplify the problem by achieving a full data cube encompassing more than 210 million of tuples from an input relation having 1 million of tuples [14]. The huge size of a data cube makes data cube computation time-consuming. Although cheap and highly-sized volume memory chips are available, it is difficult to hold the whole data cube of a large relation in the main memory. In general, the problem is due to two main reasons: the exponential number of dimensional combinations to be dealt with and, the number of attributes per dimension. In addition, data cubes are generally sparse [14], thus scarce value combinations are likely to be numerous and, when computing an entire data cube, each exception must be preserved.

Although issues related to the size of data cubes have attracted the attention of researchers, and various algorithms have been developed aiming at fast computation of large sparse data cubes [14,2] relatively fewer works concentrated on solving the complexity problem of data cube computation from its root: reducing the size of a data cube. In this work, we investigate another way of tackling the problem. First, we introduce the concept of *constrained closed non derivable data cube* and prove that the latter greatly reduces the size of a constrained data cube. Then, we propose an algorithm to efficiently compute the constrained closed non derivable cubes. Through extensive carried out experiments on synthetic and real-life datasets, we show the effectiveness of our proposal on both runtime performances and reduction of storage space.

The remainder of the paper is organized as follows. The formal background is presented in Section 2. We scrutinize, in Section 3, the related work. We define the main concepts of our representation in Section 4. We introduce the dfCLOSND algorithm in Section 5. We also discuss the encouraging results of the carried out experiments in Section 6. Finally, we conclude by summarizing our contributions and describing future research issues.

2   Formal Background

We present in this section the key settings that will be of use in the remainder.
One condensed representation of patterns is based on the concept of closure [13].

Definition 1. *Closed Pattern*
Let $\mathscr{I}$ be the set of attributes and γ be the closure operator which associates to a pattern $X \subseteq \mathscr{I}$ the maximal, w.r.t. set inclusion, superset having the same support value as X. A pattern X is said to be a closed pattern if $X = \gamma(X)$.

The concept of minimal generator [1] is defined as follows.

Definition 2. *Minimal Generator*
A pattern $g \subseteq \mathscr{I}$ is said to be a minimal generator (MG) of a closed pattern f iff $\gamma(g) = f$ and $\not\exists\, g_1 \subset g$ such that $\gamma(g_1) = f$. For a user-defined support threshold, the set of frequent minimal generators includes all generators that are frequent.

The collection of frequent non derivable patterns, denoted *NDP*, is a lossless representation of frequent patterns based on the inclusion-exclusion principle [4].

Definition 3. *Non Derivable Pattern*
Let X be a pattern and Y a proper subset of X. If $|X \backslash Y|$ is odd, then the corresponding deduction rule for an upper bound of Supp(X) is:

$$Supp(X) \leq \sum_{Y \subseteq I \subset X} (-1)^{|X \backslash I| + 1} \, Supp(I)$$

If $|X \backslash Y|$ is even, the sense of the inequality is inverted and the deduction rule gives a lower bound instead of an upper bound of the support of X. Given all subsets of X, and their supports, we obtain a set of upper and lower bounds for $Supp(X)$. In the case where the smallest upper bound is not equal to the highest lower bound, the support of X can not be derived starting from its proper subsets. Such a pattern is called *non derivable*. In the remainder, the lower and upper bounds of the support of a pattern X will respectively be denoted $X.l$ and $X.u$.

Brahmi *et al.* [3] introduced the concept of non derivable minimal generators.

Definition 4. *Non Derivable Minimal Generator*
Given a pattern $I \subseteq \mathscr{I}$, the set of non derivable minimal generator is defined as follows:
$NDMG = \{I \subseteq \mathscr{I} \mid I.l \neq I.u$ and I is a MG$\}$.

The set of frequent closed non derivable patterns has been defined in [3].

Definition 5. *Closed Non Derivable Pattern*
Given NDMG a set of frequent non derivable minimal generators. The set of frequent closed non derivable minimal generators is CNDP= $\{\gamma(X) \mid X \in NDMG\}$.

[5] provide most consistent approach to the notion of cube closures based on the cube lattice framework.

Definition 6. *Cube Lattice*
The multidimensional space of a categorical database relation R groups all the valid combinations built up by considering the value sets of dimension attributes, which are enriched with the symbolic value ALL. The multidimensional space of R is noted and defined as follows:
Space$(R) = \{X_{A \in D}(Dim(A) \cup ALL) \cup (\emptyset \dots \emptyset)\}$ *where X symbolizes the Cartesian product, and $(\emptyset \dots \emptyset)$ stands for the combination of empty values.*

Any combination belonging to the multidimensional space is a tuple and represents a multidimensional pattern. The multidimensional space of R, Space(R), is structured by the generalization/specialization order between tuples, denoted by $\preceq$. Let u and v be two tuples of the multidimensional space of R. If $u \preceq v$, we say that u is more general than v in Space(R). The two basic operators provided for tuple construction are: Sum (denoted by +) and Product (noted by ∗). The Sum of two tuples yields the most specific tuple which generalizes the two operands. The Product yields the most general tuple which specializes the two operands. The ordered set $CL(R) = (Space(R), \preceq)$ is a complete, graded lattice, called cube lattice.

Through the definition of the following lattice-isomorphism, [5] make it possible to reuse closed patterns mining algorithms in a multidimensional context.

We recall the definitions [11] of convex space, monotone/anti-monotone constraints.

Definition 7 (Convex Space). *Let $(P, \leq)$ be a partial ordered set, $C \subseteq P$ is a convex space if and only if $\forall x, y, z \in P$ such that $x \leq y \leq z$ and $x, z \in C$ then $y \in C$.*

Definition 8 (Monotone/anti-monotone constraints)

1. *A constraint Const is monotone according to the generalization order if and only if: $\forall t, u \in CL(r) : [t \preceq_g u$ and $Const(t)] \Rightarrow Const(u)$.*
2. *A constraint Const is anti-monotone according to the generalization order if and only if: $\forall t, u \in CL(r) : [t \preceq_g u$ and $Const(u)] \Rightarrow Const(t)$.*

3 Related Work

Motivated by the huge amount of results to be stored, several proposals have attempted to reduce the size of the cube representation. In order to meet such a goal, they adopt different strategies.

Approaches which do not restore the exact or complete data argue that OLAP users are interested in general trends. It takes benefit of the statistic structure of data for computing density distributions and answering OLAP queries in an approximate way. BUC [2] first proposes the concept of *iceberg cube* and employs a bottom-up computation. Iceberg cube queries only retrieve those partitions that satisfy user-specified aggregate conditions. Using the bottom-up approach, it is possible to prune off those partitions that do not satisfy the condition as early as possible. STAR-CUBING [17] exploits more opportunities in shared computation, and uses star-tree structure to integrate simultaneous aggregation into iceberg pruning. MM-CUBING [15] avoids the expensive tree operations in STAR-CUBING by partitioning the data into different subspace and using multi-way array aggregation to achieve shared computation. These approaches enforce anti-monotone constraints and partially compute data cubes to reduce both execution time and disk storage requirements.

Differential cubes [6] result from the set difference between the data cubes of two relations R_1 and R_2. They capture tuples relevant in the data cube of a relation and not existing in the cube of the other. In contrast with the previous ones, such cubes perform comparisons between two datasets. For OLAP applications, trend comparisons along time are strongly required in order to exhibit trends which are significant at a moment and then disappear (or non-existent trends which latter appear). If we consider that the original relation R_1 is stored in a data warehouse and R_2 is made of refreshment data, the differential cube shows what is new or dead.

Emerging cubes [12] capture trends which are not relevant for the users but which grow significantly or on the contrary general trends which soften but not necessarily disappear. Emergent cubes enlarge results of differential cubes and refine cube comparisons. They are of particular interest for data stream analysis because they exhibit trend reversals. For instance, in a web application, the continuous flows of received data describe in a detailed way the user navigation. Knowing the craze for (in contrast the disinterest in) such or such URL is specially important for the administrator in order to allow at best available resources according to real and fluctuating needs.

The constrained (convex) cube [11] is computed and pruned with monotone and/or anti-monotone constraints. Being a convex space [16] the constrained data cube can be represented using one of the following couple of borders (classical in data mining): (*i*) either the Lower and Upper borders; or (*ii*) the Upper$^{\sharp}$ and Upper borders. These borders are the boundaries of the solution space and can support classification tasks. The choice of borders depends on the user needs.

Nedjar et al. introduced in [11] two sound reduced representation of the constrained data cubes in order to optimize both storage space and computation time:

- The **Constrained Quotient Cube** is a summarizing structure for a constrained data cube that preserves its semantics. It can be efficiently constructed and achieves a significant reduction of the constrained cube size. The key idea behind a constrained quotient cube is to create a summary by carefully partitioning the set of cells of a cube into equivalent classes while keeping the cube ROLL-UP and DRILL-DOWN semantics and lattice structure.

- The **Constrained Closed Cube** represents a size-reduced representation of a constrained data cube when compared to the constrained quotient cube [11]. It only consists of closed cells. A cell, say c, is a closed cell if there is no cell, d, such that d is a specialization (descendant) of c, and d has the same measure value as c.

Moreover, Nedjar et al. [11] proved that the *constrained quotient cube* and the *constrained closed cube* can be computed by the application of closed patterns mining algorithms, e.g., the CLOSE algorithm [13].

Due to its usability and importance, reducing the storage space of a data cube is still a thriving and a compelling issue. In this respect, the main thrust of this work is to propose a new concise representation called *Constrained Closed Non Derivable Cube* and denoted by *CCND-Cube*. To do so, we apply a mechanism which significantly reduces the size of aggregates that have to be stored. Our aim is to compute the smallest representation of a data cube when compared to the pioneering approaches of the literature (*i.e.*, constrained closed cube, constrained quotient cube). To build up the *CCND-Cube*, we introduce a novel algorithm, called dfCLOSND (*depth-first closed non derivable patterns based on minimal generators*). This will be detailed in the following sections.

4   Structure of Constrained Closed Non Derivable Data Cubes

The idea behind our representation is to remove redundancies existing within constrained data cubes. In fact certain multidimensional tuples are built up by aggregating the very same tuples of the original relation but at different granularity levels. Thus a single tuple, the most specific of them, can stand for the whole set. The closure operator (*cf.* Definition 1) is intended for computing this representative tuple. Actually, the closed non derivable cube, including all the closed non derivable tuples, is one of the most reduced representations for the data cube [3]. Therefore, it is interesting to propose, for the constrained cube, a structure based on the concepts associated to the closed non derivable cube. Hence, we introduce a new condensed representation of the constrained data cube with a twofold objective: (*i*) defining the solution space in a compact way and deciding whether a tuple t belongs or not to this space; and (*ii*) obtain a condensed representation of the constrained cube in the presence of a conjunction of constraints according to the generalization order.

Table 1. Relation example CAR SALES

RowID	Model	City	Customer	Color	Quantity
1	Ford	Sousse	Auto	Grey	400
2	Ford	Sousse	LeMoteur	Black	100
3	Peugeot	Beja	LeMoteur	Grey	100
4	Peugeot	Sousse	Auto	Black	300
5	Peugeot	Beja	Auto	Grey	200

We take into account the monotone and anti-monotone constraints the most used in database mining. They are applied on: (1) measures of interest like pattern frequency, confidence, correlation. In these cases, only the dimensions of R are necessary; (2) aggregates computed from measures using statistic additive functions (COUNT, SUM, MIN, MAX, etc.); and (3) patterns respecting two anti-monotone constraints, namely *"to be non derivable"* (*cf.* Definition 3) and *"to be minimal generator"* (*cf.* Definition 2).

Definition 9 (Closed Non Derivable Data Cube)
Let NDMG be the set of non derivable minimal generators associated to a database relation R. The closure set of the NDMG set is a complete graded lattice, called closed non derivable data cube, structured by the generalization/specialization order between tuples, denoted by $\preceq$:
$CND\text{-}Cube(R) = (\gamma(NDMG(R)), \preceq) = (CNDP, \preceq).$

Example 1. Let us consider the relation CAR SALES (*cf.* Table 1) giving the quantities of cars sold by Model, City, Customer and Color. The constrained closed non derivable data cube is represented through Figure 1. The table on the left gives the set of constrained closed non derivable tuples.

Definition 10 (Constrained Closed Non Derivable Data Cube). *The closed non derivable cube lattice with monotone and/or anti-monotone constraints (const) is a convex space which is called Constrained Closed Non Derivable Data Cube, denoted by:*
$CCND\text{-}Cube(R) = \{t \in CND\text{-}Cube(R) \mid const(t)\}$, *such that const is a conjunction of monotone constraints, anti-monotone constraints or an hybrid conjunction of constraints. Any tuple belonging to the CCND-Cube(R) is called a constrained closed non derivable tuple.*

Example 2. With our relation example CAR SALES, Figure 1 (table on the right) gives the constrained closed non derivable tuples for the the constraints COUNT(Quantity) $\in$ [200,700]. We would like to know all the tuples for which the measure value is greater than or equal to 200. The constraint COUNT(Quantity) $\geq$ 200 is anti-monotone. If the amount of sales by Model, City and Customer is greater than 200, then the quantity satisfies this constraint at a more aggregated granularity level *e.g.* by Model and Customer (all the cities merged) or by City (all the models and customers together). In a similar way, if we aim to know all the tuples for which the quantity is lower than 700, the underlying constraint COUNT(Quantity) $\leq$ 700 is monotone.

Closed Non Derivable Tuples	COUNT (Qty)	Closed Non Derivable Tuples	COUNT (Qty)
(ALL, ALL, ALL, Grey)	700	(Peugeot, ALL, Auto, ALL)	500
(ALL, Sousse, ALL, ALL)	800	(Peugeot, Beja, ALL, Grey)	300
(ALL, ALL, LeMoteur, ALL)	200	(Peugeot, Sousse, Auto, Black)	300
(ALL, ALL, Auto, ALL)	900	(Peugeot, Beja, LeMoteur, Grey)	100
(ALL, ALL, Auto, Grey)	600	(Peugeot, Beja, Auto, Grey)	200
(ALL, Sousse, ALL, Black)	400	(Ford, Sousse, ALL, ALL)	500
(ALL, Sousse, Auto, ALL)	700	(Ford, Sousse, Auto, Grey)	400
(Peugeot, ALL, ALL, ALL)	600	(Ford, Sousse, LeMoteur, Black)	100

(Closed Non Derivable Cube)

Constrained Closed Non Derivable Tuples	COUNT (Qty)
(ALL, ALL, ALL, Grey)	700
(ALL, ALL, LeMoteur, ALL)	200
(ALL, ALL, Auto, Grey)	600
(ALL, Sousse, ALL, Black)	400
(ALL, Sousse, Auto, ALL)	700
(Peugeot, ALL, ALL, ALL)	600
(Peugeot, ALL, Auto, ALL)	500
(Peugeot, Beja, ALL, Grey)	300
(Peugeot, Beja, Auto, Grey)	200
(Peugeot, Sousse, Auto, Black)	300
(Ford, Sousse, ALL, ALL)	500
(Ford, Sousse, Auto, Grey)	400

(Constrained Closed Non Derivable Cube)

Fig. 1. Closed non derivable cube *vs.* constrained closed non derivable cube of the relation CAR SALES for the constraints $200 \leq \text{COUNT(Quantity)} \leq 700$

Definition 11 (Relative Aggregative Functions). *Let R be a database relation, $t \in CCND\text{-}Cube(R)$ a tuple, and $f \in (\text{SUM, COUNT})$ an aggregative function. We call $f_{val}(.,R)$ the relative aggregative function of f for the relation R. $f_{val}(t,R)$ the value of the aggregation function f associated to the tuple t in CCND-Cube(R). $f_{val}(t,R)$ is the ratio between the value of f for the tuple t and the value of f for the whole relation R (in other words for the tuple (ALL, ..., ALL)).*

$$f_{val}(t,R) = \frac{f(t,R)}{f_{val}((ALL,..,ALL),R)}$$

For instance, the function $\text{COUNT}_{val}(t,R)$ merely corresponds to the frequency of a multidimensional pattern t in the relation R.

Example 3. Simply, the function $\text{COUNT}_{val}(t,\text{CAR SALES}))$ is the frequency of the multidimensional pattern t in the relation CAR SALES. If we consider the sales of cars colored in "Black" in "Sousse" for all customers and models (*i.e.*, the tuple (ALL, Sousse, ALL, Black)) $\in CCND\text{-}Cube(\text{CAR SALES})$, we have: $\text{COUNT}_{val}((\text{ALL, Sousse, ALL, Black}), \text{CAR SALES}) = \frac{400}{1100} = 0.4$.

Definition 12 (Constrained Closed Non Derivable Tuple). *Let $t \in CND\text{-}Cube(R)$ be a tuple, t is a constrained closed non derivable tuple if and only if:*

(C_1) $f_{val}(t,R) \geq MinThreshold_1$;
(C_2) $f_{val}(t,R) \leq MinThreshold_2$.

where $MinThreshold_1$ et $MinThreshold_2 \in]0,1[$.

Example 4. Let us consider the relation CAR SALES (*cf.* Table 1) and the constraints $MinThreshold_1 = 0.4$ and $MinThreshold_2 = 0.7$. Among the closed non derivable tuples shown in Table 2, the tuple $t_1 = $ (Ford, Sousse, ALL, ALL) is constrained because $\text{COUNT}_{val}(t_1,R) = \frac{500}{1100} = 0.5$. In contrast, the tuple $t_2 = $ (Peugeot, Sousse, Auto, Black) is not constrained because $\text{COUNT}_{val}(t_2,R) = \frac{300}{1100} = 0.3$.

5   Computation of the CCND-Cube: dfCLOSND Algorithm

We take advantage from the conclusion drawn by Casali *et al.* [5]. The authors proved that there is a lattice isomorphism between the closed cube and the galois lattice (concept lattice) computed from a database relation R. Such an isomorphism is proved to be efficient to the computation of concise representations of a data cube using data mining algorithms as adopted for the *constrained quotient cube* and *constrained closed cube* [11]. Moreover, the approach of Casali *et al.* is based on the Birkhoff theorem [8] to bridge the concept lattice to the closed cube lattice. More precisely, starting from a database relation R, we look for extracting closed non derivable patterns by computing the closures of non derivable minimal generators. Then, based on the lattice isomorphism, we use the Birkhoff theorem to obtain the *CCND-Cube*.

In order to do so, we introduce the dfCLOSND algorithm allowing the extraction of the closed non derivable patterns set. The pseudo-code is shown by Algorithm 1. The dfCLOSND algorithm operates in two steps:

- The **first step** extracts patterns fulfilling a conjunction of monotone and/or anti-monotone constraints. This step is based on two principles:

(*a*) Besides the pruning of the unfrequent candidates based on the minimal support thresholds, dfCLOSND adopts another pruning strategy based on the estimated support. The latter is computed when a candidate J is non derivable (*cf.* lines 11-12). Thus, if the estimated support of the candidate J is equal to its real support then J is not a minimal generator. Consequently, it will be pruned (*cf.* lines 16-20). Note also that a pattern is considered as a candidate only if its upper bound is between $MinThreshold_1$ and $MinThreshold_2$ since, otherwise, it is ensured to be unfrequent (*cf.* line 11). Moreover, any frequent non derivable minimal generator J admitting a support equal to the upper bound or the lower bound will not be used to generate the candidates of a higher size (*cf.* lines 21-25). Indeed, in that case, any proper superset of J is proved to be a derivable pattern [4].

(*b*) Instead of a breadth-first exploration of the non derivable minimal generators candidates, dfCLOSND partitions the search space to make then a depth-first exploration. Hence, the algorithm begins by considering the 1-patterns[1] and examines only its conditional sub-contexts $\mathscr{D}_c$. A conditional sub-context contains only the items which occur with the 1-pattern in question. Recursively, the conditional databases $\mathscr{D}_c$ are built (*cf.* lines 3-6). Further, the choice for which type of cover[2] is not static. In dfCLOSND, this choice will be postponed to run-time. At run-time both covers are computed (*cf.* lines 13-14), and the one with minimal size is chosen (*cf.* lines 22-25). The calculation of both covers can be done with minimal overhead in the same iteration. In this way, it is guaranteed that: (*i*) the size of the covers at least halves from $\mathscr{D}$ to $\mathscr{D}_c$; and (*ii*) a compressed form of the database fits into main memory.

- The **second step** computes closures of the retained patterns from the first step, namely the non derivable minimal generators. In fact, the computation of closed non derivable patterns can be optimized if we use minimal generators. Hence, instead of computing the whole set of non derivable patterns for which the associated closures must be

---

[1] We denote by 1-pattern a pattern of size 1.
[2] The cover of a pattern X in $\mathscr{D}$ consists of the set of tuple identifiers in $\mathscr{D}$ that support X.

774   H. Brahmi and S.B. Yahia

Algorithm 1. dfCLOSND

Input:
1. $\mathscr{D}$: A dataset $\mathscr{D}$,
2. $MinThreshold_1$, $MinThreshold_2$: The support thresholds.

Output: $CNDP$: The collection of closed non derivable patterns.

1 **Begin**
2    $NDMG := \emptyset$; $CNDP := \emptyset$; /* $\mathscr{D}$ ordered descending. */
3   **Foreach** $c \in \mathscr{D}$ **do**
4   /* $c = i$ or $c = \bar{i}$ for an item i */
5      $NDMG := NDMG \cup \{(X \cup Y \cup \{i\})\}$;
6   /* Create $\mathscr{D}_c$ */ $\mathscr{D}_c := \emptyset$;
7   **Foreach** k occurring in $\mathscr{D}$ after c **do**
8   /* $k=j$ or $k=\bar{j}$ for an item j */
9   /* Let $J = X \cup Y \cup \{i,j\}$;*/
10   Count the upper bound u and lower bound u of J;
11   **If** $J.l \neq J.u$ and $MinThreshold_1 \leq J.u \leq MinThreshold_2$ **then**
12   /* J is a non derivable pattern. */
13          $C[k] := cover([c]) \cap cover([k])$;
14          $C[\bar{k}] := cover([c]) \setminus cover([k])$;
15   /* Compute $Supp(J)$, i.e $COUNT_{val}(t,\mathscr{D})$*/
16   /*Compute the estimated support*/
17          $Estimated\text{-}Supp(J) := min\{Supp(K) \mid K \subset J$ et $|K| = |J|\text{-}1\}$;
18   **If** $Estimated\text{-}Supp(J) \neq Supp(J)$ **then**
19   /* J is a non derivable minimal generator. */
20   Store J;
21   **If** $MinThreshold_1 \leq Supp(J) \leq MinThreshold_2$ and $Supp(J) \neq J.l$ and $Supp(J) \neq J.u$ **then**
22   **If** $|C[j]| \leq |C[\bar{j}]|$ **then**
23                $\mathscr{D}_c := \mathscr{D}_c \cup \{(j,C[j])\}$;
24   **else**
25                $\mathscr{D}_c := \mathscr{D}_c \cup \{(\bar{j},C[\bar{j}])\}$;

26    $NDMG := NDMG \cup$ dfCLOSND$(\mathscr{D}_c, MinThreshold_1, MinThreshold_2)$;
27    $CNDP := \{\gamma(I) \mid I \in NDMG\}$;
28   **return** $CNDP$
29 **End**

---

computed as done in [10], we can only use the set of non derivable minimal generators. The introduction of minimal generators will hence optimize both the candidate generation and closure computation steps. Indeed, the number of non derivable minimal generators is lower than that of non derivable patterns. Motivated by this idea, the dfCLOSND algorithm mines all the non derivable minimal generators. The set of the closed non derivable pattern candidates corresponds to the set of closures of the non derivable minimal generators (*cf.* line 27). For each non derivable minimal generator, its closure and its support are inserted in the set of the closed non derivable patterns.

6   Experimental Results

All experiments were carried out on a PC equipped with a 3GHz Pentium IV and 2GB of main memory running under Linux Fedora Core 6. Through these experiments, we have a twofold aim: first, we have to stress on comparing the computation time obtained by dfCLOSND *vs.* that of CLOSENDMG and FIRM [3] to compute the *CCND-Cube*. Second, we put the focus on the assessment of the compactness in storage terms of our approach *vs.* that proposed by the related approaches of the literature, namely, *constrained cube, constrained quotient cube* and *constrained closed cube*[4]. During the carried out experimentation, we used two synthetic datasets [5]: (*i*) CHESS is a dense dataset; and RETAIL is a sparse one. Moreover, we used two real datasets frequently tested for experimenting various cube algorithms [5,11], namely, COVTYPE [6] and SEP85L [7]. Table 2 sketches dataset characteristics used during our experiments.

Table 2. The considered datasets at a glance

Datasets	# Attributes	Tuples
COVTYPE	54	581012
SEP85L	7871	1015367
CHESS	75	3196
RETAIL	16470	88162

In the experiments conducted on the datasets, the parameter which varies is the minimal threshold set for the anti-monotone constraint (the measure values are above a given threshold). The minimal threshold set for the monotone constraint (the measure values are below a given threshold) stays constant.

6.1   Performance Aspect

Figure 2 plots the runtime required to generate the *CCND-Cube* for the considered datasets, using dfCLOSND, CLOSENDMG and FIRM algorithms. Clearly, *w.r.t.* efficiency terms, the dfCLOSND algorithm largely outperforms the CLOSENDMG and FIRM algorithms. Indeed, the gap between the curves tends to become wider as far as the *MinThreshold₁* values decreases.

The dfCLOSND algorithm is more efficient on dense, sparse and real datasets for all the *MinThreshold₁* values. The difference between the performances of dfCLOSND and the other algorithms reaches its maximum for the COVTYPE dataset. In fact, dfCLOSND is respectively 27 and 14 times faster than FIRM and CLOSENDMG with a *MinThreshold₁* value equals to 30%. For the CHESS dataset, the performances of dfCLOSND is widely better than those of the other algorithms. For example, with a value of

---

[3] Available at: http://www.cs.helsinki.fi/u/jomuhone/
[4] The closed Cube was extracted thanks to the CLOSE algorithm [13].
[5] Available at: http://fimi.cs.helsinki.fi/data/
[6] Available at: http://ftp.ics.uci.edu/pub/machine-learning-databases/covtype
[7] Available at: http://cdiac.esd.ornl.gov/cdiac/ndps/ndp026b.html

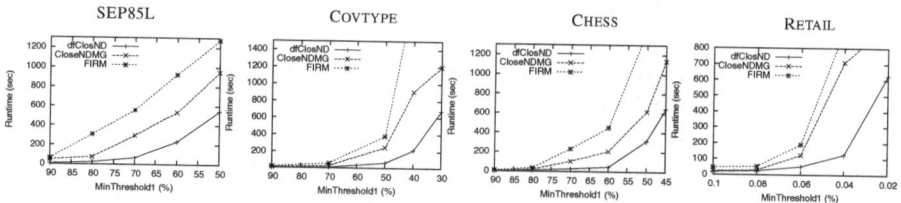

Fig. 2. Mining time of *CCND-Cubes* using the dfClosND, FIRM and ClosENDMG algorithms

MinThreshold$_1$ equals to 45%, the differences are 2584 and 487 seconds with regard to FIRM and ClosENDMG, respectively. For the sparse dataset RETAIL, dfClosND is on average 19 and 15 times faster than FIRM and ClosENDMG, respectively.

Due to the reduced number of *NDMG* to consider during the computation of the closures, dfClosND is more efficient than FIRM. The latter is hampered by redundant computations of closures. Moreover, we note that dfClosND outperforms the ClosENDMG algorithm. In fact, dfClosND avoids the bottleneck of the ClosENDMG algorithm (*i.e.* generating a prohibitive number of candidates). To do so, dfClosND is based on a depth-first search in order to apply the discovery process of the *CNDP*. This discovery process lies on a pruning strategies with respect to a conjunction of constraints.

6.2   Storage Reduction Aspect

Figure 3 presents the cardinality of tuples obtained for each data cube representation. By increasing the minimal threshold for the anti-monotone constraint *MinThreshold$_1$*, the number of relevant tuples decreases. The more the threshold increases the more the size of the concise representations decreases. Actually when the minimal threshold is high, less tuples are likely to be constrained tuples. However, the *CCND-Cube* is always more reduced than the other cubes with an appreciable gain.

For real and dense datasets, the compression rates obtained by the *CCND-Cube* are significant and by far greater than those obtained by the *constrained quotient cube* and the *constrained closed cube*. For the CHESS dataset, the size of the *CCND-Cube* is 36 and 81 times smaller than the *constrained closed cube* and the *constrained quotient cube*, respectively. Moreover, we notice that for real datasets the compression is greater

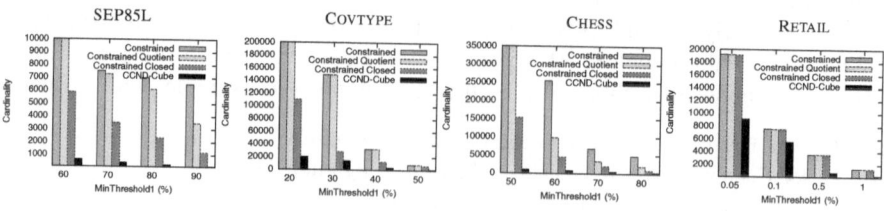

Fig. 3. Size of the data cubes generated

when using *CCND-Cube vs.* both the *constrained closed cube* and the *constrained quotient cube*. Clearly, the *CCND-Cube* is largely smaller than the *constrained cube*. On average, our condensed representation is 67 times smaller than the *constrained cube* for the COVTYPE dataset. However, the *constrained closed cube* and the *constrained quotient cube* are 16 and 4 times smaller than the *constrained cube*, for the same dataset.

As expected, the compression rates are nevertheless much more modest for the sparse dataset RETAIL. The reason behind this very weak space reduction is the following: the synthetic data is weakly correlated and thus encompasses very few redundancies [13]. However, we note that compression rates obtained for *CCND-Cube* exceed those obtained by other representations. On average, *constrained closed cubes* and *constrained quotient cubes* are 4 times smaller than the *constrained cube* while the *CCND-Cube* is 8 times smaller.

7   Conclusion and Perspectives

In this paper, we focused on concise representation of constrained data cubes using data mining algorithms to tackle the mentioned above challenges, *i.e.*, costly execution time of the data cube computation as well as a large storage space on the disk. Thus, we introduced a constrained closed non derivable cube called *CCND-Cube* based on an efficient mining algorithm called dfCLOSND. The carried out experimental results showed the effectiveness of the introduced approach and highlighted that the *CCND-Cube* outperforms the pioneering approaches fitting in the same trend. Future issues for the present work mainly concern: using the borders, classical in data mining, for the lossless representations of constrained cubes [11].

References

1. Bastide, Y., Pasquier, N., Taouil, R., Stumme, G., Lakhal, L.: Mining Minimal Non-redundant Association Rules Using Frequent Closed Itemsets. In: Palamidessi, C., Moniz Pereira, L., Lloyd, J.W., Dahl, V., Furbach, U., Kerber, M., Lau, K.-K., Sagiv, Y., Stuckey, P.J. (eds.) CL 2000. LNCS (LNAI), vol. 1861, pp. 972–986. Springer, Heidelberg (2000)
2. Beyer, K., Ramakrishnan, R.: Bottom-Up Computation of Sparse and Iceberg CUBEs. In: Proceedings of the ACM-SIGMOD International Conference on Management of Data, Pennsylvania, USA, pp. 359–370 (1999)
3. Brahmi, H., Hamrouni, T., Messaoud, R.B., Yahia, S.B.: A New Concise and Exact Representation of Data Cubes. In: Guillet, F., Ritschard, G., Zighed, D.A. (eds.) Knowledge Discovery and Management. SCI, vol. 398, pp. 27–48. Springer, Heidelberg (2012)
4. Calders, T., Goethals, B.: Non-Derivable Itemset Mining. Data Mining and Knowledge Discovery 14(1), 171–206 (2007)
5. Casali, A., Cicchetti, R., Lakhal, L.: Closed Cubes Lattices. Annals of Information Systems 3, 145–165 (2009); Special Issue on New Trends in Data Warehousing and Data Analysis
6. Casali, A.: Mining Borders of the Difference of Two Datacubes. In: Kambayashi, Y., Mohania, M., Wöß, W. (eds.) DaWaK 2004. LNCS, vol. 3181, pp. 391–400. Springer, Heidelberg (2004)
7. Chaudhuri, S., Dayal, U.: An Overview of Data Warehousing and OLAP Technology. SIGMOD Record 26(1), 65–74 (1997)

8. Ganter, B., Wille, R.: Formal Concept Analysis. Springer (1999)
9. Gray, J., Chaudhuri, S., Bosworth, A., Layman, A., Reichart, D., Venkatrao, M.: Data Cube: A Relational Aggregation Operator Generalizing Group-by, Cross-Tab, and Sub Totals. Data Mining and Knowledge Discovery 1(1), 29–53 (1997)
10. Muhonen, J., Toivonen, H.: Closed Non-derivable Itemsets. In: Fürnkranz, J., Scheffer, T., Spiliopoulou, M. (eds.) PKDD 2006. LNCS (LNAI), vol. 4213, pp. 601–608. Springer, Heidelberg (2006)
11. Cicchetti, R., Lakhal, L., Nedjar, S.: Constrained Closed and Quotient Cubes. In: Guillet, F., Ritschard, G., Zighed, D.A. (eds.) Knowledge Discovery and Management. SCI, vol. 398, pp. 3–26. Springer, Heidelberg (2012)
12. Nedjar, S., Cicchetti, R., Lakhal, L.: Extracting Semantics in OLAP Databases Using Emerging Cubes. Information Science 181, 2036–2059 (2011)
13. Pasquier, N., Bastide, Y., Taouil, R., Lakhal, L.: Efficient Mining of Association Rules Using Closed Itemset Lattices. Journal of Information Systems 24(1), 25–46 (1999)
14. Ross, K., Srivastava, D.: Fast Computation of Sparse Data Cubes. In: Proceedings of the 23rd International Conference on Very Large Databases, Athens, Greece, pp. 116–125 (1997)
15. Shao, Z., Han, J., Xin, D.: MM-Cubing: Computing Iceberg Cubes by Factorizing the Lattice Space. In: Proceedings of the 16th International Conference on Scientific and Statistical Database Management, Washington, USA, pp. 213–222 (2004)
16. Van de Vel, M.: Theory of Convex Structures. North-Holland, Amsterdam (1993)
17. Xin, D., Han, J., Li, X., Wah, B.W.: Star-Cubing: Computing Iceberg Cubes by Top-Down and Bottom-Up Integration. In: Proceedings of the 29th International Conference on Very Large Data Bases, Berlin, Germany, pp. 476–487 (2003)

VS-Cube: Analyzing Variations
of Multi-dimensional Patterns
over Data Streams[*]

Yan Tang, Hongyan Li[**], Feifei Li, and Gaoshan Miao

Key Laboratory of Machine Perception (Peking University), Ministry of Education
School of Electronics Engineering and Computer Science, Peking University
Beijing 100871, P.R. China
{tangyan,lihy,liff,miaogs}@cis.pku.edu.cn

Abstract. In many applications, patterns over time-varied data streams usually imply high domain value. The variations of patterns can often be measured from their internal structures. Traditional methods usually take each pattern as a whole to analyze data stream variations; however, few works have achieved a widely applicable resolution. This paper considers the feature of sub parts for data stream patterns and studies their variations and relationships from the perspective of multiple dimensions, to explore a comprehensive understanding for the variation history and effectively support different types of queries. This paper first decomposes patterns into different dimensions and then evaluates the variations of each dimension. After that, a data cube called VS-Cube is used to find out the variations of a single dimension as well as the relationships between different dimensions within a certain pattern. At last, the experimental results on real datasets are given to demonstrate the efficiency and effectiveness of our proposed methods.

Keywords: Data Stream, Multi-dimensional pattern, Pattern variations, OLAP.

1 Introduction

Finding and analyzing sequential patterns (e.g., frequent patterns, diverse patterns, and periodic patterns, etc.) over data streams is a challenging task in many real-life applications. Especially for patterns which are time-varied, variation analysis can be very complicated and few works have achieved a widely applicable resolution.

We observed that the variations of patterns can often be measured from their internal structures. Here, we use the term dimension to denote a component part

---

[*] This work was supported by Natural Science Foundation of China (No.60973002 and No.61170003), the National High Technology Research and Development Program of China (Grant No. 2012AA011002), National Science and Technology Major Program (Grant No. 2010ZX01042-002-002-02, 2010ZX01042-001-003-05).

[**] Corresponding author.

S. Zhou, S. Zhang, and G. Karypis (Eds.): ADMA 2012, LNAI 7713, pp. 779–792, 2012.

of the internal structure within a pattern. Fig. 1(a) shows a typical pattern of ECG (electrocardiograph). Six dimensions (P, Q, R, S, T, and U) in the ECG pattern are marked which denote different stages and actions of a human heart. In Fig. 1(c)dimension P emerges in a higher and bigger manner, which suggests that the patient may have diseases like diabetes, coronary problems, etc. Another example is a stream of the daily residential electricity consumption (Fig. 1(b)). A day can be divided into three dimensions, after-midnight dimension from 0 to 4 o'clock, day-time dimension from 4 to 16, and before-midnight from 16 to 24. Different dimensions can have different behaviors. Usually, the electricity consumption reaches its peak between 20 to 24 o'clock at night on New Year's Eve in China which is different from usual time. That is because Chinese people have the tradition of watching Spring Festival Evening at that day from 20 to 24.

As technology advances, streams of data can be massively generated in numerous applications. Patterns in many of them have multiple dimensions which may help understand them. However there has been few work on analyzing variations of patterns in perspective of dimensions.

Decomposing the variations of patterns into different internal dimensions makes it possible to find the respective variations of dimensions and the interrelationship among them, which can help us gain a more comprehensive understanding of variations of patterns. A dimension may have various kinds of variations (higher, sharper, etc.) and for some streams, a dimension may be repeated or disappear, which further increases the complexity of the variation analysis. We call it loss-or-gain constraint. For example, in Fig. 1(c), dimension T gains (two times) and U disappears. Additionally, in most cases, a pattern variation happens with the changes of several dimensions. Tracing the different variations and analyzing the interrelationships of variations help solve practical problems. Motivated by these observations, in this paper, our primary focus is designing methods for multi-dimensional analysis of pattern variations over streams, combining the OLAP operations and pattern analysis over streams. The rest of the paper is organized as follows. In section 2, related work is discussed. VS-Cube will be provided in Section 3. Section 4 discusses the queries supported by our methods. Materialization is discussed in Section 5. Experimental results are presented in Section 6. Finally, Section 7 concludes the paper with some directions for future work.

2   Related Work

In this section, we review related work of our paper from the following aspects.

2.1   Pattern Analysis over Data Streams

As the number of data stream applications grows rapidly, there is an increasing need to find various patterns on stream data. Hence, some methods have been employed to represent and find patterns in data streams and time series [2, 5, 6]. In particular, studies include Wu et al. [1] and Tang et al. [2] realized

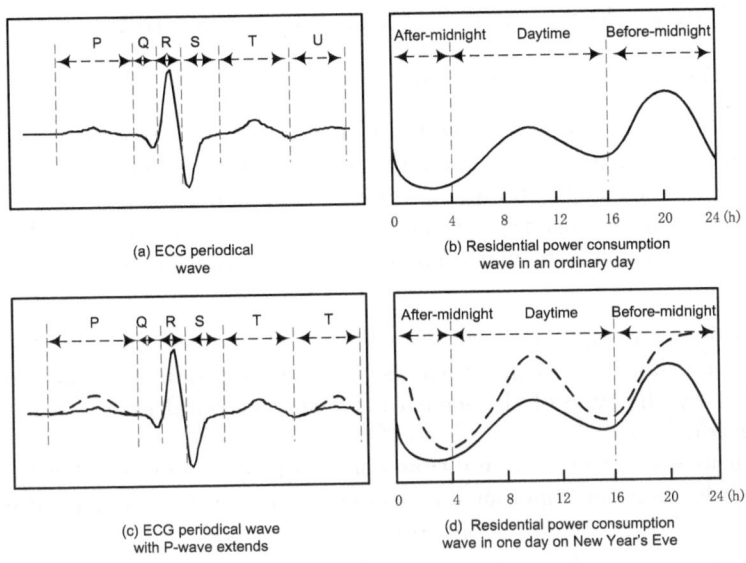

(a) ECG periodical wave

(b) Residential power consumption wave in an ordinary day

(c) ECG periodical wave with P-wave extends

(d) Residential power consumption wave in one day on New Year's Eve

Fig. 1. Multi-dimensional patterns in data streams

the significance of internal structures within a pattern. Wu et al [1] explicitly proposed internal structures within patterns, and modeled the motions of respiration using finite state automaton for the measurement of similarity. Tang et al [2] found the evolution within patterns, and proposed an automatic variation management approach. Additionally, Mueen et al. [7] detected time series motifs in data stream and realized the internal dimensions within a pattern. However, none of them paid attention to variations of each dimension and associations between them.

2.2 Multi-dimensional Analysis over Data Streams

As data stream has been studied extensively in recent years, more and more people have realized the multiple dimensions in streams. Traditional OLAP has been extended to deal with dimensions in data stream [8, 10, 11]. E-Cube [8] has proposed to facilitate computation for multi-dimensional event pattern detection by sharing results among queries. Flow Cube [9] constructs a warehouse of commodity flows based on RFID technology. It summarizes item paths, capturing the general flows trends and significant deviations from the trends. Stream Cube [10] has proposed methods for fast online multi-dimensional analysis of stream data and Regression Cubes [11] investigates methods for online multi-dimensional regression analysis of streaming data to discover dramatic changes of items. However, all of them consider neither the variation of dimensions nor the associations between them.

3   VS-Cube: Analyzing Variations of Streams

3.1   The Division of Dimensions

We define a data stream as a sequence of ordered data values. The order is decided by the arriving time. This paper employs the model of sliding window to get temporal sequences in data streams [2, 12]. Here, we are focusing on how to divide different dimensions and to measure the variations of them. We provide two general conditions here which may not cover all types of stream but can been extended.

1. Dimensions correspond to a discrete value or tuple. Streams of Sales records belong to this kind of applications. In this condition, dimensions can be decided by the type of the measures or the temporal logic of the internal dimensions.
2. Dimensions correspond to multiple continuous data points. In order to find the border between different dimensions, we introduce two typical methods which can be expanded in some other conditions.
   (a) Time-based division. In some applications, people are interested in variations in a certain time, thus we can set dimension borders according to some specific timestamps. Fig. 1(b) is an example.
   (b) Feature-based division. In most cases, the arrival time of a pattern is uncertain due to the uncertainty of data rate. So it is impossible to divide the patterns by the timestamps. In this paper, we deal with it by analyzing the features of the corresponding patterns over streams.

Here, we use ECG streams as an example to discuss the last division method which is the most complex one and is later used in our experiments. ECG streams are composed of ECG patterns, and an ECG pattern consists of six dimensions, P, Q, R, S, T and U (Fig. 1(a)). The process of division is as follows:

- Determine the baseline value X_b and the error δ. The baseline value X_b is used to decide the borders of dimensions in medical fields. X_b may have errors and sometimes the border deviates from X_b. Thus an error bound δ is set to reduce the error of border determination. X_b and δ are both specified by users with the help of machine learning and some other techniques.
- Use PLR_E [13, 14] over data in sliding window. One dimension usually contains 100-200 data points and needs to be compressed to save space and reduce computational cost. There are many existing representations for streams, SAX [15], DWT [16, 17], FFT [18], etc. We use PLR_E [13, 14] here to approximate the data, which has the best compressing effect without losing important features and can help divide the dimensions [2].
- Divide dimensions. After the second procedure, we get a set of segments. Fig. 2 shows different ways that how segments cross the base region $[X_b-\delta, X_b+\delta]$. If one segment passes the region, the point of intersection with the baseline is the border; if several segments pass the region, the nearest intersection to the baseline is the border. Later in the experiment, we will show the effectiveness of this division method.

Generally, we can decide the corresponding relationship of segments and dimensions according to the temporal logic, but as for streams which have loss-or-gain constraints, we can set some constraints over dimensions to help solve this problem. For example, in ECG streams, the timespan of dimension P should be 0.06 to 0.10s, and the height 0.22 to 0.25mV. Additionally, a normal shape of dimension can be specified by users or data mining techniques and is used to compare with the segments, and if they are similar, the segments are what we need. There are many techniques measuring the similarity between sequences, including Euclidean distance and its variations, DTW distance [19], or LCS distance [20], ERP [21], etc. Here we will not go to details of them for limited space.

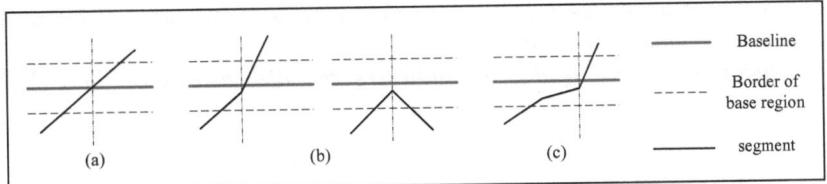

Fig. 2. (a)(b)(c) are different conditions of segments crossing the base region and the vertical dotted line denotes the boundary between dimensions. (a) One segment crosses the base region. (b) Two segments. (c) More than two segments.

For different streams, their variations vary. Here we denote the variation measures as dimensional variation attributes (we call them attributes for short). The attributes can be either user specified or expanded based on some given basic ones. Table 1 shows some attributes based on ECG applications, where a dimension is $DP^X = (x_1, t_1), , (x_m, t_m)$ after the division process, in which X is the dimension name. For different dimensions, the interested attributes are different.

Table 1. Dimensional Variation Attributes based on ECG streams

Attributes	Description	Calculation
X_{max}	Highest point	$MAX\{x_1, ..., x_m\}$
X_{min}	Lowest point	$MIN\{x_1, ..., x_m\}$
T_s	Time span	$t_m - t_1$
X_h	Height	$X_{max} - X_{min}$
X_{smooth}	Smoothness	$\sum_0^{\lfloor m/2 \rfloor} \frac{x_{i+1} - x_i}{t_{i+1} - t_i}$
$X_{displace}$	Displacement	$(x_1 + x_m)/2 - X_b$
B_{invert}	Inversion	$\begin{cases} 1 & if \|X_{max}\| < \|X_{min}\| \\ 0 & if \|X_{max}\| \geq \|X_{min}\| \end{cases}$

3.2   Multi-dimensional Variation Graph

In the section above, we have discretized dimensions and attained the dimensional variation attributes. For a deep analysis of the variations, here we introduce the data structure, Multi-dimensional Variation Graph (MVG).

Definition 1. *A **Multi-dimensional Variation Graph** (MVG) is a graph denoted as $G = (V, E, D)$.*

1. *V is a set of nodes, which represents a dimension with its attributes and the timestamp t of the pattern which it belongs to, i.e. $v_i = d_i, c_1, c_2,, c_k, t$, where k is the number of the attributes in dimension d_i. Considering loss-or-gain, we use three kinds of nodes: ordinary nodes, loss nodes which don't show in the pattern and gain nodes which repeat in the pattern.*
2. *E is a set of edges and $E = E_1, E_2$. E_1 is a set of one-way edges which connect the nodes belonging to the same dimension and the direction indicating the temporal logic, from older ones to newer ones. E_2 is a set of two-way edges connecting nodes belonging to the same pattern, that is to say, connected nodes share the same timestamp.*
3. *Dimension set $D = d_1, d_2,, d_n$ indicates the dimensions of the MVG, in which n is the number of dimensions.*

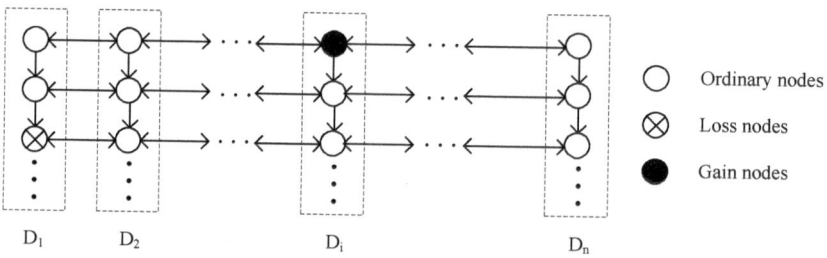

Fig. 3. An example of Multi-dimensional Variation Graph

Optimization. Fig. 3 shows an example of the structure of MVG. As time elapses and more patterns arrive, the space consumption will increases linearly and will result in excessive search costs. Therefore, we provide an optimization strategy and the details are shown in Fig. 4. In Algorithm MergeNode, line 1 and 2 merge nodes which have same attributes, which means the Euclidean distance of the two nodes is less than an error γ, and line 4 to 9 merge nodes which have the same trend. That is to say, all the attributes of the nodes being merged have to be monotonic. Let the average number of attributes in a node be m and the number of processed node be n, the constructing and optimizing methods (in Fig. 4 have a time complexity of $O(mn)$ which meets the requirement of stream processing.

Update and Maintenance. As new patterns arrive, the generated nodes are added into MVG directly and are merged according to the rules mentioned above. As the number of nodes increase, though much space is saved because of many frequent repetitions of dimensions, the structure expands.

Considering that people are more interested in recent data, tilted time frame [10] are adopted, which means that more recent data is registered with finer

```
Algorithm BuildMVG                          Algorithm MergeNode
Variables:   v, v1, CurNode: Node;          Input:    v1, v2: nodes;
             d: Dimension;                  Output:   Merged: Boolean
Output:      a MVG;                         begin
begin                                       1.if Equate(v1, v2, γ) then
1.(v,d)=fetchnextnode( );                   2.   v1.timestamp=(v1.timestamp,v2.timestamp);
2.While v!=NULL do                          3.else if v1 is a merged node then
3.   v1=Fetchlastnode(d);                   4.   if sameTrend(v1, v2)==ture then
4.   if MergeNode(v1, v)==True then         5.     for i=0 to k do
5.     AddtoLeft(CurNode, v1, d);           6.        v1.cᵢ[2]=v2.cᵢ;
6.     AddtoRight(v1, CurNode, d);          7.     endfor
7.   else                                   8.     v1.timestamp[2]=v1.timestamp;
8.     Add(v,d);                            9.     AddtoTable(v1, v2);
9.   AddtoParent(v1, v);                    10.else if v1 is a root node then
10.  AddtoLeft(CurNode, v, d);              11.   for i=0 to k do
11.  AddtoRight(v, CurNode, d);             12.      v1.cᵢ=(v1.cᵢ, v2.cᵢ);
12.endwhile                                 13.   endfor
end                                         14.   v1.timestamp=(v1.timestamp,v2.timestamp);
                                            15.   AddtoTable(v1, v2);
                                            end

          (a)                                               (b)
```

Fig. 4. Algorithms of building MVG and optimizations on MVG

granularity, while more distant data with coarser one. In this paper, we set an old time and an expired time. For data which exceeds the old time, every m nodes are merged into one node in which attributes are ranges of the corresponding ones and the attributes of the m nodes will be written into the table of the generated one. For data that exceeds the expired time, the nodes and corresponding edges are deleted, and the information can be written into files for further analysis. Thus, we can archive the variations of a stream and support analysis while controlling the time and space consumption.

3.3   VS-Cube

In spite of variations of one dimension, in many cases, people are interested in the overall changes in several dimensions. Based on Definition 1, here we propose the data structure Aggregate Multi-dimensional Variation Graph (AMVG).

Definition 2. *Given a Multi-dimensional Variation Graph $G = (V, E, D)$, and a possible division of the dimension set, $A = \{A_1, A_2, ..., A_m\}$, an Aggregate Multi-dimensional Variation Graph (AMVG) is a graph $G' = (V', E', A)$ in which m is the number of dimensions and $A_1 = \{d_1, ..., d_i\}$, $A_2 = \{d_{i+1}, ..., d_j\}$, ..., $A_m = \{d_1, ..., d_n\}$ and $\sum_{i=1}^{m} |A_i| = n$.*

For each node $v_i' = \{A_i, c_1', c_2', ..., c_k', t\}$, $c_j' = \phi(c_{j1}, ..., c_{jl})$, $j \in [i, k]$, where c_j is the j-th attribute of dimension d_i and $\phi(\cdot)$ is an aggregate function defined on variation attributes of nodes which share the same timestamp t and the dimensions of these nodes belong to the same division subset in A.

Here, we choose the aggregate function based on the type of attributes. For example, if the attribute is timespan, then the aggregate function can be $ADD(\cdot)$;

If the attribute is height, then the function can be $MAX(\cdot)$. Based on the definitions above, we give the definition of VS-Cube.

Definition 3. *Variation Stream Cube (VS-Cube) is a set of Aggregate Multi-dimensional Variation Graphs based on different divisions of the dimension set D. Each Aggregate Multi-dimensional Variation Graph is a VS-cuboid.*

Given a set of dimensions based ECG applications, $D = P, Q, R, S, T, U$, the VS-Cube is as shown in Fig. 5. Let n be the number of dimensions, the number of VS-cuboids are $\sum_{i=0}^{n-1} C(i, n-1)$. When $n = 6$, the number of VS-cuboids is 32 and not all the VS-cuboids are shown in the Fig. 5. As n increases, the number of VS-cuboids can be very large. But in reality, people usually care only some of them. For example, black nodes in Fig. 5 are more interesting to users than others.

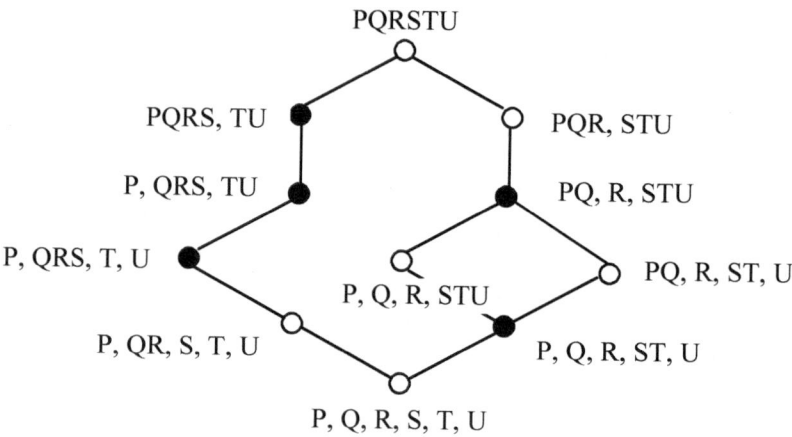

Fig. 5. VS-Cube Lattice based on ECG data stream

OLAP Operations on VS-Cube. Based on VS-Cube, we propose extended OLAP operations for users to navigate cuboids from one to another. Here, we define that if the dimension set in a cuboid G_1 is a division of that in cuboid G_2, then G_1 is an ancestor of G_2 and G_2 is a descendant of G_1 in the other way. Roll-up means going from a cuboid to one of its ancestors and drill-down from a cuboid to one of its descendants. As in Fig. 5, we can roll-up a cuboid (P, Q, R, S, T, U) to its ancestor (P, QR, S, T, U) by merging dimensions Q and R by the predefined aggregate functions. A slice-and-dice operation can be performed on cuboids by deleting some dimensions which are not currently under consideration. For example, by deleting dimensions P, T and U, we can just analyze Q, R and S without considering the other dimensions.

4   Implementation of VS-Cube

To implement a VS-Cube, we need to compute different AMVGs grouping on all possible dimension combinations. As we know, cuboids in adjacent layers share most of the dimensions, for instance cuboid (P, Q, R, S, T, U) and cuboid (P, Q, R, ST, U) share the dimensions P, Q, R, U. With a minor change of two-way pointers, we can implement these two cuboids by sharing a common version of these four dimensions. Moreover, a parent cuboid can be computed directly from its son cuboid. Based on these conditions, we selectively materialize a subset of cuboids in VS-Cube. The chosen cuboids are more probable to be accessed.

5   Queries Supported by VS-Cube

Based on VS-Cube and the OLAP operations on it, we can get variations on different abstract layers. A series of queries can be designed over VS-Cube to help discover the interrelationship between different dimensions.

As we can see, queries can be very complex. A query may correspond to a single node or a sub-graph within a dimension of MVG, and it may span several cuboids from different layers. To keep it simple, we define the basic variation bv, which expresses the change trend of an attribute with a beginning stamp $bv.bts$ and an ending stamp $bv.ets$. Based on this, we define VS-SEQ and VS-CON operators to express the relationship between variations.

Definition 4. *A VS-SEQ operator specifies an order on the timestamps in which the instances of specific dimension variations must occur to match the pattern and thus form a composite variation instance. $VS - SEQ(V_1, V_2, ..., V_n) = \{< v_1, v_2, ..., v_n > |v_1.ets < v_2.bts \wedge ... \wedge v_{n-1}.ets < v_n.bts\}, and\ v_i.bts - v_{i-1}.ets < \Delta t_{seq}\ where\ v_i = bv_j | VS - SEQ(bv_1, ..., bv_j),\ and\ bv_j\ is\ specified\ by\ users\ as basic changes. The composite instance has a beginning stamp of $v_1.bts$ and an ending stamp $v_n.ets$.*

Definition 5. *A VS-CON operator specifies variations which happen in the same time interval. $VS - CON(V_1, V_2, ..., V_n) = \{< v_1, v_2, ..., v_n > |v_1.ets \approx v_n.ets \wedge v_1.bts \approx v_n.bts\}\ where\ v_i = bv_j | VS - SEQ(bv_1, ..., bv_j)\ and\ \approx\ means the deviation is no larger than $\Delta t_c on$.*

Then queries can be expressed as a series of VS-SEQ and VS-CON operators. For example, Query 2 can be expressed as
$VS - CON(VS - SEQ(Tup, Tinvert, Tpeakhigher, Tunchange), TUshorter)$.

As we can see, the queries can be expressed as a tree. Every basic variation is a leaf node, while operators are the branch nodes and after processing the "root Node", we can get a time interval which is what we need. In this paper, we first construct a tree of the queries and then execute the algorithms which are presented in Fig. 6.

Algorithm GetQueue is to get a queue of time intervals satisfying a given basic variations on a certain attribute of a dimension which corresponds to leaf nodes, and this algorithm can be processed concurrently to reduce time cost. The corresponding cuboid is computed from its descendent in line 2 if it is not existed in memory. As for operator VS-SEQ, we use the constraints as shown in Definition 4 to prune each queue iteratively until no more changes can be detected, and the top of each queue is what satisfies the constraints (from line 5 to 13). For VS-CON, the process is similar. The time complexity is $O(xy)$, in which x is the average number of time intervals in a queue and y is the average number of nodes in a dimension, and the stamps of the nodes are less than the old time mentioned above.

```
Algorithm GetQueue
Variable: bts, ets, SameVari;
Input: Leaf: QueryNode;
Output: TQ: TimeQueue;
1.      SameVari==false;
2.      curnode=GetCoboid(layer);
3.      while(curnode.stamp!=curtime-oldtime)
4.         if(equal(curnode.trend, leaf))
5.            if(SameVari==false)
6.               bts=curnode.stamp[1];
7.               ets=curnode.stamp[2]
8.               SameVari==true;
9.            else
10.              ets=curnod.stamp[2];
11.           else
12.              if(SameVari==True)
13.                 insert(TQ, bts, ets);
14.                 SameVari==false;
15.           endif
16.           curnode=Getparent(curnode);
17.      endwhile
end
```

```
Algorithm Operator
Variable: QueueEmpty, TQSet;
Input: node: QueryNode;
Output: TQ: TimeQueue;
begin
1.      if(curnode.type==bv)
2.         return curnode.TQ;
3.      foreach childcurnode
4.         Insert(TQSet, Operator(child));
5.      if(curnode.type==VSseq)
6.         QueueEmpty==false;
7.         while(QueueEmpty==false)
8.            if(prun(TQSet, ))
9.               Insert(TQ,firstbts,lastets);
10.              pop(TQSet);
11.           else
12.              QueueEmpty==false;
13.        endwhile
14.     else //VS-CON
15.        //similar to those above
16.        //constraint is
end
```

Fig. 6. Algorithms of Queries on VS-Cube

6   Performance Study

Based on the ideas above, we implement our methods and design experiments to compare with related works in aspects of accuracy and efficiency. Specific experimental environment is as follows. OS is Microsoft Windows 7 basic with CPU Intel Core 2 3.30 GHz, and Memory is 8 GB. The algorithm development environment is MyEclipse 7.0 Enterprise Workbench and the program language we use is Java with JDK version 1.6.0.

Our experimental datasets are a real medical data flow and a residual electricity consumption (REC) stream. The medical data are downloaded from the Biomedical Signal Processing Laboratory of the Portland State University. There are 20,000,000 data points of pediatric intensive care data, including ECG, RESP, CVP and ABP streams. The residual electricity consumption stream is from Fujian, China, which is collected through a couple of observation position. The stream contains more than 120,000 data points.

6.1 Effectiveness of the Feature-Based Method

As we introduced in section 3.1, the division method based on features can divide the dimensions and maintain a low error. Here, we execute the method over ECG data and compare it with the time-based method and baseline-only method. The time-based method is based on the predefined timespan of the dimensions and the result is shown in Fig. 7(left). The feature-based method can divide the dimensions effectively while the other two give lower accuracy. That is because the timespan of dimensions change in some conditions and if one dimension is mistaken, the following are mistaken too, leading to the lowest accuracy of time-based method. In addition, as we mentioned before, the baseline may not be exact in some circumstances. That is why the baseline-only method has gradually lower effectiveness.

6.2 The Effectiveness of Variations in VS-Cube

We design a sequence of queries, which comprise different types according to Section 5 and randomly execute them on our datasets. The precision and recall of queries on VS-Cube are defined as follows:

$$R_{precision} = \frac{1}{n} \sum_{i=1}^{n} \frac{N_t^i}{N_{re}^i}; R_{recall} = \frac{1}{n} \sum_{i=1}^{n} \frac{N_t^i}{N_{ex}^i}.$$

N_t^i is the number of right answers of Query i. N_{ex}^i is the number of expected right answers of Query i which are presented by experts. N_{re}^i is the number of answers of Query i.

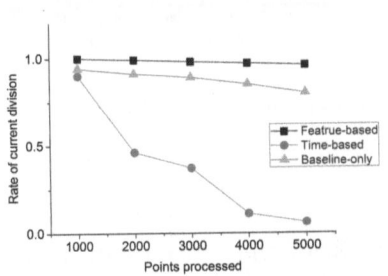
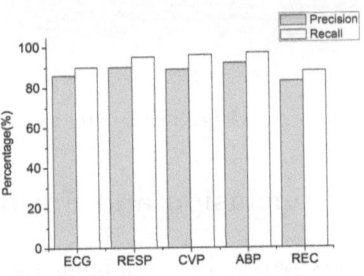

Fig. 7. Effectiveness of the feature-based division method ($X_b = 0.94$, $= 0.02$) (left) and the precision and recall of queries on VS-Cube ($= 0.04$) (right)

We implement our methods over different datasets, the result is shown in Fig. 7 (right), which demonstrates that the effectiveness of our methods meet the requirement of stream processing. Furthermore, the precision and recall are both the lowest on REC which has the most noises. And the ECG has the most complex patterns with the most dimensions, which places it on the second lowest position.

6.3   The Efficiency and Space Consumption of MVG

We compare our structure MVG with PGG [2]which stores variations over streams and takes a pattern as a whole without taking account of dimensions. In this experiment, we set the old time as 10,000s (almost 3 hours) and the expired time as 20,000s. The result is shown in Fig. 8. As we can see, PGG performs better than MVG on all datasets (Fig. 8 left). The reason is that PGG takes each pattern as a whole, but MVG decomposes each pattern into different dimensions and evaluates the variations of each dimension. However, MVG not only satisfies the requirement of applications but also can handle much more complex queries over patterns. And both the methods perform worst on ECG which is most complex than others. The result of space cost shows that MVG maintains a size of about 17KB with little changes, due to compressions of repetitions over streams. By decomposing patterns into dimensions, MVG gains more advantages of repetitions than PGG, because dimensions repeat more than patterns do. In addition, MVG does not record the whole variation history but put expired changes into files, while PGG records every pattern and scales with the data points processed. With larger old time and expired time, the space cost of MVG will increase.

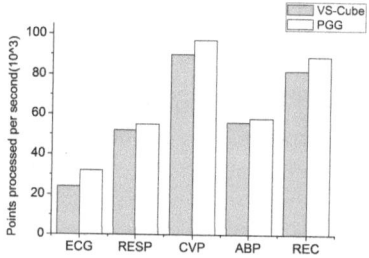 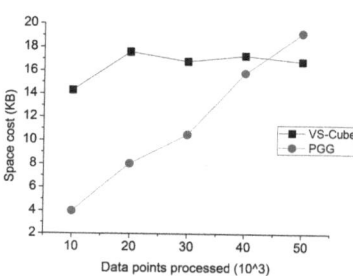

Fig. 8. The efficiency and space consumption of VS-Cube

7   Conclusion and Future Work

In this paper, by combining the OLAP technology and stream data processing methods, we proposed VS-Cube to analyze the variations of patterns over streams. Specific methods are presented to decompose a multi-dimensional pattern into different dimensions and archive the respective variations of each dimension and the interrelationships between them efficiently. Different variations are shown in different abstract layers with a series of OLAP operations. By supporting queries on VS-Cube, users can find out the variation history of a single dimension as well as the relationships between dimensions within a certain pattern. The performance study showed that VS-Cube in this paper is capable of efficiently processing different types of patterns over streams. However, the variations of patterns can be really massive and complex. And how to evaluate the variations of different dimensions remains for future study.

References

1. Wu, H., Salzberg, B., Sharp, G., Jiang, S., et al.: Subsequence Matching on Structured Time Series Data. In: Proceedings of the 2005 ACM SIGMOD International Conference on Management of Data, pp. 682–693 (2005)
2. Tang, L., Cui, B., Li, H., et al.: Effective Variation Management for Pseudo Periodical Streams. In: Proceedings of the 2007 ACM SIGMOD International Conference on Management of Data, pp. 257–268 (2007)
3. Han, J., Dong, G., Yin, Y.: Efficient mining of partial periodic patterns in time series database. In: Proceedings of the 15th International Conference on Data Engineering, p. 106 (1999)
4. Yang, J., Wang, W., Sharp, G., Yu, P.: Mining Asynchronous Periodic Patterns in Time Series Data. IEEE Transactions on Knowledge and Data Engineering (TKDE) 15(3), 275–279 (2003)
5. Lin, J., Keogh, E., Lonardi, S., Chiu, B.: A Symbolic Representation of Time Series, with Implications for Streaming Algorithms. In: Proceedings of the 8th ACM SIGMOD Workshop on Research Issues in Data Mining and Knowledge Discovery, pp. 2–11 (2003)
6. Stonebraker, M., Cetintemel, U., Zdonik, S.: The 8 Requirements of Real-time Stream Processing. ACM SIGMOD Record 34(4), 42–47 (2005)
7. Mueen, A., Koegh, E.: Online Discovery and Maintenance of Time Series Motifs. In: Proceedings of the 16th ACM SIGKDD International Conference on Knowledge Discovery and Data Mining, pp. 1089–1098 (2010)
8. Liu, M., Rundensteiner, E., Greenfield, K., et al.: E-Cube: Multi-Dimensional Event Sequence Analysis Using Hierarchical Pattern Query Sharing. In: Proceedings of the 2011 ACM SIGMOD International Conference on Management of Data, pp. 889–900 (2011)
9. Gonzalez, H., Han, J., Li, X.: FlowCube: Constructing RFID FlowCubes for Multi-Dimensional Analysis of Commodity Flows. In: Proceedings of the 32nd International Conference on Very Large Data Bases, pp. 834–845 (2006)
10. Han, J., Chen, Y., Dong, G., Pei, J., Wah, B.W., Wang, J., Cai, Y.D.: Stream Cube: An Architecture for Multi-dimensional Analysis of Data Streams. Distributed and Parallel Databases 18(2), 173–197 (2005)
11. Chen, Y., Dong, G., Han, J., et al.: Multi-Dimensional Regression Analysis of Time-Series Data Stream. In: Proceedings of the 28th International Conference on Very Large Data Bases, pp. 323–334 (2002)
12. Lim, H., Whang, K., Moon, Y.: Similar Sequence Matching Supporting Variable-length and Variable-tolerance Continuous Queries on Time-series Data Stream. Information Sciences: an International Journal 178(6), 1461–1478 (2008)
13. Keogh, E., Chu, S., Hart, D., et al.: An Online Algorithm for Segmenting Time Series. In: Proceeding IEEE International Conference on Data Mining, pp. 289–296 (2001)
14. Wang, X., Wang, Z.: A Structure-adaptive Piecewise Linear Segments Representation for Time Series. In: Proceedings of Information Reuse and Integration, pp. 433–437 (2004)
15. Lin, J., Keogh, E., Lonardi, S., Chiu, B.: A Symbolic Representation of Time Series, with Implications for Streaming Algorithms. In: Proceedings of the 8th ACM SIGMOD Workshop on Research Issues in Data Mining and Knowledge Discovery, pp. 2–11 (2003)

16. Gilbert, A.C., Kotidis, Y., Muthukrishnan, S., Strauss, M.J.: One-Pass Wavelet Decompositions of Data Streams. IEEE Transactions on Knowledge and Data Engineering 15(3), 541–554 (2003)
17. Papadimitriou, S., Brockwell, A., Faloutsos, C.: Adaptive, Unsupervised Stream Mining. The International Journal on Very Large Data Bases 13(3), 222–239 (2004)
18. Gao, L., Wang, X.S.: Continuous Similarity-Based Queries on Streaming Time Series. In: Proceedings of the 2002 ACM SIGMOD International Conference on Management of Data, vol. 17(10), pp. 370–381 (2005)
19. Gao, L., Wang, X.S.: Continuous Similarity-Based Queries on Streaming Time Series. In: Proceedings of the 2002 ACM SIGMOD International Conference on Management of Data, vol. 17(10), pp. 370–381 (2005)
20. Morse, M.D., Patel, J.M.: An Efficient and Accurate Method for Evaluating Time Series Similarity. In: Proceedings of the 2007 ACM SIGMOD International Conference on Management of Data, pp. 569–580 (2007)
21. Chen, L., Ng, R.: On the Marriage of Lp-norms and Edit Distance. In: Proceedings of the Thirtieth International Conference on Very Large Data Bases, pp. 792–803 (2004)

Author Index